Second Edition

The
Science of
Renewable
Energy

Second Edition

The
Science of
Renewable
Energy

Frank R. Spellman

CRC Press
Taylor & Francis Group
Boca Raton London New York

CRC Press is an imprint of the
Taylor & Francis Group, an **informa** business

CRC Press
Taylor & Francis Group
6000 Broken Sound Parkway NW, Suite 300
Boca Raton, FL 33487-2742

© 2016 by Taylor & Francis Group, LLC
CRC Press is an imprint of Taylor & Francis Group, an Informa business

No claim to original U.S. Government works

Printed on acid-free paper
Version Date: 20160113

International Standard Book Number-13: 978-1-4987-6047-8 (Hardback)

Library of Congress Cataloging-in-Publication Data

Names: Spellman, Frank R., author.
Title: The science of renewable energy / Frank R. Spellman.
Description: Second edition. | Boca Raton : Taylor & Francis, CRC Press, 2016. | Includes bibliographical references and index.
Identifiers: LCCN 2016000623 | ISBN 9781498760478 (alk. paper)
Subjects: LCSH: Renewable energy sources.
Classification: LCC TJ808 .S685 2016 | DDC 621.042--dc23
LC record available at http://lccn.loc.gov/2016000623

Visit the Taylor & Francis Web site at
http://www.taylorandfrancis.com

and the CRC Press Web site at
http://www.crcpress.com

Contents

Preface

Hailed on its first publication as a masterly account for the general reader, the environmentally concerned, and the serious student, *The Science of Renewable Energy, Second Edition*, continues where the first edition left off. That is, the first edition explained that studying renewable energy through science affords us the opportunity to grasp not only the pressing need for renewable energy but also the complicated task of developing viable energy sources for the future—what I consider to be a long and challenging slog. Based on critical reviews, reader input, and current developments (remember, the dynamic science of renewable energy is ever changing and advancing), it became apparent that a follow-up second edition was required to fill in some gaps left by the first edition. For example, the basic energy section in this new edition adds important energy parameters dealing with both fossil fuel supplies and renewable resources. In a science book of any type, basic parameters cannot and should not be ignored. In the second edition they are not ignored.

With regard to future energy needs, there is no question nor doubt that we are facing an international crisis of unprecedented proportion. We are ultimately headed for a train wreck when the temporarily affordable and currently accessible petroleum products are no longer so, and economic and political turmoil results. Keep in mind that we need to find replacements for liquid hydrocarbon fuels not only because the wells are almost dry (after we frack them, of course) but also because we need to clean up the environment. We need to be able to produce reliable renewable energy that will not destroy or pollute our fragile environment.

But, here is the problem. No one questions that developing renewable, sustainable energy supplies is a good thing—a very good thing—but what is the tradeoff? Most people feel strongly or are being told in no uncertain terms that renewable energy supplies are better than the filthy, polluting, disgusting fossil fuels that are destroying our environment and negatively impacting our personal health. The problem is that these same people are not being told about the problems related to renewable energy. Indeed, renewable energy is not a pure energy source devoid of any environmental or other problems. Renewable energy comes with its own set of environmental problems.

It is generally known and widely accepted that the utilization of any energy source impacts our environment. Fossil fuels (oil, coal, natural gas) do substantially more harm than renewable energy sources by almost any measure: air and water pollution, damage to public health, loss of wildlife and their habitats, significant use of water and land, and the production of emissions exacerbating global climate changes. In contrast to fossil fuels, renewable energy—wind, solar, geothermal, hydroelectric, biomass for electricity, and hydrokinetic—offers substantial benefits for our climate, our health, and our economy:

- Little to no global warming emissions
- Improved public health and environmental quality
- Vast and inexhaustible energy supply
- Jobs and other economic benefits
- Stable energy prices
- A more reliable and resilient energy system

Again, however, it is important to understand that renewable energy sources also have environmental impacts. The exact type and intensity of environmental impacts will vary depending on the specific technology used, the geographic location, and a number of other factors:

- Air quality
- Cultural resources
- Ecological resources

- Water resources
- Land use
- Soil and geologic resources
- Paleontological resources
- Transportation
- Visual resources
- Socioeconomics
- Environmental justice
- Safety and health

The important point (and the message woven in the warp and woof of this text) is that by understanding the current and potential environmental issues associated with each renewable energy source we can take steps to effectively avoid or minimize these impacts as our use of these alternative energy sources grows.

This text takes a common-sense approach and presents practical (and sometimes poetic) examples. Because this is a science text, I have adhered to scientific principles, models, and observations; however, the readers does not need to be a scientist to understand the principles and concepts presented. I go easy on the hard math and science and present the material in a user-friendly manner. What the reader really needs is an open mind, a love for the challenge of wading through the muck, an ability to decipher problems, and the patience to answer questions relevant to each topic presented. Real-life situations are interwoven throughout the text in an engaging fashion, and the material is presented in straightforward, conversational plain English to provide the facts, knowledge, and information necessary to understand the complex issues involved and to make informed decisions.

As a companion text to *The Science of Water*, *The Science of Air*, and *The Science of Environmental Pollution*, *The Science of Renewable Energy, Second Edition*, follows the same proven format used in the other three texts. But, like its forerunners, this text is not an answer book. Instead, it is designed to stimulate thought. Although solutions to specific renewable energy questions are provided, the disadvantages and hurdles that must be overcome to make renewable energy a viable alternative to fossil fuels are highlighted. The goal is to provide a framework of principles to help develop an understanding of the complexity of substituting renewables for fossil fuels. *The Science of Renewable Energy, Second Edition*, is designed to reach a wide range of diverse readers and students. The text focuses on the various forms of renewable energy derived from natural processes that can be constantly replenished: solar energy, wind energy, ocean energy, wave energy, tidal energy, geothermal energy, biomass energy, hydropower energy, biofuels, and hydrogen fuel cells. Their design and implementation should prevent pollution of the atmosphere, of surface and groundwater, and of soil (the three environmental media). Keep in mind that all of these naturally produced processes are critical to our very survival. Because renewable energy and pollution prevention are real-world issues, it logically follows that we can address these issues by using real-world methods—that's what *The Science of Renewable Energy, Second Edition*, is all about.

Note to the Reader

In 1976, energy policy analyst Amory B. Lovins coined the term *soft energy path* to describe an alternative future where energy efficiency and appropriate renewable energy sources steadily replace a centralized energy system based on fossil and nuclear fuels. In 2009, Joshua Green, then a writer for *Businessweek*, pointed out that Lovins further argued that the United States had arrived at an important crossroads and could take one of two paths. The first, supported by U.S. policy, promised a future of steadily increasing reliance on dirty fossil fuels and nuclear fission and had serious environmental risks. The alternative, which Lovins called the *soft path*, favored "benign" sources of renewable energy such as wind power, solar power, biofuels, geothermal energy, and wave and tidal power, along with a heightened commitment to energy conservation and energy efficiency.

As a lifelong student, researcher, lecturer, and ardent advocate of the development and use of renewable or alternative energy sources (eventually excluding all fossil fuel use to the extent possible), I agree with Lovins in many respects, but I take issue with those who state that renewable energy sources are "benign." In my view, the definition of the term "benign" means something that is harmless, innocent, innocuous, or inoffensive. Thus, the labeling of renewable energy sources as benign implies that the use of renewable energy sources is totally safe. The truth is the use of renewable energy sources is not totally safe.

Again, I am an advocate for the use of renewable energy. Simply, I think using renewable energy sources instead of fossil fuels is a good thing. However, with any good thing there usually comes a bad thing. Nothing made by and used by humans is absolutely harmless to the environment. Nothing. Only Mother Nature, with her ultimate plan, affects nature as we know it in beneficial ways. Even when she kills millions of us with designed orchestrations that change life as we know it, we must realize that these are simply timed mechanizations, part of her long-range plan. Remember, Mother Nature's plan is the ultimate plan. Who are we to argue otherwise?

Anyway, because I do not agree with the idea that the so-called soft path is the "benign" path, I also cannot say that the impacts of renewable energy source are necessarily bad, baneful, damaging, dangerous, deleterious, detrimental, evil, or harmful to the environment. The question is: What can I say in this book and elsewhere about renewable energy? I can say that I am biased toward the use of renewable energy, that I am for Lovins' soft path and against the hard path. But, I qualify this by also stating that renewable energy sources have impacts on the environment, both good and bad, and it is these good and bad impacts that this book addresses.

So, let's cut to the chase. My broad conclusion is that renewable energy sources are not the panacea for solving our many pollution problems that they are popularly perceived to be. In reality, their adverse environmental impacts can be just as strongly negative as the impacts of nonrenewable energy sources, but we need to eventually replace nonrenewables so renewables are the option we must pursue.

Author

Frank R. Spellman, PhD, is a retired adjunct assistant professor of environmental health at Old Dominion University, Norfolk, Virginia, and the author of more than 100 books covering topics ranging from concentrated animal feeding operations (CAFOs) to all areas of environmental science and occupational health. Many of his texts are readily available online, and several have been adopted for classroom use at major universities throughout the United States, Canada, Europe, and Russia; two have been translated into Spanish for South American markets. Dr. Spellman has been cited in more than 850 publications. He serves as a professional expert witness for three law groups and as an incident/accident investigator for the U.S. Department of Justice and a northern Virginia law firm. In addition, he consults on homeland security vulnerability assessments for critical infrastructures, including water/wastewater facilities, and conducts pre-Occupational Safety and Health Administration and Environmental Protection Agency audits throughout the country. Dr. Spellman receives frequent requests to co-author with well-recognized experts in several scientific fields; for example, he is a contributing author to the prestigious text *The Engineering Handbook*, 2nd ed. Dr. Spellman lectures on wastewater treatment, water treatment, and homeland security, as well as on safety topics, throughout the country and teaches water/wastewater operator short courses at Virginia Tech in Blacksburg. In 2011, he traced and documented the ancient water distribution system at Machu Picchu, Peru, and surveyed several drinking water resources in Amazonia, Ecuador. He has also studied and surveyed two separate potable water supplies in the Galapagos Islands, in addition to studying Darwin's finches while there. Dr. Spellman earned a BA in public administration, a BS in business management, an MBA, and both an MS and a PhD in environmental engineering.

1 Post-Oil Energy

Oil creates the illusion of a completely changed life, life without work, life for free. ... The concept of oil expresses perfectly the eternal human dream of wealth achieved through lucky accident. ... In this sense, oil is a fairy tale and like every fairy tale a bit of a lie.

—Ryszard Kapuscinski, Polish journalist and writer

To the eyes of the man of imagination, nature is imagination itself.

—William Blake, English poet and artist

Massive changes in the existence of humanity are imminent. The oil shortage? Blame it all on the Greenies; it is their entire fault. There's something guttural, something personal, about the price of gas.

Steiner (2009)

Four numbers say wind and solar can't save climate.

Bryce (2013)

THE GATHERING STORM

As I begin writing this book, several significant events and situations present themselves that warrant our attention. The promise that shifting from coal-generating power to renewable sources will create thousands of jobs has proven to be untrue. California is a good example: Promises were made that shifting to renewable energy sources would result in the creation of 11,000 new jobs but only 1700 were delivered. And, the fact is that out of the 1700 jobs created most of them went to consultants who served on a temporary, short-term basis. (*Note:* The present and projected future employment opportunities in the renewable energy field are discussed in greater detail later.)

Second on our list of significant events or situations is the current state of the U.S. economy. The real U.S. unemployment rate is at 9 to 10+%, with increasing numbers of workers losing their jobs daily. For those still working in factory jobs, their future in these same types of jobs is threatened and, in some cases, doomed—these jobs are being done away with or outsourced overseas. Many of these are jobs that are unlikely to return to the United States. Third on our list is the nominal world population clock that just posted a world population figure of 7.3 billion and counting—pointing to the increasing need for more resources, including energy.

How are these significant events and others related? What do they have to do with energy? What do they have to do with the so-called pending oil crisis (peak oil)? Finally, what do they have to do with renewable energy? I will discuss the interrelationships between oil and renewable energy and the significance of the events and situations described in the following.

Let's discuss Haiti for a moment. It is the poorest country in the Americas. Not only is it impoverished but it also is the least developed. The 2010 earthquake only exacerbated an already desperate situation in Haiti. Moreover, Haiti has consistently ranked among the most corrupt countries in the world. And, with the exception of a few agricultural crops, it has no exports. Haiti has only survived because of the foreign aid it receives, mostly from the United States and lesser amounts from other countries.

Haiti's energy supplies must be imported; it has no indigenous supply of oil-derived energy. Haiti does have extensive exposure to sunlight and wind, as well as tide and wave energy. These sources can be employed in the development and production of renewable energy, all of which would lessen the need and expense for Haiti to import expensive hydrocarbon fuel products such as gasoline. Development of renewable energy sources in Haiti could go a long way toward helping it reduce its dependence on worldwide generosity and loans. Certainly, economic development could benefit if it were tied to a decrease in corruption, a shoring up of a stable political footing, and an increase in human health and medical care.

The second event or situation that is interrelated with petroleum products and renewable energy production and the current and future state of the U.S. economy is the lack of viable jobs and lack-luster progress in technological innovation. For some time, the United States has been the leader in technological development throughout the world; however, recent reports have pointed out that from 1999 to the present the U.S. population grew by 32 million, or 1%, yet we had zero net job creation. Why? Take your pick; there are several reasons for our current economic condition. To begin with, we develop technology or technological advances in products such as computers, computer software, automobiles, farm machinery, furniture, and textiles, among others. But, the vast majority of cloth-ing, cellphones, computers, and televisions and a large percentage of our construction products (e.g., Chinese drywall) are produced in a foreign country. After we have spent large amounts of capital and time and volumes of expertise developing U.S. innovations, these are simply copied by foreign indus-tries and produced at about nine times lower cost than if they were produced in the United States. These foreign producers have much lower labor costs and no regulatory pressure, and many receive financial assistance from their government. When our technological innovations leave the country, it creates an uneven playing field that simply does not teeter-totter in a direction favorable to us.

In addition to labor, regulatory, and other related costs, the United States is not self-sustaining in oil production anymore. Our dependence on foreign energy sources is not only humiliating and shameful but also quite costly to all of us in so many different ways. For example, we have failed to grow jobs because we are no longer self-sufficient as a nation, with the exception of producing agricultural products for the country's needs. Maybe the advent of extensive hydraulic fracturing will change this ratio; maybe we will become self-sufficient. That is, until the fracked well runs dry.

Complicating any economic recovery we may achieve (at this writing, this seems doubtful and may be wishful thinking) is not only our dependence on foreign oil but also the price of oil. One thing is certain—if and when economic conditions turn around and begin to climb in the positive direction, the price of a gallon of gasoline ($2.58 and similar prices for diesel fuel at this writing) will simply increase, and increase, and increase again. The foreign energy producers are just sitting back waiting for us to recover, so they can turn the price increase lever one notch higher … and higher …, etc. These increases in the price of gasoline and diesel fuel will quickly put a lid on any economic recovery; we will descend that slippery oil-soaked slope back to our current condition of stagnant malaise, or maybe worse.

The third situation we mentioned is the continuing population growth and its impact not only on future fuel requirements but on many other resources, as well. Consider, for example, as previ-ously mentioned, that at this very moment the nominal world population clock reads 7.3 billion and counting. Population growth on this globe is something that directly affects energy availability, use, and the pollution effects associated with using hydrocarbon fuels. The U.S. Census Bureau (2008) predicts that the world's population in 2030 will be almost double that of 1980. Savinar (2010) pre-dicted that oil production in 2030 will have declined back to 1980 levels while worldwide demand for oil will significantly outpace production. Bartlett (2004) observed that the rate of oil production per capita is falling and that the decline has gone undiscussed because a politically incorrect form of population control may be implied by mitigation.

It is important to point out that one factor that has helped to somewhat ameliorate the effect of population growth on demand is the decline of the population growth rate since the 1970s, although this is offset to a degree by increasing average longevity in developed nations. In 1970, the population grew at a rate of 2.1%. By 2007, the growth rate had declined to 1.167% (CIA, 2008), but the fact remains that the world's population is still growing and is continuing to put pressure on our limited energy supplies. Energy is only one of several issues related to an increasing human population. The basic requirements of any population include air, water, food, shelter, and clothing. All of these human needs are tied directly or indirectly to energy.

With regard to the air we breathe, the air itself is ubiquitous and fills a vital need naturally for each of us; however, the air we breathe has been affected or tainted by energy use—namely, the air pollution that commonly accompanies the use of some forms of energy generation or use. At the present time, there is continuing debate on hydrocarbon energy use and its impact on global climate change. Although it is likely that we are undergoing global climate change—a general warming trend—it is not that apparent or proven that air pollution caused by energy use is the main causal factor of climate change. There can be little doubt, however, that human-caused (anthropogenic) pollution is a contributor to the current warming trend. Still, some critics argue that, in a discussion about oil, oil resources, peak oil supplies, and oil shortages, air and oil do not mix, and the use of hydrocarbons is not a primary cause of air pollution. They say there is no connection. In countering this argument, we suggest that these critics drive an automobile from any out-of-town location into downtown Los Angeles on one of those especially warm days in August or September, park the car, step out of the car, and take a deep breath of Los Angeles Basin air. We think that when this occurs the critic will get a mouthful and lungs full of more than just air and will readily come to realize that oil-derived vapors and air do make a noxious mix, making them intolerable or at least uncomfortable to inhale.

Along with good, clean air being one of those substances that we simply cannot live without is our need for water. In addition to its nutritional benefits, water used to be the primary producer of energy in the early days of humans' existence. Water is still used in hydropower applications and as such is a renewable energy source (discussed later in the text) and a major contributor to the generation of electricity. The problem with water, as with hydrocarbons, is that it is a finite natural resource. We have on Earth today the same amount of water that we have always had. We can't add to or delete from our present total water supply to any significant degree. The problem with our potable water is location; it is sparsely located or basically not available in many parts of the globe. Water is heavy and cumbersome, takes up a lot of space, and is not easily transported from location to location. Much of the freshwater of the globe is polluted, not exclusively by energy use but instead by the old mantra: We do not want it anymore so throw it away. Often the receivers of our throwaways are and have been our ponds, lakes, streams, rivers, and oceans. Beyond the use of water for the production of hydropower energy, the interface between energy and water is often ugly, unsightly, and unhealthy and results in ponds, lakes, streams, and rivers polluted by hydrocarbon-derived products such as pesticides, herbicides, rodenticides, fertilizers, and so forth. It seems rather odd that we so carelessly pollute our air and water supplies, the two substances we absolutely cannot live without. Water also interfaces with energy use and its production in that it is used as an industrial cooling medium for machinery used in the refining, production, or use of hydrocarbon products, and it is transported by truck, tanker car, or pipeline or distributed via irrigation networks through power derived (e.g., by pumps) from the use of hydrocarbon fuels such as gasoline and diesel fuel.

The connection between energy use and food has always existed and always will. Since time immemorial to the present, we have expended one form of energy or another to hunt, gather, sow, and harvest our foodstuffs. In the current era, the mechanized farm machinery we are familiar with could not function without oil-derived fuels and lubricants. Moreover, many of the chemicals used to fertilize and protect crops from insects and disease are oil derived. Simply, ours is an oil-based agriculture, and it is this oil-based agriculture that is driving the world's exploding population growth. As oil production increased, so did food production. As food production increased, so did population. In step with population growth, demand for food increased, which escalated the demand for oil. Today, almost all the energy used by the agricultural industry to produce, store, and transport food comes from oil. From tractor to combine to harvester to irrigation pumps, the energy required for storage and the trucking, railing, and flying of these foods to market depends on oil-derived products.

Because supplies of oil-derived products such as diesel fuel and gas are essential to modern agriculture techniques, a fall in global oil supplies could cause spiking food prices and unprecedented famine in the coming decades (Goodchild, 2007). Pfeiffer (2004) pointed out that current population levels are unsustainable and that to achieve a sustainable economy and avert disaster the U.S. population would have to be reduced by at least one third and the world population by two thirds.

The bottom line: Global population growth is only sustainable through our dependency on the infrastructure provided to us by oil. Whenever I make this statement to my college-level environmental students some buy the conclusion without too much thought and others give that surprised or doubtful look. It became apparent early in my environmental lecture presentations that many students and others just do not know that much about the importance of oil and its impact on our present-day society. These lectures point out that many of the more than 500,000 products we are all accustomed to using every day are manufactured using oil as a material or ingredient. Below are listed just a few of these products:

Air conditioners	Denture adhesives	Insulation	Rubber bands
Ammonia	Deodorant	Lens cleaner	Rubber boots
Antihistamine	Dice	Lip balm	Rubber cement
Antiseptics	Dishwasher liquid	Lipstick	Running shoes
Artificial turf	Dresses	Loudspeakers	Shirts (non-cotton)
Balloons	Electric blankets	Mops	Shoe polishes
Bandages	Electrician's tape	Motorcycle helmets	Shoes
Basketballs	Eyeglasses	Movie film	Shower curtains
Boats	Fertilizers	Nail polish	Solvents
Bras	Fiberglass	Oil filters	Stereos
Bubble gum	Fishing lures	Paint brushes	Synthetic clothing
Cameras	Fishing rods	Panty hose	Tape recorders
Candles	Floor wax	Parachutes	Telephones
Car batteries	Footballs	Pens	Thermos bottles
Car bodies	Glue	Perfumes	Tires
Carpet	Golf balls	Petroleum jelly	Toilet seats
Caulking	Hair coloring	Plastic chairs	Toothpaste
Chewing gum	Hair curlers	Plastic cups	Transparent tape
Cold cream	Hair dryers	Plastic forks	Trash bags
Combs/brushes	Hearing aids	Plastic toys	Umbrellas
Compact discs	Heart valves	Plastic wrap	Upholstery
Computers	House paint	Plywood adhesive	Vaporizers
Contacts	Ice chests	Purses	Volley balls
Cortisone	Inks of all types	Refrigerator seals	Water pipes
Crayons	Insect repellent	Roofing paper	Water skis

PEAK OIL

According to the U.S. Department of Energy (USDOE, 2007), the *peak oil theory* states that the world's oil production rate will reach a maximum rate and then enter terminal decline. This concept is based on the observed production rates of individual oil wells and the combined production rate of a field or related oils wells. This is not to say that the world is running out of oil, but it does signal the end of abundant and cheap oil on which all industrial nations depend (Campbell and Laherrere, 1998). Peak oil theory, also known as Hubbert peak theory, concerns the long-term rate of conventional oil (and other fossil fuel) extraction and depletion. Hubbert proposed, in a 1956 paper he presented at a meeting of the American Petroleum Institute, that oil production in the United States would peak between 1965 and 1970 (Hubbert, 1956). U.S. oil production peaked in 1971 and has been decreasing since then. Hubbert's theory is subject to continued discussion because of the potential effects of lowered oil production and because of the ongoing debate over aspects of energy policy. Opinions on the effect of passing Hubbert's peak range from faith that the market economy will produce a solution, assuming major investments in alternatives will occur before a crisis, to prediction of doomsday scenarios of a global economy unable to meet its energy needs (Gwyn, 2004). This thesis postulates that the price of oil at first will escalate and then retreat as other types of fuel and energy sources are developed and used (CERA, 2006). Pessimistic predictions of future oil production operate on the thesis that the peak has already occurred (Deffeyes, 2007) or that oil production is on the peak or will be shortly (Koppelaar, 2006).

Although many experts were skeptical of his prediction, Hubbert was proven correct when U.S. oil production peaked in 1971 (Brandt, 2007). Hubbert's prediction remained accurate even after the discovery of the Prudhoe Bay oil field (after 1971) and its great volume of oil was still not enough to bring U.S. oil production out of a long-term decline (USDOE, 2007).

Peak global oil production and its subsequent decline (based on optimistic assessments) is forecast to begin by about 2020 or later. Keep in mind that the exact date of world oil peaking is a best guess; the actual date cannot be known with any certainty, complicating the decision-making process. This decision-making process refers to deciding what or which alternative sources of energy to pursue to prevent the economic and social impacts of oil peaking from becoming chaotic. A basic problem in predicting oil peaking is uncertain and politically biased oil reserve claims from many oil-producing countries. In some instances, the veracity of such claims may be questionable (Hirsch, 2005).

If we accept predictions about peaking oil, the possible effects and consequences of peaking oil cannot be painted and displayed without using a very wide brush. Few would argue against the fact that the widespread use of fossil fuels has been one of the most important agents of economic growth and prosperity since the industrial revolution. Fossil fuels have allowed humans to participate in the consumption of energy at a greater rate than it is being replaced. Some believe that mankind is in for a momentous paradigm shift when oil production decreases. This shift will cause humans and our modern technological society to change drastically. The actual impact of peak oil will depend heavily on the rate of decline and on the development and adoption of effective alternatives.

If alternatives are not forthcoming, the products produced with oil (including fertilizers, detergents, adhesives, solvents, and most plastics) will become scarce and expensive. In 2005, the U.S. Department of Energy published a report (known as the Hirsch Report) that included this stark warning: "The peaking of world oil production presents the U.S. and the world with an unprecedented risk management problem. As peaking is approached, liquid fuel prices and price volatility will increase dramatically, and, without timely mitigation, the economic, social, and political costs will be unprecedented. Viable mitigation options exist on both the supply and demand sides, but to have substantial impact, they must be initiated more than a decade in advance of peaking" (USDOE, 2005).

HIRSCH REPORT

After careful analysis, the Hirsch Report came to a number of conclusions and three possible scenarios:

Conclusions

1. *World oil peaking is going to happen.* Some forecasters predict it will occur within a decade, others later; for example, Simmons (2006) predicted global peak oil production from 2007 to 2009, Deffeyes (2008) predicted global peaking before 2009, and Campbell and Laherrere (1998) predicted global peaking sometime between 2010 and 2020.
2. *Oil peaking could cost economies dearly,* particularly that of the United States.
3. *Oil peaking presents a unique challenge.* Previous transitions were gradual and evolutionary; oil peaking will be abrupt and revolutionary.
4. *The real problem is liquid fuels for transportation.* Motor vehicles, aircraft, trains, and ships have no readily available substitute.
5. *Mitigation efforts will require an intense effort over decades.*
6. *Both supply and demand will require attention.* Higher efficiency can reduce demand, but large amounts of substitute fuels must be produced.
7. *It is a matter of risk management.* Early mitigation will be less damaging than delayed mitigation.
8. *Government intervention will be required*; otherwise, the economic and social implications would be chaotic.
9. *Economic upheaval is not inevitable.* Without mitigation, peaking will cause major upheaval, but given enough lead time the problems are solvable.
10. *More information is needed.* Effective action requires better understanding of a number of issues.

Scenarios

1. Waiting until world oil production peaks before taking crash program action leaves the world with a significant liquid fuel deficit for more than two decades.
2. Initiating a mitigation crash program ten years before world oil peaking helps considerably but still leaves a liquid fuels shortfall roughly a decade after the time that oil would have peaked.
3. Initiating a mitigation crash program 20 years before peaking appears to offer the possibility of avoiding a worldwide liquid-fuels shortfall for the forecast period.

A MALTHUSIAN CATASTROPHE?

Obviously, the peaking of oil presents not only a number of questions but also a stark reality. For example, how will shortages and prices of oil affect you and your family? What about the effects on your community? How will oil shortages and prices affect industry and the economy? The stark reality is that the decline of oil production is not temporary. For those sitting back and waiting for some White Knight to come charging into the country on his great steed and to then suddenly sink his great platinum-clad spear into the ground to free a gigantic black stream of crude oil that shoots upward and forms a giant shaft reaching toward the clouds, it just ain't going to happen; literally, that is a pipedream of mythical proportions.

The day is fast approaching when the inability to meet demand will raise the costs of fuel. Concurrently, every sector of the economy will be affected in some way, whether it be a shortage of new materials for manufactured goods, such as those listed earlier, or increasing costs associated with transporting those goods to market, storing them in climate-controlled facilities, purchasing the machinery used in the construction of those facilities (or structures of any kind), or fertilizers and pesticides used to maintain production yields in agriculture, among other dependencies too numerous to list here. That is only the beginning of the impact that such a situation will have on our society, as the resulting domino effect will generate enormous price increases across the board for all products and services in existence, making a significant number of them economically impractical. And, while the lack of oil supplies will affect the manufacture or availability of all of these products and services, it is food that is likely to be affected the most. An adequate food supply is essential not only to maintaining world peace and preventing famine but also to ensuring the health and wellbeing of the populace—a populace that is still increasing in size and will likely continue to grow even though the liquid fuel spigots are running dry, for good. This deadly doom-and-gloom scenario has been described, in somewhat related terms and meaning, as a Malthusian catastrophe. For our purposes here, this term denotes a forced return to subsistence level conditions once energy use has outpaced energy production. The tragic irony is that if this occurs and we are not prepared, for the first time in history Third World subsistence economies will have the one-up on many of the rest of us who, in our lifetimes, have never planted, tended, or harvested any type of crop.

BLAME OR TAME?

Whenever anything goes wrong on this globe of ours, it is human tendency for someone, anyone, or all of us to point a long finger of blame at someone or anyone beside ourselves, of course. This finger pointing has certainly been seen with regard to energy shortages and energy-derived pollution problems. In particular, many critics point the finger of blame for our current and pending shortage of liquid fuels at the environmentalists; however, the U.S. Environmental Protection Agency (USEPA, 1974) suggested that a finger of blame pointed at the environmental movement is misdirected. Although it is true that environmental regulations have caused some increased energy demand and have restricted supply to some extent, it is important to recognize that other factors have been considerably more significant. As mentioned, these include rapidly escalating demand for energy, energy pricing policies, oil import quotas, lack of incentives to invest in domestic energy facilities, and depletion of domestic oil and gas reserves.

We all realize that the price of energy has increased. The summer of 2008 with its $4+ gas prices got a lot of people's attention. The higher prices for heating oil, natural gas, and other oil-derived products have also attracted our attention. The problem is we tend to blame the higher costs of petroleum-based products on those who sell the products; we think they are greedy, and that is about the extent of our concern. What we don't understand is that it is the policymakers who have failed to understand the concept of exponential growth and have failed to formulate and pass a long-term energy strategy that is sensible. The result of these failures, and others, is the creation of an escalating energy crisis as time continues to run out. Unfortunately, most people are either oblivious to this or believe that some new source of energy will magically appear—again, some White Knight will ride into town with the grand solution. We always need to blame someone or anyone for things we do not like; however, blame is never the solution to any problem. Instead, we need to get a hold of the issue and tame it. We need to think it through. Before doing that, however, we need to know the extent of the issue—the liquid fuel crisis.

WEATHERING THE STORM

With the energy crisis pending, we have three choices: We can mine for more oil, we can aggressively pursue energy conservation, or we can develop renewable energies. Because of limited fossil fuel resources, increased demand, and environmental impacts, renewable sources of energy are needed. Renewable energy is a form of energy capable of being constantly regenerated by natural processes at reasonable rates. Renewable energy technologies turn these fuels into usable forms of energy—most often electricity, but also heat, chemicals, or mechanical power. Some possibilities include the following:

- Biomass
- Wind energy
- Hydropower
- Tidal energy
- Wave energy
- Solar energy
- Geothermal power
- Hydrogen fuel cells

ALTERNATIVE AND RENEWABLE ENERGY[*]

The worldwide use of liquid fossil fuels and their decreasing availability, along with the politics involved and other economic forces, are pushing for substitute, alternative, and renewable fuel sources. This is the case, of course, because of the current and future economic problems that $4+/gal gasoline have generated (especially in the United States) and because of the perceived crisis regarding high carbon dioxide emissions, the major contributing factor of global climate change.

Before proceeding with an introductory discussion of alternative and renewable energy sources, it is important to make a clear distinction between the two terms, the current buzzwords. *Alternative energy* is an umbrella term that refers to any source of usable energy intended to replace fuel sources without the undesired consequences of the replaced fuels. The use of the term "alternative" presupposes the existence of undesirable types of energy production (for many people, fossil fuel has joined that endless list of four-letter words). Today, alternative energy is considered to be fueled energy that is an alternative to fossil fuels and that does not use up natural resources or harm the environment. Examples of "alternative" fuels throughout history, though, have included petroleum as an alternative to whale oil, coal as an alternative to wood, alcohol as an alternative to fossil fuels, and coal gasification as an alternative to petroleum. The key point in understanding the concept of alternative energy is that these fuels need not necessarily be renewable.

Renewable energy is energy generated from natural resources—sunlight, wind, water (hydro), ocean thermal, wave and tide action, biomass, geothermal heat—which are naturally replenished (and thus renewable). Renewable energy resources are virtually inexhaustible, as they are replenished at the same rate as they are used, but they are limited in the amount of energy that is available per unit time. If we have not come full circle in our cycling from renewable to nonrenewable back to renewable we are getting close to that. Consider, for example, that in 1850, about 90% of the energy consumed in the United States was from renewable energy resources (e.g., hydropower, wind, wood). Now, however, the United States is heavily reliant on nonrenewable fossil fuels: natural gas, oil, and coal. In 2014, about 7% of all energy consumed and about 8.5% of total electricity production was from renewable energy resources. Table 1.1 shows the current five primary forms of renewable energy: solar, wind, biomass, geothermal, and hydroelectric energy. Each of theses holds promise and poses challenges regarding future development.

Most renewable energy today is used for electricity generation, heat in industrial processes, heating and cooling buildings, and transportation fuels. In the United States, electricity producers (utilities, independent producers, and combined heat and power plants) consumed 22% of the world's total renewable energy in 2012 for producing electricity. Most of the rest of the remaining 78% was

---

[*] Some information in this section is from EIA, *Renewable Energy Trends*, 2004 ed., Energy Information Administration, Washington, DC, 2005 (http:www.eia.doe.gov/cneaf/solar.renewables/page/trends/rentrends04.html); EIA, *How Much Renewable Energy Do We Use?*, Energy Information Administration, Washington, DC, 2010 (http://tonto.eia.doe.gov/energy_in_brief/renewable_energy.cfm).

TABLE 1.1

U.S. Energy Consumption by Renewable Energy Source, 2014

Energy Source	Energy Consumption (quadrillion Btu)
Renewable total	9.656
Biomass (total)	4.812
Biofuels (ethanol and biodiesel)	2.067
Waste	0.516
Wood and wood-derived fuels	2.230
Geothermal	0.215
Hydroelectric power	2.475
Solar	0.421
Wind	1.733

Source: EIA, *Primary Energy Consumption by Source*, Table 10.1, Energy Information Administration, Washington, DC, 2016 (https://www.eia.gov/totalenergy/data/monthly/index.cfm#renewable).

biomass consumed for industrial applications (principally papermaking) by plants producing only heat and steam. Biomass is also used for transportation fuels (ethanol) and to provide residential and commercial space heating. In 2014, the distribution of U.S. renewable consumption by source was as follows (EIA, 2015):

- Hydropower, 26%
- Biofuels, 22%
- Biomass waste, 5%
- Biomass wood and wood-derived fuels, 23%
- Wind, 18%
- Geothermal, 2%
- Solar, 4%

The United States imports more than 50% of its oil. Replacing some of our petroleum with fuels provided from solar power or plant matter, for example, could save money and strengthen our energy security. Renewable energy is plentiful, and the technologies are improving all the time. There are many ways to use renewable energy. Our main focus should be on finding and developing a renewable source of liquid fuels (e.g., developed from biomass), because our economy runs on liquid fuels. Most of the nonliquid renewable energies will not (at the present time) provide power for airplanes and heavy trucks; that is, solar, wind, hydro, geothermal, wave, and tidal energy and

RENEWABLE ENERGY—A JEKYLL AND HYDE PERSONALITY?

Dr. Jekyll Side (Good)	Mr. Hyde Side (Bad)
Sustainable	Intermittent
Free fuel	Upfront costs
Inexhaustible	Distant from populations
Nonpolluting (to a point)	May involve environmental issues
Easily applied to off-grid/distributed power use	Large footprint

fuel cells cannot power the main transportation vehicles we use today. Note that we did not mention trains. Remember that today many trains are powered by diesel power or by diesel–electric systems. If necessary, trains could be retrofitted to steam power developed by burning coal and wood products (a step back into the past); however, we cannot do this for heavy trucks and airplanes. Most Americans still do not understand that we are running short of the fuels that we use every day, the fuels that made us what we are today. We built the world's greatest economy on oil that sold for $3 to $4 per barrel. That is no longer the case. To maintain our current level of living, the so-called good life, we must find and develop renewable energy sources to power and secure our future.

FUTURE OF RENEWABLE ENERGY AND EMPLOYMENT IN THE RENEWABLE ENERGY FIELD

Eventually the world will run out of easily obtainable fossil fuels. This will also be the case whenever the only fossil fuel left to be mined and actually available for use is coal. The United States and other countries are shifting from coal-fired or -burned energy production to natural gas or other energy sources because coal has become a target of environmentalists and others who want to prevent or lessen environmental damage from coal burning and mining processes. In 1330, King Edward I (Longshanks of Braveheart fame) decreed the death penalty for burning coal. This threat, however, did not come to fruition because there was no other alternative available. This has some relevance to a future time when no other energy source is available (including nuclear power) and renewables cannot provide the energy needed. Coal usage will experience a reverse Renaissance; that is, I believe there will be a bridge between fossil fuel and renewable energy back to coal-produced energy. This will be a Renaissance in reverse—that is, until we run out of coal.

No one doubts that the transition from fossil fuels to increased use of renewable energy sources is inevitable, but the road leading to this point is littered with potholes and sharp turns, including scams and inadequacies such as the unreliability of a source (i.e., turning on the power switch and not receiving energy when the water is not flowing, when the wind is not blowing, when the sun is not shining, and when other assorted problems and issues are present). With regard to scams related to renewable energy production, the record is crammed with examples whereby government loans for renewable energy companies were awarded to green companies that went bust. Some of the companies that went belly-up after receiving government funds are well known, but others are not; for example, most people have heard of the Solyndra fiasco where $535 million taxpayer dollars went down the drain. Other lesser known clean energy companies that did not or have yet to deliver (that we know of) include the following:

- Evergreen Solar
- SpectaWatt
- Beacon Power
- Eastern Energy
- Nevada Geothermal
- SunPower
- First Solar
- Babcock & Brown Wind Partners
- Enerl
- Amonix
- National Renewable Energy Lab
- Fisker Automotive
- Abound Solar
- Solar Trust
- A123 Systems

TABLE 1.2

Renewable Energy Job Openings Available on August 19, 2015, in the United States

Specialty	Number of Jobs Available	Specialty	Number of Jobs Available
Engineering	141	Accounting	38
Management	164	Finance	44
Sales	107	Quality control	16
Skilled labor/trades	94	Consultant	43
Information technology	39	General business	26
Marketing	34	Professional services	12
Manufacturing	27	Human resources	8
Design	25	Entry level	47
Construction	43	Purchasing	12
Business development	45	Customer services	32
Strategy/planning	56	Information technology professionals	39
Installation maintenance	30		

Note: As listed on www.careerbuilder.com. No listings with fewer than eight openings have been included, and a few vague titles are also not listed. On a personal note, several of my former students who were enrolled in a number of the environmental science, environmental health, and environmental engineering courses that I and other professors taught at Old Dominion University in Norfolk, Virginia, majored in these specific environmental areas to obtain jobs upon graduation. I have to say that, with few exceptions, over a 10-year time span (1999 to 2009) hundreds of these students either were gainfully employed in environmental job areas before graduation or obtained such positions upon graduation. To this day, I know of none of my former students who did not find employment in their chosen fields, but how many of these graduates are employed in renewable energy fields? At the present time, none that I know of. This is not to say that renewable energy jobs are not available; rather, they are available—sort of.

- William & Kelsey Solar Group
- Johnson Controls
- Schneider Electric

In addition to green company scams or their inability to deliver promised and paid-for renewable energy sources, another problem with the renewable energy field has to do with employment. As pointed out earlier, the often-stated promise that a shift to renewable energy production will produce tens of thousands of high-paying permanent jobs has proven not to be true. Table 1.2 serves as testimony to the paltry and discouraging renewable energy job creation statistics. This table shows that, for the number of job openings available on August 19, 2015 (1122 total), many of the types of jobs offered would require a 4-year degree in specific areas of study. Also, many of these jobs, such as consulting positions, are short term, temporary jobs.

Clearly, Table 1.2 provides information regarding renewable energy positions available in the United States for only one particular time frame, but it is also clear from Table 1.2 that "science" is low on the list, with only 24 positions listed. What kind of science do renewable energy employers request or require? Good question. Also note that engineering and various subfields of business are listed. This is just one list, of course, and it only lists industry positions and does not include research or teaching opportunities at the college or junior-college level. It is also important to point out that the positions listed in Table 1.2 could apply to just about any field. So, this logically leads us to ask where are the jobs specifically related to renewable energy work? For example, where are the jobs listed for wind farm managers and maintenance technicians; solar operator technicians; hydropower operators, managers, and technicians; biomass project managers, site managers, and technicians;

ocean power technicians; fuel cell designers; or geothermal specialty positions? Table 1.2 does not specifically list these jobs, as these job openings are usually advertised by the particular renewable energy specialty field employers themselves. The point is there are renewable energy jobs available and they are constantly being filled, but it is important to point out that at the present time there are not the tens of thousands renewable energy jobs available that many politicians (and others) promise.

The preceding information is not intended to discourage anyone from pursuing the study of the science of renewable energy or from pursuing employment in the field. On the contrary, the purpose is to discuss the facts, the real story, and to point out that the so-called gravy train is not always full of gravy nor is it fully fueled at the present time. There is a future in renewable energy, but the future will depend on how available and accessible fossil fuel sources are. Unfortunately, the true and ultimate push for renewable energy sources and associated workers will not occur until the lights go out or there is a lack of liquid or gaseous fuel for transportation. In the interim, solar, wind, geothermal, biomass, ocean, and fuel cell energy sources and supplies are being studied, designed, constructed, and operated on a limited scale; thus, only a limited number or renewable energy positions are available at the present time. However, the future need is real.

BOTTOM LINE: THE NEED FOR INNOVATION

Identifying and addressing the problems related to our dwindling supply of crude oil are the easy part of the fuel substitution equation; however, solving the equation is a different matter. What is needed to tackle this pending energy dilemma is good old-fashioned American innovative genius. History has shown that when humans put their minds and energies together to solve difficult problems, the problems eventually get solved. The solution to the energy crisis, in a word, is *innovation, innovation, innovation*! Consider the imaginative license we have taken in the hypothetical case study that follows that illustrates the kind of energy innovation we are talking about.

CASE STUDY: ULTRA-SUPER-SUPER SENSORS

Our select team of renewable energy experts has been given a $250 million grant to do whatever it takes to develop a super battery or fuel cell to power the cars of the future. Accompanying specifications simply state that this battery or fuel cell must be large capacity (large ampere-hour capacity) and can be continuously recharged via light striking supersensitive solar sensors. To charge any rechargeable battery, a certain amount of current or electron flow is necessary. The solar super sensors we have invented will produce substantial current flow. These super sensors are wired from the battery itself to at least four positions on the roof of our experimental automobile, where they can constantly be exposed to daylight—any level of daylight. Okay, we know what you're thinking. You're thinking that it gets dark at night, so how are you going to continuously recharge the car's battery, fuel cell, or whatever else we choose to call it? Good question, but one with a simple answer. Two additional wires and ultra-super-super light sensors (that we also invented) will run from the battery to the car's head lamps, one to each. Thus, during nighttime driving, the light from the head lamps will be sensed by the ultra-super-super sensors and sent back to the battery. Now, we know electrical engineers out there in La La Land are scratching their heads and wondering where did we got such an unworkable idea. We are not finished with our explanation yet … hold onto your brain cells. An amplification or booster device (similar to a transistor that amplifies a signal but more powerful) will be located within the wiring between the battery and the external and head lamp cell connections. We are not talking about perpetual motion here, but perpetual energy. Is this scenario possible? Anything, absolutely anything, is possible. We are Americans; we think we can do anything if we put our minds to it and history has shown this to be true. Now is the time.

Let's say that we are able to create a power-plant arrangement consisting of a powerful battery or fuel cell with a built-in recharge system like the one just described—what's next? Another good question. What's next is the final ingredient in any innovative processes: education, training, and on-the-job training. To come up with innovation and research and development grants, we must have personnel who are highly educated and trained. Unfortunately, the United States is losing (or has already lost, in some areas) our edge in higher education. The United States will not regain its preeminence in science, technology, engineering, and mathematics until we shift our focus, money, and talent to educating our youth in these critical subjects.

DISCUSSION AND REVIEW QUESTIONS

1. What is nonrenewable energy?
2. Explain the difference between alternative energy and renewable energy.
3. Is nuclear energy considered to be renewable energy?
4. What are the economic ramifications of running out of crude oil? Explain.
5. Will the crude oil crisis come about gradually or suddenly? Explain.
6. An economy based on renewable energy will help in our ongoing fight to reduce pollution. Explain.
7. Name 15 products produced by oil that we could get along without.
8. Summarize the Hirsch Report.
9. What is a Malthusian catastrophe?
10. Is the point made in the case study feasible? Explain.

REFERENCES AND RECOMMENDED READING

Bartlett, A.A. (2004). Thoughts on long-term energy supplies: scientists and the silent lie. *Physics Today*, 57(7): 53–57.

Brandt, A.R. (2007). Testing Hubbert. *Energy Policy*, 35(5): 3074–3088.

Bryce, R. (2013). Four numbers say wind and solar can't save climate. *Bloomberg View*, September 20, http://www.bloombergview.com/articles/2013-09-20/four-numbers-say-wind-and-solar-can-t-save-climate.

Campbell, C.J. (2004a). *The Essence of Oil and Gas Depletion*. Brentwood, U.K.: Multi-Science Publishing.

Campbell, C.J. (2004b). *The Coming Oil Crisis*. Brentwood, U.K.: Multi-Science Publishing.

Campbell, C.J. (2005). *Oil Crisis*. Brentwood, U.K.: Multi-Science Publishing.

Campbell, C.J. and Laherrere, J.H. (1998). The end of cheap oil. *Scientific American*, March, pp. 80–86.

Cambridge Energy Research Associates. (2006). CERA says peak oil theory is faulty. *Post Carbon Institute Energy Bulletin*, November 14 (http://www.energybulletin.net/node/22381).

CIA. (2008). *The World Factbook*. Washington, DC: Central Intelligence Agency (http://www.cia.gov/library/publications/the-world-factbook/geos/X.html).

Deffeyes, K.S. (2008). *Hubbert's Peak: The Impending World Oil Shortage*. Princeton, NJ: Princeton University Press.

Deffeyes, K.S. (2010). *Current Events: Join Us As We Watch the Crisis Unfolding*, http://www.princeton.edu/hubbert/current-events.html.

EIA. (2015). *Monthly Energy Review*. Washington, DC: Energy Information Administration (http://www.eia.gov/totalenergy/data/monthly/previous.cfm#2015).

Goodchild, P. (2007). *Peak Oil and Famine: Four Billion Deaths*, www.countercurrents.org/goodchild291007.htm.

Goodstein, D. (2005). *Out of Gas: The End of the Age of Oil*. New York: W.W. Norton.

Gwyn, R. (2004). *Demand for Oil Outstripping Supply*, http://www.commondreams.org/views04/0128-10.htm.

Hirsch, R.L. (2005). *The Mitigation of the Peaking of World Oil Production: Summary of an Analysis*. Uppsala, Sweden: Association for the Study of Peak Oil & Gas (http://www.peakoil.net/USDOE.html).

Hubbert, M.K. (1956). *Nuclear Energy and the Fossil Fuels Drilling and Production Practice*, http://www.hubbertpeak.com/hubbert/1956/1956.pdf.

Huber, P. (2005). *The Bottomless Well*. New York: Basic Books.

Koppelaar, R.H.E.M (2006). *World Oil Production & Peaking Outlook*. Amsterdam: Peak Oil Netherlands Foundation.

Leggett, J.K. (2005). *The Empty Tank: Oil, Gas, Hot Air, and the Coming Financial Catastrophe*. New York: Random House.

Pfeiffer, D.A. (2004). *Eating Fossil Fuels: Oil, Food, and the Coming Crises in Agriculture*. Gabriola Island, BC: New Society Publishers.

Rifkin, J. (2003). *The Hydrogen Economy: After Oil, Clean Energy from a Fuel-Cell-Driven Global Hydrogen Web*. New York: J.P. Tarcher.

Savinar, M. (2010). *Are We Running Out? I Thought There Was 40 Years of the Stuff Left*, http://www.lifeaftertheoilcrash.net.

Simmons, M. (2006). *Twilight in the Desert: The Coming Saudi Oil Shock and the World Economy*. New York: John Wiley & Sons.

Simon, J.I. (1998). *The Ultimate Resource*. Princeton, NJ: Princeton University Press.

Steiner, C. (2009). *$20 per Gallon: How the Inevitable Rise in the Price of Gasoline Will Change Our Lives for the Better*. New York: Grand Central Publishing.

Tertzakian, P. (2006). *A Thousand Barrels a Second*. New York: McGraw-Hill.

U.S. Census Bureau. (2008). *Total Midyear Population for the World: 1950–2050*, http://www.census.gov/compendia/statab/tables/09s1285.xls.

USDOE. (2001). *Renewable Energy: An Overview*. Washington, DC: U.S. Department of Energy.

USDOE. (2005). *Peaking of World Oil Production: Impacts, Mitigation, & Risk Management*. Washington, DC: U.S. Department of Energy.

USDOE. (2007). *Peak Oil—The Turning Point*. Washington, DC: U.S. Department of Energy.

USEPA. (1974). *EPA's Position on the Energy Crisis*. Washington, DC: U.S. Environmental Protection Agency (http://www.epa.gov/history/topics/energy/01.htm).

2 Energy Basics

Sunlight pours energy on the Earth, and the energy gets converted from one form to another, in an endless cycle of life, death, and renewal. Some of the sunlight got stored underground, which has provided us with a tremendous "savings account" of energy on which we can draw. Our civilization has developed a large thirst for the energy, as we've built billions and billions of machines large and small that all depend on fuel and electricity.

Hartmann (2004)

RENEWABLE ENERGY: PERSPECTIVE ON ENERGY

The motive force, the capacity to do the work behind the operation of just about anything and everything is energy. Energy is essential for most activities of modern society. Whether we use energy in the form of wood, fossil fuels and electricity, the goal is to make life comfortable and convenient—that is, to maintain the so-called "good life." We use electricity for our lights and fans, air-conditioner, water heater and room heaters, oven, microwave, washing machine, driers, cell phones, smartphones, computers and toasters. We use fossil fuel to run buses, trucks, trains, airplanes, ships and thus transportation uses a large percentage of all the energy used. Most people understand the importance of energy in their lives. But few understand the interface between energy and its usage and its mining or its discovery, invention, and/or innovative uses and the role of the renewable energy practitioners in all of this (heck, many environmental engineers do not understand the interface).

ENERGY

Energy can be defined in a number of ways. In the broad sense, energy means the capacity of something—a person, an animal, or a physical system (machine)—to do work and produce change. It can be used to describe someone doing energetic things such as running, talking, and acting in a lively and vigorous way. It is used in science to describe how much potential a physical system has to change. It also is used in economics to describe the part of the market where energy itself is harnessed and sold to consumers. For our purposes in this text, we simply define energy as something that can do work.

Two basic forms of energy are discussed in this text: kinetic and potential energy. Kinetic energy is energy at work or in motion; for example, a car in motion or a rotating shaft has kinetic energy. Potential energy is stored energy, like the energy stored in a coiled or stretched spring; an object stationed above a table has potential energy.

According to the law of conservation of energy, energy cannot be made or destroyed but can be made to change forms. Moreover, when energy changes from one form to another, the amount of energy stays the same. Let's consider an example of the law of conservation of energy: Measure the potential energy of something, then change the energy from potential (stored) to kinetic (moving) and back again. Measure the energy again. The energy measured initially will be the same as that measured at the end; it will always be the same. One caveat to this explanation is that we now know that matter can be converted into energy through processes such as nuclear fission and nuclear fusion. The law of conservation of energy, therefore, has been modified or amplified to become the law of conservation of matter and energy.

TYPES OF ENERGY

There are many types of energy; for example,

- Kinetic (motion) energy
- Water energy
- Potential energy
- Elastic energy
- Nuclear energy
- Chemical energy
- Sound energy
- Internal energy
- Heat/thermal energy
- Light (radiant) energy
- Electric energy

Energy sources can also be categorized as renewable or nonrenewable. When we use electricity in our homes, the electrical power was probably generated by burning coal, by a nuclear reaction, or by a hydroelectric plant at a dam. Therefore, coal, nuclear, and hydropower are called *energy sources*. When we fill up a gas tank, the source might be petroleum or ethanol made by growing and processing corn (EIA, 2009).

As just noted, energy sources can be divided into two groups—*renewable*, an energy source that can be easily replenished, and *nonrenewable*, an energy source that we are using up and cannot recreate (petroleum, for example, was formed millions of years ago from the remains of ancient sea plants and animals). In the United States, most of our energy comes from nonrenewable energy sources. Coal, petroleum, natural gas, propane, and uranium are nonrenewable energy sources. They are used to make electricity, to heat our homes, to move our cars, and to manufacture all kinds of products. Renewable and nonrenewable energy sources can be used to produce secondary energy sources, including electricity and hydrogen. Types of renewable energy sources include the following:

- Solar
- Hydro
- Wind
- Geothermal
- Ocean thermal energy conversion
- Tidal energy
- Hydrogen burning
- Biomass burning

ENERGY USE IN THE UNITED STATES

Use of energy in the United States can be broken down into four major sectors of the economy (EIA, 2013a):

- *Commercial* (19%)—Buildings such as offices, malls, stores, schools, hospitals, hotels, warehouses, restaurants, places of worship, and more
- *Industrial* (31%)—Facilities and equipment used for manufacturing, agriculture, mining, and construction
- *Residential* (22%)—Homes and apartments
- *Transportation* (28%)—Vehicles that transport people or goods, such as cars, trucks, buses, motorcycles, trains, subways, aircraft, boats, barges, and even hot air balloons

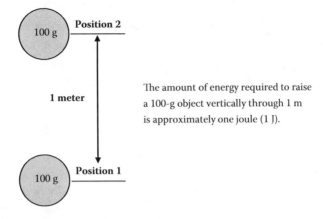

The amount of energy required to raise
a 100-g object vertically through 1 m
is approximately one joule (1 J).

FIGURE 2.1   The Joule (pronounced jewel).

MEASURING ENERGY

Energy can be measured; that is, the amount of energy a thing has can be given a number. As for
other kinds of measurements, there are measurement units. The units of measurement for measuring energy are used to make the numbers understandable and meaningful. The SI unit for both
energy and work is the joule (J) (as in "jewel," like a diamond or emerald) (Figure 2.1). It is named
after James Joule, who discovered that heat is a type of energy. One joule is equal to 1 newton meter.
In terms of SI base units, 1 J is equal to 1 kg·m^2·s^{-2}. The energy unit of measurement for electricity
is the kilowatt-hour (kWh). One kWh is equivalent to 3,600,000 J (3600 kJ or 3.6 MJ).

Energy is usually expressed in British thermal units (Btu). One Btu is the amount of heat energy
it takes to raise the temperature of 1 pound of water by 1°F at sea level:

1000 joules = 1 Btu
1000 joules = 1 kilojoule (kJ) = 1 Btu
1 therm = 100,000 Btu

Following are Btu conversion factors (weighted averages) (EIA, 2013a):

Electricity	1 kilowatt-hour (kWh) = 3412 Btu
Natural gas	1 cubic foot (ft^3) = 1028 Btu
	1 cubic foot (ft^3) = 0.01 therms
Motor gasoline	1 gallon = 124,000 Btu
Diesel fuel	1 gallon = 139,000 Btu
Heating oil	1 gallon = 139,000 Btu
Propane	1 gallon = 91,333 Btu
Wood	1 cord = 20,000,000 Btu

UNITS, STANDARDS OF MEASUREMENT, AND CONVERSIONS

A science book that does not include units and conversions is like a book without a table of contents
or index. For example, converting units from grams to slugs, centistokes to square feet per second,
or pounds per million gallons to milligrams per liter can be accomplished automatically, right away,
if you have a mind containing a library of facts and figures and conversions and data and all kinds
of that type of stuff. However, if you are normal, there are times when even the most adept, confi-
dent, competent, and brilliant engineer or engineer want-to-be must refer to some reference to find

TABLE 2.1		
Commonly Used Units		
Quantity	**SI Units**	**USCS Units**
Length	Meter	Foot (ft)
Mass	Kilogram	Pound (lb)
Temperature	Celsius	Fahrenheit (F)
Area	Square meter	Square foot (ft²)
Volume	Cubic meter	Cubic foot (ft³)
Energy	Kilojoule	British thermal unit (Btu)
Power	Watt	Btu/hr
Velocity	Meter/second	Mile/hour (mile/hr)

TABLE 2.2		
Common Prefixes		
Quantity	**Prefix**	**Symbol**
10^{-12}	Pico	p
10^{-9}	Nano	n
10^{-6}	Micro	µ
10^{-3}	Milli	m
10^{-2}	Centi	c
10^{-1}	Deci	d
10	Deca	da
10^{2}	Hecto	h
10^{3}	Kilo	k
10^{6}	Mega	M

some facts. I always hand out a printed copy of units and conversions in my college environmental classes. Most students welcome the handout; however, I have had a few who have shunned it at first, saying that they knew how to convert units and can do it from memory. I like to counter with, "Well, then, that means you know how convert abamperes to statamperes, right?" The look on their faces is always priceless. Students who do not know everything are my kind of students; they are teachable. Because not many of us are human computers with digital memories, this handbook includes a discussion of units, standards of measurement, and conversions. Moreover, because this is the age of energy consumption, renewable energy production, and hydraulic fracturing to capture more energy, energy conversion calculations are also included. By the way, does the reader know the difference, in gallons, between a U.S. fluid barrel and an oil barrel? Don't worry, if you read this chapter you will find out.

UNITS OF MEASUREMENT: THE BASICS

A basic knowledge of units of measurement and how to use them and convert them is essential. Environmental engineers should be familiar with both the U.S. Customary System (USCS), or English system, and the International System of Units (SI). Some of the important units are summarized in Table 2.1, which gives some basic SI and USCS units of measurement that will be encountered. In the study of environmental engineering math operations (and in actual practice), it is quite common to encounter both extremely large quantities and extremely small ones. The concentrations of some toxic substance may be measured in parts per million (ppm) or parts per billion (ppb), for example. To describe quantities that may take on such large or small values, it is useful to have a system of prefixes that accompany the units. Some of the more important prefixes are presented in Table 2.2.

Note: For comparative purposes, we like to say that 1 ppm is analogous to a full shotglass of water sitting in the bottom of a full standard-size swimming pool.

CONVERSION FACTORS

Sometimes we have to convert between different units. Suppose that a 60-inch piece of pipe is attached to an existing 6-foot piece of pipe. Joined together, how long are they? Obviously, we cannot find the answer to this question by adding 60 to 6, because the two lengths are given in different units. Before we can add the two lengths, we must convert one of them to the units of the other. Then, when we have two lengths in the same units, we can add them.

TABLE 2.3
SI Base Units

Quantity	Name	Symbol
Length	Meter	m
Mass	Kilogram	kg
Time	Second	s
Electric current	Ampere	A
Temperature	Kelvin	K
Amount of substance	Mole	mol
Luminous intensity	Candela	cd

To perform this conversion, we need a *conversion factor.* In this case, we have to know how many inches make up a foot: 12 inches. Knowing this, we can perform the calculation in two steps:

1. 60 in. is really $60 \div 12 = 5$ ft
2. 5 ft + 6 ft = 11 ft

From the example above, it can be seen that a conversion factor changes known quantities in one unit of measure to an equivalent quantity in another unit of measure. When making the conversion from one unit to another, we must know two things:

1. The exact number that relates the two units
2. Whether to multiply or divide by that number

Confusion over whether to multiply or divide is common; on the other hand, the number that relates the two units is usually known and thus is not a problem. Understanding the proper methodology—the "mechanics"—to use for various operations requires practice and common sense. Along with using the proper mechanics (and practice and common sense) to make conversions, probably the easiest and fastest method of converting units is to use a conversion table.

MKS System (SI Units)

The MKS system (so named because it uses the meter, kilogram, and second as base units) is comprised of SI units. All other units are derived from the base units, which are listed in Table 2.3. This is a fully consistent system; there is only one recognized unit for each variable (physical quantity).

CLEAN ENERGY PARAMETERS AND CALCULATIONS

Energy is an input fundamental to economic systems (Harris, 2006). Our current economic practice depends overwhelmingly on nonrenewable fossil fuels (90% of our energy supply), including oil, coal, and natural gas. As environmental professionals we are concerned not only with the cost of energy but also with the cost to the environment resulting from the use of nonrenewable energy supplies. Calculations related to the conversion of greenhouse gas emission numbers into different types of equivalent units and other pertinent calculations and conversions are discussed below.

> *Note:* With regard to global warming potentials (GWPs), some of the equivalencies in the calculator are reported as CO_2 equivalents (CO_2e). These are calculated using GWPs from the Intergovernmental Panel on Climate Change's Fourth Assessment report.

DID YOU KNOW?

An abampere, in electricity, is a centimeter-gram-second unit of electromagnetic current, equivalent to 10 amperes. A statampere is the electric unit of current equal to the current produced by an electromotive force of 1 statvolt acting through a resistance of 1 statohm. An abampere is multiplied by 2.99793×10^{10} to obtain a statampere.

ELECTRICITY REDUCTION (KILOWATT-HOURS)

The U.S. Environmental Protection Agency's Greenhouse Gas Equivalencies Calculator uses the Emissions & Generation Resource Integrated Database (eGRID) of U.S. annual non-baseload CO_2 output emission rates to convert reductions of kilowatt-hours into avoided units of carbon dioxide emissions. Most users of the Equivalencies Calculator who seek equivalencies for electricity-related emissions want to know equivalencies for emissions reductions due to energy efficiency or renewable energy programs. These programs are not generally assumed to affect baseload emissions (the emissions from power plants that run all the time), but rather non-baseload generation (power plants that are brought online as necessary to meet demand). For that reason, the Equivalencies Calculator uses a non-baseload emissions rate (USEPA, 2014).

EMISSION FACTOR

$$6.89551 \times 10^{-4} \text{ metric tons } CO_2/kWh$$

Note: This calculation does not include any greenhouse gases other than CO_2, and it does not include line losses. Individual subregion non-baseload emissions rates are also available on the eGRID website. To estimate indirect greenhouse gas emissions from electricity use, use Power Profiler or eGRID subregion annual output emission rates as the default emission factor.

GALLONS OF GASOLINE CONSUMED

To obtain the number of grams of carbon dioxide emitted per gallon of gasoline combusted, the heat content of the fuel per gallon is multiplied by the kilograms of CO_2 per heat content of the fuel. In the preamble to the joint USEPA, Department of Transportation, and National Highway Traffic Safety Administration rulemaking on May 7, 2010, which established the initial National Program fuel economy standards for model years 2012 to 2016, the agencies stated that they had agreed to use a common conversion factor of 8887 grams of CO_2 emissions per gallon of gasoline consumed (OFR, 2010). This value assumes that all the carbon in the gasoline is concerted to CO_2 (IPCC, 2006).

Calculation

8887 grams of CO_2 per gallon of gasoline = 8.887×10^{-3} metric tons CO_2 per gallon of gasoline.

PASSENGER VEHICLES PER YEAR

Passenger vehicles are defined as two-axle, four-tire vehicles, including passenger cars, vans, pickup trucks, and sport/utility vehicles. In 2011, the weighted average combined fuel economy of cars and light trucks was 21.4 miles per gallon (MPG) (FHWA, 2013). The average number of vehicle miles traveled in 2011 was 11,318 miles per year. In 2011, the ratio of carbon dioxide emissions to total greenhouse gas emissions (including carbon dioxide, methane, and nitrous oxide, all expressed as carbon dioxide equivalents) for passenger vehicles was 0.988 (USEPA, 2013a,b). The amount of carbon dioxide emitted per gallon of motor gasoline burned was 8.89×10^{-3} metric tons, as calculated above. To determine annual greenhouse gas emissions per passenger vehicle, the following methodology was used: The amount of vehicle miles traveled (VMT) was divided by average gas mileage to determine gallons of gasoline consumed per vehicle per year. The number of gallons of gasoline consumed was multiplied by carbon dioxide per gallon of gasoline to determine carbon dioxide emitted per vehicle per year. Carbon dioxide emissions were then divided by the ratio of carbon dioxide emissions to total vehicle greenhouse gas emissions to account for vehicle methane and nitrous oxide emissions.

Calculation

$(8.89 \times 10^{-3}$ metric tons CO_2 per gallon of gasoline$) \times (11,318$ $VMT_{car/truck\ average}) \times (1/21.4$ $MPG_{car/truck\ average}) \times [(1$ CO_2, CH_4, and $N_2O)/0.988$ $CO_2] = 4.75$ metric tons CO_2e per vehicle per year.

Miles Driven by the Average Passenger Vehicle per Year

Passenger vehicles are defined as two-axle, four-tire vehicles, including passenger cars, vans, pickup trucks, and sport/utility vehicles. In 2011, the weighted average combined fuel economy of cars and light trucks combined was 21.4 miles per gallon (FHWA, 2013). In 2011, the ratio of carbon dioxide emissions to total greenhouse gas emissions (including carbon dioxide, methane, and nitrous oxide, all expressed as carbon dioxide equivalents) for passenger vehicles was 0.988 (USEPA, 2013a,b). The amount of carbon dioxide emitted per gallon of motor gasoline burned is 8.89×10^{-3} metric tons, as calculated earlier. To determine annual greenhouse gas emission per mile, the following methodology was used: Carbon dioxide emissions per gallon of gasoline were divided by the average fuel economy of vehicles to determine carbon dioxide emitted per mile traveled by a typical passenger vehicle per year. Carbon dioxide emissions were then divided by the ratio of carbon dioxide emissions to total vehicle greenhouse gas emissions to account for vehicle methane and nitrous oxide emissions.

Calculation

$(8.89 \times 10^{-3}$ metric tons CO_2 per gallon of gasoline$) \times 1/21.4$ $MPG_{car/truck\ average} \times [(1$ CO_2, CH_4, and $N_2O)/0.988$ $CO_2] = 4.20 \times 10^{-4}$ metric tons CO_2e per mile.

Therms of Natural Gas

Carbon dioxide emissions per therm are determined by multiplying the heat content times the carbon coefficient times the fraction oxidized times the ratio of the molecular weight of carbon dioxide to that of carbon (C) (44/12). The average heat content of natural gas is 0.1 mmBtu per therm, and the average carbon coefficient of natural gas is 14.47 kg carbon per mmBtu (USEPA, 2013c). The fraction oxidized to CO_2 is 100% (IPCC, 2006).

Note: When using this equivalency, please keep in mind that it represents the CO_2 equivalency for natural gas burned as a fuel, not natural gas released to the atmosphere. Direct methane emissions released to the atmosphere (without burning) are about 21 times more powerful than CO_2 in terms of their warming effect on the atmosphere.

Calculation

$(0.1$ mmBtu/1 therm$) \times (14.46$ kg C per mmBtu$) \times (44$ g CO_2 per 12 g C$) \times (1$ metric ton/1000 kg$) = 0.005302$ metric tons CO_2 per therm.

Barrels of Oil Consumed

Carbon dioxide emissions per barrel of crude oil are determined by multiplying the heat content times the carbon coefficient times the fraction oxidized times the ratio of the molecular weight of carbon dioxide to that of carbon (44/12). The average heat content of crude oil is 5.80 mmBtu per barrel, and the average carbon coefficient of crude oil is 20.31 kg carbon per mmBtu (USEPA, 2013c). The fraction oxidized to CO_2 is 100% (IPCC, 2006).

Calculation

(5.80 mmBtu/barrel) × (20.31 kg C per mmBtu) × (44 g CO_2 per 12 g C) × (1 metric ton/1000 kg) = 0.43 metric tons CO_2 per barrel.

Tanker Trucks Filled with Gasoline

The amount of carbon dioxide emitted per gallon of motor gasoline burned is 8.89×10^{-3} metric tons, as calculated earlier. A barrel equals 42 gallons, and a typical gasoline tanker truck contains 8500 gallons (IPCC, 2006; OFR, 2010).

Calculation

(8.89×10^{-3} metric tons CO_2 per gallon) × (8500 gal per tanker truck) = 75.54 metric tons CO_2 per tanker truck.

Number of Incandescent Bulbs Switched to Compact Fluorescent Bulbs

A 13-watt compact fluorescent light (CFL) bulb produces the same light output as a 60-watt incandescent light bulb. Annual energy consumed by a light bulb is calculated by multiplying the power (60 watts) by the average daily use (3 hours/day) by the number of days per year (365). Assuming an average daily use of 3 hours per day, an incandescent bulb consumes 65.7 kWh per year, and a compact fluorescent light bulb consumes 14.2 kWh per year (USEPA, 2013d). Annual energy savings from replacing an incandescent light bulb with an equivalent compact fluorescent bulb are calculated by subtracting the annual energy consumption of the compact fluorescent light bulb (14.2 kWh) from the annual energy consumption of the incandescent bulb (65.7 kWh). Carbon dioxide emissions reduced per light bulb switched from an incandescent bulb to a compact fluorescent bulb are calculated by multiplying the annual energy savings by the national average non-baseload carbon dioxide output rate for delivered electricity. The national average non-baseload carbon dioxide output rate for generated electricity in 2010 was 1519.6 lb CO_2 per megawatt-hour (USEPA, 2014), which translates to about 1637.5 lb CO_2 per megawatt-hour for delivered electricity (assuming transmission and distribution losses at 7.2%) (EIA, 2013b; USEPA, 2014).

Calculation

47 watts × 3 hr/day × 365 days/year × 1 kWh/1000 Wh = 51.5 kWh/year/bulb replaced.
51.5 kWh/bulb/year × 1637.5 lb CO_2 per MWh delivered electricity × 1 MWh/1000 kWh × 1 metric ton/2204.6 lb = 3.82×10^{-2} metric tons CO_2 per bulb replaced.

Fluorescent Lamps and Ballasts

The Energy Policy Act of 1992, Executive Order 13123, and the Federal Acquisition Regulation, Part 23, Section 704 (48 CFR 23.704), instituted guidelines for federal agencies to purchase energy-efficient products. Lighting accounts for 20 to 25% of the United States' electricity consumption. Retrofitting with automatic controls and energy-efficient fluorescent lamps and ballasts yields paybacks within 2 to 5 years; however, the best reason for retrofitting an old lighting system—increasing the productivity of workers—is often overlooked.

Fluorescent Lighting Nomenclature

The pattern for interpreting fluorescent lamp names is FWWCCTDD, where

> F = Fluorescent lamp
> WW = Nominal power in watts (4, 5, 8, 12, 15, 20, 33, ...)
> CC = Color (W, white; CW, cool white; WW, warm white; ...)
> T = Tubular bulb.
> DD = Diameter of the tube in eighths of an inch (T8 bulb has a diameter of 1 inch, T12 bulb has a diameter of 1.5 inches, ...).

Thus, an F40T12 lamp, for example, is a 40-watt fluorescent lamp with a 1.5-inch tubular bulb.

Background on Costs

With electricity costing 8 cents per kilowatt hour, a typical 40-watt T12 fluorescent lamp will use $64 worth of electricity over its lifetime. The purchase price of the bulb ($2) accounts for just 3% of the life-cycle costs of owning and operating a lighting system, and energy accounts for 86% of the cost. Thus, the operating cost breakdown for F40T12 fluorescent lamps is energy at 86%, maintenance at 11%, and the lamp itself at 3%. These calculations readily justify the cost of more expensive lamps that produce better quality light, save energy, and increase productivity. The effect of lighting on human performance and productivity is complex. Direct effects of poor lighting include the inability to resolve detail, fatigue, and headaches. Lighting may indirectly affect someone's mood or hormonal balance.

Note: The Hawthorne effect is a phenomenon whereby workers improve or modify an aspect of their behavior in response to a change in their lighting environment.

A small change in human performance dwarfs all costs associated with lighting. The typical annual costs of 1 square foot of office space are as follows:

Heating and cooling	$2
Lighting	$0.50
Floor space	$100
Employee salary and benefits	$400

Cutting lighting consumption in half saves about 25¢ per square foot each year. A 1% increase in human productivity would save $4 per square foot each year. Costs will vary from facility to facility, but the relative magnitudes of these costs are not likely to change. The focus needs to be on providing quality lighting to meet occupants' needs; however, it is possible to improve lighting quality while reducing energy costs thanks to improvements in lighting technology.

Best Types of Fluorescent Lamps and Ballasts

The "warmness" of a light is determined by its color temperature, expressed in degrees kelvin (K); the word kelvin is not capitalized. The kelvin scale is an absolute, thermodynamic temperature scale using as its null point absolute zero, the temperature at which all normal thermal motion ceases in the classical description of thermodynamics. The kelvin is defined as the fraction of 1/273.16 of the thermodynamic temperature of the triple point of water (exactly 0.01°C or 32.0018°F). In other words, it is defined such that the triple point of water is exactly 273.16 K.

The higher the correlated color temperature, the cooler the light. Offices should use intermediate or neutral light. This light creates a friendly, yet businesslike environment. Neutral light sources have a correlated color temperature of 3500 K. The color rendering index measures the quality of light. The higher the color rendering index, the better people see for a given amount of light.

Currently available 4-foot fluorescent lamps have indexes of 70 to 98. Lamps with different correlated color temperatures and color rendering indexes should not be used in the same space. The correlated color temperature and color rendering index should be specified when ordering lamps.

The best lighting system for each operating dollar is realized with T8 fluorescent lamps that have a color rendering index of 80 or higher. Compared to standard T12 fluorescent lamps, T8 lamps have better balance between the surface area containing the phosphors that fluoresce and the arc stream that excites them. This means that T8 lamps produce more light for a given amount of energy. In Europe, T5 lamps are popular. They are more efficient than T8 lamps but cost more than twice as much. The availability of T5 lamps and fixtures is limited in the United States. T8 lamps are currently preferred.

A quick comparison of light output shows how important it is to specify ballast factor and whether the ballast is electronic or magnetic. Electronic ballasts last twice as long as magnetic ballasts, use less energy, have a lower life-cycle cost, and operate the lamp at much higher frequencies. Operating fluorescent lamps at higher frequencies improves their efficiency and eliminates the characteristic 60-cycle buzz and strobe-lighting effect associated with fluorescent lights. The 60-cycle strobe-lighting effect may cause eye fatigue and headaches. Electronic ballasts are especially desirable in shops with rotating equipment. The 60-cycle strobe-lighting effect produced by magnetic ballasts can cause rotating equipment to appear stationary. All new buildings and retrofits should use electronic ballasts.

Fluorescent Lamps and Ballast Life

Most fluorescent lamps have a rated life of 12,000 to 20,000 hours. The rated life is the time it takes for half of the bulbs to fail when they are cycled on for 3 hours and off for 20 minutes. Cycling fluorescent lamps off and on will reduce lamp life. On the other hand, turning a lamp off when it is not needed will reduce its operating hours and increase its useful life. Electricity—not lamps—accounts for the largest percentage of the operating cost of a lighting system. It is economical to turn off fluorescent lights if they are not being used. According to the Certified Ballast Manufacturers Association, the average magnetic ballast lasts about 75,000 hours, or 12 to 15 years, with normal use. The optimum economic life of a fluorescent lighting system with magnetic ballasts is usually about 15 years. At this point, dirt on reflectors and lenses has significantly reduced light output. Other factors may make it desirable to retrofit a lighting system before the end of the 12- to 15-year life cycle. Those factors include increased productivity, utility rebates, and high energy costs.

Economic Analysis

When considering the benefits of retrofitting, more lamps per existing fixture yield more energy savings per fixture and a better payback. Higher than average energy or demand of the initial installation costs or a utility rebate will also produce a faster payback. Ballast factor can be used to adjust light levels. A high ballast factor increases lumens (a measure of light output), allowing fewer lamps to provide the same amount of light. For example, when electronic ballasts with a high ballast factor are used, two-lamp fixtures will produce as much light as three-lamp fixtures. This reduces the cost of the fixtures and improves the payback. An economic analysis of retrofitting three-lamp fixtures and magnetic ballasts with two-lamp fixtures with a high-ballast-factor electronic ballast yields a payback of slightly more than 2 years. With regard to fluorescent lamp retrofit payback, a simple payback (SPB) is the time, in years, it will take for the savings (in present value) to equal the cost of the initial installation (in present value). The following calculations do not account for interest rates.

■ EXAMPLE 2.1

Problem: Compute the SPB using the following formula:

$$\text{SPB} = (\text{Cost of installed equipment} - \text{Deferred maintenance} - \text{Rebates})$$
$$\div (\text{Total energy dollar savings per year})$$

The costs to replace a T12 lamp magnetic ballast system with a T8 lamp electronic ballast system are as follows:

- New fixtures (including fixture, two T8 lamps, and electronic ballast) cost $30 per fixture.
- Installation cost is $10 per fixture.
- Deferred cost of cleaning existing fixtures is $5 per fixture.
- The power company offers a one-time $8 per fixture rebate for replacing magnetic-ballasted T12 lamps with electronic-ballasted T8 lamps.

Solution:

- Total project cost for 100 fixtures = ($30 + $10 − $5 − $8) × (100 fixtures) = $2700.
- Total energy dollar savings per year = Lighting energy savings + Cooling savings − Heating costs = $1459 + $120 − $262 = $1317 per year.
- SPB = $2700/($1317 per year) = 2.05 years.

It is obvious that retrofitting an existing lighting system that uses F40T12 lamps and magnetic ballasts with F32T8 lamps and electronic ballasts can provide a very attractive payback.

HOME ENERGY USE

In 2012, there were 113.93 million homes in the United States. On average, each home consumed 12,069 kWh of delivered electricity. Nationwide household consumption of natural gas, liquefied petroleum gas, and fuel oil totaled 4.26, 0.51, and 0.51 quadrillion Btu, respectively, in 2012 (EIA, 2013c). Averaged across households in the United States, this amounts to 52,372 cubic feet of natural gas, 70 barrels of liquefied petroleum gas, and 47 barrels of fuel oil per home. The national average carbon dioxide output rate for generated electricity in 2010 was 1232 lb CO_2 per megawatt-hour (USEPA, 2014), which translates to about 1328.0 lb CO_2 per megawatt-hour for delivered electricity (assuming transmission and distribution losses at 7.2%) (EIA, 2013c; USEPA, 2014). The average carbon dioxide coefficient of natural gas is 0.0544 kg CO_2 per cubic foot. The fraction oxidized to CO_2 is 100% (IPCC, 2006). The average carbon dioxide coefficient of distillate fuel oil is 429.61 kg CO_2 per 42-gallon barrel (USEPA, 2013c). The fraction oxidized to CO_2 is 100% (IPCC, 2006). The average carbon dioxide coefficient of liquefied petroleum gases is 219.3 kg CO_2 per 42-gallon barrel (USEPA, 2011). The fraction oxidized to CO_2 is 100% (IPCC, 2006). Total single-family home electricity, natural gas, distillate fuel oil, and liquefied petroleum gas consumption figures were converted from their various units to metric tons of CO_2 and added together to obtain total CO_2 emissions per home.

Calculation

1. *Electricity*—12,069 kWh per home × (1232.4 lb CO_2 per MWh generated) × 1/(1-0.072) MWh delivered/MWh generated × 1 MWh/1000 kWh × 1 metric ton/2204.6 lb = 7.270 metric tons CO_2 per home.
2. *Natural gas*—52,372 cubic feet per home × 0.0544 kg CO_2 per cubic foot × 1/1000 kg/metric ton = 2.85 metric tons CO_2 per home.
3. *Liquid petroleum gas*—70.4 gallons per home × 1/42 barrels/gallon × 219.3 kg CO_2 per barrel × 1/1000 kg/metric ton = 0.37 metric tons CO_2 per home.
4. *Fuel oil*—47 gallons per home × 1/42 barrels/gallon × 429.61 kg CO_2 per barrel × 1/1000 kg/metric ton = 0.48 metric tons CO_2 per home.

Total CO_2 emissions for energy use per home is equal to 7.270 metric tons CO_2 for electricity + 2.85 metric tons CO_2 for natural gas + 0.37 metric tons CO_2 for liquid petroleum gas + 0.48 metric tons CO_2 for fuel oil = 10.97 metric tons CO_2 per home per year.

Number of Tree Seedlings Grown for Ten Years

A medium-growth coniferous tree, planted in an urban setting and allowed to grow for 10 years, sequesters 23.2 lb of carbon. This estimate is based on the following assumptions:

- Medium-growth coniferous trees are raised in a nursery for 1 year until they become 1 inch in diameter at 4.5 feet above the ground (the size of tree purchased in a 15-gallon container).
- The nursery-grown trees are then planted in a suburban/urban setting; the trees are not densely planted.
- The calculation takes into account "survival factors" developed by the U.S. Department of Energy (USDOE, 1998). For example, after 5 years (1 year in the nursery and 4 in the urban setting), the probability of survival is 68%; after 10 years, the probability declines to 59%. For each year, the sequestration rate (in pounds per tree) is multiplied by the survival factor to yield a probability-weighted sequestration rate. These values are summed over the 10-year period, beginning from the time of planting, to derive the estimate of 23.2 lb of carbon per tree.

Please note the following caveats to these assumptions:

- Although most trees take 1 year in a nursery to reach the seedling stage, trees grown under different conditions and trees of certain species may take longer—up to 6 years.
- Average survival rates in urban areas are based on broad assumptions, and the rates will vary significantly depending on site conditions.
- Carbon sequestration depends on growth rate, which varies by location and other conditions.
- This method estimates only direct sequestration of carbon and does not include the energy savings that result from buildings being shaded by urban tree cover.

To convert to units of metric tons CO_2 per tree, multiply by the ratio of the molecular weight of carbon dioxide to that of carbon (44/12) and the ratio of metric tons per pound (1/2204.6).

Calculation

(23.2 lb C per tree) × (44 units CO_2 ÷ 12 units C) × (1 metric ton ÷ 2204.6 lb) = 0.039 metric ton CO_2 per urban tree planted.

Acres of U.S. Forests Storing Carbon for One Year

Growing forests accumulate and store carbon. Through the process of photosynthesis, trees remove CO_2 from the atmosphere and store it as cellulose, lignin, and other compounds. The rate of accumulation is equal to growth minus removals (i.e., harvest for the production of paper and wood) minus decomposition. In most U.S. forests, growth exceeds removals and decomposition, so the amount of carbon stored nationally is increasing overall.

Data for U.S. Forests

The *Inventory of U.S. Greenhouse Gas Emissions and Sinks: 1990–2010* (USEPA, 2012) provides data on the net greenhouse gas flux resulting from the use and changes in forest land areas. Note that the term *flux* is used here to encompass both emission of greenhouse gases to the atmosphere and removal of carbon from the atmosphere. Removal of carbon from the atmosphere is also referred to as *carbon sequestration*. Forest land in the United States includes land that is at least 10% stocked with trees of any size, or, in the case of stands dominated by certain western woodland species for which stocking parameters are not available, at least 5% crown cover by trees of any size. Timberland is defined as unreserved productive forest land producing or capable of producing crops of industrial wood. Productivity is at a minimum rate of 20 ft^3 of industrial wood per acre per year. The remaining portion of forest land is classified as "reserved forest land," which is forest withdrawn from timber use by statute or regulation, or "other forest land," which includes forests on which timber is growing at a rate less than 20 ft^3 per acre per year (Smith et al., 2010).

Calculation

Annual net change in carbon stocks per year in year n = (Carbon stocks$_{(t+1)}$ − Carbon stocks$_t$) ÷ Area of land remaining in the same land-use category.

1. Determine the carbon stock change between years by subtracting carbon stocks in year t from carbon stocks in year $(t + 1)$. (This includes carbon stocks in the above-ground biomass, below-ground biomass, dead wood, litter, and soil organic carbon pools.)
2. Determine the annual net change in carbon stocks (i.e., sequestration) per area by dividing the carbon stock change in U.S. forests from step 1 by the total area of U.S. forests remaining in forests in year $(n + 1)$ (i.e., the area of land that did not change land-use categories between the time periods).

Applying these calculations to data developed by the USDA Forest Service for the *Inventory of U.S. Greenhouse Gas Emissions and Sinks* yields a result of 150 metric tons of carbon per hectare (or 61 metric tons of carbon per acre) for the carbon stock density of U.S. forests in 2010, with an annual net change in carbon stock per area in 2010 of 0.82 metric tons of carbon sequestered per hectare per year (or 0.33 metric tons of carbon sequestered per acre per year). These values include carbon in the five forest pools of above-ground biomass, below-ground biomass, deadwood, litter, and soil organic carbon, and they are based on state-level Forest Inventory and Analysis (FIA) data. Forest carbon stocks and carbon stock change are based on the stock difference methodology and algorithms described by Smith et al. (2010).

CONVERSION FACTOR FOR CARBON SEQUESTERED ANNUALLY BY ONE ACRE OF AVERAGE U.S. FOREST

The following calculation is an estimate for "average" U.S. forests in 2010 (i.e., for U.S. forests as a whole in 2010). Significant geographical variations underlie the national estimates, and the values calculated here might not be representative of individual regions or states. To estimate carbon sequestered for additional acres in one year, simply multiply the number of acres by 1.22 metric tons CO_2 per acre per year. From 2000 to 2010, the average annual sequestration per area was 0.73 metric tons carbon per hectare per year (or 0.30 metric tons carbon per acre per year) in the United States, with a minimum value of 0.36 metric tons carbon per hectare per year (or 0.15 metric tons carbon per acre per year) in 2000, and a maximum value of 0.83 metric tons carbon per hectare per year (or 0.34 metric tons carbon per acre per year) in 2006.

Calculation

(−0.33 metric ton C per acre per year) × (44 units CO_2 ÷ 12 units C) = −1.22 metric tons CO_2 sequestered annually by 1 acre of average U.S. forest.

ACRES OF U.S. FOREST PRESERVED FROM CONVERSION TO CROPLANDS

The carbon stock density of U.S. forests in 2010 was 150 metric tons of carbon per hectare (or 61 metric tons of carbon per acre) (USEPA, 2012). This estimate is composed of the five carbon pools of above-ground biomass (52 metric tons carbon per hectare), below-ground biomass (10 metric tons carbon per hectare), dead wood (9 metric tons carbon per hectare), litter (17 metric tons carbon per hectare), and soil organic carbon (62 metric tons carbon per hectare).

The *Inventory of U.S. Greenhouse Gas Emissions and Sinks* estimates soil carbon stock changes using U.S.-specific equations and data from the USDA Natural Resource Inventory and the CENTURY biogeochemical model (USEPA, 2012). When calculating carbon stock changes in biomass due to conversion from forestland to cropland, the IPCC (2006) guidelines indicate that the average carbon stock change is equal to the carbon stock change due to removal of biomass from the outgoing land use (i.e., forestland) plus the carbon stocks from one year of growth in the incoming land use (i.e., cropland), or the carbon in biomass immediately after the conversion minus the carbon in biomass prior to the conversion plus the carbon stocks from one year of growth in the incoming land use (i.e., cropland). The carbon stock in annual cropland biomass after 1 year is 5 metric tons carbon per hectare, and the carbon content of dry above-ground biomass is 45% (IPCC, 2006). Therefore, the carbon stock in cropland after 1 year of growth is estimated to be 2.25 metric tons carbon per hectare (or 0.91 metric tons carbon per acre).

The averaged reference soil carbon stock (for high-activity clay, low-activity clay, and sandy soils for all climate regions in the United States) is 40.83 metric tons carbon per hectare (USEPA, 2012). Carbon stock change in soils is time dependent, with a default time period for transition between equilibrium soil organ carbon values of 20 years for mineral soils in cropland systems (IPCC, 2006). Consequently, it is assumed that the change in equilibrium mineral soil organic carbon will be annualized over 20 years to represent the annual flux. The IPCC (2006) guidelines indicate that there are insufficient data to provide a default approach or parameters to estimate carbon stock change from dead organic matter pools or below-ground carbon stocks in perennial cropland.

Calculation

- *Annual change in biomass carbon stocks on land converted to other land-use category:*

$$\Delta C_B = \Delta C_G + C_{Conversion} - \Delta C_L$$

where

ΔC_B = Annual change in carbon stocks in biomass due to growth on land converted to another land-use category.

ΔC_G = Annual increase in carbon stocks in biomass due to growth on land converted to another land-use category (i.e., 2.25 metric tons C per hectare).

$C_{Conversion}$ = Initial change in carbon stocks in biomass on land converted to another land-use category; the sum of the carbon stocks in above-ground, below-ground, deadwood, and litter biomass (–88.47 metric tons C per hectare). Immediately after conversion from forestland to cropland, biomass is assumed to be zero, as the land is cleared of all vegetation before planting crops.

ΔC_L = Annual decrease in biomass stocks due to losses from harvesting, fuel wood gathering, and disturbances on land converted to other land-use categories (assumed to be zero).

Thus, $\Delta C_B = \Delta C_G + C_{Conversion} - \Delta C_L = -86.22$ metric tons carbon per hectare per year of biomass carbon stocks are lost when forestland is converted to cropland.

- *Annual change in organic carbon stocks in mineral soils:*

$$\Delta C_{Mineral} = (SOC_O - SOC_{(O-T)}) \div D$$

where

$\Delta C_{Mineral}$ = Annual change in carbon stocks in mineral soils.

SOC_O = Soil organic carbon stock in last year of inventory time period (i.e., 40.83 mt C per hectare).

$SOC_{(O-T)}$ = Solid organic carbon stock at beginning of inventory time period (i.e., 62 mt C per hectare).

D = Time dependence of stock change factors which is the default time period for transition between equilibrium SOC values (i.e., 20 years for cropland systems).

Therefore, $\Delta C_{Mineral} = (SOC_O - SOC_{(O-T)}) \div D = (40.83 - 62) \div 20 = -1.06$ metric tons carbon per hectare per year of soil organic carbon are lost.

Consequently, the change in carbon density from converting forestland to cropland would be −86.22 metric tons of carbon per hectare per year of biomass plus −1.06 metric tons carbon per hectare per year of soil organic carbon, or a total loss of 87.28 metric tons carbon per hectare per year (or −35.32 metric tons carbon per acre per year).

To convert to carbon dioxide, multiply by the ratio of the molecular weight of carbon dioxide to that of carbon (44/12), to yield a value of −320.01 metric tons CO_2 per hectare per year (or −129.51 metric tons CO_2 per acre per year).

$$(-35.32 \text{ metric tons C per acre per year}) \times (44 \text{ units } CO_2 \div 12 \text{ units C})$$
$$= -129.51 \text{ metric tons } CO_2 \text{ per acre per year}$$

To estimate the amount of carbon dioxide not emitted when an acre of forest is preserved from conversion to cropland, simply multiply the number of acres of forest not converted by −129.51 mt CO_2e per acre per year. Note that this calculation method assumes that all of the forest biomass is oxidized during clearing; that is, none of the burned biomass remains as charcoal or ash. Also note that this estimate only includes mineral soil carbon stocks, as most forests in the contiguous United States are growing on mineral soils. In the case of mineral soil forests, soil carbon stocks could be replenished or even increased, depending on the starting stocks, how the agricultural lands are managed, and the time frame over which the lands are managed.

PROPANE CYLINDERS USED FOR HOME BARBECUES

Propane is 81.7% carbon, and the fraction oxidized is 100% (IPCC, 2006; USEPA, 2013c). Carbon dioxide emissions per pound of propane were determined by multiplying the weight of propane in a cylinder times the carbon content percentage times the fraction oxidized times the ratio of the molecular weight of carbon dioxide to that of carbon (44/12). Propane cylinders vary with respect to size; for the purpose of this equivalency calculation, a typical cylinder for home use was assumed to contain 18 pounds of propane.

Calculation

(18 lb propane per cylinder) × (0.817 lb C per lb propane) × (0.4536 kg/lb) × (44 kg CO_2 per 12 kg C) × (1 metric ton/1000 kg) = 0.024 metric tons CO_2 per cylinder.

RAILCARS OF COAL BURNED

The average heat content of coal consumed in the United States in 2013 was 21.48 mmBtu per metric ton (EIA, 2014). The average carbon coefficient of coal combusted for electricity generation in 2012 was 26.05 kg carbon per mmBtu (USEPA, 2013c). The fraction oxidized is 100% (IPCC, 2006). Carbon dioxide emissions per ton of coal were determined by multiplying the heat content times the carbon coefficient times the fraction oxidized times the ratio of the molecular weight of carbon dioxide to that of carbon (44/12). The amount of coal in an average railcar was assumed to be 100.19 short tons, or 90.89 metric tons (Hancock and Sreekanth, 2001).

Calculation

(21.48 mmBtu/metric ton coal) × (26.05 kg C per mmBtu) × (44 g CO_2 per 12 g C) × (90.89 metric tons coal per railcar) × (1 metric ton/1000 kg) = 186.50 metric tons CO_2 per railcar.

POUNDS OF COAL BURNED

The average heat content of coal consumed in the United States in 2013 was 21.48 mmBtu per metric ton (EIA, 2014). The average carbon coefficient of coal combusted for electricity generation in 2012 was 26.05 kg carbon per mmBtu (USEPA, 2013c). The fraction oxidized is 100% (IPCC, 2006). Carbon dioxide emissions per pound of coal were determined by multiplying the heat content times the carbon coefficient times that fraction oxidized times the ratio of the molecular weight of carbon dioxide to that of carbon (44/12).

Calculations

(21.48 mmBtu/metric ton coal) × (26.05 kg C per mmBtu) × (44 g CO_2 per 12g C) × (1 metric ton coal per 2204.6 lb coal) × (1 metric ton/1000 kg) = 9.31×10^{-4} metric tons CO_2 per pound of coal.

TONS OF WASTE RECYCLED INSTEAD OF LANDFILLED

To develop the conversion factor for recycling rather than landfilling waste, emission factors from the USEPA's Waste Reduction Model (WARM) were used (USEPA, 2013e). These emission factors were developed following a life-cycle assessment methodology using estimation techniques developed for national inventories of greenhouse gas emissions. According to WARM, the net emission reduction from recycling mixed recyclables (e.g., paper, metals, plastics), compared with a baseline in which the materials are landfilled, is 0.73 metric tons of carbon equivalent per short ton. This factor was then converted to metric tons of carbon dioxide equivalent by multiplying by 44/12, the molecular weight ratio of carbon dioxide to carbon.

Calculation

(0.76 metric tons of carbon equivalent per ton) × (44 g CO_2 per 12 g C) = 2.79 metric tons CO_2 equivalent per ton of waste recycled instead of landfilled.

GARBAGE TRUCKS OF WASTE RECYCLED INSTEAD OF LANDFILLED

The carbon dioxide equivalent emissions avoided from recycling instead of landfilling 1 ton of waste are 2.67 metric tons CO_2e per ton, as calculated in the previous section. Carbon dioxide emissions reduced per garbage truck full of waste were determined by multiplying emissions avoided from recycling instead of landfilling 1 ton of waste by the amount of waste in an average garbage truck. The amount of waste in an average garbage truck was assumed to be 7 tons (USEPA, 2002, 2013).

Calculation

(2.79 metric tons CO_2e/ton of waste recycled instead of landfilled) × (7 tons/garbage truck) = 19.51 metric tons CO_2e per garbage truck of waste recycled instead of landfilled.

COAL-FIRED POWER PLANT EMISSIONS FOR ONE YEAR

In 2010, a total of 454 power plants used coal to generate at least 95% of their electricity (USEPA, 2014). These plants emitted 1,729,127,770.8 metric tons of CO_2 in 2010. Carbon dioxide emissions per power plant were calculated by dividing the total emissions from power plants whose primary source of fuel was coal by the number of power plants.

Calculation

(1,729,127,770.8 metric tons of CO_2) × (1/454 power plants) = 3,808,651 metric tons CO_2 per coal-fired power plant.

Wind Turbines Installed

In 2012, the average nameplate capacity of wind turbines installed in the United States was 1.94 MW, and the average wind capacity factor in the United States was 31% (USDOE, 2013). Electricity generation from an average wind turbine can be determined by multiplying the average nameplate capacity of a wind turbine in the United States (1.94 MW) by the average U.S. wind capacity factor (0.31) and by the number of hours per year. It is assumed that the electricity generated from an installed wind turbine would replace marginal sources of grid electricity. The U.S. annual non-baseload CO_2 output emission rate to convert reductions of kilowatt-hours into avoided units of carbon dioxide emissions is 6.89551×10^{-4}. Carbon dioxide emissions avoided per wind turbine installed are determined by multiplying the average electricity generated per wind turbine in a year by the national average non-baseload grid electricity CO_2 output rate (USEPA, 2012).

Calculations

(1.94 MW average capacity) × (0.31) × (8760 hours/year) × (1000 kWh/MWh) × (6.89551×10^{-4} metric tons CO_2 per kWh reduced) = 3633 metric tons CO_2 per wind turbine installed.

Heat Engines

Electricity is a *secondary* energy source. The energy sources we use to make electricity can be renewable or nonrenewable, but electricity itself is neither renewable nor nonrenewable. Many of the producers of renewable energy primarily produce electricity; for example, hydropower can be converted to mechanical power via a waterwheel connected to a gear train to power a pump or other machine, but the primary purpose of hydropower is to generate electricity. Likewise, a windmill can also be connected to a gear train to power a pump or other machine, but the primary purpose of wind turbines is to produce electricity. Solar power can be used to heat water directly to produce steam which in turn powers a turbine to perform mechanical functions. Solar power can also be used to produce photovoltaic (PV) electrical energy. Even biomass liquid fuels can be used to power boilers that produce steam to power turbines connected to generators to produce electricity. In addition, waste off-gases or exhaust from various machines such as gas turbine generators can be used to heat boilers.

The use of steam to power turbines to convert energy is accomplished by machines called *heat engines*. An automobile engine converts the chemical energy of gasoline into the mechanical energy of a piston and camshaft; the turbine in an electrical generating plant converts heat energy into mechanical work to run a generator which, in turn, produces electrical power. The two heat engines of interest to us in this text are the Rankine cycle (or vapor cycle) and the Stirling cycle (or gas cycle) heat engines (Hinrichs and Kleinbach, 2006).

Because electrical energy is the main product of heat engines powered by various renewable energy sources, it is important for general readers, students, and others to have some fundamental knowledge of heat engines and electricity, electrical circuitry, electrical power generation, and electrical power uses. Accordingly, prior to beginning a detailed discussion of each major renewable energy source—solar, wind, hydropower, bioenergy, geothermal, ocean, fuel cells—we first present a basic overview of heat engines and then an even more thorough discussion of basic electricity.

A heat engine is a device that converts thermal energy to mechanical output. The thermal energy input is called *heat*, and the mechanical output is called *work*. Typically, heat engines run on a specific thermodynamic cycle. Heat engines can be open to the atmosphere (open cycle) or sealed and closed off to the outside (closed cycle). The driving agent of a heat engine is a temperature differential. That is, heat engines convert heat energy to mechanical work by exploiting the temperature gradient between a hot *source* and a cold *sink* (see Figure 2.2). Heat is transferred from the source, through the working body of the engine, to the sink, and in this process some of the heat is converted into work by exploiting the properties of a gas or liquid (the working substance). The lower the sink temperature or the higher the source temperature, the more work is available from the heat engine.

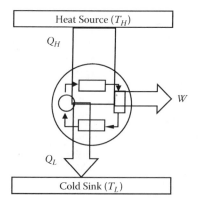

where

> Q_H = Heat energy taken form the high temperature system.
> Q_L = Heat energy delivered to the cold temperature system.
> W = Work.
> T_H = Absolute temperature of heat source.
> T_L = Absolute temperature of cold sink.

FIGURE 2.2 Heat engine.

After doing work, the working substance can either be released into the environment or sent back to the heat source to start the cycle over. If the working substance is returned to its initial state (gas or liquid), there has been no change in its total energy. Consequently, from the first law of thermodynamics, the total energy of a system can be increased by doing work on it or by adding heat, and the total work done by the system is just equal to the heat added (i.e., heat in minus heat out).

The heat source may be direct solar radiation, geothermal steam, geothermal water, ocean water heated by the sun, combustion fuel, or nuclear energy. The two types of heat engines that are most commonly associated with renewable energy processes such as solar power and geothermal energy are the Rankine and Stirling cycle heat engines.

Rankine Cycle Heat Engine

Figure 2.2 illustrates the four processes as they occur in a closed-cycle *Rankine cycle heat engine*. The expansion can be accomplished through a cylinder with a piston, as in the locomotives one still sees in the old western movies, or through a turbine. In the Rankine cycle heat engine, as described for turbine use, the working fluid changes state and is also referred to as a *vapor gas* (i.e., hot air, not to be confused with fuel natural gas) *cycle*. Rankine cycle engines are all external-combustion devices (e.g., external-combustion gas turbine engines are widely used in units for electrical generating stations). As shown in Figure 2.3, the first phase pumps the working fluid from low to high pressure; because the fluid is a liquid at this stage, the pump requires little input energy. In the

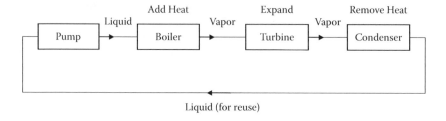

FIGURE 2.3 The four processes of the Rankine cycle.

second phase, the high-pressure liquid enters a boiler, where it is heated at constant pressure by an external heat source to become a dry saturated vapor. In the third phase, the dry saturated vapor expands through the turbine, generating power. The temperature and pressure of the vapor decrease, and some condensation may occur. In the final phase, the wet vapor enters a condenser, where it is condensed at a constant pressure and temperature to become a saturated liquid. The pressure and temperature of the condenser are fixed by the temperature of the cooling coils as the fluid undergoes a phase change.

Stirling Heat Engine

The Stirling cycle heat engine processes include isothermal compression, isometric heat addition, isothermal expansion, and isometric heat rejection. It operates by cyclic compression and expansion of air or other gas (the working fluid) at different temperature levels such that there is a net conversion of heat energy to mechanical work. Like a steam engine, heat in the Stirling heat engine flows in and out through the engine wall. The Stirling heat engine is noted for its high efficiency (resulting from heat regeneration), quiet operation, and ease with which it can use almost any heat source. This compatibility with renewable energy sources has become increasingly significant as the price of conventional fuels rises, as well as in light of concerns such as climate change and peak oil. An example of this compatibility with renewable energy sources is placing a Stirling heat engine at the focus of a parabolic mirror, where it can convert solar energy to electricity with an efficiency better than that of non-concentrated photovoltaic cells or when compared to concentrated photovoltaics.

DISCUSSION AND REVIEW QUESTIONS

1. Do you think we will run out of nonrenewable fuels in your lifetime?
2. What are the economic ramifications of shifting from crude oil to wind and solar power?
3. Are fuel cells the answer to future energy needs?
4. List 10 negative aspects of using renewable energy sources.
5. Name 15 products produced by oil that we cannot get along without.

REFERENCES AND RECOMMENDED READING

Callen, H.B. (1985). *Thermodynamics and an Introduction to Thermostatistics*, 2nd ed. New York: John Wiley & Sons.

EIA. (2009). *What Is Energy? Explained*. Washington, DC: Energy Information Administration (http://tonto.eia.doe.gov/energyexplained/print.cfm?page=about_sources_of_energy).

EIA. (2013a). *Use of Energy in the United States Explained*. Washington, DC: Energy Information Administration (http://tonto.eia.doe.gov/energyexplained/print.cfm?page=us_energy_use).

EIA. (2013b). *2013 Annual Energy Outlook*. Washington, DC: Energy Information Administration.

EIA. (2013c). *2014 Annual Energy Outlook Early Release*. Washington, DC: Energy Information Administration.

EIA. (2013d). *Form EIA-923 Detailed Data with Previous Form Data*, EIA-906/920. Washington, DC: Energy Information Administration (http://www.eia.gov/electricity/data/eia923/).

EIA. (2014). Approximate heat content of coal and coal coke. *Monthly Energy Review*, February, p. 75.

Feynman, R.P., Leighton, R.B., and Sands, M. (1963). *The Feynman Lectures on Physics*. New York: Addison-Wesley.

FHWA. (2013). *Highway Statistics 2011*. Washington, DC: Office of Highway Policy Information, Federal Highway Administration, U.S. Department of Transportation (http://www.fhwa.dot.gov/policyinformation/statistics/2011/index.cfm).

Haimes, Y.Y. (2004). *Risk Modeling, Assessment, and Management*, 2nd ed. New York: John Wiley & Sons.

Halliday, D. and Resnick, R. (1978). *Physics*, 3rd ed. New York: John Wiley & Sons.

Hancock, K. and Sreekanth, A. (2001). Conversion of weight of freight to number of railcars. *Transportation Research Record*, 1768: 1–10.

Harris, J.M. (2006). *Environmental and Natural Resource Economics*, 2nd ed. Boston: Houghton Mifflin.

Hartmann, T. (2004). *The Last Hours of Ancient Sunlight: Revised and Updated: The Fate of the World and What We Can Do Before It's Too Late*. New York: Broadway Books.

Hinrichs, R.A. and Kleinbach, M. (2006). *Energy: Its Use and the Environment*, 4th ed. Belmont, CA: Brooks & Cole.

IPCC. (2006). *2006 IPCC Guidelines for National Green House Gas Inventories*. Geneva, Switzerland: Intergovernmental Panel on Climate Change.

IPCC. (2007). *Fourth Assessment Report (AR4)*. Geneva, Switzerland: Intergovernmental Panel on Climate Change.

Kroemer, H. and Kittel, C. (1980). *Thermal Physics*, 2nd ed. New York: W.H. Freeman.

NREL. (2009a). *Energy Storage—Battery Types*. Golden, CO: National Renewable Energy Laboratory (http://www.nrel.gov/vehiclesandfuels/energystorage/battery_types.html?print).

NREL. (2009b). *Energy Storage: Batteries*. Golden, CO: National Renewable Energy Laboratory (http://www.nrel.gov/vehiclesandfuels/energystorage/batteries.html).

OFR. (2010). Light-duty vehicle greenhouse gas emission standards and corporate average fuel economic standards; final rule. *Federal Register*, 75(88): 25,323–25,728.

Organ, A.J. (1992). *Thermodynamics and Gas Dynamics of the Stirling Cycle Machine*. London: Cambridge University Press.

Organ, A.J. (1997). *The Regenerator and the Stirling Engine*. New York: John Wiley & Sons.

Smith, J.L., Heath, L., and Nichols, M. (2010). *U.S. Forest Carbon Calculation Tool User's Guide: Forestland Carbon Stocks and Net Annual Stock Change*, General Technical Report NRS-13 revised. St. Paul, MN: U.S. Department of Agriculture Forest Service, Northern Research Station.

Spellman, F.R. (1997). *A Guide to Compliance for Process Safety Management/Risk Management Planning (PSM/RMP)*. Lancaster, PA: Technomic.

Spellman, F.R. and Drinan, J.E. (2001). *Electricity*. Boca Raton, FL: CRC Press.

USDOE. (1992). *DOE Fundamentals Handbook*. Vol. 1. *Electrical Science*. Washington, DC: U.S. Department of Energy.

USDOE. (1998). *Method of Calculating Carbon Sequestration by Trees in Urban and Suburban Settings, Voluntary Reporting of Greenhouse Gases*. Washington, DC: U.S. Department of Energy.

USDOE. (2013). *2012 Wind Technologies Market Report*. Washington, DC: U.S. Department of Energy (http://www1.eere.energy.gov/wind/pdfs/2012_wind_technologies_market_report.pdf).

USEPA. (2002). *Waste Transfer Stations: A Manual for Decision-Making*. Washington, DC: U.S. Environmental Protection Agency.

USEPA. (2011). *Inventory of U.S. Greenhouse Gas Emissions and Sinks: Fast Facts 1990–2009*. Washington, DC: U.S. Environmental Protection Agency (http://epa.gov/climatechange/Downloads/ghgemissions/fastfacts.pdf).

USEPA. (2012). *Inventory of U.S. Greenhouse Gas Emissions and Sinks: 1990–2010*, EPA 430-R-12-001. Washington, DC: U.S. Environmental Protection Agency.

USEPA. (2013a). *Inventory of U.S. Greenhouse Gas Emissions and Sinks: 1990–2011*, EPA 430-R-13-001. Washington, DC: U.S. Environmental Protection Agency.

USEPA. (2013b). *Inventory of U.S. Greenhouse Gas Emissions and Sinks: 1990–2011*. Annex 6. *Additional Information*. Washington, DC: U.S. Environmental Protection Agency.

USEPA. (2013c). *Inventory of U.S. Greenhouse Gas Emissions and Sinks: 1990–2011*. Annex 2. *Methodology for Estimating CO_2 Emission from Fossil Fuel Combustion*. Washington, DC: U.S. Environmental Protection Agency.

USEPA. (2013d). *Savings Calculator for ENERGY STAR Qualified Light Bulbs*. Washington, DC: U.S. Environmental Protection Agency.

USEPA. (2013e). *Waste Reduction Model (WARM)*. Washington, DC: U.S. Environmental Protection Agency (http://epa.gov/epawaste/conserve/tools/warm/index.html).

USEPA. (2014). *eGRID*, 9th ed. Washington, DC: U.S. Environmental Protection Agency (http://www.epa.gov/cleanenergy/energy-resources/egrid/index.html).

3 Principles of Basic Electricity

There is an urgent need to stop subsidizing the fossil fuel industry, dramatically reduce wasted energy, and significantly shift our power supplies from oil, coal, and natural gas to wind, solar, geothermal, and other renewable energy sources.

—Bill McKibben, author and environmentalist

INTRODUCTION*

1. Do I need to be an electrical engineer?
2. Do I need to be an electrician?
3. Do I need electrical work experience?

When I was teaching renewable energy and other environmental courses, these three questions were among those most frequently asked by my renewable energy students, potential students, and curious wannabes. The sample curriculum sheet for a renewable energy associate of applied science degree (see Table 3.1) that I provided upon enrollment or by request always gave rise to many questions. So, now let's answer those three questions:

1. No, not exactly.
2. No, not exactly.
3. No, not exactly.

The "no" part of the answer is clear; no usually means no. But how about the "not exactly" part of the answer? I say "not exactly" because if you are an electrical engineer, an electrician, or a person with electrical work experience then you are a huge step ahead of those who desire to work in the renewable energy field. Consider, for example, Table 3.1 and its slate of electrical subject areas and then consider the energy producers or converters such as wind turbines, solar technology, fuel cells, and hydropower plants—these renewable energy sources produce electricity in one form or another. In most cases, they transform mechanical energy into electrical energy, such as the turning of wind turbines that turn generators, solar production of steam to turn turbines that turn generators, or chemical reactions that produce electricity to power an electrical motor.

Because the renewable energy practitioner must have a sound grounding or foundation in basic electrical principles, this chapter is included in this book. Again, you do not need to be an electrical engineer or an electrician or have worked in the electrical field to become a competent renewable energy practitioner but you need to have a fundamental knowledge of the principles of basic electricity.

BASIC ELECTRICAL PRINCIPLES

People living and working in modern societies generally have little difficulty in recognizing electrical equipment; electrical equipment is everywhere and (if one pays attention to his or her surroundings) is easy to spot. Despite its great importance in our daily lives, however, few of us probably stop

---

* Much of the information in this chapter is adapted from Spellman, F.R. and Drinan, J., *Electricity*, CRC Press, Boca Raton, FL, 2001.

TABLE 3.1

Renewable Energy Associate of Applied Science Curriculum (Sample)

Areas of Concentration		General Education Requirements
Introduction to Sustainability	Industrial Wiring	English Composition
Fundamentals of Electricity	Power Generation and Transmission	Physical Geography
Safe Work Practices for Green Energy Jobs	Programmable Logic Controls	Technical Mathematics
AC Current Fundamentals	Electric Motors and Control Systems	for Energy Technology
AC Circuitry	Business Concepts for Renewable Energy	Physical Science
Direct Current Fundamentals	Introduction to Wind Energy	College Mathematics
Direct Current Essentials	Introduction to Solar Energy	
Basic Wiring Applications I	Introduction to Hydropower	
Basic Wiring Applications II	Wind Turbine Requirements	
Drawing and Print Reading	Wind Turbine Systems	
Photovoltaic Systems	Turbine Troubleshooting and Repair	
Piping and Mechanical Applications	Wind Turbine Safety	
Physics with Technology Applications	Solar Cell Troubleshooting/Repair and	
Computer Fundamentals	Maintenance	
Renewable Energy Fundamentals	Process Control Basics	
Wiring, Schematics, and Blueprints	Building Materials Applications	

Note: The educational facility does not guarantee that completion of this program either ensures passage of any examination or acceptance by any state board, which may be required to work in the field.

to think what life would be like without electricity. Like air and water, we tend to take electricity for granted, but we use electricity to do many jobs for us every day—from lighting, heating, and cooling our homes to powering our televisions and computers. Then there is the workplace—can we actually perform work without electricity? The typical industrial workplace is outfitted with electrical equipment that

- Generates electricity (a generator or emergency generator)
- Stores electricity (batteries)
- Changes electricity from one level (voltage or current) to another (transformers)
- Transports or transmits and distributes electricity throughout the plant site (wiring distribution systems)
- Measures electricity (meters)
- Converts electricity into other forms of energy (mechanical energy, heat energy, light energy, chemical energy, radio energy)
- Protects other electrical equipment (fuses, circuit breakers, or relays)
- Operates and controls other electrical equipment (motor controllers)
- Converts some condition or occurrence into an electric signal (sensors)
- Converts some measured variable into a representative electrical signal (transducers or transmitters)

NATURE OF ELECTRICITY

The word "electricity" is derived from the Greek word *electron*, meaning "amber." Amber is a translucent (semitransparent), yellowish, fossilized mineral resin. The ancient Greeks used the term "electric force" to describe the mysterious forces of attraction and repulsion exhibited by amber when it was rubbed with a cloth. They did not understand the nature of this force, and they could not answer the question, "What is electricity?" This question still remains unanswered. Today, we often attempt to answer this question by describing the effect and not the force. That is, the standard answer given in physics is that electricity is the "force that moves electrons," which is about the same as defining a sail as "the force that moves a sailboat."

At the present time, we know little more about the fundamental nature of electricity than the ancient Greeks did, but we have made tremendous strides in harnessing and using it. As with many other unknown (or unexplainable) phenomena, elaborate theories concerning the nature and behavior of electricity have been advanced and have gained wide acceptance because of their apparent truth—and because they work.

Scientists have determined that electricity seems to behave in a constant and predictable manner in given situations or when subjected to given conditions. Faraday, Ohm, Lenz, and Kirchoff described the predictable characteristics of electricity and electric current in the form of certain rules. These rules are often referred to as *laws*. Thus, although electricity itself has never been clearly defined, its predictable nature and easily used nature have made it one of the most widely used power sources in modern times.

> ***The bottom line:*** We can learn about electricity by becoming familiar with the rules, or laws, that apply to the behavior of electricity and by understanding the methods of producing, controlling, and using it. Learning about electricity, then, can be accomplished without ever having determined its fundamental identity.

You are probably scratching your head, puzzled. We understand the main question running through your brain at this exact moment: "This is a section in the text about the physics of electricity and the author can't even explain what electricity is?" That is correct; I cannot. The point is, no one can definitively define electricity. Electricity is one of those subject areas where the old saying that "we don't know what we don't know about it" fits perfectly.

Only a few theories about electricity have so far stood the test of extensive analysis and much time (relatively speaking, of course). One of the oldest and the most generally accepted theories concerning electric current flow (or electricity) is known as the *electron theory*. The electron theory basically states that electricity or current flow is the result of the flow of free electrons in a conductor. Thus, electricity is the flow of free electrons or simply electron flow. This is how this text defines electricity: Electricity is the flow of free electrons. Electrons are extremely tiny particles of matter. To gain an understanding of electrons and exactly what is meant by electron flow, it is necessary to briefly review our earlier discussion about the structure of matter.

STRUCTURE OF MATTER

Matter is anything that has mass and occupies space. To study the fundamental structure or composition of any type of matter, it must be reduced to its fundamental components. All matter is made of *molecules*, or combinations of *atoms* (from the Greek word for "not able to be divided"), that are bound together to produce a given substance, such as salt, glass, or water. For example, if you keep dividing water into smaller and smaller drops, you would eventually arrive at the smallest particle that was still water. That particle is the *molecule*, which is defined as the smallest bit of a substance that retains the characteristics of that substance.

A molecule of water (H_2O) is composed of one atom of oxygen and two atoms of hydrogen. If the molecule of water was further subdivided, only unrelated atoms of oxygen and hydrogen would remain, and the water would no longer exist as such. Thus, the molecule is the smallest particle to which a substance can be reduced and still be called by the same name. This applies to all substances—solids, liquids, and gases.

> ***Note:*** Molecules are made up of atoms, which are bound together to produce a given substance.

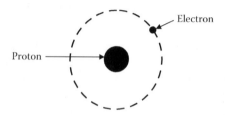

FIGURE 3.1   One proton and one electron = electrically neutral.

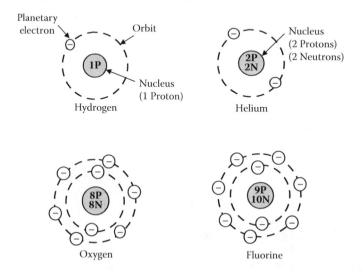

FIGURE 3.2   Atomic structure of elements.

Atoms are composed, in various combinations, of subatomic particles of *electrons*, *protons*, and *neutrons*. These particles differ in weight (a proton is much heavier than the electron) and charge. We are not concerned with the weights of particles in this text, but the *charge* is extremely important in electricity. The electron is the fundamental negative (–) charge of electricity. Electrons revolve about the nucleus or center of the atom in paths of concentric *orbits*, or shells. The proton is the fundamental positive (+) charge of electricity. Protons are found in the nucleus. The number of protons within the nucleus of any particular atom determines the atomic number of that atom; for example, the helium atom has two protons in its nucleus so the atomic number is 2. The neutron, which is the fundamental neutral charge of electricity, is also found in the nucleus.

Most of the weight of the atom is in the protons and neutrons of the nucleus. Whirling around the nucleus are one or more negatively charged electrons. Normally, there is one proton for each electron in the entire atom so the net positive charge of the nucleus is balanced by the net negative charge of the electrons rotating around the nucleus (see Figure 3.1).

> **Note:** Most batteries are marked with the symbols + and – or even with the abbreviations POS (positive) and NEG (negative). The concept of a positive or negative polarity and its importance in electricity will become clear later; however, for the moment, you need to remember that an electron has a negative charge and that a proton has a positive charge.

We stated earlier that in an atom the number of protons is usually the same as the number of electrons. This is an important point because this relationship determines the kind of element in question (the atom is the smallest particle that makes up an element; an element retains its characteristics when subdivided into atoms). Figure 3.2 provides simplified drawings of atoms of various

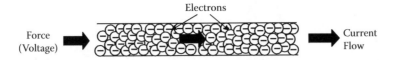

FIGURE 3.3   Electron flow in a copper wire.

materials based on the concept of electrons orbiting about the nucleus. Hydrogen, for example, has a nucleus consisting of one proton, around which rotates one electron. The helium atom has a nucleus containing two protons and two neutrons with two electrons encircling the nucleus. Both of these elements are electrically neutral (or balanced) because each has an equal number of electrons and protons. Because the negative (–) charge of each electron is equal in magnitude to the positive (+) charge of each proton, the two opposite charges cancel.

A balanced (neutral or stable) atom has a certain amount of energy that is equal to the sum of the energies of its electrons. Electrons, in turn, have different energies called *energy levels*. The energy level of an electron is proportional to its distance from the nucleus; therefore, the energy levels of electrons in shells farther from the nucleus are higher than the energy levels of electrons in shells closer to the nucleus.

When an electric force is applied to a conducting medium, such as copper wire, electrons in the outer orbits of the copper atoms are forced out of orbit (they are liberated or freed) and travel along the wire. This electrical force, which forces electrons out of orbit, can be produced in a number of ways, such as by moving a conductor through a magnetic field; by friction, as when a glass rod is rubbed with cloth (silk); or by chemical action, as in a battery.

When the electrons are forced from their orbits they are referred to as *free electrons*. Some of the electrons of certain metallic atoms are so loosely bound to the nucleus that they are relatively free to move from atom to atom. These free electrons constitute the flow of an electric current in electrical conductors. If the internal energy of an atom is raised above its normal state, the atom is said to be *excited*. Excitation may be produced by causing the atoms to collide with particles that are impelled by an electric force, as shown in Figure 3.3. In effect, energy is transferred from the electric source to the atom. The excess energy absorbed by the atom may be sufficient to cause loosely bound outer electrons (as shown in Figure 3.3) to leave the atom, resisting the force that acts to hold them within.

> *Note:* An atom that has lost or gained one or more electrons is said to be *ionized*. If the atom loses electrons it becomes positively charged and is referred to as a *positive ion*. Conversely, if the atom gains electrons, it becomes negatively charged and is referred to as a *negative ion*.

> *Note:* When an electric force is applied to a copper wire, free electrons are displaced from the copper atoms and move along the wire, producing an electric current, as shown in Figure 3.3.

CONDUCTORS, SEMICONDUCTORS, AND INSULATORS

Electric current moves easily through some materials but with greater difficulty through others. Substances that permit the free movement of a large number of electrons are *conductors*. The most widely used electrical conductor is copper because of its high conductivity and cost effectiveness. Electrical energy is transferred through a copper or other metal conductor by the movement of free electrons that migrate from atom to atom inside the conductor (see Figure 3.3). Each electron moves a very short distance to the neighboring atom, where it replaces one or more electrons by forcing them out of their orbits. The replaced electrons repeat the process in other nearby atoms until their movement is transmitted throughout the entire length of the conductor. A good conductor is said to have a low opposition, or *resistance*, to the electron (current) flow. Below are listed many of the metals commonly used as electric conductors; the best conductors appear at the top of the list, the poorer ones lower:

Silver
Copper
Gold
Aluminum
Zinc
Brass
Iron
Tin
Mercury

Note: If lots of electrons flow through a material with only a small force (voltage) applied, we refer to that material as a *conductor*. A convenient way in which to understand the purpose and function of an electrical conductor is to visualize a pipe or other conveyor of water or other liquid or gaseous substance. Just as a liquid or gas is conveyed or conducted through a pipe, so too are electrons (electricity) through a conductor. Force to push water or other liquid through a pipe is applied by a pump; in the case of electricity, voltage pushes the electrons. It is important to remember that in an electrical circuit voltage does not move; rather, voltage is the force that makes electrons move.

Note: The movement of each electron in copper wire, for example, takes a very small amount of time; it is almost instantaneous. This is an important point to keep in mind later in the book, when it might seem that events in an electrical circuit occur simultaneously.

Although electron motion is known to exist to some extent in all matter, some substances, such as rubber, glass, and dry wood, have very few free electrons. In these materials, large amounts of energy must be expended to break the electrons loose from the influence of the nucleus. Substances containing very few free electrons are called *insulators*. Insulators are important in electrical work because they prevent the current from being diverted from the wires, which in turn helps to prevent electrocutions and fires. Below are some materials that we often use as insulators in electrical circuits; these materials are listed in decreasing order of their ability to withstand high voltages without conducting:

Rubber
Mica
Wax or paraffin
Porcelain
Bakelite
Plastics
Glass
Fiberglass
Dry wood
Air

A material that is neither a good conductor nor a good insulator is called a *semiconductor*. Silicon and germanium are substances that fall into this category. Because of their peculiar crystalline structure, these materials may, under certain conditions, act as conductors and, under other conditions, as insulators. As the temperature is raised, however, a limited number of electrons become available for conduction.

Note: If the voltage is large enough, even the best insulators (rubber, plastic, wood) will break down and allow the electrons to flow.

STATIC ELECTRICITY

Electricity at rest is often referred to as *static electricity*. More specifically, when two bodies of matter with unequal charges are near one another, an electric force is exerted between them because of their unequal charges; however, because they are not in contact, their charges cannot equalize. Static electricity (or electricity at rest) will flow when given the opportunity, because static electricity is an imbalance of negative and positive charges. An example of this phenomenon is when we walk across a dry carpet and then touch a doorknob—we usually feel a slight shock and might notice a spark at our fingertips. Another familiar example is static cling; for example, when we rub an air-filled balloon against the hair on our heads and then place the balloon against a wall, the balloon will stick to the wall, defying gravity, due to static cling. In the workplace, static electricity is prevented from building up by properly bonding or grounding equipment to ground or earth.

CHARGED BODIES

The fundamental law of charged bodies states: "Like charges repel each other and unlike charges attract each other." A positive charge and a negative charge, being opposite or unlike, tend to move toward each other; that is, they are attracted to each other. In contrast, like bodies tend to repel each other. Electrons repel each other because of their like negative charges, and protons repel each other because of their like positive charges. Figure 3.4 demonstrates the law of charged bodies.

It is important to point out another significant aspect of the fundamental law of charged bodies: The force of attraction or repulsion existing between two magnetic poles decreases rapidly as the poles are separated from each other. More specifically, the force of attraction or repulsion varies directly as the product of the separate pole strengths and inversely as the square of the distance separating the magnetic poles, provided the poles are small enough to be considered as points. Let's look at an example. If you increase the distance between two magnet north poles from 2 feet to 4 feet, the force of repulsion between them is decreased to one fourth of its original value. If either pole strength is doubled and the distance remains the same, the force between the poles will be doubled.

COULOMB'S LAW

Coulomb's law states that the amount of attracting or repelling force that acts between two electrically charged bodies in free space depends on two things:

1. Their charges
2. The distance between them

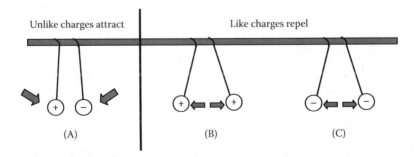

FIGURE 3.4   Reaction between two charged bodies. The opposite charges in (A) attract. The like charges in (B) and (C) repel each other.

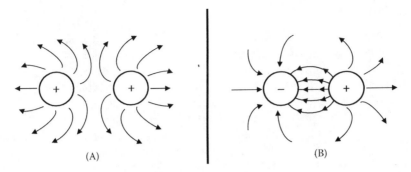

FIGURE 3.5   Electrostatic lines of force: (A) repulsion of like-charged bodies and their associated fields; (B) attraction between unlike-charged bodies and their associated fields.

Specifically, Coulomb's law states that charged bodies attract or repel each other with a force that is directly proportional to the product of their charges and is inversely proportional to the square of the distance between them.

> *Note:* The magnitude of electric charge a body possesses is determined by the number of electrons compared with the number of protons within the body. The symbol for the magnitude of electric charge is Q, expressed in units of *coulombs* (C). A charge of one positive coulomb means a body contains a charge of 6.25×10^{18}. A charge of one negative coulomb means a body contains a charge of 6.25×10^{18} more electrons than protons.

Electrostatic Fields

The fundamental characteristic of an electric charge is its ability to exert force. The space between and around charged bodies in which their influence is felt is the *electric field of force*. The electric field is always terminated on material objects and extends between positive and negative charges. This region of force can consist of air, glass, paper, or a vacuum and is referred to as an *electrostatic field*. When two objects of opposite polarity are brought near each other, the electrostatic field is concentrated in the area between them. The field is generally represented by lines referred to as *electrostatic lines of force*. These lines are imaginary and are used merely to represent the direction and strength of the field. To avoid confusion, the positive lines of force are always shown leaving, and the negative lines of force are shown entering. Figure 3.5 illustrates the use of lines to represent the field about charged bodies.

> *Note:* A charged object will retain its charge temporarily if there is no immediate transfer of electrons to or from it. In this condition, the charge is said to be *at rest*. Remember, electricity at rest is called *static* electricity.

Magnetism

Most electrical equipment depends directly or indirectly on magnetism, a phenomenon associated with magnetic fields that has the power to attract such substances as iron, steel, nickel, or cobalt (metals known as magnetic materials). A substance is said to be a *magnet* if it has the property of magnetism; for example, when a piece of iron is magnetized it becomes a magnet. When magnetized, that piece of iron will have two points, opposite each other, that most readily attract other pieces of iron. (For this discussion, we will assume the piece of iron is a flat bar 6 inches long × 1 inch wide × 0.5 inch thick—in other words, a bar magnet; see Figure 3.6.) The points of maximum

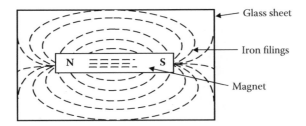

FIGURE 3.6   The magnetic field around a bar magnet. If the glass sheet is tapped gently, the filings will move into a definite pattern that describes the field of force around the magnet.

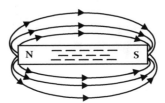

FIGURE 3.7   Magnetic field of force around a bar magnet, indicated by lines of force

attraction (one on each end) are the *magnetic poles* of the magnet: the north (N) pole and the south (S) pole. Just as like electric charges repel each other and opposite charges attract each other, like magnetic poles repel each other and unlike poles attract each other. Although invisible to the naked eye, magnetic force can be shown to exist by sprinkling small iron filings on a glass covering a bar magnet, as shown in Figure 3.6. Figure 3.7 shows how the field looks without the iron filings; it is shown as lines of force. The group of magnetic field lines, which flow from the north pole of a magnet toward the south pole, is referred to as the *magnetic flux*; the symbol for magnetic flux is the Greek lowercase letter ϕ (phi).

> **Note:** A *magnetic circuit* is a complete path through which magnetic lines of force may be established under the influence of a magnetizing force. Most magnetic circuits are composed largely of magnetic materials to contain the magnetic flux. These circuits are similar to the *electric circuit* (an important point), which is a complete path through which current is caused to flow under the influence of an electromotive force.

The three types or groups of magnets are as follows:

1. *Natural magnets* are found in the natural state in the form of the mineral magnetite (an iron compound).
2. *Permanent magnets* (artificial magnets) are hardened steel or some alloy such as Alinco bars that have been permanently magnetized. The permanent magnet most people are familiar with is the horseshoe magnet; this red U-shaped magnet is the universal symbol of magnets, recognized throughout the world (see Figure 3.8).
3. *Electromagnets* (artificial magnets) are composed of soft iron cores around which are wound coils of insulated wire. When an electric current flows through the coil, the core becomes magnetized. When the current ceases to flow, the core loses most of the magnetism.

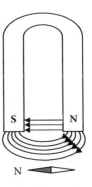

FIGURE 3.8   Horseshoe magnet.

MAGNETIC MATERIALS

Natural magnets are no longer used in electrical circuitry because more powerful and more conveniently shaped permanent magnets can be produced artificially. Commercial magnets are made from such magnetic materials as special steels and alloys. Magnetic materials are those materials that are attracted or repelled by a magnet and that can be magnetized themselves. Iron, steel, and alloy bars are the most common magnetic materials. These materials can be magnetized by inserting the material (in bar form) into a coil of insulated wire and passing a heavy direct current through the coil. The same material may also be magnetized if it is stroked with a bar magnet. It will then have the same magnetic property that the magnet used to induce the magnetism has—namely, there will be two poles of attraction, one at either end. This process produces a permanent magnet by induction; that is, the magnetism is induced in the bar by the influence of the stroking magnet. Even though they are classified as permanent magnets, it is important to point out that hardened steel and certain alloys are relatively difficult to magnetize and are said to have a *low permeability* because the magnetic lines of force do not easily permeate or distribute themselves readily through the steel.

> **Note:** *Permanent magnets* are made of hard magnetic materials (hard steel or alloys) that retain their magnetism when the magnetizing field is removed. A *temporary magnet* is one that has *no* ability to retain a magnetized state when the magnetizing field is removed.

> **Note:** *Permeability* refers to the ability of a magnetic material to concentrate magnetic flux. Any material that is easily magnetized has high permeability. A measure of permeability for different materials in comparison with air or a vacuum is the *relative* permeability, symbolized by μ (mu).

When hard steel and other alloys have been magnetized, they retain a large part of their magnetic strength and are considered to be permanent magnets. Conversely, materials that are relatively easy to magnetize—such as soft iron and annealed silicon steel—are said to have a *high permeability*. Such materials retain only a small part of their magnetism after the magnetizing force is removed and are called *temporary magnets*. The magnetism that remains in a temporary magnet after the magnetizing force is removed is called *residual magnetism*.

Early magnetic studies classified magnetic materials as being magnetic or nonmagnetic, based on the strong magnetic properties of iron; however, because even weak magnetic materials can be important in some applications, current studies classify materials into one of three groups:

- *Paramagnetic materials* include aluminum, platinum, manganese, and chromium, materials that become only slightly magnetized even under the influence of a strong magnetic field. This slight magnetization is in the same direction as the magnetizing field. Relative permeability is slightly more than 1 (i.e., they are considered to be nonmagnetic materials).
- *Diamagnetic materials* include bismuth, antimony, copper, zinc, mercury, gold, and silver, materials that can also be slightly magnetized when under the influence of a very strong field. Relative permeability is less than 1 (i.e., they are considered to be nonmagnetic materials).
- *Ferromagnetic materials* include iron, steel, nickel, cobalt, and commercial alloys, materials that comprise the most important group in applications of electricity and electronics. Ferromagnetic materials are easy to magnetize and have high permeabilities, ranging from 50 to 3000.

MAGNETIC EARTH

Earth is a huge magnet, and surrounding Earth is a magnetic field produced by Earth's magnetism. Most people would have no problem understanding or at least accepting this statement; however, they might question being told that Earth's north geographic pole is actually its south magnetic pole

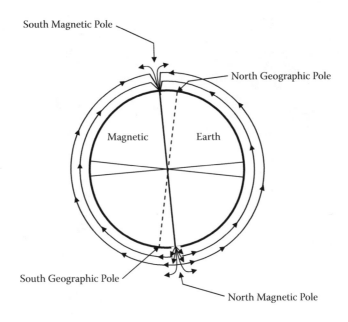

FIGURE 3.9 Earth's magnetic poles.

and that the south geographic pole is actually Earth's north magnetic pole. But, in terms of a magnet, this is true. Figure 3.9 indicates the magnetic polarities of Earth. The geographic poles are also shown at each end of the axis of rotation of Earth. Clearly, the magnetic axis does not coincide with the geographic axis; therefore, the magnetic and geographic poles are not at the same place on the surface of Earth. Recall that magnetic lines of force are assumed to emanate from the north pole of a magnet and enter the south pole as closed loops. Because Earth is a magnet, lines of force emanate from its north magnetic pole and enter the south magnetic pole. A compass needle aligns itself in such a way that Earth's lines of force enter at its south pole and leave at its north pole. Because the north pole on a compass is defined as the point where the needle points in a northerly direction, it follows that the magnetic pole in the vicinity of the north geographic pole is in reality a south magnetic pole, and *vice versa*.

DIFFERENCE IN POTENTIAL

Because of the force of its electrostatic field, an electric charge has the ability to do the work of moving another charge by attraction or repulsion. The force that causes free electrons to move in a conductor as an electric current may be referred to as

- Electromotive force (emf)
- Voltage
- Difference in potential

When a difference in potential exists between two charged bodies that are connected by a wire (conductor), electrons (current) will flow along the conductor. This flow is from the negatively charged body to the positively charged body until the two charges are equalized and the potential difference no longer exists.

> **Note:** The basic unit of potential difference is the *volt* (V). The symbol for potential difference is *V*, indicating the ability to do the work of forcing electrons (current flow) to move. Because the volt unit is used, potential difference is called *voltage*.

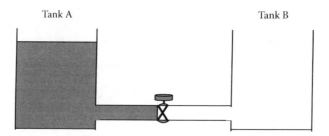

FIGURE 3.10   Water analogy of electric difference of potential.

WATER ANALOGY

When training individuals in the concepts of basic electricity, especially with regard to potential difference (voltage), current, and resistance relationships in a simple electrical circuit, it has been common practice to use what is referred to as the *water analogy*. We use the water analogy later to explain (in simple, straightforward fashion) voltage, current, and resistance and their relationships in more detail, but for now we will use the analogy to explain the basic concept of electricity: the potential difference, or voltage. Because a difference in potential causes current flow (against resistance), it is important that this concept be understood first before exploring the concept of current flow and resistance. Consider the two water tanks connected by a pipe and valve in Figure 3.10. At first, the valve is closed and all the water is in Tank A; the water pressure across the valve is at maximum. When the valve is opened, the water flows through the pipe from A to B until the water level becomes the same in both tanks. The water then stops flowing in the pipe, because there is no longer a difference in water pressure (difference in potential) between the two tanks. Just as the flow of water through the pipe in Figure 3.10 is directly proportional to the difference in water level in the two tanks, current flow through an electric circuit is directly proportional to the difference in potential across the circuit.

> **Note:** A fundamental law of current electricity is that the current is directly proportional to the applied voltage: If the voltage is increased, the current is increased. If the voltage is decreased, the current is decreased.

PRINCIPAL METHODS OF PRODUCING A VOLTAGE

There are many ways to produce electromotive force, or voltage. Some of these methods are much more widely used than others. The following is a list of the seven most common methods of producing electromotive force (USDOE, 1992):

1. *Friction* is the voltage produced by rubbing two materials together (static electricity or electrostatic force). Remember our discussion of static electricity? Let's refresh our memories. Have you ever walked across a carpet and received a shock when you touched a metal door knob? The soles of your shoes built up a charge when they rubbed on the carpet, and this charge was transferred to your body. Your body became positively charged, and when you touched the zero-charged door knob electrons were transferred to your body until both you and the door knob had equal charges.
2. *Pressure* (*piezoelectricity*) is the voltage produced by squeezing or applying pressure to crystals of certain substances (e.g., quartz, Rochelle salts, certain ceramics such as barium titanate). When pressure is applied to such substances, electrons can be driven out of orbit in the direction of the force. Electrons leave one side of the material and accumulate on the other side, building up positive and negative charges on opposite sides. When the pressure

is released, the electrons return to their orbits. Some materials will react to bending pressure, while others will respond to twisting pressure. This generation of voltage is known as the *piezoelectric effect*. If external wires are connected while pressure and voltage are present, electrons will flow and current will be produced. If the pressure is held constant, the current will flow until the potential difference is equalized. When the force is removed, the material is decompressed and immediately causes an electric force in the opposite direction. The power capacity of these materials is extremely small; however, these materials are very useful because of their extreme sensitivity to changes of mechanical force. One example is the crystal phonograph cartridge that contains a Rochelle salt crystal. A phonograph needle is attached to the crystal. As the needle moves in the grooves of a record, it swings from side to side, compressing and decompressing the crystal. The mechanical motion applied to the crystal generates a voltage signal that is used to reproduce sound.

3. *Heat (thermoelectricity)* is the voltage produced by heating the joint (junction) where two unlike metals are joined. Some materials readily give up their electrons and others readily accept electrons; for example, when two dissimilar metals such as copper and zinc are joined together, a transfer of electrons can take place. Electrons will leave the copper atoms and enter the zinc atoms. The zinc gets a surplus of electrons and becomes negatively charged. The copper loses electrons and takes on a positive charge. This creates a voltage potential across the junction of the two metals. The heat energy of normal room temperature is enough to make them release and gain electrons, causing a measurable voltage potential. As more heat energy is applied to the junction, more electrons are released, and the voltage potential becomes greater. When heat is removed and the junction cools, the charges will dissipate and the voltage potential will decrease. This process is known as thermoelectricity. A device like this is generally referred to as a *thermocouple*. Thermocouple voltage is dependent on the heat energy applied to the junction of the two dissimilar metals. Thermocouples are widely used to measure temperature and as heat-sensing devices in automatic temperature-controlled equipment. Thermocouple power capacities are very small compared to some other sources, but they are somewhat greater than those of crystals. Generally speaking, a thermocouple can be subjected to higher temperatures than ordinary mercury or alcohol thermometers.

4. *Light (photoelectricity)* is the voltage produced by light (photons) striking photosensitive (light-sensitive) substances. When the photons in a light beam strike the surface of a material, they release their energy and transfer it to the atomic electrons of the material. This energy transfer may dislodge electrons from their orbits around the surface of the substance. Upon losing electrons, the photosensitive (light-sensitive) material becomes positively charged and an electric force is created. This phenomenon is called the *photoelectric effect* and has wide applications in electronics (e.g., photoelectric cells, photovoltaic cells, optical couplers, television camera tubes). Three uses of the photoelectric effect are described below.

    • *Photovoltaic*—The light energy in one of two plates that are joined together causes one plate to release electrons to the other; the plates build up opposite charges, as in a battery.

    • *Photoemission*—The photon energy from a beam of light can cause a surface to release electrons in a vacuum tube, and a plate then collects the electrons.

    • *Photoconduction*—The light energy applied to some materials that are normally poor conductors causes free electrons to be produced in the materials so they become better conductors.

5. *Chemical action* is the voltage produced by chemical reaction in a battery cell; an example is a voltaic chemical cell, in which a chemical reaction produces and maintains opposite charges on two dissimilar metals that serve as positive and negative terminals. The metals are in contact with an electrolyte solution. Connecting more than one of these cells will produce a battery.

6. *Magnetism* is the voltage produced in a conductor when the conductor moves through a magnetic field or a magnetic field moves through the conductor in such a manner as to cut the magnetic lines of force of the field. A generator is a machine that converts mechanical energy into electrical energy by using the principle of *magnetic induction*. Magnetism is used to produce vast quantities of electric power.

7. *Thermionic emission* is produced by a thermionic energy converter, which consists of two electrodes placed near one another in a vacuum. One electrode is normally the cathode, or emitter, and the other is the anode, or plate. Ordinarily, electrons in the cathode are prevented from escaping from the surface by a potential-energy barrier. When an electron starts to move way from the surface, it induces a corresponding positive charge in the material, which tends to pull it back into the surface. To escape, the electron must somehow acquire enough energy to overcome this energy barrier. At ordinary temperatures, almost none of the electrons can acquire enough energy to escape; however, when the cathode is very hot, the electron energies are greatly increased by thermal motion. At sufficiently high temperatures, a considerable number of electrons are able to escape. The liberation of electrons from a hot surface is thermionic emission. The electrons that have escaped from the hot cathode form a cloud of negative charges near it called a *space charge*. If the plate is maintained positive with respect to the cathode by a battery, the electrons in the cloud are attracted to it. As long as the potential difference between the electrodes is maintained, there will be a steady current flow from the cathode to the plate. The simplest example of a thermionic device is a vacuum tube diode, in which the only electrodes are the cathode and plate, or anode. The diode can be used to convert alternating current (AC) flow to a pulsating direct current (DC) flow.

In the study of the basic electricity related to renewable energy production, we are most concerned with magnetism (e.g., generators powered by hydropower), light (photoelectricity produced by solar cells), and chemistry (chemical energy converted to electricity in batteries) as the means to produce voltage. Friction has little practical application, although we discussed it earlier with regard to static electricity. Pressure and heat do have useful applications, but we do not need to consider them in this text. Magnetism used in generators, electricity produced by solar light, and the chemistry involved in storing electricity in batteries, on the other hand, are the principal sources of voltage and are discussed at length in this text.

ELECTRIC CURRENT

The movement or the flow of electrons is called *current*. To produce current, the electrons must be moved by a potential difference or pressure (voltage).

> *Note:* The terms *current, current flow, electron flow,* and *electron current* all describe the same phenomenon.

For our purposes in this text, electron flow, or current, in an electric circuit is from a region of less negative potential to a region of more positive potential—from negative to positive.

> *Note:* Current is represented by the letter *I*. The basic unit in which current is measured is the *ampere*, or *amp* (A). One ampere of current is defined as the movement of one coulomb past any point of a conductor during one second of time.

Recall that we used the water analogy to help us understand potential difference. We can also use the water analogy to help us understand current flow through a simple electric circuit. Consider Figure 3.11, which shows a water tank connected via a pipe to a pump with a discharge pipe. If the water tank contains an amount of water above the level of the pipe opening to the pump, the water

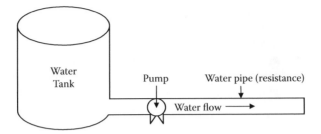

FIGURE 3.11   Water analogy: current flow.

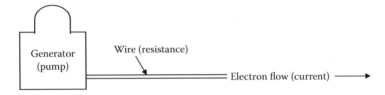

FIGURE 3.12   Simple electric circuit with current flow.

exerts pressure (a difference in potential) against the pump. When sufficient water is available for pumping with the pump, water flows through the pipe against the resistance of the pump and pipe. The analogy should be clear—in an electric circuit, if a difference in potential exists, current will flow in the circuit. Another simple way of looking at this analogy is to consider Figure 3.12, where the water tank has been replaced with a generator, the pipe with a conductor (wire), and water flow with the flow of electric current. Again, the key point illustrated by Figures 3.11 and 3.12 is that, to produce current, the electrons must be moved by a potential difference.

Electric current is generally classified into two general types:

• Direct current (DC)
• Alternating current (AC)

Direct current is current that moves through a conductor or circuit in one direction only. *Alternating current* periodically reverses direction.

RESISTANCE

Earlier we pointed out that free electrons, or electric current, could move easily through a good conductor, such as copper, but that an insulator, such as glass, was an obstacle to current flow. In the water analogy shown in Figure 3.11 and the simple electric circuit shown in Figure 3.12, resistance is indicated by either the pipe or the conductor. Every material offers some resistance, or opposition, to the flow of electric current through it. Good conductors such as copper, silver, and aluminum offer very little resistance. Poor conductors, or insulators, such as glass, wood, and paper, offer a high resistance to current flow.

Note: The amount of current that flows in a given circuit depends on two factors: voltage and resistance.

Note: Resistance is represented by the letter R. The basic unit in which resistance is measured is the *ohm* (Ω). One ohm is the resistance of a circuit element, or circuit, that permits a steady current of 1 ampere (1 coulomb per second) to flow when a steady electromotive force (emf) of 1 volt is applied to the circuit. Manufactured circuit parts containing definite amounts of resistance are called *resistors*.

The size and type of material of the wires in an electric circuit are chosen so as to keep the electrical resistance as low as possible. In this way, current can flow easily through the conductors, just as water flows through the pipe between the tanks in Figure 3.10. If the water pressure remains constant, the flow of water in the pipe will depend on how far the valve is opened. The smaller the opening, the greater the opposition (resistance) to the flow and the smaller will be the rate of flow in gallons per second. In the electric circuit shown in Figure 3.12, the larger the diameter of the wire, the lower will be its electrical resistance (opposition) to the flow of current through it. In the water analogy, pipe friction opposes the flow of water between the tanks. This friction is similar to electrical resistance. The resistance of a pipe to the flow of water through it depends on (1) the length of the pipe, (2) the diameter of the pipe, and (3) the nature of the inside walls (rough or smooth). Similarly, the electrical resistance of the conductors depends on (1) the length of the wires, (2) the diameter of the wires, and (3) the material of the wires (e.g., copper, silver). It is important to note that temperature also affects the resistance of electrical conductors to some extent. In most conductors (e.g., copper, aluminum) the resistance increases with temperature. Carbon is an exception. In carbon, the resistance decreases as temperature increases.

> **Note:** Electricity is a study that is frequently explained in terms of opposites. The term that is exactly the opposite of resistance is *conductance*. Conductance (*G*) is the ability of a material to pass electrons. The SI unit of conductance is the *siemens*. The commonly used unit of conductance is the *mho*, which is ohm spelled backward. The relationship that exists between resistance and conductance is the *reciprocal*. A reciprocal of a number is obtained by dividing the number into one. If the resistance of a material is known, dividing its value into one will give its conductance. Similarly, if the conductance is known, dividing its value into one will give its resistance.

BATTERY-SUPPLIED ELECTRICITY

Battery-supplied direct current electricity has many applications and is widely used in household, commercial, and industrial operations. Applications include providing electrical energy for industrial vehicles, emergency diesel generators, material handling equipment (forklifts), portable electric/electronic equipment, backup emergency power for light packs, hazard warning signal lights, flashlights, and standby power supplies or uninterruptible power supplies (UPS) for computer systems. In some instances, they are used as the only source of power, whereas in others they are used as a secondary or standby power supply. In renewable energy applications, batteries are used to store electrical energy; for example, a battery pack can store electricity produced by a solar electric system. Today, they are commonly used in specialized applications such as plug-in hybrid electric vehicle energy storage systems. Batteries are often used in hybrid wind systems to store excess wind energy and then provide supplementary energy when the wind cannot generate sufficient power to meet the electric load. Batteries are also used in roadside solar-charging systems; that is, solar energy is used to charge a battery pack that supplies electrical power to some type of electrical signaling device, navigation buoys, or other remote application (e.g., emergency telephone) where it would be impractical and costly to run miles of electrical supply cable.

BATTERY TERMINOLOGY

- A *voltaic cell* is a combination of materials used to convert chemical energy into electrical energy.
- A *battery* is a group of two or more connected voltaic cells.
- An *electrode* is a metallic compound, or metal, that has an abundance of electrons (negative electrode) or an abundance of positive charges (positive electrode).
- An *electrolyte* is a solution capable of conducting an electric current.

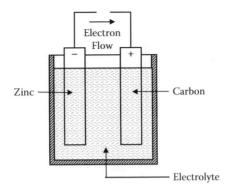

FIGURE 3.13   Simple voltaic cell.

- *Specific gravity* is defined as the ratio comparing the weight of any liquid to the weight of an equal volume of water.
- An *ampere-hour* is defined as a current of 1 ampere flowing for 1 hour.

Voltaic Cell

The simplest cell (a device that transforms chemical energy into electrical energy) is the voltaic (or galvanic) cell (see Figure 3.13). It consists of a piece of carbon (C) and a piece of zinc (Zn) suspended in a jar that contains a solution of water (H_2O) and sulfuric acid (H_2SO_4).

> *Note:* A simple cell consists of two strips, or *electrodes*, placed in a container that holds the *electrolyte*. A battery is formed when two or more cells are connected.

The electrodes are the conductors by which the current leaves or returns to the electrolyte. In the simple cell described above, they are carbon and zinc strips placed in the electrolyte. Zinc contains an abundance of negatively charged atoms, and carbon has an abundance of positively charged atoms. When the plates of these materials are immersed in an electrolyte, chemical action between the two begins. In the *dry cell* (Figure 3.14), the electrodes are the carbon rod in the center and the zinc container in which the cell is assembled. The electrolyte is the solution that acts upon the electrodes that are placed in it. The electrolyte may be a salt, an acid, or an alkaline solution. In the simple voltaic cell and in the automobile storage battery, the electrolyte is in a liquid form; in the dry cell, the electrolyte is a moist paste.

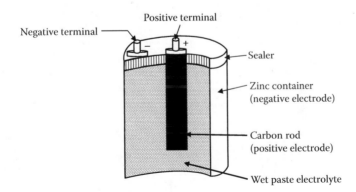

FIGURE 3.14   Dry cell (cross-sectional view).

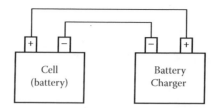

FIGURE 3.15   Hookup for charging a secondary cell with a battery charger.

PRIMARY AND SECONDARY CELLS

Primary cells are normally those that cannot be recharged or returned to good condition after their voltage drops too low. Dry cells in flashlights and transistor radios are examples of primary cells. Some primary cells have been developed to the state where they can be recharged. A *secondary cell* is one in which the electrodes and the electrolyte are altered by the chemical action that takes place when the cell delivers current. These cells are rechargeable. During recharging, the chemicals that provide electric energy are restored to their original condition. Recharging is accomplished by forcing an electric current through them in the opposite direction of the discharge. Figure 3.15 shows how a secondary cell is recharged. Some battery chargers have a voltmeter and an ammeter that indicate the charging voltage and current. The automobile storage battery is the most common example of the secondary cell.

BATTERY

As stated previously, a cell is an electrochemical device capable of supplying the energy that results from an internal chemical reaction to an external electrical circuit. A battery consists of two or more cells placed in a common container. The cells are connected in series, in parallel, or in some combination of series and parallel, depending on the amount of voltage and current required of the battery. The connection of cells in a battery is discussed in more detail later.

Battery Operation

The chemical reaction within a battery provides the voltage. This occurs when a conductor is connected externally to the electrodes of a cell which causes electrons to flow under the influence of a difference in potential across the electrodes from the zinc (negative) through the external conductor to the carbon (positive), returning within the solution to the zinc. After a short period of time, the zinc will begin to waste away because of the acid. The voltage across the electrodes depends on the materials from which the electrodes are made and the composition of the solution. The difference of potential between the carbon and zinc electrodes in a dilute solution of sulfuric acid and water is about 1.5 volts. The current that a primary cell may deliver depends on the resistance of the entire circuit, including that of the cell itself. The internal resistance of the primary cell depends on the size of the electrodes, the distance between them in the solution, and the resistance of the solution. The larger the electrodes and the closer together they are in solution (without touching), the lower the internal resistance of the primary cell and the more current it is capable of supplying to the load.

   Note: When current flows through a cell, the zinc gradually dissolves in the solution and the acid is neutralized.

Combining Cells

In many operations, battery-powered devices may require more electrical energy than one cell can provide. Various devices may require either a higher voltage or more current, and in some cases both. Under such conditions it is necessary to combine, or interconnect, a sufficient number of cells

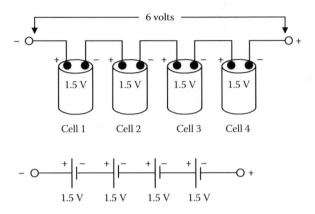

FIGURE 3.16 Cells in series (schematic representation).

to meet the higher requirements. Cells connected in series provide a higher voltage, and cells connected in parallel provide a higher current capacity. To provide adequate power when both voltage and current requirements are greater than the capacity of one cell, a series–parallel network of cells must be interconnected.

When cells are connected in *series* (Figure 3.16), the total voltage across the battery of cells is equal to the sum of the voltage of each of the individual cells. In Figure 3.16, the four 1.5-V cells in series provide a total battery voltage of 6 V. When cells are placed in series, the positive terminal of one cell is connected to the negative terminal of the other cell. The positive electrode of the first cell and negative electrode of the last cell then serve as the power takeoff terminals of the battery. The current flowing through such a battery of series cells is the same as from one cell because the same current flows through all of the series cells.

To obtain a greater current, cells can be connected in *parallel*, as shown in Figure 3.17. In this parallel connection, all of the positive electrodes are connected to one line, and all negative electrodes are connected to the other. Any point on the positive side can serve as the positive terminal of the battery and any point on the negative side can be the negative terminal. The total voltage output

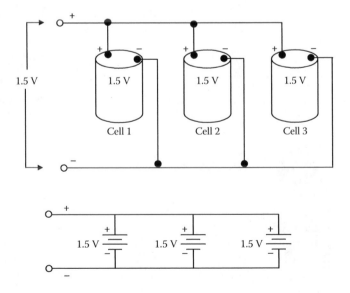

FIGURE 3.17 Cells in parallel (schematic representation).

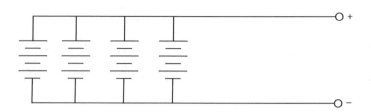

FIGURE 3.18 Series–parallel connected cells.

of a battery of three parallel cells is the same as that for a single cell (Figure 3.17), but the available current is three times that of one cell; that is, the current capacity has been increased. Identical cells in parallel all supply equal parts of the current to the load; for example, of three different parallel cells producing a load current of 210 milliamperes (mA), each cell contributes 70 mA. Figure 3.18 is a schematic of a *series–parallel* battery network supplying power to a load requiring both a voltage and current greater than one cell can provide. To provide the required increased voltage, groups of three 1.5-V cells are connected in series. To provide the required increased amperage, four series groups are connected in parallel.

Types of Batteries

In the past 30 years, several different types of batteries have been developed. In this text, we briefly discuss the dry cell battery and those batteries that are currently used to store electrical energy in a reversible chemical reaction as applied to renewable energy production. The renewable energy source (solar, wind, or hydro) produces the energy, and the battery stores it for times of low or no renewable energy production. The types of batteries used in this application and discussed here include the lead–acid, alkaline, nickel–cadmium, mercury, nickel–metal hydride, lithium-ion, and lithium-ion polymer batteries. Keep in mind that a battery does not *create* energy; rather, it *stores* energy. For most renewable energy applications, the preferred battery type is the deep-cycle battery. A deep-cycle battery is designed to deliver a constant voltage as the battery discharges. A car-starting battery, in contrast, is designed to deliver sporadic current spikes. Battery-driven vehicles, such as forklifts, golf carts, and floor sweepers, commonly use deep-cycle batteries. Deep-cycle batteries can be charged with a lower current than regular batteries. Following are descriptions of the various types of batteries (National Renewable Energy Laboratory, 2009):

- *Dry cell*—In the dry cell, or carbon–zinc cell, the electrolyte is not in a liquid state but is a moist paste. A carbon rod placed in the center of the cell is the positive terminal. The case of the cell is made of zinc and is the negative terminal (see Figure 3.14). Between the carbon electrode and the zinc case is the electrolyte, a moist, chemical, paste-like mixture. The cell is sealed to prevent the liquid in the paste from evaporating. The voltage of a cell of this type is about 1.5 V.
- *Lead–acid battery*—The lead–acid battery is a secondary cell (e.g., storage battery or rechargeable battery) that stores chemical energy until it is released as electrical energy. The lead–acid battery differs from a primary cell battery mainly in that it may be recharged, whereas most primary cells cannot be recharged. As the name implies, the lead–acid battery consists of a number of lead–acid cells immersed in a dilute solution of sulfuric acid. Each cell has two groups of lead plates; one set is the positive terminal and the other is the negative terminal. Active materials within the battery (lead plates and sulfuric acid electrolyte) react chemically to produce a flow of direct current whenever current-consuming devices are connected to the battery terminal posts. This current is produced by a chemical reaction between the active material of the plates (electrodes) and the electrolyte (sulfuric acid). This type of cell produces slightly more than 2 V. Most automobile batteries contain

six cells connected in series so the output voltage from the battery is slightly more than 12 V. In addition to being rechargeable, the main advantage of the lead–acid battery over the dry cell battery is that it can supply current for a much longer time than the average dry cell. Lead–acid batteries can be designed to be high power and are inexpensive, safe, and reliable; also, a recycling infrastructure is in place for them. But, low specific energy, poor cold-temperature performance, and short calendar and cycle life are still impediments to their use. Advanced high-power, deep-cycle lead–acid batteries are being developed for hybrid electric vehicle (HEV) applications; however, lead–acid batteries are used for residential solar electric systems because of their low maintenance requirements and cost.

Safety Note: Whenever a lead–acid storage battery is charging, the chemical action produces dangerous hydrogen gas; thus, the charging operation should take place only in a well-ventilated area.

- *Alkaline cell*—The alkaline cell is a secondary cell that gets its name from its alkaline electrolyte, potassium hydroxide. The alkaline battery has a negative electrode of zinc and a positive electrode of manganese dioxide. It generates 1.5 V.
- *Nickel–cadmium cell*—The nickel–cadmium cell, or NiCad cell, is the only dry cell that is a true storage battery with a reversible chemical reaction, allowing recharging many times. In the secondary nickel–cadmium dry cell, the electrolyte is potassium hydroxide, the negative electrode is nickel hydroxide, and the positive electrode is cadmium oxide. The operating voltage is 1.25 V. Because of its rugged characteristics (stands up well to shock, vibration, and temperature changes) and availability in a variety of shapes and sizes, it is ideally suited for powering portable communication equipment. NiCad batteries are very expensive. Moreover, although nickel–cadmium batteries are used in many electronic consumer products and have a higher specific energy and better life cycle than lead–acid batteries, they are low efficiency (65 to 80%). They do not deliver sufficient power and are not being considered for HEV applications. Cadmium is a heavy metal that is toxic and is very expensive to dispose of which reduces its desirability for use in hybrid applications.
- *Mercury cell*—The mercury cell was developed for space exploration activities, for which small transceivers and miniaturized equipment required a small power source. In addition to reduced size, the mercury cells have a good shelf life and are very rugged; they also produce a constant output voltage under different load conditions. Two types of mercury cells are available. One is a flat cell shaped like a button; the other is a cylindrical cell that looks like a standard flashlight cell. The advantage of the button-type cell is that several of them can be stacked inside one container to form a battery. A mercury cell produces 1.35 V.
- *Nickel–metal hydride battery*—Nickel–metal hydride batteries, used routinely in computer and medical equipment, offer reasonable specific energy and specific power capabilities. Their components are recyclable, but a recycling structure is not yet in place. Nickel–metal hydride batteries have a much longer life cycle than lead–acid batteries; they are safe and tolerate abuse. These batteries have been used successfully in production electric vehicles and recently in low-volume production of HEVs. The main challenges with nickel–metal hydride batteries are their high cost, high rate of self-discharge, very high gassing/waste consumption, heat generation at high temperatures, the need to control hydrogen loss, and their low cell efficiency (may be as low as 50%, typically 60 to 65%).
- *Lithium-ion battery*—Lithium-ion batteries are rapidly penetrating the laptop and cell phone markets because of their high specific energy. They also have high specific power, high energy efficiency, good high-temperature performance, and low self-discharge rates. Components of lithium-ion batteries can be recycled. These characteristics make lithium-ion batteries suitable for HEV applications; however, to make them commercially viable for HEVs, further development is necessary, including improvements in their calendar and cycle life, a greater degree of cell and battery safety, improved abuse tolerance, and acceptable costs.

- *Lithium-ion polymer battery*—Lithium-ion polymer batteries with high specific energy (i.e., high energy per unit mass), initially developed for cell phone applications, also have the potential to provide high specific power for HEV applications. Other key characteristics of the lithium-ion polymer battery are safety and a good cycle life. The battery could be commercially viable if the costs are lowered and even higher specific power batteries are developed.

Battery Characteristics

Batteries are generally classified by their various characteristics. Parameters such as internal resistance, specific gravity, capacity, shelf life, C rate, mid-point voltage (MPV), gravimetric energy density, volumetric energy density, constant-voltage charge, constant-current charge, and specific power are used to describe and classify batteries by operation and type.

When discussing *internal resistance*, it is important to keep in mind that a battery is a DC voltage generator. As such, the battery has internal resistance or equivalent series resistance (ESR). In a chemical cell, the resistance of the electrolyte between the electrodes is responsible for most of the internal resistance of the cell. Because any current in the battery must flow through the internal resistance, this resistance is in series with the generated voltage. With no current, the voltage drop across the resistance is zero, so fully generated voltage develops across the output terminals. This is the open-circuit voltage, or no-load voltage. If a load resistance is connected across the battery, the load resistance is in series with internal resistance. When current flows in this circuit, the internal voltage drop reduces the terminal voltage of the battery.

The ratio of the weight of a certain volume of liquid to the weight of the same volume of water is the *specific gravity* of the liquid. Pure sulfuric acid has a specific gravity of 1.835 because it weighs 1.835 times as much as water per unit volume. The specific gravity of a mixture of sulfuric acid and water varies with the strength of the solution from 1.000 to 1.830.

The specific gravity of the electrolyte solution in a lead–acid cell ranges from 1.210 to 1.300 for new, fully charged batteries. The higher the specific gravity, the less internal resistance of the cell and the higher the possible load current. As the cell discharges, the water formed dilutes the acid and the specific gravity gradually decreases to about 1.150, at which time the cell is considered to be fully discharged.

The specific gravity of the electrolyte is measured with a *hydrometer*, which has a compressible rubber bulb at the top, a glass barrel, and a rubber hose at the bottom of the barrel. When taking readings with a hydrometer, the decimal point is usually omitted; for example, a specific gravity of 1.260 is read simply as "twelve-sixty." A hydrometer reading of 1210 to 1300 indicates a full charge; about 1250, half-charge; and 1150 to 1200, complete discharge.

The *capacity* of a battery is measured in ampere-hours (Ah). The ampere-hour capacity is equal to the product of the current in amperes and the time in hours during which the battery is supplying this current; it is defined as the amount of current that a battery can deliver for 1 hour before the battery voltage reaches the end-of-life point. In other words, 1 ampere-hour is the equivalent of drawing 1 amp steadily for 1 hour or 2 amps steadily for 1/2 an hour. A typical 12-volt system may have 800 ampere-hours of battery capacity; that is, the battery can draw 100 amps for 8 hours if fully discharged and starting from a fully charged state. This is equivalent to 1200 watts for 8 hours (power in watts = amps × volts). The ampere-hour capacity varies inversely with the discharge current. The size of a cell is determined generally by its ampere-hour capacity. The capacity of a storage battery determines how long it will operate at a given discharge rate and depends on many factors; the most important of these are as follows:

- The area of the plates in contact with the electrolyte
- The quantity and specific gravity of the electrolyte
- The type of separators

- The general condition of the battery (e.g., degree of sulfating, buckled plates, warped separators, sediment in bottom of cells)
- The final limiting voltage

The *shelf life* of a cell is that period of time during which the cell can be stored without losing more than approximately 10% of its original capacity. The loss of capacity of a stored cell is due primarily to the electrolyte in a wet cell drying out and to chemical actions that alter the materials within the cell. The shelf life of a cell can be extended by keeping it in a cool, dry place.

The *C rate* of a current is numerically equal to the ampere-hour rating of the cell. Charge and discharge currents are typically expressed in fractions of multiples of the C rate. The *mid-point voltage (MPV)* is the nominal voltage of the cell, the voltage that is measured when the battery has discharged 50% of its total energy. The *gravimetric energy density* of a battery is a measure of how much energy a battery contains in comparison to its weight. The *volumetric energy density* of a battery is a measure of how much energy a battery contains in comparison to its volume.

A *constant-voltage charger* is a circuit that recharges a battery by sourcing only enough current to force the battery voltage to a fixed value. A *constant-current charger* is a circuit that charges a battery by sourcing a fixed current into the battery, regardless of battery voltage. In batteries, *specific power* usually refers to the power-to-weight ratio, measured in kilowatts per kilogram (kW/kg).

DC CIRCUITS

An electric circuit includes an *energy source* (the source of the electromotive force or voltage, such as a battery or generator), a conductor (wire), a load, and a means of control (see Figure 3.19). The energy source could be a battery, as shown in Figure 3.19, or some other means of producing a voltage. The *load* that dissipates the energy could be a lamp, a resistor, or some other device that does useful work, such as an electric toaster, a power drill, a radio, a soldering iron, a laptop computer, or a battery pack that stores energy and serves as a backup to a renewable energy source such as solar or wind. Again, *conductors* are wires that offer low resistance to current; they connect all the loads in the circuit to the voltage source. No electrical device dissipates energy unless current flows through it. Because conductors, or wires, are not perfect conductors, they heat up (dissipate energy), so they are actually part of the load. For simplicity, however, we usually think of the connecting wiring as having no resistance, as it would be tedious to assign a very low resistance value to the wires every time we wanted to solve a problem. *Control devices* might be switches, variable resistors, circuit breakers, fuses, or relays.

A complete pathway for current flow, or *closed circuit* (Figure 3.19), is an unbroken path for current from the electromotive force, through a load, and back to the source. An *open circuit* (see Figure 3.20) has a break in the circuit (e.g., open switch) that does not provide a complete path for current.

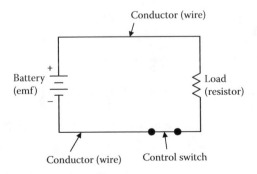

FIGURE 3.19 Simple closed circuit.

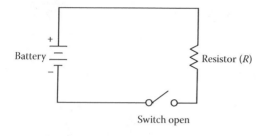

Switch open

FIGURE 3.20   Open circuit.

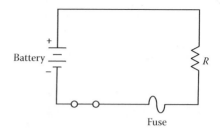

Fuse

FIGURE 3.21   Simple fused circuit.

> *Note:* Current flows from the negative (–) terminal of the battery (Figures 3.19 and 3.20), through the load to the positive (+) battery terminal, and continues through the battery from the positive (+) terminal back to the negative (–) terminal. As long as this pathway is unbroken, it is a closed circuit and current will flow; however, if the path is broken at *any* point, it is an open circuit and no current flows.

To protect a circuit, a *fuse* is placed directly into the circuit (Figure 3.21). A fuse will open the circuit whenever a dangerous large current begins to flow; a short-circuit condition occurs when an accidental connection between two points in a circuit offers very little resistance. A fuse will permit currents smaller than the fuse value to flow but will melt and therefore break or open the circuit if a larger current flows.

Schematic Representation

The simple circuits shown in Figures 3.19, 3.20, and 3.21 are displayed in *schematic* form. A schematic is a simplified drawing that represents the electrical, not the physical, situation in a circuit. The symbols used in schematic diagrams are the electrician's shorthand; they make the diagrams easier to draw and easier to understand. Consider the symbol shown in Figure 3.22 that is used to represent a battery power supply. The symbol is rather simple and straightforward, but it is very important. By convention, the shorter line in the symbol for a battery represents the *negative* terminal. It is important to remember this, because sometimes when you examine the schematic it is necessary to note the direction of current flow, which is from negative to positive. The battery symbol shown in Figure 3.22 represents a single cell, so only one short and one long line are used. The number of lines used to represent a battery vary (and they are not necessarily equivalent to the number of cells), but they are always in pairs, with long and short lines alternating. In the circuit shown in Figure 3.21, the current would flow in a *counterclockwise* direction. If the long and short lines of the battery symbol (see Figure 3.22) were reversed, the current in the circuit shown in Figure 3.21 would flow *clockwise*.

> *Note:* When studying electricity and electronics, many circuits are analyzed that consist mainly of specially designed resistive components. These components are *resistors*. Throughout the remaining analysis of the basic circuit, the resistive component will be a physical resistor; however, resistive components could be any of several electrical devices.

Keep in mind that, in the simple circuits shown in the figures to this point, we have only illustrated and discussed a few of the many symbols used in schematics to represent circuit components. (Other symbols will be introduced as we need them.) It is also important to keep in mind that a closed loop of wire (conductor) is not necessarily a circuit. A source of voltage must be included to make it an electric circuit. In any electric circuit where electrons move around a closed loop, current, voltage, and resistance are present. The

FIGURE 3.22   Schematic symbol for a battery.

physical pathway for current flow is actually the circuit. By knowing any two of the three quantities, such as voltage and current, the third (resistance) may be determined. This is done using Ohm's law, which is the foundation on which electrical theory is based.

OHM'S LAW

Simply put, *Ohm's law* defines the relationship among current, voltage, and resistance in electric circuits. Ohm's law can be expressed mathematically in three ways:

1. The *current* (I) in a circuit is equal to the voltage applied to the circuit divided by the resistance of the circuit. Stated another way, the current in a circuit is *directly* proportional to the applied voltage and *inversely* proportional to the circuit resistance. Ohm's law may be expressed by the equation:

$$I = E/R \tag{3.1}$$

where I = current in amps, E = voltage in volts, and R = resistance in ohms.
2. The *resistance* (R) of a circuit is equal to the voltage applied to the circuit divided by the current in the circuit:

$$R = E/I \tag{3.2}$$

3. The applied *voltage* (E) to a circuit is equal to the product of the current and the resistance of the circuit:

$$E = I \times R = IR \tag{3.3}$$

If any two of the quantities in Equations 3.1, 3.2, or 3.3 are known, the third may be easily found. Let's look at an example.

■ EXAMPLE 3.1

Problem: Figure 3.23 shows a circuit containing a resistance of 6 ohms and a source voltage of 3 volts. How much current flows in the circuit?
Solution:

Given:
  R = 6 ohms
  E = 3 volts
  I = ?

$$I = E/R = 3/6 = 0.5 \text{ amp}$$

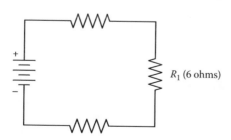

FIGURE 3.23 Determining current in a simple circuit.

To observe the effect of source voltage on circuit current, we can use the circuit shown in Figure 3.23 but double the voltage to 6 volts.

■ EXAMPLE 3.2

Problem: How much current is flowing when $E = 6$ volts and $R = 6$ ohms?
Solution:

$$I = E/R = 6/6 = 1 \text{ amp}$$

Notice that as the source of voltage doubles, the circuit current also doubles. To verify that current is inversely proportional to resistance, assume that the resistor in Figure 3.23 has a value of 12 ohms.

Note: Circuit current is directly proportional to the applied voltage and will change by the same factor that the voltage changes.

■ EXAMPLE 3.3

Problem: Given $E = 3$ volts and $R = 12$ ohms, what is I?
Solution:

$$I = E/R = 3/12 = 0.25 \text{ amp}$$

Comparing this current of 0.25 amp for the 12-ohm resistor to the 0.5-amp of current obtained with the 6-ohm resistor shows that doubling the resistance will reduce the current to one half the original value.

Note: Circuit current is inversely proportional to the circuit resistance.

Recall that if you know any two quantities of E, I, and R, you can calculate the third. In many circuit applications, current is known and either the voltage or the resistance will be the unknown quantity. To solve a problem in which current and resistance are known, the basic formula for Ohm's law must be transposed to solve for E, for I, or for R; however, the Ohm's law equations can be memorized and practiced effectively by using an Ohm's law circle (see Figure 3.24).

To find the equation for E, I, or R when two quantities are known, cover the unknown third quantity with your finger, as shown in Figure 3.25:

$$I = E/R \qquad R = E/I \qquad E = I \times R$$

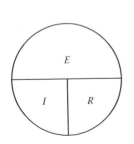

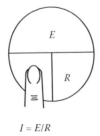
$I = E/R$

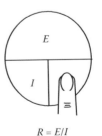
$R = E/I$

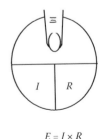
$E = I \times R$

FIGURE 3.24 Ohm's law circle. **FIGURE 3.25** Putting the Ohm's law circle to work.

■ EXAMPLE 3.4

Problem: Find *I* when *E* = 120 V and *R* = 40 ohms.
Solution: Place your finger on *I*, as shown in the figure. Use Equation 3.1 to find the unknown *I*:

$$I = E/R = 120/40 = 3 \text{ A}$$

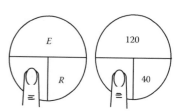

■ EXAMPLE 3.5

Problem: Find *R* when *E* = 220 V and *I* = 10 A.
Solution: Place your finger on *R*, as shown in the figure:

$$R = E/I = 220/10 = 22 \text{ ohms}$$

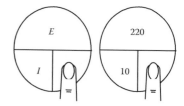

■ EXAMPLE 3.6

Problem: Find *E* when *I* = 2.5 A and *R* = 25 ohms.
Solution: Place your finger on *E*, as shown in the figure.

$$E = I \times R = 2.5 \times 25 = 62.5 \text{ V}$$

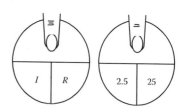

Note: In the previous examples, we have demonstrated how we can use the Ohm's law circle to help us solve simple voltage, current, and amperage problems. Beginning students, however, are cautioned not to rely entirely on the use of this circle when transposing simple formulas but rather to use it to supplement their knowledge of the algebraic method. Algebra is a basic tool in the solution of electrical problems, and the importance of knowing how to use it should not be underemphasized or bypassed after the operator has learned a shortcut method such as the one shown in this circle.

■ EXAMPLE 3.7

Problem: An electric light bulb draws 0.5 A when operating on a 120-V DC circuit. What is the resistance of the bulb?
Solution: The first step in solving a circuit problem is to sketch a schematic diagram of the circuit itself, labeling each of the parts and showing the known values (see Figure 3.26). Because *I* and *E* are known, we use Equation 3.2 to solve for *R*:

$$R = E/I = 120/0.5 = 240 \text{ ohms}$$

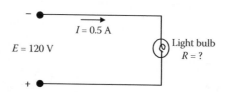

FIGURE 3.26   Circuit for Example 3.7.

ELECTRIC POWER

Power, whether electrical or mechanical, pertains to the rate at which work is being done, so the power consumption in your plant is related to current flow. A large electric motor or air dryer consumes more power (and draws more current) in a given length of time than, for example, an indicating light on a motor controller. *Work* is done whenever a force causes motion. If a mechanical force is used to lift or move a weight, work is done; however, force exerted *without* causing motion, such as the force of a compressed spring acting between two fixed objects, does not constitute work.

Note: Power is the rate at which work is done.

ELECTRICAL POWER CALCULATIONS

The electric power (P) used in any part of a circuit is equal to the current (I) in that part multiplied by the voltage (E) across that part of the circuit. In equation form:

$$P = E \times I \tag{3.4}$$

where
P = Power (watts, W).
E = Voltage (volts, V).
I = Current (amps, A).

If we know the current (I) and the resistance (R) but not the voltage, we can find the power (P) by using Ohm's law for voltage, so by substituting Equation 3.3,

$$E = I \times R$$

into Equation 3.4, we obtain

$$P = (I \times R) \times I = I^2 \times R \tag{3.5}$$

In the same manner, if we know the voltage (V) and the resistance (R) but not the current (I), we can find P by using Ohm's law for current, so by substituting Equation 3.1:

$$I = E/R$$

into Equation 3.4, we obtain

$$P = E \times (E/R) = E^2/R \tag{3.6}$$

Note: If we know any two quantities, we can calculate the third.

■ EXAMPLE 3.8

Problem: The current through a 200-ohm resistor to be used in a circuit is 0.25 A. Find the power rating of the resistor.
Solution: Because I and R are known, use Equation 3.5 to find P:

$$P = I^2 \times R = (0.25)^2 \times 200 = 0.0625 \times 200 = 12.5 \text{ W}$$

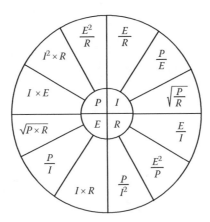

FIGURE 3.27 Ohm's law circle: summary of basic formulas.

Note: The power rating of any resistor used in a circuit should be twice the wattage calculated by the power equation to prevent the resistor from burning out. Thus, the resistor used in Example 3.8 should have a power rating of 25 watts.

■ EXAMPLE 3.9

Problem: How many kilowatts of power are delivered to a circuit by a 220-V generator that supplies 30 A to the circuit?
Solution:

$$P = E \times I = 220 \times 30 = 6600 \text{ W} = 6.6 \text{ kW}$$

■ EXAMPLE 3.10

Problem: If the voltage across a 30,000-ohm resistor is 450 V, what is the power dissipated in the resistor?
Solution:

$$P = E^2/R = (450)^2/30,000 = 202,500/30,000 = 6.75 \text{ W}$$

In this section, *P* was expressed in terms of various pairs of the other three basic quantities *E*, *I*, and *R*. In practice, you should be able to express any one of the three basic quantities, as well as *P*, in terms of any two of the others. Figure 3.27 is a summary of 12 basic formulas you should know. The four quantities *E*, *I*, *R*, and *P*, are at the center of the figure. Adjacent to each quantity are three segments. Note that in each segment the basic quantity is expressed in terms of two other basic quantities, and no two segments are alike.

ELECTRIC ENERGY

Energy (the mechanical definition) is defined as the ability to do work (energy and time are essentially the same and are expressed in identical units). Energy is expended when work is done, because it takes energy to maintain a force when that force acts through a distance. The total energy expended to do a certain amount of work is equal to the working force multiplied by the distance through which the force moved to do the work. In electricity, total energy expended is equal to the *rate* at which work is done multiplied by the length of time the rate is measured. Essentially, energy (watts) is equal to the power (P) times time (t). The kilowatt-hour (kWh) is a unit commonly used for large amounts of electric energy or work. The amount of kilowatt-hours is calculated as the product of the power in kilowatts (kW) and the time in hours (h) during which the power is used:

$$kWh = kW \times h \tag{3.7}$$

■ EXAMPLE 3.11

Problem: How much energy is delivered in 4 hours by a generator supplying 12 kW?
Solution:

$$kWh = kW \times h = 12 \times 4 = 48$$

The energy delivered is 48 kWh.

SERIES DC CIRCUIT CHARACTERISTICS

As previously mentioned, an electric circuit is made up of a voltage source, the necessary connecting conductors, and the effective load. If the circuit is arranged so the electrons have only *one* possible path, the circuit is a *series circuit*. A series circuit, then, is defined as a circuit that contains only one path for current flow. Figure 3.28 shows a series circuit having several loads (resistors).

Note: A *series circuit* is a circuit having only one path for current to flow along.

RESISTANCE

To follow its electrical path, the current in a series circuit must flow through resistors inserted in the circuit (Figure 3.28). Thus, each additional resistor offers added resistance. In a series circuit, the total circuit resistance (R_T) is equal to the sum of the individual resistances:

$$R_T = R_1 + R_2 + R_3 \dots R_n \tag{3.8}$$

where
  R_T = Total resistance (ohms).
  R_1, R_2, R_3 = Resistance in series (ohms).
  R_n = Any number of additional resistors in the equation.

■ EXAMPLE 3.12

Problem: Three resistors of 10 ohms, 12 ohms, and 25 ohms are connected in series across a battery whose emf is 110 volts (Figure 3.29). What is the total resistance?

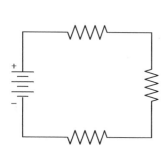

FIGURE 3.28 Series circuit.

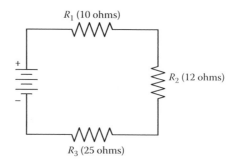

FIGURE 3.29 Solving for total resistance in a series circuit.

Solution:

   Given:
$$R_1 = 10 \text{ ohms}$$
$$R_2 = 12 \text{ ohms}$$
$$R_3 = 25 \text{ ohms}$$
$$R_T = ?$$

$$R_T = R_1 + R_2 + R_3 = 10 + 12 + 25 = 47 \text{ ohms}$$

Transposition can be used in some circuit applications where the total resistance is known but the value of a circuit resistor has to be determined.

■ EXAMPLE 3.13

Problem: The total resistance of a circuit containing three resistors is 50 ohms (see Figure 3.30). Two of the circuit resistors are 12 ohms each. Calculate the value of the third resistor.
Solution:

   Given:
$$R_T = 50 \text{ ohms}$$
$$R_1 = 12 \text{ ohms}$$
$$R_2 = 12 \text{ ohms}$$
$$R_3 = ?$$

$$R_T = R_1 + R_2 + R_3$$

Subtracting $(R_1 + R_2)$ from both sides of the equation, we obtain:

$$R_3 = R_T - R_1 - R_2$$

$$R_3 = 50 - 12 - 12$$

$$R_3 = 50 - 24 = 26 \text{ ohms}$$

Note: When resistances are connected in series, the total resistance in the circuit is equal to the sum of the resistances of all parts of the circuit.

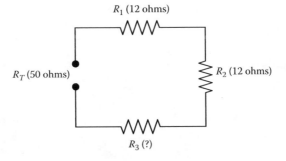

FIGURE 3.30   Calculating the value of one resistance in a series circuit.

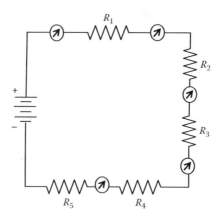

FIGURE 3.31 DC current in a series circuit.

CURRENT

Because there is but one path for current in a series circuit, the same current (*I*) must flow through each part of the circuit. To determine the current throughout a series circuit, only the current through one of the parts must be known. The fact that the same current flows through each part of a series circuit can be verified by inserting ammeters into the circuit at various points as shown in Figure 3.31. As indicated in Figure 3.31, each meter indicates the same value of current.

> **Note:** In a series circuit, the same current flows in every part of the circuit. *Do not* add the currents in each part of the circuit to obtain *I*.

VOLTAGE

The voltage drop across the resistor in the basic DC circuit is the total voltage across the circuit and is equal to the applied voltage. The total voltage across a series circuit is also equal to the applied voltage but consists of the sum of two or more individual voltage drops. This statement can be proven by an examination of the circuit shown in Figure 3.32. In this circuit, a source potential (E_T) of 30 volts is impressed across a series circuit consisting of two 6-ohm resistors. The total resistance of the circuit is equal to the sum of the two individual resistances, or 12 ohms. Using Ohm's law, the circuit current may be calculated as follows:

$$I = E_T/R_T = 30/12 = 2.5 \text{ amps}$$

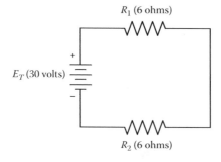

FIGURE 3.32 Calculating total resistance in a series circuit.

Because we know that the value of the resistors is 6 ohms each, and the DC current through the resistors is 2.5 amps, the voltage drops across the resistors can be calculated. The voltage (E_1) across R_1 is, therefore:

$$E_1 = I \times R_1$$

$$E_1 = 2.5 \text{ amps} \times 6 \text{ ohms} = 15 \text{ volts}$$

Because R_2 is the same ohmic value as R_1 and carries the same current, the voltage drop across R_2 is also equal to 15 volts. Adding these two 15-volt drops together gives a total drop of 30 volts, exactly equal to the applied voltage. For a series circuit, then:

$$E_T = E_1 + E_2 + E_3 + \ldots + E_n \tag{3.9}$$

where
  E_T = Total voltage (V).
  E_1 = Voltage across resistance R_1 (V).
  E_2 = Voltage across resistance R_2 (V).
  E_3 = Voltage across resistance R_3 (V).

■ EXAMPLE 3.14

Problem: A series DC circuit consists of three resistors having values of 10, 20, and 40 ohms. Find the applied voltage if the current through the 20-ohm resistor is 2.5 amps.
Solution: To solve this problem, first draw a circuit diagram and label it as shown in Figure 3.33.

  Given:
    R_1 = 10 ohms
    R_2 = 20 ohms
    R_3 = 40 ohms
    I = 2.5 amps

Because the circuit involved is a DC series circuit, the same 2.5 amps of current flows through each resistor. Using Ohm's law, the voltage drops across each of the three resistors can be calculated:

E_1 = 25 V
E_2 = 50 V
E_3 = 100 V

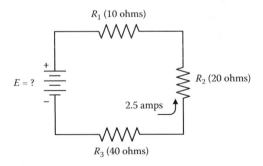

FIGURE 3.33   Solving for applied voltage in a series circuit.

When the individual drops are known they can be added to find the total or applied voltage:

$$E_T = E_1 + E_2 + E_3$$

$$E_T = 25 \text{ V} + 50 \text{ V} + 100 \text{ V} = 175 \text{ V}$$

Note: The total voltage (E_T) across a DC series circuit is equal to the sum of the voltages across each resistance of the circuit.

Note: The voltage drops that occur in a series circuit are in direct proportions to the resistance across which they appear. This is the result of having the same current flow through each resistor. Thus, the larger the resistor, the larger will be the voltage drop across it.

POWER

Each resistor in a DC series circuit consumes power. This power is dissipated in the form of heat. Because this power must come from the source, the total power must be equal in amount to the power consumed by the circuit resistances. In a series circuit, the total power is equal to the sum of the powers dissipated by the individual resistors. Total power (P_T), then, is equal to

$$P_T = P_1 + P_2 + P_3 + \dots + P_n \tag{3.10}$$

where
    P_T = Total power (W).
    P_1 = Power used in the first part (W).
    P_2 = Power used in the second part (W).
    P_3 = Power used in the third part (W).
    P_n = Power used in the nth part (W).

■ EXAMPLE 3.15

Problem: A DC series circuit consists of three resistors having values of 5, 15, and 20 ohms. Find the total power dissipation when 120 volts is applied to the circuit (see Figure 3.34).
Solution:

    Given:
        $R_1 = 5$ ohms
        $R_2 = 15$ ohms
        $R_3 = 20$ ohms
        $E = 120$ volts

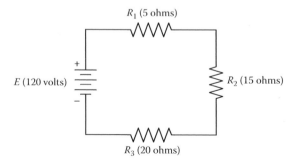

FIGURE 3.34   Solving for total power in a series circuit.

First find the total resistance:

$$R_T = R_1 + R_2 + R_3 = 5 + 15 + 20 = 40 \text{ ohms}$$

Using total resistance and the applied voltage, calculate the circuit current:

$$I = E_T/R_T = 120/40 = 3 \text{ amps}$$

Using the power formula, calculate the individual power dissipations:

For resistor R_1:

$$P_1 = I^2 \times R_1 = (3)^2 \times 5 = 45 \text{ watts}$$

For resistor R_2:

$$P_2 = I^2 \times R_2 = (3)^2 \times 15 = 135 \text{ watts}$$

For resistor R_3:

$$P_3 = I^2 \times R_3 = (3)^2 \times 20 = 180 \text{ watts}$$

To obtain total power:

$$P_T = P_1 + P_2 + P_3 = 45 + 135 + 180 = 360 \text{ watts}$$

To check the answer, calculate the total power delivered by the source:

$$P = E \times I = 120 \text{ volts} \times 3 \text{ amps} = 360 \text{ watts}$$

Thus, the total power is equal to the sum of the individual power dissipations.

We found that Ohm's law can be used for total values in a DC series circuit as well as for individual parts of the circuit. Similarly, the formula for power may be used for total values:

$$P_T = E_T \times I \tag{3.11}$$

SUMMARY OF THE RULES FOR SERIES DC CIRCUITS

To this point, we have covered many of the important factors governing the operation of basic DC series circuits. In essence, what we have really done is to lay a strong foundation to build upon in preparation for more advanced circuit theory that follows. Following is a summary of the important factors governing the operation of a DC series circuit:

- The same current flows through each part of a series circuit.
- Total resistance of a series circuit is equal to the sum of the individual resistances.
- Total voltage across a series circuit is equal to the sum of the individual voltage drops.
- Voltage drop across a resistor in a series circuit is proportional to the size of the resistor.
- Total power dissipated in a series circuit is equal to the sum of the individual dissipations.

General DC Series Circuit Analysis

Now that we have discussed the pieces required to put together the puzzle of DC series circuit analysis, we now move on to the next step in the process: series circuit analysis in total.

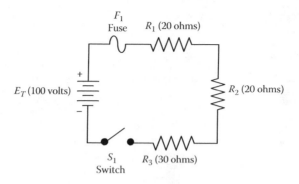

FIGURE 3.35   Solving for various values in a series circuit.

■ EXAMPLE 3.16

Problem: Three resistors of 20, 20, and 30 ohms are connected across a battery supply rated at 100 volts terminal voltage. Completely solve the circuit shown in Figure 3.35.

   Note: When current is known, the voltage drops and power dissipations can be calculated.

Solution: The total resistance is

$$R_T = R_1 + R_2 + R_3 = 20 \text{ ohms} + 20 \text{ ohms} + 30 \text{ ohms} = 70 \text{ ohms}$$

By Ohm's law, the current is

$$I = E/R_T = 100/70 = 1.43 \text{ amps (rounded)}$$

The voltage (E_1) across R_1 is

$$E_1 = I \times R_1 = 1.43 \text{ amps} \times 20 \text{ ohms} = 28.6 \text{ volts}$$

The voltage (E_2) across R_2 is

$$E_2 = I \times R_2 = 1.43 \text{ amps} \times 20 \text{ ohms} = 28.6 \text{ volts}$$

The voltage (E_3) across R_3 is

$$E_3 = I \times R_3 = 1.43 \text{ amps} \times 30 \text{ ohms} = 42.9 \text{ volts}$$

The power dissipated by R_1 is

$$P_1 = E_1 \times I = 28.6 \text{ volts} \times 1.43 \text{ amps} = 40.9 \text{ watts}$$

The power dissipated by R_2 is

$$P_2 = E_2 \times I = 28.6 \text{ volts} \times 1.43 \text{ amps} = 40.9 \text{ watts}$$

The power dissipated by R_3 is

$$P_3 = E_3 \times I = 42.9 \text{ volts} \times 1.43 \text{ amps} = 61.3 \text{ watts}$$

The total power dissipated is

$$P_T = E_T \times I = 100 \text{ volts} \times 1.43 \text{ amps} = 143 \text{ watts}$$

Note: When applying Ohm's law to a DC series circuit, consider whether the values used are component values or total values. When the information available enables the use of Ohm's law to find total resistance, total voltage, and total current, total values must be inserted into the formula.

To find total resistance:

$$R_T = E_T / I_T$$

To find total voltage:

$$E_T = I_T \times R_T$$

To find total current:

$$I_T = E_T / R_T$$

KIRCHOFF'S VOLTAGE LAW

Kirchoff's voltage law states that the voltage applied to a closed circuit equals the sum of the voltage drops in that circuit. It should be obvious that this fact was used in the study of series circuits to this point. It was expressed as follows:

Voltage applied = Sum of voltage drops

$$E_A = E_1 + E_2 + E_3 \tag{3.12}$$

where E_A is the applied voltage and E_1, E_2, and E_3 are voltage drops.

Another way of stating Kirchoff's law is that the algebraic sum of the instantaneous emf values and voltage drops around any closed circuit is zero. Kirchoff's law can be used to solve circuit problems that would be difficult and often impossible to solve with only knowledge of Ohm's law. When Kirchoff's law is properly applied, an equation can be set up for a closed loop and the unknown circuit values may be calculated.

POLARITY OF VOLTAGE DROPS

When there is a voltage drop across a resistance, one end must be more positive or more negative than the other end. The polarity of the voltage drop is determined by the direction of current flow. In the circuit shown in Figure 3.36 the current is seen to be flowing in a counterclockwise direction due to the arrangement of the battery source (E). Notice that the end of resistor R_1 into which the current flows is marked negative (–). The end of R_1 at which the current leaves is marked positive (+). These polarity markings are used to show that the end of R_1 into which the current flows is at a higher negative potential than is the end of the resistor at which the current leaves. Point A is thus more negative than point B. Point C, which is at the same potential as point B, is labeled negative to indicate that point C, though positive with respect to point A, is more negative than point D. To say a point is positive (or negative), without stating what it is positive *with respect to* has no meaning. Kirchoff's voltage law written as an equation is

$$E_a + E_b + E_c + \ldots + E_n = 0 \tag{3.13}$$

where E_a, E_b, etc. are the voltage drops and emf values around any closed-circuit loop.

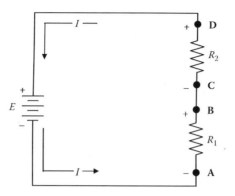

FIGURE 3.36   Polarity of voltage drops.

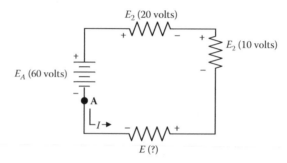

FIGURE 3.37   Determining unknown voltage in a series circuit.

■ EXAMPLE 3.17

Problem: Three resistors are connected across a 60-volt source. What is the voltage across the third resistor if the voltage drops across the first two resistors are 10 volts and 20 volts?

Solution: Draw a diagram like the one in Figure 3.37 and assume a direction of current as shown. Using this current, place the polarity markings at each end of each resistor and on the terminals of the source. Starting at point A, trace around the circuit in the direction of current flow, recording the voltage and polarity of each component. Starting at point A, these voltages would be as follows:

$$E_a + E_b + E_c + \ldots + E_n = 0$$

From the circuit:

$$(+E_?) + (+E_2) + (+E_3) - (E_A) = 0$$

Substituting values from circuit:

$$E_? + 10 + 20 - 60 = 0$$

$$E_? - 30 = 0$$

$$E_? = 30 \text{ volts}$$

Note: In much the same way, a problem can be solved in which the current is the unknown quantity.

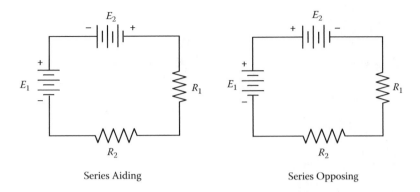

FIGURE 3.38 Aiding and opposing sources.

SERIES AIDING AND OPPOSING SOURCES

Sources of voltage that cause current to flow in the same direction are considered to be *series aiding* and their voltages add. Sources of voltage that would tend to force current in opposite directions are said to be *series opposing*, and the effective source voltage is the difference between the opposing voltages. When two opposing sources are inserted into a circuit, current flow would be in a direction determined by the larger source. Examples of series aiding and opposing sources are shown in Figure 3.38.

KIRCHOFF'S LAW AND MULTIPLE SOURCE SOLUTIONS

Kirchoff's law can be used to solve multiple-source circuit problems. When applying this method, the exact same procedure is used for multiple-source circuits as for single-source circuits. This is demonstrated by the following example.

■ EXAMPLE 3.18

Problem: Find the amount of current in the circuit shown in Figure 3.39.
Solution: Start at point A:

$$E_a + E_b + E_c + \ldots + E_n = 0$$

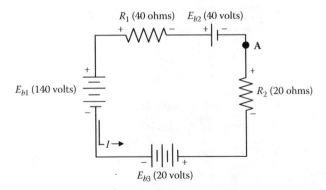

FIGURE 3.39 Solving for circuit current in a multiple source circuit.

From the circuit:

$$E_{b2} + E_1 - E_{b1} + E_{b3} + E_2 = 0$$

$$40 + 40I - 140 + 20 + 20I = 0$$

Combining like terms:

$$60I - 80 = 0$$

$$60I = 80$$

$$I = 1.33 \text{ amps}$$

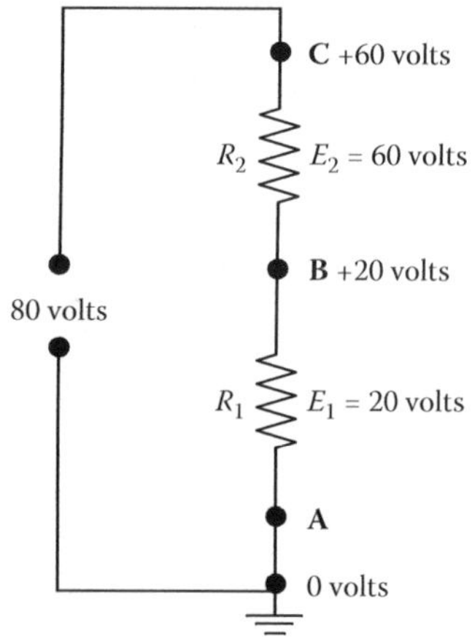

GROUND

The term *ground* is used to denote a common electrical point of zero potential. The reference point of a circuit is always considered to be at zero potential. The earth (ground) is said to be at zero potential. In Figure 3.40, point A is the zero reference or ground and is symbolized as such. Point C is 60 volts positive and point B is 20 volts positive with respect to ground. The common ground for much electrical/electronics equipment is the metal chassis. The value of ground is noted when considering its contribution to economy, simplification of schematics, and ease of measurement. When completing each electrical circuit, common points of a circuit at zero potential are connected directly to the metal chassis, thereby eliminating a large amount of connecting wire. An example of a grounded circuit is illustrated in Figure 3.41.

FIGURE 3.40   Use of ground symbols.

> **Note:** Most voltage measurements used to check proper circuit operation in electronic equipment are taken with respect to ground. One meter lead is attached to ground and the other meter lead is moved to various test points.

OPEN AND SHORT CIRCUITS

A circuit is *open* if a break in the circuit does not provide a complete path for DC or AC current. Figure 3.42 shows an open circuit, because the fuse is blown. To protect a circuit, a fuse is placed directly into the circuit (see Figure 3.42). A fuse will open the circuit whenever a dangerously large current begins to flow. A fuse will permit currents smaller than the fuse value to flow but will melt and therefore break or open the circuit if a larger current flows. A dangerously large current will flow when a *short circuit* occurs. A short circuit is usually caused by an accidental connection between two points in a circuit that offers very little resistance and passes an abnormal amount of current (see Figure 3.43). A short circuit often occurs as a result of improper wiring or broken insulation.

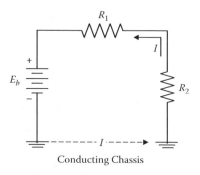

Conducting Chassis

FIGURE 3.41   Ground used as a conductor.

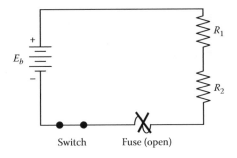

FIGURE 3.42 Open circuit with fuse blown.

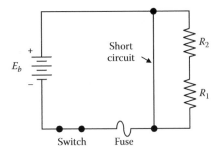

FIGURE 3.43 Short circuit.

PARALLEL DC CIRCUITS

The principles we applied to solving simple DC series circuit calculations for determining the reactions of such quantities as voltage, current, and resistance also can be used in DC parallel and series–parallel circuits.

PARALLEL CIRCUIT CHARACTERISTICS

A *parallel circuit* is defined as one having two or more components connected across the same voltage source (see Figure 3.44). Recall that a series circuit has only one path for current flow. As additional loads (e.g., resistors) are added to the circuit, the total resistance increases and the total current decreases. This is *not* the case for a parallel circuit. In a parallel circuit, each load (or branch) is connected directly across the voltage source. In Figure 3.44, commencing at the voltage source (E_b) and tracing counterclockwise around the circuit, two complete and separate paths can be identified in which current can flow. One path is traced from the source through resistance R_1 and back to the source; the other, from the source through resistance R_2 and back to the source.

VOLTAGE IN PARALLEL CIRCUITS

Recall that in a series circuit the source voltage divides proportionately across each resistor in the circuit. In a parallel circuit (see Figure 3.44), the same voltage is present across all the resistors of a parallel group. This voltage is equal to the applied voltage (E_b) and can be expressed in equation form as

$$E_b = E_{R1} = E_{R2} = E_{Rn} \tag{3.14}$$

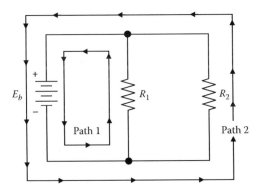

FIGURE 3.44 Basic parallel circuit.

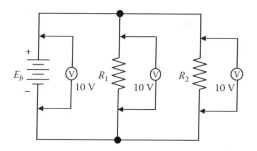

FIGURE 3.45   Voltage comparison in a parallel circuit.

We can verify Equation 3.14 by taking voltage measurements across the resistors of a parallel circuit, as illustrated in Figure 3.45. Notice that each voltmeter indicates the same amount of voltage; that is, the voltage across each resistor is the same as the applied voltage.

Note: In a parallel circuit, the voltage remains the same throughout the circuit.

■ EXAMPLE 3.19

Problem: Assume that the current through a resistor of a parallel circuit is known to be 4 milliamperes (mA) and the value of the resistor is 40,000 ohms. Determine the potential (voltage) across the resistor. The circuit is shown in Figure 3.46.
Solution: First find E_{R2} and then E_b.

Given:
 $R_2 = 40,000$ ohms
 $I_{R2} = 4.0$ ma

Select the proper equation:

$$E = I \times R$$

Substitute known values:

$$E_{R2} = I_{R2} \times R_2$$

$$E_{R2} = 4.0 \text{ mA} \times 40,000 \text{ ohms}$$

Use powers of ten:

$$E_{R2} = (4.0 \times 10^{-3}) \times (40 \times 10^3)$$

$$E_{R2} = 4.0 \times 40 = 160 \text{ V}$$

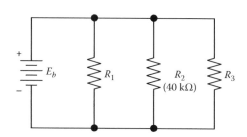

FIGURE 3.46   Example 3.19.

Therefore,

$$E_b = 160 \text{ V}$$

Note: Ohm's law states that the current in a circuit is inversely proportional to the circuit resistance. This fact, important as a basic building block of electrical theory, obviously, is also important in the following explanation of current flow in parallel circuits.

CURRENT IN PARALLEL CIRCUITS

In a series circuit, a single current flows. Its value is determined in part by the total resistance of the circuit; however, the source current in a parallel circuit divides among the available paths in relation to the value of the resistors in the circuit. Ohm's law remains unchanged. For a given voltage, current varies inversely with resistance. The behavior of current in a parallel circuit is best illustrated by example. In Figure 3.47, the resistors R_1, R_2, and R_3 are in parallel with each other and with the battery. Each parallel path is then a branch with its own individual current. When total current I_T leaves voltage source E, part I_1 of current I_T will flow through R_1, part I_2 will flow through R_2, and I_3 will flow through R_3. Branch currents I_1, I_2, and I_3 can be different; however, if a voltmeter (used for measuring the voltage of a circuit) is connected across R_1, R_2, and R_3, the respective voltages E_1, E_2, and E_3 will be equal. Therefore,

$$E = E_1 = E_2 = E_3$$

Total current I_T is equal to the sum of all branch currents:

$$I_T = I_1 = I_2 = I_3 \tag{3.15}$$

This formula applies for any number of parallel branches, whether the resistances are equal or unequal.

By Ohm's law, each branch current equals the applied voltage divided by the resistance between the two points where the voltage is applied. Hence, for each branch we have the following equations (see Figure 3.47):

Branch 1

$$I_1 = E_1/R_1 = V/R_1$$

Branch 2

$$I_2 = E_2/R_2 = V/R_2$$

Branch 3

$$I_3 = E_3/R_3 = V/R_3$$

With the same applied voltage, any branch that has less resistance allows more current through it than a branch with higher resistance.

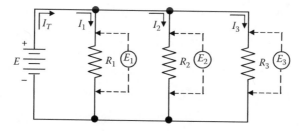

FIGURE 3.47 Parallel circuit.

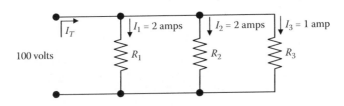

FIGURE 3.48   Example 3.20.

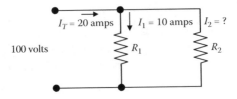

FIGURE 3.49   Example 3.21.

■ EXAMPLE 3.20

Problem: Two resistors, each drawing 2 amps, and a third resistor drawing 1 amp are connected in parallel across a 100-V line (see Figure 3.48). What is the total current?
Solution: The formula for total current is

$$I_T = I_1 + I_2 + I_3 = 2 + 2 + 1 = 5 \text{ amps}$$

The total current is 5 amps.

■ EXAMPLE 3.21

Problem: Two branches, R_1 and R_2, across a 100-V power line draw a total line current of 20 A (Figure 3.49). Branch R_1 takes 10 A. What is current I_2 in branch R_2?
Solution: Starting with Equation 3.15, transpose to find I_2 and then substitute the given values:

$$I_T = I_1 + I_2$$
$$I_2 = I_T - I_1 = 20 - 10 = 10 \text{ A}$$

The current in branch R_2 is 10 A.

■ EXAMPLE 3.22

Problem: A parallel circuit consists of two 15-ohm and one 12-ohm resistors across a 120-V line (see Figure 3.50). What current will flow in each branch of the circuit and what is the total current drawn by all the resistors?
Solution: There is a 120-V potential across each resistor, so

$$I_1 = V/R_1 = 120/15 = 8 \text{ amps}$$
$$I_2 = V/R_2 = 120/15 = 8 \text{ amps}$$
$$I_3 = V/R_3 = 120/12 = 10 \text{ amps}$$

Now find total current:

$$I_T = I_1 + I_2 + I_3 = 8 + 8 + 10 = 26 \text{ A}$$

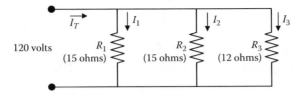

FIGURE 3.50 Example 3.22.

PARALLEL CIRCUITS AND KIRCHOFF'S CURRENT LAW

The division of current in a parallel network follows a definite pattern. This pattern is described by *Kirchoff's current law*: The algebraic sum of the currents entering and leaving any junction of conductors is equal to zero. This can be stated mathematically as

$$I_a + I_b + \ldots + I_n = 0 \tag{3.16}$$

where I_a, I_b, etc. are the currents entering and leaving the junction. Currents entering the junction are assumed to be positive; currents leaving the junction, negative. When solving a problem using Equation 3.16, the currents must be placed into the equation with the proper polarity.

■ EXAMPLE 3.23

Problem: Solve for the value of I_3 in Figure 3.51.
Solution: First, give the currents the proper signs:

$$I_1 = +10 \text{ amps} \qquad I_3 = ? \text{ amps}$$
$$I_2 = -3 \text{ amps} \qquad I_4 = -5 \text{ amps}$$

Then, place these currents into the equation with the proper signs, as follows:

$$I_a + I_b + \ldots + I_n = 0$$
$$I_1 + I_2 + I_3 + I_4 = 0$$
$$(+10) + (-3) + (I_3) + (-5) = 0$$

Combining like terms:

$$I_3 + 2 = 0$$
$$I_3 = -2 \text{ amps}$$

Thus, I_3 has a value of 2 amps, and the negative sign shows that it is a current leaving the junction.

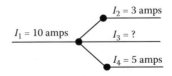

FIGURE 3.51 Example 3.23.

PARALLEL RESISTANCE

Unlike series circuits where total resistance (R_T) is the sum of the individual resistances, in a parallel circuit the total resistance is *not* the sum of the individual resistances. In a parallel circuit, we can use Ohm's law to find total resistance. We use the equation

$$R = E/I \qquad \text{or} \qquad R_T = E_s/I_T$$

where R_T is the total resistance of all the parallel branches across the voltage source E_S, and I_T is the sum of all the branch currents.

■ EXAMPLE 3.24

Problem: What is the total resistance of the circuit shown in Figure 3.52?
Solution:

Given:
$E_s = 120$ V
$I_T = 26$ A

$$R_T = E_s/I_T = 120/26 = 4.62 \text{ ohms}$$

Note: Notice that R_T is smaller than any of the three resistances shown in Figure 3.52. This fact may surprise you—it may seem strange that the total circuit resistance is *less* than that of the smallest resistor (R_3, 12 ohms). However, if we refer back to the water analogy we have used previously, it makes sense. Consider water pressure and water pipes, and assume it is possible to keep the water pressure constant. A small pipe offers more resistance to the flow of water than a larger pipe, but, if we add another pipe in parallel, one of even smaller diameter, the total resistance to water flow is *decreased*. In an electrical circuit, even a larger resistor in another parallel branch provides an additional path for current flow, so the *total* resistance is less. Remember, if we add one more branch to a parallel circuit, the total resistance decreases and the total current increases.

What we essentially demonstrated in working this particular problem is that the total load connected to the 120-V line is the same as the single equivalent resistance of 4.62 ohms connected across the line. It is probably more accurate to call this total resistance the "equivalent resistance," but by convention we use R_T, or total resistance, although the terms are often used interchangeably. We illustrate the equivalent resistance in the equivalent circuit shown in Figure 3.53.

Other methods can be used to determine the equivalent resistance of parallel circuits. The most appropriate method for a particular circuit depends on the number and value of the resistors; for example, consider the parallel circuit shown in Figure 3.54. For this circuit, the following simple equation is used:

$$R_{eq} = R/N \qquad\qquad (3.17)$$

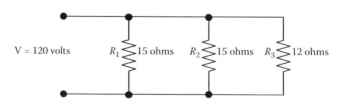

FIGURE 3.52   Example 3.24.

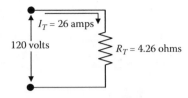

FIGURE 3.53   Circuit equivalent to that of Figure 3.52.

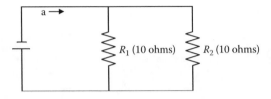

FIGURE 3.54   Two equal resistors connected in parallel.

where

  R_{eq} = Equivalent parallel resistance.
  R = Ohmic value of one resistor.
  N = Number of resistors.

Thus, R_{eq} = 10 ohms/2 = 5 ohms

When two equal value resistors are connected in parallel, they present a total resistance equivalent to a single resistor of one-half the value of either of the original resistors.

Note: Equation 3.17 is valid for any number of equal value parallel resistors.

■ EXAMPLE 3.25

Problem: When five 50-ohm resistors are connected in parallel, what is the equivalent circuit resistance?
Solution: Using Equation 3.17:

$$R_{eq} = R/N = 50 \text{ ohms}/5 = 10 \text{ ohms}$$

What about parallel circuits containing resistances of unequal value? How is equivalent resistance determined? Example 3.26 demonstrates how this is accomplished.

■ EXAMPLE 3.26

Problem: Refer to Figure 3.55 and determine R_{eq}.

  Given:
    R_1 = 3 ohms
    R_2 = 6 ohms
    E_a = 30 volts

  Known:
    I_1 = 10 amps
    I_2 = 5 amps
    I_T = 15 amps

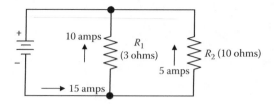

FIGURE 3.55 Example 3.26.

Solution:

$$R_{eq} = E_a/I_T = 30 \text{ ohms}/15 = 2 \text{ ohms}$$

In Example 3.26, the equivalent resistance of 2 ohms is less than the value of either branch resistor. Remember, in parallel circuits the equivalent resistance will always be smaller than the resistance of any branch.

RECIPROCAL METHOD

When circuits are encountered in which resistors of unequal value are connected in parallel, the equivalent resistance may be computed by using the *reciprocal method*.

> **Note:** A *reciprocal* is an inverted fraction; the reciprocal of the fraction 3/4, for example, is 4/3. We consider a whole number to be a fraction with 1 as the denominator, so the reciprocal of a whole number is that number divided into 1; for example, the reciprocal of R_T is $1/R_T$. The equivalent resistance in parallel is given by

$$\frac{1}{R_T} = \frac{1}{R_1} + \frac{1}{R_2} + \frac{1}{R_3} + \ldots + \frac{1}{R_n} \tag{3.18}$$

where R_T is the total resistance in parallel and R_1, R_2, R_3, and R_n are the branch resistances.

■ EXAMPLE 3.27

Problem: Find the total resistance of 2-, 4-, and 8-ohm resistors in parallel (Figure 3.56).
Solution: Write the formula for three resistors in parallel:

$$\frac{1}{R_T} = \frac{1}{R_1} + \frac{1}{R_2} + \frac{1}{R_3}$$

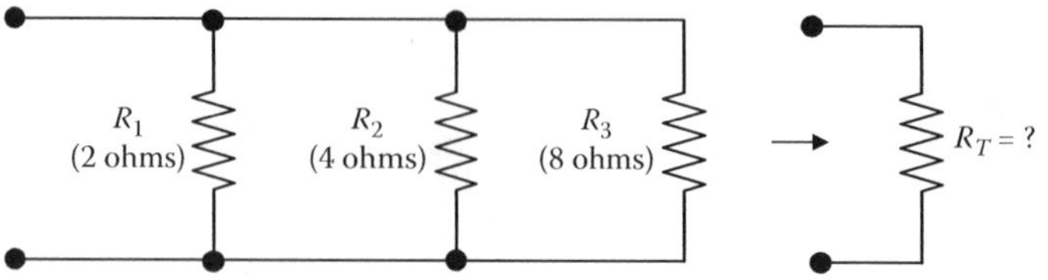

FIGURE 3.56 Example 3.27.

Substitute the resistance values:

$$\frac{1}{R_T} = \frac{1}{2} + \frac{1}{4} + \frac{1}{8}$$

Add fractions:

$$\frac{1}{R_T} = \frac{4}{8} + \frac{2}{8} + \frac{1}{8} = \frac{7}{8}$$

Invert both sides of the equation:

$$R_T = \frac{8}{7} = 1.14 \text{ ohms}$$

Note: When resistances are connected in parallel, the total resistance is always *less* than the smallest resistance of any single branch.

PRODUCT-OVER-SUM METHOD

When any two unequal resistors are in parallel, it is often easier to calculate the total resistance by multiplying the two resistances and then dividing the product by the sum of the resistances:

$$R_T = \frac{R_1 \times R_2}{R_1 + R_2} \tag{3.19}$$

where R_T is the total resistance in parallel, and R_1 and R_2 are the two resistors in parallel.

■ EXAMPLE 3.28

Problem: What is the equivalent resistance of a 20-ohm and a 30-ohm resistor connected in parallel?
Solution:

Given:
  R_1 = 20 ohms
  R_2 = 30 ohms

$$R_T = \frac{R_1 \times R_2}{R_1 + R_2} = \frac{20 \times 30}{20 + 30} = \frac{600}{50} = 12 \text{ ohms}$$

REDUCTION TO AN EQUIVALENT CIRCUIT

In the study of basic electricity, it is often necessary to resolve a complex circuit into a simpler form. Any complex circuit consisting of resistances can be reduced to a basic equivalent circuit containing the source and total resistance. This process is called *reduction to an equivalent circuit*. An example of circuit reduction was demonstrated in Example 3.28 and is illustrated in Figure 3.57. The circuit shown in Figure 3.57A is reduced to the simple circuit shown in Figure 3.57B.

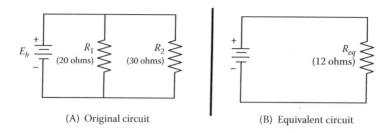

(A) Original circuit   (B) Equivalent circuit

FIGURE 3.57   Parallel circuit with equivalent circuit.

POWER IN PARALLEL CIRCUITS

As in the series circuit, the total *power* consumed in a parallel circuit is equal to the sum of the power consumed in the individual resistors:

$$P_T = P_1 + P_2 + P_3 + \dots + P_n \qquad (3.20)$$

where P_T is the total power and P_1, P_2, P_3, and P_n are the branch powers.

Note: Because power dissipation in resistors consists of a heat loss, power dissipations are additive regardless of how the resistors are connected in the circuit.

Total power can also be calculated by

$$P_T = E \times I_T \qquad (3.21)$$

where
  P_T = Total power.
  E = Voltage source across all parallel branches.
  I_T = Total current.

The power dissipated in each branch is equal to EI and to V^2/R.

Note: In both parallel and series arrangements, the sum of the individual values of power dissipated in the circuit equals the total power generated by the source. The circuit arrangements cannot change the fact that all power in the circuit comes from the source.

■ EXAMPLE 3.29

Problem: Find the total power consumed by the circuit in Figure 3.58.
Solution:

$$P_{R1} = E_b \times I_{R1} = 50 \times 5 = 250 \text{ watts}$$

$$P_{R2} = E_b \times I_{R2} = 50 \times 2 = 100 \text{ watts}$$

$$P_{R3} = E_b \times I_{R3} = 50 \times 1 = 50 \text{ watts}$$

$$P_T = P_1 + P_2 + P_3 = 250 + 100 + 50 = 400 \text{ watts}$$

Note: The power dissipated in the branch circuits in Figure 3.58 is determined in the same manner as the power dissipated by individual resistors in a series circuit. The total power (P_t) is then obtained by summing up the powers dissipated in the branch resistors using Equation 3.21.

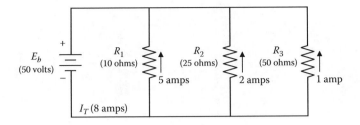

FIGURE 3.58 Example 3.29.

Because, in the example shown in Figure 3.58, the total current is known, we could determine the total power by the following method:

$$P_T = E_b \times I_T = 50 \text{ V} \times 8 \text{ A} = 400 \text{ W}$$

RULES FOR SOLVING PARALLEL DC CIRCUITS

Problems involving the determination of resistance, voltage, current, and power in a parallel circuit are solved as simply as in a series circuit. The procedure is basically the same: (1) draw a circuit diagram, (2) state the values given and the values to be found, (3) state the applicable equations, and (4) substitute the given values and solve for the unknown. Along with following this problem-solving procedure, it is also important to remember to apply the rules for solving parallel DC circuits:

1. The same voltage exists across each branch of a parallel circuit and is equal to the source voltage.
2. The current through a branch of a parallel network is inversely proportional to the amount of resistance of the branch.
3. The total current of a parallel circuit is equal to the sum of the currents of the individual branches of the circuit.
4. The total resistance of a parallel circuit is equal to the reciprocal of the sum of the reciprocals of the individual resistances of the circuit.
5. The total power consumed in a parallel circuit is equal to the sum of the power consumption of the individual resistances.

SERIES–PARALLEL CIRCUITS

To this point we have discussed series and parallel DC circuits; however, the maintenance operator will seldom encounter a circuit that consists solely of either type of circuit. Most circuits consist of both series and parallel elements. A circuit of this type is referred to as a *series–parallel circuit*, or as a *combination circuit*. Solving a series–parallel (combination) circuit is simply a matter of applying the laws and rules discussed up to this point.

SOLVING A SERIES–PARALLEL CIRCUIT

At least three resistors are required to form a series–parallel circuit. An example of a series–parallel circuit is shown in Figure 3.59, where two parallel resistors, R_2 and R_3, are connected in series with resistor R_1 and voltage source E. In a circuit of this type, current I_T divides after it flows through R_1; part flows through R_2, and part flows through R_3. The current then joins at the junction of the two resistors and flows back to the positive terminal of the voltage source (E) and through the voltage source to the positive terminal.

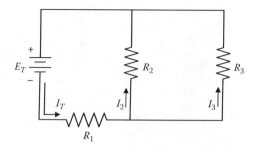

FIGURE 3.59   Series–parallel circuit.

When solving for values in a series–parallel circuit (current, voltage, and resistance), follow the rules that apply to a series circuit for the series part of the circuit and follow the rules that apply to a parallel circuit for the parallel part of the circuit. Solving series–parallel circuits is simplified if all parallel and series groups are first reduced to single equivalent resistances and the circuits are redrawn in simplified form. Recall that the redrawn circuit is called an *equivalent circuit*. The procedure for developing an equivalent circuit is shown in Figure 3.60.

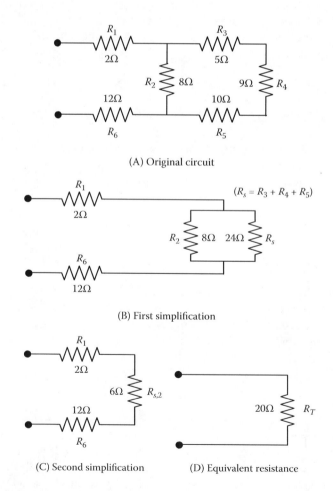

FIGURE 3.60   Developing an equivalent circuit.

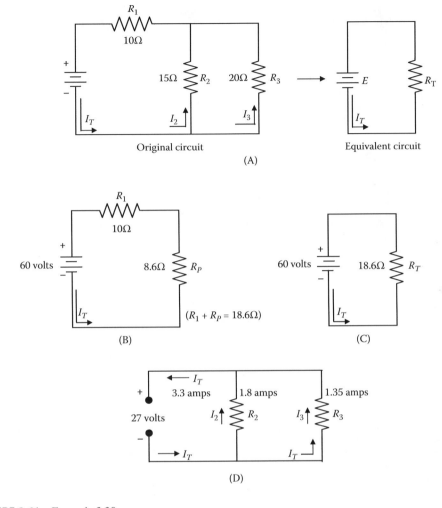

FIGURE 3.61 Example 3.30.

Note: There are no general formulas for the solution of series–parallel circuits because there are so many different forms of these circuits.

■ EXAMPLE 3.30

Problem: Find the total resistance (R_T), total circuit current (I_T), and branch currents of the circuit shown in Figure 3.61A.

Solution: Find the equivalent resistance of the parallel branch:

$$R_P = \frac{R_2 \times R_3}{R_2 + R_3} = \frac{15 \times 20}{15 + 20} = \frac{300}{35} = 8.6 \text{ ohms}$$

The equivalent circuit reduces to a series circuit (Figure 3.61B). Find the resistance of the equivalent series circuit:

$$R_T = R_1 + R_P = 10 + 8.6 = 18.6 \text{ ohms}$$

The equivalent circuit reduces to a single voltage source and a single resistance (Figure 3.61C). Find the actual current being supplied in the original series–parallel circuit (I_T):

$$I_T = V/R_T = 60/18.6 = 3.3 \text{ amps}$$

Find I_2 and I_3. The voltage across R_2 and R_3 is equal to the applied voltage (E) less the voltage drop across R_1 (Figure 3.61D):

$$V_2 = V_3 = V - (I_T \times R_1) = 60 - (3.3 \times 10) = 27 \text{ volts}$$

Then,

$$I_2 = V_2/R_2 = 27/15 = 1.8 \text{ amps}$$

$$I_3 = V_3/R_3 = 27/20 = 1.35 \text{ amps}$$

Note: The total current in the series–parallel circuit depends on the effective resistance of the parallel portion *and* on the other resistances.

■ EXAMPLE 3.31

Problem: Answer the following questions related to Figure 3.62:

1. The figure shows a circuit in which two resistors in series (R_2 and R_3) form one branch of a parallel circuit. The total current (I_T) flows into the parallel circuit, splitting into two branches at point a. At what point do the branch currents rejoin to form I_T?
   Answer: At point c.

2. Source voltage E_S (30 V) drops between points a and c. The largest voltage drop is across which resistor? Why?
   Answer: R_1, because the 30 V must be divided between R_2 and R_3.

3. What is the resistance of the top branch of the circuit in Figure 3.62?
   Answer: $R_2 + R_3 = 14$ ohms.

4. What is the value of R_T?
   Answer:
   $$R_T = (8 \times 14)/(8 + 14) = 112/22 = 5.1 \text{ ohms}$$

5. What is the current through the top branch ($I_{2,3}$)?
   Answer:
   $$I = E/R = 30/14 = 2.14 \text{ amps}$$

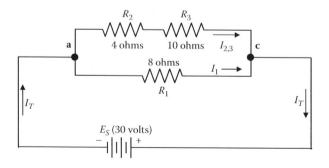

FIGURE 3.62   Example 3.31.

6. What is the value of P_T?
   Answer:

$$P_T = (E_s)^2/R_T = 30^2/5.1 = 900/5.1 = 176.5 \text{ watts}$$

CONDUCTORS

Recall that we pointed out earlier that electric current moves easily through some materials but with greater difficulty through others. Three good electrical conductors are copper, silver, and aluminum (generally, we can say that most metals are good conductors). At the present time, copper is the material of choice used in electrical conductors. Under special conditions, certain gases are also used as conductors; for example, neon gas, mercury vapor, and sodium vapor are used in various kinds of lamps. The function of the wire conductor is to connect a source of applied voltage to a load resistance with a minimum *IR* voltage drop in the conductor so most of the applied voltage can produce current in the load resistance. Ideally, a conductor must have a very low resistance; a typical value for a conductor such as copper is less than 1 ohm per 10 feet. Because all electrical circuits utilize conductors of one type or another, in this section we discuss the basic features and electrical characteristics of the most common types of conductors. Moreover, because conductor splices and connections (and insulation of such connections) are also an essential part of any electric circuit, they are also discussed.

Unit Size of Conductors

A standard (or unit size) of a conductor has been established to compare the resistance and size of one conductor with another. The unit of linear measurement used with regard to the diameter of a piece of wire is the *mil* (0.001 of an inch). A convenient unit of wire length used is the *foot*. Thus, the standard unit of size in most cases is the *mil-foot*; a wire will have unit size if it has a diameter of 1 mil and a length of 1 foot. The resistance in ohms of a unit conductor or a given substance is called the *resistivity* (or *specific resistance*) of the substance. As a further convenience, *gauge* numbers are also used to compare the diameter of wires. The Brown & Sharpe (B&S) gauge was used in the past; today, the *American wire gauge* (AWG) is more commonly used.

Square Mil

Figure 3.63 shows a square mil, which is a convenient unit of cross-sectional area for square or rectangular conductors. As shown in the figure, a square mil is the area of a square, the sides of which are 1 mil. To obtain the cross-sectional area in square mils of a square conductor, square one side measured in mils. To obtain the cross-sectional area in square mils of a rectangular conductor, multiply the length of one side by that of the other, each length being expressed in mils.

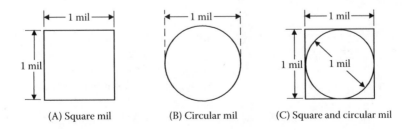

(A) Square mil   (B) Circular mil   (C) Square and circular mil

FIGURE 3.63   (A) Square mil; (B) circular mil; (C) comparison of circular to square mil.

■ **EXAMPLE 3.32**

Problem: Find the cross-sectional area of a large rectangular conductor 5/8 inch thick and 5 inches wide.

Solution: The thickness may be expressed in mils as $0.625 \times 1000 = 625$ mils and the width as $5 \times 1000 = 5000$ mils. The cross-sectional area is 625×5000, or 3,125,000 square mils.

CIRCULAR MIL

The *circular mil* is the standard unit of wire cross-sectional area used in most wire tables. To avoid the use of decimals (because most wires used to conduct electricity may be only a small fraction of an inch), it is convenient to express these diameters in mils. For example, the diameter of a wire is expressed as 25 mils instead of 0.025 inch. A circular mil is the area of a circle having a diameter of 1 mil, as shown in Figure 3.63B. The area in circular mils of a round conductor is obtained by squaring the diameter measured in mils; thus, a wire having a diameter of 25 mils has an area of 25^2, or 625, circular mils. By way of comparison, the basic formula for the area of a circle is

$$A = \pi \times R^2 \tag{3.22}$$

In this example, the area in square inches (in.^{2}) is

$$A = \pi \times R^2 = 3.14 \times (0.0125)^2 = 0.00049 \text{ in.}^2$$

If D is the diameter of a wire in mils, the area in square mils can be determined using the following equation:

$$A = \pi(D/2)^2 \tag{3.23}$$

which translates to

$$A = 3.14/4D^2 = 0.785 \text{ } D^2 \text{ square mils}$$

Thus, a wire 1 mil in diameter has an area of

$$A = 0.785 \times 1^2 = 0.785 \text{ square mils}$$

which is equivalent to 1 circular mil. The cross-sectional area of a wire in circular mils is therefore determined as

$$A = (0.785 \times D^2)/0.785 = D^2 \text{ circular mils}$$

where D is the diameter in mils; thus, the constant $\pi/4$ is eliminated from the calculation. Note that when comparing square and round conductors that the circular mil is a smaller unit of area than the square mil, so there are more circular mils than square mils in any given area. The comparison is shown in Figure 3.63C. The area of a circular mil is equal to 0.785 of a square mil.

> *Note:* To determine the circular-mil area when the square-mil area is given, divide the area in square mils by 0.785. Conversely, to determine the square-mil area when the circular-mil area is given, multiply the area in circular mils by 0.785.

■ **EXAMPLE 3.33**

Problem: A No. 12 wire has a diameter of 80.81 mils. (1) What is its area in circular mils? (2) What is its area in square mils?
Solution:

(1) Area = D^2 = $(80.81)^2$ = 6530 circular mils
(2) Area = 0.785 × 6530 = 5126 square mils

■ **EXAMPLE 3.34**

Problem: A rectangular conductor is 1.5 inches wide and 0.25 inch thick. (1) What is its area in square mils? (2) What size of round conductor in circular mils is necessary to carry the same current as the rectangular bar?
Solution:

(1) 1.5 in. = 1.5 × 1000 = 1500 mils
0.25 in. = 0.25 × 1000 = 250 mils
Area = 1500 × 250 = 375,000 square mils

(2) To carry the same current, the cross-sectional area of the rectangular bar and the cross-sectional area of the round conductor must be equal. There are more circular mils than square mils in this area; therefore,

$$A = 375,000/0.785 = 477,700 \text{ circular mils}$$

Note: Many electric cables are composed of stranded wires. The strands are usually single wires twisted together in sufficient numbers to make up the necessary cross-sectional area of the cable. The total area in circular mils is determined by multiplying the area of one strand in circular mils by the number of strands in the cable.

Circular-Mil-Foot

As shown in Figure 3.64, a *circular-mil-foot* is actually a unit of volume. More specifically, it is a unit conductor 1 foot in length and having a cross-sectional area of 1 circular mil. Because it is considered a unit conductor, the circular-mil-foot is useful in making comparisons between wires that are made of different metals. For example, a comparison of the *resistivity* of various substances can be made by determining the resistance of a circular-mil-foot of each of the substances.

Note: It is sometimes more convenient to employ a different unit of volume when working with certain substances. Accordingly, unit volume may also be expressed in cubic centimeters. Cubic inches may also be used. The unit of volume employed is given in tables of specific resistances.

Resistivity

All materials differ in their atomic structure and therefore in their ability to resist the flow of an electric current. As we have discussed, the measure of the ability of a specific material to resist the flow of electricity is its *resistivity*, or specific resistance—the resistance in ohms offered by the unit

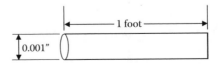

FIGURE 3.64   Circular-mil-foot.

volume (the circular-mil-foot) of a substance to the flow of electric current. Resistivity is the reciprocal of conductivity (i.e., the ease by which current flows in a conductor). A substance that has a high resistivity will have a low conductivity, and *vice versa*. The resistance of a given length, for any conductor, depends on the resistivity of the material, the length of the wire, and the cross-sectional area of the wire according to the equation:

$$R = \rho \times (L/A) \tag{3.24}$$

where
  R = Resistance of the conductor (ohms).
  ρ = Specific resistance or resistivity (cm × ohms/ft).
  L = Length of the wire (ft).
  A = Cross-sectional area of the wire (cm).

The factor ρ (Greek letter *rho*) permits different materials to be compared for resistance according to their nature without regard to different lengths or areas. Higher values of ρ mean more resistance.

Note: The resistivity of a substance is the resistance of a unit volume of that substance.

Many tables of resistivity are based on the resistance in ohms of a volume of the substance 1 foot long and 1 circular mil in cross-sectional area. The temperature at which the resistance measurement is made is also specified. If the kind of metal of which the conductor is made is known, the resistivity of the metal may be obtained from a table. The resistivity, or specific resistance, values of some common substances are given in Table 3.2.

Note: Because silver, copper, gold, and aluminum have the lowest values of resistivity, they are the best conductors. Tungsten and iron have much higher resistivity values.

■ EXAMPLE 3.35

Problem: What is the resistance of 1000 feet of copper wire having a cross-sectional area of 10,400 circular mils (No. 10 wire)? The wire temperature is 20°C.
Solution: Table 3.2 indicates that the resistivity (specific resistance) is 10.37. Substituting the known values in Equation 3.24, resistance R is determined as

$$R = \rho \times L/A = 10.37 \times (1000/10,400) = 1 \text{ ohm (approximately)}$$

TABLE 3.2
Resistivity (Specific Resistance)

Substance	Specific Resistance at 20°C ($\Omega\cdot$cmil/ft)
Silver	9.8
Copper (drawn)	10.37
Gold	14.7
Aluminum	17.02
Tungsten	33.2
Brass	42.1
Steel (soft)	95.8
Nichrome	660.0

TABLE 3.3
Copper Wire Table

Gauge	Diameter	Circular mils	Ohms/1000 ft at 25°C	Gauge	Diameter	Circular mils	Ohms/1000 ft at 25°C
1	289.0	83,700.0	0.126	20	32.0	1020.0	10.4
2	258.0	66,400.0	0.159	21	28.5	810.0	13.1
3	229.0	52,600.0	0.201	22	25.3	642.0	16.5
4	204.0	41,700.0	0.253	23	22.6	509.0	20.8
5	182.0	33,100.0	0.319	24	20.1	404.0	26.4
6	162.0	26,300.0	0.403	25	17.9	320.0	33.0
7	144.0	20,800.0	0.508	26	15.9	254.0	41.6
8	128.0	16,500.0	0.641	27	14.2	202.0	52.5
9	114.0	13,100.0	0.808	28	12.6	160.0	66.2
10	102.0	10,400.0	1.02	29	11.3	127.0	83.4
11	91.0	8230.0	1.28	30	10.0	101.0	105.0
12	81.0	6530.0	1.62	31	8.9	79.7	133.0
13	72.0	5180.0	2.04	32	8.0	63.2	167.0
14	64.0	4110.0	2.58	33	7.1	50.1	211.0
15	57.0	3260.0	3.25	34	6.3	39.8	266.0
16	51.0	2580.0	4.09	35	5.6	31.5	335.0
17	45.0	2050.0	5.16	36	5.0	25.0	423.0
18	40.0	1620.0	6.51	37	4.5	19.8	533.0
19	36.0	1290.0	8.21	38	4.0	15.7	673.0
20	32.0	1020.0	10.4	39	3.5	12.5	848.0
21	28.5	810.0	13.1	40	3.1	9.9	1070.0

WIRE MEASUREMENT

Wires are manufactured in sizes numbered according to a table known as the American wire gauge (AWG). Table 3.3 lists the standard wire sizes that correspond to the AWG. The gauge numbers specify the size of round wire in terms of its diameter and cross-sectional area. Note the following:

- As the gauge numbers increase from 1 to 40, the diameter and circular area decrease. Higher gauge numbers indicate smaller wire sizes; thus, No. 12 wire is a smaller wire than No. 4 wire.
- The circular area doubles for every three gauge sizes; for example, No 12 wire has about twice the area of No. 15 wire.
- The higher the gauge number and the smaller the wire, the greater the resistance of the wire for any given length; therefore, 1000 ft of No. 12 wire has a resistance of 1.62 ohms while 1000 ft of No. 4 wire has a resistance of 0.253 ohm.

FACTORS GOVERNING SELECTION OF WIRE SIZE

Several factors must be considered when selecting the size of wire to be used for transmitting and distributing electric power. These factors include allowable power loss in the line, permissible voltage drop in the line, current-carrying capacity of the line, and ambient temperatures in which the wire is to be used:

- Allowable power loss (I^2R) in the line—This loss represents electrical energy converted into heat. The use of large conductors will reduce the resistance and therefore the I^2R loss; however, large conductors are heavier and require more substantial supports, so they are more expensive initially than small ones.
- Permissible voltage drop (IR drop) in the line—If the source maintains a constant voltage at the input to the line, any variation in the load on the line will cause a variation in line current and a consequent variation in the IR drop in the line. A wide variation in the IR drop in the line causes poor voltage regulation at the load.
- Current-carrying capacity of the line—When current is drawn through the line, heat is generated. The temperature of the line will rise until the heat radiated, or otherwise dissipated, is equal to the heat generated by the passage of current through the line. If the conductor is insulated, the heat generated in the conductor is not so readily removed as it would be if the conductor were not insulated.
- Conductors installed in relatively high ambient temperatures—When installed in such surroundings, the heat generated by external sources constitutes an appreciable part of the total conductor heating. Due allowance must be made for the influence of external heating on the allowable conductor current, and each case has its own specific limitations.

COPPER VS. OTHER METAL CONDUCTORS

If it were not cost prohibitive, silver, the best conductor of electron flow (electricity), would be the conductor of choice in electrical systems. Instead, silver is used only in special circuits where a substance with high conductivity is required. The two most generally used conductors are copper and aluminum. Each has characteristics that make its use advantageous under certain circumstances. Likewise, each has certain disadvantages, or limitations. Copper has a higher conductivity and is more ductile (able to be drawn out into wire); it has relatively high tensile strength and can be easily soldered. It is more expensive and heavier than aluminum. Aluminum has only about 60% of the conductivity of copper, but its lightness makes possible long spans, and its relatively large diameter for a given conductivity reduces corona, which is the discharge of electricity from the wire when it has a high potential. The discharge is greater when smaller diameter wire is used than when larger diameter wire is used; however, aluminum conductors are not easily soldered, and aluminum's relatively large size for a given conductance does not permit the economical use of an insulation covering. A comparison of some of the characteristics of copper and aluminum is given in Table 3.4.

Note: Recent practice involves using copper wiring instead of aluminum wiring in homes and some industrial applications. Aluminum connections are not as easily made as they are with copper, and, over the years, many fires have been started because of improperly connected aluminum wiring; poor connections result in high-resistance connections and excessive heat generation.

TABLE 3.4
Characteristics of Copper and Aluminum

Characteristic	Copper	Aluminum
Tensile strength (lb/in.^{2})	55,000	25,000
Tensile strength for same conductivity (lb)	55,000	40,000
Weight for same conductivity (lb)	100	48
Cross-section for same conductivity (cm)	100	160
Specific resistance (Ω/mil-ft)	10.6	17

TABLE 3.5

Properties of Conducting Materials (Approximate)

Material	Temperature Coefficient (α/°C)	Material	Temperature Coefficient (α/°C)
Aluminum	0.004	Iron	0.006
Carbon	−0.0003	Nichrome	0.0002
Constantan	0 (average)	Nickel	0.005
Copper	0.004	Silver	0.004
Gold	0.004	Tungsten	0.005

WIRE TEMPERATURE COEFFICIENTS

The resistance of pure metals (e.g., silver, copper, aluminum) increases as the temperature increases. The *temperature coefficient* of resistance, α (*alpha*), indicates how much the resistance changes for a change in temperature. A positive value for α means R increases with temperature, a negative α means R decreases, and a zero α means R is constant, not varying with changes in temperature. Typical values of α are listed in Table 3.5. The amount of increase in the resistance of a 1-ohm sample of the copper conductor per degree rise in temperature (i.e., the temperature coefficient of resistance) is approximately 0.004. For pure metals, the temperature coefficient of resistance ranges between 0.004 and 0.006 ohm. Thus, a copper wire having a resistance of 50 ohms at an initial temperature of 0°C will have an increase in resistance of 50 × 0.004, or 0.2 ohms (approximate) for the entire length of wire for each degree of temperature rise above 0°C. At 20°C the increase in resistance is approximately 20 × 0.2, or 4 ohms. The total resistance at 20°C is 50 + 4, or 54 ohms.

Note: As shown in Table 3.5, carbon has a negative temperature coefficient. In general, α is negative for all semiconductors such as germanium and silicon. A negative value for α means less resistance at higher temperatures; therefore, the resistance of semiconductor diodes and transistors can be reduced considerably when they become hot with normal load current. Observe, also, that constantan has a value of zero for α. Thus, it can be used for precision wire-wound resistors that do not change resistance when the temperature increases.

CONDUCTOR INSULATION

Electric current must be contained; it must be channeled from the power source to a useful load safely. To accomplish this, electric current must be forced to flow only where it is needed. Moreover, current-carrying conductors must not be allowed (generally) to come in contact with one another, their supporting hardware, or personnel working near them. To accomplish this, conductors are coated or wrapped with various materials. These materials have such a high resistance that they are, for all practical purposes, nonconductors. They are generally referred to as *insulators* or *insulating materials.*

Numerous types of insulated conductors are available to meet the requirements of any job; however, only the necessary minimum of insulation is applied for any particular type of cable designed to do a specific job because insulation is expensive and has a stiffening effect. Also, it is required to withstand a great variety of physical and electrical conditions. Two fundamental but distinctly different properties of insulation materials (e.g., rubber, glass, asbestos, plastics) are insulation resistance and dielectric strength:

- *Insulation resistance* is the resistance to current leakage through and over the surface of insulation materials.
- *Dielectric strength* is the ability of the insulator to withstand potential difference and is usually expressed in terms of the voltage at which the insulation fails because of the electrostatic stress.

Various types of materials are used to provide insulation for electric conductors, including rubber, plastics, varnished cloth, paper, silk, cotton, and enamel.

CONDUCTOR SPLICES AND TERMINAL CONNECTIONS

When conductors join each other, or connect to a load, splices or terminals must be used. It is important that they be properly made, as any electric circuit is only as good as its weakest connection. The basic requirement of any splice or connection is that it be both mechanically and electrically as strong as the conductor or device with which it is used. High-quality workmanship and materials must be employed to ensure lasting electrical contact, physical strength, and insulation (if required).

Note: Conductor splices and connections are essential parts of any electric circuit.

SOLDERING OPERATIONS

Soldering operations are a vital part of electrical and electronics maintenance procedures. Soldering is a manual skill that must be learned by all personnel who work in the field of electricity. Obviously, practice is required to develop proficiency in the techniques of soldering. Both the solder and the material to be soldered (e.g., electric wire or terminal lugs) must be heated to a temperature that allows the solder to flow. If either is heated inadequately, cold solder joints result (i.e., high-resistance connections are created). Such joints do not provide either the physical strength or the electrical conductivity required. Moreover, it is necessary to select a solder that will flow at a temperature low enough to avoid damage to the part being soldered or to any other part or material in the immediate vicinity.

SOLDERLESS CONNECTORS

Generally, terminal lugs and splicers that do not require solder are more widely used than those that do require solder because they are easier to mount correctly. Solderless connectors—made in a wide variety of sizes and shapes—are attached to their conductors by means of several different devices, but the principle of each is essentially the same. They are all crimped (squeezed) tightly onto their conductors. They offer adequate electrical contact, plus great mechanical strength.

INSULATION TAPE

The carpenter has his saw, the dentist his pliers, the plumber his wrench, and the electrician his insulation tape. Accordingly, one of the first things the rookie maintenance operator learns (a rookie who is also learning proper and safe techniques for performing electrical work) is the value of electrical insulation tape. Normally, the use of electrical insulating tape comes into play as the final step in completing a splice or joint, to place insulation over the bare wire at the connection point. Typically, the insulation tape used should be the same basic substance as the original insulation, usually a rubber-splicing compound. When using rubber (latex) tape as the splicing compound where the original insulation was rubber, it should be applied to the splice with a light tension so each layer presses tightly against the one underneath it. In addition to the rubber tape application (which restores the insulation to original form), restoring with friction tape is also often necessary.

In recent years, plastic electrical tape has come into wide use. It has certain advantages over rubber and friction tape; for example, it will withstand higher voltages for a given thickness. Single, thin layers of certain commercially available plastic tape will tolerate several thousand volts without breaking down.

> *Note:* Be advised that, although plastic electrical tape is widely used in industrial applications, it must be applied in more layers to ensure an extra margin of safety because it is thinner than rubber or friction tape.

ELECTROMAGNETISM

Earlier, we discussed the fundamental theories concerning simple magnets and magnetism. That discussion dealt mainly with forms of magnetism that were not related directly to electricity—permanent magnets, for instance. Further, only brief mention was made of those forms of magnetism having a direct relation to electricity (e.g., producing electricity with magnetism). In medicine, anatomy and physiology are so closely related that the medical student cannot study one at length without involving the other. A similar relationship holds for the electrical field; that is, magnetism and basic electricity are so closely related that one cannot be studied at length without involving the other. This close fundamental relationship is continually borne out in the study of generators, transformers, battery packs, and motors. To be proficient in electricity, we must become familiar with the general relationships that exist between magnetism and electricity:

- Electric current flow will always produce some form of magnetism.
- Magnetism is by far the most commonly used means for producing or using electricity.
- The occasional peculiar behavior of electricity is caused by magnetic influences.

MAGNETIC FIELD AROUND A SINGLE CONDUCTOR

In 1819, Hans Christian Oersted, a Danish scientist, discovered that a field of magnetic force exists around a single wire conductor carrying an electric current. In Figure 3.65, a wire is passed through a piece of cardboard and connected through a switch to a dry cell. When the switch is open (no current flowing), iron filings sprinkled on the cardboard will fall back haphazardly when tapped. If we close the switch, current will begin to flow in the wire. This time, when we tap the cardboard the magnetic effect of the current in the wire will cause the filings to fall back into a definite pattern of concentric circles, with the wire as the center of the circles. Every section of the wire has this field of force around it in a plane perpendicular to the wire, as shown in Figure 3.66.

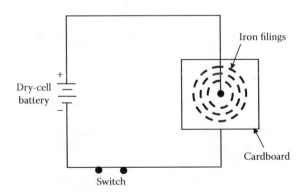

FIGURE 3.65   A circular patterns of magnetic force exists around a wire carrying an electric current.

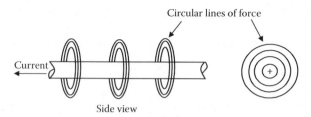

FIGURE 3.66   The circular fields of force around a wire carrying a current are in planes that are perpendicular to the wire.

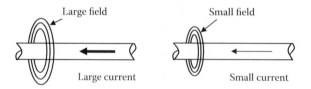

FIGURE 3.67   The strength of the magnetic field around a wire carrying a current depends on the amount of current.

The ability of the magnetic field to attract bits of iron (as demonstrated in Figure 3.45) depends on the number of lines of force present. The strength of the magnetic field around a wire carrying a current depends on the current, as it is the current that produces the field. The greater the current, the greater the strength of the field. A large current will produce many lines of force extending far from the wire, while a small current will produce only a few lines close to the wire, as shown in Figure 3.67.

POLARITY OF A SINGLE CONDUCTOR

The relation between the direction of the magnetic lines of force around a conductor and the direction of current flow along the conductor may be determined by means of the *left-hand rule for a conductor*. If the conductor is grasped in the left hand with the thumb extended in the direction of electron flow (– to +), the fingers will point in the direction of the magnetic lines of force. This is the same direction that the north pole of a compass would point to if the compass were placed in the magnetic field.

> *Note:* Arrows are generally used in electric diagrams to denote the direction of current flow along the length of wire. Where cross-sections of wire are shown, a special view of the arrow is used. A cross-sectional view of a conductor that is carrying current toward the observer is illustrated in Figure 3.68A. The direction of current is indicated by a dot, which represents the head of the arrow. A conductor that is carrying current away from the observer is illustrated in Figure 3.68B. The direction of current is indicated by a cross, which represents the tail of the arrow.

MAGNETIC FIELD AROUND TWO PARALLEL CONDUCTORS

When two parallel conductors carry current in the same direction, the magnetic fields tend to encircle both conductors, drawing them together with a force of attraction, as shown in Figure 3.69A. Two parallel conductors carrying currents in opposite directions are shown in Figure 3.69B. The field around one conductor is opposite in direction to the field around the other conductor. The resulting lines of force are crowded together in the space between the wires and tend to push the wires apart; two parallel adjacent conductors carrying currents in the same direction attract each other, and two parallel conductors carrying currents in opposite directions repel each other.

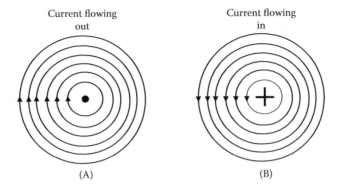

FIGURE 3.68   Magnetic field around a current-carrying conductor.

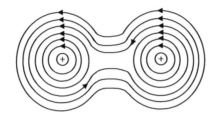

(A) Current flowing in the same direction

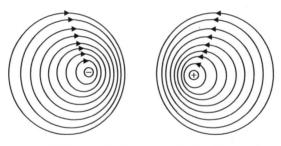

(B) Currents flowing in opposite directions

FIGURE 3.69   Magnetic field around two parallel conductors.

MAGNETIC FIELD OF A COIL

The magnetic field around a current-carrying wire exists at all points along its length. Bending the current-carrying wire into the form of a single loop has two results. First, the magnetic field consists of more dense concentric circles in a plane perpendicular to the wire (see Figure 3.66), although the total number of lines is the same as for the straight conductor. Second, all the lines inside the loop are in the same direction. When this straight wire is wound around a core, as is shown in Figure 3.70, it becomes a coil and the magnetic field assumes a different shape. When current is passed through the coiled conductor, the magnetic field of each turn of wire links with the fields of adjacent turns. The combined influence of all the turns produces a two-pole field similar to that of a simple bar magnet. One end of the coil will be a north pole and the other end will be a south pole.

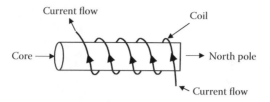

FIGURE 3.70 Current-carrying coil.

POLARITY OF AN ELECTROMAGNETIC COIL

In Figure 3.67, it was shown that the direction of the magnetic field around a straight conductor depends on the direction of current flow through that conductor; thus, a reversal of current flow through a conductor causes a reversal in the direction of the magnetic field that is produced. It follows that a reversal of the current flow through a coil also causes a reversal of its two-pole field. This occurs because that field is the product of the linkage between the individual turns of wire on the coil; therefore, if the field of each turn is reversed, it follows that the total field (coils' field) is also reversed. When the direction of electron flow through a coil is known, its polarity may be determined by use of the *left-hand rule for coils*. This rule is illustrated in Figure 3.70 and can be stated as follows: When the coil is grasped in the left hand, with the fingers wrapped around in the direction of electron flow, the thumb will point toward the north pole.

STRENGTH OF AN ELECTROMAGNETIC FIELD

The strength, or intensity, of the magnetic field of a coil depends on a number of factors:

- Number of turns of the conductor
- Amount of current flow through the coil
- Ratio of the coil length to its width
- Type of material in the core

MAGNETIC UNITS

The law of current flow in the electric circuit is similar to the law for establishing flux in the magnetic circuit. The *magnetic flux*, ϕ (*phi*), is similar to current in the Ohm's law formula and is comprised of the total number of lines of force existing in the magnetic circuit. The maxwell is the unit of flux; 1 line of force is equal to 1 maxwell.

Note: The maxwell is often referred to as simply a line of force, line of induction, or line.

The *strength* of a magnetic field in a coil of wire depends on how much current flows in the turns of the coil. The more current, the stronger the magnetic field. Also, the more turns, the more concentrated are the lines of force. The *force* that produces the flux in the magnetic circuit (comparable to electromotive force in Ohm's law) is known as *magnetomotive force* (mmf). The practical unit of magnetomotive force is the ampere-turn (At). In equation form:

$$F \text{ (ampere-turns)} = N \times I \tag{3.25}$$

where
  F = Magnetomotive force (At).
  N = Number of turns.
  I = Current (A).

■ EXAMPLE 3.36

Problem: Calculate the ampere-turns for a coil with 2000 turns and a 5-mA current.
Solution: Use Equation 3.25 and substitute $N = 2000$ and $I = 5 \times 10^{-3}$ A:

$$N \times I = 2000 \times (5 \times 10^{-3}) = 10 \text{ At}$$

The unit of *intensity* of magnetizing force per unit of length is designated as H and is sometimes expressed as Gilberts per centimeter of length. Expressed as an equation:

$$H = (N \times I)/L \tag{3.26}$$

where
H = Magnetic field intensity (ampere-turns per meter, At/m).
$N \times I$ = Number of turns × current (ampere-turns, At).
L = Length between poles of the coil (meters, m).

Note: Equation 3.26 is for a solenoid, and H is the intensity of an air core. With an iron core, H is the intensity through the entire core, and L is the length of or distance between poles of the iron core.

PROPERTIES OF MAGNETIC MATERIALS

In this section, we discuss two important properties of magnetic materials: permeability and hysteresis.

Permeability

When the core of an electromagnet is made of annealed sheet steel it produces a stronger magnet than if a cast iron core is used because annealed sheet steel is more readily acted upon by the magnetizing force of the coil than is the hard cast iron. Simply put, soft sheet steel is said to have greater permeability because of the greater ease with which magnetic lines are established in it. Recall that permeability is the relative ease with which a substance conducts magnetic lines of force. The permeability of air is arbitrarily set at 1. The permeability of other substances is the ratio of their ability to conduct magnetic lines compared to that of air. The permeability of nonmagnetic materials, such as aluminum, copper, wood, and brass, is essentially unity, or the same as for air.

Note: The permeability of magnetic materials varies with the degree of magnetization, being smaller for high values of flux density. *Reluctance*, which is analogous to resistance and is the opposition to the production of flux in a material, is inversely proportional to permeability. Iron has high permeability and, therefore, low reluctance. Air has low permeability and hence high reluctance.

Hysteresis

When the current in a coil of wire reverses thousands of times per second, a considerable loss of energy can occur. This loss of energy is caused by *hysteresis*. Hysteresis means "a lagging behind"; that is, the magnetic flux in an iron core lags behind the increases or decreases of the magnetizing force. The simplest method of illustrating the property of hysteresis is by graphical means, such as the hysteresis loop for a magnetic material shown in Figure 3.71. The hysteresis loop is a series of curves that show the characteristics of a magnetic material. Opposite directions of current result in the opposite directions of +H and –H for field intensity. Similarly, the range of flux density is indicated by +B and –B. The current starts at the center 0 (zero) when the material is unmagnetized. Positive H values increase B to saturation at $+B_{max}$. Next H decreases to zero, but B drops only to the value of B_r because of hysteresis. The current that produced the original magnetization now is reversed so H becomes negative. B drops to zero and continues to $-B_{max}$. As the –H values decrease,

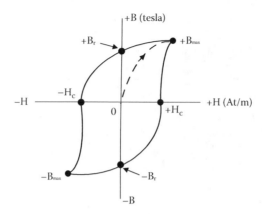

FIGURE 3.71   Hysteresis loop.

B is reduced to $-B_r$ when H is zero. Now, with a positive swing of current, H becomes positive, producing saturation at $+B_{max}$ again. The hysteresis loop is now complete. The curve does not return to zero at the center because of hysteresis.

Electromagnets

An *electromagnet* is composed of a coil of wire wound around a core that is normally soft iron, because of its high permeability and low hysteresis. When direct current flows through the coil, the core will become magnetized with the same polarity that the coil would have without the core. If the current is reversed, the polarities of the coil and core are reversed. The electromagnet is of great importance in electricity simply because the magnetism can be turned on or turned off at will. The starter solenoid (an electromagnet) in automobiles and power boats is a good example. In an automobile or boat, an electromagnet is part of a relay that connects the battery to the induction coil and generates the very high voltage required to start the engine. The starter solenoid isolates this high voltage from the ignition switch. When no current flows in the coil, it is an air core, but when the coil is energized, a movable soft-iron core does two things. First, the magnetic flux is increased because the soft-iron core is more permeable than the air core. Second, the flux is more highly concentrated. All of this concentration of magnetic lines of force in the soft-iron core results in a very good magnet when current flows in the coil, but soft iron loses its magnetism quickly when the current is shut off. The effect of the soft iron is, of course, the same whether it is movable, as in some solenoids, or permanently installed in the coil. An electromagnet, then, consists basically of a coil and a core; it becomes a magnet when current flows through the coil. The ability to control the action of magnetic force makes an electromagnet very useful in many circuit applications. Many applications of electromagnets are discussed throughout this manual.

AC THEORY

Because voltage is induced in a conductor when lines of force are cut, the amount of the induced electromotive force (emf) depends on the number of lines cut in a unit time. To induce an emf of 1 volt, a conductor must cut 100,000,000 lines of force per second. To obtain this great number of cuttings, the conductor is formed into a loop and rotated on an axis at great speed (see Figure 3.72). The two sides of the loop become individual conductors in series, each side of the loop cutting lines of force and inducing twice the voltage than a single conductor would induce. In commercial generators, the number of cuttings and the resulting emf are increased by (1) increasing the number of lines of force by using more magnets or stronger electromagnets, (2) using more conductors or loops, or (3) rotating the loops faster. (Both AC and DC generators are covered later.)

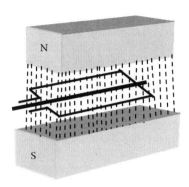

FIGURE 3.72 AC voltage produced by a loop rotating in a magnetic field.

How an AC generator operates to produce an AC voltage and current is a basic concept taught in elementary and middle-school science classes. Of course, today we accept technological advances as commonplace—we surf the Internet, text our friends, watch cable television, use our cell phones, and take outer space flight as a given. We consider the production of electricity, which makes all of these technologies possible, to be a right. These technologies are bottom shelf to us today; they are readily available to us so we use them—no big deal, right? Not worth thinking about. This point of view, though, surely was not typical of those who broke ground in developing technology and electricity.

During the groundbreaking years of electric technology development, the geniuses of the science of electricity (including George Simon Ohm) performed their technological breakthroughs in faltering steps. We tend to forget that those first faltering steps of scientific achievement in the field of electricity were achieved with crude and, for the most part, homemade apparatus. (Does this sound something like the more contemporary garage and basement inventors who came up with the first basic user-friendly microcomputer and software packages and perhaps will come up with the renewable energy innovations of tomorrow?)

Indeed, the innovators of electricity had to fabricate nearly all of the laboratory equipment used in their experiments. At the time, the only convenient source of electrical energy available to these early scientists was the voltaic cell, invented some years earlier. Because of the fact that cells and batteries were the only sources of power available, some of the early electrical devices were designed to operate from *direct current* (DC); thus, direct current was used extensively at the time. When the use of electricity became widespread, certain disadvantages in the use of direct current became apparent. In a direct-current system, the supply voltage must be generated at the level required by the load. To operate a 240-volt lamp, for example, the generator must deliver 240 volts. A 120-volt lamp could not be operated from the same generator by any convenient means. A resistor could be placed in series with the 120-volt lamp to drop the extra 120 volts, but the resistor would waste an amount of power equal to that consumed by the lamp.

Another disadvantage of direct-current systems is the large amount of power lost due to the resistance of the transmission wires used to carry current from the generating station to the consumer. This loss could be greatly reduced by operating the transmission line at a very high voltage and low current. This is not a practical solution in a DC system, however, because the load would also have to operate at high voltage. As a result of the difficulties encountered with direct current, practically all modern power distribution systems use alternating current (AC).

Unlike DC voltage, AC voltage can be stepped up or down by a device called a *transformer.* Transformers allow the transmission lines to be operated at high voltage and low current for maximum efficiency. Then, at the consumer end, the voltage is stepped down to whatever value the load requires by using a transformer. Due to its inherent advantages and versatility, alternating current has replaced direct current in all but a few commercial power distribution systems.

Basic AC Generator

As shown in Figure 3.72, an AC voltage and current can be produced when a conductor loop rotates through a magnetic field and cuts lines of force to generate an induced AC voltage across its terminals. This describes the basic principle of operation of an alternating current generator, or *alternator*. An alternator converts mechanical energy into electrical energy. It does this by utilizing the principle of *electromagnetic induction*. The basic components of an alternator are an armature, around which many turns of conductor are wound and which rotates in a magnetic field, as well as some means of delivering the resulting alternating current to an external circuit. (We will cover generator construction in more detail later; in this section, we concentrate on the theory of operation.)

Cycle

An AC voltage is one that continually changes in magnitude and periodically reverses in polarity (see Figure 3.73). The zero axis is a horizontal line across the center. The vertical variations on the voltage wave show the changes in magnitude. The voltages above the horizontal axis have positive (+) polarity, while voltages below the horizontal axis have negative (–) polarity. Figure 3.74 shows a suspended loop of wire (conductor or armature) being rotated (moved) in a counterclockwise direction through the magnetic field between the poles of a permanent magnet. For ease of explanation, the loop has been divided into a thick and thin half. Notice that in Figure 3.74A the thick half is moving along (parallel to) the lines of force; consequently, it is cutting none of these lines. The same is true of the thin half, moving in the opposite direction. Because the conductors are not cutting any lines of force, no emf is induced. As the loop rotates toward the position shown in Figure 3.74B, it cuts more and more lines of force per second because it is cutting more directly across the field (lines of force) as it approaches the position shown in Figure 3.74B. At the position shown in Figure 3.74B, the induced voltage is greatest because the conductor is cutting directly across the field.

As the loop continues to be rotated toward the position shown in Figure 3.74C, it cuts fewer and fewer lines of force per second. The induced voltage decreases from its peak value. Eventually, the loop is once again moving in a plane parallel to the magnetic field, and no voltage (zero voltage) is induced. The loop has now been rotated through half a circle (one alternation, or 180°). The sine curve shown in the lower part of Figure 3.74 shows the induced voltage at every instant of rotation of the loop. Notice that this curve contains 360°, or two alternations. In Figure 3.74, if the loop is rotated at a steady rate and if the strength of the magnetic field is uniform, the number of cycles per second (cps), or *hertz*, and the voltage will remain at fixed values. Continuous rotation will produce a series of sine-wave voltage cycles or, in other words, an AC voltage. In this way, mechanical energy is converted into electrical energy.

Note: Two complete alternations in a period of time is called a *cycle*.

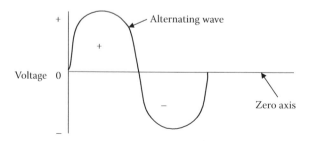

FIGURE 3.73   AC voltage waveform.

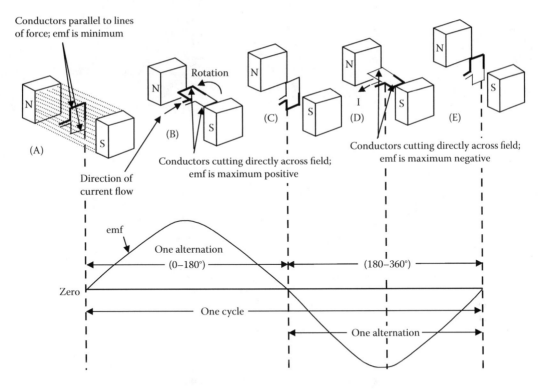

FIGURE 3.74   Basic alternating current generator.

FREQUENCY, PERIOD, AND WAVELENGTH

The *frequency* of an alternating voltage or current is the number of complete cycles occurring in each second of time. It is indicated by the symbol f and is expressed in hertz (Hz). One cycle per second equals 1 hertz. Thus, 60 cycles per second (cps) equals 60 Hz. A frequency of 2 Hz (Figure 3.75B) is twice the frequency of 1 Hz (Figure 3.75A). The amount of time for the completion of 1 cycle is the *period*. It is indicated by the symbol T for time and is expressed in seconds. Frequency and period are reciprocals of each other:

$$f = 1/T \tag{3.27}$$

$$T = 1/f \tag{3.28}$$

Note: The higher the frequency, the shorter the period.

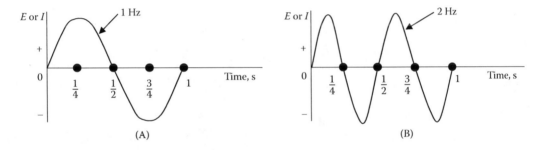

FIGURE 3.75   Comparison of frequencies.

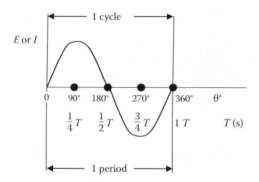

FIGURE 3.76 Relationship between electrical degrees and time.

The angle of 360° represents the time for 1 cycle, or the period T. So, we can show the horizontal axis of the sine wave in units of either electrical degrees or seconds (see Figure 3.76). The *wavelength* is the length of one complete wave or cycle. It depends upon the frequency of the periodic variation and its velocity of transmission. It is indicated by the symbol λ (Greek lowercase lambda). Expressed as a formula:

$$\lambda = \text{Velocity/Frequency} \tag{3.29}$$

CHARACTERISTIC VALUES OF AC VOLTAGE AND CURRENT

Because an AC sine wave voltage or current has many instantaneous values throughout the cycle, it is convenient to specify magnitudes to compare one wave with another. The peak, average, or root-mean-square (RMS) value can be specified (see Figure 3.77). These values apply to current or voltage.

PEAK AMPLITUDE

One of the most frequently measured characteristics of a sine wave is its amplitude. Unlike DC measurement, the amount of alternating current or voltage present in a circuit can be measured in various ways. In one method of measurement, the maximum amplitude of either the positive or the negative alternation is measured. The value of current or voltage obtained is called the *peak voltage* or the *peak current*. An oscilloscope is used to measure the peak value of current or voltage. The peak value is illustrated in Figure 3.77.

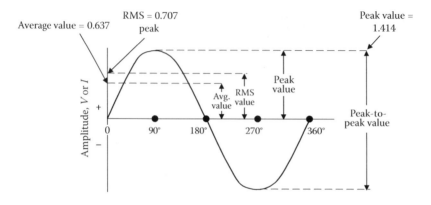

FIGURE 3.77 Amplitude values for AC sine wave.

PEAK-TO-PEAK AMPLITUDE

A second method of indicating the amplitude of a sine wave consists of determining the total voltage or current between the positive and negative peaks. This value of current or voltage is the peak-to-peak value (see Figure 3.77). Because both alternations of a pure sine wave are identical, the peak-to-peak value is twice the peak value. Peak-to-peak voltage is usually measured with an oscilloscope, although some voltmeters have a special scale calibrated in peak-to-peak volts.

INSTANTANEOUS AMPLITUDE

The *instantaneous value* of a sine wave of voltage for any angle of rotation is expressed by the formula:

$$e = E_m \times \sin(\theta) \tag{3.30}$$

where
    e = Instantaneous voltage.
    E_m = Maximum or peak voltage.
    $\sin(\theta)$ = Sine of the angle at which e is desired.

Similarly, the equation for the instantaneous value of a sine wave of current would be

$$i = I_m \times \sin(\theta) \tag{3.31}$$

where
    i = Instantaneous current.
    I_m = Maximum or peak current.
    $\sin(\theta)$ = Sine of the angle at which i is desired.

Note: The instantaneous value of voltage constantly changes as the armature of an alternator moves through a complete rotation. Because current varies directly with voltage, according to Ohm's law, the instantaneous changes in current also result in a sine wave for which the positive and negative peaks and intermediate values can be plotted exactly as we plotted the voltage sine wave. However, instantaneous values are not useful in solving most AC problems, so an *effective* value is used.

EFFECTIVE OR RMS VALUE

The effective value of an AC voltage or current of sine waveform is defined in terms of an equivalent heating effect of a direct current. Heating effect is independent of the direction of current flow. The alternating current of a sine waveform having a maximum value of 14.14 amps produces the same amount of heat in a circuit having a resistance of 1 ohm as a direct current of 10 amps. Knowing this, we can work out a constant value for converting any peak value to a corresponding effective value. This constant is represented by x in the simple equation below. Solve for x to three decimal places:

$$14.14x = 10$$

$$x = 0.707$$

Note: Because all instantaneous values of induced voltage are somewhere between zero and E_m (maximum or peak voltage), the effective value of a sine wave voltage or current must be greater than zero and less than E_m.

The effective value is also called the *root-mean-square* (RMS) value because it is the square root of the average of the squared values between zero and maximum. The effective value of an alternating current is stated in terms of an equivalent direct current. The phenomenon that is used as the standard comparison is the heating effect of the current. In many instances, it is necessary to convert from effective to peak or *vice versa* using a standard equation. Figure 3.77 shows that the peak value of a sine wave is 1.414 times the effective value; therefore, the equation we use is

Note: Anytime an AC voltage or current is stated without any qualifications, it is assumed to be an effective value.

$$E_m = E \times 1.414 \qquad (3.32)$$

where
E_m = Maximum or peak voltage.
E = Effective or RMS voltage.

and

$$I_m = I \times 1.414 \qquad (3.33)$$

where
I_m = Maximum or peak current.
I = Effective or RMS current.

Occasionally it is necessary to convert a peak value of current or voltage to an effective value. This is accomplished by using the following equations:

$$E = E_m \times 0.707 \qquad (3.34)$$

where
E = Effective voltage.
E_m = Maximum or peak voltage.

$$I = I_m \times 0.707 \qquad (3.35)$$

where
I = Effective current.
I_m = Maximum or peak current.

Average Value

Because the positive alternation is identical to the negative alternation, the average value of a complete cycle of a sine wave is zero. In certain types of circuits, however, it is necessary to compute the average value of one alternation. Figure 3.77 shows that the average value of a sine wave is 0.637 × peak value; therefore,

$$\text{Average value} = 0.637 \times \text{peak value} \qquad (3.36)$$

or

$$E_{avg} = E_m \times 0.637$$

where
E_{avg} = Average voltage of one alternation.
E_m = Maximum or peak voltage.

TABLE 3.6
AC Sine Wave Conversion Table

Multiply the Value	by	To Get the Value
Peak	2	Peak-to-peak
Peak-to-peak	0.5	Peak
Peak	0.637	Average
Average	1.637	Peak
Peak	0.707	Root-mean-square (RMS, effective)
Root-mean-square (RMS, effective)	1.414	Peak
Average	1.110	Root-mean-square (RMS, effective)
Root-mean-square (RMS, effective)	0.901	Average

Similarly,

$$I_{avg} = I_m \times 0.637 \tag{3.37}$$

where

I_{avg} = Average current in one alternation.
I_m = Maximum or peak current.

Table 3.6 lists various sine wave amplitude values used to convert AC sine wave voltage and current.

RESISTANCE IN AC CIRCUITS

If a sine wave of voltage is applied to a resistance, the resulting current will also be a sine wave. This follows Ohm's law, which states that the current is directly proportional to the applied voltage. Figure 3.78 shows a sine wave of voltage and the resulting sine wave of current superimposed on the same time axis. Notice that as the voltage increases in a positive direction the current increases along with it. When the voltage reverses direction, the current reverses direction. At all times the voltage and current pass through the same relative parts of their respective cycles at the same time. When two waves, such as those shown in Figure 3.78, are precisely in step with one another they are said to be *in phase*. To be in phase, the two waves must go through their maximum and minimum points at the same time and in the same direction. In some circuits, several sine waves can be in phase with each other. Thus, it is possible to have two or more voltage drops in phase with each other and in phase with the circuit current.

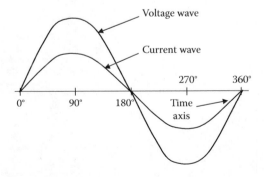

FIGURE 3.78   Voltage and current waves in phase.

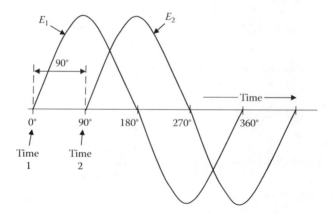

FIGURE 3.79 Voltage waves 90° out of phase.

Note: It is important to remember that Ohm's law for DC circuits is applicable to AC circuits with resistance only.

Voltage waves are not always in phase; for example, Figure 3.79 shows voltage wave E_1, which starts at 0° (time 1). As voltage wave E_1 reaches its positive peak, a second voltage wave, E_2, begins to rise (time 2). Because these waves do not go through their maximum and minimum points at the same instant of time, a phase difference exists between the two waves, and the two waves are said to be *out of phase*. For the two waves in Figure 3.79, this phase difference is 90°.

PHASE RELATIONSHIPS

In the preceding section, we discussed the important concepts of being in phase and phase difference. Another important phase concept is *phase angle*. The phase angle between two waveforms of the same frequency is the angular difference at a given instant of time. As an example, the phase angle between waves B and A (see Figure 3.80) is 90°. Take the instant of time at 90°. The horizontal axis is shown in angular units of time. Wave B starts at maximum value and reduces to zero value at 90°, while wave A starts at zero and increases to maximum value at 90°. Wave B reaches its maximum value 90° ahead of wave A, so wave B *leads* wave A by 90° (and wave A *lags* wave B by 90°).

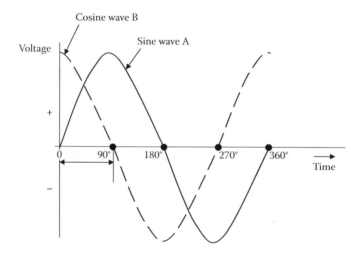

FIGURE 3.80 Wave B leads wave A by a phase angle of 90°.

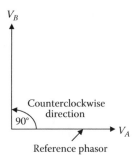

FIGURE 3.81   Phasor diagram.

This 90° phase angle between waves B and A is maintained throughout the complete cycle and all successive cycles. At any instant of time, wave B has the value that wave A will have 90° later. Wave B is a cosine wave because it is displaced 90° from wave A, which is a sine wave.

Note: The amount by which one wave leads or lags another is measured in degrees.

To compare phase angles or phases of alternating voltages or currents, it is more convenient to use vector diagrams corresponding to the voltage and current waveforms. A *vector* is a straight line used to denote the magnitude and direction of a given quantity. Magnitude is denoted by the length of the line drawn to scale, and the direction is indicated by the arrow at one end of the line, together with the angle that the vector makes with a horizontal reference vector.

Note: In electricity, because different directions really represent *time* expressed as a phase relationship, an electrical vector is called a *phasor*. In an AC circuit containing only resistance, the voltage and current occur at the same time; that is, they are in phase. To indicate this condition by means of phasors all that is necessary is to draw the phasors for the voltage and current in the same direction. The value of each is indicated by the length of the phasor.

A vector, or phasor, diagram is shown in Figure 3.81, where vector V_B is vertical to show the phase angle of 90° with respect to vector V_A, which is the reference. Because lead angles are shown in the counterclockwise direction from the reference vector, V_B leads V_A by 90°.

INDUCTANCE

To this point we have learned the following key points about magnetic fields:

- A field of force exists around a wire carrying a current.
- This field has the form of concentric circles around the wire, in planes perpendicular to the wire and with the wire at the center of the circles.
- The strength of the field depends on the current. Large currents produce large fields; small currents produce small fields.
- When lines of force cut across a conductor, a voltage is induced in the conductor.

Moreover, to this point we have studied circuits that have been *resistive* (i.e., resistors presented the only opposition to current flow). Two other phenomena, inductance and capacitance, exist in DC circuits to some extent, but they are major players in AC circuits. Both inductance and capacitance present a kind of opposition to current flow that is called *reactance*, which we will cover later. Before we examine reactance, however, we must first study *inductance* and *capacitance*.

What Is Inductance?

Inductance is the characteristic of an electrical circuit that makes itself evident by opposing the starting, stopping, or changing of current flow. A simple analogy can be used to explain inductance. We are all familiar with how difficult it is to push a heavy load (a cart full of heavy items, for example). It takes more work to start the load moving than it does to keep it moving. This is because the load possesses the property of *inertia*. Inertia is the characteristic of mass that opposes a change in velocity; therefore, inertia can hinder us in some ways and help us in others. Inductance exhibits the same effect on current in an electric circuit as inertia does on velocity of a mechanical object. The effects of inductance are sometimes desirable, sometimes undesirable.

Note: Simply put, inductance is the characteristic of an electrical conductor that opposes a change in current flow.

Because inductance is the property of an electric circuit that opposes any change in the current through that circuit, if the current increases then a self-induced voltage opposes this change and delays the increase. On the other hand, if the current decreases then a self-induced voltage tends to aid (or prolong) the current flow, delaying the decrease. Thus, current can neither increase nor decrease as fast in an inductive circuit as it can in a purely resistive circuit. In AC circuits, this effect becomes very important because it affects the phase relationships between voltage and current. When inductance is a factor in a circuit, the voltage and current generated by the same armature are out of phase. We will examine these phase relationships later. Our objective now is to understand the nature and effects of inductance in an electric circuit.

UNIT OF INDUCTANCE

The unit for measuring inductance (*L*) is the *henry* (named for the American physicist Joseph Henry), which is abbreviated as h. Figure 3.82 shows the schematic symbol for an inductor. An inductor has an inductance of 1 henry if an emf of 1 volt is induced in the inductor when the current through the inductor is changing at the rate of 1 ampere per second. The relation among the induced voltage, inductance, and rate of change of current with respect to time can be stated mathematically as

$$E = L \times \Delta I / \Delta t \qquad (3.38)$$

where
  E = Induced emf (volts).
  L = Inductance (henrys).
  Δ*I* = Change in amperes occurring in Δ*t* seconds.

Note: The symbol Δ (Delta) means "change in."

The henry is a large unit of inductance that is used with relatively large inductors. The unit employed with small inductors is the millihenry (mh). For still smaller inductors, the unit of inductance is the microhenry (μh).

SELF-INDUCTANCE

As previously explained, current flow in a conductor always produces a magnetic field surrounding, or linking with, the conductor. When the current changes, the magnetic field changes, and an emf is induced in the conductor. This emf is called a *self-induced emf* because it is induced in the conductor carrying the current.

Note: Even a perfectly straight length of conductor has some inductance.

FIGURE 3.82   Schematic symbol for an inductor.

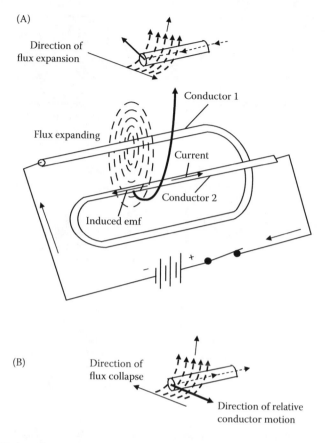

FIGURE 3.83 Self-inductance.

The direction of the induced emf has a definite relation to the direction in which the field that induces the emf varies. When the current in a circuit is increasing, the flux linking with the circuit is increasing. This flux cuts across the conductor and induces an emf in the conductor in such a direction as to oppose the increase in current and flux. This emf is sometimes referred to as *counterelectromotive force* (cemf). The two terms are used synonymously throughout this manual. Likewise, when the current is decreasing, an emf is induced in the opposite direction and opposes the decrease in current.

Note: The effects described here are summarized by *Lenz's law,* which states that the induced emf in any circuit is always in a direction opposite that of the effect that produced it.

Shaping a conductor so the electromagnetic field around each portion of the conductor cuts across some other portion of the same conductor increases the inductance. This is shown in its simplest form in Figure 3.83A. A loop of conductor is looped so two portions of the conductor lie adjacent and parallel to one another. These portions are labeled Conductor 1 and Conductor 2. When the switch is closed, electron flow through the conductor establishes a typical concentric field around all portions of the conductor. The field is shown in a single plane (for simplicity) that is perpendicular to both conductors. Although the field originates simultaneously in both conductors, it is considered as originating in Conductor 1, and its effect on Conductor 2 will be noted. With increasing current, the field expands outward, cutting across a portion of Conductor 2. The resultant induced emf in Conductor 2 is shown by the dashed arrow. Note that it is in opposition to the battery current and

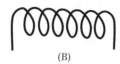

(A)   (B)

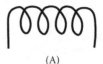

(A)   (B)

FIGURE 3.84   (A) Few turns, low inductance; (B) more turns, higher inductance.

FIGURE 3.85   (A) Wide spacing between turns, low inductance; (B) close spacing between turns, higher inductance.

voltage, according to Lenz's law. In Figure 3.83B, the same section of Conductor 2 is shown, but with the switch open and the flux collapsing. Four major factors affect the self-inductance of a conductor, or circuit:

Note: From Figure 3.83, the important point to note is that the voltage of self-induction opposes both changes in current. It delays the initial buildup of current by opposing the battery voltage and delays the breakdown of current by exerting an induced voltage in the same direction in which the battery voltage acted.

1. *Number of turns*—Inductance depends on the number of wire turns. Wind more turns to increase inductance. Take turns off to decrease the inductance. Figure 3.84 compares the inductance of two coils made with different numbers of turns.
2. *Spacing between turns*—Inductance depends on the spacing between turns, or the length of the inductor. Figure 3.85 shows two inductors with the same number of turns. The turns of the first inductor have a wide spacing. The turns of the second inductor are closer together. The second coil, though shorter, has a larger inductance value because of its close spacing between turns.
3. *Coil diameter*—Coil diameter, or cross-sectional area, is highlighted in Figure 3.86. The larger diameter inductor has higher inductance. Both coils shown have the same number of turns, and the spacing between turns is the same, but the first inductor has a smaller diameter than the second one. The inductance of the second inductor is greater than that for the first inductor.
4. *Type of core material*—Permeability, as pointed out earlier, is a measure of how easily a magnetic field goes through a material. Permeability also tells us how much stronger the magnetic field will be with the material inside the coil. Figure 3.87 shows three identical coils. One has an air core, one has a powdered iron core in the center, and the other has a

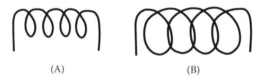

(A)   (B)

FIGURE 3.86   (A) Small diameter, low inductance; (B) larger diameter, higher inductance.

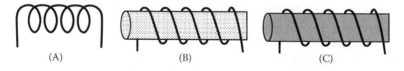

(A)   (B)   (C)

FIGURE 3.87   (A) air core, low inductance; (B) powdered iron core, higher inductance; (C) soft iron core, highest inductance

soft iron core. This figure illustrates the effects of core material on inductance. The inductance of a coil is affected by the magnitude of current when the core is a magnetic material.
When the core is air, the inductance is independent of the current.

Note: The inductance of a coil increases very rapidly as the number of turns is increased. It also
increases as the coil is made shorter, the cross-sectional area is made larger, or the permeability
of the core is increased.

GROWTH AND DECAY OF CURRENT IN AN *RL* SERIES CIRCUIT

If a battery is connected across a pure inductance, the current builds up to its final value at a rate that
is determined by the battery voltage and the internal resistance of the battery. The current buildup
is gradual because of the counter emf (cemf) generated by the self-inductance of the coil. When the
current starts to flow, the magnetic lines of force move out, cut the turns of wire on the inductor, and
build up a cemf that opposes the emf of the battery. This opposition causes a delay in the time it takes
the current to build up to a steady value. When the battery is disconnected, the lines of force collapse,
again cutting the turns of the inductor and building up an emf that tends to prolong the current flow.

Although the analogy is not exact, electrical inductance is somewhat like mechanical inertia. A
boat begins to move on the surface of water at the instant a constant force is applied to it. At this
instant, its rate of change of speed (acceleration) is greatest, and all the applied force is used to
overcome the inertia of the boat. After a while, the speed of the boat increases but its acceleration
decreases; the applied force is used up overcoming the friction of the water against the hull. As the
speed levels off and acceleration drops to zero, the applied force equals the opposing friction force
at this speed and the inertia effect disappears. In the case of inductance, it is electrical inertia that
must be overcome.

Figure 3.88 shows a circuit that includes two switches, a battery, and a voltage divider containing
a resistor (*R*) and an inductor (*L*). Switches S_1 and S_2 are mechanically interlocked ("ganged"), as
indicated by the dashed line; when one closes, the other opens at exactly the same instant. Such an
arrangement is an *RL* (resistive–inductive) series circuit. The source voltage of the battery is applied
across the resistor and inductor when S_1 is closed. As S_1 is closed, as shown in Figure 3.88, a voltage
(*E*) appears across the resistor and inductor. A current attempts to flow, but the inductor opposes this
current by building up a cemf. At the first instant S_1 is closed, the cemf exactly equals the battery
emf, and its polarity is opposite. Under this condition, no current will flow in resistor *R*. Because no
current can flow when the cemf is exactly equal to the battery voltage, no voltage is dropped across
R. As time goes on, more of the battery voltage appears across the resistor and less across the inductor. The rate of change of current is less and the induced emf is less. As the steady-state condition
of the current flow is approached, the drop across the inductor approaches zero, and all the battery
voltage is used to overcome the resistance of the circuit.

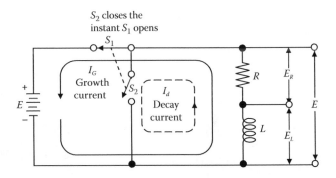

FIGURE 3.88   Growth and decay of current in an *RL* circuit.

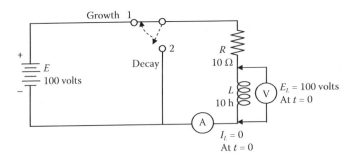

FIGURE 3.89 *L/R* time constant.

When S_2 is closed (source voltage E is removed from the circuit), the flux that has been established around L collapses through the windings and induces voltage E_L in L which has a polarity opposite that of E and is essentially equal to it in magnitude. The induced voltage, E_L, causes current I_d to flow through R in the same direction that it was flowing when S_1 was closed. Voltage E_R, which is initially equal to E, is developed across R. It rapidly falls to zero as voltage E_L across L falls to zero due to the collapsing flux.

Note: When a switch is first closed to complete a circuit, inductance opposes the buildup of current in the circuit.

Note: In many electronic circuits, the time required for the growth or decay of current is important, but these applications are beyond the scope of this manual; however, you need to learn the fundamentals of the *L/R* time constant, which are discussed in the following section.

L/R TIME CONSTANT

The time required for the current through an inductor to increase to 63.2% (63%) of the maximum current or to decrease to 36.7% (37%) is known as the *time constant* of the circuit. An *RL* circuit is shown in Figure 3.89. The value of the time constant in seconds is equal to the inductance in henrys divided by the circuit resistance in ohms. One set of values is given in Figure 3.89. *L/R* is the symbol used for this time constant. If L is in henrys and R is in ohms, t (time) is in seconds. If L is in microhenrys and R is in ohms, t is in microseconds. If L is in millihenrys and R is in ohms, t is in milliseconds. R in the *L/R* equation is always in ohms, and the time constant is on the same order of magnitude as L. Two useful relations used in calculating *L/R* time constants are as follows:

$$L \text{ (henrys)}/R \text{ (ohms)} = t \text{ (seconds)} \tag{3.39}$$

$$L \text{ (microhenrys)}/R \text{ (ohms)} = t \text{ (microseconds)} \tag{3.40}$$

Note: The time constant of an *RL* circuit is always expressed as a ratio between inductance (L) and resistance (R).

MUTUAL INDUCTANCE

When the current in a conductor or coil changes, the varying flux can cut across any other conductor or coil located nearby, thus inducing voltages in both. A varying current in L_1, therefore, induces voltage across L_1 and across L_2 (Figure 3.90); see Figure 3.91 for the schematic symbol for two coils with mutual inductance. When the induced voltage (E_{L2}) produces current in L_2, its varying magnetic field induces voltage in L_1; hence, the two coils L_1 and L_2 have *mutual inductance* because

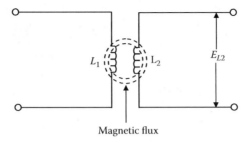

FIGURE 3.90   Mutual inductance between L_1 and L_2.

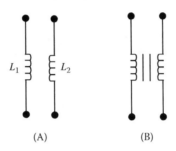

(A)   (B)

FIGURE 3.91   (A) Schematic symbol for two coils (air core) with mutual inductance; (B) two coils (iron core) with mutual inductance.

current change in one coil can induce voltage in the other. The unit of mutual inductance is the henry, and the symbol is L_M. Two coils have a mutual inductance of 1 henry when a current change of 1 A/sec in one coil induces 1 volt in the other coil. The factors affecting the mutual inductance of two adjacent coils are dependent on

- Physical dimensions of the two coils
- Number of turns in each coil
- Distance between the two coils
- Relative positions of the axes of the two coils
- Permeability of the cores

Note: The amount of mutual inductance depends on the relative position of the two coils. If the coils are separated a considerable distance, the amount of flux common to both coils is small and the mutual inductance is low. Conversely, if the coils are close together so that nearly all the flow of one coil links the turns of the other, then mutual inductance is high. The mutual inductance can be increased greatly by mounting the coils on a common iron core.

CALCULATION OF TOTAL INDUCTANCE

In the study of advanced electrical theory, it is necessary to know the effect of mutual inductance in solving for total inductance in both series and parallel circuits; however, for our purposes in this manual, we do not attempt to make these calculations. Instead, we discuss the basic total inductance calculations that the maintenance operator should be familiar with. If inductors in series are located far enough apart, or well shielded to make the effects of mutual inductance negligible, the total inductance is calculated in the same manner as for resistances in series; we merely add them:

$$L_T = L_1 + L_2 + L_3 + \ldots + L_n \qquad (3.41)$$

■ **EXAMPLE 3.37**

Problem: If a series circuit contains three inductors with values of 40, 50, and 20 μh, what is the total inductance?
Solution:

$$L_T = 40 + 50 + 20 = 110 \text{ μh}$$

In a parallel circuit containing inductors (without mutual inductance), the total inductance is calculated in the same manner as for resistances in parallel:

$$\frac{1}{L_T} = \frac{1}{L_1} + \frac{1}{L_2} + \frac{1}{L_3} + \dots \frac{1}{L_n} \tag{3.42}$$

■ **EXAMPLE 3.38**

Problem: A circuit contains three totally shielded inductors in parallel. The values of the three inductances are 4, 5, and 10 mh. What is the total inductance?
Solution:

$$\frac{1}{L_T} = \frac{1}{4} + \frac{1}{5} + \frac{1}{10} = 0.25 + 0.2 + 0.1 = 0.55$$

$$L_T = \frac{1}{0.55} = 1.8 \text{ mh}$$

CAPACITANCE

No matter how complex the electrical circuit, it is composed of no more than three basic electrical properties: resistance, inductance, and capacitance. Accordingly, gaining a thorough understanding of these three basic properties is a necessary step toward the understanding of electrical equipment. We have covered resistance and inductance, and the last of the basic three, capacitance, is covered in this section. Earlier, we learned that inductance opposes any change in current. Capacitance is the property of an electric circuit that opposes any change of *voltage* in a circuit. If applied voltage is increased, capacitance opposes the change and delays the voltage increase across the circuit. If applied voltage is decreased, capacitance tends to maintain the higher original voltage across the circuit, thus delaying the decrease. Capacitance is also defined as that property of a circuit that enables energy to be stored in an electric field. Natural capacitance exists in many electric circuits; however, in this manual, we are concerned only with the capacitance that is designed into the circuit by means of devices called *capacitors*.

> **Note:** The most noticeable effect of capacitance in a circuit is that voltage can neither increase nor decrease as rapidly in a capacitive circuit as it can in a circuit that does not include capacitance.

CAPACITORS

A capacitor, or condenser, is a manufactured electrical device that consists of two conducting plates of metal separated by an insulating material called a *dielectric* (see Figure 3.92). (The prefix *di-* means "through" or "across.") When a capacitor is connected to a voltage source, there is a short current pulse. A capacitor stores this electric charge in the dielectric (it can be charged and

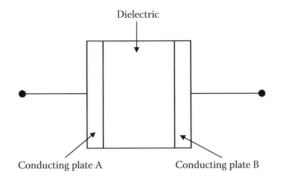

FIGURE 3.92   Capacitor.

FIGURE 3.93   (A) Schematic for a fixed capacitor; (B) variable capacitor.

discharged, as we shall see later). To form a capacitor of any appreciable value, however, the area of the metal pieces must be quite large and the thickness of the dielectric must be quite small. The symbol used to designate a capacitor is C; the unit of capacitance is the farad (F). The farad is that capacitance that will store 1 coulomb of charge in the dielectric when the voltage applied across the capacitor terminals is 1 volt. The schematic symbols for capacitors are shown in Figure 3.93.

Note: A capacitor is essentially a device that stores electrical energy.

The capacitor is used in a number of ways in electrical circuits. It may block DC portions of a circuit, as it is effectively a barrier to direct current (but not to alternating current). It may be part of a tuned circuit—one such application is in the tuning of a radio to a particular station. It may be used to filter AC out of a DC circuit. Most of these are advanced applications that are beyond the scope of this presentation; however, a basic understanding of capacitance is necessary to the fundamentals of AC theory.

Note: A capacitor does not conduct direct current. The insulation between the capacitor plates blocks the flow of electrons. We learned earlier that there is a short current pulse when we first connect the capacitor to a voltage source. The capacitor quickly charges to the supply voltage, and then the current stops.

The two plates of the capacitor shown in Figure 3.94 are electrically neutral because each plate has as many protons (positive charge) as electrons (negative charge); thus, the capacitor has no charge. Now connect a battery across the plates (Figure 3.95A). When the switch is closed (Figure 3.95B), the negative charge on plate A is attracted to the positive terminal of the battery. This movement of charges will continue until the difference in charge between plates A and B is equal to the electromotive force (voltage) of the battery. The capacitor is now charged. Because almost none of the charge can cross the space between the plates, the capacitor will remain in this condition even if the battery is removed (Figure 3.96A); however, if a conductor is placed across the plates (Figure 3.96B), the electrons find a path back to plate A and the charges on each plate are again neutralized. The capacitor is now discharged.

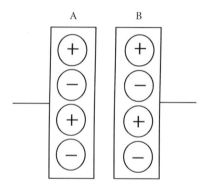

FIGURE 3.94   Two plates of a capacitor with a neutral charge.

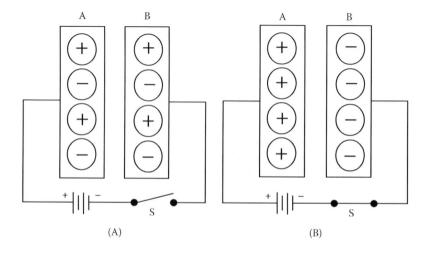

FIGURE 3.95   (A) Neutral capacitor; (B) charged capacitor.

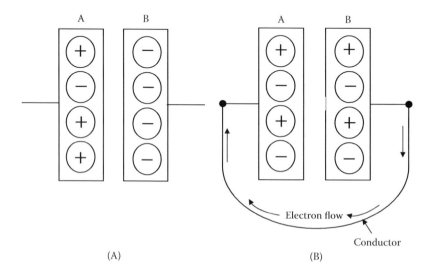

FIGURE 3.96   (A) Charged capacitor; (B) discharging a capacitor.

TABLE 3.7
Dielectric Constants

Material	Constant	Material	Constant
Vacuum	1.0000	Rubber	2.5–35
Air	1.0006	Wood	2.5–8
Paraffin paper	3.5	Porcelain	5.1–5.9
Glass	5–10	Glycerin (15°C)	56
Quartz	3.8	Petroleum	2
Mica	3–6	Pure water	81

Note: In a capacitor, electrons cannot flow through the dielectric, because it is an insulator. Because it takes a definite quantity of electrons to charge a capacitor, it is said to have *capacity*. This characteristic is referred to as *capacitance*.

DIELECTRIC MATERIALS

Somewhat similar to the phenomenon of permeability in magnetic circuits, various materials differ in their ability to support electric flux (lines of force) or to serve as dielectric material for capacitors. Materials are rated in their ability to support electric flux in terms of a number called a *dielectric constant*. Other factors being equal, the higher the value of the dielectric constant, the better the dielectric material. Dry air is the reference against which all other materials are rated. Dielectric constants for some common materials are given in Table 3.7.

Note: From Table 3.7 it is obvious that pure water is the best dielectric. Keep in mind that the key word here is "pure." Water capacitors are used today in some high-energy applications, in which differences in potential are measured in thousands of volts.

UNIT OF CAPACITANCE

Capacitance is equal to the amount of charge that can be stored in a capacitor divided by the voltage applied across the plates:

$$C = Q/E \qquad (3.43)$$

where
  C = Capacitance (farads, F).
  Q = Amount of charge (coulombs, C).
  E = Voltage (volts, V).

■ EXAMPLE 3.39

Problem: What is the capacitance of two metal plates separated by 1 centimeter of air if 0.001 coulomb of charge is stored when a potential of 300 volts is applied to the capacitor?
Solution: Given that $Q = 0.001$ coulomb, $E = 200$ volts, and

$$C = Q/E$$

Convert to the power of ten:

$$C = (10 \times 10^{-4})/(2 \times 10^2) = 5 \times 10^{-6} = 0.000005 \text{ farads}$$

Note: Although the capacitance value obtained in Example 3.39 appears small, many electronic circuits require much smaller capacitors. The farad is a cumbersome unit that is far too large for many applications. The microfarad (μf) is one millionth of a farad (1×10^{-6} farad) and is a more convenient unit.

Equation 3.43 can be rewritten as follows:

$$Q = C \times E \qquad (3.44)$$

$$E = Q/C \qquad (3.45)$$

Note: From Equation 3.44, do not get the mistaken idea that capacitance is dependent on charge and voltage. Capacitance is determined entirely by physical factors, which are covered later.

FACTORS AFFECTING VALUE OF CAPACITANCE

The capacitance of a capacitor depends on three main factors: plate surface area, distance between plates, and dielectric constant of the insulating material:

- *Plate surface area*—Capacitance varies directly with plate surface area. We can double the capacitance value by doubling the plate surface area of the capacitor. Figure 3.97 shows a capacitor with a small surface area and another one with a large surface area. Adding more capacitor plates can increase the plate surface area. Figure 3.98 shows alternate plates connecting to opposite capacitor terminals.
- *Distance between plates*—Capacitance varies inversely with the distance between plate surfaces. The capacitance increases when the plates are closer together. Figure 3.99 shows capacitors with the same plate surface area but different spacing.

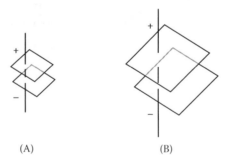

(A)   (B)

FIGURE 3.97 (A) Small plates, small capacitance; (B) larger plates, higher capacitance.

FIGURE 3.98 Several sets of plates connected to produce a capacitor with greater surface area.

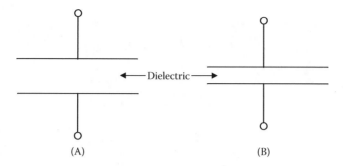

FIGURE 3.99 (A) Wide plate spacing, small capacitance; (B) narrow plate spacing, larger capacitance.

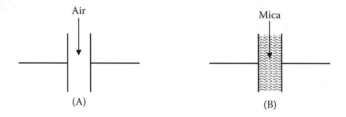

FIGURE 3.100 (A) Low capacitance; (B) higher capacitance.

- *Dielectric constant of the insulating material*—An insulating material with a higher dielectric constant produces a higher capacitance rating. Figure 3.100 shows two capacitors. Both have the same plate surface area and spacing. Air is the dielectric in the first capacitor, and mica is the dielectric in the second one. The dielectric constant of mica is 5.4 times greater than the dielectric constant of air, so the mica capacitor has 5.4 times more capacitance than the other capacitor.

Voltage Rating of Capacitors

There is a limit to the voltage that may be applied across any capacitor. If too large a voltage is applied, it will overcome the resistance of the dielectric and a current will be forced through it from one plate to the other, sometimes burning a hole in the dielectric. In this event, a short circuit exists and the capacitor must be discarded. The maximum voltage that may be applied to a capacitor is known as the *working voltage* and must never be exceeded. The working voltage of a capacitor depends on (1) the type of material used as the dielectric, and (2) the thickness of the dielectric. As a margin of safety, the capacitor should be selected so its working voltage is at least 50% greater than the highest voltage to be applied to it; for example, if a capacitor is expected to have a maximum of 200 volts applied to it, its working voltage should be at least 300 volts.

Charge and Discharge of an *RC* Series Circuit

According to Ohm's law, the voltage across a resistance is equal to the current through it times the value of the resistance. This means that a voltage will be developed across a resistance *only when current flows through it*. As previously stated, a capacitor is capable of storing or holding a charge of electrons. When uncharged, both plates contain the same number of free electrons. When charged, one plate contains more free electrons than the other. The difference in the number of electrons is a measure of the charge on the capacitor. The accumulation of this charge builds up a voltage across the terminals of the capacitor, and the charge continues to increase until this voltage equals the

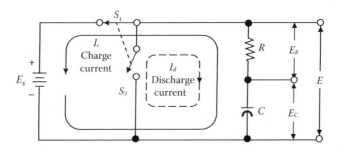

FIGURE 3.101   Charge and discharge of an *RC* series circuit.

applied voltage. The greater the voltage, the greater the charge on the capacitor. Unless a discharge path is provided, a capacitor keeps its charge indefinitely. Any practical capacitor, however, has some leakage through the dielectric so the voltage will gradually leak off. A voltage divider with resistance and capacitance may be connected in a circuit by means of a switch, as shown in Figure 3.101. Such a series arrangement is called an *RC* series circuit.

If S_1 is closed, electrons flow counterclockwise around the circuit containing the battery, capacitor, and resistor. This flow of electrons ceases when *C* is charged to the battery voltage. At the instant current begins to flow, there is no voltage on the capacitor and the drop across *R* is equal to the battery voltage. The initial charging current (*I*) is therefore equal to E_S/R. The current flowing in the circuit soon charges the capacitor. Because the voltage on the capacitor is proportional to its charge, a voltage (E_C) will appear across the capacitor. This voltage opposes the battery voltage—that is, these two voltages buck each other. As a result, voltage E_R across the resistor is equal to $E_S - E_C$, which is equal to the voltage drop ($I_C R$) across the resistor. Because E_S is fixed, I_C decreases as E_C increases. The charging process continues until the capacitor is fully charged and the voltage across it is equal to the battery voltage. At this instant, the voltage across *R* is zero and no current flows through it. In Figure 3.101, if S_2 is closed (S_1 opened), a discharge current (I_D) will discharge the capacitor. Because I_D is opposite in direction to I_C, the voltage across the resistor will have a polarity opposite to the polarity during the charging time; however, this voltage will have the same magnitude and will vary in the same manner. During discharge the voltage across the capacitor is equal and opposite to the drop across the resistor. The voltage drops rapidly from its initial value and then approaches zero slowly. The actual time it takes to charge or discharge is important in advanced electricity and electronics. Because the charge or discharge time depends on the values of resistance and capacitance, an *RC* circuit can be designed for the proper timing of certain electrical events. The *RC* time constant is covered in the next section.

RC TIME CONSTANT

The time required to charge a capacitor to 63% of maximum voltage or to discharge it to 37% of its final voltage is known as the *time constant* of the current. An *RC* circuit is shown in Figure 3.102. The time constant *T* for an *RC* circuit is

$$T = R \times C \qquad (3.46)$$

The time constant of an *RC* circuit is usually very short because the capacitance of a circuit may be only a few microfarads or even picofarads.

Note: An *RC* time constant expresses the charge and discharge times for a capacitor.

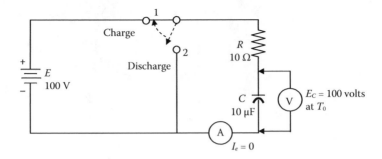

FIGURE 3.102 *RC* circuit.

CAPACITORS IN SERIES AND PARALLEL

Like resistors or inductors, capacitors may be connected in series, in parallel, or in a series–parallel combination. Unlike resistors or inductors, however, total capacitance in series, parallel, or series–parallel combinations is found in a different manner. Simply put, the rules are not the same for the calculation of total capacitance. This difference is explained as follows: Parallel capacitance is calculated like series resistance, and series capacitance is calculated like parallel resistance:

- When capacitors are connected in *series* (see Figure 3.103), the total capacitance (C_T) is

$$\frac{1}{C_T} = \frac{1}{C_1} + \frac{1}{C_2} + \frac{1}{C_3} + \dots \frac{1}{C_n} \tag{3.47}$$

■ EXAMPLE 3.40

Problem: Find the total capacitance of a 3-µF, a 5-µF, and a 15-µF capacitor in series.
Solution: Write Equation 3.47 for three capacitors in series:

$$\frac{1}{C_T} = \frac{1}{C_1} + \frac{1}{C_2} + \frac{1}{C_3} = \frac{1}{3} + \frac{1}{5} + \frac{1}{15} = \frac{9}{15} = \frac{3}{5}; \quad C_T = \frac{5}{3} = 1.7\,\mu\text{F}$$

- When capacitors are connected in *parallel* (see Figure 3.104), the total capacitance (C_T) is the sum of the individual capacitances:

$$C_T = C_1 + C_2 + C_3 + \dots + C_n \tag{3.48}$$

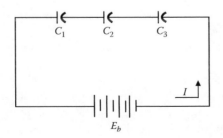

FIGURE 3.103 Series capacitive circuit.

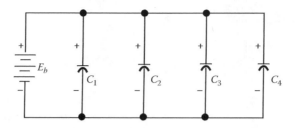

FIGURE 3.104 Parallel capacitive circuit.

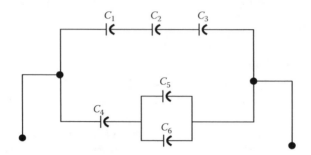

FIGURE 3.105 Series–parallel capacitance configuration.

■ EXAMPLE 3.41

Problem: Determine the total capacitance in a parallel capacitive circuit, given:

$C_1 = 2\ \mu F$
$C_2 = 3\ \mu F$
$C_3 = 0.25\ \mu F$

Solution: Write Equation 3.48 for three capacitors in parallel:

$$C_T = C_1 + C_2 + C_3 = 2 + 3 + 0.25 = 5.25\ \mu F$$

- If capacitors are connected in a combination of *series and parallel* (see Figure 3.105), the total capacitance is found by applying Equations 3.47 and 3.48 to the individual branches.

TYPES OF CAPACITORS

Capacitors used for commercial applications are divided into two major groups—fixed and variable—and are named according to their dielectric. Most common are air, mica, paper, and ceramic capacitors, plus the electrolytic type. These types are compared in Table 3.8. The fixed capacitor has a set value of capacitances that is determined by its construction. The construction of the variable capacitor allows a range of capacitances. Within this range, the desired value of capacitance is obtained by some mechanical means, such as by turning a shaft (as in turning a radio tuner knob, for example) or adjusting a screw to adjust the distance between the plates. The electrolytic capacitor consists of two metal plates separated by an electrolyte. The electrolyte, either paste or liquid, is in contact with the negative terminal, and this combination forms the negative electrode. The dielectric is a very thin film of oxide deposited on the positive electrode, which is aluminum sheet. Electrolytic capacitors are polarity sensitive (i.e., they must be connected in a circuit according to their polarity markings) and are used where a large amount of capacitance is required.

TABLE 3.8

Comparison of Capacitor Types

Dielectric	Construction	Capacitance Range
Air	Meshed plates	10–400 pF
Mica	Stacked plates	10–5000 pF
Paper	Rolled foil	0.001–1 μF
Ceramic	Tubular	0.5–1600 pF
	Disk	0.002–0.1 μF
Electrolytic	Aluminum	5–1000 μF
	Tantalum	0.01–300 μF

ULTRACAPACITORS

Like batteries, ultracapacitors, also known as supercapacitors, pseudocapacitors, electric double-layer capacitors, or electrochemical double-layer capacitors (EDLCs), are energy storage devices (NREL, 2009). To meet the power, energy, and voltage requirements for a wide range of applications, they use electrolytes and various-sized cells configured into modules. As storage devices, however, they differ from batteries in that ultracapacitors (which are true capacitors in that energy is stored via charge separation at the electrode–electrolyte interface) store energy electrostatically, but batteries store energy chemically. Ultracapacitors provide quick bursts of energy and also offer an improvement of about two to three orders of magnitude in capacitance (as compared to an average capacitor), but with a lower working voltage. Moreover, they can withstand hundreds of thousands of charge/discharge cycles without degrading. As an alternative energy source, ultracapacitors have proven themselves as reliable energy storage components that can be used to power a variety of electronic and portable devices such as radios, flashlights, cell phones, and emergency kits. As ultracapacitor technology matures they are being developed to function as batteries; for example, the vehicle industry is deploying ultracapacitors as a replacement for chemical batteries.

Ultracapacitor Operation

An ultracapacitor polarizes an electrolytic solution to store energy electrostatically. Though it is an electrochemical device, no chemical reactions are involved in its energy storage mechanism. This mechanism is highly reversible and allows the ultracapacitor to be charged and discharged hundreds of thousands of times. As shown in Figure 3.106, an ultracapacitor is comprised of two nonreactive porous plates, or collectors, suspended within an electrolyte, with a voltage potential applied across

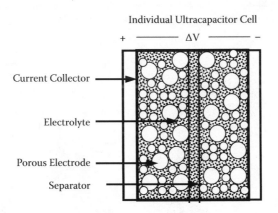

FIGURE 3.106 Ultracapacitor.

the collectors. In an individual ultracapacitor cell, the applied potential on the positive electrode attracts the negative ions in the electrolyte, while the potential on the negative electrode attracts the positive ions. A dielectric separator between the two electrodes prevents the charge from moving between the two electrodes. Once the ultracapacitor is charged and energy stored, a load can use this energy. The amount of energy stored is very large compared to a standard capacitor because of the enormous surface area created by the porous carbon electrodes and the small separation (10 angstroms) created by the dielectric separate; however, it stores a much smaller amount of energy than does a battery. Because the rates of charge and discharge are determined solely by is physical properties, the ultracapacitor can release energy much faster (with more power) than a battery that relies on slow chemical reactions.

INDUCTIVE AND CAPACITIVE REACTANCE

Earlier, we learned that the inductance of a circuit acts to oppose any change of current flow in that circuit and that capacitance acts to oppose any change of voltage. In DC circuits these reactions are not important, because they are momentary and occur only when a circuit is first closed or opened. In AC circuits, these effects become very important because the direction of current flow is reversed many times each second, and the opposition presented by inductance and capacitance is, for practical purposes, constant. In purely resistive circuits, either DC or AC, the term for opposition to current flow is *resistance*. When the effects of capacitance or inductance are present, as they often are in AC circuits, the opposition to current flow is called *reactance*. The total opposition to current flow in circuits that have both resistance and reactance is called *impedance*. In this section, we cover the calculation of inductive and capacitive reactance and impedance; the phase relationships of resistance, inductive, and capacitive circuits; and power in reactive circuits.

INDUCTIVE REACTANCE

In order to gain an understanding of the reactance of a typical coil, we need to review exactly what occurs when AC voltage is impressed across the coil. The AC voltage produces an alternating current. When a current flows in a wire, lines of force are produced around the wire. Large currents produce many lines of force; small currents produce only a few lines of force. As the current changes, the number of lines of force will change. The field of force will seem to expand and contract as the current increases and decreases, as shown in Figure 3.107. As the field expands and contracts, the lines of force must cut across the wires that form the turns of the coil.

These cuttings induce an emf in the coil. This emf acts in the direction so as to oppose the original voltage and is called a *counter* (or back) *emf*. The effect of this counter emf is to reduce the original voltage impressed on the coil. The net effect will be to reduce the current below that which would flow if there were no cuttings or counter emf. In this sense, the counter emf is acting as a

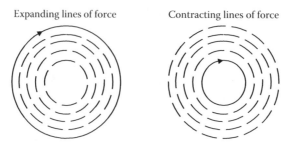

FIGURE 3.107   AC current producing a moving (expanding and collapsing) field; in a coil, this moving field cuts the wires of the coil.

resistance in reducing the current. Although it would be more convenient to consider the current-reducing effect of a counter emf as a number of ohms of effective resistance, we don't do this. Instead, because a counter emf is not actually a resistance but merely *acts* as a resistance, we use the term *reactance* to describe this effect.

> **Note:** The *reactance* of a coil is the number of ohms of resistance that the coil seems to offer as a result of a counter emf induced in it. Its symbol is X to differentiate it from the DC resistance R.

The inductive reactance of a coil depends primarily on (1) the inductance of the coil, and (2) the frequency of the current flowing through the coil. The value of the reactance of a coil is therefore proportional to its inductance and the frequency of the AC circuit in which it is used. The formula for inductive reactance is

$$X_L = 2\pi f L \tag{3.49}$$

Because $2\pi = 2 \times 3.14 = 6.28$, Equation 3.49 becomes

$$X_L = 6.28fL$$

where
X_L = Inductive reactance (ohms).
f = Frequency (Hz).
L = Inductance (h).

If any two quantities are known in Equation 3.49, the third can be found:

$$L = X_L/6.28f \tag{3.50}$$

$$f = X_L/6.28L \tag{3.51}$$

■ EXAMPLE 3.42

Problem: The frequency of a circuit is 60 Hz and the inductance is 20 mh. What is X_L?
Solution:

$$X_L = 2\pi f L = 6.28 \times 60 \times 0.02 = 7.5 \text{ ohms}$$

■ EXAMPLE 3.43

Problem: A 30-mh coil is in a circuit operating at a frequency of 1400 kHz. Find its inductive reactance.
Solution: Given that $L = 30$ mh and $f = 1400$ kHz, find X_L. First, change the units of measurement:

$$30 \text{ mh} = 30 \times 10^{-3} \text{ h}$$

$$1400 \text{ kHz} = 1400 \times 10^3$$

Now find the inductive reactance:

$$X_L = 6.28fL$$

$$X_L = 6.28 \times (1400 \times 10^3) \times (30 \times 10^{-3}) = 263,760 \text{ ohms}$$

■ **EXAMPLE 3.44**

Problem: Given $L = 400$ µh and $f = 1500$ Hz, find X_L.
Solution:

$$X_L = 2\pi fL = 6.28 \times 1500 \times 0.0004 = 3.78 \text{ ohms}$$

Note: If frequency or inductance varies, inductive reactance must also vary. The inductance of a coil does not vary appreciably after the coil is manufactured, unless it is designed as a variable inductor; thus, frequency is generally the only variable factor affecting the inductive reactance of a coil. The inductive reactance of the coil will vary directly with the applied frequency.

CAPACITIVE REACTANCE

Previously, we learned that as a capacitor is charged electrons are drawn from one plate and deposited on the other. As more and more electrons accumulate on the second plate, they begin to act as an opposing voltage, which attempts to stop the flow of electrons just as a resistor would do. This opposing effect is called the *reactance* of the capacitor and is measured in ohms. As noted earlier, the basic symbol for reactance is X, and the subscript defines the type of reactance. Earlier, we used X_L to represent inductive reactance; the subscript L refers to inductance. Following the same pattern, we use X_C to represent capacitive reactance. The two factors affecting capacitive reactance (X_C) are

*   Size of the capacitor
*   Frequency

Note: Capacitive reactance, X_C, is the opposition to the flow of AC current due to capacitance in the circuit.

The larger the capacitor, the greater the number of electrons that may be accumulated on its plates. When the plate area is large, the electrons do not accumulate in one spot but spread out over the entire area of the plate and do not impede the flow of new electrons onto the plate; therefore, a large capacitor offers a small reactance. In a capacitor with a small plate area, the electrons cannot spread out, and they attempt to stop the flow of electrons coming onto the plate; therefore, a small capacitor offers a large reactance. The reactance, therefore, is *inversely* proportional to the capacitance.

If an AC voltage is impressed across the capacitor, electrons are accumulated first on one plate and then on the other. If the frequency of the changes in polarity is low, the time available to accumulate electrons is large. This means that a large number of electrons will be able to accumulate, which will result in a large opposing effect, or a large reactance. If the frequency is high, the time available to accumulate electrons will be small. This means that only a few electrons will accumulate on the plates, which will result in a small opposing effect, or a small reactance. The reactance, therefore, is *inversely* proportional to the frequency. The formula for capacitive reactance is

$$X_C = 1/(2\pi fC) \tag{3.52}$$

where
  X_C = Capacitive reactance (ohms).
  f = Frequency (Hz).
  C = Capacitance (F).

■ **EXAMPLE 3.45**

Problem: What is the capacitive reactance of a circuit operating at a frequency of 60 Hz, if the total capacitance is 130 µf?
Solution:

$$X_C = 1/(2\pi fC) = 1/(6.28 \times 60 \times 0.00013) = 20.4 \text{ ohms}$$

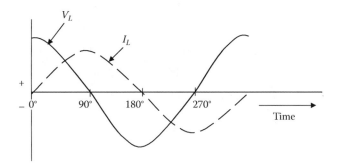

FIGURE 3.108   Inductive circuit: voltage leads current by 90°.

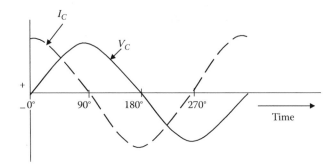

FIGURE 3.109   Capacitive circuit: current leads voltage by 90°.

PHASE RELATIONSHIPS FOR *R*, *L*, AND *C* CIRCUITS

Unlike a purely resistive circuit, where current rises and falls with the voltage (i.e., it neither leads nor lags; current and voltage are in phase), current and voltage are not in phase in inductive and capacitive circuits. This is the case, of course, because events are not quite instantaneous in circuits that have either inductive or capacitive components. In the case of an inductor, voltage is first applied to the circuit, then the magnetic field begins to expand, and self-induction causes a countercurrent to flow in the circuit, opposing the original circuit current. In this case, *voltage leads current* by 90° (Figure 3.108). When a circuit includes a capacitor, a charge current begins to flow and then a difference in potential appears between the plates of the capacitor. In this case, *current leads voltage* by 90° (Figure 3.109).

> **Note:** In an inductive circuit, voltage leads current by 90°; in a capacitive circuit, current leads voltage by 90°.

IMPEDANCE

Impedance is the total opposition to the flow of alternating current in a circuit that contains resistance and reactance. In the case of pure inductance, inductive reactance (X_L) is the total opposition to the flow of current through it. In the case of pure resistance, R represents the total opposition. The combined opposition of R and X_L in series or in parallel to current flow is the impedance. The symbol for impedance is Z. The impedance of resistance in series with inductance is

$$\frac{1}{R_T} = \frac{1}{R_1} + \frac{1}{R_2} + \frac{1}{R_3} + \ldots + \frac{1}{R_n} \tag{3.53}$$

where

    Z = Impedance (ohms).
    R = Resistance (ohms).
    X_L = Inductive reactance (ohms).

The impedance of resistance in series with capacitance is

$$Z = \sqrt{R^2 + X_L^2} \qquad (3.54)$$

where

    Z = Impedance (ohms).
    R = Resistance (ohms).
    X_C = Inductive capacitance (ohms).

When the impedance of a circuit includes R, X_L, and X_C, both resistance and net reactance must be taken into account. The equation for impedance that includes both X_L and X_C is

$$Z = \sqrt{R^2 + X_C^2} \qquad (3.55)$$

POWER IN REACTIVE CIRCUITS

The power in a DC circuit is equal to the product of volts and amps, but in an AC circuit this is true only when the load is resistive and has no reactance. In a circuit possessing inductance only, the true power is zero. The current lags the applied voltage by 90°. The *true power* in a capacitive circuit is also zero. True power is the *average power* actually consumed by the circuit, the average being taken over one complete cycle of alternating current. The *apparent power* is the product of the RMS volts and RMS amps. The ratio of true power to apparent power in an AC circuit is called the *power factor*. It may be expressed as a percent or as a decimal.

AC CIRCUIT THEORY

To this point, we have pointed out how combinations of inductance and resistance and then capacitance and resistance behave in an AC circuit, and we saw how *RL* and *RC* combinations affect the current, voltages, power, and power factor of a circuit:

1. The voltage drop across a resistor is *in phase* with the current through it.
2. The voltage drop across an inductor *leads* the current through it by 90°.
3. The voltage drop across a capacitor *lags* the current through it by 90°.
4. The voltage drops across inductors and capacitors are 180° *out of phase*.

Solving AC problems is complicated by the fact that current varies with time as the AC output of an alternator goes through a complete cycle. The various voltage drops in the circuit vary in phase—they are not at their maximum or minimum values at the same time. AC circuits frequently include all three circuit elements: resistance, inductance, and capacitance. In this section, all three of these fundamental circuit parameters are combined and their effect on circuit values studied.

SERIES *RLC* CIRCUIT

Figure 3.110 shows both the sine waveforms and the vectors for purely resistive, inductive, and capacitive circuits. Only the vectors show the direction, because the magnitudes are dependent on the values chosen for a given circuit. (We are only interested in the *effective* root-mean-square

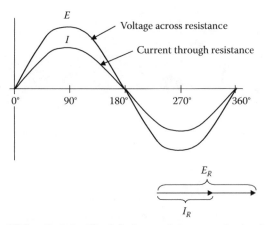

(A) Pure Resistive Circuit (voltage and current are in phase)

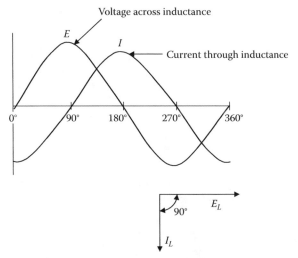

(B) Pure Inductive Circuit (voltage leads current by 90°).

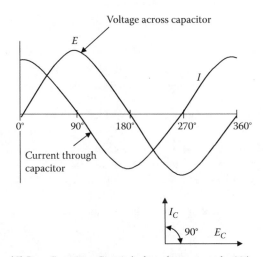

(C) Pure Capacitive Circuit (voltage lags current by 90°).

FIGURE 3.110   Sine waveforms and vectorial representation of R, L, and C circuits.

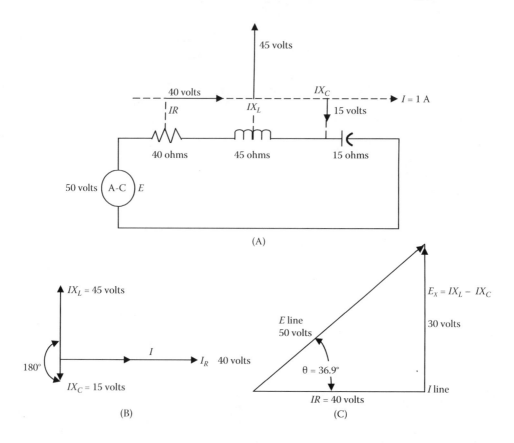

FIGURE 3.111   Resistance, inductance, and capacitance connected in a series.

values.) If the individual resistances and reactances are known, Ohm's law may be applied to find the voltage drops. For example, we know that $E_R = I \times R$ and $E_C = I \times X_C$. Then, according to Ohm's law, $E_L = I \times X_L$. In AC circuits, current varies with time, and the voltage drops across the various elements also vary with time; however, the same variation is not always present in each at the same time (except in purely resistive circuits) because current and voltage are not in phase. We are primarily concerned, in practical terms, with effective values of current and voltage; however, to understand basic AC theory, we need to know what occurs from instant to instant. In Figure 3.111, note first that current is the common reference for all three element voltages, because there is only one current in a series circuit, and it is common to all elements. The dashed line in Figure 3.111A represents the common series current. Voltage vectors for each element are provided to demonstrate their individual relationships to the common current. The total source voltage (E) is the vector sum of the individual voltages of IR, IX_L, and IX_C. The arrangement of these three voltages is summarized in Figure 3.111B. Because IX_L and IX_C are each 90° away from I, they are 180° from each other. Vectors in direct opposition (180° out of phase) may be subtracted directly. The total reactive voltage (E_X) is the difference of IX_L and IX_C; for example, $E_X = IX_L - IX_C = 45 - 15 = 30$ volts. The final relationship of line voltage and current, as seen from the source, is shown in Figure 3.111C. Had X_C been larger than X_L, the voltage would lag, rather than lead. When X_C and X_L are of equal value, the line voltage and current will be in phase.

Note: In a resistive circuit, the phase difference between voltage and current is zero.

Note: The voltage across a single reactive element in a series circuit can have a greater effective value than that of the applied voltage.

Note: One of the most important characteristics of an *RLC* circuit is that it can be made to respond most effectively to a single given frequency. When operated in this condition, the circuit is said to be in *resonance* with or *resonant* to the operating frequency. A circuit is at resonance when the inductive reactance (X_L) is equal to the capacitive reactance (X_C). At resonance, impedance (Z) equals resistance (R).

In summary, the series *RLC* circuit illustrates three important points:

- The current in a series *RLC* circuit either leads or lags the applied voltage, depending on whether X_C is greater or less than X_L.
- A capacitive voltage drop in a series circuit always subtracts directly from an inductive voltage drop.
- The voltage across a single reactive element in a series circuit can have a greater effective value than that of the applied voltage.

PARALLEL *RLC* CIRCUITS

The *true power* of a circuit is $P = E \times I \times \cos(\theta)$. For any given amount of power to be transmitted, the current (I) varies inversely with the power factor, $\cos(\theta)$. The addition of capacitance in parallel with inductance will, under the proper conditions, improve the power factor of the circuit and make possible the transmission of electric power with reduced line loss and improved voltage regulation. Figure 3.112A shows a three-branch, parallel AC circuit with a resistance in one branch, inductance in the second branch, and capacitance in the third branch. The voltage is the same across each parallel branch, so $V_T = V_R = V_L = V_C$. The applied voltage (V_T) is used as the reference line to measure phase angle θ. The total current (I_T) is the vector sum of I_R, I_L, and I_C. The current in the resistance (I_R) is in phase with the applied voltage (V_T) (Figure 3.112B). The current in the capacitor (I_C) leads total voltage V_T by 90°. I_L and I_C are exactly 180° out of phase and thus acting in opposite directions (Figure 3.112B). When $I_L > I_C$, I_T lags V_T (Figure 3.112C), so the parallel *RLC* circuit is considered inductive.

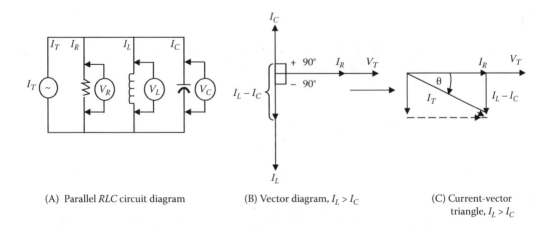

(A) Parallel *RLC* circuit diagram

(B) Vector diagram, $I_L > I_C$

(C) Current-vector triangle, $I_L > I_C$

FIGURE 3.112 R, X_L, and X_C in parallel.

POWER IN AC CIRCUITS

In circuits that have only resistance, but no reactance, the amount of power absorbed in the circuit is easily calculated by $P = I^2 \times R$. However, in dealing with circuits that include inductance and capacitance (or both), which is often the case in AC circuits, the calculation of power is a more complicated process. Earlier, we explained that power is a measure of the rate at which work is done. The work of a resistor is to limit current flow to the correct, safe level. In accomplishing this, the resistor dissipates heat, and we say that power is consumed or absorbed by the resistor. Inductors and capacitors also oppose current flow, but they do so by producing current that opposes the line current. In either inductive or capacitive circuits, instantaneous values of power may be very large, but the power actually absorbed is essentially zero, since only resistance dissipates heat (absorbs power). Both inductance and capacitance return the power to the source.

Any component that has resistance, such as a resistor or the wiring of an inductor, consumes power. Such power is not returned to the source, because it is dissipated as heat. Previously, we stated that power consumed in the circuit is called *true power*, or *average power*. The two terms are interchangeable, but we generally use the term average power, because the overall value is more meaningful than the instantaneous values of power appearing in the circuit during a complete cycle. Not all apparent power is consumed by the circuit; however, because the alternator does deliver the power, it must be considered in the design. The average power consumption may be small, but instantaneous values of voltage and current are often very large. Apparent power is an important design consideration, especially in assessing the amount of insulation necessary. In an AC circuit that includes both reactance and resistance, some power is consumed by the load and some is returned to the source. How much of each depends on the phase angle, because current normally leads or lags voltage by some angle.

Note: In terms of the dissipation of power as heat in a circuit, *average power* is the power that is dissipated as heat, but *apparent power* includes both power that is dissipated as heat and power that is returned to the source.

Note: Recall that in a purely reactive circuit, current and voltage are 90° out of phase.

In an *RLC* circuit, the R/Z ratio is the cosine of the phase angle, θ; therefore, it is easy to calculate average power in an *RLC* circuit:

$$P = E \times I \times \cos(\theta) \tag{3.56}$$

where
  P = Average power absorbed by the circuit.
  E = Effective value of the voltage across the circuit.
  I = Effective value of current in the circuit.
  θ = Phase angle between voltage and current.

Note: Recall that the equation for average power in a purely resistive circuit is $P = E \times I$. In a resistive circuit, $P = E \times I$ because the $\cos(\theta)$ is 1 and need not be considered. In most cases, the phase angle will be neither 90° nor zero, but somewhere between those extremes.

■ EXAMPLE 3.46

Problem: An *RLC* circuit has a source voltage of 500 volts, line current is 2 amps, and current leads voltage by 60°. What is the average power?
Solution: The cosine of 60° = 0.5, so

Average power = 500 V × 2 A × 0.5 = 500 W

■ **EXAMPLE 3.47**

Problem: An *RLC* circuit has a source voltage of 300 volts, line current is 2 amps, and current lags voltage by 31.8°. What is the average power?
Solution: The cosine of 31.8° = 0.8499, so

$$\text{Average power} = 300 \text{ V} \times 2 \text{ A} \times 0.8499 = 509.9 \text{ W}$$

■ **EXAMPLE 3.48**

Problem: Given E = 100 volts, I = 4 amps, and θ = 58.4°, what is the average power?
Solution: The cosine of 58.4° = 0.5240, so

$$\text{Average power} = 100 \text{ V} \times 4 \text{ A} \times 0.5240 = 209.6 \text{ W}$$

GENERATORS

DC GENERATORS

A DC generator is a rotating machine that converts mechanical energy into electrical energy. This conversion is accomplished by rotating an armature, which carries conductors, in a magnetic field, thus inducing an emf in the conductors. As stated previously, in order for an emf to be induced in the conductors, a relative motion must always exist between the conductors and the magnetic field in such a manner that conductors cut through the field. In most DC generators, the armature is the rotating member and the field is the stationary member. A mechanical force is applied to the shaft of the rotating member to cause the relative motion. Thus, when mechanical energy is put into the machine in the form of a mechanical force or twist on the shaft, causing the shaft to turn at a certain speed, electrical energy in the form of voltage and current is delivered to the external load circuit.

> **Note:** Mechanical power must be applied to the shaft constantly as long as the generator is supplying electrical energy to the external load circuit.

To gain a basic understanding of the operation of a DC generator, consider the following explanation. A simple DC generator consists of an armature coil with a single turn of wire (Figure 3.113A,B). (The armature coils used in large DC machines are usually wound in their final shape before being put on the armature, and the sides of the preformed coil are placed in the slots of the laminated armature core.) This armature coil cuts across the magnetic field to produce voltage. If a complete

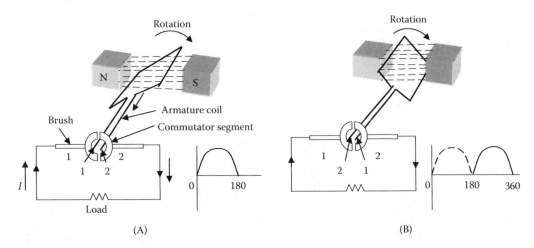

FIGURE 3.113   Basic operation of a DC generator.

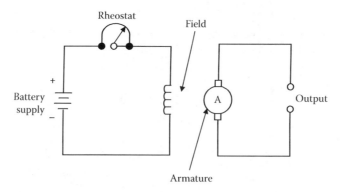

FIGURE 3.114   Separately excited DC generator.

path is present, current will move through the circuit in the direction shown by the arrows in Figure 3.113A. In this position of the coil, commutator segment 1 is in contact with brush 1, while commutator segment 2 is in contact with brush 2. As the armature rotates a half turn in a clockwise direction, the contacts between the commutator segments and the brushes are reversed (Figure 3.113B). At this moment, segment 1 is in contact with brush 2 and segment 2 is in contact with brush 1. Because of this commutator action, that side of the armature coil that is in contact with either of the brushes is always cutting across the magnetic field in the same direction. Thus, brushes 1 and 2 have constant polarity, and a *pulsating DC current* is delivered to the external load circuit.

> **Note:** In DC generators, voltage induced in individual conductors is AC. It is converted to DC (rectified) by the commutator, which rotates in contact with carbon brushes so current generated is in one direction (i.e., direct current).

The several different types of DC generators take their names from the type of field excitation used; that is, they are classified according to the manner in which the field windings are connected to the armature circuit. When the field of the generator is excited (or supplied) from a separate DC source (such as a battery) other than its own armature, it is called a *separately excited DC generator* (Figure 3.114). The field windings of a *shunt generator* (self-excited) are connected in series with a rheostat, across the armature in shunt with the load, as shown in Figure 3.115; the shunt generator is widely used in industry. The field windings of a *series generator* (self-excited) are connected in series with the armature and load, as shown in Figure 3.116. Series generators are seldom used. *Compound generators* (self-excited) contain both series and shunt field windings, as shown in Figure 3.117; compound generators are widely used in industry.

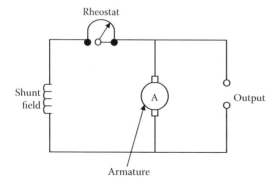

FIGURE 3.115   DC shunt generator.

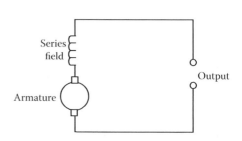

FIGURE 3.116   DC series generator.

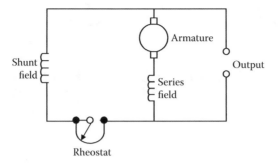

FIGURE 3.117   DC compound generator.

Note: As central generating stations increased in size along with number and power distribution distances, DC generating systems, because of the high power losses in long DC transmission lines, were replaced by AC generating systems to reduce power transmission costs.

AC GENERATORS

Most electric power utilized today is generated by *alternating-current generators* (also called *alternators*). They are made in many different sizes, depending on their intended use. Regardless of size, however, all generators operate on the same basic principle—a magnetic field cutting through conductors or conductors passing through a magnetic field. They include (1) a group of conductors in which the output voltage is generated, and (2) a second group of conductors through which direct current is passed to obtain an electromagnetic field of fixed polarity. The conductors in which the electromagnetic field originates are always referred to as the *field windings*.

In addition to the armature and field, there must also be motion between the two. To provide this motion, AC generators are built in two major assemblies, the *stator* and the *rotor*. The rotor rotates inside the stator. The revolving-field AC generator (see Figure 3.118) is the most widely used type. In this type of generator, direct current from a separate source is passed through windings on the rotor by means of slip rings and brushes. (Slip rings and brushes are adequate for the DC field supply because the power level in the field is much smaller than in the armature circuit.) This maintains a rotating electromagnetic field of fixed polarity. The rotating magnetic field, following the rotor, extends outward and cuts through the armature windings imbedded in the surrounding stator. As the rotor turns, AC voltages are induced in the windings because magnetic fields of first one polarity and then the other cut through them. The output power is taken from the

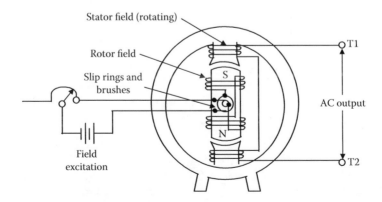

FIGURE 3.118   Essential parts of a rotating field AC generator.

stationary windings and may be connected through fixed output terminals (T1 and T2 in Figure 3.118). This is advantageous, in that there are no sliding contacts and the entire output circuit is continuously insulated.

> **Note:** In AC generators, frequency and electromagnetic wave cycles per second depend on how fast the rotor turns and the number of electromagnetic field poles. The voltage generated depends on the rotor speed, number of coils in the armature, and strength of the magnetic field.

ELECTRIC MOTORS

There is an almost endless variety of tasks that electric motors perform in the operation of industrial and domestic equipment. An *electric motor* is a machine used to change electrical energy to mechanical energy to do the work. (Recall that a generator does just the opposite; that is, a generator changes mechanical energy to electrical energy.) Previously, we pointed out that when a current passes through a wire a magnetic field is produced around the wire. If this magnetic field passes through a stationary magnetic field, the fields either repel or attract, depending on their relative polarity. If both are positive or negative, they repel. If they have opposite polarities, they attract. Applying this basic information to motor design, an electromagnetic coil, the armature, rotates on a shaft. The armature and shaft assembly are called the *rotor*. The rotor is assembled between the poles of a permanent magnet and each end of the rotor coil (armature) is connected to a commutator also mounted on the shaft. A commutator is composed of copper segments insulated from the shaft and from each other by an insulting material. As like poles of the electromagnet in the rotating armature pass the stationary permanent magnet poles, they are repelled, continuing the motion. As the opposite poles near each other, they attract, continuing the motion.

DC MOTORS

The construction of a DC motor is essentially the same as that of a DC generator; however, it is important to remember that the DC generator converts mechanical energy into the electrical energy back into mechanical energy. A DC generator may be made to function as a motor by applying a suitable source of DC voltage across the normal output electrical terminals. The various types of DC motors differ in the way their field coils are connected. Each has characteristics that are advantageous under given load conditions. *Shunt motors* (see Figure 3.119) have field coils connected in parallel with the armature circuit. This type of motor, with constant potential applied, develops variable torque at an essentially constant speed, even under changing load conditions. Such loads are found in machine-shop equipment, such as lathes, shapes, drills, milling machines, and so forth. *Series motors* (see Figure 3.120) have field coils connected in series with the armature circuit. This

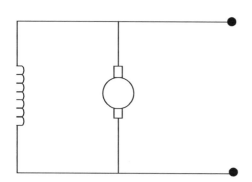

FIGURE 3.119   DC shunt motor.

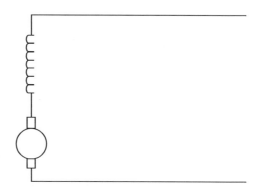

FIGURE 3.120   DC series motor.

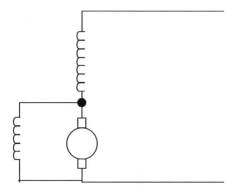

FIGURE 3.121 DC compound motor.

type of motor, with constant potential applied, develops variable torque but its speed varies widely under changing load conditions. The speed is low under heavy loads but becomes excessively high under light loads. Series motors are commonly used to drive electric hoists, winches, cranes, and certain types of vehicles (e.g., electric trucks). Also, series motors are used extensively to start internal combustion engines. *Compound motors* (see Figure 3.121) have one set of field coils in parallel with the armature circuit and another set of field coils in series with the armature circuit. This type of motor is a compromise between shunt and series motors. It develops an increased starting torque over that of the shunt motor and has less variation in speed than the series motor. The speed of a DC motor is variable. It is increased or decreased by a rheostat connected in series with the field or in parallel with the rotor. Interchanging either the rotor or field winding connections reverses direction.

Brushless DC Motor

Brushless DC motors can be deployed in any area currently utilizing brushed DC motors. In renewable energy applications, the brushless DC motor can be found in such high-power applications as electric vehicles, hybrid vehicles, and some industrial machinery. The Segway® scooter and the Vectrix™ maxi-scooter use brushless DC motors. They are more commonly found in consumer devices, such as computer hard drives, CD/DVD players, and PC cooling fans. The brushless DC motor is a synchronous motor powered by DC current that has permanent magnets (usually rare-earth-type magnets) mounted on the spinning rotor; instead of a mechanical commutation system based on brushes, it has a commutation system that is electronically controlled by a solid-state controller. The commutator and brush system form a set of electrical switches, each opening and closing in sequence, such that electrical power always flows through the armature coil closest to the stationary stator.

AC Motors

Alternating current voltage can easily be transformed from low voltages to high voltages, or *vice versa*, and can be moved over a much greater distance without too much loss in efficiency. Most of the power-generating systems today, therefore, produce alternating current. Thus, it logically follows that a great majority of the electrical motors utilized today are designed to operate on alternating current. AC motors offer other advantages, though, in addition to the wide availability of AC power. In general, AC motors are less expensive than DC motors. Most types of AC motors do not employ brushes and commutators which eliminates many problems of maintenance and wear and eliminates dangerous sparking. AC motors are manufactured in many different sizes, shapes, and ratings for use on an even greater number of jobs. They are designed for use with either polyphase or single-phase power systems. This text cannot possibly cover all aspects of the subject of AC motors; consequently, it deals mainly with the operating principles of the two most common types—the induction motor and the synchronous motor.

Induction Motors (Polyphase)

The induction motor is the most common AC motor because of its simple, rugged construction and good operating characteristics. It consists of two parts: the *stator* (stationary part) and the *rotor* (rotating part). The most important type of polyphase induction motor is the three-phase motor.

Note: A three-phase system (3θ) is a combination of three single-phase (1θ) systems. In a three-phase, balanced system, the power comes from an AC generator that produces three separate but equal voltages, each of which is out of phase with the other voltages by 120°. Although single-phase circuits are widely used in electrical systems, most generation and distribution of AC current are three phase.

The driving torque of both DC and AC motors is derived from the reaction of current-carrying conductors in a magnetic field. In the DC motor, the magnetic field is stationary and the armature, with its current-carrying conductors, rotates. The current is supplied to the armature through a commutator and brushes. In induction motors, the rotor currents are supplied by electromagnet induction. The stator windings, connected to the AC supply, contain two or more out-of-time-phase currents, which produce corresponding magnetomotive force (mmf), establishing a rotating magnetic field across the air gap. This magnetic field rotates continuously at constant speed regardless of the load on the motor. The stator winding corresponds to the armature winding of a DC motor or to the primary winding of a transformer. The rotor is not connected electrically to the power supply. The induction motor derives its name from the fact that mutual induction (or transformer action) takes place between the stator and the rotor under operating conditions. The magnetic revolving field produced by the stator cuts across the rotor conductors, inducing a voltage in the conductors. This induced voltage causes rotor current to flow. Hence, motor torque is developed by the interaction of the rotor current and the magnetic revolving field.

Synchronous Motors

Like induction motors, synchronous motors have stator windings that produce a rotating magnetic field, but, unlike the induction motor, the synchronous motor requires a separate source of direct current from the field. It also requires special starting components, including a salient-pole field with starting grid winding. The rotor of the conventional type synchronous motor is essentially the same as that of the salient-pole AC generator. The stator windings of induction and synchronous motors are essentially the same. In operation, the synchronous motor rotor locks into step with the rotating magnetic field and rotates at the same speed. If the rotor is pulled out of step with the rotating stator field, no torque is developed and the motor stops. Because a synchronous motor develops torque only when running at synchronous speed, it is not self-starting and hence needs some device to bring the rotor to synchronous speed. A synchronous motor may be started by a DC motor on a common shaft. After the motor is brought to synchronous speed, AC current is applied to the stator windings. The DC starting motor now acts as a DC generator, which supplies DC field excitation for the rotor. The load then can be coupled to the motor.

Single-Phase Motors

Single-phase motors are so called because their field windings are connected directly to a single-phase source. These motors are used extensively in fractional horsepower sizes in commercial and domestic applications. The advantages of using single-phase motors in small sizes are that they are less expensive to manufacture than other types, and they eliminate the need for three-phase AC lines. Single-phase motors are used in fans, refrigerators, portable drills, grinders, and so forth. A single-phase induction motor with only one stator winding and a cage rotor is like a three-phase induction motor with a cage rotor except that the single-phase motor has no magnetic revolving field at start and hence no starting torque. However, if the rotor is brought up to speed by external means, the induced currents in the rotor will cooperate with the stator currents to produce a revolving field, which causes the rotor to continue to run in the direction in which it was started. Several methods are used to provide the single-phase induction motor with starting torque. These methods categorize the motor as split-phase, capacitor, shaded-pole, or repulsion-start. Another class of single-phase motors is the AC series (universal) type.

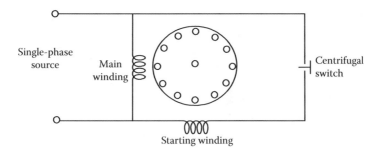

FIGURE 3.122   Split-phase motor.

Split-Phase Motor

The split-phase motor (Figure 3.122), has a stator composed of slotted lamination that contains a starting winding and a running winding. The starting winding has fewer turns and smaller wire than the running winding, hence higher resistance and less reactance. The main winding occupies the lower half of the slots and the starting winding occupies the upper half. When the same voltage is applied to both windings, the current in the main winding lags behind the current in the starting winding. The angle (θ) between the main winding and the starting winding is enough phase difference to provide a weak rotating magnetic field that produces a starting torque. When the motor reaches a predetermined speed, usually 75% of synchronous speed, a centrifugal switch mounted on the motor shaft opens, thereby disconnecting the starting winding. Because of their low starting torque, fractional-horsepower, split-phase motors are used in a wide variety of equipment, such as washers, oil burners, ventilating fans, and woodworking machines. The direction of rotation of the split-phase motor can be reversed by interchanging the starting winding leads.

> *Note:* If two stator windings of unequal impedance are spaced 90° apart but connected in parallel to a single-phase source, the field produced will appear to rotate. This is the principle of *phase splitting*.

Capacitor Motor

The capacitor motor is a modified form of split-phase motor. Its capacitor is in series with the starting winding (Figure 3.123). The capacitor motor operates with an auxiliary winding and series capacitor permanently connected to the line (Figure 3.123). The capacitance in series may be of one value for starting and another value for running. As the motor approaches synchronous speed, the centrifugal switch disconnects one section of the capacitor. If the starting winding is cut out after the motor has increased in speed, the motor is a *capacitor-start motor*. If the starting winding and capacitor are designed to be left in the circuit continuously, the motor is a *capacitor-run motor*. Capacitor motors are used to drive grinders, drill presses, refrigerator compressors, and other loads that require relatively high starting torque. The direction of rotation of the capacitor motor can be reversed by interchanging the starting winding leads.

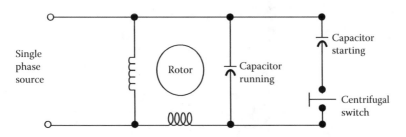

FIGURE 3.123   Capacitor motor.

Shaded-Pole Motor

A shaded-pole motor employs a salient-pole stator and a cage rotor. The projecting poles on the stator resemble those of DC machines except that the entire magnetic circuit is laminated and a portion of each pole is split to accommodate a short-circuited coil called a *shading coil* (Figure 3.124). The coil is usually a single band or strap of copper. The effect of the coil is to produce a small sweeping motion of the field flux from one side of the pole piece to the other as the field pulsates. This slight shift in the magnetic field produces a small starting torque. Thus, shaded-pole motors are self-starting. This motor is generally manufactured in very small sizes, up to 1/20 horse-power, for driving small fans, small appliances, and clocks.

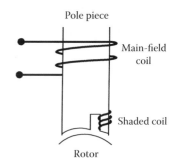

FIGURE 3.124   Shaded pole.

In operation, during that part of the cycle when the main pole flux is increasing, the shading coil is cut by the flux, and the resulting induced emf and current in the shading coil tend to prevent the flux from rising readily through it. Thus, the greater portion of the flux rises in that portion of the pole that is not in the vicinity of the shading coil. When the flux reaches its maximum value, the rate of change of flux is zero, and the voltage and current in the shading coil are also zero. At this time, the flux is distributed more uniformly over the entire pole face. Then, as the main flux decreases toward zero, the induced voltage and current in the shading coil reverse their polarity, and the resulting mmf tends to prevent the flux from collapsing through the iron in the region of the shading coil. The result is that the main flux first rises in the unshaded portion of the pole and later in the shaded portion. This action is equivalent to a sweeping movement of the field across the pole face in the direction of the shaded pole. The rotor conductors are cut by this moving field, and the force exerted on them causes the rotor to turn in the direction of the sweeping field. The shaded-pole method of starting is used in very small motors, up to about 1/25 hp, for driving small fans, small appliances, and clocks.

Repulsion-Start Motor

Like a DC motor, the repulsion-start motor has a form-wound rotor with commutator and brushes. The stator is laminated and contains a distributed single-phase winding. In its simplest form, the stator resembles that of the single-phase motor. In addition, the motor has a centrifugal device that removes the brushes from the commutator and places a short-circuiting ring around the commutator. This action occurs at about 75% of synchronous speed. Thereafter, the motor operates with the characteristics of the single-phase induction motor. This type of motor is made in sizes ranging from 1/2 to 15 hp and is used in applications requiring a high starting torque.

AC Series Motor

The AC series motor operates on either AC or DC circuits. When an ordinary DC series motor is connected to an AC supply, the current drawn by the motor is low due to the high series-field impedance. The result is low running torque. To reduce the field reactance to a minimum, AC series motors are built with as few turns as possible. Armature reaction is overcome by using *compensating windings* (see Figure 3.125) in the pole pieces. As in DC series motors, the speed in AC series

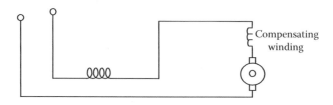

FIGURE 3.125   AC series motor.

motors increases to a high value with a decrease in load. The torque is high for high armature currents, so the motor has a good starting torque. AC series motors operate more efficiently at low frequencies. Fractional horsepower AC series motors are called *universal motors*. They do not have compensating windings. They are used extensively to operate fans and portable tools, such as drills, grinders, and saws.

TRANSFORMERS

A transformer is an electric control device (with no moving parts) that raises or lowers voltage or current in an electric distribution system. The basic transformer consists of two coils electrically insulated from each other and wound on a common core (Figure 3.126). Magnetic coupling is used to transfer electric energy from one coil to another. The coil that receives energy from AC sources is the primary coil. The coil that delivers energy to AC loads is the secondary coil. The core of transformers used at low frequencies is generally made of magnetic material, usually laminated sheet steel. Cores of transformers used at higher frequencies are made of powdered iron and ceramics, or nonmagnetic materials. Some coils are simply wound on nonmagnetic hollow forms such as cardboard or plastic so the core material is actually air.

In operation, an alternating current will flow when an AC voltage is applied to the primary coil of a transformer. This current produces a field of force that changes as the current changes. The changing magnetic field is carried by the magnetic core to the secondary coil, where it cuts across the turns of that coil. In this way, an AC voltage in one coil is transferred to another coil, even though there is no electrical connection between them. The number of lines of force available in the primary is determined by the primary voltage and the number of turns on the primary, with each turn producing a given number of lines. If there are many turns on the secondary, each line of force will cut many turns of wire and induce a high voltage. If the secondary contains only a few turns, there will be few cuttings and low induced voltage. The secondary voltage, then, depends on the number of secondary turns as compared with the number of primary turns. If the secondary has twice as many turns as the primary, the secondary voltage will be twice as large as the primary voltage. If the secondary has half as many turns as the primary, the secondary voltage will be one-half as large as the primary voltage. A voltage ratio of 1:4 means that for each volt on the primary there are 4 volts on the secondary. This is called a *step-up* transformer. A step-up transformer receives a low voltage on the primary coil and delivers a high voltage from the secondary coil. In contrast, a voltage ratio of 4:1 means that for 4 volts on the primary coil there is only 1 volt on the secondary. This is called a *step-down transformer*. A step-down transformer receives a high voltage on the primary coil and delivers a low voltage from the secondary.

Note: The voltage on the coils of a transformer is directly proportional to the number of turns on the coils.

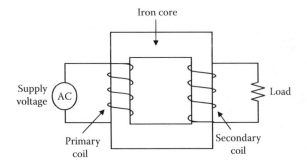

FIGURE 3.126   Basic transformer.

POWER DISTRIBUTION SYSTEM PROTECTION

Interruptions are very rare in a power distribution system that has been properly designed. Still, protective devices are necessary because of the load diversity. Most installations are quite complex. In addition, externally caused variations might overload them or endanger personnel: Figure 3.127 shows the general relationship between protective devices and different components of a complete system. Each part of the circuit has its own protective device or devices that protect not only the load but also the wiring and control devices themselves. These disconnect and protective devices are described in the following sections.

FUSES

The passage of an electric current produces heat. The larger the current, the more heat is produced. To prevent large currents from accidentally flowing through expensive apparatus and burning it up, a fuse is placed directly into the circuit, as in Figure 3.127, so as to form a part of the circuit through which all the current must flow. The fuse will permit currents smaller than the fuse value to flow but will melt and therefore break the circuit if a larger, dangerous current ever appears; for example, a dangerously large current will flow when a short circuit occurs. A short circuit is usually caused by an accidental connection between two points in a circuit which offer very little resistance to the flow of electrons. If the resistance is small, there will be nothing to stop the flow of the current, and the current will increase enormously. The resulting heat generated might cause a fire. If the circuit is protected by a fuse, the heat caused by the short-circuit current will melt the fuse wire, thus breaking the circuit and reducing the current to zero.

Note: A fuse is a thin strip of easily melted material. It protects a circuit from large currents by melting quickly, thereby breaking the circuit.

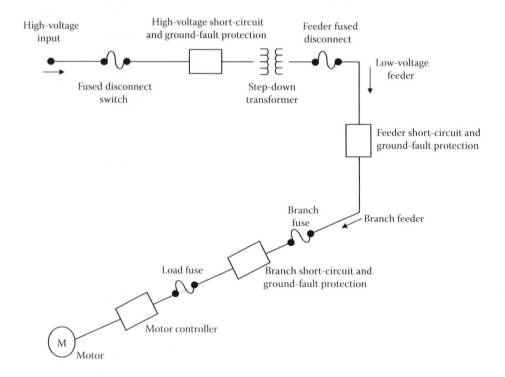

FIGURE 3.127   Motor power distribution system.

Fuses are rated by the number of amps of current that can flow through them before they melt and break the circuit; thus, a fuse can be, for example, 10, 15, 20, or 30 amps. We must be sure that any fuse inserted in a circuit is rated low enough to melt, or "blow," before the apparatus is damaged. In a plant building wired to carry a current of 10 A, it is best to use a fuse no larger than 10 A so a current larger than 10 A could never flow. Some equipment, such as an electric motor, requires more current during starting than for normal running; fast-time or medium-time fuses that provide running protection might blow during the initial period when high starting current is required. *Delayed-action* fuses are used to handle these situations.

CIRCUIT BREAKERS

Circuit breakers are protective devices that open automatically at a preset ampere rating to interrupt an overload or short circuit. Unlike fuses, they do not require replacement when they are activated. They are simply reset to restore power after the overload has been cleared. Circuit breakers are made in both plug-in and bolt-on designs. Plug-in breakers are used in load centers. Bolt-on breakers are used in panelboards and exclusively for high interrupting current applications. Circuit breakers are rated according to current and voltage, as well as short-circuit-interrupting current. A single handle opens or closes contacts between two or more conductors. Breakers are single pole, but single-pole units can be ganged to form double- or triple-pole devices opened with a single handle.

> **Note:** A circuit breaker is designed to break the circuit and stop the current flow when the current exceeds a predetermined value.

Several types of circuit breakers are commonly used. They may be thermal or magnetic, or a combination of the two. Thermal breakers are tripped when the temperature rises because of heat created by the overcurrent condition. Bimetallic strips provide the time delay for overload protection. Magnetic breakers operate on the principle that a sudden current rise will create enough magnetic field to turn an armature, tripping the breaker and opening the circuit. Magnetic breakers provide the instantaneous action needed for short-circuit protection. Magnetic breakers are also used where ambient temperatures might adversely affect the action of a thermal breaker. Thermal–magnetic breakers combine features of both types. An important feature of the circuit breaker is its arc chutes, which allow the breaker to extinguish very hot arcs harmlessly. Some circuit breakers must be reset by hand, while others reset themselves automatically. If the overload condition still exists when the circuit breaker is reset, the circuit breaker will trip again to prevent damage to the circuit.

CONTROL DEVICES

Control devices are electrical accessories (switches and relays) that govern the power delivered to any electrical load. In its simplest form, the control applies voltage to, or removes it from, a single load. In more complex control systems, the initial switch may set into action other control devices (relays) that govern motor speeds, servomechanisms, temperatures, and numerous other pieces of equipment. In fact, all electrical systems and equipment are controlled in some manner by one or more controls. A controller is a device or group of devices that serves to govern, in some predetermined manner, the device to which it is connected. In large electrical systems, it is necessary to have a variety of controls for operation of the equipment. These controls range from simple push buttons to heavy-duty contactors that are designed to control the operation of large motors. The push button is manually operated; a contactor is electrically operated.

ELECTRICAL DRAWINGS

Working drawings for the fabrication and troubleshooting of electrical machinery, switching devices, and chassis for electronic equipment, cabinets, housings, and other mechanical elements associated with electrical equipment are based on the same principles as given earlier. Those qualified to operate, maintain, and repair electrical equipment must understand electrical systems. The electrician or student of electricity must be able to read electrical drawings to understand the system and to determine what is wrong when electrical equipment fails to run properly. This section introduces electrical drawings, and the functions of important electrical components and how they are shown on drawings are explained.

ELECTRICAL SYMBOLS

Figure 3.128 shows some of the most common symbols used on electrical drawings. It is not necessary to memorize these symbols, but the maintenance operator should be familiar with them as an aid to reading electrical drawings.

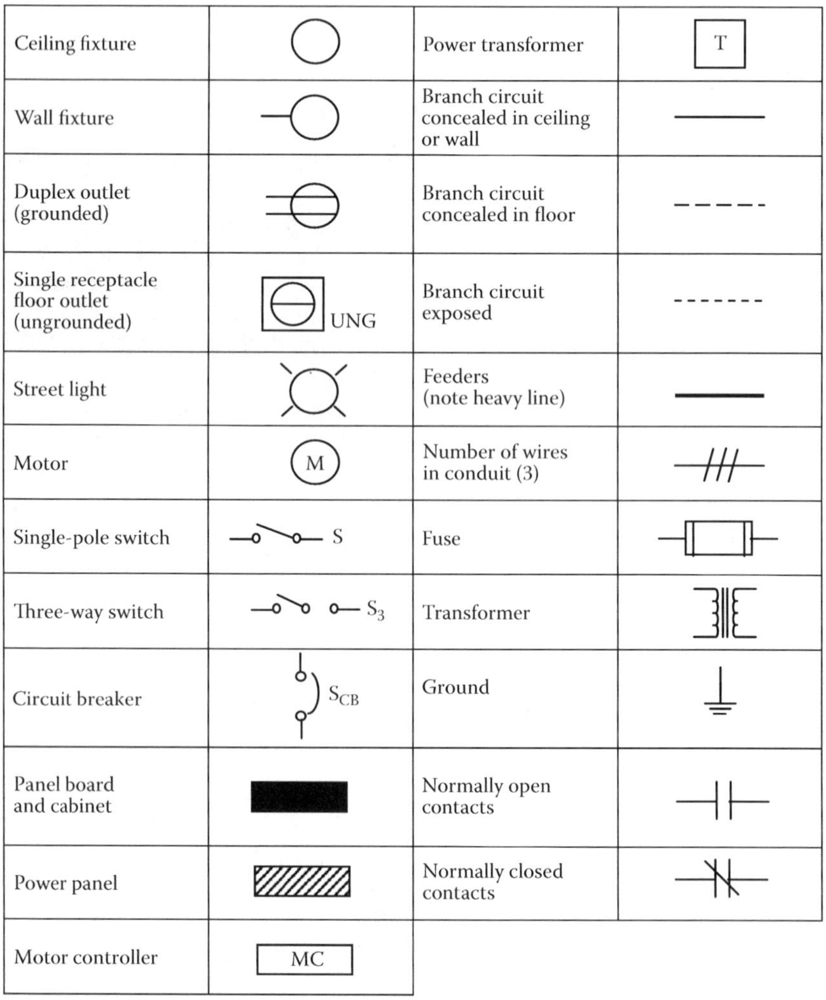

FIGURE 3.128   Common electrical symbols.

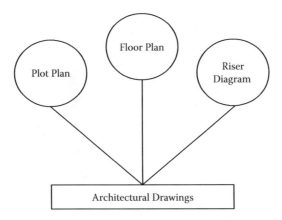

FIGURE 3.129   Architectural drawings.

TYPES OF ELECTRICAL DRAWINGS

The two kinds of electrical drawings used for troubleshooting operations are *architectural drawings* and *circuit drawings*. An architectural drawing shows the physical locations of the electric lines in a plant building or between buildings. A circuit drawing shows the electrical loads served by each circuit.

> *Note:* A circuit drawing does not indicate the physical location of any load or circuit.

Architectural Drawings

Figure 3.129 shows three types of architectural drawings. The *plot plan* shows the electric distribution to all the plant buildings. The *floor plan* shows where branch circuits are located in one building or pumping station, where equipment is located, and where outside and inside tie-ins to water, heat, and electric power are located. The *riser diagram* shows how the wiring goes to each floor of the building.

Circuit Drawings

A circuit drawing shows how a single circuit distributes electricity to various loads (e.g., pump motors, grinders, bar screens, mixers). Unlike an architectural drawing, a circuit drawing does not show the location of these loads. Figure 3.130 depicts a typical single-line circuit drawing and shows the power distribution to 11 loads. The number in each circle indicates the power rating of the loads in horsepower.

> *Note:* Electrical loads in all plants can be divided into two categories: critical and noncritical. Critical loads are essential to the operation of the plant and cannot be turned off (e.g., critical unit processes). Noncritical loads include pieces of equipment that would not disrupt the operation of the plant or pumping station or compromise safety if they were turned off for a short period of time (e.g., air conditioners, fan systems, electric water heaters, certain lighting systems).

The numbers in the rectangles show the current ratings of circuit breakers. The upper number is the current in amps the circuit breaker will allow as a momentary surge.

> *Note:* Circuit breakers are typically equipped with surge protection for three-phase motors and other devices. When a three-phase motor is started, current demand is six to ten times normal value. After start, current flow decreases to its normal rated value. Surge protection is also provided to allow slight increases in current flow when the load varies or increases slightly. The lower number is the maximum current the circuit breaker will allow to flow continuously. Most circuit breakers are also equipped with an instantaneous trip value for protection against short circuits.

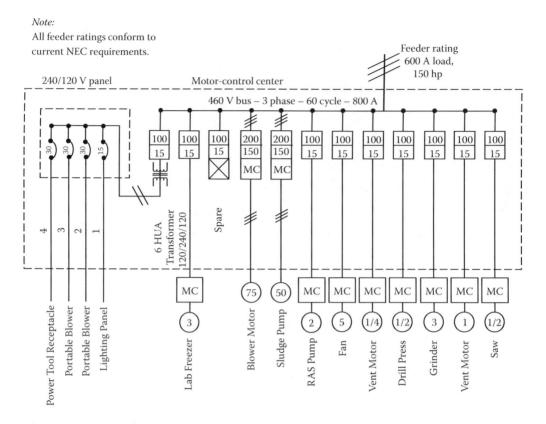

FIGURE 3.130   Single-line circuit diagram.

Ladder Drawing

A ladder drawing is a type of schematic diagram that shows a control circuit. The parts of the control circuit lie on horizontal lines, like the rungs of a ladder. Figure 3.131 is an example of a ladder diagram. The purpose of a ladder drawing, such as the one shown in Figure 3.131, is to cut maintenance and troubleshooting time. This is accomplished when the designer follows certain guidelines when making electrical drawings and layouts. Let's take a closer look at the ladder drawing for a control circuit shown in Figure 3.131. Note the numbering of elementary circuit lines. Normally, closed contacts are indicated by a bar under the line number. Moreover, note that the line numbers are enclosed in a geometric figure to prevent mistaking the line numbers for circuit numbers. All contacts and the conductors connected to them are properly numbered. Typically, numbering is carried throughout the entire electrical system. This may involve going through one or more terminal blocks. The incoming and outgoing conductors as well as the terminal blocks carry the proper electrical circuit numbers. When possible, connections to all electrical components are taken back to one common checkpoint. All electrical elements on a machine should be correctly identified with the same markings as shown on the ladder drawing in Figure 3.131; for example, if a given solenoid is marked "Solenoid A2" on the drawing, the actual solenoid on the machine should carry the marking "Solenoid A2."

> **Note:** The size of electrical drawings is important. This can be understood first in the problem of storing drawings. If every size and shape were allowed, the task of systematic and protective filing of drawings could be tremendous. Pages that are 8-1/2 × 11 or 9 × 12 inches and multiples thereof are generally accepted. The drawing size can also be a problem for the troubleshooter or maintenance operator. If the drawing is too large, it is unwieldy to handle at the machine. If it is too small, it is difficult to read the schematic.

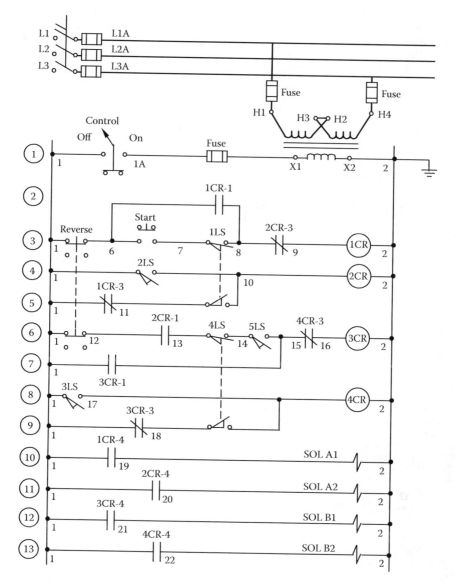

FIGURE 3.131 Typical ladder drawing.

DISCUSSION AND REVIEW QUESTIONS

1. Explain the difference between a series and a parallel circuit.
2. What is the sum of all voltages in a series circuit equal to?
3. For any total voltage rise in a circuit, there must be an equal total _____.

Refer to the figure to the right for questions 4 through 8.

4. Is the direction of current flow clockwise or counterclockwise?
5. $I_T = ?$
6. E dropped across $R_1 = ?$
7. What is the power absorbed by R_2?
8. $P_T = ?$

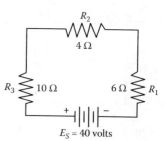

9. The equivalent resistance (R_T) of parallel branches is _____ than the smallest branch resistance because all of the branches must take _____ current from the source than any one branch.
10. A short circuit has _____ resistance, resulting in _____ current.
11. What is the total resistance of four 30-ohm resistors connected in series?
12. There is only _____ voltage across all components in parallel.
13. The sum of the _____ values of power dissipated in parallel resistances equals the _____ power produced by the source.

Refer to the figure to the right for questions 14 through 16.

14. $R_T = ?$
15. $I_T = ?$
16. $P_T = ?$

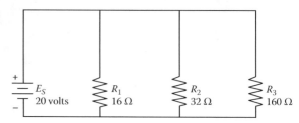

Refer to the figure to the right for questions 17 through 20.

17. What is the total resistance (R_T) of the circuit?
18. What is the total current (I_T) in the circuit?
19. What is the voltage across R_3?
20. What is the total power consumed in the circuit?

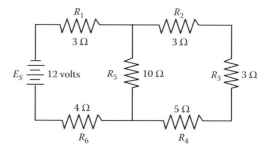

Answers

1. A series circuit has only one path for current flow, whereas a parallel circuit has more than one path.
2. Source voltage
3. Voltage drop
4. Counterclockwise
5. 2 amps
6. 12 volts
7. 16 volts
8. 80 watts
9. Less, more
10. Zero, excessive
11. 120 ohms
12. One
13. Individual, total
14. $R_T = 10$ ohms
15. $I_T = 2$ amps
16. $P_T = 40$ watts
17. 12 ohms
18. 1 amp (rounded)
19. 1.5 volts
20. 12 watts (rounded)

REFERENCES AND RECOMMENDED READING

Callen, H.B. (1985). *Thermodynamics and an Introduction to Thermostatistics*, 2nd ed. New York: John Wiley & Sons.

EIA. (2009). *What Is Energy? Explained*. Washington, DC: Energy Information Administration (http://tonto.eia.doe.gov/energyexplained/print.cfm?page=about_sources_of_energy).

EIA. (2013). *Use of Energy in the United States Explained*. Washington, DC: Energy Information Administration (http://tonto.eia.doe.gov/energyexplained/print.cfm?page=us_energy_use).

Feynman, R.P., Leighton, R.B., and Sands, M. (1963). *The Feynman Lectures on Physics*. New York: Addison-Wesley.

Halliday, D. and Resnick, R. (1978). *Physics*, 3rd ed. New York: John Wiley & Sons.

Hinrichs, R.A. and Kleinbach, M. (2006). *Energy: Its Use and the Environment*, 4th ed. Belmont, CA: Brooks & Cole.

Kroemer, H. and Kittel, C. (1980). *Thermal Physics*, 2nd ed. New York: W.H. Freeman.

NREL. (2009). *Energy Storage: Batteries*. Washington, DC: National Renewable Energy Laboratory (http://www.nrel.gov/vehiclesandfuels/energystorage/batteries.html).

Organ, A.J. (1992). *Thermodynamics and Gas Dynamics of the Stirling Cycle Machine*. London: Cambridge University Press.

Organ, A.J. (1997). *The Regenerator and the Stirling Engine*. New York: John Wiley & Sons.

Spellman, F.R. and Drinan, J.E. (2001). *Electricity*. Boca Raton, FL: CRC Press.

USDOE. (1992). *DOE Fundamentals Handbook: Electrical Science*, Vol. 1. Washington, DC: U.S. Department of Energy.

4 Solar Energy

One generation passes away, and another generation cometh: but the Earth abides forever. The sun also rises.

—Ecclesiastes 1:4–5

Busy old fool, unruly Sun,
Why dost thou thus,
Through windows and through curtains, call on us?

—John Dunne, "The Sun Rising"

The Sun, with all the planets revolving around it and dependent on it, can still ripen a bunch of grapes as though it had nothing else in the universe to do.

—Galileo Galilei

INTRODUCTION

It is fitting to begin our discussion of the various kinds of renewable energy with the sun—the medium-sized star that symbolizes life, power, strength, force, clarity, and, yes, energy. In 1931, not long before he died, the inventor Thomas Edison told his friends Henry Ford and Harvey Firestone, "I'd put my money on the sun and solar energy. What a source of power! I hope we don't have to wait until oil and coal run out before we tackle that" (Newton, 1989). The sun nourishes our planet. Solar energy is a renewable resource because it is continuously supplied to the Earth by the sun. When we consider the sun and solar energy, we quickly realize that there is nothing new about renewable energy. The sun was the first energy source; it has been around for 4.5 billion years, as long as anything else we are familiar with. On Earth, without the sun there would be nothing—absolutely nothing, zilch. The old 5-5-5 rule sums up the essence of life on Earth: We can survive 5 minutes without oxygen, 5 days without water, and 5 weeks without food. Without the sun, though, the 5-5-5-rule has no validity. Without the sun, nothing related to life has validity. Without the sun, there is nothing: no air, no water, no food ... no life. The sun, the ultimate energy source, provided light and heat to the first humans. During daylight, people searched for food. They hunted and gathered and probably stayed together for their safety. When nightfall arrived, we can only imagine that they huddled together for warmth and reassurance under the light of the stars and moon, waiting for the sun and its live-giving and life-sustaining light to return.

Solar energy (a term used interchangeably with solar power) uses various technologies to take advantage of the power of the sun to produce energy (Figure 4.1). Solar energy is one of the best renewable energy sources available because it is one the cleanest sources of energy. Direct solar radiation absorbed in solar collectors can provide space heating and hot water. Passive solar can be used to enhance the solar energy used in buildings for space heating and lighting requirements. Solar energy can also be used to produce electricity, and this is the type of renewable energy that we focus on in this chapter. Radiant energy from the sun, in the form of photons, strikes the surface of the Earth with the average equivalent of about 168 kWh of energy (equal to 575,000 Btu thermal energy) per square foot per year; it varies, of course, with location, cloud cover, and orientation with the surface (Hanson, 2004). The question, then, becomes one of how much of this energy can be used by consumers in the United States. Table 4.1 shows the energy consumption by energy source for 2014; in that year, solar energy accounted for 0.307 quadrillion Btu, a figure that is expected to steadily increase, which should be reflected in later figures when they are released.

(A)

(B)

FIGURE 4.1   (A) Nevada Solar One solar thermal plant, Boulder City, Nevada. (B) Solar panel used for lighting in a remote restroom, Zion National Park, Utah. (Photographs by Frank R. Spellman.)

TABLE 4.1

U.S. Energy Consumption by Renewable Energy Source, 2014

Energy Source	Energy Consumption (quadrillion Btu)
Renewable total	9.656
Biomass (total)	4.812
Biofuels (ethanol and biodiesel)	2.067
Waste	0.516
Wood and wood-derived fuels	2.230
Geothermal	0.215
Hydroelectric power	2.475
Solar	**0.421**
Wind	1.733

Source: EIA, *Primary Energy Consumption by Source*, Table 10.1, Energy Information Administration, Washington, DC, 2016 (https://www.eia.gov/totalenergy/data/monthly/index.cfm#renewable).

With regard to the use and overall environmental impact of solar energy (along with the other sources of renewable energy discussed later), it can be characterized as a double-edged sword. On one edge of the sword, solar technologies diversify the energy supply and also reduce the country's dependence on imported fuel. Moreover, solar energy technologies provide a beneficial environmental impact by helping to improve air quality and offset greenhouse gas emissions. An additional benefit of a growing solar technology industry (that has yet to be proven or realized on a large scale) is that it stimulates our economy by creating jobs in solar manufacturing and installation. On the other edge of the sword, consider that all energy-generating technologies, including solar technologies, affect the environment in many ways. Again, solar energy has some obvious advantages in that the source is free; however, the initial investment in operating equipment is not. Probably the greatest downside of solar energy use is that in areas without direct sunlight during certain times of the year solar panels cannot capture enough energy to provide heat for homes or offices. Geographically speaking, the higher latitudes do not receive as much direct sunlight as tropical areas. Because of the position of the sun in the sky, solar panels must be placed in sun-friendly locations such as the U.S. desert southwest, which could cause land use and siting problems. Moreover, solar energy also has a water footprint, which is discussed in detail later. A solar panel by itself or a solar farm consisting of thousands of solar panels is not hazardous after it is up and running (except for birds), but during the manufacturing and ultimate disposal processes waste is produced. And there are the birds to consider … not the malicious *Birds* of Alfred Hitchcock fame but instead those innocent and mostly harmless feathered friends who simply fly into the wrong areas at the wrong time (i.e., during daylight). Recall that in Greek mythology Icarus attempted to escape from Crete by means of wings that his father constructed from feathers and wax. Icarus ignored instructions not to fly too close to the sun, and the melting wax caused him to fall into the sea where he drowned. At the Ivanpah Solar Electric Generating System in the Mojave Desert, near the California/Nevada border, amid miles of rock and scrub in a windy stretch once roamed by tortoises and coyotes, is a vast array of 7- by-10-foot solar mirrors. From above, these mirrors create the image of an ethereal lake shimmering atop the desert floor. A modern Icarus-like horror is played out at this site whenever birds fly over and end up with scorched wings and bodies or when they dive into this faux lake and are literally fried by the 1000°F heat flux radiated by the mirrors.

SOLAR HISTORICAL TIMELINE

Solar technology isn't new nor is humans' use of solar energy. The history and usage of solar energy spans from the 7th century BC to today. Humans started out using solar energy by concentrating the sun's heat with glass and mirrors to light fires. Today, humans use it for everything from solar-powered buildings to solar-powered vehicles. The following historical timeline (USDOE, 2003) presents the milestones in the development of solar technology, century by century and year by year. This timeline lists the milestones in the historical development of solar technology beginning in the 7th century BC.

7th Century BC

Magnifying glass was used to concentrate the sun's rays to make fire and to burn ants.

3rd Century BC

Greeks and Romans used burning mirrors to light torches for religious purposes.

2nd Century BC

As early as 212 BC, the Greek scientist Archimedes used the reflective properties of bronze shields to focus sunlight and to set fire to wooden ships from the Roman Empire that were besieging Syracuse. Although no proof of such a feat exists, the Greek navy recreated the experiment in 1973 and successfully set fire to a wooden boat at a distance of 50 meters (54.7 yards or 164 feet).

20 AD

The Chinese documented the use of burning mirrors to light torches for religious purposes.

1st to 4th Century AD

The famous Roman bathhouses in the first to fourth centuries AD had large south-facing windows to let in the sun's warmth.

6th Century AD

Sunrooms on houses and public buildings were so common that the Justinian Code initiated "sun rights" to ensure individual access to the sun.

1200

In North America, the Anasazi, ancestors of Pueblo Indians, lived in south-facing cliff dwellings that captured the winter sun.

1767

Swiss scientist Horace de Saussure was credited with building the world's first solar collector, later used by Sir John Herschel to cook food during his South Africa expedition in the 1830s.

1816

On September 27, 1816, Robert Stirling applied for a patent for his economizer at the Chancery in Edinburgh, Scotland. By trade, Stirling was a minister in the Church of Scotland, and he continued to give services until he was 86 years old, but in his spare time he built heat engines in his home workshop. Lord Kelvin used one of the working models during some of his university classes. This engine was later used in the dish/Stirling system, a solar thermal electric technology that concentrates the sun's thermal energy to produce power.

1839

French scientist Edmond Becquerel discovered the photovoltaic effect while experimenting with an electrolytic cell made up of two metal electrodes placed in an electricity-conducting solution—electricity generation increased when exposed to light.

1860s

French mathematician Augustin Mouchot proposed an idea for solar-powered steam engines. In the following two decades, he and his assistant, Able Pifre, constructed the first solar powered engines and used them for a variety of applications. These engines became the predecessors of modern parabolic dish collectors. Later, in 1878, Mouchot observed that, "One must not believe, despite the silence of modern writings, that the idea of using solar heat for mechanical operations is recent. On the contrary, one must recognize that this idea is very ancient and its slow development across the centuries has given birth to various curious devices."

1873

Willoughby Smith, an English electrical engineer, discovered the photoconductivity of selenium.

1876

William Grylls Adams and Richard Evans Day discovered that the selenium and platinum junction produces electricity when exposed to light. Although selenium solar cells failed to convert enough sunlight to power electrical equipment, Adams and Day proved that solid material could change light into electricity without heat or moving parts.

1880

Samuel P. Langley invented the bolometer, which is used to measure light from the faintest stars and the sun's heat rays. It consists of a fine wire connected to an electric circuit. When radiation falls on the wire, it becomes very slightly warmer. This increases the electrical resistance of the wire.

1883

Charles Fritts, an American inventor, described the first solar cells made from selenium wafers.

1887

Heinrich Hertz discovered that ultraviolet light altered the lowest voltage capable of causing a spark to jump between two metal electrodes.

1891

Baltimore inventor Clarence Kemp patented the first commercial solar water heater.

1904

Wilhelm Hallwachs discovered that a combination of copper and cuprous oxide is photosensitive.

1905

Albert Einstein published his paper on the photoelectric effect (along with a paper on his theory of relativity).

1908

William J. Bailey of the Carnegie Steel Company invented a solar collector with copper coils and an insulated box—roughly its current design.

1914

The existence of a barrier layer in photovoltaic devices was noted.

1916

Robert Millikan provided experimental proof of the photoelectric effect.

1918

Polish scientist Jan Czochralski developed a way to grow single-crystal silicon.

1921

Albert Einstein won the Nobel Prize for his theories regarding the photoelectric effect.

1932

Audobert and Stora discovered the photovoltaic effect in cadmium sulfide (CdS).

1947

Passive solar buildings in the United States were in such demand, as a result of scarce energy during the prolonged World War II, that Libbey-Owens-Ford Glass Company published a book entitled *Your Solar House*, which profiled 49 of the nation's greatest solar architects.

1953

Dr. Dan Trivich, at Wayne State University, made the first theoretical calculations regarding the efficiencies of various materials of different band gap widths based on the spectrum of the sun.

1954

Photovoltaic technology was born in the United States when Daryl Chapin, Calvin Fuller, and Gerald Pearson developed the silicon photovoltaic (PV) cell at Bell Labs—the first solar cell capable of converting enough of the sun's energy into power to run everyday electrical equipment. Bell Telephone Laboratories produced a silicon solar cell with 4% efficiency and later achieved 11% efficiency.

Mid-1950s

Architect Frank Bridgers designed the world's first commercial office building using solar water heating and passive design. This solar system has been in continuous operation since that time, and the Bridgers–Paxton Building is now listed in the National Historic Register as the world's first solar-heated office building.

1955

Western Electric began to sell commercial licenses for silicon photovoltaic technologies. Early successful products included photovoltaic-powered dollar bill changers and devices that decoded computer punch cards and tape.

1956

William Cherry, of U.S. Signal Corps Laboratories, approached RCA Laboratories' Paul Rappaport and Joseph Loferski about developing photovoltaic cells for proposed orbiting Earth satellites.

1957

Hoffman Electronics achieved 8% efficient photovoltaic cells.

1958

T. Mandelkorn, of U.S. Signal Corps Laboratories, fabricated n-on-p silicon photovoltaic cells, which were critically important for space cells, as they are more resistant to radiation. Hoffman Electronics achieved 9% efficient photovoltaic cells. The *Vanguard I* space satellite used a small (less than 1 watt) array to power its radios. Later that year, *Explorer 3*, *Vanguard 2*, and *Sputnik 3* were launched with PV-powered systems on board. Despite faltering attempts to commercialize the silicon solar cell in the 1950s and 1960s, it was used successfully in powering satellites. It became the accepted energy source for space applications and remains so today.

1959

Hoffman Electronics achieved 10% efficient, commercially available photovoltaic cells. Hoffman also learned to use a grid contact to reduce the series resistance significantly. On August 7, the *Explorer 6* satellite was launched with a photovoltaic array of 9600 cells (1 cm × 2 cm each). Then, on October 13, the *Explorer 7* satellite was launched.

1960

Hoffman Electronics achieved 14% efficient photovoltaic cells. Silicon Sensors, Inc., of Dodgeville, Wisconsin, was founded and began producing selenium and silicon photovoltaic cells.

1962

Bell Telephone Laboratories launched the first telecommunications satellite, the *Telstar* (initial power of 14 watts).

1963

Sharp Corporation succeeded in producing practical silicon photovoltaic modules. Japan installed a 242-watt, photovoltaic array on a lighthouse, the world's largest array at that time.

1964

NASA launched the first *Nimbus* spacecraft, a satellite powered by a 470-watt photovoltaic array.

1965

Peter Glaser conceived the idea of a satellite solar power station.

1966

NASA launched the first Orbiting Astronomical Observatory, which was powered by a 1-kilowatt photovoltaic array, to provide astronomical data in the ultraviolet and x-ray wavelengths filtered out by the Earth's atmosphere.

1969

The Odeillo solar furnace, located in Odeillo, France, was constructed. It featured an eight-story parabolic mirror.

1970s

Dr. Elliot Berman, with help from Exxon Corporation, designed a significantly less costly solar cell, bringing the price down from $100 per watt to $20 per watt. Solar cells begin to power navigation warning lights and horns on many offshore gas and oil rigs, lighthouses, and railroad crossings, and domestic solar applications began to be viewed as sensible applications in remote locations where grid-connected utilities could not exist affordably.

1972

The French installed a cadmium sulfide (CdS) photovoltaic system to operate an educational television at a village school in Niger. The Institute of Energy Conversion was established at the University of Delaware to perform research and development on thin-film photovoltaic and solar thermal systems, becoming the world's first laboratory dedicated to photovoltaic research and development.

1973

The University of Delaware built "Solar One," one of the world's first photovoltaic-powered residences. The system was a photovoltaic/thermal hybrid. The roof-integrated arrays fed surplus power through a special meter to the utility during the day and purchased power from the utility at night. In addition to electricity, the arrays acted as flat-plate thermal collectors, with fans blowing the warm air from over the array to phase-change heat-storage bins.

1976

NASA's Lewis Research Center began installing 83 photovoltaic power systems on every continent except Australia. These systems provide such diverse applications as vaccine refrigeration, room lighting, medical clinic lighting, telecommunications, water pumping, grain milling, and classroom television. The Center completed the project in 1995, having worked on it from 1976 to 1985 and then again from 1992 to 1995.

1976

David Carlson and Christopher Wronski, of RCA Laboratories, fabricated the first amorphous silicon photovoltaic cells.

1977

The U.S. Department of Energy launched the Solar Energy Research Institute, a federal facility dedicated to harnessing power from the sun. Total photovoltaic manufacturing production exceeded 500 kilowatts.

1978

NASA's Lewis Research Center dedicated a 3.5-kilowatt photovoltaic system it had installed on the Papago Indian Reservation located in southern Arizona. It was the world's first village photovoltaic system. The system was used to provide for water pumping and residential electricity in 15 homes until 1983, when grid power reached the village. The PV system was then dedicated to pumping water from a community well.

1980

ARCO Solar became the first company to produce more than 1 megawatt of photovoltaic modules in one year. At the University of Delaware, the first thin-film solar cell exceeded 10% efficiency using copper sulfide/cadmium sulfide.

1981

Paul MacCready built the first solar-powered aircraft—the *Solar Challenger*—and flew it from France to England across the English Channel. The aircraft had over 16,000 solar cells mounted on its wings, which produced 3000 watts of power.

1982

The first, photovoltaic megawatt-scale power station went online in Hisperia, California. Its 1-megawatt capacity system was developed by ARCO Solar and featured modules on 108 dual-axis trackers. Australian Hans Tholstrup drove the first solar-powered car ("Quiet Achiever") almost 2800 miles between Sydney and Perth in 20 days—10 days faster than the first gasoline-powered car to do so. The U.S. Department of Energy, along with an industry consortium, began operation of Solar One, a 10-megawatt central-receiver demonstration project. The project established the feasibility of power-tower systems, a solar–thermal electric or concentrating solar power technology. In 1988, the final year of operation, the system could be dispatched 96% of the time. Volkswagen of Germany began testing photovoltaic arrays mounted on the roofs of Dasher station wagons, generating 160 watts for the ignition system. Worldwide photovoltaic production exceeded 9.3 megawatts.

1983

ARCO Solar dedicated a 6-megawatt photovoltaic substation in central California. The 120-acre, unmanned facility supplied the Pacific Gas & Electrical Company's utility grid with enough power for 2000 to 2500 homes. Solar Design Associates completed a stand-alone, 4-kilowatt powered home in the Hudson River Valley. Worldwide photovoltaic production exceeded 21.3 megawatts, with sales of more than $250 million.

1984

The Sacramento Municipal Utility District commissioned its first 1-megawatt photovoltaic electricity generating facility.

1985

University of South Wales broke the 20% efficiency barrier for silicon solar cells under 1-sun conditions.

1986

The world's largest solar thermal facility, located in Kramer Junction, California, was commissioned. The solar field contained rows of mirrors that concentrated the sun's energy onto a system of pipes circulating a heat-transfer fluid. The heat-transfer fluid was used to produce steam, which powered a conventional turbine to generate electricity.

1986

ARCO Solar released the G-4000, the world's first commercial thin-film power module.

1988

Dr. Alvin Marks received patents for two solar power technologies he developed: Lepcon and Lumeloid. Lepcon consists of glass panels covered with a vast array of millions of aluminum or copper strips, each less than a micron or thousandth of a millimeter wide. As sunlight hits the metal strips, the energy in the light is transferred to electrons in the metal, which escape at one end in the form of electricity. Lumeloid uses a similar approach but substitutes less expensive, film-like sheets of plastic for the glass panels and covers the plastic with conductive polymers, long chains of molecular plastic units.

1991

President George Bush redesignated the U.S. Department of Energy's Solar Energy Research Institute as the National Renewable Energy Laboratory.

1992

The University of South Florida developed a 15.9% efficient thin-film photovoltaic cell made of cadmium telluride, breaking the 15% barrier for the first time for this technology. A 7.5-kilowatt prototype dish system using an advanced stretched-membrane concentrator became operational.

1993

Pacific Gas & Electric completed installation of the first grid-supported photovoltaic system in Herman, California. The 500-kilowatt system was the first "distributed power" effort.

1994

The National Renewable Energy Laboratory completed construction of its Solar Energy Research Facility, which was recognized as the most energy-efficient of all U.S. government buildings world-wide. It features not only a solar electric system but also a passive solar design. The first solar dish generator using a free-piston Stirling engine was tied to a utility grid. The National Renewable Energy Laboratory developed a solar cell made from gallium indium phosphide and gallium arsenide that became the first to exceed 30% conversion efficiency.

1996

The world's most advanced solar-powered airplane, the *Icare*, flew over Germany. The wings and tail surfaces of the *Icare* were covered by 3000 super-efficient solar cells. The U.S. Department of Energy, along with an industry consortium, began operating Solar Two and an upgrade of its Solar One concentrating solar power-tower project. Operated until 1999, Solar Two demonstrated how solar energy can be stored efficiently and economically so that power can be produced even when the sun is not shining. It also fostered commercial interest in power towers.

1998

The remote-controlled, solar-powered aircraft *Pathfinder* set an altitude record (80,000 feet) on its 39th consecutive flight on August 6 in Monrovia, California. This altitude was higher than any prop-driven aircraft thus far. Subhendu Guha, a scientist noted for his pioneering work in amorphous silicon, led the invention of flexible solar shingles, a roofing material, and state-of-the-art technology for converting sunlight to electricity.

1999

Construction was completed on 4 Times Square, the tallest skyscraper built in the 1990s in New York City. It incorporates more energy-efficient building technologies than any other commercial skyscraper and also includes building-integrated photovoltaic (BIPV) panels on the 37th through 43rd floors on the south- and west-facing facades that produce a portion of the building's power.

1999

Spectrolab, Inc., and the National Renewable Energy Laboratory developed a photovoltaic solar cell able to convert 32.3% of the sunlight hitting it into electricity. The high conversion efficiency was achieved by combining three layers of photovoltaic materials into a single solar cell. The cell performed most efficiently when it received sunlight concentrated to 50 times normal. To use such cells in practical applications, the cell is mounted in a device that uses lenses or mirrors to concentrate sunlight onto the cell. Such concentrator systems are mounted on tracking systems that keep them pointed toward the sun. The National Renewable Energy Laboratory achieved a new efficiency record for thin-film photovoltaic solar cells. The measurement of 18.8% efficiency for the prototype solar cell topped the previous record by more than 1%. Cumulative worldwide installed photovoltaic capacity reached 1000 megawatts.

2000

First Solar began production in Perrysburg, Ohio, at the world's largest photovoltaic manufacturing plant with an estimated capacity of producing enough solar panels each year to generate 100 megawatts of power. At the International Space Station, astronauts began installing solar panels on what would be the largest solar power array deployed in space. Each wing of the array consisted of 32,800 solar cells. Sandia National Laboratories developed a new inverter for solar electric systems that will increase the safety of the systems during a power outage. Inverters convert the direct current (DC) electrical output from solar systems into alternating current (AC), which is the standard current for household wiring and for the power lines that supply electricity to homes. Two new thin-film solar modules developed by BP Solarex broke previous performance records. The company's 0.5-m^2 module achieved 10.8% conversion efficiency, the highest in the world for thin-film modules of its kind; its 0.9-m^2 module achieved 10.6% conversion efficiency and a power output of 91.5 watts, the highest power output for any thin-film module in the world. A family in Morrison, Colorado, installed a 12-kilowatt solar electric system on its home that was recognized as the largest residential installation in the United States to be registered with the U.S. Department of Energy's "Million Solar Roofs" program. The system provided most of the electricity for the 6000-ft^2 home and family of eight.

2001

Home Depot began selling residential solar power systems in three of its stores in San Diego, California. A year later it expanded sales to include 61 stores nationwide. The National Space Development Agency of Japan (NASDA) announced plans to develop a satellite-based solar power system that would beam energy back to Earth. A satellite carrying large solar panels would use a laser to transmit the power to an airship at an altitude of about 12 miles, which would then transmit the power to Earth. TerraSun LLC developed a unique method of using holographic films to concentrate sunlight onto a solar cell. Concentrating solar cells typically use Fresnel lenses or mirrors to concentrate sunlight. TerraSun claimed that the use of holographic optics allows more selective use of the sunlight, allowing light not needed for power production to pass through the transparent modules. This capability allows the modules to be integrated into buildings as skylights. In Hawaii, PowerLight Corporation placed online the world's largest hybrid system that combined the power from both wind and solar energy. The grid-connected system was unusual in that its solar energy capacity of 175 kilowatts was actually larger than its wind energy capacity of 50 kilowatts. Such hybrid power systems combine the strengths of both energy systems to maximize the available power. British Petroleum (BP) and BP Solar announced the opening of a service station in Indianapolis that featured a solar-electric canopy. The Indianapolis station was the first U.S. "BP Connect" store, a model that BP intended to use for all new or significantly revamped BP service stations. The canopy was built using translucent photovoltaic modules made of thin films of silicon deposited onto glass.

2002

NASA successfully conducted two tests of a solar-powered, remote-controlled aircraft called *Pathfinder Plus*. In the first test in July, researchers demonstrated the aircraft's use as a high-altitude platform for telecommunications technologies. Then, in September, a test demonstrated its use as an aerial imaging system for coffee growers. Union Pacific railroad installed 350 blue-signal rail yard lanterns that incorporated energy-saving light-emitting diode (LED) technology with solar cells at its North Platt, Nebraska, rail yard—the largest rail yard in the United States. Automation Tooling Systems, Inc., in Canada began to commercialize an innovative method of producing solar cells, called Spheral Solar technology. The technology is based on tiny silicon beads bonded between two sheets of aluminum foil and promises lower costs due to its greatly reduced use of silicone relative to conventional multicrystalline silicon solar cells. The technology was not new. It was championed by Texas Instruments (TI) in the early 1990s, but despite U.S Department of Energy (DOE) funding TI dropped the initiative. The largest solar power facility in the Northwest—the 38.7-kilowatt White Bluffs Solar Station—went online in Richland, Washington.

CONCENTRATING SOLAR POWER[*]

Concentrating solar power (CSP) offers a utility-scale, reliable, firm, dispatchable, renewable energy option that can help meet a nation's demand for electricity. Nine trough plants producing more than 400 megawatts (MW) of electricity have been operating reliably in the California Mojave Desert since the 1980s (USDOE, 2008). CSP plants produce power by first using mirrors to concentrate and focus sunlight onto a thermal receiver, similar to a boiler tube. The receiver absorbs and converts sunlight into heat. Ultimately, this high-temperature fluid is used to spin a turbine or power an engine that drives a generator that produces electricity. Concentrating solar power systems can be classified by how they collect solar energy: linear concentrators, dish/engine systems, or power tower systems (NREL, 2014). All three are typically engineered with tracking devices for following the sun, both seasonally and throughout the day, to maximize the electrical output of the system (Chiras, 2002).

LINEAR CONCENTRATORS

Linear CSP collectors capture or collect the sun's energy using long, rectangular, curved (U-shaped) mirrors. The mirrors, tilted toward the sun, focus sunlight on tubes (or receivers) that run the length of the mirrors. The reflected sunlight heats a fluid flowing through the tubes. The hot fluid is then used to boil water to create superheated steam that spins a conventional steam-turbine generator to produce electricity. Alternatively, steam can be generated directly in the solar field, eliminating the need for costly heat exchangers. Linear concentrating collector fields consist of a large number of collectors in parallel rows that are typically aligned in a north–south orientation to maximize both annual and summertime energy collection. A single-axis sun-tracking system allows the mirrors to track the sun from east to west during the day, ensuring that the sun reflects continuously onto the receiver tubes (USDOE, 2006b). The two major types of linear concentrator systems are *parabolic trough systems*, where receiver tubes are positioned along the focal line of each parabolic mirror, and *linear Fresnel reflector systems*, where one receiver tube is positioned above several mirrors to allow the mirrors greater mobility in tracking the sun.

Parabolic Trough Systems

The predominant CSP systems currently in operation in the United States are linear concentrators using parabolic trough collectors (USDOE, 2006b). Parabolic trough systems have long, curved arrays of mirrors (usually coated silver or polished aluminum). All rays of light that enter parallel to

---

[*] Adapted from EERE, *Concentrating Solar Power: Energy from Mirrors*, DOE/GO-102001-1147, Office of Energy Efficiency & Renewable Energy, U.S. Department of Energy, Washington, DC, 2001.

the axis of a parabolic-shaped mirror will be reflected to one point, the focus; hence, it is possible to concentrate virtually all of the radiation incident upon the mirror in a relatively small area in the focus (Perez-Blanco, 2009). The receiver tube (Dewar tube) runs its length and is positioned along the focal line of each parabola-shaped reflector (Duffie and Beckman, 1991; Patel, 1999). The tube is fixed to the mirror structure, and the heated fluid—either a heat-transfer fluid (usually oil) or water or steam—flows through and out of the field of solar mirrors to where it will be used to create steam; in the case of a water/steam receiver, the flow is sent directly to the turbine (i.e., the turbine is the prime mover of the electrical generator that produces electrical current flow in an accompanying electrical distribution system). Temperatures in these systems range from 150 to 750°F (80 to 400°C). Currently, the largest individual trough systems can generate 80 MW of electricity; however, individual systems being developed will generate 250 MW. In addition, individual systems can be collocated in power parks. The potential capacity of such power parks would be constrained only by transmission capacity and the availability of contiguous land area. Trough designs can incorporate thermal storage (discussed in more detail later). In thermal storage systems, the collector field is oversized to heat a storage system during the day that can be used in the evening or during cloudy weather to generate additional steam to produce electricity.

Parabolic trough plants can be designed as hybrids, meaning that they use fossil fuel to supplement the solar output during periods of low solar radiation. In such a design, a natural-gas-fired heater or gas–steam boiler/reheater is used. In the future, troughs may be integrated with existing or new combined-cycle, natural gas, and coal-fired plants (USDOE, 2006b).

Fresnel Reflector Systems

Another type of linear concentrator technology is the linear Fresnel reflector system. Flat or slightly curved mirrors mounted on trackers on the ground are configured to reflect sunlight onto a water-filled receiver tube fixed in space above these mirrors. A small parabolic mirror is sometimes added atop the receiver to further focus the sunlight. The key advantage of Fresnel reflectors systems is their simplicity as compared to other systems.

Dish/Engine Systems

Dish/engine systems use a mirrored, dish-shaped, parabolic mirror similar to a very large satellite dish. The dish-shaped surface directs and concentrates sunlight onto a thermal receiver (the interface between the dish and the engine or generator), which absorbs and collects the heat and transfers it to the engine generator. Solar dish/engine systems convert the energy from the sun into electricity at a very high efficiency. The power conversion unit includes the thermal receiver and the engine or generator. It absorbs the concentrated beams of solar energy, converts them to heat, and transfers the heat to the engine or generator. A thermal receiver can be a bank of tubes with a cooling fluid—usually hydrogen or helium—that typically is the heat-transfer medium and also the working fluid for an engine. Alternative thermal receivers are heat pipes, where the process of boiling and condensing an intermediate fluid transfers the heat to the engine. The most common type of heat engine used today in dish/engine systems is the Stirling engine (conceived in 1816). This system uses the fluid heated by the receiver to move pistons and create mechanical power. The mechanical power is then used to run a generator or alternator to produce electricity (USDOE, 1998). Disk/engine systems use dual-axis collectors to track the sun. As mentioned, the ideal concentrator shape is parabolic, created with a single reflective surface or multiple reflectors, or facets. Many options exist for receiver and engine type, including the Stirling cycle, microturbine, and concentration photovoltaic modules. Because of the high concentration ratios achievable with parabolic dishes and the small size of the receiver, solar dishes are efficient (as high as 30%) at collecting solar energy at very high temperatures (900 to 2700°F, or 500 to 1500°C). Solar dish/engine systems have environmental, operational, and potential economical advantages over more conventional power generation options because (USDOE, 1998):

DID YOU KNOW?

Smaller CSP systems can be located directly where the power is needed; for example, a single dish/engine system can produce 3 to 25 kW of power and is well suited for such distributed applications. Larger, utility-scale CSP applications provide hundreds of megawatts of electricity for the power grid. Both linear concentrator and power tower systems can be easily integrated with thermal storage, helping to generate electricity during cloudy periods or at night. Alternatively, these systems can be combined with natural gas, and the resulting hybrid power plants can provide high-value, dispatchable power throughout the day.

- They produce zero emissions when operating on solar energy.
- They operate more quietly than diesel or gasoline engines.
- They are easier to operate and maintain than conventional engines.
- They start up and shut down automatically.
- They operate for long periods with minimal maintenance.

Solar dish/engine systems are well suited and often used for nontraditional power generation because of their size and durability. Individual units range in size from 10 to 25 kW. They can operate independently of power grids in remote sunny locations for uses such as pumping water and providing power to people living in isolated locations.

The high temperatures produced by dish/engine systems can be used to produce steam that can be used for either electricity production or various high-temperature industrial processes (Chiras, 2002). Dish/engine systems also can be linked together to provide utility-scale power to a transmission grid. Such systems could be located near consumers, substantially reducing the need for building or upgrading transmission capacity. Largely because of their high efficiency, the cost of these systems is expected to be lower than that of other solar systems for these applications (USDOE, 1998).

POWER TOWER SYSTEMS

The power tower or the central receiver system consists of a large field of flat, sun-tracking mirrors known as *heliostats* (silver-laminated acrylic membranes) focused to concentrate sunlight onto a receiver on the top of a tower. A heat-transfer fluid heated in the receiver is used to generate steam, which, in turn, is used in a conventional turbine generator to produce electricity. Some power towers use water or steam as the heat-transfer fluid. Other advanced designs are experimenting with molten (liquefied) nitrate salt because of its superior heat-transfer and energy-storage (heat-retention) capabilities, allowing continued electricity production for several consecutive cloudy days (NREL, 2010). The energy-storage capability, or thermal storage, allows the system to continue to dispatch electricity during cloudy weather or at night. A power tower system is composed of five main components: heliostats, receiver, heat transport and exchange, thermal storage, and controls (Figure 4.2). Individual commercial plants can be sized to produce up to 200 MW of electricity.

DID YOU KNOW?

Concentrated solar power plants use water not only for lens cleaning but also for steam cycle production. Water consumption is an issue because these plants require water but they are most effective in locations where the sun is most intense, which in turn often corresponds to places like the Mohave Desert where there is little water (USDOE, 2008).

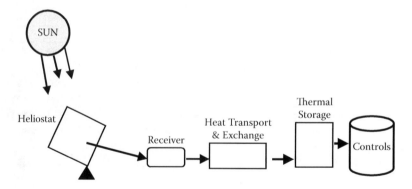

FIGURE 4.2   Solar power tower main components.

THERMAL STORAGE

Thermal energy storage (TES) has become a critical aspect of any concentrating solar power system deployed today (USDOE, 2013). One challenge facing the widespread use of solar energy is the reduced or curtailed energy production when the sun sets or is blocked by clouds. Thermal energy storage provides a workable solution to this challenge. In a CSP system, the sun's rays are reflected onto a receiver, creating heat that is then used to generate electricity. If the receiver contains oil or molten salt as the heat-transfer medium, then the thermal energy can be stored for later use. This allows CSP systems to be a cost-competitive option for providing clean, renewable energy. Current steam-based receivers cannot store thermal energy for later use.

An important criterion in selecting a storage medium, such as the oil or molten salt referred to above, is the *specific heat* of the substance. Specific heat (*c*) is the measure of the heat energy required to increase the temperature of a unit quantity of a substance by a unit of temperature. For example, at a temperature of 15°C, the heat energy required to raise the temperature of water 1 Kelvin (equal to 1°C) is 4186 joules per kilogram (J/kg). Different materials absorb different amounts of heat when undergoing the same temperature increase. The relationship between a temperature change (ΔT) and the amount of heat (Q) added (or subtracted) is given by

$$Q = mc\Delta T \tag{4.1}$$

where *c* is the specific heat and *m* the mass of the substance. Table 4.2 lists some specific heats for different materials per pound and per cubic foot. Another important parameter related to specific

TABLE 4.2
Thermal Energy Storage Materials

Material	Specific Heat (Btu/lb °F)	Density (lb/ft³)	Density (kg/m³)	Heat Capacity (Btu/ft³ °F)	Heat Capacity (kJ/m³ °C)
Water	1.00	62	1000	62	4186
Iron	0.12	490	7860	59	3521
Copper	0.09	555	8920	50	3420
Aluminum	0.22	170	2700	37	2430
Concrete	0.23	140	2250	32	2160
Stone	0.21	170	2700	36	2270
White pine	0.67	2	435	18	1220
Sand	0.19	95	1530	18	1540
Air	0.24	0.075	1.29	0.02	1.3

heat and thermal energy storage materials is *heat capacity*, which is defined as the ratio of the heat energy absorbed by a substance to the substance's increase in temperature. It is given by

$$\text{Heat capacity} = \text{Specific heat} \times \text{Density} \tag{4.2}$$

Other media for thermal storage are *phase-change materials*, which are classified as latent heat storage units. Specifically, a phase-change material is a substance with a high heat of fusion, which is the amount of thermal energy that must be absorbed or evolved for 1 mole of a substance to change states from a solid to a liquid or *vice versa* without a temperature change. A phase-change material melts and solidifies at a certain temperature and is capable of storing and releasing large amounts of energy. Heat is absorbed or released when the material changes from solid to liquid and *vice versa*. One common group of substances being used as phase-change materials for solar applications (and elsewhere) are eutectic salts, such as sodium sulfate decahydrate, also known as Glauber's salt. This salt melts at 91°F with the addition of 108 Btu/lb. Conversely, when the temperature drops below 91°F, 108 Btu/lb of heat energy is released as the salt solidifies (Hinrichs and Kleinbach, 2006). Several TES technologies have been tested and implemented since 1985. These include the two-tank direct system, two-tank indirect system, and single-tank thermocline system.

Two-Tank Direct System

Solar thermal energy in this system is stored in the same fluid used to collect it. The fluid is stored in two tanks—one at high temperature and the other at low temperature. Fluid from the low-temperature tank flows through the solar collector or receiver, where solar energy heats it to the high temperature; it then flows back to the high-temperature tank for storage. Fluid from the high-temperature tank flows through a heat exchanger, where it generates steam for electricity production. The fluid exits the heat exchanger at the low temperature and returns to the low-temperature tank. Two-tank direct storage was used in early parabolic trough power plants and at the Solar Two power tower in California. The trough plants used mineral oil as the heat-transfer and storage fluid, and the Solar Two power tower used molten salt.

Two-Tank Indirect System

The two-tank indirect system functions in the same way as the two-tank direct system, except that different fluids are used for the heat-transfer fluid and for the storage fluid. This system is used in plants where the heat-transfer fluid is too expensive or not suited for use as the storage fluid. The storage fluid from the low-temperature tank flows through an extra heat exchanger, where it is heated by the high-temperature heat-transfer fluid. The high-temperature storage fluid then flows back to the high-temperature storage tank. The fluid exits this heat exchanger at a low temperature and returns to the solar collector or receiver, where it is heated back to the high temperature. Storage fluid from the high-temperature tank is used to generate steam in the same manner as the two-tank direct system. The indirect system requires an extra heat exchanger, which adds cost to the system. This system will be used in many of the parabolic power plants in Spain and has also been proposed for several U.S. parabolic plants. The plants will use organic oil as the heat-transfer fluid and molten salt as the storage fluid.

Single-Tank Thermocline System

This system stores thermal energy in a solid medium—most commonly silica sand—located in a single tank. At any time during operation, a portion of the medium is at high temperature and a portion is at lower temperature. The hot- and cold-temperature regions are separated by a temperature gradient, or *thermocline* (see Figure 4.3). High-temperature heat-transfer fluid flows into the top of the thermocline and exits the bottom at low temperature. This process moves the thermocline downward and adds thermal energy to the system for storage. Reversing the flow moves the thermocline

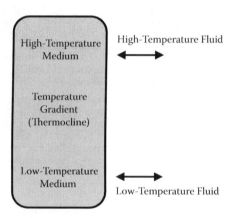

FIGURE 4.3   Single-tank thermocline thermal energy storage system.

upward and removes thermal energy from the system to generate steam and electricity. Buoyancy effects create thermal stratification of the fluid within the tank, which helps to stabilize and maintain the thermocline.

> *Note:* Using a solid storage medium and only requiring one tank can reduce the cost of the single-tank thermocline system relative to two-tank systems.

PHOTOVOLTAICS

Photovoltaics (PV) has been and will continue to be one of the more glamorous technologies in the energy field (Hinrichs and Kleinbach, 2006). Photovoltaic (*photo* from the Greek word for "light" and *volt* for electricity pioneer Alessandro Volta) technology makes use of the abundant energy in the sun, and it has little impact on our environment. Photovoltaics is the direct conversion of light (photons) into electricity (voltage) at the atomic level. Some materials exhibit a property known as the *photoelectric effect* (discovered and described by Becquerel in 1839) that causes them to absorb photons of light and release electrons. When these free electrons are captured, an electric current (flow of free electrons) results, which can be used as electricity. The first photovoltaic module (billed as a solar battery) was built by Bell laboratories in 1954. In the 1960s, the space program began to make the first serious use of the technology to provide power aboard spacecraft. Use of this technology in the space program produced giant advancements in its reliability and helped to lower costs; however, it was the oil embargo of the 1970s (the so-called energy crisis) that focused attention on using photovoltaic technology for applications other than the space program. Photovoltaics can be used in a wide range of products, from small consumer items to large commercial solar electric systems.

Figure 4.4 illustrates the photoelectric effect when light shines on a negative plate; electrons are emitted with an amount of kinetic energy inversely proportional to the wavelength of the incident light. Figure 4.5 illustrates the operation of a basic *photovoltaic cell*, also called a *solar cell*. Solar cells are made of silicon and other semiconductor materials such as germanium, gallium arsenide, and silicon carbide that are used in the microelectronics industry. For solar cells, a thin semiconductor wafer is specially treated to form an electric field, positive on one side and negative on the other. When light energy strikes the solar cell, electrons are jarred loose from the atoms in the semiconductor material (see Figures 4.4 and 4.5). If electrical conductors are attached to the positive and negative sides, forming an electrical circuit, the electrons can be captured in the form of an electrical current (recall that electron flow is electricity). This electricity can then be used to power a load, such as for a light, tool, or toaster, for example.

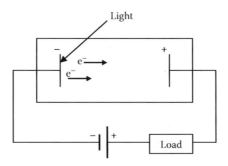

FIGURE 4.4 Photoelectric effect.

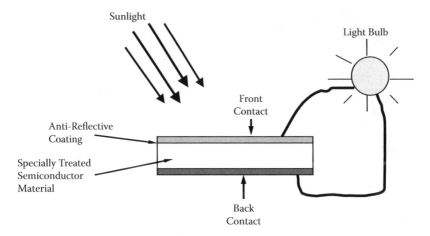

FIGURE 4.5 Operation of basic photovoltaic cell.

A *photovoltaic module* is comprised of a number of solar cells electrically connected to each other and mounted on a support panel or frame (see Figure 4.6). Solar panels used to power homes and businesses are typically made from solar cells combined into modules that hold about 40 cells. Modules are designed to supply electricity at a certain voltage (e.g., 12 volts). The current produced

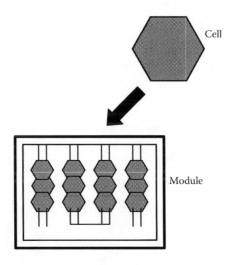

FIGURE 4.6 Single solar cell and solar cell module.

is directly dependent on how much light strikes the module. Multiple modules can be wired together to form an array. In general, the larger the area of a module or array, the more electricity that will be produced. Photovoltaic modules and arrays produce direct-current (DC) electricity. They can be connected in both series and parallel electrical arrangements to produce any required voltage and current combination.

DOMESTIC AND INDUSTRIAL USE OF SOLAR ENERGY

When thinking about a potential renewable energy source, we first want to know if it is reliable, clean, and affordable. This is especially the case for industry owners contemplating solar applications for industrial use. Until recently, solar energy has been used for both domestic and industrial applications, but only on a somewhat limited basis. This trend is changing, especially in those locations where sunlight is prevalent throughout the year. With the soaring costs of fossil fuel supplies and their pending decreased availability, solar power is beginning to receive more attention. Solar technologies for domestic use include photovoltaics, passive heating, window and structure daylighting, and water heating. Industrial, commercial, and municipal treatment facilities may use the same solar technologies that are used for residential buildings, in addition to solar energy technologies that would be impractical for a home, such as ventilation air preheating, solar process heating, and solar cooling.

SOLAR HOT WATER

Just as the sun heats the surface layers of exposed bodies of water—ponds, lakes, streams, and oceans—it can also heat water used in buildings and swimming pools. Harnessing the ability of the sun to heat water requires a solar collector and a storage tank. The most common collector is the *flat-plate collector* (Figure 4.6). The flat-plate collector system is the most commonly used collector and is ideal for applications that require water temperatures under 140°F (60°C). Mounted on the roof, the system consists of a thin, flat, rectangular, insulated box with a transparent glass cover that faces the sun. Small copper tubes run through the box and carry the fluid (water or other fluid such as an antifreeze solution) to be heated. The tubes, typically arranged in series (i.e., water flows in one end and out the other end), are attached to an absorber plate, which, along with the tubes, is painted black to absorb the heat. As heat builds up in the collector, it heats the fluid (water or propylene glycol) passing through the tubes. The motive force for passing fluid through the tubes can be accomplished by either active or passive means. In an active system, the most common type of system, water heaters rely on an electric pump and controller to circulate water or other heat-transfer fluids through the collectors. Passive solar water-heater systems rely on gravity and the tendency for water to naturally circulate as it is heated. Because passive systems have no moving parts, they are more reliable, are easier to maintain, and often have a longer life than active systems. The storage tank holds the hot liquid (see Figure 4.7). It is usually a large well-insulated tank, but modified water heaters can also be used to store the hot liquid. When the fluid used is other than water, the water is heated by passing it through a coil of tubing in the tank, which is full of hot fluid. Active solar systems usually include a storage tank along with a conventional water heater. In two-tank systems, the solar water heater preheats water before it enters the conventional water heater.

SPACE HEATING

In order to maintain indoor air quality (IAQ), many large buildings require ventilated air. Although all that fresh air can be great for indoor air quality, heating this air in cold climates can use large amounts of energy and can be very expensive. An elegantly simple solar ventilation system can preheat the air, though, saving both energy and money. This type of system typically uses a transpired air collector. Transpired air collector systems essentially consist of a dark-colored, perforated

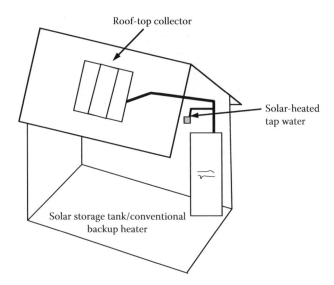

FIGURE 4.7 Simplified representation of a home solar water-heating system.

facade installed on the south-facing wall of a building. A fan or the existing ventilation system draws ventilation air into the building through the perforated absorber plate on the facade and up the plenum (the air space between the absorber and the south wall). Solar energy absorbed by the dark absorber and transferred to the air flowing through it can preheat the intake air by as much as 40°F (22°C). Reduced heating costs will pay for the systems in 3 to 12 years (USDOE, 2006a).

In addition to meeting a portion of a building's heating load with clean, free solar energy, the transpired collector helps save energy and money in other ways. It recaptures heat loss through a building's south-facing wall, as heat that escapes through the south wall is captured in the air space between the structural wall and the transpired collector and returned to the interior. Also, by introducing make-up air though ceiling-mounted ducts, the system eliminates the wasteful air stratification that often plagues high-ceiling buildings (USDOE, 2006a).

To illustrate an active, real-world example of transpired air collectors used as solar preheaters for outdoor ventilation air, it is probably best to use an actual example of a working system in an industrial application. The following case study is such an example; it has been monitored extensively by experts.

General Motors Battery Plant[*]

The General Motors battery plant in Oshawa, Canada, is a 100,000-ft^2 facility in which automotive batteries are manufactured. The plant was built in the 1970s and consists of an open shop floor and a 28-foot-high ceiling. GM operates two full-time production shifts within the plant and conducts maintenance activities at night and on weekends, so the building is continuously occupied.

Until the early 1990s, GM relied solely on a steam-operated fan coil system for space heating, but the system was incapable of providing the necessary quantities of heated outdoor air. As a result, the plant was not being adequately ventilated. In 1991, plant management installed a transpired collector to correct the ventilation problems. Over the next two years the transpired collector system was modified slightly to further improve airflow; the original fans and motors were replaced with vane axial fans and high-efficiency motors, and the original ducting was replaced with upgraded fabric ducting.

---

[*] Adapted from USDOE, *Transpired Collectors (Solar Preheaters for Outdoor Ventilation Air)*, Federal Technology Alert, U.S. Department of Energy, Washington, DC, 1998.

The GM plant collector is comprised of 4520 ft^2 of absorber sheeting. The lower 21 feet of the transpired collector is black, perforated, aluminum wall cladding with 1.6-mm holes totaling 2% porosity. The average depth of the plenum between the transpired collector and the plant's structural wall is 6 inches. The canopy at the top of the wall acts as both a manifold for air flow and a solar heat collection device. The canopy face is made of perforated plate with 1% porosity. The transpired collector covers about 50% of the total area of the plant's south-facing wall; the remainder of the south facade has shipping doors and other obstructions that make it unsuitable for mounting collector cladding.

The GM transpired collector has two fan/distribution systems, each consisting of a constant-speed fan, a recirculation damper system, and a fabric distribution duct. The total airflow delivered by the system's fans is 40,000 cfm. Both recirculated air and air drawn through the solar collectors make up this flow; the percentages of each depend on the temperature of the air coming from the collector.

The GM battery plant's transpired collector has been monitored extensively since it was installed. The data in this case study reflect the performance of the system during the 1993–1994 heating season. An in-depth report on the monitoring program is available (Enermodal Engineering, Ltd., 1995). The data show that the annual energy savings for the 4520-ft^2 collector were 940 million Btu/year: 678 MBtu resulted from the thermal energy gained directly from the outside air as it passed through the absorber, and 262 MBtu resulted from heat loss recaptured by the wall from inside the building. Other possible energy-saving mechanisms—such as destratification and heat recapture—likely contributed to improved system performance; however, these effects are highly structure specific and have not been incorporated into the savings reported here.

The cost of the transpired collector system at the GM plant was $66,530, or $14.72/ft^2 of installed collector. The cost per square foot is higher than typical installations for two reasons: (1) this system was installed soon after the technology was introduced, before design and installation procedures had been streamlined; and (2) the cost includes the fan and ducting modifications that were implemented during the first 2 years of operation.

It is important to point out that the GM transpired collector experienced a number of operational problems. After the system was initially installed, employees complained about fan noise and feeling cold drafts, and employees occasionally disabled the system. The fan and duct upgrades described previously eliminated the problems on one of the fan systems; the other fan continues to generate noise, and employees still disable it when working in the immediate vicinity. The manufacturer has addressed these complaints by specifying smaller, but more numerous, fans in subsequent installations.

Also, both bypass dampers and a recirculation damper required additional maintenance. The recirculation damper became stuck in the full recirculation mode, and a new modulating motor was installed to fix the problem. The bypass dampers occasionally became bound, which led to unacceptably high leakage rates. These dampers were kept closed manually through the 1993–1994 heating season.

WATER HEATING

Solar water-heating systems are designed to provide large quantities of hot water for nonresidential buildings. A typical system includes solar collectors that work along with a pump, heat exchanger, and one or more large storage tanks. The two main types of solar collectors used for nonresidential buildings—evacuated-tube collector and linear concentrator—can operate at high temperatures with high efficiency. An evacuated-tube collector is a set of many double-walled glass tubes and reflectors to heat the fluid inside the tubes. A vacuum between the two walls insulates the inner tube, retaining the heat. Linear concentrators use long, rectangular, curved (U-shaped) mirrors tilted to focus sunlight on tubes that run along the length of the mirrors. The concentrated sunlight heats the fluid within the tubes.

SPACE COOLING

Space cooling can be accomplished using a thermally activated cooling system (TACS) driven by solar energy. Because of a high initial cost, the TACS is not widespread. The two systems currently in operation are solar absorption systems and solar desiccant systems. Solar absorption systems use thermal energy to evaporate a refrigerant fluid to cool the air. In contrast, solar desiccant systems use thermal energy to regenerate desiccants that dry the air, thereby cooling the air. These systems also work well with evaporative coolers in more humid climates.

LET THERE BE NATURAL LIGHT*

Lighting is an essential element of any building. Proper lighting improves the aesthetics of indoor spaces and provides illumination for tasks and activities. An efficient lighting strategy, including natural daylighting, can provide proper levels of illumination and reduce energy costs. *Daylighting* is a passive solar design used to illuminate a living space or an industrial or commercial working space. In the Northern Hemisphere, southern exposures receive the greatest amount of sunlight, while the opposite is true in the Southern Hemisphere. In other words, installing large windows in a south-facing wall or roof allows natural light to penetrate and reduces the use of expensive electrical lighting. Because the proper use of daylighting requires the integration of natural and artificial lighting sources early in the building design process, it is important to have an understanding of the fundamentals of lighting and lighting technology. Accordingly, in this section, we first present a discussion of lighting technology before discussing the principles and practices associated with daylighting.

Lighting technologies include the following:

- *Lamps*—Lighting sources, such as fluorescent and incandescent light bulbs, as well as solid-state lighting
- *Solid-state lighting (SSL)*—Light-emitting diode (LED) lighting, which has the potential to save energy, enhance the quality of building environments, and contribute to energy and climate change solutions
- *Ballasts*—Components of electric-discharge lamps such as fluorescent lamps that transform and control electrical power to the light
- *Luminaires (fixtures)*—Complete lighting units that contain the bulbs and, if necessary, the ballasts
- *Lighting controls*—Devices such as timers and sensors that can save energy by turning lights off when not needed
- *Daylighting*—Use of natural light in a building

LAMPS

Commonly called light bulbs, lamps produce light. When comparing lamps, it is important to understand the following performance characteristics:

- *Color rendering index (CRI)* is a measurement of the accuracy of a light source in rendering colors compared to a reference light source with the same correlated color temperature. The highest attainable CRI is 100. Lamps with CRI values above 70 are typically used in office and living environments.

---

* Adapted from EERE, *Commercial Building Initiative*, Office of Energy Efficiency & Renewable Energy, U.S. Department of Energy, Washington, DC, 2010 (www1.eere.energy.gov/buildings/commercial/printable_versions/lighting.html).

- *Correlated color temperature (CCT)* is a measurement on the Kelvin (K) scale that indicates the warmth or coolness of the color of a lamp. The higher the color temperature, the cooler the color. Typically, a CCT rating below 3200 K is considered warm, whereas a rating above 4000 K is considered cool.
- *Efficacy* is the ratio of the light output to the power, measured in lumens per watt (lm/W). The higher the efficacy, the more efficient the lamp.

Incandescent Lamps

A standard incandescent lamp consists of a fairly large, thin, frosted glass envelope. Inside the glass is an inert gas such as argon or nitrogen. At the center of the lamp is a tungsten filament. Electricity heats the filament. The heated tungsten emits visible light through a process called *incandescence.* Most standard light bulbs are incandescent lamps. They have a CRI value of 100 and CCT values between 2600 and 3000, making them attractive lighting sources for many applications; however, these bulbs are typically inefficient (10 to 15 lm/W). They convert only about 10% of the energy into light while transforming the rest into heat.

Another type of incandescent lamp is the halogen lamp. Halogen lamps also have a CRI value of 100, but they are slightly more energy efficient, and they maintain their light output over time. A halogen lamp also uses a tungsten filament, but the filament is encased inside a much smaller quartz envelope. The gas inside the envelope is from the halogen group. If the temperature is high enough, the halogen gas will combine with tungsten atoms as they evaporate and redeposit them on the filament. This recycling process means the filament lasts much longer. In addition, it is now possible to run the filament hotter, which means that more light per unit of energy can be obtained. Because the quartz envelope is so close to the filament, it becomes about four times hotter than a standard incandescent lamp. As a result of this wasted heat energy, halogen lamps, which are popular in torchieres (i.e., portable electric lamps with a reflector bowl that directs light upward to given indirect illumination), are not very energy efficient. The exposed heat from halogen torchieres can also pose a serious fire risk, especially near flammable objects. Because of this inefficiency and risk, manufacturers have developed torchieres that can use other types of lamps, such as compact fluorescent lamps.

Fluorescent Lamps

A fluorescent lamp (70 to 100 lm/W) is a sealed glass tube containing a small amount of mercury and an inert gas, such as argon, kept under very low pressure. In these electric-discharge lamps, a fluorescing coating (phosphor powder) on the glass transforms some of the ultraviolet energy generated into light. Fluorescent lamps require a ballast to start and maintain their operation. Early fluorescent lamps were sometimes criticized for not producing warm enough colors, making them appear too white or not complimentary to skin tones; a cool white fluorescent lamp had a CRI value of 62. Today, some fluorescent lamps have CRI values of 80 and above, and they simulate natural daylighting and incandescent light. They also are available in a variety of CCT values, ranging from 2900 to 7000. The "T" designation for fluorescent lamps stands for tubular, the shape of the lamp. The number after the "T" gives the diameter of the lamp in eighths of an inch. The T8 lamp

DID YOU KNOW?

Cold cathode fluorescent lamps are one of the latest technological advances in fluorescent technology. The "cold" in cold cathode means there is no heating filament in the lamp to heat up the gas. This makes them more efficient. Also, because there is no filament to break, they are ideal for use in rough service environments where a regular lamp may fail. They are often used as backlights in LCLD monitors. They are also used in exit signs.

(available straight or U-shaped) has become the standard for new construction. It is also commonly serves as a retrofit replacement for 40-watt T12 lamps, improving efficacy, CRI, and efficiency. In some cases, T10 lamps offer advantages over both T12 and T8 lamps, including higher efficiency, higher CRI values, a wider selection of CCT values, and compatibility with several ballast types. Another lamp type is the T5FT fluorescent lamp. These lamps produce maximum light output at higher ambient temperatures than those that are linear or U-shaped. Linear fluorescent lamps often are less expensive than compact fluorescent lamps. They can also produce more light, are easier to dim, and last longer.

Compact Fluorescent Lamps

Compact fluorescent lamps (CFLs) are small-diameter fluorescent lamps folded for compactness, with an efficacy of 50 to 75 lm/W for 27 to 40 watts. There are several styles of CFLs: two-, four-, and six-tube lamps, as well as circular lamps. Some CFLs have permanently connected tubes and ballasts; others have separate tubes and ballasts. Some CFLs feature a round adaptor, allowing them to screw into common electrical sockets and making them ideal replacements for incandescent lamps. They last up to 10 times longer than incandescent lamps, and they use about one-fourth the energy, producing 90% less heat. Typical 60- to 100-watt incandescent lamps are no more than 5.3 inches long, but standard CFLs are longer than 6 inches. For this reason, sub-CFLs have been developed that are no more than 4.5 inches long and fit into most incandescent fixtures.

High-Intensity Discharge Lamps

Compared to fluorescent and incandescent lamps, high-intensity discharge (HID) lamps produce a large quantity of light in a small package. HID lamps produce light by striking an electrical arc across tungsten electrodes housed inside a specially designed inner glass tube. This tube is filled with both gas and metals. The gas aids in the starting of the lamps; the metals then produce the light when they are heated to a point of evaporation. Like fluorescent lamps, HID lamps require a ballast to start and maintain their operation. Types of HID lamps include mercury vapor (CRI range, 15 to 55), metal halide (CRI range, 65 to 80), and high-pressure sodium (CRI range, 22 to 75). Mercury vapor lamps (25 to 45 lm/W), which originally produced a bluish-green light, were the first commercially available HID lamps. Today, they are also available as a color-corrected, whiter light, but they are still often being replaced by the newer, more efficient high-pressure sodium and metal halide lamps. Standard high-pressure sodium lamps have the highest efficacy of all HID lamps, but they produce a yellowish light. High-pressure sodium lamps that produce a whiter light are now available, but efficiency is somewhat sacrificed. Metal halide lamps are less efficient but produce an even whiter, more natural light. Colored metal halide lamps are also available.

DID YOU KNOW?

One way to create white light with LEDs is to mix the three primary colors of light (red, green, and blue, or RBG). Phosphor-conversion white light can be produced by blue, violet, or near-ultraviolet LEDs coated with yellow or multichromatic phosphors; the combined light emission appears white.

Low-Pressure Sodium Lamps

Low-pressure sodium lamps have the highest efficacy of all commercially available lighting sources, producing up to 180 lm/W. Even though they emit a yellow light, a low-pressure sodium lamp should not be confused with a standard high-pressure sodium lamp—a high-intensity discharge lamp. Low-pressure sodium lamps operate much like a fluorescent lamp and require a ballast. The lamps are also physically large (about 4 feet long for the 180-watt size), so light distribution from fixtures is less controllable. The lamp requires a brief warm-up period to reach full brightness. With a CRI value of 0, low-pressure sodium lamps are used where color rendition is not important but energy efficiency is. They are commonly used for outdoor, roadway, parking lot, and pathway lighting. Low-pressure sodium lamps are preferred around astronomical observatories because the yellow light can be filtered out of the random light surrounding the telescope.

SOLID-STATE LIGHTING

Compared to incandescent and fluorescent lamps, solid-state lighting (SSL) creates light with less directed heat. A semiconducting material converts electricity directly into light, which makes the light very energy efficient. Solid-state lighting includes a variety of light-producing semiconductor devices, such as light-emitting diodes (LEDs) and organic light-emitting diodes (OLEDs). Warm white LEDs have an efficacy of 50 lm/W, while cool white LEDs can achieve efficacies up to 100 lm/W. Until recently, LEDs—basically tiny light bulbs that fit easily into an electrical circuit—were used as simple indicator lamps in electronics and toys, but recent research has achieved efficiencies equal to those of fluorescent lamps. Also, the costs of semiconductor materials, which used to be quite expensive, are lower today, making LEDs a more cost-effective lighting option.

Ongoing research shows that LEDs have great potential as energy-efficient lighting for residential and commercial building use. New uses for LEDs include small-area lighting (e.g., task lighting and under-shelf fixtures), decorative lighting, and pathway and step marking. As LEDs become more powerful and effective, they will be used in more general illumination applications, perhaps with entire walls and ceilings becoming the lighting system. They are already being used successfully in many general illumination applications, such as traffic signals and exit signs. OLEDs currently are used in very thin, flat display screens, such as those in portable televisions; in some vehicle dashboard readouts; and in postage-stamp-sized data screens built into pilots' helmet visors. Because OLEDs emit their own light and can be incorporated into arrays on very thin, flexible materials, they also could be used to fashion large, extremely thin panels for light sources in buildings.

BALLASTS

Ballasts consume, transform, and control electrical power for electric-discharge lamps, providing the necessary circuit conditions for starting and operating them. Electric-discharge lamps include fluorescent, high-intensity discharge, and low-pressure sodium. When comparing ballasts, it is important to understand the following performance characteristics:

- *Ballast factor (BF)*—The ratio of the light output of a lamp or lamps operated by a specific ballast to the light output of the same lamps operated by a reference ballast. It can be used to calculate the actual light output of that specific lamp–ballast combination. The BF is typically different for each lamp type. Ballasts with extremely high BF values could reduce lamp life and accelerate lumen deficiency because of high lamp current. Extremely low BF values also could reduce lamp life because they reduce lamp current.
- *Ballast efficacy factor (BEF)*—The ratio of ballast factor (as a percentage) to power (in watts). BEF comparisons should be made only among ballasts operating the same type and number of lamps.
- *System efficacy*—The ratio of the light output to the power. Efficacy is measured in lumens per watt (lm/W) for a particular lamp ballast system.

The three basic types of ballasts are magnetic, hybrid, and electronic:

- *Magnetic ballasts* contain a magnetic core of several laminated steel plates wrapped with copper windings. These ballasts operate lamps at line frequency (60 Hz in North America). Of all ballasts, magnetic ones are the least expensive and also the least efficient. They have greater power losses than electronic ballasts, but magnetic ballasts manufactured today are 10% more efficient than the older, high-loss magnetic ballasts that used aluminum windings. Magnetic ballasts are available with dimming capability; however, they cannot be dimmed below 20% and still use more electricity than electronic ballasts.
- *Hybrid ballasts* (cathode-disconnect ballasts) use a magnetic core-and-coil transformer and an electronic switch for the electrode heating circuit. Like magnetic ballasts, they operate lamps at line frequency (60 Hz in North America). After they start the lamp, these ballasts disconnect the electrode heating circuit. Hybrid ballasts cost more than magnetic ballasts, but they are more energy efficient.
- *Electronic ballasts* appeared in the early 1980s, when manufacturers began to replace the core-and-coil transformer with solid-state, electronic compounds that could operate lamps at 20 to 50 kHz. These electronic ballasts experience half the power loss of magnetic ballasts. Also, lamp efficacy is increased by approximately 10 to 15% compared to 60-Hz operation. Electronic ballasts are the most expensive, but they are also the most efficient. Operating lamps with electronic ballasts reduces electricity use by 10 to 15% over magnetic ballasts of the same light output. They are quieter and lighter, and they virtually eliminate lamp flicker. Electronic ballasts are also available as dimming ballasts. These ballasts allow the light level to be controlled between 1 and 100%. Several types of electronic ballasts are available for use with fluorescent lamps. Electronic ballasts have been successfully used with lower watt high-intensity discharge (HID) lamps (primarily 35 to 100 W). These ballasts provide an energy savings over magnetic ballasts of 8 to 20%. Their lighter weight also helps in some HID applications, such as track lighting.

LUMINAIRES (LIGHTING FIXTURES)

A lamp or lamp–ballast combination may produce light very efficiently, but if it is installed in an inefficient luminaire the overall system efficiency may still be poor. The best luminaire manufacturers will design their fixtures around specific lamps to optimize the amount of light delivered to the work area; for example, a luminaire designed specifically for a compact fluorescent lamp can deliver almost 10 times as much illumination as an incandescent fixture fitted with the same compact fluorescent lamp. Luminaire components include reflectors, diffusers (which absorb some of the light from a lamp), and polarizing panels. Reflectors can be used to direct more of the light produced by the lamp out of the luminaire onto the work area. Polarizing panels can sometimes increase the contrast of a visual task.

DID YOU KNOW?

The most efficient light source technology for exit signs is light-emitting diodes (LEDs). The most popular parking lot luminaires use energy-efficient, high-intensity discharge lamps or low-pressure sodium lamps. The most efficient light source technology for outdoor use, though, is outdoor photovoltaic lighting.

When comparing luminaires, it is important to understand the following luminaire performance characteristics:

- *Illuminance*—The amount of light that reaches a surface; it is measured in footcandles (lumens per square foot) or lux (lumens per square meter)
- *Luminaire efficacy/efficiency rating (LER)*—The light output (lumens) per watt (W) of electricity use

Before selecting luminaires or lighting fixtures for an office building, factory, warehouse, or even a parking lot, it is a good idea to consult a certified lighting designer. A lighting designer will not only help find the most energy-efficient luminaires but also provide lighting that makes for a comfortable and more productive work environment. Today, energy-efficient commercial lighting design includes more than just the ambient or general lighting of a workspace, such as the use of ceiling luminaires. When designing or retrofitting the lighting, the general illuminance can be reduced if task lighting is implemented properly into the overall design. Task lighting can result in significant energy savings and improved visibility for workers.

LIGHTING CONTROLS

Lighting controls help conserve energy and make a lighting system more flexible. The most common light control is the on/off switch. Other types of light control technologies include manual dimming, photosensors, occupancy sensors, clock switches or timers, and centralized controls:

- *Manual dimming*—These controls allow occupants of a space to adjust the light output or illuminance. This can result in energy savings through reductions in input power, as well as reductions in peak power demand and enhanced lighting flexibility. Slider switches allow the occupant to change the lighting over the complete output range. Preset scene controls change the dimming setting for various lights all at once with the press of a button; it is possible to have different settings for the morning, afternoon, and evening. Remote control dimming is also available; this type of technology is well suited for retrofit projects, where it is useful to minimize rewiring.
- *Photosensors*—These devices automatically adjust the light output of a lighting system based on detected illuminance. The technology behind photosensors is the photocell. A photocell is a light-responding silicon chip that converts incident radiant energy into electrical current. Whereas some photosensors just turn lights off and on, others can also dim lights. Automatic dimming can help with lumen maintenance. Lumen maintenance involves dimming luminaires when they are new, which minimizes the wasteful effects of over-design. The power supplied to them is gradually increased to compensate for light loss over the life of the lamp.
- *Occupancy sensors*—These sensors turn lights on and off based on their detection of motion within a space. Some sensors can also be used in conjunction with dimming controls to keep the lights from turning completely off when a space is unoccupied. This

DID YOU KNOW?

Nearly all photosensors are used to decrease the electric power for lighting. In addition to lowering the electric power demand, dimming the lights also reduces the heat load on the cooling system of a building. Any solar heat gain that occurs in a building during the day must be taken into account for a whole building energy usage analysis.

control scheme may be appropriate when occupancy sensors control separate zones in a large space, such as in a laboratory or in an open office area. In these situations, the lights can be dimmed to a predetermined level when the space is unoccupied. Sensors can also be used to enhance the efficiency of centralized controls by switching off lights in unoccupied areas during normal working hours as well as after hours. The three basic types of occupancy sensors are listed below.

- *Passive infrared (PIR) sensors*—These sensors react to the movement of a heat-emitting body through their field of view. Wall-box-type PIR occupancy sensors are best suited for small, enclosed spaces such as private offices, where the sensor replaces the light switch on the wall and no extra wiring is required. They should not be used where walls, partitions, or other objects might block the ability of the sensor to detect motion.
- *Ultrasonic sensors*—These sensors emit an inaudible sound pattern and read the reflection. They react to changes in this reflected sound pattern. These sensors detect very minor motion better than most infrared sensors; therefore, they are good to use in spaces such as restrooms with stalls, which can block the field of view, because the hard surfaces will reflect the sound pattern.
- *Dual-technology (hybrid) sensors*—These occupancy sensors use both passive infrared and ultrasonic technologies to minimize the risk of false triggering (lights coming on when the space is unoccupied). They also tend to be more inexpensive.
- *Clock switches or timers*—These control lighting for a preset period of time. They come equipped with an internal mechanical or digital clock, which will automatically adjust for the time of year. The user determines when the light should be turned on and when they should be turned off. Clock switches can be used in conjunction with photosensors.
- *Centralized controls*—These are building controls or building automation systems that can be used to automatically turn on, turn off, or dim electrical lights around a building. In the morning, the centralized control system can be used to turn on the lights before employees arrive. During the day, a central control system can be used to dim the lights during periods of high power demand. And, at the end of the day, the lights can be turned off automatically. A centralized lighting control system can significantly reduce energy use in buildings where lights are left on when not needed.

DAYLIGHTING

As mentioned earlier, daylighting is the practice of placing windows or other openings and reflective surfaces so that during the day natural light provides effective internal lighting. When properly designed and effectively integrated with the electric lighting system, daylighting can offer significant energy savings by offsetting a portion of the electric lighting load. A related benefit is the reduction in cooling capacity and use due to lowering a significant component of internal gains. In addition to energy savings, daylighting generally improves occupant satisfaction and comfort. Windows also provide visual relief, contact with nature, time orientation, the possibility of ventilation, and emergency egress.

Daylight Zone

High daylight potential is particularly found in those spaces that are predominantly occupied in the daytime. Site solar analyses should assess the access to daylight by considering what can be viewed from the various potential window orientations. What proportion of the sky is seen from typical task locations in the room? What are the exterior obstructions and glare sources? Is the building design going to shade a neighboring building or landscape feature that is dependent on daylight or solar access? It is important to establish which spaces will most benefit from daylight and which spaces have little or no need for daylight. Within the spaces that can use daylight, place the most critical visual tasks in positions near the window. Try to group tasks by similar lighting requirements and occupancy patterns. Avoid placing the windows in the direct line of sight of the occupant as this can cause extreme contrast and glare. It is best to orient the occupant at 90° from the window. Where privacy is not a major concern, consider interior glazing (known as *relights* or *borrow lights*), which allows light from one space to be shared with another. This can be achieved with transom lights, vision glass, or translucent panels if privacy is required. The floor plan configuration should maximize the perimeter daylight zone. This may result in a building with a higher skin-to-volume ratio than a typical compact building design. A standard window can produce useful illumination to a depth of about 1.5 times the height of the window. With light shelves or other reflector systems this can be increased to 2.0 times or more. As a general rule of thumb, the higher the window is placed on the wall, the deeper the daylight penetration.

Window Design Considerations

The daylight that arrives at a work surface comes from three sources:

1. *Exterior reflected component*—This includes ground surfaces, pavement, adjacent buildings, wide windowsills, and objects. Remember that excessive rough reflectance will result in glare.
2. *Direct sun/sky component*—Typically, the direct sun component is blocked from occupied spaces because of heat gain, glare, and ultraviolet degradation issues. The sky dome then becomes an important contribution to daylighting the place.
3. *Internal reflected component*—Once the daylight enters the room, the surrounding wall, ceiling, and floor surfaces are important light reflectors. Using high-reflectance surfaces will better bounce the daylight around the room, and it will reduce extreme brightness contrast. Window-frame materials should be light colored to reduce contrast with the view and should have a nonspecular finish to eliminate glare spots. The window jambs and sills can be beneficial light reflectors. Deep jambs should be splayed (angled toward the interior) to reduce the contrast around the perimeter of the window.

Major room furnishings such as office cubicles or partitions can have a significant impact on reflected light so select light-colored materials. Suggested reflectance levels for various room surfaces are

---

DID YOU KNOW?

The most important interior light-reflecting surface is the ceiling. High-reflectance paints and ceiling tiles are now available with 90% or higher reflectance values. Tilting the ceiling plane toward the daylight source increases the daylight that is reflected from this surface. The rear wall of a small room should also have a highly reflective finish. Side walls, followed by the floor, have less impact on the reflected daylight in the space.

- Ceilings, >80%
- Walls, 50 to 70%
- Floors, 20 to 40%
- Furnishings, 25 to 45%

Because light essentially has no scale for architectural purposes, the proportions of a room are more important than the dimensions. A room that has a higher ceiling compared to the room depth will have deeper penetration of daylight, whether from side lighting (windows) or top lighting (skylights and clerestories). Raising the window head height will result in deeper penetration and more even illumination in the room. Punched window openings, such as small, square windows separated by wall area, result in uneven illumination and harsh contrast between the window and adjacent wall surfaces. A more even distribution is achieved with horizontal strip windows.

Effective Aperture

One method of assessing the relationship between visible light and the size of the window is the effective aperture method. The *effective aperture* (EA) is defined as the product of the *visible transmittance* (VT) and the *window-to-wall ratio* (WWR). The window-to-wall ratio is the proportion of window area compared to the total wall area where the window is located; for example, if a window covers 25 square feet in a 100-square-foot wall, then the WWR is 25/100, or 0.25. A good starting target for the EA is in the range of 0.20 to 0.30. For a given EA number, a higher WWR (larger window) results in a lower visible transmittance. For example, a WWR = 0.5 (half the wall is glazing) has a VT = 0.6 and EA = 0.3. For WWR = 0.75, VT = 0.4 for the same EA of 0.3.

Light Shelves

A light shelf is an effective horizontal light-reflecting overhang placed above eye level with a transom window placed above it. Light shelves enhance the lighting from windows of the equator-facing side of a building. Exterior shelves are more effective shading devices than interior shelves. A combination of exterior and interior will work best in providing an even illumination gradient.

Toplighting Strategies

Large single-level floor areas and the top floors of multiple-story buildings can benefit from toplighting. Types of toplighting include skylights, clerestories, monitors, and sawtooth roofs. Horizontal skylights can be an energy problem because they tend to receive maximum solar gain at the peak of the day; the daylight contribution peaks at midday and falls off severely in the morning and afternoon. Some high-performance skylight designs incorporate reflectors or prismatic lenses that

DID YOU KNOW?

A building designed for daylighting but without an integrated electrical lighting system will be a net energy loser because of the increased thermal loads. Only when the electric lighting load is reduced will we realize more than offsetting savings in electrical and cooling loads. The benefits from daylighting are maximized when both occupancy and lighting sensors are used to control the electric lighting system.

reduce peak daylight and heat gain while increasing early and late afternoon daylight contributions. Another option is light pipes, high-reflectance ducts that channel light from a skylight down to a diffusing lens in the room. These may be advantageous in deep roof constructions.

Clerestory windows are vertical glazings located high on an interior wall. South-facing clerestories can be effectively shaded from direct sunlight by a properly designed horizontal overhang. In this design, the interior north wall can be sloped to better reflect the light down into the room. Use light-colored overhangs and adjacent roof surfaces to improve the reflected component. If exterior shading is not possible, consider interior vertical baffles to better diffuse the light. A south-facing clerestory will produce higher daylight illumination than a northern-facing clerestory. East- and west-facing clerestories present the same problems as east and west windows: difficult shading and potentially high heat gains. A roof monitor consists of a flat roof section raised above the adjacent roof with vertical glazing on all sides. This design often results in an excessive glazing area, which results in higher heat losses and gains than a clerestory design. The multiple orientations of the glazing can also create shading problems. The sawtooth roof is an old design often seen in industrial buildings. Typically, one sloped surface is opaque and the other is glazed. A contemporary sawtooth roof may have solar collectors or photovoltaic cells on the south-facing slope and daylight glazing on the north-facing slope. Unprotected glazing on the south-facing sawtooth surface may result in high heat gains. In these applications, an insulated diffusing panel may be a good choice.

OTHER SOLAR ENERGY APPLICATIONS

If one is tasked with putting together a list of potential solar energy applications, a full sheet of paper that can be written on both sides would probably be needed. The point is that it is only a lack of imagination or a lack of desire for ingenuity and for innovation that limits us in our quest to employ solar energy for useful purposes. Although we use solar energy for space heating and cooling and for swimming pool heating, there are other applications of solar energy that many people are not aware of, such as the use of solar energy for crop drying, water purification, cooking, and electrifying fences on farms.

CROP DRYING

Agriculture was an early adapter of solar energy as a remote energy source, and many of those initial applications are still cost effective today due to low maintenance costs and the high cost of extending electricity to remote locations. One of the primary uses of solar energy in agriculture is crop drying. If agricultural products (e.g., prunes, walnuts, pecans, herbs, grains) are not dried within a few days of harvest and prior to transport, insects and fungi will make them unusable. Most traditional commercial dryers on the market run on oil, natural, gas, propane, or steam and often produce higher temperatures than necessary. Transpired solar collector panels (e.g., SolarWall®) are well suited for heating large volumes of air. Usually, the panels are placed on the roofs or walls of buildings housing existing dryers, and they either heat or preheat the air entering the fan and dryer. The best candidates for solar crop drying are those that dry all year long, as they allow longer utilization of the equipment and offer a quicker payback time to recover capital investment.

WATER PURIFICATION

Unsanitary drinking water is the source of many serious diseases. Solar water purification uses solar energy to make biologically contaminated (e.g., bacteria, viruses, protozoa, worms) water safe to drink. Probably the simplest way to employ this solar energy technology to disinfect small quantities of water is the solar water disinfection method known as SODIS, which requires only a clean transparent bottle and sunshine. In this method, a sealed polyethylene terephthalate (PET) plastic bottle containing contaminated water (but not cloudy water) is simply exposed to the sun for about 6 hours, allowing ultraviolet radiation to kill pathogens.

SOLAR COOKING

The basic box-type solar cooker is very popular in areas where national and local governments, to prevent massive deforestation and desertification, have put into effect stringent laws and regulations about gathering and using forest products for cooking purposes. The basic cooker has insulated sides and a transparent cover on top and can reach temperatures up to 212°F (100°C). More sophisticated and expensive solar energy models can attain temperatures as high as 680°F (360°C).

ENVIRONMENTAL IMPACTS OF SOLAR ENERGY

All energy-generating technologies, including solar technologies, affect the environment in many ways. Solar energy has some obvious advantages in that the source is free; however, the initial investment in operating equipment is not. Solar energy is also environmentally friendly, requires almost no maintenance, and reduces our dependence on foreign energy supplies. Probably the greatest downside of solar energy use is that in areas without direct sunlight during certain times of the year solar panels cannot capture enough energy to provide heat for homes or offices. Geographically speaking, the higher latitudes do not receive as much direct sunlight as tropical areas. Because of the position of the sun in the sky, solar panels must be placed in sun-friendly locations such as the U.S. desert southwest and the Sahara region of northern Africa.

LAND USE AND SITING

Most of the land used for larger utility-scale solar facilities, depending on their location, can raise concerns about land degradation and habitat loss. This is the case even if abandoned industrial, fallow agricultural, or former mining sites are used. Total land area requirements vary depending on the technology, the topography of the site, and the intensity of the solar resource. Estimates for utility-scale PV systems range from 3.5 to 10 acres per megawatt, while estimates for CSP facilities are between 4 and 16.5 acres per megawatt (UCS, 2013). Unlike wind facilities, there is less opportunity for solar projects to share land with agricultural uses; however, land impacts for utility-scale solar systems can be minimized by siting them at lower-quality locations such as brownfields, existing transportation and transmission corridors, or abandoned mining lands (AMLs) (Hand et al., 2012; USEPA, 2011).

DID YOU KNOW?

Abandoned mine lands (AMLs) are those lands, waters, and surrounding watersheds where extraction, beneficiation, or processing of ores and minerals (excluding coal) has occurred. These lands also include areas where mining or processing activity is temporarily inactive.

WATER RESOURCES

Just as with the production of energy from all other sources, in one way or another solar energy production has a *water footprint*, defined as the total volume of freshwater used to produce energy and the services consumed by the production process. Water consumption for solar generation varies by technology and location. For our purposes in this text, water consumption is defined as the amount of water that is "evaporated, transpired, incorporated into products or crops, consumed by humans or livestock, or otherwise removed from the immediate water environment" (Kenny et al., 2009). Water consumption is distinct from water withdrawal. Water withdrawal is the total amount of "water removed from the ground or diverted from a surface-water source for use" (Kenny et al., 2009) but which may be returned to the sources. Both water withdrawal and consumption are important metrics, but consumption is a very useful metric for water-scarce regions, especially in the context of future resource development, because consumption effectively removes water from the system so it is not available for other uses (e.g., agriculture, drinking).

Table 4.3 shows estimated water-consumption ranges for solar deployment in 2030 and 2050 under a U.S. Department of Energy scenario. These values represent estimates of gross water consumption for deployed solar technologies only; that is, they do not consider the amount of water consumption avoided due to replacement of other electricity-generating technologies by solar. Table 4.4 gives water use estimates for solar, wind, fossil fuel, and nuclear generating technologies. Biomass and co-fired biomass power plants have cooling/generating water consumption similar to that of comparable coal plants, but water consumption related to growing biomass fuel is highly variable (Gerbens-Leenes et al., 2009; Macknick et al., 2011). As Table 4.3 shows, many solar configurations can reduce water consumption dramatically compared with conventional technologies that use evaporative cooling systems (i.e., cooling towers).

TABLE 4.3

U.S. Solar-Related Water Consumption for Solar Technology Deployment in 2030 and 2050 under USDOE (2012) Scenario

	Solar Generation in 2030 (TWh)	Solar-Related Water Consumption in 2030 (billion gal)	Solar Generation in 2050 (TWh)	Solar-Related Water Consumption in 2050 (billion gal)
Rooftop PV	164	0–0.8	318	0–1.6
Utility-scale PV	341	0–1.7	718	0–3.6
CSP[a]	137	14–75	412	42–227
Total	642	14–78	1448	42–232

Source: USDOE, *SunShot Vision Study*, U.S. Department of Energy, Washington, DC, 2012.

[a] The concentrating solar power (CSP) water-use values shown here reflect a range of trough/tower water-use estimates. The low number reflects trough/tower technology with 90% use of dry cooling and 10% use of wet cooling, with per-megawatt-hour consumption at the low end of the trough/tower ranges. The high number reflects trough/tower technology with 50% use of wet cooling and 50% use of dry cooling, with per-megawatt-hour consumption at the high end of the trough/tower ranges. The USDOE scenario assumes 100% dry cooling as a conservative estimate of costs, but it is likely that the mix would consist of various technologies. Thus, the values given in this table are meant to illustrate a range of possible scenarios of CSP deployment. Dish/engine CSP technologies were used in these calculations because substantially more data are available for them, but, assuming dish/engine technologies meet the price and performance characteristics envisioned in the USDOE scenario, widespread deployment of these technologies could help reduce CSP-related water use.

TABLE 4.4
Water Intensity of Electricity Generation by Fuel Source and Technology[a]

Generation Technology	Cooling System	Water Consumed for Cooling (gal/MWh)	Other Water Consumed in Generation (gal/MWh)	Water Consumed in Producing Fuel (gal/MWh)
CSP trough or tower (wet-cooled)[b]	Closed-loop cooling tower	710–960	40–60	0
CSP trough or tower (dry-cooled)[c]	Dry air cooling	0	30–80	0
CSP dish/engine[d]	Dry air cooling	0	4–6	0
PV[e]	None	0	0–5	0
Wind[f]	None	0	0	0
Pulverized coal[f,g]	Closed-loop cooling tower	360–590	60–120	5–74
Pulverized coal with CO_2 capture[f,h]	Closed-loop cooling tower	700–770	150–180	5–74
Integrated gasification combined cycle (IGCC)[f,i]	Closed-loop cooling tower	250–370	40–70	5–74
IGCC with CO_2 capture[f,j]	Closed-loop cooling tower	390–410	130–150	5–74
Natural gas combined cycle (CC)[f,k]	Closed-loop cooling tower	180–280	2	11
Nuclear[f,l]	Closed-loop cooling tower	580–850	30	45–150

Source: USDOE, *SunShot Vision Study*, U.S. Department of Energy, Washington, DC, 2012.

[a] The table does not account for water consumption in system manufacturing or construction of any of the technologies. Water consumption for fuel extraction is considered for fossil and nuclear. All wet-cooled Rankine power cycles are assumed to use closed-loop cooling towers with four cycles of concentration and blowdown water discharge to an onsite evaporation pond. Water consumption values for wet-cooled Rankine power cycles using once-through cooling systems are not shown because their large water withdrawal requirements make them infeasible for the Southwest. Dry cooling is possible with all Rankine cycles, although it is explicitly shown for concentrating solar power (CSP) only.

[b] From Cohen et al. (1999) and Viebahn et al. (2008). Other water consumed for trough and tower technologies includes water for washing mirrors and steam-cycle blowdown and makeup. Mirror soiling rates/washing rates are site and developer specific. Towers will be at the lower end of the cooling-water range and troughs at the higher end due to thermal efficiency differences.

[c] From Brightsource Energy (2007) and Kelly (2006). Other water consumed for trough and tower technologies includes water for washing mirrors and steam-cycle blowdown and make-up. Mirror soiling rates/washing rates are site and developer specific. There is more uncertainty in other water consumed for dry-cooled trough/tower technologies than for wet-cooled technologies because fewer dry-cooled plants have been built.

[d] Dish/engine washing rates and other water use are not well documented and vary by site and developer. The estimate of 4 to 6 gal/MWh is based on Leitner (2002) and CEC (2010b), as well as on industry knowledge.

[e] Utility-scale PV washing rates and other water use are not well documented and vary by site and developer. The estimate of 0 to 5 gal/MWh is based on Aspen Environmental Group (2011) and on industry knowledge.

[f] From USDOE (2006b).

[g] From NETL (2007, 2010). Cooling and other-generation values are for new subcritical and supercritical coal plants.

[h] From NETL (2010). Cooling and other-generation values are for new subcritical and supercritical coal plants.

[i] From NETL (2007, 2010).

[j] From NETL (2010).

[k] From EPRI (2002) and NETL (2007).

[l] From Gleick (1993) and Gerdes and Nichols (2009).

Solar cooling towers regulate temperature by dissipating heat from recirculating water used to cool process equipment. Heat is rejected from the tower primarily through evaporation; therefore, by design, cooling towers consume significant amounts of water (see Table 4.4). The Achilles heel of solar cooling towers is twofold: (1) equipment failure and (2) operator and/or management error. The thermal efficiency and longevity of the cooling tower and equipment used to cool depend on the proper operation and management of water recirculated through the tower. Water leaves a cooling tower system in any one of four ways (EERE, 2013a):

1. *Evaporation*—This is the primary function of the tower and is the method that transfers heat from the cooling tower system to the environment. The quantity of evaporation is not a subject for water efficiency efforts (although improving the energy efficiency of the systems being cooled will reduce the evaporative load on the tower).
2. *Drift*—A small quantity of water may be carried from the tower as mist or small droplets. Drift loss is small compared to evaporation, and blowdown and is controlled with baffles and drift eliminators.
3. *Blowdown or bleed-off*—When water evaporates from the tower, dissolved solids (such as calcium, magnesium, chloride, and silica) are left behind. As more water evaporates, the concentration of these dissolved solids increases. If the concentration gets too high, the solids can cause scale to form within the system or the dissolved solids can lead to corrosion problems. The concentration of dissolved solids is controlled by blowdown. Carefully monitoring and controlling the quantity of blowdown provides the most significant opportunity to conserve water in cooling tower operations.
4. *Basin leaks or overflows*—Properly operated towers should not have leaks or overflows. Float control equipment must be checked to ensure that the basin level is being maintained properly, and the system valves must be checked to make sure that there are no unaccounted for losses. The sum of water that is lost from the tower must be replaced by make-up water:

$$\text{Make-up water} = \text{Evaporation} + \text{Blowdown} + \text{Drift}$$

A key parameter used to evaluate cooling tower operation is *cycles of concentration*. This is calculated as the ratio of the concentration of dissolved solids (or conductivity) in the blowdown water compared to the make-up water. Because dissolved solids enter the system in the make-up water and exit the system in the blowdown water, the cycles of concentration are also approximately equal to the ratio of volume of make-up to blowdown water. From a water efficiency standpoint, cycles of concentration should be maximized, which will minimize blowdown water quantity and reduce make-up water demand. However, this can only be done within the constraints of make-up water availability and cooling tower chemistry. Dissolved solids increase as cycles of concentration increase, which can cause scale and corrosion problems unless carefully controlled.

Other cooling types (e.g., once-through and pond systems) may have different water consumption and withdrawal rates, but these technologies are generally not feasible in arid regions due to their higher withdrawal rates. Photovoltaics consume little, if any, water during operation; some PV operators wash panels to maintain optimal performance, whereas others do not. Concentrating solar technologies, including concentrating photovoltaics (CPV) and CSP, require water for rinsing panels, mirrors, and reflectors to ensure maximum energy production. Manufacturing solar technologies also consumes water. For a trough-based CSP facility with 6 hours of two-tank indirect thermal energy storage (TES), Burkhardt et al. (2010) estimated that about 120 gal/MWh are consumed, mainly in the production of solar collector assemblies, nitrate salts, and heat-transfer fluid (HTF). Although water-consumption values for PV manufacturing have not been established, Fthenakis and Kim (2010) provided some information about water withdrawals related to PV manufacturing (i.e., water used in the PV manufacturing process but not entirely consumed, with some of the water processed and returned to the immediate water environment). Water consumed to extract, process, and transport fuels can be significant for fossil fuel and nuclear technologies but is not required for solar and wind technologies (see Table 4.4).

The largest water consumption associated with solar-electricity production is for cooling CSP trough and tower plants. The amount of water a CSP system consumes for cooling depends on the technology, cooling system, location, climate, and water availability. Three types of CSP cooling systems can be deployed: wet (once-through or evaporative cooling using cooling towers), dry, or hybrid (combination wet/dry). *Wet cooling* (using cooling towers—evaporative water cooling) is the most common cooling method for new power plants and currently offers the highest performance at the lowest overall cost (Turchi et al., 2010), but it also consumes the largest amount of water. *Dry cooling* cuts operational water consumption by as much as 97% compared with wet cooling, but it increases capital costs and reduces efficiency on hot days (Turchi, 2010). In addition, dry-cooling technology is significantly less effective at temperatures above 100°F. The cost of electricity from a dry-cooled parabolic-trough plant in the Mojave Desert is about 7% higher than from a similar wet-cooled plant (Turchi, 2010; USDOE, 2006b). Dish/engine CSP plants are dry cooled.

To overcome the cost and performance penalty associated with dry cooling, some developers are considering *hybrid systems* that employ dry cooling when temperatures are below 38°C (100°F) and wet cooling for hotter periods. Hybrid systems can consume 40 to 90% less water than a wet-cooled system while maintaining 97 to 99% of the performance (USDOE, 2006b); however, hybrid systems currently have a higher life-cycle cost than wet-cooled systems (Turchi et al., 2010).

In addition to consuming water for cooling, trough and tower CSP systems consume a relatively small amount of water to produce steam for electricity generation. In a typical Rankine cycle steam turbine, water in a closed loop is heated to produce steam and spin a turbine, then cooled, re-condensed, and used again. A relatively small amount of water—compared with the water consumed in an evaporative cooling system—is drained to remove particulates and salts (the blowdown process) in the boiler and cooling systems. The amount of blowdown water depends on the quality of the source water; more is required when using degraded water sources. Dish/engine CSP plants with Stirling engines do not use a water–steam cycle; the movement of a gas is used to produce electricity in these systems.

The distribution of solar water consumption will not be uniform across the United States; it will be highest in the arid Southwest, where CSP development will be concentrated. Unless dry cooling is used, siting CSP in arid areas presents a potentially insurmountable deployment challenge because of water constraints in these areas (Carter and Campbell, 2009). The West accounted for half of all U.S. population growth from 1990 to 2000, creating additional demand for water (Anderson and Woosley, 2005). Water resources in arid regions may also decline with climate change, and the Southwest has experienced the most rapid warming in the United States (U.S. Global Change Research Program, 2009). Water consumption per unit of area for PV and CSP is less intensive than for a number of other activities. Thus, although water consumption is likely to be an issue of contention in the Southwest going forward, it is possible that solar developers will be able to obtain water rights from existing water-rights holders, sometimes resulting in less intensive water consumption.

Hazardous Waste

Like all other technologies, solar technologies require proper waste management and recycling. PV is associated with a few particular waste management and recycling issues, whereas CSP shares issues with other technologies that use common materials such as concrete, glass, and steel. Waste management and recycling issues for each technology are discussed below, with a focus on the issues surrounding PV. The PV cell manufacturing process includes a number of hazardous materials, such as compounds of cadmium (Cd), selenium (Se), and lead (Pb), and there are concerns about potential emissions at the end of a module's useful life. Managing the disposal and/or recycling of these materials to avoid groundwater contamination (via landfills) and air pollution (via incinerators) is an important environmental consideration. Another important consideration is that in creating millions of solar panels each year millions of pounds of polluted sludge and contaminated water are produced. To dispose of the material properly, the producers must transport it by truck or rail far from their own plants to waste facilities hundreds or, in some cases, thousands of miles away. The fossil fuels used to transport that waste are not typically considered in calculating solar's carbon footprint, giving scientists and consumers who use the measurement to gauge a product's impact on global climate change the impression that solar is cleaner than it is (Anon., 2013).

In addition to materials contained within the completed module, a number of chemicals may be used during PV manufacturing. For crystalline silicon modules, feedstock materials are made through a purification process, the byproducts of which typically include silicon tetrachloride ($SiCl_4$). To reduce costs and protect the environment, most of today's manufacturing plants use a closed-loop process that greatly minimizes waste products by converting, separating, and reusing trichlorosilane from the $SiCl_4$ byproduct. Silicon nitride (SiN_4) is used as an antireflective-coating material and is generally deposited via chemical vapor deposition. This process requires the safe handling and management of pyrophoric silane gas—a gas that can ignite spontaneously when exposed to air. Silane is also the major feedstock in thin-film amorphous silicon (a-Si) PV and is often used as a coupling agent to adhere fibers to polymer materials. The a-Si/thin-film tandem segment of the PV industry also uses nitrogen trifluoride (NF_3) for reactor cleaning, which has a global warming potential 17,000 times greater than CO_2. The controlled use and production of NF_3 have been proven for specific production and end-use systems (for example, in the liquid crystal display industry), and its use in the a-Si/microcrystalline silicon PV industry will not alter the environmental benefits of PV replacing fossil fuels if best practices are adopted globally (Fthenakis et al., 2010).

The greatest concern surrounding thin-film cadmium telluride (CdTe) and copper indium gallium selenide (CIGS) PV is potential exposure to Cd, which the U.S. Environmental Protection Agency defines as a Class B1 carcinogen (i.e., probable human carcinogen). Typical CdTe PV material contains 5 g of Cd per m^2 of module, whereas typical CIGS material (which can contain cadmium sulfide) contains less than 1 g of Cd per m^2 of module (Fthenakis and Zweibel, 2003). Although Cd is not emitted during normal module operation, small emissions could occur during manufacturing or accidental fires. However, the life-cycle Cd emissions of CdTe and CIGS PV are orders of magnitude lower than Cd emissions from the operation of fossil fuel power plants (Fthenakis, 2004; Fthenakis et al., 2005, 2008).

Not all the news about PV production and subsequent PV waste is negative. Recycling can help resolve end-of-life (cradle-to-grave) PV module issues, and the PV industry is proactively engaged in building recycling infrastructure. The technical and economic feasibility of recycling the semiconductor materials, metals, and glass from manufacturing scrap and spent PV modules has been established (Fthenakis, 2000). Furthermore, recycling can provide a significant secondary source of materials that may be used in the production of future PV technologies, such as tellurium, indium, and germanium (Fthenakis, 2009). First Solar® (Tempe, AZ), which manufactures thin-film CdTe PV, established the industry's first comprehensive pre-funded module collection and recycling program, which the company claims will result in recycling 90% of the weight of each recovered First Solar PV module. In Europe, the PV industry has established PV Cycle, a voluntary program to recycle PV modules (PV Cycle, 2014). The United States could adopt this type of industrywide approach to manage the large-scale recycling and management of PV materials.

The major constituents of CSP plants include glass, steel, and concrete. In addition, some CSP plants will contain a significant quantity of nitrate salt and organic heat transfer oil. All of these materials are recyclable (EERE, 2012).

ECOLOGICAL IMPACTS OF SOLAR ENERGY

All development creates ecological and other land-use impacts. The primary impacts of solar development relate to land used for utility-scale PV and CSP. Even with the most careful land selection, the projected utility-scale solar development may have significant local land-use impacts, especially on portions of the southern United States. Solar development should be consistent with national and local land-use priorities. With regard to direct ecological impacts of solar development, these include soil disturbance, habitat fragmentation, and noise. Indirect impacts include changes in surface water quality because of soil erosion at the construction site. The specific impacts of utility-scale solar development will depend on the project location, solar technology employed, size of the development, and proximity to existing roads and transmission lines.

The potential ecological impacts in the southwestern United States are particularly important because of the large scale of solar development envisioned for this area. The Southwest supports a wide variety of plant communities and habitats, including arid and semiarid desert-scrub and shrub land, grasslands, woodlands, and savannas. The wildlife in these areas includes diverse species of amphibians, reptiles, birds, and small and large mammals. Government agencies and conservation groups have identified a significant list of species that may be affected by solar development (USDOE, 2010). Altering plant communities with development can strain wildlife living in or near these communities, making it more difficult to find shelter, hunt, forage, and reproduce. Fenced-in power plants can add further strain by affecting terrestrial and avian migration patterns. Aquatic species also can be affected—as can terrestrial and avian species that rely on aquatic habitats—if the water requirements of solar development result in substantial diversion of local water sources. Large areas covered by solar collectors also may affect plants and animals by interfering with natural sunlight, rainfall, and drainage. Solar equipment may provide perches for birds of prey that could affect bird and prey populations.

The potential impacts of solar development are not limited to ecological impacts. Solar development could affect a variety of activities that take place on public and private land. For example, conflicts may arise if development impacts cultural sites, or interferes with U.S. Department of Defense (DOD) activities. In addition, loss of forage base could result in reduced grazing, which would disrupt the longstanding economic and cultural characteristics of ranching operations. Potential direct impacts include conversion of land to provide support services and housing for people who move to the region to support the solar development, with associated increases in roads, traffic, and penetration into previously remote areas. The additional transmission infrastructure associated with solar development could create various impacts as well.

These are merely examples of the types of impacts that may be associated with solar development. For an exhaustive discussion, see USDOE (2010) and other detailed environmental-impact studies. Less well-studied impacts are also important and must be evaluated as solar development progresses. For example, the local and global climate effects of changes in albedo due to widespread PV and CSP deployment are not well studied. *Albedo* is the ratio between the light reflected from a surface and the total light falling on it. Albedo is a surface phenomenon—basically a radiation reflector. Albedo always has a value less than or equal to 1. An object with a high albedo, near 1, is very bright, while a body with a low albedo, near 0, is dark. For example, freshly fallen snow typically has an albedo that is between 75% and 90%; that is, 75% to 95% of the solar radiation that is incident on snow is reflected. At the other extreme, the albedo of a rough, dark surface, such as a green forest, may be as low as 5%. The albedo values of some common surfaces are listed in Table 4.5. The portion of insolation not reflected is absorbed by the Earth's surface, warming it. This means that Earth's albedo plays an important role in the Earth's radiation balance and influences the mean annual temperature and the climate on both local and global scales. One study evaluated the net balance between the greenhouse gas emissions reduction resulting from PV replacing fossil-fuel-based power generation (with PV growing to meet 50% of world energy demand in 2100) and a decrease in desert albedo due to PV module covering, concluding that the PV albedo effect would have little impact on global warming (Nemet, 2009).

TABLE 4.5
Albedo of Some Surface Types

Surface	Albedo (% Reflected)
Water (low sun)	10–100
Water (high sun)	3–10
Grass	16–26
Glacier ice	20–40
Deciduous forest	15–20
Coniferous forest	5–15
Old snow	40–70
Fresh snow	75–95
Sea ice	30–40
Blacktopped tarmac	5–10
Desert	25–30
Crops	15–25

With regard to solar energy production and the possible impact on global warming (climate change), note that there are no global climate change emissions associated with generating electricity from solar energy. However, there are emissions associated with other stages of the solar life-cycle, including manufacturing, materials transportation, installation, maintenance, and decommissioning and dismantlement. Most estimates of life-cycle emissions for photovoltaic systems are between 0.07 and 0.18 pounds of carbon dioxide equivalent per kilowatt-hour.

MOJAVE DESERT TORTOISES AND SOLAR ENERGY

It seems the American standard has become let me sue you before you sue me. In the case of siting solar facilities, we urge rapid development of renewables (solar and wind energy in particular), but sue when a project is fast tracked. With respect to siting solar energy production facilities in southern Nevada and southeastern California, many issues have been presented by plaintiffs and appellants for litigation in defense of preserving desert landscapes (USFWS, 2014). One of these issues concerns *Gopherus agassizii* and *Gopherus morafkai*, species of the tortoise (commonly called the desert tortoise; see Figure 4.8) native to the Mojave desert and Sonoran desert of the southwestern United States.

FIGURE 4.8   Desert tortoise. (Photograph by U.S. Fish and Wildlife Service.)

Under natural conditions (undisturbed by humans), the desert tortoises live approximately 50 to 80 years; they grow slowly and generally have low reproductive rates. They spend most of the time in burrows, rocks shelters, and pallets to regulate body temperature and reduce water loss. In addition to destruction of the desert tortoise by humans and their associated activities, red-tailed hawks, burrowing owls, golden eagles, loggerhead shrikes, mountain lions, ravens, Gila monsters, kit foxes, badgers, roadrunners, coyotes, and fire ants are all natural predators of the desert tortoise. However, the most significant threats to tortoises include urbanization, disease, habitat destruction and fragmentation, illegal collection and vandalism by humans, and habitat conversion from invasive plant species. Of course, our concern here is with the construction of solar facilities and their impact—namely, causing habitat loss, fragmentation, and degradation of the desert tortoise population.

Since the 1800s, portions of the desert southwest occupied by desert tortoises have been subjected to a variety of impacts that cause habitat loss, fragmentation, and degradation, thereby threatening the long-term survival of the species. Some of the most apparent threats are those that result in mortality and permanent habitat loss across large areas.

Conflicts between energy development projects, which can cover thousands of acres, and the desert tortoise have been recognized since at least 1986. Desert tortoises may be killed during exploration, construction, and ongoing operations, as well as by maintenance activities associated with energy facilities and transmission corridors. Project sites are typically contoured and fenced, resulting in direct mortality and habitat loss. Ground-disturbing activities that may cause negative impacts to the desert tortoise can increase soil erosion, increase establishment of invasive plant species, reduce cryptobiotic soil crusts (biological crusts composed of living bacteria, algae, fungi, lichens, and/or mosses stabilize the soil), and alter drainage patterns that impact plant communities downstream from the project footprint.

The bottom-line question is how do we install solar facilities in desert tortoise habitat areas without killing off the tortoises? Unfortunately, the desert tortoise displays high fidelity to its home area and would be difficult to persuade to depart the solar construction and operation areas on its own. Therefore, the only plausible fix to this problem is to move the tortoises wholesale to distant locations within the same type of area. One thing in their favor in the southwest is the almost endless regions of untamed terrain, much of it suitable for desert tortoises. At least this is the case at the present time.

SOLAR ENERGY JOB HAZARDS*

FATALITIES AND INCIDENTS

Solar energy workers are exposed to hazards that can result in fatalities and serious injuries. Many incidents involving falls, electrocution, severe burns from electrical shocks, and arc flashes/fires have been reported to OSHA. Some examples are given below:

- *Solar panel installer dies when he falls off a roof.* A 30-year-old solar panel installer died after he fell 45 feet off the roof of a three-story apartment building. He was part of a three-man crew working to install solar panels on a sloped roof. The worker walked backward and stepped off the roof while checking the position of some brackets. No one was wearing personal fall protection equipment and no other fall protection system was in place.
- *Solar energy technician is electrocuted.* A 34-year-old solar energy technician died after being injured at work. He was bringing a metal brace to a rooftop work site. The technician was standing on a scaffold and lifting the brace when it reached the top of the scaffold. The other end of the brace swung into a nearby high-voltage power line. The technician was shocked by the electrical current and fell 35 feet to the ground. He died the next day of injuries from the electrocution and fall.

---

* Adapted from OSHA, *Green Job Hazards: Solar Energy*, http://osha.gov/dep/greenjobs/solar.html.

- *Deadly skylights! Solar energy and warehouse workers killed.* A 46-year-old electrical worker and a 56-year-old warehouse worker both died after falling through skylights while working on a roof. The electrician was carrying a solar panel and tripped on the ledge of the skylight and fell backward through the skylight. The warehouse worker was on the warehouse roof repairing a broken air conditioner. It is not known exactly how the warehouse worker fell through the skylight.

Workers in the solar energy industry are potentially exposed to many of the same hazards as those who work with or around wind turbines. In addition, solar energy workers can be exposed to a variety of serious hazards, such as arc flashes (which include arc flash burn and blast hazards), electric shock, and thermal burn hazards that can cause injury and death. Solar energy employers (connecting to grid) are covered by the Electric Power Generation, Transmission, and Distribution standard (29 CFR 1910.269) and therefore may be required to implement the safe work practices and worker training requirements of the standard. Although solar energy is a growing industry, the hazards are not unique, and OSHA has many standards that cover them. The hazards and controls that workers in the solar energy industry may encounter are provided below.

HAZARDS AND CONTROLS

- Falls
- Lockout/tagout
- Crane and hoist safety
- Electrical
- Thermal stress
- Personal protective equipment (PPE)

Those charged with managing workers in the solar energy production fields are responsible to ensure compliance with OSHA, EPA, CDC, and state and local safety and health requirements. Each of the hazards and controls listed above must be addressed and appropriate measures taken to ensure worker health and safety.

FUTURE DIRECTION FOR SOLAR ENERGY

Because of our future energy needs and because of our growing knowledge that traditional fossil-fuel sources of energy will not last forever, a bright future for solar energy production and its use can be projected. Moreover, it is not a leap of faith—that is, it is safe to say—that the future for all forms of renewable energy sources is bright and growing in intensity. It is not a stretch, for example, to predict that all buildings will be built to combine energy-efficient design and construction practices and renewable energy technologies for a net zero energy building. In effect, such buildings will conserve and produce enough energy to create a new generation of cost-effective buildings that have zero net annual need for nonrenewable energy. Moreover, photovoltaics research and development will continue to attract intense interest in new materials, cells, designs, and novel approaches to solar material and product development. It is a future where the clothes we wear and our modes of transportation can produce power that is clean and safe. It is also not a stretch to predict that the future research and development in concentrating solar power (CSP) will lead to full competitiveness with conventional power generation technologies within a decade or so. The potential of solar power in the southwestern Untied States is comparable in scale to the hydropower resources of the northwest. A desert area 10 miles by 15 miles could provide 20,000 megawatts of power, whereas the electricity needs of the entire United States could theoretically be met by a photovoltaic array within an area 100 miles on a side.

Concentrating solar power, or solar thermal electricity, could harness the sun's heat energy to provide large-scale, domestically secure, and environmentally friendly electricity (USDOE, 2003). Concentrating solar power, or solar thermal electricity, could harness the sun's heat energy to provide large-scale, domestically secure, and environmentally friendly electricity. Many feel that the price of photovoltaic power will be competitive with traditional sources of electricity within 10 years. Fuel cell technology is discussed later in the book, but for now it is important to point out that we could use solar energy as an inexpensive means of electrolyzing water, producing hydrogen for fuel cells for transportation and buildings. Remember that solar energy as a primary source of energy is great while the sun is shining and is unobstructed by clouds or by darkness, but our supply of seawater is endless, almost infinite. The hydrogen contained within this endless supply is just floating there, and no matter the weather or darkness it just sits there waiting to be utilized for fuel cells.

BOTTOM LINE ON SOLAR ENERGY

Solar energy has some obvious advantages in that the source is free; however, the initial investment in operating equipment is not free or inexpensive. Solar energy is also environmentally friendly (to a degree), requires almost no maintenance, and reduces our dependence on foreign energy supplies. Probably the greatest downside of solar energy use is that in areas without direct sunlight during certain times of the year solar panels cannot capture enough energy to provide heat for homes or offices. Geographically speaking, the higher latitudes do not receive as much direct sunlight as tropical areas. Because of the position of the sun in the sky, solar panels must be placed in sun-friendly locations such as the desert southwest in the United States and the Sahara region of northern Africa. Another downside of solar energy is the efficiency of the system; it may be seriously affected by how well it was installed. The real bottom line on solar energy is obvious: Its advantages outnumber its disadvantages.

DISCUSSION AND REVIEW QUESTIONS

1. The sun has been described in many ways. How would you describe the sun?
2. Is there a difference between solar energy and solar power? Explain.
3. What is the 5-5-5 rule? Does it make sense? Explain.
4. Do you feel that solar energy can eventually replace nonrenewable energy sources?
5. Do you agree or disagree with the author's view that the best use of solar energy may be to electrolyze water, producing hydrogen for fuel cells?

REFERENCES AND RECOMMENDED READING

Anderson, M.T. and Woosley, Jr., L.H. (2005). *Water Availability for the Western United States—Key Scientific Challenges*, Circular 51. Washington, DC: U.S. Geological Survey.

Anon. (2013). Solar power boom fuels increase in hazardous waste sent to dumps. *The Blade*, February 11 (http://www.toledoblade.com/Energy/2013/02/11/Solar-power-boom-fuels-increase-in-hazardous-waste-sent-to-dumps.html).

Argonne National Laboratory. (2015). Solar Energy Development Programmatic Environmental Impact Statement Information Center, http://solareis.anl.gov/.

Aspen Environmental Group. (2011). *California Valley Solar Ranch Conditional Use Permit, and Twisselman Reclamation Plan and Conditional Use Permit: Final Environmental Impact Report (DRC2008-00097, DRC 2009-00004)*. Prepared for County of San Luis Obispo Department of Planning and Building by Aspen Environmental Group, San Francisco, CA (http://www.sloplanning.org/EIRs/CaliforniaValleySolarRanch/index.htm).

Baker, M.S. (1990). Modeling complex daylighting with DOE 2.1-C, *DOE-2 User News*, 11(1).

Baylon, D. and Storm, P. (2008). Comparison of commercial LEED buildings and non-LEED buildings within the 2002–2004 Pacific Northwest commercial building stock. In: *Proceedings of ACEEE Summer Study on Energy Efficiency in Buildings*. Washington, DC: American Council for an Energy-Efficient Economy.

Bierman, A. (2007). Photosensors: dimming and switching systems for daylight harvesting, *NLPIP Specifier Reports*, 11(1), 1–54.

Birt, B. and Newsham, G.R. (2009). Post-occupancy evaluation of energy and indoor environment quality in green building: a review. In: *Proceedings of 3rd International Conference on Smart and Sustainable Built Environments*, Delft, The Netherlands, June 15–19.

Brightsource Energy. (2007). *Application for Certification, Volumes I and II, for the Ivanpah Solar Electric Generating System*. Submitted to California Energy Commission Docket Unit, Application for Certification (07-AFC-5), August 31.

Brookfield, H.C. (1989). *Sensitivity to Global Change: A New Task for Old/New Geographers*. Reading, U.K.: Norma Wilkinson Memorial Lecture, University of Reading.

Brown, M.H. and Sedano, R.P. (2004). *Electricity Transmission: A Primer*. Washington, DC: National Council on Energy Policy.

Brundtland, G.H. (1987). *Our Common Future*. New York: Oxford University Press.

Brunger, A. and Hollands, K. (1996). Back-of-Plate Heat Transfer in Unglazed Perforated Collectors Operated Under Non-Uniform Air Flow Conditions, paper presented at the 22nd Annual Conference of the Solar Energy Society of Canada, Inc., Orillia, Ontario, Canada, June 9–10.

Burkhardt, J.J., Heath, G., and Turchi, C. (2010). Life Cycle Assessment of a Parabolic Trough Concentrating Solar Power Plant and the Impacts of Key Design Alternatives, paper presented at National Renewable Energy Laboratory Conference, Granada, Spain, March 23–25.

Caldeira, K. and Wickett, M.E. (2003). Anthropogenic carbon and ocean pH. *Nature*, 425(6956): 365.

Calder, W.A. (1996). *Size, Function and Life History*. Mineola, NY: Dover.

Carter, N.T. and Campbell, R.J. (2009). *Water Issues of Concentrating Solar Power (CSP) in the U.S. Southwest*, Order No. R40631. Washington, DC: Congressional Research Service.

CEC. (2009). Renewable Energy Transmission Initiative website, http://www.energy.ca.gov/reti/index.html.

CEC. (2010a). *Notice of Meeting for the Renewable Energy Policy Group*. Sacramento: California Energy Commission.

CEC. (2010b). *Calico Solar Power Project: Commission Decision*, CEC-800-2010-012-CMF. Sacramento: California Energy Commission (http://www.energy.ca.gov/2010publications/CEC-800-2010-012/CEC-800-2010-012-CMF.PDF).

Chiras, D.D. (2002). *The Solar House: Passive Heating and Cooling*. White River Junction, VT: Chelsea Green Publishing.

Christensen, C. (1982). As quoted in *Making It Happen: A Positive Guide to the Future*, Richardson, J., Ed. Warwick, NY: Roundtable Press.

Christensen, C., Hancock, E., Barker, G., and Kutscher, C. (1990). Cost and performance predictions for advanced active solar concepts. In: *Proceedings of the American Solar Energy Society Annual Meeting*, Austin, TX, March 19–22.

Cohen, G., Kearney, D., Drive, C., Mar, D., and Kolb, G. (1999). *Final Report on the Operation and Maintenance Improvement Program for Concentrating Solar Plants*, SAD99-1290. Albuquerque, NM: Sandia National Laboratories.

CPUC and BLM. (2008). *Final Environmental Impact Report/Environmental Impact Statement and Proposed Land Use Plan Amendments for the Sunrise Powerlink Project of the Proposed Sunrise Powerlink Transmission Line Project*. San Francisco: California Public Utility Commission and Bureau of Land Management.

Debres, K. (2005). Burgers for Britain: a cultural geography of McDonald's UK. *Journal of Cultural Geography*, 22(2): 115–139.

Dengler, J. and Wittwear, V. (1994). Glazings with granular aerogels, *SPIE*, 255: 718–727.

Duffie, J. and Beckman, W. (1991). *Solar Engineering of Thermal Processes*. New York: John Wiley & Sons.

Dumortier, D. (1997). Evaluation of luminous efficacy models according to sky types and atmospheric conditions. In: *Proceedings of Lux Europa '97*, Ecole Nationale des Travaux Publics de l'Etat, Vaulx-en-Velin, France, May 11–14.

EERE. (2006). *Solar Energy Technologies Program: Overview and Highlights*. Washington, DC: Office of Energy Efficiency & Renewable Energy, U.S. Department of Energy.

EERE. (2011). *History of Hydropower*. Washington, DC: Office of Energy Efficiency & Renewable Energy, U.S. Department of Energy (http://www1.eere.energy.gov/water/hydro_history.html).

EERE. (2012). *SunShot Initiative*. Washington, DC: Office of Energy Efficiency & Renewable Energy, U.S. Department of Energy (http://www1.eere.Energy.gov/solar/printable_versions/about.html).

EERE. (2013a). *Best Management Practice: Cooling Tower Management*. Washington, DC: Office of Energy Efficiency & Renewable Energy, U.S. Department of Energy (http://energy.gov/eere/femp/best-management-practice-cooling-tower -management).

EERE. (2013b). *Linear Concentrator System Basics for Concentrating Solar Power*. Washington, DC: Office of Energy Efficiency & Renewable Energy, U.S. Department of Energy (http://energy.gov/eere/energybasics/articles/linear-concentrator-system-basics-concentrating-solar-power).

Enermodal Engineering, Ltd. (1995). *Performance of the Perforated-Plate/Canopy Solarwall at GM Canada, Oshawa*. Ottawa, ON: Energy Technology Branch, Department of Natural Resources.

EPRI. (2002). *Water and Sustainability*. Vol. 2. *An Assessment of Water Demand, Supply, and Quality in the U.S.—The Next Half Century*, Technical Report 1006785. Palo Alto, CA: Electric Power Research Institute.

Fenchel, T. (1974). Intrinsic rate of natural increase: the relationship with body size. *Oecologia*, 14: 317–326.

Fontoynont, M., Place, W., and Bauman, P. (1984). Impact of electric lighting efficiency on the energy saving potential of daylighting from roof monitors. *Energy and Buildings*, 6(4): 375–386.

Fthenakis, V.M. (2000). End-of-life management and recycling of PV modules. *Energy Policy*, 28: 1050–1058.

Fthenakis, V.M. (2004). Life cycle impact analysis of cadmium in CdTe PV production. *Renewable and Sustainable Energy Reviews*, 8: 303–334.

Fthenakis, V.M. (2009). Sustainability of photovoltaics: the case for thin-film solar cells. *Renewable and Sustainable Energy Reviews*, 13: 2746–2750.

Fthenakis, V. and Alsema, E. (2006). PV energy payback times, greenhouse gas emissions and external costs. *Progress in Photovoltaics: Research and Applications*, 14: 275–280.

Fthenakis, V. and Kim, H.C. (2007). Greenhouse gas emissions from solar electric and nuclear power: a life cycle study. *Energy Policy*, 35: 2549–2557.

Fthenakis, V. and Kim, H.C. (2009). Land use and electricity generation: a life cycle analysis. *Renewable and Sustainable Energy Review*, 13: 1465–1474.

Fthenakis, V. and Kim, H.C. (2010). Life cycle uses of water in U.S. electricity generation. *Renewable and Sustainable Energy Reviews*, 14: 2039–2048.

Fthenakis, V. and Zweibel, K. (2003). CdTe PV: Real and Perceived EHS Risks, paper presented at National Center for Photovoltaics and Solar Program Review Meeting, Denver, CO, March 24–26 (http://www.nrel.gov/docs/fy03osti/33561.pdf).

Fthenakis, V., Fuhrmann, M, Heiser, J., Lanzirotti, A., Fitts, J., and Wang, W. (2005). Emissions and encapsulation of cadmium in CdTe PV modules during fires. *Progress in Photovoltaics: Research and Application*, 13: 713–723.

Fthenakis, V., Kim, H.C., and Alsema, E. (2008). Emissions from photovoltaic life cycles. *Environmental Science & Technology*, 42(6): 2168–2174.

Fthenakis, V., Raugei, M., Held, M., Kim, H.C., and Krones, J. (2009). An Update of Energy Payback Times and Greenhouse Gas Emissions in the Life Cycle of Photovoltaics, paper presented at 24th European Photovoltaic Solar Energy Conference, Hamburg, Germany, September 21–25.

Fthenakis, V., Clark, C., Moalem, M., Chandler, P., Ridgeway, R., Hulbert, F., Cooper, D., and Maroulls, P. (2010). Life-cycle nitrogen trifluoride emissions from photovoltaics. *Environmental Science & Technology*, 44(22): 8750–8757.

Fthenakis, V., Green, T., Blunden, J., and Krueger, L. (2011). Large Photovoltaic Power Plants: Wildlife Impacts and Benefits, paper presented at 37th IEEE Photovoltaic Specialists Conference, Seattle, WA, June 19–24.

Galasiu, AD, Atif, M.R., and MacDonald, R.A. (2004). Impact of window blinds on daylight-linked dimming and automatic on/off lighting controls. *Solar Energy*, 76(5): 523–544.

Galasiu, AD, Newsham, G.R., Suvagau, C., and Sander, D.M. (2007). Energy saving lighting control system for open-plan offices: a field study. *Leukos*, 4(1): 7–29.

Gerbens-Leenes, W., Hoekstra, A.Y., and van Der Meer, T.H. (2009). The water footprint of bioenergy. *Proceedings of the National Academy of Sciences of the United States of America*, 106: 10219–10223.

Gerdes, K. and Nichols, C. (2009). *Water Requirements of Existing and Emerging Thermoelectric Plant Technologies*, DOE/NETL-402/080108. Morgantown, WV: National Energy Technology Laboratory.

Gleick, P. (1993). *Water in Crisis: A Guide to the World's Fresh Water Resources*. New York: Oxford University Press.

Gunnewick, L. (1994). An Investigation of the Flow Distribution Through Unglazed Transpired-Plate Solar Air Heaters, master's thesis, Department of Mechanical Engineering, University of Waterloo, Canada.

Hand, M.M., Baldwin, S., DeMeo, E., Reilly, J.M., Mai, T., Arent, D., Porro, G., Meshek, M., and Sandlor, D., Eds. (2012). *Renewable Electricity Futures Study*, NREL/TP-6A20-52409. Golden, CO: National Renewable Energy Laboratory.

Hanson, B.J. (2004). *Energy Power Shift: Benefiting from Today's New Technologies*. Maple, WI: Lakota Scientific Press.

Hardin, G. (1986). Cultural carrying capacity: a biological approach to human problems. *BioScience*, 36: 599–606.

Heschong, L. and McHugh, J. (2003). *Integrated Design of Commercial Building Ceiling Systems Project*. Sacramento: Public Interest Energy Research (PIER), California Energy Commission.

Heschong, L. and McHugh, J. (2006). Skylights: calculating illumination levels and energy impacts. *Journal of Illuminating Engineering Society*, 29(1): 90–100.

Hinrichs, R.A. and Kleinback, M.H. (2006). *Energy: Its Use and the Environment*, 4th ed. New York: Thomson Learning.

Hirsch, J.J. and Associates. (2004). *DOE 2.2 Building Energy Use and Cost Analysis Program*. Vol. 3. *Topics*. Berkeley: Lawrence Berkeley National Laboratory, University of California.

HMG. (2006). *Sidelighting Photocontrols Field Study*. Sacramento, CA: Heschong Mahone Group (http://www.h-m-g.com/Projects/Photocontrols/sidelighting_photocontrols_field.htm).

Holdren, J.P. (1991). Population and the energy problem. *Population and the Environment*, 12: 231–255.

Howlett, O. et al. (2006). Sidelighting photocontrols field study. In: *Proceedings of ACEEE Summer Study on Energy Efficiency in Buildings*, Pacific Grove, CA, August 13–18, pp. 3-148–3-159.

Hubbell, S.P. and Johnson, L.K. (1977). Competition and nest spacing in a tropical stingless bee community. *Ecology*, 58: 949–963.

Janak, M. (1997). Coupling building energy and lighting simulation. In: *Proceedings of the Fifth International IBPSA Conference*, Prague, September 8–10.

Janak, M. and Macdonald, I. (1999). Current state-of-the-art of integrated thermal and lighting simulation and future issues. In: *Proceedings of the Sixth International IBPSA Conference*, Kyoto, Japan, September 13–15.

Jeffrey, L.W. (1993). Sizing up skylights. *Home Energy*, November/December (http://www.homeenergy.org/show/article/nav/windows/page/4/id/998).

Kelly, B. (2006). *Nexant Parabolic Trough Solar Power Plant Systems Analysis. Task 2: Comparison of Wet and Dry Rankine Cycle Heat Rejection*, NREL/SR-550-40163. Golden, CO: National Renewable Energy Laboratory (http://www.nrel.gov/docs/fy06osti/40163.pdf).

Kenny, J.G., Barber, N.L., Hutson, S.S., Linsey, K.S., Lovelace, J.K., and Maupin, M.A. (2009). *Estimated Use of Water in the United States in 2005*, USGS Circular 1344. Reston, VA: U.S. Geological Survey.

Krebs, R.E. (2001). *Scientific Laws, Principles and Theories*. Greenwood Press, Westport, CT.

Kuhlken, R. (2002). Intensive agricultural landscapes of Oceania. *Journal of Cultural Geography*, 19(2): 161–195.

Kutscher, C.F. (1992). An Investigation of Heat Transfer for Air Flow Through Low Porosity Perforated Plates, doctoral thesis, University of Colorado, Boulder.

Kutscher, C.F., Christensen, C., and Barker, G. (1993). Unglazed transpired solar collectors: heat loss theory. *ASME Journal of Solar Energy Engineering*, 115: 182–188.

Lauoadi, A. and Aresnault, C. (2004). *Validation of Skyvision*, IRC-RR-167. Washington, DC: National Research Council.

Lauoadi, A. and Atif, M.R. (1999). Predicting optical and thermal characteristics of transparent single-glazed domed skylights. *ASHRAE Transactions*, 105(2): 325–333.

Lederer, E.M. (2008). UN says half the world's population will live in urban areas by end of 2008, *Associated Press*, February 26 (http://www.iht.com/aritcles/ap/2008/02/26/news/UN-GEN-UB-Growing-Cities.php).

Leitner, A. (2002). *Fuel from the Sky: Solar Power's Potential for Western Energy Supply*, NREL/SR-550-32160. Golden, CO: National Renewable Energy Laboratory (www.nrel.gov/docs/fy02osti/32160.pdf).

LOC. (2009). *The World's First Hydroelectric Power Plant*. Washington, DC: Library of Congress (http://www.americaslibrary.gov/jb/gilded/jb_gilded_hydro_1.html).

Lund, J.W. (2007). Characteristics, development and utilization of geothermal resources. *Institute of Technology*, 28(2): 1–9.

Macknick, J., Newmark, R., Heath, G., and Hallett, K. (2011). *A Review of Operational Water Consumption and Withdrawal Factors for Electricity Generating Technologies*, NREL/TP-6A20-50900. Golden, CO: National Renewable Energy Laboratory (http://www.cwatershedalliance.com/pdf/SolarDoc01.pdf).

Masters, G.M. (1991). *Introduction to Environmental Engineering and Science*. Englewood Cliffs, NJ: Prentice Hall.

Miller, G.T. (1988). *Environmental Science: An Introduction*. Belmont, CA: Wadsworth.

Molburg, J.C., Kavicky, J.A., and Picel, K.C. (2007). *The Design, Construction and Operation of Long-Distance High-Voltage Electricity Transmission Technologies*. Lemont, IL: Argonne National Laboratory.

Moore, F. (1991). *Concepts and Practice of Architectural Daylighting*. New York: John Wiley & Sons.

Nemet, G.F. (2009). Net radiative forcing from widespread deployment of photovoltaics. *Environmental Science & Technology*, 43(6): 2173–2178.

NETL. (2007). *Power Plant Water Usage and Loss Study*. Pittsburgh, PA: National Energy Technology Laboratory.

NETL. (2010). *Cost and Performance Baseline for Fossil Energy Plants*. Vol. 1. *Bituminous Coal and Natural Gas to Electricity, Revision 2*, DOE/NETL-2010/1397. Pittsburgh, PA: National Energy Technology Laboratory.

Newsham, G.R. (2004). *American National Standard Practice for Office Lighting*, RP-1. Washington, DC: ANSI/IESNA.

Newton, J. (1989). *Uncommon Friends: Life with Thomas Edison, Henry Ford, Harvey Firestone, Alexis Carrel, and Charles Lindbergh*. New York: Mariner Books.

NREL. (2010). *Concentrating Solar Power Fact Sheet*. Golden, CO: National Renewable Energy Laboratory (www.nrel.gov/csp/pdfs/48658.pdf).

NREL. (2014). *Learning About Renewable Energy: Concentrating Solar Power Basics*. Golden, CO: National Renewable Energy Laboratory (http://www.nrel.gov/learning/re_csp.html).

O'Connor, J., Lee, E., Rubinstein, L.E., and Selkowitz, S. (1997). *Tips for Daylighting with Windows*. Berkeley: Lawrence Berkeley National Laboratory, University of California.

Patel, M. (1999). *Wind and Solar Power Systems*. Boca Raton, FL: CRC Press.

Perez-Blanco, H. (2009). *The Dynamics of Energy: Supply, Conversion, and Utilization*. Boca Raton, FL: CRC Press.

Place, W. et al. (1984). The predicted impact of roof aperture design on the energy performance of office buildings. *Energy and Building*, 6(4): 361–373.

PV Cycle. (2014). PV Cycle website, http://www.pvcycle.org.uk/.

Reid, W. et al. (1988). *Bankrolling Successes*. Washington, DC: Environmental Policy Institute and National Wildlife Federation.

Rubinstein, F., Ward, G., and Verderber, R. (1989). Improving the performance of photo-electrically controlled lighting systems. *Journal of the Illuminating Engineering Society*, 18(1): 70–94.

Searchinger, T. et al. (2008). Use of U.S. croplands for biofuels increases greenhouse gases through emissions from land-use change. *Science*, 319(5867): 1238–1240.

Simon, J.L. (1980). Resources, population, environment: an oversupply of false bad news, *Science*, 208: 1431–1437.

Turchi, C. (2010). *Parabolic Trough Reference Plant for Cost Modeling with the Solar Advisor Model (SAM)*, NREL/TP-550-47605. Golden, CO: NREL (http://www.nrel.gov/docs/fy10osti/47605.pdf).

Turchi, C., Wagner, M.J., and Kutscher, C.F. (2010). *Water Use in Parabolic Trough Power Plants: Summary Results from Worley Parsons' Analyses*, NREL/TP-5500-49468. Golden, CO: National Renewable Energy Laboratory (www.nrel.gov/docs/fy11osti/ 49468.pdf).

UCS. (2008). *Land-Use Changes and Biofuels*, Fact Sheet. Cambridge, MA: Union of Concerned Scientists.

UCS. (2013). *Environmental Impacts of Solar Power*. Cambridge, MA: Union of Concerned Scientists (http://www.ucsusa.org/clean_energy/our-energy-choices/renewable-energy/environmental-impacts-solar-power.html).

U.S. Global Change Research Program. (2009). *Global Climate Change Impacts in the United States, Regional Highlights, Southwest*. Washington, DC: U.S. Global Change Research Program (www.globalchange.gov/usimpacts).

USDOE. (1998). *Solar Dish/Engine Systems*. Washington, DC: U.S. Department of Energy.

USDOE. (2003). *The History of Solar Energy*. Washington, DC: U.S. Department of Energy.

USDOE. (2006a). *Solar Buildings: Transpired Air Collectors*. Washington, DC: U.S. Department of Energy (http://www.nrel.gov/docs/fy06osti/29913.pdf).

USDOE. (2006b). *DOE Solar Energy Technologies Program: Overview and Highlights*. Washington, DC: U.S. Department of Energy (https://www1.eere.energy.gov/solar/pdfs/39081.pdf).

USDOE. (2008). *Concentrating Solar Power Commercial Application Study: Reducing Water Consumption of Concentrating Solar Power Electricity Generation*. Washington, DC: U.S. Department of Energy (www1.eere.energy.gov/solar/pdfs/csp_water_study.pdf).

USDOE. (2010). *Draft Programmatic Environmental Impact Statement for Solar Energy Development on BLM-Administered Lands in the Southwestern United States*. Washington, DC: U.S. Department of Energy (http://energy.gov/nepa/downloads/eis-0403-draft-programmatic-environmental-impact-statement).

USDOE. (2013). *Concentrating Solar Power Thermal Storage System Basics.* Washington, DC: U.S. Department of Energy (http://energy.gov/eere/energybasics/articles/concentrating-solar-power-thermal -storage-system-basics).

USEPA. (2011). *Shining Light on a Bright Opportunity.* Washington, DC: U.S. Environmental Protection Agency.

USFWS. (2014). *Mojave Desert Tortoises.* Washington, DC: U.S. Fish and Wildlife Service (http://www.fws. gov/nevada/desert_tortoise/dt/dt_threats.html).

Viebahn, P., Kronshage, S., Trieb, F., and Lechone, Y. (2008). *Final Report on Technical Data, Costs, and life Cycle Inventories of Solar Thermal Power Plants*, Project No. 502687. Rome: New Energy Externalities Developments for Sustainability (NEEDS).

Vitousek, P.M., Ehrlich, A.H., and Matson, P.A. (1986). Human appropriation of the products of photosynthesis. *BioScience*, 36: 368–373.

Wilson, E.O. (2001). *The Future of Life.* New York: Knopf.

Winkelman, F. and Selkowitz, S. (1985). Daylighting simulation in the DOE-2 building energy analysis program. *Energy and Buildings*, 8(4): 271–286.

5 Wind Energy

The wind goeth toward the south, and turneth about unto the north; it whirleth about continually, and the wind returneth again.

—Ecclesiastes 1:6

We rushed into renewable energy without any thought. The schemes are largely hopelessly inefficient and unpleasant. I personally can't stand windmills at any price.

—James Lovelock, environmentalist

The Good, Bad, and Ugly of Wind Energy

Good: As long as Earth exists, the wind will always exist. The energy in the winds that blow across the United States each year could produce more than 16 billion GJ of electricity—more than one and one-half times the electricity consumed in the United States in 2000.

Bad: Turbines are expensive. Wind doesn't blow all the time, so they have to be part of a larger plan. Turbines make noise. Turbine blades kill birds.

Ugly: Some look upon giant wind turbine blades cutting through the air as grotesque scars on the landscape, as visible polluters.

The bottom line: Do not expect Don Quixote, mounted in armor on his old nag, Rocinate, with or without Sancho Panza, to lead the charge to build those windmills. Instead, expect—you can count on it, bet on it, and rely on it—that the charge will be made by the rest of us to satisfy our growing, inexorable need for renewable energy. What other choice do we have?

Whenever I teach undergraduate/graduate college students renewable energy courses and whenever I shift lectures from solar to wind energy, invariably one of the students asks, "Why do we not call wind turbines *windmills*, as they are actually nothing more than sophisticated windmills?" I generally reply as follows: In the study of renewable energy, the ancient term *windmill* is rarely used because of its connotation of a device used to mill grain. When I look out in this classroom at all of you I do not see grain. Instead, I see energy, and when you get right down to it turbines are all about energy—wind power harnessed for useful purposes. So, it is your energy that I want to mill.

INTRODUCTION[*]

Most of the differences between solar- and wind-produced energy are apparent and obvious. Even though the potential energy produced by solar and wind energy is expressed in the same units for both (watts per square meter, W/m^2), it is important to point out (and for the reader to remember) that the relevant area for solar energy is a square meter of the Earth's surface, whereas for wind it is a vertical square meter perpendicular to the wind flow.

Obviously, wind energy or power is all about wind. In simple terms, wind is the response of the atmosphere to uneven heating conditions. Earth's atmosphere is constantly in motion. Anyone observing the constant cloud movement and weather changes around them is well aware of this phenomenon. Although its physical manifestations are obvious, the importance of the dynamic state of our atmosphere is much less obvious. The constant motion of Earth's atmosphere is both horizontal (*wind*) and vertical (*air currents*). This air movement is the result of thermal energy produced from

---

[*] Adapted from Spellman, F.R., *Environmental Science and Technology*, 3rd ed., Bernan Press, Lanham, MD, 2016.

DID YOU KNOW?

Wind speed is generally measured in meters per second (m/s), but Americans usually think in terms of miles per hour (mph). To convert m/s to mph, a good rule of thumb is to double the value in m/s and add 10%.

heating of the surface of the Earth and the air molecules above. Because of differential heating of the surface of the planet, energy flows from the equator poleward. The energy resources contained in the wind in the United States are well known and mapped in detail (Hanson, 2004). It is clear that air movement plays a critical role in transporting the energy of the lower atmosphere, bringing the warming influences of spring and summer and the cold chill of winter, and wind and air currents are fundamental to how nature functions. Still, though, the effects of air movements on our environment are often overlooked. All life on Earth has evolved or has been sustained with mechanisms dependent on air movement; for example, pollen is carried by the winds for plant reproduction, animals sniff the wind for essential information, and wind power was the motive force during the earliest stages of the Industrial Revolution. We can also see other effects of winds. Wind causes weathering (erosion) of the Earth's surface, wind influences ocean currents, and the wind carries air pollutants and contaminants such as radioactive particles that impact our environment.

KEY TERMS FOR WIND ENERGY DEVELOPMENT

To better prepare the reader for the material contained in this chapter, terms relevant to wind turbines are defined here (BPWENA, 2011). These terms and their definitions apply to modern wind power plant construction in the United States. Other general terms are defined as they appear or are listed in the glossary at the end of the text.

Access road—Provides primary access to large blocks of land and connects with, or is an extension of, the public road system. The access roads carry the large turbine components and construction and operations personnel from the public road system to the initial point of entry and inspection on the project site.

Ancillary facilities—Wind turbine generator (WTG) support components (e.g., transformers, electrical collection system, substations, O&M) that allow the electricity produced by the WTG to be connected to the existing electrical grid.

Anemometer—One of the components of a meteorological tower, the anemometer is a sensor that measures wind speed and direction.

Clearing and grubbing—Process of removing the top layer of soil and vegetation within the areas indicated on the design drawings; it involves cutting and removing all brush, shrubs, debris, and vegetation to approximately flush with the ground surface or 3 to 6 inches below surface.

Concrete batch plant—A manufacturing plant where cement is mixed before being transported to a construction site, ready to be poured. Equipment and materials including batchers, mixers, sand, aggregate, and cement are required for batching and mixing concrete.

Electrical collection system—Consists of underground and overhead cables that carry electricity from and within groups of wind turbines and transmit it to a collection substation and point of interconnection switchyard, which transfers the electricity generated by the project to the regional power grid.

Electromagnetic fields (EMFs)—A combination of invisible electric and magnetic fields of force; they can occur both naturally or due to human constructs.

Environmental Impact Statement (EIS) corridor—Potential area of impact resulting from the proposed construction activities.

Gearbox—A protective casing for a system of gears that converts the wind into mechanical energy.

Generator—A device for converting mechanical energy to electrical energy that is located in the nacelle.

Grid (power grid, utility grid)—A common term referring to an electricity transmission and distribution system.

Interior roads—Provide lower volume secondary road access and serve a smaller area than access roads; they connect to access roads. They are typically low-volume spur roads that provide point access and connect to the access roads. Interior roads are roads along the turbine line corridors used during construction to construct the concrete turbine foundations and to deliver turbine towers, components, and operation personnel to the turbine during operations.

Main lay-down area (staging area)—A designated secure area or space adjacent to the construction site where construction equipment and material/supplies in transit are temporarily stored, assembled, or processed as part of a construction operation. In addition, temporary construction trailers and vehicles may be parked within the boundary limits of this secure space.

Megawatt—A unit used to measure power, equal to one million watts.

Meteorological mast—One of the components of a meteorological tower, the meteorological mast supports the anemometers and data logger.

Meteorological towers—Wind measurement systems that can be of steel tube or lattice construction and can be freestanding or guyed; they are equipped with sensors to measure wind speed and direction, temperature, and pressure.

Nacelle—The cover for the gearbox, drive train, and generator of a wind turbine that converts the energy of the wind into electrical energy. The nacelle can rotate a full 360 degrees at the top of the tower to capture the prevailing wind, and it can weigh as much as 50 tons.

Operations and maintenance (O&M) facilities—Facilities used to store equipment and supplies required during operation. Some maintenance facilities include control functions such as supervisory control and data acquisition (SCADA) to provide two-way communication with each wind turbine.

Overhead transmission line—Potential route for overhead electrical lines connecting from the project substation to a determined point of interconnection at the existing system.

Power grid (utility grid)—A common term referring to an electricity transmission and distribution system.

Rotor—The blades and other rotating components of a wind energy conversion turbine.

SCADA (Supervisory Control and Data Acquisition)—System that collects data throughout the wind farm to monitor and provide control form a remote location.

Sedimentation—Deposition of sediment into water bodies and wetlands.

Shadow flicker—The effect caused by the sun's casting shadows from moving wind turbine blades.

Soil erosion—A natural process in which soil particles are detached and removed by wind or water.

Staging area (main lay-down area)—A designated secure area or space, adjacent to the construction site, where construction equipment and materials/supplies in transit are temporarily stored, assembled, or processed as part of a construction operation.

Switching station—A particular type of substation where energy, of the same voltage, is routed either from different sources or to different customers. Switching stations often contain circuit breakers, switches, and other automated mechanisms that switch or divide their output between different distribution lines when system faults occur or shut down transmission altogether in the event of a serious problem.

Transmission/interconnection facilities—A collection substation terminates collection feeder cables and steps up the voltage to that of the transmission system to which the project ultimately connects.

Turbine—A term used for a wind energy conversion device that produces electricity. See also *wind turbine*.

Utility grid—A common term referring to an electricity transmission and distribution system. See also *power grid*.

Wind energy—Power generated by converting the mechanical energy of the wind into electrical energy through the use of a wind generator. See also *wind power*.

Wind generator—A wind energy conversion system designed to produce electricity.

Wind load—The lateral pressure on a structure, in pounds per square foot, due to wind blowing in any direction.

Wind power—Power generated by converting the mechanical energy of the wind into electrical energy through the use of a wind generator. See also *wind energy*.

Wind power plant—A group of wind turbines interconnected to a common utility system.

Wind project—Can vary in size, from small projects of one to a few turbines (known as "behind the meter" or "distributed wind systems") serving individual customers to large projects ("utility" or "commercial-scale") designed to provide wholesale electricity to utilities or an electricity market.

Wind turbine—A wind energy conversion device that produces electricity; it typically consists of three major mechanical components: tower, nacelle, and rotor.

Wind turbine lay-down area—An area adjacent to the wind turbine foundation where wind turbine components are temporarily stored, assembled, or processed as part of the wind turbine assembly operation.

AIR IN MOTION

In all dynamic situations, forces are necessary to produce motion and changes in motion. The atmosphere, which is made up of various gases, is subject to two primary forces: gravity and pressure differences from temperature variations. *Gravity* (gravitational forces) holds the atmosphere close to the Earth's surface. Newton's law of universal gravitation states that every body in the universe attracts another body with a force equal to

$$F = G\left(\frac{m_1 m_2}{r^2}\right)$$

where F is the magnitude of the gravitational force between the two bodies, G is the gravitational constant $\approx 6.67 \times 10^{-11}$ N (m²/kg²), m_1 and m_2 are the masses of the two bodies, and r is the distance between the two bodies.

The force of gravity decreases as an inverse square of the distance between the two bodies. Thermal conditions affect density, which in turn affects vertical air motion and planetary air circulation (and how air pollution is naturally removed from the atmosphere). Although forces acting in other directions can overrule gravitational force, gravity constantly acts vertically downward, on every gas molecule, which accounts for the greater density of air near the Earth.

Atmospheric air is a mixture of gases, so the gas laws and other physical principles govern its behavior. Pressure is equal to the force per unit area (pressure = force/area). Because the pressure of a gas is directly proportional to its temperature, temperature variations in the air give rise to differences in pressure or force. These differences in pressure cause air movement, both large and small scale, from a high-pressure region to a low-pressure region.

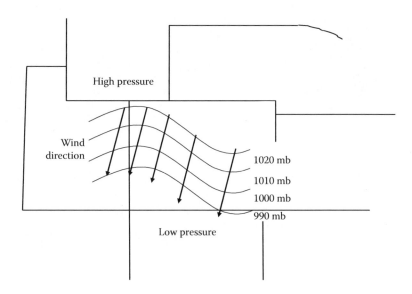

FIGURE 5.1 Isobars drawn through locations having equal atmospheric pressures. The air motion, or wind direction, is at right angles to the isobars and moves from a region of high pressure to a region of low pressure.

Horizontal air movements (*advective winds*) result from temperature gradients, which give rise to density gradients and, subsequently, pressure gradients. The force associated with these pressure variations (*pressure gradient force*) is directed perpendicular to lines of equal pressure (called *isobars*) and is directed from high to low pressure. In Figure 5.1, the pressures over a region are mapped by taking barometric readings at different locations. Lines drawn through the points (locations) of equal pressure are the isobars. All points on an isobar are of equal pressure, indicating no air movement along the isobar. The wind direction is at right angles to the isobar in the direction of the lower pressure. In Figure 5.1, notice that air moves down a pressure gradient toward a lower isobar like a ball rolls down a hill. If the isobars are close together, the pressure gradient force is large, and such areas are characterized by high wind speeds. If isobars are widely spaced, the winds are light because the pressure gradient is small.

Localized air circulation gives rise to *thermal circulation* (a result of the relationship based on a law of physics whereby the pressure and volume of a gas are directly related to its temperature). A change in temperature causes a change in the pressure and volume of a gas. With a change in volume comes a change in density (density = mass/volume), so regions of the atmosphere with different temperatures may have different air pressures and densities. As a result, localized heating sets up air motion and gives rise to *thermal circulation*. To gain an understanding of this phenomenon, consider Figure 5.2.

Once the air has been set into motion, secondary forces (velocity-dependent forces) act. These secondary forces are caused by Earth's rotation (*Coriolis force*) and contact with the rotating Earth (friction). The Coriolis force, named after its discoverer, French mathematician Gaspard Coriolis (1772–1843), is the effect of rotation on the atmosphere and on all objects on the Earth's surface. In the Northern Hemisphere, it causes moving objects and currents to be deflected to the right; in the Southern Hemisphere, it causes deflection to the left because of the Earth's rotation. Air, in large-scale north or south movements, appears to be deflected from its expected path. That is, air moving poleward in the Northern Hemisphere appears to be deflected toward the east; air moving southward appears to be deflected toward the west. The Coriolis effect on a propelled particle is analogous to the apparent affect of an air mass flowing from one point to another point.

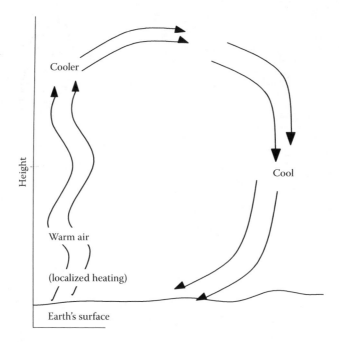

FIGURE 5.2 Thermal circulation of air. Localized heating, which causes air in the region to rise, initiates the circulation. As the warm air rise and cools, cool air near the surface moves horizontally into the region vacated by the rising air. The upper, still cooler, air then descends to occupy the region vacated by the cool air.

Friction (drag) can also cause the deflection of air movements. This friction (resistance) is both internal and external. The friction of molecules generates internal friction, and external friction is caused by contact with terrestrial surfaces. The magnitude of the frictional force along a surface is dependent on the air's magnitude and speed, and the opposing frictional force is in the opposite direction of the air motion.

WIND ENERGY[*]

Wind energy is power produced by the movement of air. Since early recorded history, people have been harnessing the energy of the wind to, for example, mill grain and pump water. Wind energy propelled boats along the Nile River as early as 5000 BC. By 200 BC, simple windmills in China were pumping water, and windmills with woven reed sails were grinding grain in Persia and the Middle East. The use of wind energy spread around the world, and by the 11th century people in the Middle East were using windmills extensively for food production; returning merchants and crusaders carried this idea back to Europe. The Dutch refined the windmill and adapted it for draining lakes and marshes in the Rhine River delta. When settlers took this technology to the New World in the later 19th century, they used windmills to pump water for farms and ranches and, later, to generate electricity for homes and industry.

The first known wind turbine designed to produce electricity was built in 1888 by Charles F. Brush, in Cleveland, Ohio; it was a 12-kW unit that charged batteries in the cellar of a mansion. The first wind turbine used to generate electricity outside of the United States was built in Denmark in 1891 by Poul la Cour, who used electricity from his wind turbines to electrolyze water to make hydrogen for the gas lights at the local schoolhouse. By the 1930s and 1940s, hundreds of thousands

[*] Much of the information in this section is from EERE, *History of Wind Energy*, Office of Energy Efficiency & Renewable Energy, U.S. Department of Energy, Washington, DC, 2005 (http://www1.eere.energy.gov/windandhydro/wind_history. html).

TABLE 5.1

U.S. Energy Consumption by Renewable Energy Source, 2014

Energy Source	Energy Consumption (quadrillion Btu)
Renewable total	9.656
Biomass (total)	4.812
Biofuels (ethanol and biodiesel)	2.067
Waste	0.516
Wood and wood-derived fuels	2.230
Geothermal	0.215
Hydroelectric power	2.475
Solar	0.421
Wind	**1.733**

Source: EIA, *Primary Energy Consumption by Source*, Table 10.1, Energy Information Administration, Washington, DC, 2016 (https://www.eia.gov/totalenergy/data/monthly/index.cfm#renewable).

of wind turbines were used in rural areas of the United States that were not yet served by the grid. The oil crisis in the 1970s created a renewed interest in wind, until the U.S. government stopped giving tax credits.

Today, several hundred thousand windmills are in operation around the world, many of which are used for pumping water. The use of wind energy as a pollution-free means of generating electricity on a significant scale is attracting the most interest in the subject today. As a matter of fact, due to current and pending shortages and high costs of fossil fuels to generate electricity, as well as the green movement toward the use of cleaner fuels, wind energy is the world's fastest-growing energy source and could power industry, businesses, and homes with clean, renewable electricity for many years to come. In the United States, wind-based electricity generating capacity has increased markedly since the 1970s. Today, though, it still represents only a small fraction of total electric capacity and consumption (see Table 5.1), despite highly variable gasoline prices, increases in the cost of electricity, high heating and cooling costs, and worldwide political unrest or uncertainty in oil-supplying countries. Traveling the wind corridors of the United States (primarily Arizona, New Mexico, Texas, Missouri, and north through the Great Plains to the Pembina Escarpment and Turtle Mountains of North Dakota) gives some indication of the considerable activity and seemingly exponential increase in wind energy development and wind turbine installations.

WIND POWER BASICS

The terms *wind energy* and *wind power* reflect the process by which the wind is used to generate mechanical power or electricity. Wind turbines convert the kinetic energy in the wind into mechanical power. This mechanical power can be used for specific tasks (such as grinding grain or pumping water) or a generator can convert this mechanical power into electricity (EERE, 2006a). We have been harnessing the energy of the wind for hundreds of years. From old Holland to farms in the

DID YOU KNOW?

Whenever wind energy is considered as a possible source of renewable energy, it is important to consider the amount of land area required, accessibility to generators, and aesthetics.

DID YOU KNOW?

We can classify wind energy as a form of solar energy. Winds are caused by uneven heating of the atmosphere by the sun, irregularities of the Earth's surface, and rotation of the Earth. As a result, winds are strongly influenced and modified by local terrain, bodies of water, weather patterns, and vegetative cover, among other factors. The energy of winds harvested by wind turbines can be used to generate electricity.

United States, windmills have been used for pumping water or grinding grain; today, the modern equivalent of a windmill—a wind turbine—can use the energy of the wind to generate electricity. The blades of a wind turbine spin like aircraft propeller blades. Wind turns the blades, which in turn, spin a shaft connected to a generator and produce electricity (Wind Energy EIS, 2009). Unlike fans, which use electricity to make wind, wind turbines use wind to make electricity.

WIND TURBINE TYPES

Whether referred to as *wind-driven generators, wind generators, wind turbines, wind-turbine generators*, or *wind energy conversion systems*, modern wind turbines fall into two basic groups: the horizontal-axis wind turbine (HAWT), like the traditional farm windmills used for water pumping (see Figure 5.3), and the vertical-axis wind turbine (VAWT), like the eggbeater-style, Darrieus rotor model, named after its French inventor, the only vertical-axis machine with any commercial success. Wind hitting the vertical blades, called *aerofoils*, generates lift to create rotation. No yaw (rotation about vertical axis) control is needed to keep them facing into the wind. The heavy machinery in the nacelle (cowling) is located on the ground. Blades are closer to the ground, where wind speeds are lower. Most large modern wind turbines are horizontal-axis turbines; therefore, they are highlighted and described in detail in this text.

HORIZONTAL-AXIS WIND TURBINES

Wind turbines are available in a variety of sizes and power ratings. Utility-scale turbines range in size from 100 kW to as large as several megawatts. Horizontal-axis wind turbines typically have two or three blades. Downwind horizontal-axis wind turbines have a turbine with the blades behind

FIGURE 5.3 Typical farm windmill used for pumping water.

(downwind from) the tower. No yaw control is needed because they naturally orient themselves in line with the wind; however, these downwind HAWTs experience a shadowing effect, in that when a blade swings behind the tower the wind it encounters is briefly reduced and the blade flexes. Upwind HAWTs usually have three blades in front (upwind) of the tower. These upwind wind turbines require a somewhat complex yaw control to keep them facing into the wind. They operate more smoothly and deliver more power and thus are the most commonly used modern wind turbines. The largest machine has blades that span more than the length of a football field, stands 20 building stories high, and produces enough electricity to power 1400 homes.

Inside the HAWT

Basically, a horizontal-axis wind turbine consists of three main parts: a turbine, a nacelle, and a tower. Several other important parts are contained within the tower and nacelle, including anemometer, blades, brake, controller, gearbox, generator, high-speed shaft, low-speed shaft, pitch, rotor, tower, wind direction, wind vane, yaw drive, and yaw motor (see Figure 5.5) (Anon., 2011).

FIGURE 5.4   Wind turbines at Tehachapi Pass Wind Farm. Wind development in the Tehachapi Pass began in the early 1980s, and it is one of the first large-scale wind farms installed in the United States.

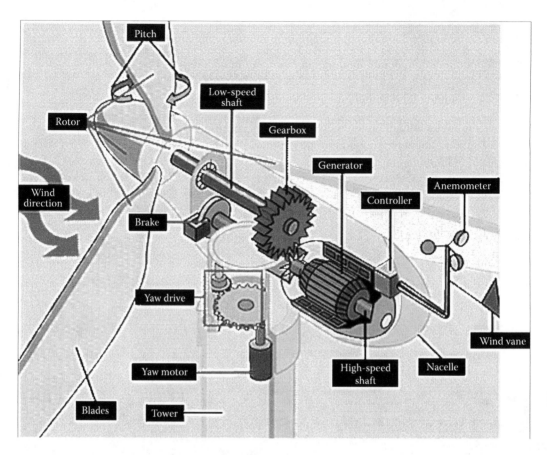

FIGURE 5.5   Horizontal-axis wind turbine components.

WIND TURBINE COMPONENTS

Anemometer—An anemometer is a device that measures the wind speed and transmits wind
   speed data to the controller.

Blades—Wind blowing over the two or three blades causes the blades to lift and rotate; they
   capture the kinetic energy of the wind and help the turbine rotate. It is interesting to note
   that most wind turbines have three blades because that design offers the greatest efficiency
   and less vibration, and designers think they are more symmetric looking. The novice to
   wind energy production might wonder why most wind turbines are limited to only three
   blades. Would it not be more efficient to fill all the spaces between the blades with more
   blades? No. The so-called wasted space between blades is necessary and has a designed
   engineering purpose. The blades affect the entire airstream that passes through the rotor,
   and the entire airstream in turn reacts back on the blades, causing them to turn.

Brake—A disc brake can be applied mechanically, electrically, or hydraulically to stop the
   rotor in emergencies.

Controller—The controller starts up the machine at wind speeds of about 8 to 16 mph and
   shuts off the machine at about 55 mph. Turbines do not operate at wind speeds above about
   55 mph because they might be damaged by the high winds.

Gearbox—Gears connect the low-speed shaft to the high-speed shaft and increase the rotational speeds from about 30 to 60 rotations per minute (rpm) to about 1000 to 1800 rpm,
   the rotational speed required by most generators to produce electricity. The gearbox is a

costly (and heavy) part of the wind turbine, and engineers are exploring direct-drive generators that operate at lower rotational speeds and do not require gearboxes.

Generator—An off-the-shelf induction generator produces 60-hertz AC electricity.

High-speed shaft—This shaft drives the generator.

Low-speed shaft—This shaft turns at about 30 to 60 rotations per minute.

Nacelle—The nacelle unit sits atop the tower and contains the gearbox, generator, low- and high-speed shafts, controller, and brake.

Pitch—Bales are turned, or pitched, out of the wind to control the rotor speed and keep the rotor turning in winds that are too high or too low to produce electricity.

Rotor—The blades and the hub together are called the rotor.

Tower—The tower is made from tubular steel, concrete, or steel lattice. Wind speed increases with height, so turbines on taller towers capture more energy and generate more electricity.

Wind direction—An upwind turbine operates facing into the wind; other turbines are designed to run downwind, facing away from the wind.

Wind vane—The wind vane measures wind direction and communicates with the yaw drive to orient the turbine properly with respect to the wind.

Yaw drive—The yaw drive of upwind turbines is used to keep the entire nacelle and thus the rotor facing into the wind as the wind direction changes. Downwind turbines do not require a yaw drive, as the wind blows the rotor downwind.

Yaw motor—The yaw motor powers the yaw drive.

WIND ENERGY AND POWER CALCULATIONS

A wind turbine is a machine that converts the kinetic energy of wind into the mechanical energy of a shaft. Calculating the energy and power available in the wind relies on a knowledge of basic physics and geometry. The kinetic energy of an object is the extra energy it possesses because of its motion. It is defined as the work necessary to accelerate a body of a given mass from rest to its current velocity. Once in motion, a body maintains it kinetic energy unless its speed changes. The kinetic energy of a body is given by

$$\text{Kinetic energy} = 0.5 \times m \times v^2 \tag{5.1}$$

where

    m = Mass.

    v = Velocity.

■ EXAMPLE 5.1. DETERMINING POWER IN THE WIND

For the purpose of determining the kinetic energy of moving air (wind), let's say we have a large packet of wind (i.e., a geometrical package of air passing through the plane of a wind turbine's blades, which sweep out at cross-sectional area A and have thickness D), passing through the plane over a given time. (See Figure 5.6.)

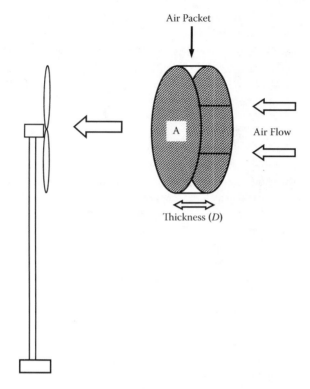

FIGURE 5.6 A packet of air passing through the plane of a wind turbine's blades, with thickness D, passing through the plane over a given time.

Step 1. To determine the power in the wind, we must first consider the kinetic energy of this packet of air, along with its mass (m) and velocity (v), as shown in Equation 5.1. We then divide by time to get power:

$$\text{Power through area } A = \frac{1}{2}\left(\frac{m \text{ passing through } A}{t}\right)v^2 \tag{5.2}$$

Step 2. The mass flow rate can be expressed as follows (where ρ is air density, the mass per unit volume of Earth's atmosphere):

$$m = \left(\frac{m \text{ passing through } A}{t}\right)\rho A v \tag{5.3}$$

Step 3. Combining Equations 5.2 and 5.3, we obtain

$$\text{Power through area } A = 1/2(\rho A v)v^2$$
$$\text{Power in the wind} = P_W = 1/2(\rho A v^3) \tag{5.4}$$

where
   P_W = Power in the wind (watts).
   ρ = Air density (2.225 kg/m^3 at 15°C and 1 atm).
   A = Cross-sectional area that wind passes through perpendicular to the wind (m^2).
   v = Wind speed normal to A (m/s; 1 m/s = 2.237 mph).

From Equation 5.4, it is evident that wind power is a function of the cube of the wind speed; that is, doubling the wind speed increases the power by eight. As an example, the energy produced during 1 hour of 20-mph winds is the same as the energy produced during 8 hours of 10-mph winds. When we speak of power (W/m^2) and wind speed (mph), we cannot use average wind speed because the relationship is nonlinear. Power in the wind is also proportional to A. For a conventional horizontal-axis wind turbine, $A = (\pi/4)D^2$, so wind power is proportional to the blade diameter squared. Because cost is approximately proportional to blade diameter, larger wind turbines are more cost effective.

AIR-DENSITY CORRECTION FACTORS

Earlier we pointed out that air density is affected by different temperature and pressure. Air-density correction factors can correct air density for temperature and altitude. Correction factors for both temperature and altitude correction can be found in standardized tables. Equation 5.5 is used to determine air density (ρ) for various temperatures and pressures:

$$\rho = \frac{P \times MW \times 10^{-3}}{RT} \tag{5.5}$$

where
  P = Absolute gas pressure (atm).
  MW = Molecular weight of air (= 28.97 g/mol).
  R = Ideal gas constant (= 8.2056×10^{-5} $m^3 \cdot atm \cdot K^{-1} \cdot mol^{-1}$).
  T = Absolute temperature (degrees Kelvin = 273 + °C).

ELEVATION AND EARTH'S ROUGHNESS

The speed of wind is affected by its elevation above the Earth and the roughness of Earth. Because power increases like the cube of wind speed, we can expect a significant economic impact from even a moderate increase in wind speed; thus, in the operation of wind turbines, wind speed is a very important parameter. The surface features of Earth cannot be ignored when deciding where to place wind turbines and calculating their output productivity. Natural obstructions such as mountains and forests and human-made obstructions such as buildings cause friction as winds flow over them. Generally, a lot of friction is encountered in the first few hundred meters above ground (although smooth surfaces such as water offer little resistance). For this reason, taller wind turbine towers are better.

When actual measurements are not possible or available, it is possible to characterize or approximate the impact of rough surfaces and height on wind speed:

$$\frac{v}{v_0} = \left(\frac{H}{H_0}\right)^{\alpha} \tag{5.6}$$

> **DID YOU KNOW?**
>
> The energy in wind is a cubed function of wind speed, which means that if the wind speed doubles there is eight times as much available energy, not twice as much, as one might expect.

where

α = Friction coefficient, obtained from standardized tables (typical values of α are 0.10 for calm water or smooth, hard ground; 0.15 for open terrain; and 0.40 for a large city with tall buildings).

v = Wind speed at height H.

v_0 = Wind speed at height H_0 (H_0 is usually 10 m).

WIND TURBINE ROTOR EFFICIENCY

Generally, when we think or talk about efficiency we think about input vs. output and know that if we put 100% into something and get 100% output then we have a very efficient machine, operation, or process. In engineering, we can approximate efficiency (input vs. output) by performing mass balance calculations. We know that, according to the laws of conservation, mass cannot disappear or be destroyed; thus, it must continue to exist in one form or another. With regard to this input vs. output concept, we should address a couple views on maximum rotor efficiency that are misinformed and make no sense but are stated here to point out what really does make sense. The two wrong assumptions are that (1) it can be assumed that the downwind velocity is zero and the turbine extracts 100% of the power from the wind, or (2) downwind velocity is the same as the upwind velocity and the turbine has not reduced the wind speed at all. Wrong! Albert Betz set the record straight in 1919 when he proposed that there exists a maximum theoretical efficiency for extracting kinetic energy from the wind.

Derivation of Betz's Law

To understand Betz's law, we must first understand the constraints on the ability of a wind turbine to convert kinetic energy in the wind into mechanical power. Visualize wind passing through a turbine (see Figure 5.7)—it slows down and the pressure is reduced so it expands. Equation 5.7 is used to determine the power extracted by the blades:

$$P_b = 1/2\, m\left(v^2 - v_d^2\right) \tag{5.7}$$

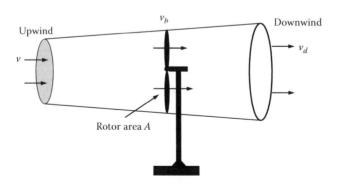

FIGURE 5.7   Wind passing through a turbine.

where

m = Mass flow rate of air within stream tube.
v = Upwind undisturbed wind speed.
v_d = Downwind wind speed.

Determining the mass flow rate is the next step. In making the determination, it is easiest to use the cross-sectional area A at the plane of the rotor because we know what this value is; thus, the mass flow rate is

$$m = p \times A \times v_b \tag{5.8}$$

Assume that the velocity through the rotor (v_b) is the average of the upwind velocity (v) and the downwind velocity (v_d):

$$v_b = \left(\frac{v + v_d}{2}\right) \rightarrow m = \rho A \left(\frac{v + v_d}{2}\right)$$

Equation 5.7 becomes

$$P_b = 1/2 \rho A \left(\frac{V + v_d}{2}\right)\left(v^2 - v_d^2\right) \tag{5.9}$$

Before moving on in the derivation process, it is important that we define operating speed ratio (λ):

$$\lambda = \frac{v_d}{v} \tag{5.10}$$

We can now rewrite Equation 5.9 as

$$P_b = 1/2 \rho A \left(\frac{v + \lambda v}{2}\right)\left(v^2 - \lambda^2 v^2\right) \tag{5.11}$$

$$\downarrow$$

$$\left(\frac{v + \lambda v}{2}\right)\left(v^2 - \lambda^2 v^2\right) = \frac{v^3}{2} - \frac{\lambda^2 v^3}{2} + \frac{\lambda v^3}{2} - \frac{\lambda^3 v^3}{2} = \frac{v^3}{2}\left[(1+\lambda) - \lambda^2(1+\lambda)\right] = \frac{v^3}{2}\left[(1+\lambda)(1-\lambda^2)\right]$$

$$P_b = \left(1/2 \rho A v^3\right) \qquad \times \qquad 1/2\left[(1+\lambda)(1-\lambda^2)\right]$$

$$\downarrow \qquad\qquad\qquad\qquad \downarrow$$

$$\left(P_W = \text{Power in the wind}\right) \qquad \left(C_P = \text{Rotor efficiency}\right)$$

The next step is to find the operating speed ratio (λ) that maximizes the rotor efficiency (C_P):

$$C_P = 1/2\left[(1+\lambda)(1-\lambda^2)\right] = \frac{1}{2} - \frac{\lambda^2}{2} + \frac{\lambda}{2} - \frac{\lambda^3}{2}$$

Set the derivative of rotor efficiency to zero and solve for λ:

$$\frac{\partial C_P}{\partial \lambda} = -2\lambda + 1 - 3\lambda^2 = 0$$

$$\frac{\partial C_P}{\partial \lambda} = 3\lambda^2 + 2\lambda - 1 = 0$$

$$\frac{\partial C_P}{\partial \lambda} = (3\lambda - 1)(\lambda + 1) = 0$$

$$\vdots$$

$$\lambda = 1/3$$

We plug the optimal value for λ back into the equation for C_P to find the maximum rotor efficiency:

$$C_P = 1/2\left[\left(1 + \frac{1}{3}\right)\left(1 - \frac{1}{3}\right)\right] = \frac{16}{27} = 59.3\% \qquad (5.12)$$

The maximum efficiency of 59.3% occurs when air is slowed to 1/3 of its upstream rate. This value is called the *Betz efficiency* (Betz, 1966). In plain English, Betz's law states that all wind power cannot be captured by the rotor; otherwise, air would be completely still behind the rotor and not allow more wind to pass through. For illustrative purposes, in Table 5.2, we list wind speed, power of the wind, and power of the wind based on the Betz limit (59.3%).

TABLE 5.2
Betz Limit for 80-M Rotor Turbines

Wind Speed (mph)	Wind Speed (m/s)	Wind Power (kW)	Power Based on 59.3% Betz Limit (kW)
5	2.2	36	21
10	4.5	285	169
15	6.7	962	570
20	8.9	2280	1352
25	11.2	4453	2641
28	12.5	6257	3710
30	13.4	7695	4563
35	15.6	12,220	7246
40	17.9	18,241	10,817
45	20.1	25,972	15,401
50	22.4	35,626	21,126
55	24.6	47,419	28,119
56 (cutoff speed)	25.0	50,053	29,681
60	26.8	61,563	36,507

Source: Adapted from Devlin, L., *Wind Turbine Efficiency*, 2007 (http://k0lee.com/turbineeff.htm).

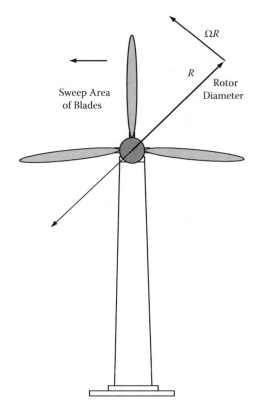

FIGURE 5.8 Tip speed ratio (TSR).

TIP SPEED RATIO

Efficiency is a function of how fast the rotor turns. The tip speed ratio (TSR), an extremely important factor in wind turbine design, is the ratio of the speed of the rotating blade tip to the speed of the free stream wind (see Figure 5.8). Stated differently, TSR is the speed of the outer tip of the blade divided by wind speed. There is an optimum angle of attack that creates the highest lift-to-drag ratio. If the rotor of the wind turbine spins too slowly, most of the wind will pass straight through the gap between the blades, thus giving it no power. But, if the rotor spins too fast, the blades will blur and act like a solid wall to the wind. Moreover, rotor blades create turbulence as they spin through the air. If the next blade arrives too quickly, it will hit that turbulent air. Thus, it is actually better to slow down the blades. Because the angle of attack is dependent on wind speed, there is an optimum tip speed ratio:

$$\text{TSR} = \Omega R / V \qquad (5.13)$$

where
   Ω = Rotational speed (rad/s).
   R = Rotor radius.
   V = Wind "free stream" velocity.

SMALL-SCALE WIND POWER

To meet the typical domestic user's need for electricity, a small-scale wind machine may be the answer. Such a wind turbine has rotors between 8 and 25 feet in diameter, stands around 30 feet tall, and can supply the power needs of an all-electric home or small business. Utility-scale turbines

DID YOU KNOW?

The difference between power and energy is that power (kilowatts) is the rate at which electricity is consumed and energy (kilowatt-hours) is the quantity consumed.

range in size from 50 to 750 kW. Single small turbines below 50 kW in size are used for homes, telecommunication dishes, or water pumping. Potential users of small wind turbines should ask many questions before installing such a system (EERE, 2015). Let's ask the obvious question first: "What are the benefits to homeowners from using wind turbines?" Wind energy systems provide a cushion against electricity price increases. Wind energy systems reduce U.S. dependence on fossil fuels, and they don't emit greenhouse gases. If you are building a home in a remote location, a small wind energy system can help you avoid the high cost of extending utility power lines to your site.

Although wind energy systems involve a significant initial investment, they can be competitive with conventional energy sources when you account for a lifetime of reduced or altogether avoided utility costs. The length of payback time—the time before the savings resulting from your system equal the system cost—depends on the system you choose, the wind resources at your site, the electric utility rates in your area, and how you use your wind system.

Another frequently asked question is "Is wind power practical for me?" Small wind energy systems can be used in connection with an electricity transmission and distribution system (*grid-connected systems*) or in stand-alone applications that are not connected to the utility grid. A grid-connected wind turbine can reduce the consumption of utility-supplied electricity for lighting, appliances, and electric heat. If the turbine cannot deliver the amount of energy required, the utility makes up the difference. When the wind system produces more electricity than the household requires, the excess can be sold to the utility. With the interconnections available today, switching takes place automatically. Stand-alone wind energy systems can be appropriate for homes, farms, or even entire communities that are far from the nearest utility lines.

Stand-alone systems (systems not connected to the utility gird) require batteries to store excess power generated for use when the wind is calm. They also must have a charge controller to keep the batteries from overcharging. Deep-cycle batteries, such as those used for golf carts, can discharge and recharge 80% of their capacity hundreds of times, which makes them a good option for remote renewable energy systems. Automotive batteries are shallow-cycle batteries and should not be used in renewable energy systems because of their short life for deep-cycling operations.

Small wind turbines generate direct current (DC) electricity. In very small systems, DC appliances operate directly off the batteries. To use standard appliances that use conventional household alternating current (AC), an inverter must be installed to convert DC electricity from the batteries to AC (see Figure 5.9). Although the inverter slightly lowers the overall efficiency of the system, it allows the home to be wired for AC, a definite plus with lenders, electrical code officials, and future homebuyers. In grid-connected (or interactive) systems, the only additional equipment required is a power condition unit (inverter) that makes the turbine electrical output compatible with the utility grid. Usually, batteries are not needed. Either type of system can be practical if the following conditions exist:

SAFETY NOTE

For safety, batteries should be isolated from living areas and electronics because they contain corrosive and explosive substances. Lead–acid batteries also require protection from temperature extremes.

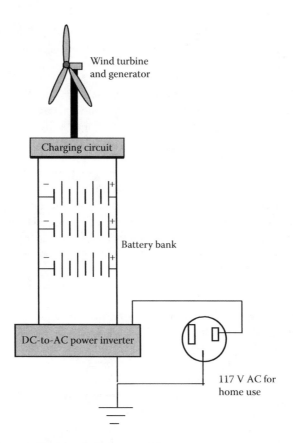

FIGURE 5.9   Stand-alone small-scale wind-power system.

- *Conditions for stand-alone systems*
  You live in an area with average annual wind speeds of at least 4.0 mps (9 mph).
  A grid connection is not available or can only be made through an expensive extension. The cost of running a power line to a remote site to connect with the utility grid can be prohibitive, ranging from $15,000 to more than $50,000 per mile, depending on the terrain.
  You have an interest in gaining energy independence from the utility.
  You would like to reduce the environmental impact of electricity production.
  You acknowledge the intermittent nature of wind power and have a strategy for using intermittent resources to meet your power needs.

- *Conditions for grid-connected systems*
  You live in an area with average annual wind speeds of at least 4.5 mps (10 mph).
  Utility-supplied electricity is expensive in your area (about 10 to 15¢ per kilowatt-hour).
  The utility's requirements for connecting the system to its grid are not prohibitively expensive.
  Local building codes or covenants allow you to erect a wind turbine on your property.
  You are comfortable with long-term investments.

After comparing stand-alone systems and grid-connected systems and determining which is best-suited for your particular circumstance, the next question to consider is whether your location is the appropriate site for installing a small-scale wind turbine system. Is it legal to install the system on your property? Are there any environmental or economic issues? Does the wind blow frequently and hard enough to make a small wind turbine system economically worthwhile? This is a key question that is not always easily answered. The wind resource can vary significantly over an area of just a few miles because of local terrain influences on the wind flow; yet, there are steps you can take that will go a long way toward answering the above question.

Wind resource maps like the ones included in the U.S. Department of Energy's *Wind Energy Resource Atlas of the United States* (Elliott et al., 1987) can be used to estimate the wind resources in your region. The highest average wind speeds in the United States are generally found along seacoasts, on ridge lines, and on the Great Plains; however, many other areas have wind resources strong enough to power a small wind turbine economically. The wind resource estimates provided by the *Wind Energy Resource Atlas* generally apply to terrain features that are well exposed to the wind, such as plains, hilltops, and ridge crests. Local terrain features may cause the wind resources at specific sites to differ considerably from these estimates.

Average wind speed information can be obtained from a nearby airport; however, caution should be used because local terrain and other factors may cause the wind speed to differ from that recorded at an airport. Airport wind data are generally measured at heights about 20 to 33 ft (6 to 10 m) above ground. Also, average wind speeds increase with height and may be 15 to 25% greater at a typical wind turbine hub height of 80 ft (24 m) than those measured at airport anemometer heights. The *Wind Energy Resource Atlas* contains data from airports in the United States and makes wind data summaries available.

Again, it is important to have site-specific data to determine the wind resource at your exact location. If wind speed data for a particular site are not available, it may be necessary to measure wind speeds at that location for a year. A recording anemometer generally costs $500 to $1500. The most accurate readings are taken at hub height (i.e., the elevation at the top of the wind turbine tower). This requires placing the anemometer high enough to avoid turbulence created by trees, buildings, and other obstructions. The standard wind sensor height used to obtain data for the Department of Energy maps is 10 meters (33 feet).

Within the same property it is not unusual to have varied wind resources. Those living in complex terrains should take care when selecting an installation site. A wind turbine installed on the top of a hill or on the windy side of a hill, for example, will have greater access to prevailing winds than if it were located in a gully or on the leeward (sheltered) side of a hill on the same property. Consider existing obstacles and plan for future obstructions, including trees and buildings, that could block the wind. Also recall that the power in the wind is proportional to its speed cubed. This means that the amount of power provided by a generator increases exponentially as the wind speed increases; for example, a site that has an annual average wind speed of about 5.6 mps (12.6 mph) has twice the energy available as a site with an average of 4.5 mps (10 mph) ($12.6/10^3$). Another useful indirect measurement of the wind resource is observation of an area's vegetation. Trees, especially conifers or evergreens, can be permanently deformed by strong winds. This deformity, known as *flagging*, has been used to estimate the average wind speed for an area (see Figure 5.10).

In addition to ensuring the proper siting of a small wind turbine system, other legal, environmental, and economic issues must be addressed:

- Research potential legal and environmental obstacles.
- Obtain cost and performance information from manufacturers.
- Perform a complete economic analysis that accounts for a multitude of factors.
- Understand the basics of a small wind system.
- Review possibilities for combining the system with other energy sources, backups, and energy efficiency improvements.

With regard to economic issues, because energy efficiency is usually less expensive than energy production, making your house more energy efficient first will likely result in spending less money on a wind turbine, because a smaller one may meet your needs. *A word of caution:* Before investing in a wind turbine, research potential legal and environmental obstacles to installing one. Some

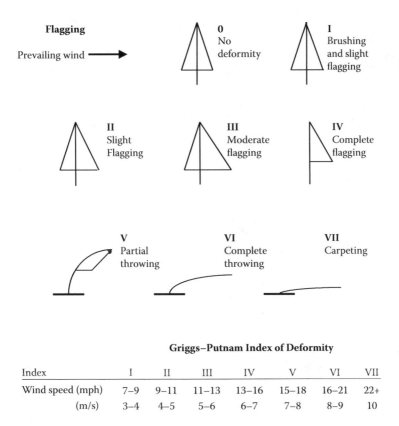

Griggs–Putnam Index of Deformity

Index	I	II	III	IV	V	VI	VII
Wind speed (mph)	7–9	9–11	11–13	13–16	15–18	16–21	22+
(m/s)	3–4	4–5	5–6	6–7	7–8	8–9	10

FIGURE 5.10 A crude method of approximating average annual wind speed from the deformation of trees (and other foliage).

jurisdictions, for example, restrict the height of the structures permitted in residentially zoned areas, although variances are often obtainable. Neighbors might object to a wind machine that blocks their view, or they might be concerned about noise. Consider obstacles that might block the wind in the future (large planned developments or saplings, for example). Saplings that will grow into large trees can be a problem in the future. As mentioned, trees can affect wind speed (see Figure 5.10). If you plan to connect the wind generator to your local utility company's grid, find out its requirements for interconnections and buying electricity from small independent power producers. When you are convinced that a small wind turbine is what you want and there are no obstructions restricting its installation, approach buying a wind system as you would any major purchase.

ENVIRONMENTAL IMPACTS OF WIND ENERGY

RECENT HEADLINES ON WIND TURBINES

> "Wind Turbine Health Effects: Are Wind Turbines Really Unhealthy for Humans?" (Casey, 2012a)
> "How Noisy Is a Wind Turbine?" (Casey, 2012b)
> "U.S. to Allow Eagle Deaths—To Aid Wind Power" (Cappiello, 2013a)
> "Eagle Deaths at Wind Farms Glossed Over" (Cappiello, 2013b)

The potential impacts of wind energy during site development and evaluation, site construction, and site operations and maintenance are discussed in the following sections.

WIND ENERGY SITE EVALUATION IMPACTS

Site evaluation phase activities, such as monitoring and testing, are temporary and are conducted at a smaller scale than those at the construction and operation phases. Potential impacts of these activities are presented below by type of affected resource. The impacts described are for typical site evaluation and exploration activities, such as ground clearing (removal of vegetative cover), vehicular and pedestrian traffic, borings for geotechnical surveys and guy wire installation, and positioning of equipment, such as meteorological towers. If excavation of road construction is necessary during this phase, potential impacts would be similar in character to those for the construction phase, but generally of smaller magnitude.

Air Quality

Impacts on air quality during monitoring and testing activities would be limited to temporary and local generation of vehicle emissions and fugitive dust. These impacts are unlikely to cause an exceedance of air quality standards or to impact climate change.

Cultural Resources

Surface disturbance is minimal during the site evaluation phase, and cultural resources buried below the surface are unlikely to be affected. Cultural material present on the surface could be disturbed by vehicular traffic, ground clearing, and pedestrian activity (including collection of artifacts). Monitoring and testing activities could affect areas of interest to Native Americans depending on the physical placement or level of visual intrusion. Surveys conducted during this phase to evaluate the presence or significance of cultural resources in the area would assist developers in designing the project to avoid or minimize impacts on these resources.

Ecological Resources

Impacts on vegetation, wildlife habitat, and aquatic habitat would be minimal during site monitoring and testing because of the limited nature of activities. The introduction and spread of invasive vegetation could occur as a result of vehicular traffic. Surveys conducted during this phase to evaluate the presence and/or significance of ecological resources in the area would assist developers in designing the project to avoid or minimize impacts on these resources.

Water Resources

There likely would have minimal impact on water resources, local water quality, water flows, and surface water/groundwater interactions. Very little water would likely be used during the site evaluation phase. Any water required could be trucked in from offsite.

Land Use

Monitoring and testing activities would likely result in temporary and localized impacts on land use. These activities could create a temporary disturbance to wildlife and cattle in the immediate vicinity of the monitoring/testing site while workers are present; however, monitoring equipment is unlikely to change land-use patterns over a longer period of time. Although a buffer area may be established around equipment to protect the public, wildlife, and the equipment, access to the area for continued recreational use would not be affected. There could be visual impacts, though, from the presence of equipment and access roads, potentially impacting the recreational experience. Monitoring and testing activities are unlikely to affect mining activities, military operations, or aviation.

Soils and Geologic Resources

Surface disturbance and use of geologic materials are minimal during the site evaluation phase, and soils and geologic resources are unlikely to be affected. These activities would also be unlikely to activate geological hazards or increase soil erosion. Borings for soil testing and geotechnical surveys provide useful site-specific data on these resources. Surface effects from pedestrian and vehicular traffic could occur in areas that contain special (e.g., cryptobiotic) soils.

Paleontological Resources

Surface disturbance is minimal during the site evaluation phase, and paleontological resources buried below the surface are unlikely to be affected. Fossil material present on the surface could be disturbed by vehicular traffic, ground clearing, and pedestrian activity (including collection of fossils). Surveys conducted during this phase to evaluate the presence and/or significance of paleontological resources in the area would assist developers in designing the project to avoid or minimize impacts on these resources.

Transportation

No impacts on transportation are anticipated during the site evaluation phase. Transportation activities would be temporary, intermittent, and limited to low volumes of heavy- and medium-duty pickup trucks and personal vehicles.

Visual Resources

Monitoring and testing activities would have temporary and minor visual effects caused by the presence of equipment.

Socioeconomics

Site evaluation and exploration activities are temporary and limited and would not result in socio-economic impacts on employment, local services, or property values.

Environmental Justice

Site evaluation and exploration activities are limited and would not result in significant high and adverse impacts in any resource area; therefore, environmental justice impacts are not expected during this phase.

Hazardous Materials and Waste Management

Impacts from use, storage, and disposal of hazardous materials and waste would be minimal to nonexistent if appropriate management practices are followed.

Acoustics (Noise)

Activities associated with site monitoring and testing would generate low levels of temporary and intermittent noise; however, it is important to point out that wind turbine noise due to construction, operation, and maintenance activities is a major environmental issue and complaint. Because noise associated with wind turbines is a major environmental impact issue, an in-depth discussion about noise is provided below. Keep in mind that the noise generated and discussed here refers to that generated during all phases of wind turbine operation. The following text goes into more detail on the subject of noise.

DID YOU KNOW?

In 1983, the Occupational Safety and Health Administration (OSHA) adopted a Hearing Conservation Amendment to 29 CFR 1910.95 requiring employers to implement *hearing conservation programs* in any work setting where employees are exposed to an 8-hour, time-weighted average of 85 dB and above (LaBar, 1989). Employers must monitor all employees whose noise exposure is equivalent to or greater than a noise exposure received in 8 hours where the noise level is constantly 85 dB. The exposure measurement must include all continuous, intermittent, and impulsive noise within an 80-dB to 130-dB range and must be taken during a typical work situation. This requirement is performance oriented because it allows employers to choose the monitoring method that best suits each individual situation (OSHA, 2002).

NOISE

Noise is commonly defined as any unwanted sound. Based on the author's observation of the operation of wind farms in Washington, Oregon, California, Indiana, West Virginia, North Dakota, South Dakota, New Mexico, and Wyoming, turbine-generated noise can be characterized as ranging from the swooshing sound of rotating rotor blades to a deep, bass-like hum produced by a single operating wind turbine. Using a calibrated sound pressure level (SPL) decibel (dB) measuring device, the author determined that the wind turbine-generated noise monitored and measured varied depending on the size of the turbines, their location, and distance away from the turbine or wind farm. To better comprehend the material presented in this section, it is important to have a fundamental knowledge of noise, sound, and the properties involved. Thus, in the following section a basic introduction of noise and sound properties is presented.

High Noise Levels

High noise levels are a hazard to anyone within hearing distance. High noise levels are a physical stress that may produce psychological effects by annoying, startling, or disrupting concentration, which can lead to accidents. High levels can also result in damage to hearing, resulting in hearing loss. When wind farms are taken into consideration, the noise produced has the potential to escalate into a severe nuisance for small, local populations. This impact alone can have detrimental effects on a wide range of related aspects including health and property values.

Noise and Hearing Loss Terminology

Many specialized terms are used to express concepts in noise, noise control, and hearing loss prevention. The individual or group concerned with wind turbine noise pollution and its potential impacts should be familiar with these terms. The National Institute for Occupational Safety and Health (NIOSH, 2005) definitions below were written in as nontechnical a fashion as possible.

Acoustic trauma—A single incident that produces an abrupt hearing loss. Welding sparks (to the eardrum), blows to the head, and blast noise are examples of events capable of providing acoustic trauma.

Action level—The sound level that, when reached or exceeded, necessitates implementation of activities to reduce the risk of noise-induce hearing loss. OSHA currently uses an 8-hour, time-weighted average of 85 dBA as the criterion for implementing an effective hearing conservation program.

Aerodynamic noise—Generated noise; noise of aerodynamic origin in a moving fluid arising from flow instabilities.

Attenuate—To reduce the amplitude of sound pressure (noise).

Attenuation, real ear attenuation at threshold (REAT)—A standardized procedure for conducting psychoacoustic tests on human subjects, it is designed to measure the sound protection features of hearing protective devices. Typically, these measures are obtained in a calibrated sound field and represent the difference between subjects' hearing thresholds when wearing a hearing protector vs. not wearing the protector.

Attenuation, real-world—Estimated sound protection provided by hearing protective devices as worn in "real-world" environments.

Audible range—The frequency range over which individuals with normal hearing acuity hear (approximately 20 to 20,000 Hz).

Audiogram—A chart, graph, or table resulting from an audiometric test showing an individual's hearing threshold levels as a function of frequency.

Audiologist—A professional specializing in the study and rehabilitation of hearing and who is certified by the American Speech–Language–Hearing Association or licensed by a state board of examiners.

Background noise—Noise coming from sources other than the particular noise sources being monitored.

Baseline audiogram—A valid audiogram against which subsequent audiograms are compared to determine if hearing thresholds have changed. The baseline audiogram is preceded by a quiet period so as to obtain the best estimate of the person's hearing at that time.

Continuous noise—Noise of a constant level as measured over at least one second using the "slow" setting on a sound level meter. Note that a noise that is intermittent (e.g., on for over a second and then off for a period) would be both variable and continuous.

Controls, administrative—Efforts, usually by management, to limit workers' noise exposure by modifying workers' schedules or location or by modifying the operating schedule of noisy machinery.

Controls, engineering—Any use of engineering methods to reduce or control the sound level of a noise source by modifying or replacing equipment, making any physical changes at the noise source or along the transmission path (with the exception of hearing protectors).

dB (decibel)—The unit used to express the intensity of sound. The decibel was named after Alexander Graham Bell. The decibel scale is a logarithmic scale in which 0 dB approximates the threshold of hearing in the mid-frequencies for young adults and in which the threshold of discomfort is between 85 and 95 dB SPL and the threshold for pain is between 120 and 140 dB SPL.

Dosimeter—When applied to noise, refers to an instrument that measures sound levels over a specified interval; stores the measures; calculates the sound as a function of sound level and sound duration; and describes the results in terms of, dose, time-weighted average, and (perhaps) other parameters such as peak level, equivalent sound level, or sound exposure level.

Double hearing protection—A combination of both ear-plug and ear-muff types of hearing protection devices is required for employees who have demonstrated temporary threshold shift during audiometric examination and for those who have been advised to wear double protection by a medical doctor in work areas that exceed 104 dBA.

Equal-energy rule—The relationship between sound level and sound duration based on a 3-dB exchange rate; that is, the sound energy resulting from doubling or halving a noise exposure's duration is equivalent to increasing or decreasing the sound level by 3 dB, respectively.

Exchange rate—The relationship between intensity and dose. OSHA uses a 5-dB exchange rate; thus, if the intensity of an exposure increases by 5 dB, the dose doubles. This is also referred to as the *doubling rate*. The U.S. Navy uses a 4-dB exchange rate, the U.S. Army and Air Force use a 3-dB exchange rate, and NIOSH recommends a 3-dB exchange rate. Note that the equal-energy rule is based on a 3-dB exchange rate.

Frequency—Rate at which pressure oscillations are produced; measured in hertz (Hz).

Hazardous noise—Any sound for which any combination of frequency, intensity, or duration is capable of causing permanent hearing loss in a specified population.

Hearing handicap—A specified amount of permanent hearing loss usually averaged across several frequencies that negatively impacts employment and/or social activities. Handicap is often related to an impaired ability to communicate. The degree of handicap will also be related to whether the hearing loss is in one or both ears, and whether the better ear has normal or impartial hearing.

Hearing loss—Often characterized by the area of the auditory system responsible for the loss; for example, when injury or a medical condition affects the outer ear or middle ear (i.e., from the pinna, ear canal, and ear drum to the cavity behind the ear drum, including the ossicles) the resulting hearing loss is referred to as a *conductive* loss. When an injury or medical condition affects the inner ear or the auditory nerve that connects the inner ear to the brain (i.e., the cochlea and the VIIIth cranial nerve) the resulting hearing loss is referred to as a *sensorineural* loss. Thus, when a welder's spark damages the ear drum that would be considered conductive hearing loss. Because noise can damage the tiny hair cells located in the cochlea, it causes a sensorineural hearing loss.

Hearing threshold level (HTL)—The hearing level, above a reference value, at which a specified sound or tone is heard by an ear in a specified fraction of the trials. Hearing threshold levels have been established so that 0 dB HTL reflects the best hearing of a group of persons.

Hertz (Hz)—The unit of measurement for audio frequencies. The frequency range for human hearing lies between 20 Hz and approximately 20,000 Hz. The sensitivity of the human ear drops off sharply below about 500 Hz and above 4000 Hz.

Impulsive noise—A term used to generally characterize impact or impulse noise, which is typified by a sound that rapidly rises to a sharp peak and then quickly fades. The sound may or may not have a "ringing" quality (such as a striking a hammer on a metal plate or a gunshot in a reverberant room). Impulsive noise may be repetitive, or it may be a single event (as with a sonic boom). *Note:* If impulses occurring in very rapid succession (such as with some jack hammers), the noise would not be described as impulsive.

Loudness—The subjective attribute of a sound by which it would be characterized along a continuum from "soft" to "loud." Although this is a subjective attribute, it depends primarily upon sound pressure level and, to a lesser extent, the frequency characteristics and duration of the sound.

Material hearing impairment—As defined by OSHA, an average hearing threshold level of 25 dB HTL at the frequencies of 1000, 2000, and 3000 Hz.

Mechanical noise—Noise generated as a result of moving parts or components such as gears, shafts, or reciprocating parts.

Medical pathology—A disorder or disease; with regard to noise, a condition or disease affecting the ear, which a physician specialist should treat.

Noise—Any unwanted sound.

Noise dose—The noise exposure expressed as a percentage of the allowable daily exposure. For OSHA, a 100% dose would equal an 8-hour exposure to a continuous 90-dBA noise. A 50% dose would equal an 8-hour exposure to an 85-dBA noise or a 4-hour exposure to a 90-dBA noise. If 85 dBA is the maximum permissible level, then an 8-hour exposure to a continuous 85-dBA noise would equal a 100% dose. If a 3-dB exchange rate is used in conjunction with an 85-dBA maximum permissible level, a 50% dose would equal a 2-hour exposure to 88 dBA or an 8-hour exposure to 82 dBA.

Noise dosimeter—An instrument that integrates a function of sound pressure over a period of time to directly indicate a noise dose.

Noise hazard area—Any area where noise levels are equal to or exceed 85 dBA. OSHA requires employers to designate work areas as a "noise hazard area," post warning signs, and warn employees when work practices exceed 90 dBA. Hearing protection must be worn whenever 90 dBA is reached or exceeded.

Noise hazard work practice—Performing or observing work where 90 dBA is equaled or exceeded. Some work practices, however, will be specified as a "rule of thumb." Whenever attempting to hold a normal conversation with someone who is 1 foot away and shouting must be employed to be heard, one can assume that a 90-dBA noise level or greater exists and hearing protection is required. Typical examples of work practices where hearing protection is required are jackhammering, heavy grinding, heavy equipment operations, and similar activities.

Noise-induced hearing loss—A sensorineural hearing loss that is attributed to noise and for which no other etiology can be determined.

Noise-level measurement—Total sound level within an area; includes workplace measurements indicating the combined sound levels of tool noise (from ventilation systems, cooling compressors, circulation pumps, etc.).

Noise reduction rating (NRR)—The NRR is a single-number rating method that attempts to describe a hearing protector based on how much the overall noise level is reduced by the hearing protector. When estimating A-weighted noise exposures, it is important to remember to first subtract 7 dB from the NRR and then subtract the remainder from the A-weighted noise level. The NRR theoretically provides an estimate of the protection that should be met or exceeded by 98% of the wearers of a given device. In practice, this does not prove to be the case, so a variety of methods for "de-rating" the NRR have been discussed.

Ototoxic—A term typically associated with the sensorineural hearing loss resulting from therapeutic administration of certain prescription drugs.

Ototraumatic—A broader term than ototoxic. As used in hearing loss prevention, it refers to any agent (e.g., noise, drug, industrial chemical) that has the potential to cause permanent hearing loss subsequent to acute or prolonged exposure.

Presbycusis—The gradual increase in hearing loss that is attributable to the effects of aging and not related to medical causes or noise exposure.

Sensorineural hearing loss—Hearing loss resulting from damage to the inner ear (from any source).

Sociacusis—Hearing loss related to non-occupational noise exposure.

Sound intensity (I)—At a specific location, it is the average rate at which sound energy is transmitted through a unit area normal to the direction of sound propagation.

Sound level meter (SLM)—Device that measures sound and provides a readout of the resulting measurement. Some SLMs provide only A-weighted measurements, others provide A- and C-weighted measurements, and some provide weighted, linear, and octave (or narrower) band measurements. Some SLMs are also capable of providing time-integrated measurements.

Sound power—The total sound energy radiated by a source per unit time. Sound power cannot be measured directly.

Sound pressure level (SPL)—A measure of the ratio of the pressure of a sound wave relative to a reference sound pressure. Sound pressure level in decibels is typically referenced to 20 mPa. When used alone (e.g., 90 dB SPL), a given decibel level implies an unweighted sound pressure level.

Standard threshold shift (STS)—(1) OSHA uses the term to describe a change in hearing threshold relative to the baseline audiogram of an average of 10 dB or more at 2000, 3000, and 4000 Hz in either ear. It is used by OSHA to trigger additional audiometric testing and related follow-up. (2) NIOSH uses this term to describe a change of 15 dB or more at any frequency, 5000 through 6000 Hz, from baseline levels that is present on an immediate retest in the same ear and at the same frequency. NIOSH recommends a confirmation audiogram within 30 days, with the confirmation audiogram being preceded by a quiet period of at least 14 hours.

Threshold shift—Audiometric monitoring programs will encounter two types of changes in hearing sensitivity (i.e., threshold shifts): permanent threshold shift (PTS) and temporary threshold shift (TTS). As the names imply, any change in hearing sensitivity that is persistent is considered a PTS. Persistence may be assumed if the change is observed on a 30-day follow-up exam. Exposure to loud noise may cause a temporary worsening in hearing sensitivity (i.e., a TTS) that may persist for 14 hours (or even longer in cases where the exposure duration exceeded 12 to 16 hours). Hearing health professionals need to recognize that not all threshold shifts represent decreased sensitivity, and not all temporary or permanent threshold shifts are due to noise exposure. When a permanent threshold shift can be attributable to noise exposure, it may be referred to as a *noise-induced permanent threshold shift* (NIPTS).

Velocity—The speed at which the regions of sound producing pressure changes move away from the sound source.

Wavelength (λ)—The distance required for one complete pressure cycle to be completed (1 wavelength), measured in feet or meters.

Weighted measurements—Two weighting curves are commonly applied to measures of sound levels to account for the way the ear perceives the "loudness" of sounds.

> *A-weighting*—A measurement scale that approximates the "loudness" of tones relative to a 40-db SPL, 1000-Hz reference tone. A-weighting has the added advantage of being correlated with annoyance measures and is most responsive to the mid-frequencies, 500 to 4000 Hz.

> *C-weighting*—A measurement scale that approximates the "loudness" of tones relative to a 90-dB SPL, 1000-Hz reference tone. C-weighting has the added advantage of providing a relatively flat measurement scale that includes very low frequencies.

Noise Exposure

Noise literally surrounds us every day and is with us just about everywhere we go; however, the noise we are concerned with here is that produced by wind turbines. Excessive amounts of noise in the wind farm environment (and outside of it) cause many problems for people, including increased stress levels, interference with communication, disrupted concentration, and, most importantly, varying degrees of hearing loss. Exposure to high noise levels also adversely affects quality of life and increases accident rates.

One of the major problems with attempting to protect an inhabitant's hearing acuity is the tendency of many people to ignore the dangers of noise. Because hearing loss, like cancer, is insidious (you do not always know that you have cancer and you do not always know that you have lost hearing), it's easy to ignore. It sort of sneaks up on you slowly and is not apparent (in many cases) until after the damage is done. Alarmingly, hearing loss from noise exposure has been well documented since the 18th century; yet, since the advent of the industrial revolution, the number of exposed people has greatly increased.

Determining Noise Levels

The unit of measurement for sound is the decibel. Decibels are the preferred unit for measuring sound; the name is derived from the word *bel*, a unit of measure in electrical communications engineering. The decibel is a dimensionless unit used to express the logarithm of the ratio of a measured quantity to a reference quantity. Determination of noise exposure is accomplished by conducting a noise-level survey of the site or area of concern. Sound measuring instruments are used to make this determination. These include noise dosimeters, sound level meters, and octave-band analyzers. The uses and limitations of each kind of instrument are discussed below.

Noise Dosimeter

The noise dosimeters used by OSHA meet the American National Standards Institute (ANSI) Standard S1.25-1978 (Specifications for Personal Noise Dosimeter), which sets performance and accuracy tolerances. For OSHA use, the dosimeter must have a 5-dB exchange rate, use a 90-dBA criterion level, be set at slow response, and use either an 80-dBA or a 90-dBA threshold gate, or a dosimeter can be used that has both capabilities, whichever is appropriate for evaluation.

Sound Level Meter

When conducting a noise-level survey, the operator should use an ANSI-approved sound level meter (SLM)—a device used most commonly to measure sound pressure. The SLM measures in decibels. One decibel is 1/10 of a bel and is the minimum difference in loudness that is usually perceptible. The sound level meter consists of a microphone, an amplifier, and an indicating meter, which responds to noise in the audible frequency range of about 20 to 20,000 Hz. Sound level meters usually contain weighting networks designated A, B, or C. Some meters have only one weighting network; others are equipped with all three. The A network approximates the equal loudness curves at low sound pressure levels, the B network is used for medium sound pressure levels, and the C network is used for high levels. When conducting a routine sound level survey, using the A-weighted network (dBA) in the assessment of the overall noise hazard has become common practice. The A-weighted network is the preferred choice because it is thought to provide a rating of industrial noises that indicates the injurious effects such noise has on the human ear (i.e., gives a frequency response similar to that of the human ear at relatively low sound pressure levels). With an approved and freshly calibrated (always calibrate test equipment prior to use) sound level meter in hand, the user is ready to begin the sound level survey. In doing so, the user is primarily interested in answering the following questions:

1. What is the noise level in each wind turbine and/or wind farm area?
2. What equipment or process is generating the highest/lowest level of noise?
3. Which local residents are exposed to the noise?
4. How long are nearby residents exposed to the noise?

When addressing these questions, the monitors record their findings as they move from location to location, following a logical step-by-step procedure. The first step involves using the sound level meter set in A-scale slow response mode to measure a wind farm grid area. When making such measurements, restrict the size of the area (grid) being measured to less than 1000 square feet. If the maximum sound level does not exceed 80 dBA, it can be assumed that all residents or visitors in this grid are in a relatively safe or satisfactory noise level.

The next step depends on the readings recorded when the entire work area was measured. For example, if the measurements indicate sound levels greater than 80 dBA, then another set of measurements must be taken in each grid. The purpose here, of course, is to determine two things: which wind turbine or group of turbines is making noise above acceptable levels (i.e., >80 dBA) and which residents and visitors are exposed to these levels.

TABLE 5.3

Permissible Noise Exposures per 29 CFR 1910.95

Duration per Day (hours)	Sound Level dBA Slow Response
8	90
6	92
4	95
3	97
2	100
1.5	102
1	105
0.5	110
0.25 or less	115

Note: When the daily noise exposure is composed of two or more periods of noise exposure of different levels, their combined effect should be considered, rather than the individual effect of each. If the sum of the following fractions $C_1/T_1 + C_2/T_2 + C_n/T_n$ exceeds unity, then the mixed exposure should be considered to exceed the limit value. C_n indicates the total time of exposure at a specified noise level, and T_n indicates the total time of exposure permitted at that level. Exposure to impulsive or impact noise should not exceed the 140-dB peak sound pressure level.

If grid measurements indicate readings that exceed the 85-dBA level, another step must be performed. This step involves determining the length of time of exposure for residents and/or visitors (do not forget wind turbine service personnel). The easiest, most practical way to make this determination is to have these persons wear a noise dosimeter, which records the noise energy to which the person was exposed during an 8-hour period (or longer).

What happens next? It is necessary to determine if anyone is or can be exposed to noise levels that exceed the permissible noise exposure levels listed in Table 5.3. The key point to remember is that the findings must be based on a time-weighted average (TWA); for example, in Table 5.3, notice that a noise level of 95 dBA is allowed up to 4 hours per day. Note that this parameter assumes that the exposed person has good hearing acuity with no loss. If the exposed person has documented hearing loss, exposure to 95 dBA or higher, without proper hearing protection, may be unacceptable under any circumstances.

Octave-Band Noise Analyzers

Several Type 1 sound level meters have built-in octave-band analysis capability. These devices can be used to determine the feasibility of controls for individual noise sources for abatement purposes and to evaluate hearing protectors. Octave-band analyzers segment noise into its component parts. The octave-band filter sets provide filters with the following center frequencies: 31.5, 63, 125, 250, 500, 1000, 2000, 4000, 8000, and 16,000 Hz. The special signature of a given noise can be obtained by taking sound level meter readings at each of these settings (assuming that the noise is fairly constant over time). The results may indicate those octave bands that contain the majority of the total radiated sound power. Octave-band noise analyzers can assist users in determining the adequacy of various types of frequency-dependent noise controls. They also can be used to select hearing protectors because they can measure the amount of attenuation offered by the protectors in the octave bands responsible for most of the sound energy in a given situation.

DID YOU KNOW?

- Adding together two identical noise sources will increase the total sound-power level by 3 dB ($10\log_2$).
- *Sound power* and *sound power level* are often used to specify the noise of sound emitted from technical equipment such as fans, pumps, or other machines.
- Sound measured with sensors or microphones is *sound pressure*.
- SPL = sound pressure level.
- Decibel (dB) = unit of sound production level (logarithmic scale).
- dBA = A-weighted scale decibel (levels weighted according to sound frequency).
- L_{eq} = equivalent continuous sound level.
- SEL = sound equivalent level integrated per 1 second.
- Hertz (Hz) = cycles per second (measure of sound frequency).
- 1 kilohertz (kHz) = 1000 hertz.
- ROW = right of way (pertaining to highway noise).

Noise Units, Relationships, and Equations

A number of noise units, relationships, and equations that are involved with controlling noise hazards are discussed below.

Sound Power

Sound power of a source is the total sound energy radiated by the source per unit time. It is expressed in terms of the sound power level (L_w) in decibels referenced to 10^{-12} watts (w_0). The relationship to decibels is shown below:

$$L_w = 10 \log (w/w_0) \tag{5.14}$$

where

L_w = Sound power level (decibels).
10 log = Logarithm to the base 10.
w = Sound power (watts).
w_0 = Reference power (10^{-12} watts).

Sound Pressure

Units used to describe sound pressures are

$$1 \; \mu bar = 1 \; dyne/cm^2 = 0.1 \; N/cm^2 = 0.1 \; Pa \tag{5.15}$$

Sound Pressure Level

$$SPL \; (decibels) = 10 \log (p^2/p_0^2) \tag{5.16}$$

where

p = Measured RMS sound pressure (N/m², μbars).[*]
p_0 = Reference RMS sound pressure (20 μPa, N/m², μbars).

---

[*] The root-mean-square (RMS) value of a changing quantity, such as sound pressure, is the square root of the mean of the squares of the instantaneous values of the quantity.

Speed of Sound

$$c = f_1 \qquad (5.17)$$

Wavelength

$$\lambda = c/f \qquad (5.18)$$

Octave Band Frequencies

$$\text{Upper frequency band} = f_2 = 2f_1 \qquad (5.19)$$

where
  f_2 = Upper frequency band.
  f_1 = Lower frequency band.

$$\text{One-half octave band} = f_2 = \sqrt{2(f_1)} \qquad (5.20)$$

where
  f_2 = 1/2 octave band.
  f_1 = Lower frequency band.

$$\text{One-third octave band} = f_2 = \sqrt[3]{2(f_1)} \qquad (5.21)$$

where
  f_2 = 1/3 octave band.
  f_1 = Lower frequency band.

Adding Noise Sources
When sound power is known:

$$L_w = 10 \log\left(\frac{w_1 + w_2}{w_0 + w_0} \right) \qquad (5.22)$$

where
  L_w = Sound power (watts).
  w_1 = Sound power of noise source 1 (watts).
  w_2 = Sound power of noise source 2 (watts).
  w_0 = Reference sound power (reference 10^{-12}) (watts).

Sound Pressure Additions
When sound pressure is known:

$$\text{SPL (decibels)} = 10 \log (p^2/p_0^2) \qquad (5.23)$$

where
  $p^2/p_0^2 = 10^{SPL/10}$.
  p = Measured root-mean-square (RMS) sound pressure (N/m², μbars).
  p_0 = Reference RMS sound pressure (20 μPa, N/m², μbars).

For three sources, the equation becomes

$$SPL = 10 \ \log\left[\left(10^{SPL_1/10}\right) + \left(10^{SPL_2/10}\right) + \left(10^{SPL_3/10}\right) \right] \qquad (5.24)$$

When adding any number of sources, whether or not the sources are identical, the equation becomes

$$SPL = 10 \ \log\left[\left(10^{SPL_1/10}\right)+\cdots+\left(10^{SPL_n/10}\right)\right] \tag{5.25}$$

To determine the sound pressure level from multiple identical sources, use the following equation:

$$SPL_f = SPL_i + 10 \ \log(n) \tag{5.26}$$

where
   SPL_f = Total sound pressure level (dB).
   SPL_i = Individual sound pressure level (dB).
   n = Number of identical sources.

Noise Levels in a Free Field

$$SPL = L_w - 20 \ \log(r) - 0.5 \tag{5.27}$$

where
   SPL = Sound pressure (reference 0.00002 N/m^2).
   L_w = Sound power (reference 10^{-12} watts).
   r = Distance (feet).

Noise Levels with Directional Characteristics

$$SPL = L_w - 20 \ \log(r) - 0.5 + \log(Q) \tag{5.28}$$

where
   SPL = Sound pressure (reference 0.00002 N/m^2).
   L_w = Sound power (reference 10^{-12} watts).
   r = Distance (feet).
   Q = Directivity factor.
      = 2 for one reflecting plane.
      = 4 for two reflecting planes.
      = 8 for three reflecting planes.

Noise Level at a New Distance from the Noise Source

$$SPL = SPL_1 + 20\log\left(\frac{d_1}{d_2}\right) \tag{5.29}$$

where
   SPL = Sound pressure level at new distance (d_2).
   SPL_1 = Sound pressure level at d_1.
   d_n = Distance from source.

Daily Noise Dose
The following formula combines the effects of different sound pressure levels and allowable exposure times:

$$\text{Daily noise dose} = \frac{C_1 + C_2 + C_3 + \cdots + C_n}{T_1 + T_2 + T_3 + \cdots + T_n} \tag{5.30}$$

where
   C_i = Number of hours exposed at given SPL_i.
   T_i = Number of hours exposed at given SPL_i.

OSHA Permissible Noise Levels

$$T_{SPL} = 8/2^{(SPL-90)/5}$$ (5.31)

where
   T_{SPL} = Time at given SPL (hours).
   SPL = Sound pressure level (dBA).

Equivalent Eight-Hour TWA

$$TWA_{eq} = 16.61 \log\left(\frac{D}{100}\right) + 90$$ (5.32)

where
   TWA_{eq} = Eight-hour equivalent TWA (dBA).
   D = Noise dosimeter reading (%).

WIND ENERGY CONSTRUCTION IMPACTS

Typical activities during the wind energy facility construction phase include ground clearing (removal of vegetative cover), grading, excavation, blasting, trenching, vehicular and pedestrian traffic, and drilling. Activities conducted in locations other than the facility site include excavation/blasting for construction materials such as sands and gravels, as well as access road construction (TEEIC, 2015).

AIR QUALITY

Emissions resulting from construction activities include vehicle emissions, diesel emissions from large construction equipment and generators; volatile organic compound (VOC) releases from storage and transfer of vehicle or equipment fuels; small amounts of carbon monoxide, nitrogen oxides, and particulates from blasting activities; and fugitive dust. Fugitive dust would be caused by

- Disturbing and moving soils (clearing, grading, excavation, trenching, backfilling, dumping, and truck and equipment traffic)
- Mixing concrete and associated storage piles
- Drilling and pile driving

A construction permit is needed from the state or local air agency to control or mitigate these emissions; therefore, these emissions would not likely cause an exceedance of air quality standards or have an impact on climate change.

CULTURAL RESOURCES

Direct impacts on cultural resources could occur from construction activities, and indirect impacts might be caused by soil erosion and increased accessibility to possible site locations. Potential impacts include the following:

DID YOU KNOW?

Wind energy is underutilized as of now and holds tremendous potential for the future. Although there has been a 25% increase in wind turbine use in the last 12 years, wind energy still provides only a small percentage of the energy of the world.

- Complete destruction of the resource could occur if present in areas undergoing surface disturbance or excavation.
- Degradation or destruction of near-surface cultural resources on- and offsite could result from topographic or hydrological pattern changes or from soil movement (removal, erosion, sedimentation). (Note that the accumulation of sediment could protect some localities by increasing the amount of protective cover.)
- Unauthorized removal of artifacts or vandalism at the site could occur as a result of increase in human access to previously inaccessible areas, if significant cultural resources are present.
- Visual impacts could result from vegetation clearing, increases in dust, and the presence of large-scale equipment, machinery, and vehicles (if the resources have an associated landscape component that contributes to their significance, such as a sacred landscape or historic trail).

ECOLOGICAL RESOURCES

Ecological resources that could be affected include vegetation, fish, and wildlife, as well as their habitats. Adverse ecological effects during construction could be caused by the following:

- Erosion and runoff
- Fugitive dust
- Noise
- Introduction and spread of invasive vegetation
- Modification, fragmentation, and reduction of habitat
- Mortality of biota (i.e., death of plants and animals)
- Exposure to contaminants
- Interference with behavioral activities

Site clearing and grading, along with construction of access roads, towers, and support facilities, could reduce, fragment, or dramatically alter existing habitat in the disturbed portions of the project area. Ecological resources would be most affected during construction by the disturbance of habitat in areas near turbines, support facilities, and access roads. Wildlife in surrounding habitats might also be affected if the construction activity (and associated noise) disturbs normal behaviors, such as feeding and reproduction.

WATER RESOURCES

Water Use

Water would be used for dust control when clearing vegetation and grading and for road traffic; for making concrete for foundations of towers, substations, and other buildings; and for consumptive use by the construction crew. Water could be trucked in from offsite or obtained from local groundwater wells or nearby surface water bodies, depending on availability.

Water Quality

Water quality could be affected by

- Activities that cause soil erosion
- Weathering of newly exposed soils that could cause leaching and oxidation, thereby releasing chemicals into the water
- Discharge of waste or sanitary water
- Pesticide applications

Flow Alteration

Surface and groundwater flow systems could be affected by withdrawals made for water use, wastewater and stormwater discharges, and the diversions of surface water flow for access road construction of stormwater control systems. Excavation activities and the extraction of geological materials could affect surface and groundwater flow. The interaction between surface water and groundwater could also be affected if the surface water and groundwater were hydrologically connected, potentially resulting in unwanted dewatering or recharging of water resources.

LAND USE

This section provides an analysis of the land use associated with modern, large wind power plants (defined as greater than 20 megawatts and constructed after 2000). Impacts to land use could occur during construction if there were conflicts with existing land use plans and community goals; conflicts with existing recreational, educational, religious, scientific, or other use areas: or conversion of the existing commercial land use for the area (e.g., mineral extraction). During construction, impacts to most land uses would be temporary, such as removal of livestock from grazing areas during blasting or heavy equipment operations, or temporary effects on the character of a recreation area because of construction noise, dust, and visual intrusions. Long-term land use impacts would occur if existing land uses are not compatible with wind energy development, such as remote recreational experiences; however, those uses could potentially be resumed if the land is reclaimed to pre-development conditions.

Quantifying the area of a wind power plant is challenging given the discontinuous nature of its configuration. "Area" includes not only land directly disturbed by installation of the turbines but also the surrounding area that potentially may be impacted. There are two general types of "areas" considered. The first is the *direct surface area impact* (i.e., disturbed land) due to plant construction and infrastructure. The second is more vaguely defined but is associated with the *total area of the wind power plant* as a whole. Figure 5.11 provides a simplified illustration of the two types of areas, which are vastly different in both quantity and quality of impacts as discussed in subsequent sections (Denholm et al., 2009).

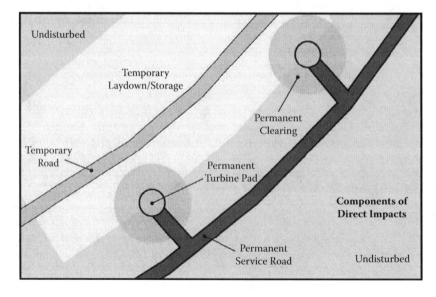

FIGURE 5.11   Illustration of the two types of wind plant use: total area and direct impact area (including permanent and temporary). (From USDOE, *Environmental Assessment for Spring Canyon Wind Project, Logan County, Colorado*, DOE/EA-1521, U.S. Department of Energy, Washington, DC, 2005.)

Direct Impact Area

Development of a wind power plant results in a variety of temporary and permanent (lasting the life of the project) disturbances (Denholm et al., 2009). These disturbances include land occupied by wind turbine pads, access roads, substations, service buildings, and other infrastructure which physically occupy land area or create impermeable surfaces. Additional direct impacts are associated with development in forested areas, where additional land must be cleared around each turbine. Although land cleared around a turbine pad does not result in impervious surfaces, this modification represents a potentially significant degradation in ecosystem quality. In addition to permanent impacts, which last the life of the facility, there are temporary impacts from plant construction. These impacts are associated with temporary construction-access roads, storage, and lay-down. After plant construction is completed, these areas will eventually return to their previous state. The amount of time required to return to the pre-disturbance condition is estimated to be 2 to 3 years for grasslands and decades for desert environments (Arnett et al., 2007).

Total Wind Plant Area

Although the area and impacts associated with physical infrastructure as described earlier may be the easiest to quantify, the more commonly cited land-use metric associated with wind power plants is the footprint of the project as a whole. However, unlike the area occupied by roads and pads, the total area is more challenging to define and subjective in nature. Generally, the total area of a wind power plant consists of the area within a perimeter surrounding all of the turbines in the project; however, the perimeter is highly dependent on terrain, turbine site, current land use, and other considerations such as setback regulations. There is no uniform definition of the perimeter or boundary surrounding a wind power plant—in fact, the total area of a wind power plant could have a number of definitions. The boundary could be defined based on the required turbine spacing as a function of rotor diameter, or a standardized setback from turbines at the edge of a project could be used (Denholm et al., 2009).

When wind farm construction spreads, local opposition to the mass towers (some over 400 feet tall) impacts landowners' opinions, land values, and state regulations; however, this local opposition is relative in the sense that opinions are based on the receipt of or nonreceipt of monetary rewards for use of the land. That is, opposition is based on whether the wind farms constructed on personal property accrue a monetary reward or usage fee for property owners who have allowed the presence of turbines on their land. This view is typically different for residents who do not reap an economical benefit from the presence of wind turbines in their backyards.

Impacts on aviation could be possible if the project is located within 20,000 feet (6100 meters) or less of an existing public or military airport, or if the proposed construction involves objects that are greater than 200 feet (61 meters) in height. The Federal Aviation Administration (FAA) must be notified if either of these two conditions occurs, and the FAA would be responsible for determining if the project would adversely affect commercial, military, or personal air navigation safety. Similarly, impacts on military operations could occur if a project is located near a military facility if that facility conducts low-altitude military testing and training activities.

SOILS AND GEOLOGIC RESOURCES

Sands, gravels, and quarry stone would be excavated for construction access roads; for concrete for buildings, substations, transformer pads, foundations, and other ancillary structures; and for improving ground surfaces for laydown areas and crane staging areas. Possible geological hazards such as landslides could be activated by excavation and blasting for raw materials, increasing slopes during site grading and construction of access roads, altering natural drainage patterns, and toe-cutting bases of slopes. Altering drainage patterns could also accelerate erosion and create slope instability. Surface disturbance, heavy equipment traffic, and changes to surface runoff patterns could cause soil erosion and impacts on special soils (e.g., cryptobiotic soils). Impacts of soil erosion could include soil nutrient loss and reduced water quality in nearby surface water bodies.

PALEONTOLOGICAL RESOURCES

Impacts on paleontological resources could occur directly from the construction activities or indirectly from soil erosion and increase accessibility to fossil locations. Potential impacts include the following:

- Complete destruction of the resource could occur if present in areas undergoing surface disturbance or excavation.
- Degradation or destruction of near-surface fossil resources on- and offsite could result from changes in topography, changes in hydrological patterns, and soil movement (removal, erosion, sedimentation). (Note that the accumulation of sediment could serve to protect some locations by increasing the amount of protective cover.)
- Unauthorized removal of fossil resources or vandalism to the site could occur as a result of increased human access to previously inaccessible areas if significant paleontological resources are present.

TRANSPORTATION

Short-term increases in the use of local roadways would occur during the construction period. Heavy equipment likely would remain at the site. Shipments of materials are unlikely to affect primary or secondary road networks significantly, but this would depend on the location of the project site relative to material source. Oversized loads could cause temporary transportation disruptions and could require some modifications to roads or bridges (such as fortifying bridges to accommodate the size or weight). Shipment weight might also affect the design of access roads for grade determinations and turning clearance requirements.

VISUAL RESOURCES

Although many of us consider wind turbines to be visually acceptable and in some cases even pleasing to look at, wind turbines disturb the visual area of other people by creating negative changes in the natural environment. The test of whether a wind turbine or wind farm is a visual pollutant is to ask this question: How many of us would seriously like one or more wind turbines a few hundred feet or meters from our homes? It is important to remember that wind turbines can be anywhere from a few meters to a hundred meters high. Having a wind turbine tower over one's home is the last thing many residents want. Having said this, the possible sources of visual impacts during construction include the following:

- Road development (e.g., new roads or expansion of existing roads) and parking areas could introduce strong visual contrasts in the landscape, depending on the route relative to surface contours and the width, length, and surface treatment of the roads.
- Conspicuous and frequent small-vehicle traffic for worker access and frequent large-equipment (e.g., trucks, graders, excavators, cranes) traffic for road construction, site preparation, and turbine installation could produce visible activity and dust in dry soils. Suspension and visibility of dust would be influenced by vehicle speeds and road surface materials.
- Site development could be intermittent, staged, or phased, giving the appearance that work starts and stops. Depending on the length of time required for development, the project site could appear to be under construction for an extended period. This could give rise to perceptions of lost benefit and productivity, like those alleged for the equipment. Timing and duration concerns may result.
- There would be a temporary presence of large cranes or other large machines to assemble towers, nacelles, and rotors. This equipment would also produce emissions while operating and could create visible exhaust plumes. Support facilities and fencing associated with the construction work would also be visible.

- Ground disturbance and vegetation removal could result in visual impacts that produce contrasts of color, form, texture, and line. Excavation for turbine foundations and ancillary structures, trenching to bury electrical distribution systems, grading and surfacing roads, cleaning and leveling staging areas, and stockpiling soil and spoils (if not removed) would (1) damage or remove vegetation, (2) expose bare soil, and (3) suspend dust. Soil scars and exposed slope faces would result from excavation, leveling, and equipment movement. Invasive species could colonize disturbed and stockpiled soils and compacted areas.

SIDEBAR 2.1. VISUAL CONSERVATION: BANANA VS. NIMBY, OR BOTH

Renewable Energy Paradox

Ask an environmentalist if renewable energy is a good idea. It's a no-brainer, they'll say. Hell, yes. If you tell them that you will build a wind farm or a series of electrical power towers in their backyard, though, they will scream, "No way, Jose!"

Just as they are with fire or waterfall gazing, some drivers are also mesmerized whenever they drive Interstate 40 through Oklahoma and Texas or Interstate 15 through Tehachapi Pass, California, on clear days and witness a host of hundreds of whirling, swooshing, slashing wind turbines. Many of these folks do not need a road-to-Damascus change of view to accept these massive turbines. Moreover, they are not necessarily advocates for renewable energy; instead, they are simply captivated by the human-made machines as they stand tall against a backdrop of plains and deserts and hills and mountains and blue or cloud-filled nothingness.

This somewhat romantic view of wind farms is held by a few people here and there, but it is safe to say, without qualification, that many other people have a different view of whirling, swooshing wind turbines scattered helter-skelter across the U.S. landscape. Even though very few of these people dispute the environmental benefits of wind energy (or solar and other renewable energy producers), many feel that the construction of wind farms (anywhere) ruins or spoils the otherwise pristine landscape.

The possible or potential construction of wind farms (or solar or other renewable energy sources) often elicits the "not in my back yard" (NIMBY) phenomenon—even more so, in some cases, for the associated electric power transmission towers and lines. Generally, NIMBY opponents acknowledge the need for the wind turbines and transmission lines, while arguing that they just don't want them nearby to them. Most people understand the need for them, but hardly anyone wants to live within sight of them, usually because they look "ugly" or for personal safety concerns.

The opposition generated by the NIMBY phenomenon is one thing, but quite another is the opposition generated by the "build absolutely nothing anywhere near anything" (BANANA) phenomenon, which protests, unlike NIMBY, the overall necessity of any development. The opponents are often environmentalists, in which case the argument is generally we don't need new wind turbine farms and associated transmission lines and power stations. Simply, BANANA activists argue against any more of *whatever* is being planned. With regard to construction of new wind turbines and associated equipment, the BANANA argument is that instead we need to use power more wisely, not generate more.

Whether in reference to the NIMBY or the BANANA phenomenon, what we are talking about is complaints about visual pollution. Visual pollution is an aesthetic issue regarding the impacts of pollution impairing one's ability to enjoy a view or vista. Visual pollution disturbs coveted views by creating what are perceived as negative changes in the natural environment; that is, visual pollution intrudes on, ruins, spoils, and mars the natural landscape and

FIGURE SB5.1.1 High-voltage power lines.

can best be described as an eyesore. The most common negative changes or forms of visual pollution are buildings, automobiles, trash dumps, space debris, telephone and electric towers and wires, and electrical substations (see Figures SB5.1.1 and SB5.1.2).

One thing is certain; you can't hide wind turbines from view. They can't be hidden nor can they be camouflaged like cell phone towers and satellite dishes (Komanoff, 2003; Pasqualetti, 2002; Righter, 2002; Saito, 2004). Because of their visibility (and for other reasons), siting and establishing wind-energy production farms and transmission infrastructure present unique challenges.

With regard to transmission siting, the remote location of much of the utility-scale wind power capacity requires the construction of new high-voltage transmission lines to transport electricity to population centers. Because transmission lines can cross private, public (state and federal), and tribal lands, the process of planning, permitting, and building new lines is highly visible and implicates many diverse interests—and it can be costly, time consuming, and controversial.

FIGURE SB5.1.2 Electrical substation.

FIGURE SB5.1.3   Power line right of way (ROW).

Another major consideration for the installation of transmission lines from remote locations (and any other location) is right of way (ROW). The right of way for a transmission corridor includes land set aside for the transmission line and associated facilities, land needed to facilitate maintenance, and land needed to avoid risk of fires and other accidents. It provides a safety margin between the high-voltage lines and surrounding structures and vegetation (see Figure SB5.1.3). Some vegetation clearing may be needed for safety or access reasons. A ROW generally consists of native vegetation or plants selected for favorable growth patterns (slow growth and low mature heights). However, in some cases, access roads constitute a portion of the ROW and provide more convenient access for repair and inspection vehicles. Vegetation clearing or recontouring of land may be required for access road construction. The width of a ROW varies depending on the voltage rating on the line, ranging from 50 feet to approximately 175 feet or more for 500-kV lines.

Before approval for new transmission is granted, the regulatory authority must determine that the project is necessary. Non-transmission alternatives must often be considered, including energy conservation, energy efficiency, distributed generation, and fully utilizing unused capacity on existing transmission lines. When new transmission lines are deemed necessary, developers and utilities must find the best routes to the greatest concentrations of renewable energy and build with the least possible impact on the environment. Transmission lines from tall transmission towers carry high-voltage energy (115 to 500 kV) over long distances to a substation. Both transmission and distribution lines (from substations) carry enough energy to harm or kill both people and birds. Kill birds? Yes, actually, contrary to popular belief, some birds are electrocuted by electrical power lines. Small birds do not usually get electrocuted because they fail to complete a circuit either by touching a grounded wire or structure, or another energized wire, so the electricity stays in the line. Larger birds, however, such as the California condor, which has a wingspan of up to 9.5 feet, are more likely to touch a power line and ground wire, another energized wire, or a pole at the same time, giving electricity a path to the ground. In both situations, the birds are electrocuted and killed, a fuse is blown, power fails, and everyone is impacted.

Birds also fly into power lines. It is generally believed that birds collide with power lines because the lines are invisible to them or because they do not see the line before it is too late to avoid it. Birds' limited ability to judge distance makes power lines especially difficult to see, even as they are flying closer to them. Large birds are especially vulnerable because they are not always quick enough to change their direction before it is too late. Poor weather conditions, such as fog, rain or snow, as well as darkness, may make the lines even more difficult to see.

When birds collide with power lines they are either killed outright by the impact or injured by contact with electrical lines, resulting in crippling that is likely fatal. Electrocutions can also start wildfires and cause power outages. An estimated 5 to 15% of all power outages can be attributed to bird collisions with power lines (USFWS, 2005).

In addition to increasing wildlife mortality due to collisions and electrocutions, as well as by serving as perches for predators, transmission lines can fragment and interfere with wildlife habitats and corridors (WGA, 2009). Again, there are also concerns about the visual impacts of transmission lines. Moreover, many people feel that living or working near transmissions lines is hazardous to their health. Burying transmission lines can help avoid many of the environmental and aesthetic issues, but burying lines may also have negative impacts on soil, vegetation, and other resources (Molburg et al., 2007), and underground lines are typically four times as expensive as overhead lines (Brown and Sedano, 2004). Also, although high-voltage direct current (DC) lines can be buried, there is a limit on the maximum voltage and length of alternating current (AC) lines that can be buried.

In all, constructing major new transmission infrastructure can require 7 to 10 years from planning to operation: 1 year for final engineering, 1 to 2 years for construction, and the rest of the time for planning and permitting. Substantial time and controversy are added to the process when environmental and related concerns are addressed at the end instead of at the beginning. The specific environmental impact concerns are discussed in the following sections.

SOCIOECONOMICS

Direct impacts would include the creation of new jobs for workers (approximately two workers per megawatt) at wind energy development projects and the associated income and taxes paid. Indirect impacts would occur as a result of the new economic development and would include new jobs at businesses that support the expanded workforce or provide project materials and the associated income taxes. Wind energy development activities could also potentially affect property value, either positively from increased employment effects or the image of clean energy or negatively from proximity to the wind farm and any associated or perceived environmental effect (noise, visual, etc.). Adverse impacts could occur if a large migrant workforce, culturally different from the local indigenous group, is brought in during construction. This influx of migrant workers could strain the existing community infrastructure and social services.

ENVIRONMENTAL JUSTICE

If significant impacts occurred in any resource areas, and the impact disproportionately affected minority or low-income populations, then there could be environmental justice concerns. Potential issues during construction are noise, dust, and visual impacts from the construction site and possible impacts associated with the construction of new access roads.

HAZARDOUS MATERIALS AND WASTE MANAGEMENT

Solid and industrial waste would be generated during construction activities. The solid waste would likely be nonhazardous and consist mostly of containers, packing material, and wastes from equipment assembly and construction crews. Industrial wastes would include minor amounts of paints,

coatings, and spent solvents. Hazardous materials stored onsite for vehicle and equipment maintenance would include petroleum fluids (lubricating oils, hydraulic fluid, fuels), coolants, and battery
electrolytes. Oils, transmission fluids, and dielectric fluids would be brought to the site to fill turbine
components and other large electrical devices. Also, compressed gases would be used for welding,
cutting, brazing, etc. These materials would be transported offsite for disposal, but impacts could
result if the wastes are not properly handled and are released to the environment.

WIND ENERGY OPERATIONS IMPACTS

Typical activities during the wind energy facility operations phase include turbine operation, power
generation, and associated maintenance activities that would require vehicular access and heavy
equipment operation when large components are being replaced. Potential impacts from these activities are presented below, by the type of affected resource.

AIR QUALITY

There are no direct air emissions from operating a wind turbine. Minor volatile organic compound
(VOC) emissions are possible during routine maintenance activities of applying lubricants, cooling
fluids, and greases. Minor amounts of carbon monoxide and nitrogen oxides would be produced
during periodic operation of diesel emergency generators as part of preventative maintenance.
Vehicular traffic would continue to produce small amounts of fugitive dust and tailpipe emissions
during the operations phase. These emissions would not likely exceed air quality standards or have
any impact on climate change.

CULTURAL RESOURCES

Impacts during the operations phase would be limited to the unauthorized collection of artifacts and
visual impacts. The threat of unauthorized collection would be present once access roads are constructed during the site evaluation or construction phase, making remote lands accessible to the public. Visual impacts resulting from the presence of large wind turbines and associated facilities and
transmission lines could affect some cultural resources, such as sacred landscapes or historic trails.

ECOLOGICAL RESOURCES

During operation, adverse ecological effects could occur from (1) disturbance of wildlife by turbine noise and human activity, (2) site maintenance (e.g., mowing), (3) exposure of biota to contaminants, and (4) mortality of birds and bats that collide with the turbines and meteorological
towers. During the operation of a wind facility, plant and animal habitats could still be affected by
habitat fragmentation due to the presence of turbines, support facilities, and access roads. In addition, the presence of an energy development project and its associated access roads may increase
human use of surrounding areas, which could in turn impact ecological resources in the surrounding areas through

1. Introduction and spread of invasive vegetation
2. Fragmentation of habitat
3. Disturbance of biota
4. Increased potential for fire

As discussed in detail later, the presence of a wind energy project (and its associated infrastructure)
could also interfere with migratory and other behaviors of some wildlife.

WATER RESOURCES

Impacts on water use and quality and flow systems during the operation phase would be limited to possible degradation of water quality resulting from vehicular traffic and pesticide application, if conducted improperly.

LAND USE

Impacts on land use would be minimal, as many activities can continue to occur among the operating turbines, such as agriculture and grazing. It might be possible to collocate other forms of energy development, provided the necessary facilities could be installed without interfering with operation and maintenance of the wind farm. Collocation of other forms of energy development could include directionally drilled oil and gas wells, underground mining, and geothermal or solar energy development. Recreation activities (e.g., off-highway vehicle use, hunting) are also possible, but activities centered on solitude and scenic beauty could be affected. Military operations and aviation could be affected by radar interference associated with the operating turbines, and low-altitude activities could be affected by the presence of turbines over 200 feet high.

SOILS AND GEOLOGIC RESOURCES

Following construction, disturbed portions of the site would be revegetated and the soil and geologic conditions would stabilize. Impacts during the operations phase would be limited largely to soil erosion impacts caused by vehicular traffic for operations maintenance.

PALEONTOLOGICAL RESOURCES

Impacts during the operations phase would be limited to the unauthorized collection of fossils. This threat is present once the access roads are constituted in the site evaluation or construction phases, making remote land accessible to the public.

TRANSPORTATION

No noticeable impacts on transportation are likely during operations. Low volumes of heavy- and medium-duty pickup trucks and personal vehicles are expected for routine maintenance and monitoring. Infrequent but routine shipments of component replacements during maintenance procedures are likely over the period of operation.

VISUAL RESOURCES

Wind energy development projects would be highly visible in rural or natural landscapes, many of which have few other comparable structures. The artificial appearance of wind turbines may have visually incongruous industrial-type associations for some, particularly in a predominantly natural landscape; however, other viewers may find wind turbines visually pleasing and consider them a positive visual impact. Visual evidence of wind turbines cannot easily be avoided, reduced, or concealed, due to their size and exposed location; therefore, effective mitigation is often limited. Additional issues of concern are shadow flicker (strobe-like effects from flickering shadows cast by the moving rotors), blade glint from the sun reflecting off moving blades, visual contrasts from support facilities, and light pollution from the lighting on facilities and towers (which are required safety features). Additional visual impacts from vehicular traffic would occur during maintenance and as towers, nacelles, and rotors are upgraded or replaced. When replacing turbines and other facility components, the opportunity and pressures to break uniformity of spacing between turbines and uniformity of size, shape, and color among facility components could increase visual contrast and visual clutter. Infrequent outages, disassembly, and repair of equipment may occur, producing the appearance of idle or missing rotors, "headless" towers (when nacelles are removed), and lowered towers, as well as negative visual perceptions of lost benefits (e.g., loss of wind power) and "bone yards" (storage areas).

SOCIOECONOMICS

Direct impacts would include the creation of approximately one new job for every three megawatts of installed capacity for operations and maintenance workers at wind energy development projects, as well as the associated income and taxes paid. Indirect impacts would occur from new economic development, including new jobs at businesses that support the expanded workforce or that provide project materials and the associated income and taxes. Wind energy development activities could also potentially affect property values, either positively from increased employment effects or the image of clean energy or negatively from proximity to the wind farm and any associated or perceived environmental effects (e.g., noise, visual).

ENVIRONMENTAL JUSTICE

Possible environmental justice impacts during operation include the alteration of scenic quality in areas of traditional or cultural significance to minority or low-income populations. Noise impacts and health and safety impacts are also possible sources of disproportionate effects.

HAZARDOUS MATERIALS AND WASTE MANAGEMENT

Industrial and sanitary wastes are generated during routine operations (e.g., lubricating oils, hydraulic fluids, coolants, solvents, cleaning agent, sanitary wastewaters). These wastes are typically put in containers, characterized and labeled, possibly stored briefly, and transported by a licensed hauler to an appropriate permitted offsite disposal facility as a standard practice. Impacts could result if these wastes were not properly handled and were released to the environment. Releases could also occur if individual turbine components or electrical equipment were to fail.

WIND ENERGY IMPACTS ON WILDLIFE

> He clasps the crag with crooked hands;
> Close to the sun in lonely lands,
> Ringed with the azure world, he stands.
> The wrinkled sea beneath him crawls;
> He watches from his mountain walls,
> And like a thunderbolt he falls.
>
> **—Alfred Lord Tennyson, "The Eagle"**

Alfred Lord Tennyson, in his classic poem above, wants us to see the eagle as both a swift predator and a powerful bird who is nonetheless susceptible to defeat by other forces (most likely by humans). Human-made and installed wind turbines are responsible for bird deaths. This has been a less documented impact of wind turbines and has mainly been argued by wildlife groups. Noise standards, for example, for wind turbines developed by countries such as Sweden and New Zealand and some specific site-level standards implemented in the United States focus primarily on sleep disturbance and annoyance to humans (USFWS, 2014b). Noise standards do not generally exist for wildlife, except in a few instances where federally listed species may be impacted. Findings from recent research clearly indicate the need to better address noise–wildlife issues. As such, noise impacts on wildlife should clearly be included as a factor in wind turbine siting, construction, and operation.

A detailed description of eagle conservation guidance with regard to land-based wind energy is presented later. For now it is important to point out some of the key issues, which include (1) how wind facilities affect background noise levels; (2) how and what fragmentation, including acoustical fragmentation, occurs, especially to species sensitive to habitat fragmentation; (3) comparison of turbine noise levels at lower valley sites—where it may be quieter—to turbines placed on ridge lines above rolling terrain where significant topographic sound shadowing can occur with the potential to significantly elevate sound levels above ambient conditions; and (4) correction and accounting of a 15-dB underestimate from daytime wind turbine noise readings used to estimate nighttime turbine noise levels (Barber et al., 2010; Van den Berg, 2004). The sensitivities of various groups of wildlife can be summarized as follows:

- Birds (more uniform than mammals)—100 Hz to 8–10 kHz; sensitivity at 0.10 dB
- Mammals—<10 Hz to 150 kHz; sensitivity to –20 dB
- Reptiles (poorer than birds)—50 Hz to 2 kHz; sensitivity at 40 to 50 dB
- Amphibians—100 Hz to 2 kHz; sensitivity from 10 to 60 dB

Turbine blades at normal operating speeds can generate significant levels of noise. How much noise? Based on a propagation model of an industrial-scale, 1.5-MW wind turbine at 263-foot hub height, positioned approximately 1000 feet apart from neighboring turbines, the following decibel levels were determined for peak sound production. At a distance 300 feet from the blades, 45 to 50 dBA were detected; at 2000 feet, 40 dBA; and at 1 mile, 30 to 35 dBA (Kaliski, 2009). Declines in the densities of woodland and grassland bird species have been shown to occur at noise thresholds between 45 and 48 dB, respectively, whereas the most sensitive woodland and grassland species showed declines between 35 and 43 dB, respectively. Songbirds specifically appear to be sensitive to very low sound levels equivalent to those in a library reading room (~30 dBA) (Foreman and Alexander, 1998). Given this knowledge, it is possible that effects to sensitive species may be occurring at ≥1 mile from the center of a wind facility at periods of peak sound production.

Noise does not have to be loud to have negative effects. Very low-frequency sounds including infrasound (sound lower in frequency than 20 Hz) are also being investigated for their possible effects on both humans and wildlife. Wind turbine noise results in a high infrasound component (Salt and Hullar, 2010). Infrasound is inaudible to the human ear, but this unheard sound can cause human annoyance, sensitivity, disturbance, and disorientation (Anon., 2010). For birds, bats, and other wildlife, the effects may be more profound. Noise from traffic, wind, and operating turbine blades produces low-frequency sounds (<1 to 2 kHz) (Dooling, 2002; Lohr et al., 2003). Bird vocalizations are generally within the frequency range of 2 to 5 kHz (Dooling and Popper, 2007), and birds hear best between 1 and 5 kHz (Dooling, 2002). Although traffic noise generally falls below the frequency of bird communication and hearing, several studies have documented that traffic noise can have significant negative impacts on bird behavior, communication, and ultimately on avian health and survival (Barber et al, 2010; Lengagne, 2008; Lohr et al., 2003). Whether these effects are attributable to infrasound effects or to a combination of other noise factors is not yet fully understood. The fact is that little is known about the combination effect of traffic noise and wind turbine noise. However, given that wind-generated noise, including blade turbine noise, produces a fairly persistent, low-frequency sound similar to that generated by traffic noise (Dooling, 2002; Lohr et al., 2003), it is plausible that wildlife effects from these two sounds could be similar. It is also plausible that the effects from these two sounds combined could be detrimental to wildlife of all kinds. Based on experience, this book supports this view.

Some may feel that the combination of road noise and wind turbine noise causing wildlife effects is plainly a stretch. Although the author has studied, observed, measured, and monitored this phenomenon, the truth is little is known about the effects of noise related to road noise and wind turbines combined; moreover, at present, little to nothing on the subject is reported in the peer-reviewed literature.

Let's get back to why some people feel the statement that a combination of road noise and wind turbine noise has a wildlife effect is plainly a stretch. This point of view seems to be prevalent for those who do not travel the east–west U.S. interstates from East Coast areas to the Southwest or Pacific Northwest, or the north–south route through Indiana on I-65 and others. On many roads it is not uncommon to drive in any direction and see hundreds of tall wind turbines off in the distance far ahead and to the right and left of the highway. Although some of these wind turbines are very close to the shoulder of the roads, many wind turbine farms can only be seen off in the vast, far distance as one travels the highways. It is this perception of distance on some outlying, remote hillside or ridge in some unoccupied wilderness or high plains area that gives the unknowing viewer the misconception about noise from road traffic and wind turbines not having a combined effect. After all, if the wind turbines are off in some remote corner of nowhere, how can road noise add to normal wind turbine noise? Very easily, actually. Remember, service roads built to perform maintenance and preventive maintenance, such as inspection of components, servicing items on a regular basis (e.g., retorquing bolts), and replacing consumable items at or before a specified age (e.g., replacing filters, changing the oil in the gearbox), are utilized by light and heavy trucks and other vehicles on a routine basis, no matter the location. The larger the wind farm, the greater the access—the more traffic, the more noise. All access vehicles, including helicopters used to transport parts and personnel, produce noise, in some cases a lot of noise.

The counter argument, of course, is that wind turbines run on their own and require little or no operations and maintenance (O&M). If this were true (and it is not), then there would be very little traffic on the associated service roads and thus little noise added to wind turbine noise. Wind turbines, though, do not operate by themselves; they are normally operated remotely from a plant operations room by a human operator, but they can also be operated within the turbine nacelle. Their operation is also monitored by a supervisory control and data acquisition (SCADA) communication system designed to alert appropriate service personnel through computer warnings and automated telephone calls.

The major causes of wind turbine downtime have been operations and maintenance (O&M) work and faults. The occurrence of O&M work and faults has been variable. Some sites have downtown of no more than 43 hours per month, while others have had as much as 127 hours (NREL, 2000). O&M downtime includes all troubleshooting, inspections, adjustments, retrofits, and repairs performed on the turbines. Faults generally require no more than a reset and most can be performed remotely. Increased downtime for O&M work and fault reasons means more traffic to and from the turbine farm.

Fast-forwarding to wildlife effects resulting from wind turbine noise alone, it is important to point out that a bird's inability to detect turbine noise at a close range may also be problematic. The threshold for hearing in birds is higher than that for humans at all frequencies, and the overlap in the discernible frequencies between species indicates that birds do not filter out other species simply by being unable to detect theme (i.e., birds can hear songs of other species). In their environment, birds must be able to discriminate their own vocalizations and those of other species apart from any background noise (Dooling, 1982). Calls are important in the isolation of species, pair bond formation, precopulatory display, territorial defense, danger, advertisement of food sources, and flock cohesion (Knight, 1974).

For the average bird, within a signal frequency range of 1 to 4 kHz, noise must be 24 to 30 dB above the ambient noise level in order for a bird to detect it. Turbine blade and wind noise frequencies generally fall below the optimal hearing frequency of birds. Additionally, by the inverse square law, the sound pressure level decreases by 6 dB with every doubling of distance. Therefore, although the sound level of the blade may be significantly above the ambient wind noise level and detectable by birds at the source, as the distance from the source increases and the blade noise level decreases toward the ambient wind noise level, a bird may lose its ability to detect the blade and risk colliding with the moving blade.

Some researchers have attempted to blame avian collisions with turbine blades on birds' inability to divide their attention between surveying the ground for prey and monitoring the horizon and above for obstacles; that is, the birds are so busy searching the ground that they do not notice the turbines. This hypothesis derives from substituting our knowledge of human vision for that of avian vision. Humans are foveate animals; that is, we search the visual world with a small area of the retina known as the fovea, which is our area of sharpest vision, like someone searching a dark room with a narrow-beam searchlight. This results from our very low ratio (approximately 1:1) of photoreceptors to ganglion cells in the macular region of the retina. Outside the macular region, the ratio of receptors to ganglion cells increases progressively to 50:1 to 100:1, and our visual acuity drops sharply. Birds and many other animals, on the other hand, have universal macularity, which means that they have a low ratio of receptors to ganglion cells (4:1 to 8:1) out to the periphery of the retina. They maintain good acuity even in peripheral vision. In addition, raptors possess the specialization of two foveal regions: one for frontal vision and one for looking at the ground. Moreover, birds have various optical methods for keeping objects at different distances simultaneously in focus on the retina. Because of these considerations, failure to divide attention seems like an unlikely hypothesis.

SIDEBAR 5.2. BATS GET A BAD RAP

Ask the average, unknowing person what they think about bats and they may give you that look that says: "Are you nuts or just plain batty? Bats? Hmmmm, you mean those vampire things—those blind, filthy things that hang around Dracula types and suck everyone's blood? Or are you referring to those Halloween make-believe goblin things?" The respondent may simply summarize his or her feelings about bats by stating, "I hate bats. They are scary."

Hate bats? Scary? Experienced conservationists and environmentalists know better. Bats are not something to be hated or frightened of; instead, if we could, we should thank them, embrace them, hug them … well, at least respect them. Why? Because we need them. Yes, we need them, but before explaining why we need them let's debunk some of myths that give bats a bad rap.

- Bats are not blind.
- Bats do not get tangled in your hair.
- Few bats are rabid, and they do not attack humans.
- Vampire bats do not use their stiletto-like teeth to suck blood; instead, they lap it up.
- Bats can digest their food in about 20 minutes; they have a great metabolism.
- Bats are mammals.
- Bats are the only mammals that can fly.
- Bats use echolocation to get around in the dark.
- Bats make up a quarter of all mammals.
- Many bat species are endangered.
- An anticoagulant found in vampire bats may be a medical breakthrough substance to fight heart disease.
- The average bat can live to be 30 years old.
- Bats are cleanliness nuts. They are constantly washing themselves, and they are not filthy!
- A single bat can catch and consume about 1200 mosquito-size insects in an hour.

Along with being one of Mother Nature's creations and thus having a designed purpose on Earth that may be beyond our comprehension, it is the final item on the list above that is important here. Bats are beneficial consumers of agricultural insect pests, and migratory species of bats provide free (and safe) pest-control services across international borders and ecosystems.

FIGURE SB5.2.1 Hoary bat (*Lasiurus cinereus*) found dead beneath a wind turbine, the apparent victim of a blade strike or near-contact barotrauma (lung failure from severe and abrupt pressure chance caused by the spinning blades). (From USGS, *USGS Multimedia Gallery*, U.S. Geological Survey, Washington, DC, 2014, http://gallery.usgs.gov/photos/10_19_2009_s84Aq11PPk_10_19_2009_5#.VpZ0X_EZ0mU. Photograph by Paul Cryan.)

We need bats.

The reader is probably asking what does all this information about bats have to do with wind turbines? Dead or badly injured bats are being found beneath wind turbines all over the world (see Figure SB5.2.1). Bat fatalities have now been documented at most wind facilities in the United States and Canada, and it is estimated that tens to hundreds of thousands die at wind turbines in North America each year. Needless to say, because of the incalculable value of bats to agriculture this unanticipated issue has moved to the forefront of conservation and management efforts directed toward this poorly understood group of mammals. The mystery of why bats die at industrial wind turbines remains unsolved. We do not know what we do not know about bat fatalities.

Let's take a closer look at today's wind turbines before we speak further about bat fatalities caused by wind turbines. Early turbines were mounted on towers 60 to 80 feet in height and had rotors 50 to 60 feet in diameter that turned 60 to 80 revolutions per minute (rpm). Today's land-based turbines are mounted on towers 200 to 250 feet in height with rotors 150 to 260 feet in diameter, resulting in blade tips that can reach over 425 feet above ground level. Rotor-swept areas now exceed 1 acre and are expected to reach nearly 1.5 acres within the next several years. Even though the speed of rotor revolution has significantly decreased to 11 to 28 rpm, blade tip speeds have remained about the same; under normal operating conditions, blade tip speeds range from 138 to 182 mph. Wider and longer blades produce greater vortices and turbulence in their wake as they rotate, posing a potential problem for bats. Because large turbines are more efficient, most modern wind developments for a given number of megawatts (MW; 1 MW equals 1 million watts) have fewer machines with wider spacing. Still, larger turbines are being developed (USDOE, 2010).

Wind energy's ability to generate electricity without many of the environmental impacts associated with other energy sources (e.g., air pollution, water pollution, mercury emissions, climate impact) could benefit bats and many other plant and animal species. However, possible adverse impacts of wind facilities on bats and their habitats have been documented and continue to be an issue. Populations of many bat species are experiencing long-term declines, due in part to habitat loss and fragmentation, invasive species, and numerous anthropogenic impacts, increasing the concern over the potential effects of energy development (USDOE, 2010).

Many questions and theories exist about the bat casualty/wind turbine contact interface. Is it a simple case of flying in the wrong place at the wrong time? Are bats attracted to the spinning turbine blades? Probably the most puzzling question is why are so many bats colliding with turbines compared to the infrequent crashes with other tall, human-made structures? Then there is the ongoing and active research question: Are there ways to predict and minimize risk to bats before turbines are built? (USGS, 2015). Although these questions remain mostly unanswered, potential clues can be found in the patterns of fatalities. Foremost, the majority of bat fatalities at industrial turbines involve species that migrate long distances and rely on trees as roosts throughout the year. Some of these bats migrate long distances and are what we refer to as *tree bats*. Tree bats represent more than three quarters of the bat fatalities observed at wind energy sites. The other striking pattern is that the vast majority of bat fatalities at wind turbines occur during later summer and autumn. This seasonal peak in fatalities coincides with periods of both autumn migration and mating behavior of tree bats. Seasonal involvement of species with shared behavior indicates that behavior plays a key role in the susceptibility of bats to wind turbines, and that migratory tree bats might actually be attracted to turbines (USGS, 2015).

Research also points to weather patterns that may influence bat fatalities (USDOE, 2010). Studies show that fatalities occur primarily on nights with low wind speed and typically increase immediately before and after the passage of storm fronts. Weather patterns therefore may be a predictor of bat activity and fatalities, and mitigation efforts that focus on these high-risk periods may reduce bat fatalities substantially (Arnett et al., 2008). Other research findings indicate that more adults and more male bats tend to be killed by wind turbines. Although this pattern has been documented at a number of facilities, it may represent an idiosyncrasy of those species most commonly killed during their fall migration in North America. Furthermore, the pattern of adult fatalities may not necessarily reflect increased susceptibility of adults but rather a preponderance of adults in the populations. There are notable exceptions, and some studies have reported female and juvenile bias among bat fatalities (e.g., Brown and Hamilton, 2006a,b; Fielder, 2004; Fiedler et al., 2007). It has recently been hypothesized that migratory tree bats (e.g., hoary and eastern red bats) may exhibit lek mating systems (i.e., a gathering of males, of certain animals species, for the purposes of competitive mating display), so the males may be congregating around turbines during autumn in an effort to attract females (Cryan and Brown, 2007).

The bottom line on bat fatalities and wind turbine collisions highlights the importance of ongoing research into causal factors of bat/turbine collisions and the need to increase general knowledge on the importance of bats in the environment. At the present time there is little doubt that bats get a bad rap.

We need bats.

Motion Smear

A bird approaching a moving blade under high wind conditions may be unable to see the blade due to motion smear—reduced visibility of the blades, especially at the tips—and may not hear the blade until it is very close—if it is able at all to hear it at all (Dooling, 2002). As an object moves across the retina with increasing speed, it becomes progressively blurred; this phenomenon is known as *motion smear, motion blur,* or *motion transparency* and is well known in human psychophysical research. It results because the human visual system is sluggish in its response to temporal stimulation; that is, the visual system in humans summates signals over periods of about 120 ms in daylight.

The phenomenon of motion smear is apparent at the tips of wind turbine rotor blades as the observer (bird, human, or camera) approaches the turbine. Motion smear is not apparent in the central regions of the rotors. Even though the central regions and the tips are rotating at the same number of revolutions per minute (RPM), the absolute velocity of the blades is much higher at the peripheral regions. The higher velocity of the bade tip has placed it in the temporal-summation zone, where the retina is sluggish in its ability to resolve temporarily separated stimuli. In contrast, the lower velocities of the more central portions are below the transition point between blur and non-blur, so the individual blades can be seen more or less clearly. Moreover, the absolute velocity of the blade in the visual world is not critical; rather, it is the absolute velocity of the image of the blade that sweeps across the retina that is the critical variable. For reasons that will be explained later, as the observer approaches the turbine, the retinal image of the blades increases in velocity until the retina can no longer process the information. This results in motion smear or motion transparency—the blade becomes transparent to the viewer. A solution to avian collisions with wind turbines must take into account the causes of motion smear and consider whether blade patterns could minimize this effect.

Theory of Motion Smear

One of the characteristics of motion smear is that it eliminates the high spatial frequencies from visual patterns, which is why they appear to go out of focus and become virtually transparent. High spatial frequencies are those Fourier components of a visual object that are found at edges and corners and in the fine details. The print on this page, for example, is made up mainly of high spatial frequencies. If they are removed by optical blur or refractive error, the text becomes transparent and, in the worst case, virtually disappears. Motion smear causing bird collisions with seemingly slow-moving turbines seems paradoxical given the acute vision that most birds, especially raptors, possess; however, as the eye approaches the rotating blades, the retinal image of the blade increases in velocity until it is moving so fast that the retina cannot keep up with it. At this point the retinal image becomes a transparent blur that the bird probably interprets as a safe area to fly through, with disastrous consequences (NREL, 2000).

Law of the Visual Angle

Figure 5.12 shows how objects of different sizes and different distances can form the same size image within the eye. The angle (A′) inside the eye is the same as the angle (A) from the eye to each of the objects. These angles, called *visual angles*, are the conventional units used to describe object size because they are directly related to retinal-image size, which is the only relevant variable for these purposes. Thus, a small object close to the eye can cast the same size retinal image as a large object seen from a much farther distance.

AMBIENT NOISE

Another concern involves the effect of ambient noise on communication distance and an animal's ability to detect calls. For effects to birds, this can mean (1) behavioral and/or physiological effects, (2) damage to hearing from acoustic over-exposure, and (3) masking of communication signals and

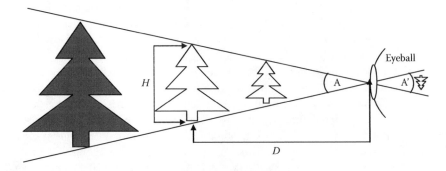

FIGURE 5.12   The law of the visual angle. Objects of different sizes and distances that subtend the same angle will cast the same size image on the retina. Angles A and A′ are the same.

other biologically relevant sounds (Dooling and Popper, 2007). Based on the 49 bird species whose behavioral audibility curves or physiological recordings have been determined, Dooling and Popper (2007) developed a conceptual model for estimating the masking effects of noise on birds. Based on the distance between birds and the spectrum level, bird communication was predicted to be "at risk" (~755-ft distance where noise was 20 dB), "difficult" (~755-ft distance where noise was 25 dB), and "impossible" (~755-ft distance where noise was 30 dB). Although clearly there is variation among species and there is no single noise level where one size fits all, this masking effect of turbine blades is of concern and should be considered as part of the cumulative impacts analysis of a wind facility on wildlife. It must be recognized that noise in the frequency region of avian vocalizations will be most effective in masking these vocalizations (Dooling, 2007).

Barber et al. (2010) assessed the threats of chronic noise exposure, focusing on grouse communication calls, urban bird calls, and other songbird communications. They determined that although some birds were able to shift their vocalizations to reduce the masking effects of noise, when shifts did not occur or were insignificant masking could prove detrimental to the health and survival of wildlife. Although much is still unknown in the real world about the masking effects of noise on wildlife, the results of a physical model analyzing the impacts of transportation noise on listening area (i.e., the active space of vocalization in which animals search for sounds) of animals resulted in some significant findings. With a noise increase of just 3 dB—a noise level identified as "just perceptible to humans"—this increase corresponded to a 50% loss of listening area for wildlife (Barber et al., 2010). Other data suggest that noise increases of 3 to 10 dB correspond to 30 to 90% reductions in alerting distances (i.e., the maximum distance at which a signal can be heard by an animal, particularly important for detecting threats) for wildlife (Barber et al., 2010). Impacts of noise could thus be putting species at risk by impairing signaling and listening capabilities necessary for successful communication and survival.

Swaddle and Page (2007) tested the effects of environmental noise on pair preference selection of zebra finches. They noted a significant decrease in females' preference for their pair-bonded males under high environmental noise conditions. Bayne et al. (2008) found that areas near noise-less energy facilities had total passerine (i.e., birds of the order Passeriformes which includes more than half of all bird species) density 1.5 times greater than areas near noise-producing energy facilities. Specifically, white-throated sparrows, yellow-rumped warblers, and red-eyed vireos were less dense in noisy areas. Habib et al. (2007) found a significant reduction in ovenbird pairing success at compressor sites (averaging 77% success) compared to noiseless well pads (92%). Quinn et al. (2006) found that noise increases perceived predation risk in chaffinches, leading to increased vigilance and reduced food intake rates, a behavior that could over time result in reduced fitness. Francis et al. (2009) showed that noise alone reduced nesting species richness and led to a different composition of avian communities. Although they found that noise disturbance ranged from positive to negative, responses were predominantly negative.

Schaub et al. (2008) investigated the influence of background noise on the foraging efficiency and foraging success of the greater mouse-eared bat, a model selected because it represents an especially vulnerable group of gleaning bats (predators of herbivorous insects) that rely on their capacity to listen for prey rustling sounds to locate food. Their study clearly found that traffic noise, and other sources of intense, broadband noise, deterred bats from foraging in areas where these noises were present, presumably because these sounds masked relevant sounds or echoes the bats use to locate food.

Although there are few studies specifically focused on the noise effects of wind energy facilities on birds, bats, and other wildlife, scientific evidence regarding the effects of other noise sources (e.g., transportation) is widely documented. The results show, as documented in the various examples above, that varying sources and levels of noise can affect both the sending and receiving of important acoustic signaling and sounds. This can also cause behavioral modifications in certain species of birds and bats such as decreased foraging and mating success and overall avoidance of noisy areas. The inaudible frequencies of sound may also have negative impacts on wildlife.

To this point we have focused on wildlife effects of wind turbines and wind turbine farms on birds. Little is known about the effects of noise related to wind turbines on invertebrates, fish, reptiles and amphibians, and various mammals. As stated earlier, some correlation or association has been made between road noise and wind turbine noise in producing wildlife effects. Additional extrapolation in making the connection between road-generated and wind-turbine-generated noise is avoided here because not only is the jury still out but the fact is the jury has yet to be formed.

Given the mounting evidence regarding the negative impacts of noise (specifically, the low-frequency levels of noise such as those created by wind turbines) on birds, bats, and other wildlife, it is important to take precautionary measures to ensure that noise impacts at wind facilities are thoroughly investigated prior to development. Noise impacts on wildlife must be considered during the landscape site evaluation and constructions processes. In an attempt to meet this need, the U.S. Fish and Wildlife Service (USFWS, 2013) developed the Eagle Conservation Plan Guidance (ECPG). Few would argue against the point that, of all of America's wildlife, eagles hold perhaps the most revered place in our national history and culture. The United States has long established special protections for its bald and golden eagle populations. Now, as the nation seeks to increase its production of domestic energy, wind energy developers and wildlife agencies have recognized a need for specific guidance to help make wind energy facilities compatible with eagle conservation and the laws and regulations that protect eagles. The ECPG provides specific in-depth guidance for conserving bald and golden eagles in the course of siting, construction, and operating wind energy facilities. As research specific to noise effects from wind turbines further evolves, these findings should be added to ECPG guidelines and utilized to develop technologies and measures to further minimize noise impacts on wildlife (USFWS, 2013).

WIND ENERGY IMPACTS ON HUMAN HEALTH

Anyone looking at an operational wind turbine or a massive wind farm would have a particular perception about the view. For any number of reasons, the observer's perception might be positive or it might be negative—it's all in the eyes of the beholder (literally, the perception is subjective). But, the eyes are simply the windows to receive neuron-transmitted signals to our brain cells and thus to our thought process. One thought that such observers might have is that a wind turbine could not affect their health, but they would be wrong. What is health, exactly? For our purposes in this text, let's use the World Health Organization's definition of health: "Health is a state of complete physical, mental, and social well-being and not merely the absence of disease or infirmity."

Although from a practical and well-documented viewpoint, the operation of a wind turbine or wind turbine farm does not directly impact human health, it is also true that factors such as stress and loss of sleep contribute to health problems for some residents living close to these installations. Stress can be generated by frustrated residents having to put up with noise pollution and the adverse visual impact and loss of land value; also, loss of sleep can be experienced by people living close to wind turbines as a result of noise pollution.

Observers have complained about turbine noise annoyance and unpleasant sounds that include rhythmic modulation of low-frequency noise (which may be more annoying than steady noise) and increasing sound pressure levels, resulting in increased levels of annoyance. Interestingly, such annoyance was reported more frequently when turbines were visible and when the observer reported a negative impact on surrounding landscape.

It has been widely reported that wind turbines are creating sounds and vibrations that can be sensed by people up to 10 miles away. Magee (2014, p. 227) noted that, "Nina Pierpont, MD, PhD, is reporting that people who live within 2 kilometers of wind turbines are reporting sickness that can be traced to the presence of these. Low frequency noise and infrasound (sound that is less than 20 Hz) appear to be the problem." The problem that Pierpont and others have reported is commonly called *wind turbine syndrome*, which is the disruption or abnormal stimulation of the inner ear's vestibular system caused by turbine infrasound and low-frequency noise. Symptoms of wind turbine syndrome include the following:

- Sleep problems
- Headaches
- Dizziness
- Exhaustion, anxiety, anger, irritability, and depression
- Problems with concentration and learning
- Tinnitus (ringing in the ears)

Along with the turbine noise annoyance generated by mechanical and aerodynamic factors—the feeling of resentment displeasure, discomfort, dissatisfaction, or offense that occurs when noise interferes with someone's thoughts, feeling, or daily activities (Concha-Barrientos et al., 2005)—there have been complaints about rhythmic light flicker causing intermittent shadows known as *shadow flicker* or *flickering shadows*. Schworm and Filipov (2013) reported that a woman from Kingston, Massachusetts, stated that the problem with shadow flicker begins in the late afternoon. Stripes of shadow whip across her living room, kitchen, and bedroom, a pulse of flashing light and dark that can continue more than an hour and makes her think she is losing her mind. "You can't stay in your room. You get a headache," the woman said. "You can't live your life."

Another increasing complaint being heard concerning wind turbine noise generation is related to high levels of low-frequency noise over years of exposure. This problem is called *vibroacoustic disease* (VAD). The clinical progression is insidious, and lesions are found in many systems throughout the body. This disease or syndrome is commonly classified as mild, moderate, or severe as described below:

- *Mild* (1–4 years)—Slight mood swings, indigestion, heartburn, mouth/throat infections, bronchitis
- *Moderate* (4–10 years)—Chest pain, definite mood swings, back pain, fatigue, skin infections (fungal, viral, and parasitic), inflammation of the stomach lining, painful urination and blood in the urine, conjunctivitis, allergies
- *Severe* (>10 years)—Psychiatric disturbances, hemorrhages (nasal, digestive, conjunctiva mucosa), varicose veins, hemorrhoids, duodenal ulcers, spastic colitis, decrease in visual activity, headaches, severe joint pain, intense muscular pain, neurological disturbances

POWER TRANSMISSION LINES[*]

In the preceding paragraphs we have set the stage for the more detailed discussion that follows regarding the environmental impact of power transmission lines. Basically, energy transmission involves three stages of implementation: site evaluation, project construction, and transmission operation.

ENERGY TRANSMISSION SITE EVALUATION IMPACTS

Energy site evaluation phase activities are generally temporary and conducted at a smaller magnitude than those during the construction and operation phases. Potential impacts from these activities are presented by the type of affected resource and are described for typical site evaluation activities, such as limited ground clearing, vehicular and pedestrian traffic, borings for geotechnical surveys, and positioning of equipment. If excavation or access road construction is necessary at this stage, impacts on resources would be similar in character, but lesser in magnitude, to those for the construction phase. Route and access road selection that avoids major environmental impacts is ideal; therefore,

---

[*] Adapted from Tribal Energy and Environmental Information Clearinghouse website, http://teeic.indianaffairs.gov/.

additional activities that could occur during this phase are field surveys for recording significant resources present in the potential project area (e.g., threatened species and endangered species, wetlands, archaeological sites). These surveys are typical of the short and limited disturbance that occurs during the site evaluation phase. The following impacts may result from site evaluation activities.

Air Quality

Impacts on air quality during surveying and testing activities would be limited to temporary and local generation of vehicle and boring equipment emissions and fugitive dust and would not likely cause an exceedance of air quality standards nor have any impact on climate change.

Cultural Resources

The amount of surface and subsurface disturbance is minimal during the site evaluation phase. Cultural resources buried below the surface are unlikely to be affected, but material present on the surface could be disturbed by vehicular traffic, ground clearing, and pedestrian activity (including collection of artifacts). Surveying and testing activities could affect areas of interest to Native Americans depending on the placement of equipment or level of visual intrusion. Surveys conducted during this phase to evaluate the presence or significance of cultural resources in the area would assist developers in routing and designing the project to avoid or minimize impacts on these resources.

Ecological Resources

Impacts on ecological resources (vegetation, wildlife, aquatic, biota, special status species, and their habitats) would be minimal and localized during surveying and testing because of the limited nature of the activities. The introduction or spread of some nonnative invasive vegetation could occur as a result of vehicular traffic, but this would be relatively limited in extent. Surveys conducted during this phase to evaluate the presence or significance of ecological resources in the area would assist developers in routing and designing the project to avoid or minimize impacts on these resources (e.g., wetlands, migratory birds, threatened and endangered species).

Water Resources

Minimal impact on water resources, local water quality, water flows, and surface water/groundwater interaction is anticipated. Very little water would likely be used or generated during the site evaluation phase. Any water needed could be trucked in from offsite.

Land Use

Site evaluation activities would likely result in temporary and localized impacts on land use. These activities could create a temporary disturbance in the immediate vicinity of a surveying or monitoring site (e.g., disturb recreational activities or livestock grazing). Site evaluation activities are unlikely to affect mining activities, military operations, or aviation.

Soils and Geologic Resources

The amount of surface disturbance and the use of geologic materials are minimal during the site evaluation phase, and soils and geologic resources are unlikely to be affected. Surveying and testing activities would be unlikely to activate geological hazards or increase soil erosion. Borings for soil testing and geotechnical surveys provide useful site-specific data on these resources. Surface effects from pedestrian and vehicular traffic could occur in areas that contain special (e.g., cryptobiotic) soils.

Paleontological Resources

The amount of subsurface disturbance is minimal during the site evaluation phase, and paleontological resources buried below the surface are unlikely to be affected. Fossil material present on the surface could be disturbed by vehicular traffic, ground clearing, and pedestrian activity (including collection of fossils). Surveys conducted during this phase to evaluate the presence or significance of paleontological resources in the area would assist developers in routing and designing the project to avoid or minimize impacts on these resources.

Transportation

No impacts on transportation are anticipated during the site evaluation phase. Transportation activities would be temporary and intermittent and limited to low volumes of heavy- and medium-duty construction vehicles (e.g., pickup trucks) and personal vehicles.

Visual Resources

Surveying and testing activities would have temporary and minor visual effects caused by the presence of workers, vehicles, and equipment.

Socioeconomics

The activities during the site evaluation phase are temporary and limited and would not result in socioeconomic impacts on employment, local services, or property values.

Environmental Justice

Site evaluation phase activities are limited and would not result in significant adverse impacts in any resource area; therefore, environmental justice is not expected to be an issue during this phase.

ENERGY TRANSMISSION CONSTRUCTION IMPACTS

Typical activities during the construction phase of an energy transmission project include ground clearing and removal of vegetative cover, grading, excavation, blasting, trenching, drilling, vehicular and pedestrian traffic, and project component construction and installation. Activities conducted in locations other than within the project right of way (ROW) include excavation/blasting for construction materials (such as sands and gravels), access road and staging area construction, and construction of other ancillary facilities such as compressor stations or pump stations.

Air Quality

Emissions generated during the construction phase include vehicle emissions; diesel emissions from large construction equipment and generators; VOC emissions from storage and transfer of fuels for construction equipment; small amounts of carbon monoxide, nitrogen oxides, and particulates from blasting activities; and fugitive dust from many sources such as disturbing and moving soils (clearing, grading, excavating, trenching, backfilling, dumping, and truck and equipment traffic), mixing concrete, storage of unvegetated soil piles, and drilling and pile driving. Air quality impacts could also occur if cleared vegetation is burned.

Cultural Resources

Potential impacts on cultural resources include the following:

- Complete destruction of the resource could occur if present in areas undergoing surface disturbance or excavation.
- Degradation or destruction of near-surface cultural resources on- and offsite could result from changing the topography, changing the hydrological patterns, and soil movement (removal, erosion, sedimentation).

- Unauthorized removal of artifacts or vandalism could occur as a result of human access to previously inaccessible areas.
- Visual impacts could result from large areas of exposed surface, increases in dust, and the presence of large-scale equipment, machinery, and vehicles (if the resources have an associated landscape component that contributes to their significance, such as a sacred landscape or historic trail). Note that the accumulation of sediment mentioned above could serve to protect some buried resources by increasing the amount of protective cover.

Ecological Resources

Adverse impacts on ecological resources could occur during construction due to the following:

- Erosion and runoff
- Fugitive dust
- Noise
- Introduction and spread of invasive nonnative vegetation
- Modification, fragmentation, and reduction of habitat
- Mortality of biota
- Exposure to contaminants
- Interference with behavioral activities

Site clearing and grading, coupled with construction of access roads, towers, and support facilities, could reduce, fragment, or dramatically alter existing habitats in the disturbed portions of the project area. Wildlife in surrounding habitats might also be affected if construction activities (and associated noise) disturb normal behaviors, such as feeding and reproduction.

Water Resources

Water would be required for dust control, making concrete, and consumptive use by the construction crew. Depending on availability, it may be trucked in from offsite or obtained from local groundwater wells or nearby surface water bodies. Water quality can be affected by the following:

- Activities that cause soil erosion
- Weathering of newly exposed soils causing leaching and oxidation that can release chemicals into the water
- Discharges of waste or sanitary water
- Herbicide applications
- Contaminant spills, especially oil

Applying sand and gravel for road construction, layout areas, foundations, etc., can alter the drainage near where the material is used. The size of the area affected can range from a few hundred square feet (for a support tower foundation) to a few hundred acres (for an access road). Surface and groundwater flow systems could be affected by withdrawals made for water use, wastewater and stormwater discharges, and the diversion of surface water flow for access road construction or stormwater control systems. Excavation activities and the extraction of geological materials may affect surface and groundwater flow. The interaction between surface water and groundwater may also be affected if the two are hydrologically connected, potentially resulting in unwanted dewatering or recharging.

Land Use

Impacts on land use could occur during construction if there are conflicts with existing land use plans and community goals; conflicts with existing recreation, education, religious, scientific, or other use areas; or conversion or cessation of the existing commercial land use of the area (e.g., mineral extraction). During construction, most land use impacts would be temporary, such as removal of livestock from grazing areas during periods of blasting or heavy equipment operations; curtailing hunting near work crews; or temporary effects on the character of a recreation area because of construction noise, dust, and visual intrusions. Long-term land use impacts would occur if existing land uses are not compatible with an energy transmission project, such as remote recreational experiences. Within forested areas, ROW clearing could result in the long-term loss of timber production.

Impacts on aviation could be possible if the project is located within 20,000 feet (6100 meters) or less of an existing public or military airport or if proposed construction involves objects greater than 200 feet (61 meters) in height. The Federal Aviation Administration (FAA) must be notified if either of these two conditions occurs and would be responsible for determining if the project would adversely affect commercial, military, or personal air navigation safety. Similarly, impacts on military operations may occur if the project is located near a military facility and that facility conducts military testing and training activities that occur at low altitudes.

Soils and Geologic Resources

Surface disturbance, heavy equipment traffic, and changes to surface runoff patterns can cause soil erosion. Impacts of soil erosion include soil nutrient loss and reduced water quality in nearby surface water bodies. Impacts on special soils (e.g., cryptobiotic soils) could also occur. Sands, gravels, and quarry stone would be excavated for use in the construction of access roads; for use in making concrete for foundations and ancillary structures; for improving the ground surface for lay-down areas and crane staging areas; and, as necessary, for use as backfill in pipeline trenches. Mining operations would disturb the ground surface, and runoff would erode fine-grained soils, increasing the sediment load farther down in streams and rivers. Mining on steep slopes or on unstable terrain without appropriate engineering measures increases the landslide potential in the mining areas. Possible geological hazards (earthquakes, landslides, avalanches, forest fires, geomagnetic storms, ice jams, mudflows, rock falls, flash floods, volcanic eruptions, geyser deposits, ground settlement, sand dune migration, thermals springs, etc.) can be activated by excavation and blasting for raw material, increasing slopes during site grading and construction of access roads, altering natural drainage patterns, and toe-cutting bases of slopes. Altering drainage patterns accelerates erosion and creates slope instability.

Paleontological Resources

Potential impacts on paleontological resources during construction include (1) complete destruction of the resource if present in areas undergoing surface disturbance or excavation; (2) degradation or destruction of near-surface fossil resources on- and offsite resulting from changing the topography, changing the hydrological patterns, and soil movement (removal, erosion, sedimentation); and (3) unauthorized removal of fossil resources or vandalism to the locality as a result of human access to previous inaccessible areas. The accumulation of sediment mentioned above could serve to protect some localities by increasing the amount of protective cover.

Transportation

Short-term increases in the use of local roadways would occur during the construction period. Heavy equipment would need to be continuously moved as construction progresses along the linear project. Shipments of materials are not expected to significantly affect primary or secondary road networks but would depend on the ever-changing location of the construction site area relative to material source. Overweight and oversized loads could cause temporary disruptions and could require some modification to roads or bridges (such as fortifying bridges to accommodate the size or weight). The weight of shipments is also a parameter in the design of access roads for grade determinants and turning clearance requirements.

Visual Resources

Potential sources of visual impacts during construction include visual constraints in the landscape due to access roads and staging areas, conspicuous and frequent small-vehicle traffic for worker access, and frequent large-equipment (trucks, graders, excavators, cranes, and, possibly helicopters) traffic for project and access road construction. Project component installation would produce visible activity and dust in dry soils. Project construction may be progressive, persisting over a significant period of time. Ground disturbance (e.g., trenching and grading) would result in visual impacts that produce contrasts of color, form, texture, and line. Soil scars and exposed slope faces could result from excavation, leveling, and equipment movement.

Socioeconomics

Direct impacts would include the creation of new jobs for construction workers and the associated income and taxes generated by the project. As an example, the number of construction workers required for a 150-mile (241-kilometer) length of pipeline is only about 230 annual direct workers (fewer than 175 for an equivalent length of a transmission line or a petroleum pipeline). Indirect impacts are those impacts that would occur as a result of the new economic development and would include things such as new jobs at businesses that support the expanded workforce or that provide project materials and the associated income and taxes. Construction of an energy transmission project may affect the value of residential properties located adjacent to the ROW (there are conflicting reports as to whether these would be adverse, beneficial, or neutral).

Environmental Justice

If significant impacts were to occur in any of the resource areas and were to disproportionately affect minority or low-income populations, there could be an environmental justice impact. Issues that could be of concern are noise, dust, visual impacts, and habitat destruction from construction activities and possible impacts associated with new access roads.

ENERGY TRANSMISSION OPERATIONS IMPACTS

Typical activities during the operation and maintenance phase include operation of compressor stations or pump stations, ROW inspections, ROW vegetation clearing, and maintenance and replacement of facility components. Environmental impacts that could occur during the operation and maintenance phase would mostly occur from long-term habitat change with the ROW, maintenance activities (e.g., ROW vegetation clearing, facility component maintenance or replacement), noise (e.g., compressor station, corona discharge), the presence of workers, and potential spills (e.g., oil spills).

Air Quality

Vehicular traffic and machinery would continue to produce small amounts of fugitive dust and exhaust emissions during the operation and maintenance phase. These emissions would not likely cause an exceedance of air quality standards nor have any impact on climate change. Trace amounts of ozone would be produced by corona effects from transmission lines (e.g., less than 1.0 part per billion, which is considerably less than air quality standards). Routine venting of pipelines and breakout tanks (for liquid petroleum products and crude oil) would also cause localized air quality impacts.

Cultural Resources

Impacts during the operations and maintenance phase could include damage to cultural resources during vegetation management and other maintenance activities, unauthorized collection of artifacts, and visual impacts. This threat is present when the access roads have been constructed and the ROW has been established, making remote areas more accessible to the public. Visual impacts resulting from the presence of the aboveground portion of a pipeline, transmission lines, and associated facilities could impact cultural resources that have an associated landscape component that contributes to their significance, such as a sacred landscape or historic trails.

Ecological Resources

During operations and maintenance, adverse impacts on ecological resources could occur from the following:

- Disturbance of wildlife from noise and human activity
- ROW maintenance (e.g., vegetation removal)
- Exposure or biota to contaminants
- Mortality of biota caused by collisions with transmission lines or aboveground pipeline components

Ecological resources may continue to be affected by the reduction in habitat quality associated with habitat fragmentation due to the presence of the ROW, support facilities, and access roads. In addition, the presence of an energy transmission project and its associated access roads may increase human use of surrounding areas, which in turn could impact ecological resources in the surrounding areas through

- Introduction and spread of invasive nonnative vegetation
- Fragmentation of habitat
- Disturbance of biota
- Collisions with or electrocution of birds
- Increased potential for fire

Water Resources

Impacts on water resources during the operation and maintenance phase would be limited to possible minor degradation of water quality resulting from vehicular traffic and machinery operation during maintenance (e.g., erosion and sedimentation) or herbicide contamination during vegetation management (e.g., from accentual sills). However, a large oil pipeline spill could potentially cause extensive degradation of surface waters or shallow groundwater.

Land Use

Land use impacts would be minimal, as many activities could continue with the ROW (e.g., agriculture and grazing). Other industrial and energy projects would likely be excluded within the ROW. In addition, construction of facilities (such as houses and other structures) would be precluded within the ROW, and roads would only be allowed to cross ROWs, not run along their length. Recreation activities (e.g., off-highway vehicle use, hunting) are also possible, although restrictions may exist for the use of guns, especially for aboveground pipelines or transmission lines. The ROW and access roads may make some areas more accessible for recreation activities. Activities centered on solitude and scenic beauty would potentially be affected. Military operations and aviation could be affected by the presence of transmission lines; for example, transmission lines could affect military training and testing operations that may occur at low altitudes (e.g., military training routes).

Soils and Geologic Resources

Following construction, disturbed portions of the site would be revegetated and the soil and geologic conditions would stabilize. Impacts during the operation phase would be limited largely to soil erosion impacts caused by vehicular traffic and machinery operation during maintenance activities. Any excavations required for pipeline maintenance would cause impacts similar to those from construction, but to a lesser spatial and temporal extent. Herbicides would likely be used for ROW maintenance. The accidental spills of herbicides or pipeline product would likely cause soil contamination. Except in the case of a large oil spill, soil contamination would be localized and limited in extent and magnitude.

Paleontological Resources

Impacts during the operations phase would be limited to unauthorized collection of fossils. This threat is present once the access roads are constructed and the ROW is established, making remote areas more accessible to the public.

Transportation

No noticeable impacts on transportation are likely during the operation and maintenance phase. Low volumes of heavy- and medium-duty pickup trucks, personal vehicles, and other machinery are expected to be used during this phase. Infrequent, but routine, shipments of component replacements during maintenance procedures are likely over the period of operation.

Visual Resources

The aboveground portions of energy transmission projects would be highly visible in rural or natural landscapes, many of which have few other comparable structures. The artificial appearance of a transmission line or pipeline may have visually incongruous industrial-type associations for some, particularly in a predominantly natural landscape. Visual evidence of these projects cannot be completely avoided, reduced, or concealed. Additional visual impacts would occur during maintenance from vehicular traffic, aircraft, and workers. Maintenance, replacement, or upgrades of project components would repeat the initial visual impacts of the construction phase, although at a more localized scale.

Socioeconomics

Direct impacts would include the creation of new jobs for operation and maintenance workers and the associated income and taxes paid. Indirect impacts are those impacts that would occur as result of the new economic development and would include things such as new jobs at businesses that support the expanded workforce or that provide project materials and the associated income and taxes. The number of project personnel required during the operation and maintenance phase would be about an order of magnitude less than during construction; therefore, socioeconomic impacts related directly to jobs would be minimal. Potential impacts on the value of residential properties located adjacent to an energy transmission project would continue during this phase.

Environmental Justice

Possible environmental justice impacts during operation include the alteration of scenic quality in areas of traditional or cultural significance to minority or low-income populations. Habitat modification, noise impacts, and health and safety impacts are also possible sources of environmental justice impacts.

WIND TURBINE OPERATIONS AND MAINTENANCE PERSONNEL SAFETY CONCERNS

WIND ENERGY FATALITIES/INCIDENTS

Wind energy workers are exposed to hazards that can result in fatalities and serious injuries. Many incidents involving falls, severe burns from electrical shocks and arc flashes/fires, and crushing injuries have been reported to OSHA. Some examples are given below:

- On August 29, 2009, at 8:30 a.m., a 33-year-old male lineman was shocked as he grasped a trailer ramp attached to a low-boy trailer containing an excavator. The excavator was being operated in anticipation of being offloaded from the trailer. The trailer was parked on a rural aggregate road adjacent to an access road for a wind turbine generator. The excavator operator rotated the upper works of the machine prior to moving the machine from the trailer. During the rotation the boom contacted a 7200-volt primary rural power line. The power line was approximately 12 feet from the road, and the trailer was parked approximately 2 feet from the road edge. The injured worker had entry wounds in his hands and exit wounds in his feet. He was transported by ambulance to a local hospital, where he was treated and admitted for observation. He was discharged approximately 24 hours later and returned to work the following day.
- On May 10, 2009, a worker was in the bottom power cabinet of a wind turbine to check the electrical connections. He came into contact with a bus bar, and an arc flash erupted, causing him injury. The victim was being taken to a hospital by a technician when they were met by an ambulance on the way. After arriving at the hospital, the victim was later transferred by medi-vac to another hospital in Oklahoma City where he was treated for injuries. On June 12, the company was notified by a representative of the hospital that the victim had died.
- On November 11, 2005, a man and two coworkers were removing and replacing a broken bolt in the nacelle assembly of a wind turbine tower that was approximately 200 feet above the ground. They were heating the bolt with an oxygen–acetylene torch when a fire started. The man retreated to the rear of the nacelle, away from the ladder access area. Although his two coworkers were able to descend the tower, the man fell approximately 200 feet to the ground, struck an electrical transformer box, and was killed.
- At approximately 11:40 a.m. on June 17, 1992, a worker attempted to descend an 80-foot ladder that accessed a wind turbine generator. The worker slipped and fell from the ladder and was killed. The victim was wearing his company-furnished safety belt, but the safety lanyards were not attached. Both lanyards were later discovered attached to their tie-off connection at the top of the turbine generator.
- A site foreman was replacing a 480-volt circuit breaker serving a wind turbine. He turned a rotary switch to what he thought was the open position in order to isolate the circuit breaker; however, the worker did not test the circuit to ensure that it was de-energized. The worker had placed the rotary switch in a closed position, and the circuit breaker remained energized by back feed from a transformer. Using two plastic-handled screwdrivers, he shorted two contacts on the breaker to discharge static voltage buildup. This caused a fault and the resultant electric arc caused deep flash burns to the worker's face and arms and ignited his shirt. The worker was hospitalized in a burn unit for 4 days.

Case Study 5.1. Wind Turbine Fatality[*]

The Oregon Department of Consumer and Business Services, Occupational Safety and Health Division (Oregon OSHA), fined Siemens Power Generation, Inc., a total of $10,500 for safety violations related to an August 25, 2007, wind turbine collapse that killed one worker and injured another. "The investigation found no structural problems with the tower," said Michael Wood, Oregon OSHA administrator. "This tragedy was the result of a system that allowed the operator to restart the turbine after service while the blades were locked in a hazardous position. Siemens has made changes to the tower's engineering controls to ensure it does not happen again."

---

[*] Adapted from Department of Consumer Business & Services, Oregon OSHA Releases Findings in Wind Turbine Collapse [news release], Oregon Occupational Safety and Health Administration, Salem, OR, February 26, 2008.

The event took place at the Klondike III Wind Farm near Wasco, where three wind technicians were performing maintenance on a wind turbine tower. After applying a service brake to stop the blades from moving, one of the workers entered the hub of the turbine. He then positioned all three blades to the maximum wind-resistance position and closed all three energy isolation devices on the blades. The devices are designed to control the mechanism that directs the blade pitch so workers do not get injured while they are working in the hub. Before leaving the confined space, the worker did not return the energy isolation devices to the operational position.

As a result, when he released the service brake, wind energy on the out-of-position blades created an "overspeed" condition, causing one of the blades to strike the tower and the tower to collapse, the Oregon OSHA investigation found. Chadd Mitchell, who was working at the top of the tower, died in the collapse. William Trossen, who was on his way down a ladder in that tower when it collapsed, was injured. The third worker was outside the tower and unharmed.

During the investigation, Oregon OSHA found several violations of safety rules:

- Workers were not properly instructed and supervised in the safe operation of machinery, tool, equipment, process, or practice they were authorized to use or apply. The technicians working on the turbine each had less than two months' experience, and there was no supervisor onsite. The workers were unaware of the potential for catastrophic failure of the turbine that could occur as a result of not restoring energy isolation devices to the operational position.
- The company's procedures for controlling potentially hazardous energy during service or maintenance activities did not fully comply with Oregon OSHA regulations. Oregon OSHA requirements include developing, documenting, and using detailed procedures and applying lockout or tagout devices to secure hazardous energy in a "safe" or "off" position during service or maintenance. Several energy isolation devices in the towers, such as valves and lock pins, were not designed to hold a lockout device, and energy control procedures in place at the time of the accident did not include the application and removal of tagout devices.
- Employees who were required to enter the hub (a permit-required confined space) or act as attendants to employees entering the hub had not been trained in emergency rescue procedures from the hub.

WIND TURBINE HAZARDS AND APPLICABLE OHSA STANDARDS AND CONTROLS

As shown in Case Study 5.1, wind energy employers need to protect their workers from workplace hazards. Workers should be engaged in workplace safety and health, and they need to understand how to protect themselves from these hazards. Even though the wind energy industry is a growing industry, the hazards are not unique, and OSHA has many standards that cover various worker on-the-job activities and exposures. The hazards (along with controls) that workers in wind energy may face are provided in this section.

Working around, with, or on wind turbines and associated equipment presents many hazards to the installers, operators, and maintenance personnel. The hazards along with applicable OSHA standards and controls include the following:

- Falls
- Confined space entry
- Fire
- Lockout/tagout
- Medical and first aid
- Crane, derrick, and hoist safety

- Electrical safety
- Machine guarding
- Respiratory protection

Those charged with managing workers in the wind energy production fields are responsible to ensure compliance with OSHA, EPA, CDC, state, and local safety and health requirements. Each of the hazards and controls listed above must be addressed and appropriate measures taken to ensure worker health and safety.

BOTTOM LINE ON WIND POWER

Technology is much more advanced today in utilizing our wind resource, and the United States is home to one of the best wind resource areas in the world: the Midwest states of North and South Dakota, Nebraska, Kansas, Montana, Iowa, and Oklahoma (Archer and Jacobson, 2004). However, as with any other source of energy, nonrenewable or renewable, there are advantages and disadvantages associated with their use. On the positive side, it should be noted that wind energy is a free, renewable resource, so no matter how much is used today there will still be the same supply in the future. Wind energy is also a source of clean, non-polluting electricity. Wind turbines can be installed on farms or ranches, thus benefiting the economy in rural areas, where most of the best wind sites are found. Moreover, farmers and ranchers can continue to work the land because the wind turbines use only a fraction of the land—the height and distance between turbines mean that land used for wind turbines can also be used for agriculture and grazing; only about 5% of the land in a wind farm is actually occupied by the turbines themselves. One huge advantage of wind energy is that it is a domestic source of energy; in the United States, the wind supply is abundant.

On the other side of the coin, wind energy does have a few negatives—wind projects face opposition. Wind power must compete with conventional generation sources on a cost basis. Even though the cost of wind power has decreased dramatically in the past 10 years, the technology requires a higher initial investment than for fossil-fueled generators. The challenge to using wind as a source of power is that the wind is intermittent and does not always blow when electricity is needed. Wind energy cannot be stored (unless batteries are being used), and not all winds can be harnessed to meet the timing of electricity demands. Another problem is that good sites are often located in remote locations, far from cities where the electricity is needed. Moreover, wind resource development may compete with other uses for the land, and those alternative uses may be more highly valued than electricity generation. Finally, with regard to the environment, wind power plants have relatively little impact on the environment compared to other conventional power plants, but there is some concern over the noise produced by the rotor blades (most experts agree that wind turbine noise is generally not a major concern beyond a half mile) and aesthetic (visual) impacts. Sometimes birds are killed by flying into the rotors. Most of these problems have been resolved or greatly reduced through technological development or by properly siting wind plants, but the NIMBY point of view is still alive and strong in the United States and has succeeded in killing many projects, even renewable, clean energy projects.

Wind energy jobs are the proverbial double-edged sword. The development of wind energy provides benefits related to moving us toward energy independence, creates new jobs, and improves environmental conditions. However, wind energy jobs expose workers to a number of safety hazards. Workers may be exposed to the same conventional hazards found in most work-places—such as slips, trips, and falls, confined spaces, electrical, fire, and other similar hazards. Additionally, though, wind energy workers may be exposed to new hazards, many of which have yet to be identified.

DISCUSSION AND REVIEW QUESTIONS

1. What factors must be taken into account when considering wind energy as an alternative renewable energy source? Explain.
2. In your opinion, is wind power a viable source of renewable energy?
3. When opponents of wind farms argue against them for aesthetic reasons, the counter argument has been offered that the wind farm has to proceed based on their ecological benefits alone, along the lines of "eat your spinach" mode of persuasion. Does this argument make sense? Explain.
4. If you were assigned to design wind turbine farms to be aesthetically pleasing, how would you do it?
5. How would you protect wildlife from wind turbine operation?
6. If a hillside was filled with oil rigs instead of wind turbines, would you then prefer the wind turbines instead?

REFERENCES AND RECOMMENDED READING

Anon. (2010). Measuring wind turbine noise. Are decibel levels the most important metric for determining impact? *RenewableEnergyWorld.com*, November 22 (http://www.renewableenergyworld.com/rea/news/article/2010/11/measuring-wind-turbine-noise).

Anon. (2011). Wind turbines—kinetic wind energy generator technology. *Alternative Energy News*, November 14 (http://www.alternative-energy-news.info/technology/wind-power/wind-turbines/).

Anon. (2014). *dB: What Is a Decibel?* Sydney, Australia: University of New South Wales School of Physics (http://www.animations.physics.unsw.edu.au/jw/dB.htm).

Archer, C. and Jacobson, M.Z. (2004). *Evaluation of Global Wind Power.* Stanford, CA: Department of Civil and Environmental Engineering, Stanford University.

Arnett, E.B. et al. (2007). *Impacts of Wind Energy Facilities on Wildlife and Wildlife Habitat*, Technical Review 07-2. Bethesda, MD: The Wildlife Society.

Arnett, E.B. et al. (2008). Behavior of bats at wind turbines. *PNAS*, 111: 42.

Barber, J.R., Cooks, K.R., and Fristrup, K. (2010). The costs of chronic noise exposure for terrestrial organisms. *Trends in Ecology & Evolution*, 25(3): 180–189.

Bayne, E.M., Habib, L., and Boutin. S. (2008). Impacts of chronic anthropogenic noise from energy-sector activity on abundance of songbirds in the boreal forest. *Conservation Biology*, 22(5): 1186–1193.

Betz, A. (1966). *Introduction to the Theory of Flow Machines* (D.G. Randall, trans.). Oxford: Pergamon Press.

BPWENA. (2011). *Mohave County Wind Farm Project.* Houston, TX: BP Wind Energy North America.

Brown, M.H, and Sedano, R.P. (2004). *Electricity Transmission—A Primer.* Washington, DC: Regulatory Assistance Project, National Council on Energy Policy.

Brown, W.K. and Hamilton, B.L. (2006a). *Bird and Bat Interactions with Wind Turbines: Castle River Wind Farm, Alberta (2001–2002).* Calgary, AB: Vision Quest Windelectric.

Brown, W.K. and Hamilton, B.L. (2006b). *Monitoring of Bird and Bat Collisions with Wind Turbines at the Summerview Wind Power Project, Alberta (2005–2006).* Calgary, AB: Vision Quest Windelectric.

Burton, T., Sharpe, D., Jenkins, N., and Bossanyi, E. (2001). *Wind Energy Handbook.* New York: John Wiley & Sons.

Cappiello, D. (2013a). U.S. to allow eagle deaths—to aid wind power. *The Big Story*, December 6 (http://bigstory.ap.org/article/wind-power-us-extends-permit-eagle-deaths).

Cappiello, D. (2013b). Eagle deaths at wind farms glossed over. *Virginian-Pilot (Norfolk, VA)*, May 15.

Casey, Z. (2012a). Wind turbine health effects: are wind turbines really unhealthy for humans? *Green Chip Stocks*, November 9 (http://www.greenchipstocks.com/articles/wind-turbine-health-effects/2154).

Casey, Z. (2012b). How noisy is a wind turbine? *EWEA Blog*, November 13 (http://www.ewea.org/blog/2012/11/how-noisy-is-a-wind-turbine/).

CDC. (2005). *Noise and Hearing Loss Prevention.* Atlanta, GA: Centers for Disease Control and Prevention (http://www.cdc.gov/niosh/topics/noise/faq.html).

CDOT. (1998). *Technical Noise Supplement: A Technical Supplement to the Traffic Noise Analysis Protocol.* Sacramento: California Department of Transportation (http://www.dot.ca.gov/hq/env/noise/pub/Technical%20Noise%20Supplement.pdf).

Concha-Barrientos, M., Campbell-Lendrum, D., and Steenland, K. (2005). *Occupational Noise: Assessing the Burden of Disease from Work-Related Hearing Impairment at National and Local Levels*. Geneva, Switzerland: World Health Organization (http://www.who.int/quantifying_ehimpacts/publications/en/ebd9.pdf).

Cryan, P.M. (2008). Mating behavior as a possible cause of bat fatalities at wind turbines. *Journal of Wildlife Management*, 72: 845–849.

Cryan, P.M. and Brown, A.C. (2007). Migration of bats past a remote island offers clues toward the problem of bat fatalities at wind turbines. *Biological Conservation*, 139: 1–11.

Denholm, P., Hand, M., Jackson, M., and Ong, S. (2009). *Land-Use Requirements of Modern Wind Power Plants in the United States*, Technical Report NAREL/TP-6A2-45834. Washington, DC: National Renewable Energy Laboratory.

Dooling, R.J. (1982). Auditory perception in birds. In: *Acoustic Communications in Birds*, Vol. 1, Kroodsma, D.E. and Miller, E.H., Eds., pp. 95–130. New York: Academic Press.

Dooling, R.J. (2002). *Avian Hearing and the Avoidance of Wind Turbines*, NREL/TP-500-30844. Golden, CO: National Renewable Energy Laboratory (http://www.nrel.gov/wind/pdfs/30844.pdf).

Dooling, R.J. and Popper, A.N. (2007). *The Effects of Highway Noise on Birds*, report to the California Department of Transportation, Contract 43A0139. Sacramento: California Department of Transportation, Division of Environmental Analysis.

EERE. (2006a). *Wind & Water Power Program*. Washington, DC: Office of Energy Efficiency & Renewable Energy, U.S. Department of Energy (http://www1.eere.energy.gov/windandhydro/).

EERE. (2006b). *How Wind Turbines Work*. Washington, DC: Office of Energy Efficiency & Renewable Energy, U.S. Department of Energy (http://www1.eere.energy.gov/Windandhydro/wind_how.html).

EERE. (2015). *Frequently Asked Questions on Small Wind Systems*. Washington, DC: U.S. Office of Energy Efficiency & Renewable Energy, U.S. Department of Energy (http://energy.gov/eere/wind/frequently-asked-questions-small-wind-systems).

Elliott, D.L., Holladay, C.G., Barchet, W.R., Foote, H.P., and Sandusky, W.F. (1987). *Wind Energy Resource Atlas of the U.S.* Golden, CO: Solar Technical Information Program, Solar Energy Institute (rredc.nrel.gov/wind/pubs/atlas/).

Fiedler, J.K. (2004). Assessment of Bat Mortality and Activity at Buffalo Mountain Windfarm, Eastern Tennessee, master's thesis, University of Tennessee, Knoxville.

Fiedler, J.K., Henry, T.H., Tankersley, R.D., and Nicholson, C.P. (2007). *Results of Bat and Bird Mortality Monitoring at the Expanded Buffalo Mountain Windfarm, 2005*. Knoxville: Tennessee Valley Authority.

Foreman, R.T.T. and Alexander, L.E. (1998). Roads and their major ecological effects. *Annual Review of Ecological Systems*, 29: 207–231.

Francis, C.D., Ortega, C.P., and Cruz, A. (2009). Noise pollution changes avian communities and species interactions. *Current Biology*, 19: 1415–1419.

Gasaway, D.C. (1985). *Hearing Conservation: A Practical Manual and Guide*. Englewood Cliffs, NJ: Prentice-Hall.

Habib, L., Bayne, E.M., and Boutin, S. (2007). Chronic industrial noise affects pairing success and age structure of ovenbirds *Seiurus aurocapilla*. *Journal of Applied Ecology*, 44: 176–184.

Hanson, B.J. (2004). *Energy Power Shift*. Maple, WI: Lakota Scientific Press.

Kaliski, K. (2009). *Calibrating Sound Propagation Models for Wind Power Projects, State of the Art in Wind Siting Seminar*. Washington, DC: National Wind Coordinating Collaborative.

Khaligh, A. and Onar, O.C. (2010). *Energy Harvesting*. Boca Raton, FL: Taylor & Francis.

Knight, T.A. (1974). A review of hearing and song in birds with comments on the significance of song in display. *Emu*, 74: 5–8.

Komanoff, C. (2003). Even wind power can't be invisible. *The Providence Journal*, June 6.

LaBar, G. (1989). Sound policies for protecting workers' hearing. *Occupational Hazards*, 51(7): 46.

Lengagne, T. (2008). Traffic noise affects communication behavior in a breeding anuran, *Hyla arborea*. *Biological Conservation*, 141: 2023–2031.

Lohr, B., Wright, T.F., and Dooling, R.J. (2003). Detection and discrimination of natural calls in masking noise by birds: estimating the active space of a signal. *Animal Behavior*, 65: 763–777.

Magee, S. (2014). *Health Forensics*. Seattle, WA: CreateSpace Independent Publishing Platform.

Molburg, J.C., Kavicky, J.A., and Picel, K.C. (2007). *The Design, Construction and Operation of Long-Distance High-Voltage Electricity Transmission Technologies*. Argonne, IL: Argonne National Laboratory.

NIOSH. (1973). *The Industrial Environment: Its Evaluation and Control*, Publ. No. 74-117. Cincinnati, OH: National Institute for Occupational Safety and Health.

NIOSH. (2005). *Common Hearing Loss Prevention Terms*. Cincinnati, OH: National Institute for Occupational Safety and Health.

NREL. (2000). *Review of Operation and Maintenance Experience in the DOE-EPR Wind Turbine Verification Program*. Golden, CO: National Renewable Energy Laboratory.

OSHA. (2002). *Hearing Conservation*. Washington, DC: Occupational Safety and Health Administration.

Pasqualetti, M.J. (2002). Living with wind power in a hostile landscape. In: *Wind Power in View*, Pasqualetti, M.J., Gipe, P., and Righter, R.W., Eds., pp. 153–172. San Diego, CA: Academic Press.

Quinn, J.L., Whittingham, M.J., Butler, S.J., and Cresswell, W. (2006). Noise, predation risk compensation and vigilance in the chaffinch *Fringilla coelebs*. *Journal of Avian Biology*, 37: 601–608.

Rabin, L.A., Coss, R.G., and Owings. D.H. (2006). The effects of wind turbines on antipredator behavior in California ground squirrels (*Spermophilus beechyl*). *Biological Conservation*, 131: 410–420.

Righter, R.W. (2002). *Exoskeletal Outer-Space Creations*. Atlanta, GA: Academic Press.

Saito, Y. (2004). Machines in the ocean: the aesthetics of wind farms. *Contemporary Aesthetics*, 2 (http://www.contempaesthetics.org/newvolume/pages/article.php?articleID=247).

Salt, A.N. and Hullar, T.E. (2010). Responses of the ear to low frequency sounds, infrasound and wind turbines. *Hearing Research*, 268: 12–21.

Schaub, A., Ostwald, J., and Siemers, B.M. (2008). Foraging bats avoid noise. *Journal of Experimental Biology*, 211: 3174–3180.

Schworm, P. and Filipov, D. (2013). Flickering shadows from wind turbines draw complaints. *Boston Globe*, April 5.

Seltenrich, N. (2014). Wind turbines: a different breed of noise. *Environmental Health Perspectives*, 122(1): A20–A25.

Serway, R.A. and Jewett, J.W. (2004). *Physics for Scientists and Engineers*, 6th ed. New York: Thomson/Brooks/Cole.

Siemens. (2014). *Sustainable Energy in the U.S.* Munich: Siemens (http://www.usa.siemens.com/sustainable-energy/?stc=usccc025186).

Swaddle, J.P. and Page. L.C. (2007). High levels of environmental noise erode pair preferences in zebra finches: implications for noise pollution. *Animal Behaviour*, 74: 363–368.

TEEIC. (2015). *Wind Energy Construction Impacts*. Washington, DC: Tribal Energy and Environmental Information.

Tripler, P. (2004). *Physics for Scientists and Engineers: Mechanics, Oscillations and Wave, Thermodynamics*, 5th ed. New York: W.H. Freeman.

Tripler, P. and Llewellyn, R. (2002). *Modern Physics*, 4th ed. New York: W.H. Freeman.

USDOE. (2005). *Environmental Assessment for Spring Canyon Wind Project, Logan County, Colorado*, DOE/EA-1521. Washington, DC: U.S. Department of Energy.

USDOE. (2010). *Wind Turbine Interactions with Birds, Bats, and Their Habitats*. Washington, DC: U.S. Department of Energy (https://www1.eere.energy.gov/wind/pdfs/birds_and_bats_fact_sheet.pdf).

USFWS. (2005). *A Fine Line for Birds: A Guide to Bird Collisions at Power Lines*. Washington, DC: U.S. Fish and Wildlife Service (http://www.fws.gov/birds/documents/powerlines.pdf).

USFWS. (2013). *Eagle Conservation Plan Guidance: Module 1—Land-Based Wind Energy*. Washington, DC: U.S. Fish and Wildlife Service (http://www.fws.gov/migratorybirds/Eagle_Conservation_Plan_Guidance-Module%201.pdf).

USFWS. (2014a). *Mojave Desert Tortoise*. Washington, DC: U.S. Fish and Wildlife Service (http://www.fws.gov/nevada/desert_tortoise/dt/dt_threats.html).

USFWS. (2014b). *The Effects of Noise on Wildlife*. Washington, DC: U.S. Fish and Wildlife Service (www.fws.gov/windenergy/docs/noise.pdf).

USGS. (2015). *Bat Fatalities at Turbines: Investigating the Causes and Consequences*. Washington, DC: U.S. Geological Survey (https://www.fort.usgs.gov/science-feature/96).

Van den Berg, G.P. (2004). Effects of the wind profile at night on wind turbine sound. *Journal of Sound and Vibration*, 277: 955–970.

Vanderheiden, G. (2004). *About Decibels (dB)*. Madison: Trace R&D Center, University of Wisconsin (http://trace.wisc.edu/docs/2004-About-dB/).

WGA. (2009). *Western Renewable Energy Zones Phase 1 Report*. Denver, CO: Western Renewable Energy Zones Initiative.

White, F.M. (1988). *Fluid Mechanics*, 2nd ed. Singapore: McGraw-Hill.

Wind Energy EIS. (2009). *Wind Energy Basics*. Washington, DC: Wind Energy Development Programmatic EIS Information Center (http://windeis.anl.gov/guide/basics/index.cfm).

6 Hydropower

Natural powers, principally those of steam and falling water, are subsidized and taken into human employment. Spinning-machines, power-looms, and all the mechanical devices, acting, among other operatives, in the factories and work-shops, are but so many laborers. They are usually denominated labor-saving machines, but it would be more just to call them labor-doing machines. They are made to be active agents; to have motion, and to produce effect; and though without intelligence, they are guided by laws of science, which are exact and perfect, and they produce results, therefore, in general, more accurate than the human hand is capable of producing.

—Daniel Webster, American statesman

Bureaus build roads into new hinterlands, then buy more hinterlands to absorb the exodus accelerated by the roads. A gadget industry pads the bumps against nature-in-the-raw; woodcraft becomes the art of using gadgets. And now, to cap the pyramid of banalities, the trailer. To him who seeks in the woods and mountains only those things obtainable from travel or golf, the present situation is intolerable. But to him who seeks something more, recreation has become a self-destructive process of seeking but never quite finding, a major frustration of mechanized society.

Leopold (1949)

When we speak of water and its many manifestations we are speaking of that endless quintessential cycle that predates all other cycles. Water is our most precious natural resource; we can't survive without it. There is no more water today than there was yesterday; that is, there is no more this calendar year than 100 million years ago. The water present today is the same water used by all the animals that ever lived ... by cave dwellers, Caesar, Cleopatra, Christ, Leonardo da Vinci, John Snow, Teddy Roosevelt, and the rest of us. Again, there is not one drop more nor one drop less of water than there has always been. Although it is a unique blend of thermal and mechanical aspects, the life-giving water cycle is dependent on solar energy and gravity (see Figure 6.1). Nothing on Earth is truly infinite in supply, but the energy available from water sources, in practical terms, comes closest to that ideal.

INTRODUCTION

When we look at rushing waterfalls and rivers, we may not immediately think of electricity, but water-powered (hydroelectric) power plants are responsible for lighting many of our homes and neighborhoods. Hydropower is the harnessing of water to perform work. The power of falling water has been used in industry for thousands of years (see Table 6.1). The Greeks used water wheels to grind wheat into flour more than 2000 years ago. In addition to grinding flour, the power of water has been used to saw wood and power textile mills and manufacturing plants.

The technology for using falling water to create hydroelectricity has existed for more than a century. The evolution of the modern hydropower turbine began in the mid-1700s when a French hydraulic and military engineer, Bernard Forest de Belidor, wrote a four-volume work describing the use of a vertical-axis vs. horizontal-axis machine. Water turbine development continued into the 1800s. A Brush arc light dynamo driven by a water turbine provided theater and storefront lighting in Grand Rapids, Michigan, in 1880, and in 1881 a Brush dynamo connected to a turbine in a flour mill provided street lighting at Niagara Falls, New York. These two projects used direct current (DC) technology.

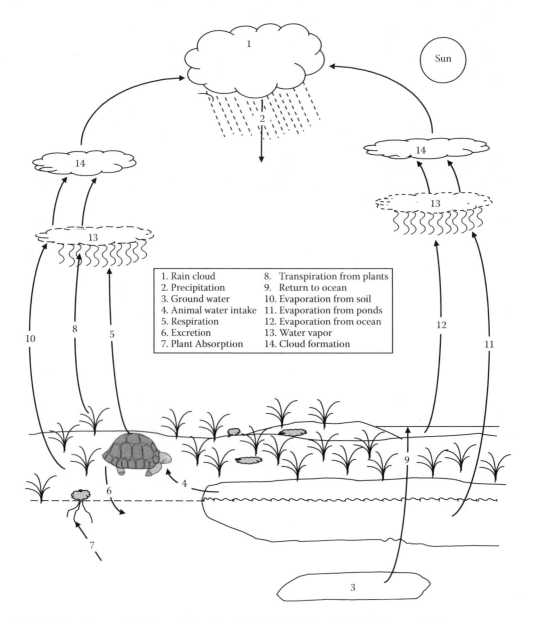

1. Rain cloud   8. Transpiration from plants
2. Precipitation   9. Return to ocean
3. Ground water   10. Evaporation from soil
4. Animal water intake   11. Evaporation from ponds
5. Respiration   12. Evaporation from ocean
6. Excretion   13. Water vapor
7. Plant Absorption   14. Cloud formation

FIGURE 6.1 Water cycle. (Adapted from Carolina Biological Supply Co., Burlington, NC.)

Alternating current (AC) is used today. That breakthrough came when the electric generator was coupled to the turbine in 1882, which resulted in the world's first hydroelectric plant, located on the Fox River in Appleton, Wisconsin. The Appleton hydroelectric power plant is considered to be one of the major accomplishments of the Gilded Age (1878–1889) (Library of Congress, 2009). People across the United States soon enjoyed electricity in their homes, schools, and offices, where they were able to read by electric lamp instead of candlelight or kerosene. Today, we take electricity for granted and cannot imagine life without it.

Table 6.2 shows the energy consumption by energy source for 2014; in that year, hydropower power accounted for 2.475 quadrillion Btu. Ranging in size from small systems (100 kW to 30 MW) for a home or village to large projects (capacity greater than 30 MW) that produce electricity for

TABLE 6.1
History of Hydropower

Date	Hydropower Event
BCE	Hydropower was used by the Greeks more than 2000 years ago to turn water wheels to grind wheat into flour.
Mid-1770s	French hydraulic and military engineer Bernard Forest de Belidor wrote a four-volume work describing vertical- and horizontal-axis machines.
1775	U.S. Army Corps of Engineers was founded, with establishment of a Chief Engineer for the Continental Army.
1880	Michigan's Grand Rapids Electric Light and Power Company, generating electricity by a dynamo belted to a water turbine at the Wolverine Chair Factory, lit up 16 Brush-arc lamps.
1881	Niagara Falls city street lamps were powered by hydropower.
1882	World's first hydroelectric power plant began operation on the Fox River in Appleton, Wisconsin.
1886	About 45 water-powered electric plants were operating in the United States and Canada.
1887	First hydroelectric plant in the west was opened in San Bernardino, California.
1889	Two hundred electric plants in the United States used water power for some or all of their power generation.
1901	First Federal Water Power Act was passed.
1902	Bureau of Reclamation was established.
1907	Hydropower provided 15% of U.S. electrical generation.
1920	Hydropower provided 25% of U.S. electrical generation. The Federal Power Act established the Federal Power Commission with authority to issue licenses for hydro development on public lands.
1933	Tennessee Valley Authority was established.
1935	Federal Power Commission authority was extended to all hydroelectric projects built by utilities engaged in interstate commerce.
1937	Bonneville Dam, the first federal dam, began operation on the Columbia River in Oregon and Washington. The federal Bonneville Power Administration was established in the Pacific Northwest.
1940	Hydropower provided 40% of electrical generation. Conventional capacity had tripled in the United States since 1920.
1980	Conventional capacity had nearly tripled in the United States since 1900.
2003	About 10% of U.S. electricity came from hydropower. Capacity was about 80,000 MW of conventional capacity and 18,000 MW of pumped storage.

Source: EERE, *History of Hydropower*, Office of Energy Efficiency & Renewable Energy, U.S. Department of Energy, Washington, DC, 2008 (http://www1.eere.energy.gov/windandhydro/hydro_history.html).

TABLE 6.2
U.S. Energy Consumption by Renewable Energy Source, 2014

Energy Source	Energy Consumption (quadrillion Btu)
Renewable total	9.656
Biomass (total)	4.812
Biofuels (ethanol and biodiesel)	2.067
Waste	0.516
Wood and wood-derived fuels	2.230
Geothermal	0.215
Hydroelectric power	**2.475**
Solar	0.421
Wind	1.733

Source: EIA, *Primary Energy Consumption by Source*, Table 10.1, Energy Information Administration, Washington, DC, 2016 (https://www.eia.gov/totalenergy/data/monthly/index.cfm#renewable).

utilities, hydropower plants are of three types: impoundment, diversion, and pumped storage. Some hydropower plants use dams, and some do not. Many dams were built for other purposes, and the hydropower function was added later. The United States has about 80,000 dams, of which only 2400 produce power. The other dams are for recreation, stock/farm ponds, flood control, water supply, and irrigation. The types of hydropower plants are described below.

IMPOUNDMENT

The most common type of hydroelectric power plant is the impoundment facility, which typically is a large hydropower system that uses a dam to store river water in a reservoir. This type of facility works best in mountainous or hilly terrains where high dams can be built and deep reservoirs can be maintained. Potential energy available in a reservoir depends on the mass of water contained in it, as well as on overall depth of the water. Water released from the reservoir flows through a turbine and makes it spin, which in turn activates a generator that produces electricity. The water may be released either to meet changing electricity needs or to maintain a constant reservoir level.

DIVERSION

A diversion (sometimes called *run-of-river*) facility channels all or a portion of the flow of a river from its natural course through a canal or penstock, and the current is used to drive turbines. This approach may not require the use of a dam. This type of system is best suited for locations where a river drops considerably per unit of horizontal distance. The ideal location is near a natural waterfall or rapids. The chief advantage of a diversion system is the fact that, lacking a dam, it has far less impact on the environment than an impoundment facility (Gibilisco, 2007).

PUMPED STORAGE

When the demand for electricity is low, a pumped storage facility stores energy by pumping water from a lower reservoir to an upper reservoir. During periods of high electrical demand, the water is released back to the lower reservoir to generate electricity.

KEY TERMS

To assist in understanding material presented in this chapter, the key components that make up hydro turbines and hydropower plants are defined in the following:

DID YOU KNOW?

The potential energy in a specific slug or parcel of water is expressed in newton-meters (N·m). The newton is the standard unit of force, equivalent to meters per second squared (1 m/s^2). The potential energy is the product of the mass of the slug (kg), acceleration due to gravity (about 9.8 m/s^2), and elevation of the parcel in meters (vertical distance it falls as its energy is harnessed). The equivalent kinetic energy unit is the joule (J), which is, in effect, equal to a watt-second (W·s).

Alternating current (AC)—Electric current that reverses direction many times per second.

Ancillary services—Capacity and energy services provided by power plants that are able to respond on short notice, such as hydropower plants, and are used to ensure stable electricity delivery and optimized grid reliability.

Cavitation—The phase changes that occur due to pressure changes in a fluid that cause bubbles to form, resulting in noise or vibration in the water column. The implosion of these bubbles against a solid surface, such as a hydraulic turbine, may cause erosion and lead to reductions in capacity and efficiency pressure.

Control gate—A barrier that regulates water released from a reservoir to the power generation unit.

Direct current (DC)—Electric current that flows in one direction.

Diversion—A facility that channels a portion of a river through a canal or penstock.

Draft tube—A water conduit, which can be straight or curved depending on the turbine installation and which maintains a column of water from the turbine outlet and the downstream water level.

Efficiency—A percentage obtained by dividing the actual power or energy by the theoretical power or energy. It represents how well the hydropower plant converts the potential energy of water into electrical energy.

Fish ladder—A transport structure that provides safe upstream fish passage around hydropower projects.

Flow—Volume of water, expressed as cubic feet or cubic meters per second, passing a point in a given amount of time.

Generator—Device that converts the rotational energy from a turbine to electrical energy.

Head—Vertical change in elevation, expressed in feet or meters, between the head (reservoir) water level and the tailwater (downstream) level.

Headwater—The water level above the powerhouse or at the upstream face of a dam.

Hydropower—The harnessing of flowing water using a dam or other type of diversion structure to create energy that can be captured via a turbine to generate electricity.

Impoundment—A body of water formed by damming a river or stream, commonly known as a reservoir.

Low head—Head of 66 feet or less.

Microhydro—Hydropower projects that generate up to 100 kilowatts.

Penstock—A closed conduit or pipe for conducting water to the powerhouse.

Power house—The structure that houses generators and turbines.

Pumped storage—A type of hydropower that works like a battery, pumping water from a lower reservoir to an upper reservoir for storage and later generation.

Runner—The rotating part of the turbine that converts the energy of falling water into mechanical energy.

Scroll case—A spiral-shaped steel intake guiding the flow in the wicket gates located just prior to the turbine.

Small hydro—Hydropower projects that generate 10 MW or less of power.

Spillway—A structure used to provide the release of flows from a dam into a downstream area. Some spillways are designed like an inverted bell so that water can enter all around the perimeter. These uncontrolled spillway devices are also called *morning glory*, *glory hole* (see Figure 6.2), or *bell-mouth* spillways.

Tailrace—The channel that carries water away from a dam.

Tailwater—The water downstream of the powerhouse or dam.

Transformer—Device that takes power from the generator and converts it to higher-voltage current.

FIGURE 6.2 Hungry Horse Dam glory hole, Hungry Horse, Montana. The dam is equipped with a glory hole emergency spillway for high water; it is the highest spillway or glory hole structure in the world. (Photograph by Frank R. Spellman.)

Turbine—A machine that produces continuous power in which a wheel or rotor revolves by a
   fast-moving flow of water.
Ultra low head—Head of 10 feet or less.
Wicket gates—Adjustable elements that control the flow of water to the turbine.

HYDROPOWER BASIC CONCEPTS[*]

Air pressure (at sea level) = 14.7 pounds per square inch (psi)

The relationship shown above is important because our study of hydropower basics begins with air. A blanket of air, many miles thick surrounds the Earth. The weight of this blanket on a given square inch of the Earth's surface will vary according to the thickness of the atmospheric blanket above that point. As shown above, at sea level the pressure exerted is 14.7 pounds per square inch (psi). On a mountaintop, air pressure decreases because the blanket is not as thick.

1 cubic foot (ft^3) of water = 62.4 lb

This relationship is also important; note that both cubic feet and pounds are used to describe a volume of water. A defined relationship exists between these two methods of measurement. The specific weight of water is defined relative to a cubic foot. One cubic foot of water weighs 62.4 lb. This relationship is true only at a temperature of 4°C and at a pressure of 1 atmosphere, conditions referred to as standard temperature and pressure (STP). One atmosphere equals 14.7 psi at sea level and 1 ft^3 of water contains 7.48 gal. The weight varies so little that, for practical purposes, this weight is used for temperatures ranging from 0 to 100°C. One cubic inch of water weighs 0.0362 lb. Water 1 ft deep will exert a pressure of 0.43 psi on the bottom area (12 in. × 0.0362 lb/in.^{3}). A column of water 2 ft high exerts 0.86 psi (2 ft × 0.43 psi/ft), one 10 ft high exerts 4.3 psi (10 ft × 0.43 psi/ft), and one 55 ft high exerts 23.65 psi (55 ft × 0.43 psi/ft). A column of water 2.31 ft high will exert 1.0 psi. To produce a pressure of 50 psi requires a 115.5-ft water column (50 psi × 2.31 ft/psi).

---

[*] Much of the information in this section is from Spellman, F.R., *The Science of Water*, CRC Press, Boca Raton, FL, 2014.

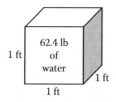

FIGURE 6.3 One cubic foot of water weighs 62.4 lb.

Two important points are being made here:

1. 1 ft^3 of water = 62.4 lb (see Figure 6.3).
2. A column of water 2.31 ft high will exert 1.0 psi.

Another relationship is also important. As noted above, at standard temperature and pressure, 1 ft^3 of water contains 7.48 gal and weighs 62.4 lb. Thus, we can determine the weight of 1 gal of water:

$$\text{Weight of 1 gal water} = 62.4 \text{ lb/ft}^3 \div 7.48 \text{ gal/ft}^3 = 8.34 \text{ lb/gal}$$

Thus,

$$1 \text{ gal water} = 8.34 \text{ lb}$$

Note: Convert cubic feet to gallons simply by multiplying the number of cubic feet by 7.48 gal/ft^3.

■ EXAMPLE 6.1

Problem: Find the number of gallons in a reservoir that has a volume of 855.5 ft^3.
Solution:

$$855.5 \text{ ft}^3 \times 7.48 \text{ gal/ft}^3 = 6399 \text{ gal (rounded)}$$

Note: The term *head* is used to designate water pressure in terms of the height of a column of water in feet; for example, a 10-ft column of water exerts 4.3 psi. This can be referred to as 4.3-psi pressure or 10 feet of head.

STEVIN'S LAW

Stevin's law deals with water at rest. Specifically, it states: "The pressure at any point in a fluid at rest depends on the distance measured vertically to the free surface and the density of the fluid." Stated as a formula, this becomes

$$p = w \times h \tag{6.1}$$

where
p = Pressure in pounds per square foot (lb/ft^2, or psf).
w = Density in pounds per cubic foot (lb/ft^3).
h = Vertical distance in feet.

■ EXAMPLE 6.2

Problem: What is the pressure at a point 18 ft below the surface of a reservoir?
Solution: To calculate this, we must know that the density (w) of the water is 62.4 lb/ft^3.

$$p = w \times h = 62.4 \text{ lb/ft}^3 \times 18 \text{ ft} = 1123 \text{ lb/ft}^2 \text{ (psf)}$$

Water practitioners generally measure pressure in pounds per square *inch* rather than pounds per square *foot*; to convert, divide by 144 in.^{2}/ft^2 (12 in. × 12 in. = 144 in.^{2}):

$$p = 1123 \text{ lb/ft}^2 \div 144 \text{ in.}^2/\text{ft}^2 = 7.8 \text{ lb/in.}^2$$

TABLE 6.3

Water Properties (Temperature, Specific Weight, and Density)

Temperature (°F)	Specific Weight (lb/ft³)	Density (slugs/ft³)	Temperature (°F)	Specific Weight (lb/ft³)	Density (slugs/ft³)
32	62.4	1.94	130	61.5	1.91
40	62.4	1.94	140	61.4	1.91
50	62.4	1.94	150	61.2	1.90
60	62.4	1.94	160	61.0	1.90
70	62.3	1.94	170	60.8	1.89
80	62.2	1.93	180	60.6	1.88
90	62.1	1.93	190	60.4	1.88
100	62.0	1.93	200	60.1	1.87
110	61.9	1.92	210	59.8	1.86
120	61.7	1.92			

DENSITY AND SPECIFIC GRAVITY

Table 6.3 shows the relationship among temperature, specific weight, and density of water. When we say that iron is heavier than aluminum, we say that iron has greater density than aluminum. In practice, what we are really saying is that a given volume of iron is heavier than the same volume of aluminum.

> *Note:* What is density? *Density* is the *mass per unit volume* of a substance. Specific gravity is the weight (or density) of a substance compared to the weight (or density) of an equal volume of water. The specific gravity of water is 1. A cubic foot of water, for example, weighs 62.4 lb, and a cubic foot of aluminum weighs 178 lb; thus, aluminum is 2.7 times heavier than water.

Suppose you had a tub of lard and a large box of cold cereal, each having a mass of 600 grams. The density of the cereal would be much less than the density of the lard because the cereal occupies a much larger volume than the lard occupies. The density of an object can be calculated by using the formula:

$$\text{Density} = \frac{\text{Mass}}{\text{Volume}} \tag{6.2}$$

Perhaps the most common measures of density are pounds per cubic foot (lb/ft³) and pounds per gallon (lb/gal):

- 1 ft³ of water weighs 62.4 lb; density = 62.4 lb/ft³.
- 1 gal of water weighs 8.34 lb; density = 8.34 lb/gal.

The density of dry material, such as cereal, lime, soda, or sand, is usually expressed in pounds per cubic foot. The density of a liquid, such as liquid alum, liquid chlorine, or water, can be expressed either as pounds per cubic foot or as pounds per gallon. The density of a gas, such as chlorine gas, methane, carbon dioxide, or air, is usually expressed in pounds per cubic foot. As shown in Table 6.3, the density of a substance such as water changes slightly as the temperature of the substance changes. This occurs because substances usually increase in volume by expanding as they become warmer. Because of this expansion with warming, the same weight is spread over a larger volume, so the density is lower when a substance is warm rather than when it is cold.

It is not that difficult to find the specific gravity of a piece of metal. All you have to do is to weigh the metal in air, then weigh it under water. Its loss of weight is the weight of an equal volume of water. To find the specific gravity, divide the weight of the metal by its loss of weight in water:

$$\text{Specific gravity} = \frac{\text{Weight of substance}}{\text{Weight of equal volume of water}} \tag{6.3}$$

■ **EXAMPLE 6.3**

Problem: Suppose a piece of metal weighs 150 lb in air and 85 lb under water. What is the specific gravity?
Solution:

$$150 \text{ lb} - 85 \text{ lb} = 65 \text{ lb loss of weight in water}$$

$$\text{Specific gravity} = 150 \div 65 = 2.3$$

The specific gravity of water is 1, which is the standard, the reference against which all other liquid or solid substances are compared. Specifically, any object that has a specific gravity greater than 1 will sink in water (e.g., rocks, steel, iron, grit, floc, sludge). Substances with a specific gravity of less than 1 will float (e.g., wood, scum, gasoline). Considering the total weight and volume of a ship, its specific gravity is less than 1; therefore, it can float. The most common use of specific gravity in water operations is in gallon-to-pound conversions. In many cases, the liquids being handled have a specific gravity of 1 or very nearly 1 (between 0.98 and 1.02), so a value of 1 may be used in the calculations without introducing significant error. For calculations involving a liquid with a specific gravity of less than 0.98 or greater than 1.02, however, the conversions from gallons to pounds must consider specific gravity. The technique is illustrated in the following example.

■ **EXAMPLE 6.4**

Problem: A basin contains 1455 gal of a liquid. If the specific gravity of the liquid is 0.94, how many pounds of liquid are in the basin?
Solution: Normally, for a conversion from gallons to pounds, we would use the factor 8.34 lb/gal (the density of water) if the specific gravity of the substance is between 0.98 and 1.02; in this instance, however, the substance has a specific gravity outside this range, so the 8.34 factor must be adjusted. Multiply 8.34 lb/gal by the specific gravity to obtain the adjusted factor:

$$8.34 \text{ lb/gal} \times 0.94 = 7.84 \text{ lb/gal (rounded)}$$

Then convert 1455 gal to pounds using the correction factor:

$$1455 \text{ gal} \times 7.84 \text{ lb/gal} = 11,407 \text{ lb (rounded)}$$

FORCE AND PRESSURE

Water exerts force and pressure against the walls of its container, whether it is stored in a tank or flowing in a pipeline. Force and pressure are different, although they are closely related. Force and pressure are defined below. Force is the push or pull influence that causes motion. In the English system, force and weight are often used in the same way. The weight of 1 ft³ of water is 62.4 lb. The force exerted on the bottom of a 1-ft cube is 62.4 lb (see Figure 6.3). If we stack two 1-ft cubes on top of one another, the force on the bottom will be 124.8 lb. Pressure is the force per unit of area. In equation form, this can be expressed as

$$P = \frac{F}{A} \tag{6.4}$$

where
  P = Pressure.
  F = Force.
  A = Area over which the force is distributed.

Pounds per square inch (lb/in.² or psi) or pounds per square foot (lb/ft² or psf) are commonly used to express pressure. The pressure on the bottom of our 1-ft cube is 62.4 lb/ft² (see Figure 6.3). It is normal to express pressure in pounds per square inch. This is easily accomplished by determining the weight of 1 in.² of a 1-ft cube. If we have a cube that is 12 in. on each side, the number of square inches on the bottom surface of the cube is $12 \times 12 = 144$ in.². Dividing the weight by the number of square inches determines the weight on each square inch:

$$\text{Pounds per square inch (psi)} = \frac{62.4 \text{ lb}}{144 \text{ in.}^2} = 0.433 \text{ psi}$$

This is the weight of a column of water 1 in. square and 1 ft tall. If the column of water were 2 ft tall, the pressure would be 2 ft × 0.433 psi/ft = 0.866.

Note: 1 ft of water = 0.433 psi.

With this information, feet of head can be converted to psi by multiplying the feet of head times 0.433 psi/ft.

■ EXAMPLE 6.5

Problem: A tank is mounted at a height of 90 ft. Find the pressure at the bottom of the tank.
Solution:

$$90 \text{ ft} \times 0.433 \text{ psi/ft} = 39 \text{ psi (rounded)}$$

Note: To convert psi to feet, divide the psi by 0.433 psi/ft.

■ EXAMPLE 6.6

Problem: Find the height of water in a tank if the pressure at the bottom of the tank is 22 psi.
Solution:

$$\text{Height} = 22 \text{ psi} \div 0.433 \text{ psi/ft} = 51 \text{ ft (rounded)}$$

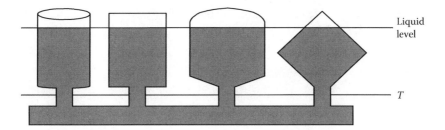

FIGURE 6.4   Hydrostatic pressure.

HYDROSTATIC PRESSURE

Figure 6.4 shows a number of differently shaped, connected, open containers of water. Note that the water level is the same in each container, regardless of the shape or size of the container. This occurs because pressure is developed, within water (or any other liquid), by the weight of the water above. If the water level in any one container is momentarily higher than that in any of the other containers, the higher pressure at the bottom of this container would cause some water to flow into the container having the lower liquid level. In addition, the pressure of the water at any level (such as line T) is the same in each of the containers. Pressure increases because of the weight of the water. The farther down from the surface, the more pressure is created. This illustrates that the *weight*, not the volume, of water contained in a vessel determines the pressure at the bottom of the vessel. Some important principles that always apply for hydrostatic pressure include the following (Nathanson, 1997):

1. The pressure depends only on the depth of water above the point in question (not on the water surface area).
2. The pressure increases in direct proportion to the depth.
3. The pressure in a continuous volume of water is the same at all points that are at the same depth.
4. The pressure at any point in the water acts in all directions at the same depth.

HEAD

Head is defined as the vertical distance the water must be lifted from the supply tank to the discharge, or as the height a column of water would rise due to the pressure at its base. A perfect vacuum plus atmospheric pressure of 14.7 psi would lift the water 34 ft. If the top of the sealed tube is open to the atmosphere and the reservoir is enclosed, the pressure in the reservoir is increased; the water will rise in the tube. Because atmospheric pressure is essentially universal, we usually ignore the first 14.7 psi of actual pressure measurements and measure only the difference between the water pressure and the atmospheric pressure; we call this *gauge pressure*. Water in an open reservoir, for example, is subjected to the 14.7 psi of atmospheric pressure, but subtracting this 14.7

DID YOU KNOW?

One of the problems encountered in a hydraulic system is storing the liquid. Unlike air, which is readily compressible and is capable of being stored in large quantities in relatively small containers, a liquid such as water cannot be compressed. It is not possible to store a large amount of water in a small tank; 62.4 lb of water occupies a volume of 1 cubic foot, regardless of the pressure applied to it.

psi leaves a gauge pressure of 0 psi. This shows that the water would rise 0 ft above the reservoir surface. If the gauge pressure in a water main is 120 psi, the water would rise in a tube connected to the main:

$$120 \text{ psi} \times 2.31 \text{ ft/psi} = 277 \text{ ft (rounded)}$$

Total Dynamic (System) Head

The total dynamic head includes the vertical distance the liquid must be lifted (*static head*), the loss to friction (*friction head*), and the energy required to maintain the desired velocity (*velocity head*):

$$\text{Total dynamic head} = \text{Static head} + \text{Friction head} + \text{Velocity head} \qquad (6.5)$$

Static Head

Static head is the actual vertical distance the liquid must be lifted:

$$\text{Static head} = \text{Discharge elevation} - \text{Supply elevation} \qquad (6.6)$$

■ EXAMPLE 6.7

Problem: A supply tank is located at elevation 118 ft. The discharge point is at elevation 215 ft. What is the static head in feet?
Solution:

$$\text{Static head} = 215 \text{ ft} - 118 \text{ ft} = 97 \text{ ft}$$

Friction Head

Friction head is the equivalent distance of the energy that must be supplied to overcome friction. Engineering references include tables showing the equivalent vertical distances for various sizes and types of pipes, fittings, and valves. The total friction head is the sum of the equivalent vertical distances for each component:

$$\text{Friction head (ft)} = \text{Energy losses due to friction} \qquad (6.7)$$

Velocity Head

Velocity head is the equivalent distance of the energy consumed in achieving and maintaining the desired velocity in the system:

$$\text{Velocity head (ft)} = \text{Energy losses to maintain velocity} \qquad (6.8)$$

Pressure and Head

The pressure exerted by water is directly proportional to its depth or head in the pipe, tank, or channel. If the pressure is known, the equivalent head can be calculated.

$$\text{Head (ft)} = \text{Pressure (psi)} \times 2.31 \text{ (ft/psi)} \qquad (6.9)$$

■ EXAMPLE 6.8

Problem: The pressure gauge on the discharge line from the influent pump reads 72.3 psi. What is the equivalent head in feet?
Solution:

$$\text{Head} = 72.3 \times 2.31 \text{ ft/psi} = 167 \text{ ft}$$

Head and Pressure

If the head is known, the equivalent pressure can be calculated as follows:

$$\text{Pressure (psi)} = \frac{\text{Head (ft)}}{2.31 \text{ ft/psi}} \tag{6.10}$$

■ EXAMPLE 6.9

Problem: A tank is 22 ft deep. What is the pressure in psi at the bottom of the tank when it is filled with water?
Solution:

$$\text{Pressure} = \frac{22 \text{ ft}}{2.31 \text{ ft/psi}} = 9.52 \text{ psi (rounded)}$$

FLOW AND DISCHARGE RATES: WATER IN MOTION

The study of fluid flow is much more complicated than that of fluids at rest, but it is important to have an understanding of these principles because the water used in hydropower applications is nearly always in motion (e.g., the water is used to propel turbine blades). *Discharge* (or flow) is the quantity of water passing a given point in a pipe or channel during a given period. Stated another way for open channels, the flow rate through an open channel is directly related to the velocity of the liquid and the cross-sectional area of the liquid in the channel:

$$Q = A \times V \tag{6.11}$$

where
  Q = Flow, or discharge in cubic feet per second (cfs).
  A = Cross-sectional area of the pipe or channel (ft^2).
  V = Water velocity in feet per second (fps or ft/s).

■ EXAMPLE 6.10

Problem: A channel is 6 ft wide, and the water depth is 3 ft. The velocity in the channel is 4 fps. What is the discharge or flow rate in cubic feet per second?
Solution:

$$\text{Flow} = 6 \text{ ft} \times 3 \text{ ft} \times 4 \text{ ft/s} = 72 \text{ cfs}$$

Discharge or flow can be recorded as gal/day (gpd), gal/min (gpm), or cubic feet per second (cfs). Flows treated by many hydropower systems are large and are often expressed in million gallons per day (MGD). The discharge or flow rate can be converted from cfs to other units such as gpm or MGD by using appropriate conversion factors.

■ EXAMPLE 6.11

Problem: A 12-in.-diameter pipe has water flowing through it at 10 fps. What is the discharge in (a) cfs, (b) gpm, and (c) MGD?

Solution: Before we can use the basic formula, we must determine the area (A) of the pipe. The formula for the area of a circle is

$$\text{Area } (A) = \pi \times (D^2/4) = \pi \times r^2 \tag{6.12}$$

where

π = Constant value 3.14159, or simply 3.14.
D = Diameter of the circle (ft).
r = Radius of the circle (ft).

Therefore, the area of the pipe is

$$A = \pi \times (D^2/4) = 3.14 \times (1 \text{ ft}^2/4) = 0.785 \text{ ft}^2$$

(a) Now we can determine the discharge in cfs:

$$Q = V \times A = 10 \text{ fps} \times 0.785 \text{ ft}^2 = 7.85 \text{ ft}^3/\text{s (cfs)}$$

(b) We need to know that 1 cfs is 449 gpm, so 7.85 cfs × 449 gpm/cfs = 3525 gpm (rounded).

(c) Finally, 1 million gallons per day is 1.55 cfs, so

$$7.85 \text{ cfs} \div 1.55 \text{ cfs/MGD} = 5.06 \text{ MGD}$$

AREA AND VELOCITY

The *law of continuity* states that the discharge at each point in a pipe or channel is the same as the discharge at any other point (if water does not leave or enter the pipe or channel). That is, under the assumption of steady-state flow, the flow that enters the pipe or channel is the same flow that exits the pipe or channel. In equation form, this becomes:

$$Q_1 = Q_2 \quad \text{or} \quad A_1 V_1 = A_2 V_2 \tag{6.13}$$

■ EXAMPLE 6.12

Problem: A pipe 12 inches in diameter is connected to a 6-in.-diameter pipe. The velocity of the water in the 12-in. pipe is 3 fps. What is the velocity in the 6-in. pipe?

Solution: Using the equation $A_1 V_1 = A_2 V_2$, we need to determine the area of each pipe:

$$\text{Area } (A) = \pi \times (D^2/4)$$

$$\text{12-in. pipe area} = 3.14 \times (1 \text{ ft}^2/4) = 0.785 \text{ ft}^2$$

$$\text{6-in. pipe area} = 3.14 \times (0.5 \text{ ft}^2/4) = 0.196 \text{ ft}^2$$

The continuity equation now becomes

$$0.785 \text{ ft}^2 \times 3 \text{ ft/s} = 0.196 \text{ ft}^2 \times V_2$$

Solving for V_2:

$$V_2 = \frac{0.785 \text{ ft}^2 \times 3 \text{ ft/s}}{0.196 \text{ ft}^2} = 12 \text{ ft/s (fps)}$$

PRESSURE AND VELOCITY

In a closed pipe flowing full (under pressure), the pressure is indirectly related to the velocity of the liquid:

$$\text{Velocity}_1 \times \text{Pressure}_1 = \text{Velocity}_2 \times \text{Pressure}_2 \quad \text{or} \quad V_1 P_1 = V_2 P_2 \qquad (6.14)$$

CONSERVATION OF ENERGY

The volume of water flowing past any given point in the pipe or channel per unit time is called the *flow rate* or *discharge*—or just *flow*. The *continuity of flow* and the *continuity equation* have been discussed (see Equation 6.13). Along with the continuity of flow principle and continuity equation, the law of conservation of energy, piezometric surface, and Bernoulli's theorem (or principle) are also important to our study of water hydraulics. Many of the principles of physics are important to the study of hydraulics. When applied to problems involving the flow of water, few of the principles of physical science are more important and useful to us than the *law of conservation of energy*. Simply, the law of conservation of energy states that energy can be neither created nor destroyed, but it can be converted from one form to another. In a given closed system, the total energy is constant.

ENERGY HEAD

Hydraulic systems have two types of energy—kinetic and potential—and they have three forms of mechanical energy—potential energy due to elevation, potential energy due to pressure, and kinetic energy due to velocity. Energy has the units of foot pounds (ft-lb). It is convenient to express hydraulic energy in terms of *energy head* in feet of water. This is equivalent to foot-pounds per pound of water (ft-lb/lb = ft).

ENERGY AVAILABLE

Energy available is directly proportional to flow rate and to the hydraulic head. As mentioned, the head is equivalent to stored potential energy. This is shown as

$$\text{Head} = m \times g \times h$$

where
   m = Mass of water.
   g = Acceleration due to gravity (can be taken as 10 m/s^2 in most applications).
   h = Head difference.

The diameter of a pipe must be large enough to handle the volume of water flowing. Friction in the pipes will reduce the effective head of water so larger diameters are used, although cost is a factor. Ideally, the pipes should narrow as they proceed downhill; however, friction losses are highest where the velocity is highest, so usually the pipe diameter changes very little. Friction losses in piping are classified as either major head loss or minor head loss (Tovey, 2005).

MAJOR HEAD LOSS

Major head loss consists of pressure decreases along the length of pipe caused by friction created as water encounters the surfaces of the pipe. It typically accounts for most of the pressure drop in a pressurized or dynamic water system. The components that contribute to major head loss are roughness, length, diameter, and velocity:

DID YOU KNOW?

For the same diameter pipe, when flow increases, head loss increases.

- *Roughness*—Even when new, the interior surfaces of pipes are rough. The roughness varies, of course, depending on pipe material, corrosion (tuberculation and pitting), and age. Because normal flow in a water pipe is turbulent, the turbulence increases with pipe roughness, which, in turn, causes pressure to drop over the length of the pipe.
- *Pipe length*—With every foot of pipe length, friction losses occur. The longer the pipe, the greater the head loss. Friction loss because of pipe length must be factored into head loss calculations.
- *Pipe diameter*—Generally, small-diameter pipes have more head loss than large-diameter pipes. In large-diameter pipes, less of the water actually touches the interior surfaces of the pipe (thus encountering less friction) than in a small-diameter pipe.
- *Water velocity*—Turbulence in a water pipe is directly proportional to the speed (or velocity) of the flow; thus, the velocity head also contributes to head loss.

Calculating Major Head Loss

Darcy, Weisbach, and others developed the first practical equation used to determine pipe friction in about 1850. The formula now known as the *Darcy–Weisbach* equation for circular pipes is

$$h_f = f\left(\frac{LV^2}{D^2 g}\right) \tag{6.15}$$

In terms of the flow rate (Q), the equation becomes

$$h_f = \frac{8fLQ^2}{\pi^2 g D^5} \tag{6.16}$$

where
  h_f = Head loss (ft).
  f = Coefficient of friction.
  L = Length of pipe (ft).
  Q = Flow rate (ft³/s).
  D = Diameter of pipe (ft).
  g = Acceleration due to gravity (32.2 ft/s²).

The Darcy–Weisbach formula was meant to apply to the flow of any fluid, and into this friction factor was incorporated the degree of roughness and an element called the *Reynold's number*, which is based on the viscosity of the fluid and the degree of turbulence of flow. The Darcy–Weisbach formula is used primarily for determining head loss calculations in pipes. For open channels, the *Manning equation* was developed during the latter part of the 19th century. Later, this equation was used for both open channels and closed conduits.

DID YOU KNOW?

An alternative to calculating the Hazen–Williams formula, called an *alignment chart*, has become quite popular for fieldwork. The alignment chart can be used with reasonable accuracy.

TABLE 6.4
C Factors

Type of Pipe	C Factor	Type of Pipe	C Factor
Asbestos cement	140	Galvanized iron	120
Brass	140	Glass	140
Brick sewer	100	Lead	130
Cast iron		Masonry conduit	130
10 years old	110	Plastic	150
20 years old	90	Steel	
Ductile iron (cement lined)	140	Coal-tar enamel lined	150
Concrete or concrete lined		New unlined	140
Smooth, steel forms	140	Riveted	110
Wooden forms	120	Tin	130
Rough	110	Vitrified	120
Copper	140	Wood stave	120
Fire hose (rubber lined)	135		

Source: Adapted from Lindeburg, M.R., *Civil Engineering Reference Manual*, 4th ed., Professional Publications, San Carlos, CA, 1986.

In the early 1900s, a more practical equation, the *Hazen–Williams* equation, was developed for use in making calculations related to water pipes and wastewater force mains:

$$Q = 0.435 \times C \times D^{2.63} \times S^{0.54} \qquad (6.17)$$

where

Q = Flow rate (ft³/s).
C = Coefficient of roughness (C decreases with roughness).
D = Hydraulic radius R (ft).
S = Slope of energy grade line (ft/ft).

C Factor

The C factor, as used in the Hazen–Williams formula, designates the coefficient of roughness. C does not vary appreciably with velocity, and by comparing pipe types and ages it includes only the concept of roughness, ignoring fluid viscosity and Reynold's number. C factor values have been established for various types of pipe (see Table 6.4). Generally, the C factor decreases by one with each year of pipe age. Flow for a newly designed system is often calculated with a C factor of 100, based on averaging it over the life of the pipe system.

Slope

Slope is defined as the head loss per foot. In open channels, where the water flows by gravity, slope is the amount of incline of the pipe and is calculated as feet of drop per foot of pipe length (ft/ft). Slope is designed to be just enough to overcome frictional losses, so the velocity remains constant, the water keeps flowing, and solids will not settle in the conduit. In piped systems, where pressure loss for every foot of pipe is experienced, slope is not provided by slanting the pipe but instead by pressure added to overcome friction.

DID YOU KNOW?

A high C factor means a smooth pipe. A low C factor means a rough pipe.

DID YOU KNOW?

In practice, if minor head loss is less than 5% of the total head loss, it is usually ignored.

MINOR HEAD LOSS

In addition to the head loss caused by friction between the fluid and the pipe wall, losses also are caused by turbulence created by obstructions (i.e., valves and fittings of all types) in the line, changes in direction, and changes in flow area.

RESERVOIR STORED ENERGY

A major component of a hydroelectric dam is the area behind the dam, its reservoir (Figure 6.5). The water temporarily stored there is called *gravitational potential energy*. The water is stored above the rest of the dam facility so gravity can carry the water down to the turbines. Because this higher altitude is not where the water would naturally be, the water is considered to be at an altered equilibrium. The result is stored energy of position, or gravitational potential energy. The water has the potential to do work because of the position it is in. As shown in Figure 6.6, gravity will force the water to fall to a lower position through the intake and the control gate. When the control gate is opened, the water from the reservoir moves through the intake and becomes translational kinetic energy as it falls through the next main part of the system, the penstock. *Translational kinetic energy* is the energy due to motion from one location to another. The water moves (falls) from the reservoir through the long shaft of the penstock toward the turbines, where the kinetic energy becomes mechanical energy. The force of the water is used to turn the turbines, which turn the generator shaft. The generators convert the energy of water into electricity and then step-up transformers increase the voltage produced to higher voltage levels.

Because the potential energy stored in the reservoir is converted into kinetic energy at the inlet to the water turbine, we can equate:

$$m \times g \times h = mV^2 \tag{6.18}$$

FIGURE 6.5 Wells Dam reservoir, Columbia River, Washington. (Photograph by Frank R. Spellman.)

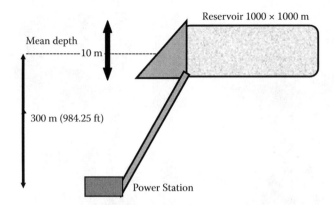

FIGURE 6.6 Schematic representation of a hydropower station. (Adapted from Tovey, N.K., *ENV-2E02 Energy Resources (2004–2005)*, University of East Anglia, Norwich, UK, 2005.)

where

m = Mass of water.

g = Acceleration due to gravity (can be taken as 10 m/s^2 in most applications).

h = Head difference.

V = Velocity of water at the inlet.

■ EXAMPLE 6.13[*]

Problem: A reservoir has an area of 1 km^2, and the difference between the crest of the dam and the inlet to a hydro station is 10 m (see Figure 6.6). The station runs at an overall efficiency of 80% and is situated 305 m below the crest of the dam. The rainfall is 1000 mm per annum, the catchment area of the reservoir is 10 times the area of the reservoir, and the run is 50%. What should the rated output of the turbine be if its maximum output is designed to be 5 times the mean output at the site? What is the maximum time the station could operate at full power during a sustained drought?

Solution: Mean head between maximum and minimum levels = 305 − 10/2 = 300 m. Average annual flow into the reservoir is equal to 50% of 10 times the area multiplied by the rainfall:

1. Mean head between the maximum and minimum levels = 305 − 10/2 = 300 m.
2. Average annual flow into the reservoir = 50% of 10 × area × rainfall = 0.5 × 10 × (1000 m × 1000 m) × 1 m = 5,000,000 m^3.
3. Mean energy generated per annum at 80% efficiency = $m \times g \times h \times 0.8$ = 5,000,000 × 1000 × 10 × 300 × 0.8 = 12,000,000 MJ.
4. Rated output (mean power) = 12,000,000/(60 × 60 × 24 × 365) = 0.381 MW.
5. So, maximum power out = 5 × 0.381 = 190 MW, and the time at maximum power assuming that the reservoir falls by 10 m is

$$\text{Days} = \frac{\text{Area} \times \text{Depth} \times \text{Density} \times g \times h \times 0.8}{\text{Maximum power}}$$

$$= \frac{1000 \times 1000 \times 10 \times 1000 \times 300 \times 0.8}{1,900,000 \times 60 \times 60 \times 24}$$

$$= 146.2 \text{ days}$$

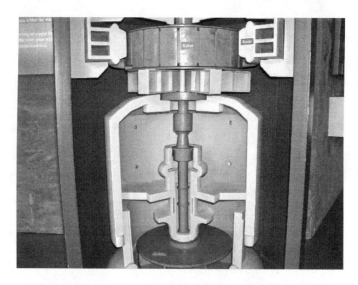

FIGURE 6.7   Cutaway view of a hydroturbine, shafting, and generator (Grand Coulee Dam, Columbia River, Washington). (Photograph by Frank R. Spellman.)

WATER TURBINES

The two main types of hydroturbines are *impulse* and *reaction* (EERE, 2008). The type of hydropower turbine chosen for a project is based on the height of standing water—referred to as *head*—and the flow, or volume of water, at the site. Cost, efficiency desired, and how deep the turbine must be set are other deciding factors. A cutaway view of the internal workings of a turbine, shafting, generator connection, and generator of a general type is provided in Figure 6.7, and Figure 6.8 shows the generator end of installed turbines.

FIGURE 6.8   Tops of generator end of hydroturbine assembly, Rocky Reach Dam, Columbia River, Washington. (Photograph by Frank R. Spellman.)

IMPULSE TURBINE

The impulse turbine uses the velocity of the water to move the runner and discharges to atmospheric pressure. The water stream hits each bucket on the runner. There is no suction on the down side of the turbine, and water flows out of the bottom of the turbine housing after hitting the runner. An impulse turbine is generally suitable for high-head, low-flow applications.

REACTION TURBINE

A reaction turbine develops power from the combined action of pressure and moving water. The runner is placed directly in the water stream flowing over the blades rather than striking each individually. Reaction turbines are generally used for sites with lower head and higher flows than compared with impulse turbines.

ADVANCED HYDROPOWER TECHNOLOGY

The U.S. Department of Energy (USDOE, 2008) and its associated technical activities support the development of technologies that will enable existing hydropower facilities to generate more electricity with less environmental impact. This will be done by (1) developing new turbine systems that have improved overall performances; (2) developing new methods to optimize hydropower operations at the unit, plant, and reservoir system levels; and (3) conducting research to improve the effectiveness of the environmental mitigation practices required at hydropower projects.

The main objective of its research into advanced hydropower technology is to develop new system designs and operation modes that will enable both better environmental performance and competitive generation of electricity. The products of USDOE research will allow hydropower projects to generate cleaner electricity, and USDOE-sponsored projects will develop new equipment and operational techniques to optimize water-use efficiency, increase generation, and improve environmental performance and mitigation practices at existing plants. Ongoing research efforts contributing to the success of these objectives will enable up to a 10% increase in the hydropower generation at existing dams; these objectives include the following:

- Test a new generation of large turbines in the field to demonstrate that these turbines are commercially viable, compatible with today's environmental standards, and capable of balancing environmental, technical, operational, and cost considerations.
- Develop new tools to improve water use efficiency and operations optimization within hydropower units, plants, and river systems with multiple hydropower facilities.
- Identify improved practices that can be applied at hydropower plants to mitigate for environmental effects of hydro development and operation.

ENVIRONMENTAL AND ECOLOGICAL IMPACTS OF HYDROPOWER

DISSOLVED OXYGEN CONCERNS

The benefits derived from the use of hydropower include the following: (1) it is a clean fuel source; (2) it is a fuel source that is domestically supplied; (3) it relies on the water cycle and thus is a renewable power source; (4) it is generally available as needed; (5) it creates reservoirs that offer a variety of recreational opportunities, such as fishing, swimming, and boating; and (6) it creates a supply of water where needed and assists in flood control. Many of these benefits are well known and often taken for granted.

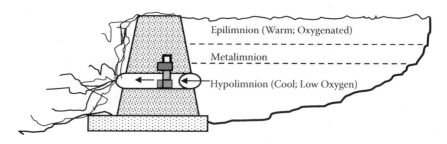

FIGURE 6.9 Thermal stratification of a hydropower reservoir.

Coins are two sided, of course; that is, the good side of anything is generally accompanied by a bad side. Many view this to be the case with hydropower. The bad side or disadvantages of hydropower include the impact on fish populations, such as salmon, when the fish cannot migrate upstream past impoundment dams to their spawning grounds or migrate downstream to the ocean. Hydropower can also be impacted by drought, because when water is not available the plant cannot produce electricity. Hydropower plants also compete with other uses for the land.

Other lesser known negatives of hydropower plants concern their impact on water flow and quality; hydropower plants can cause low water levels that impact riparian habitats. Water quality is also affected by hydropower plants. Low dissolved oxygen (DO) levels in the water, which are harmful to riparian habitats, can result when reservoirs stratify; that is, they develop layers of water of different temperatures (see Figure 6.9). Stratification can affect the water temperature, thus causing changes in dissolved oxygen levels, nutrient levels, productivity, and the bioavailability of heavy metals. During the summer, the natural process of stratification can divide a reservoir into distinct vertical strata, such as a warm, well-mixed upper layer (epilimnion) over a cooler, relativity stagnant lower layer (hypolimnion). Plant and animal respiration, bacterial decomposition of organic matter, and chemical oxidation can all act to progressively remove dissolved oxygen from hypolimnetic waters. This decrease in hypolimnetic dissolved oxygen is not generally offset by the renewal mechanisms of atmospheric diffusion, circulation, and photosynthesis that operate in the epilmnion (Spellman, 1996, 2008). In temperate regions, the decline in hypolimnetic dissolved oxygen concentrations begins at the onset of stratification (spring or summer) and continues until either anaerobic conditions predominate or reoxygenation occurs during the fall turnover of the water body.

Numerous structural, operational, and regulatory techniques are available to resolve low dissolved oxygen issues. Levels of dissolved oxygen can be increased through modifications in dam operations that include such techniques as fluctuating the timing and duration of flow releases, spilling or sluicing water, increasing minimum flows, flow mixing, and turbine aeration; at some sites, injection of air or oxygen for weir aeration has proven effective. The most effective strategy for addressing the dissolved oxygen problem depends on the site.

During operation, adverse ecological effects could result from disturbance of wildlife by equipment noise, site inspection and maintenance activities, exposure of biota to contaminants, and mortality of birds due to collisions with the project facilities and/or electrocution by transmission lines. During operation, wildlife could still be affected by habitat fragmentation or the presence of barriers in fenced areas, canals, or aboveground pipelines, utility rights of way (ROWs), and access roads. In addition, the presence of the hydropower project and its associated access roads and ROWs may increase the human use of surrounding areas, which could, in turn, impact ecological resources in the surrounding areas through

- Introduction and spread of invasive vegetation
- Disturbance of habitats
- Mortality of wildlife from vehicles
- Increase in hunting (including poaching)

- Increased potential for fire
- Physical barriers to fish migration
- Flow alteration and fluctuations
- Biological impacts of flow fluctuations

In the discussion that follows, for the purposes of this text, we concentrate on the last three impacts listed above related to the Pacific Northwest region of the United States.

Physical Barrier to Fish Migration

The presence of a dam can impose a physical barrier to fish migration. The following sidebar account is provided here to illustrate this issue and to explain mitigation procedures.

SIDEBAR 6.1. THE RACHEL RIVER*

The Rachel River, a hypothetical river system in the northwestern United States, courses its way through an area that includes a Native American reservation. The river system outfalls to the Pacific Ocean, and the headwaters begin deep and high within the Cascade Range in the state of Washington. For untold centuries, this river system provided a natural spawning area for salmon. The salmon fry thrived in the river and eventually grew the characteristic dark blotches on their bodies and transformed from fry to parr. When the time came to make their way to the sea, their bodies larger and covered with silver pigment, the salmon, now called smolt, inexorably migrated to the ocean, where they thrived until time to return to the river and spawn, about 4 years later. In spawning season, the salmon instinctively headed toward the odor generated by the Rachel River (their homing signal) and up the river to their home waters, as their life-cycle instincts demanded.

Before non-Native American settlers arrived in this pristine wilderness region, nature, humans, and salmon lived in harmony and provided for each other. Nature gave the salmon the perfect habitat, the salmon provided Native Americans with sustenance, and the Native Americans gave both their natural world and the salmon the respect they deserved.

After the settlers came to the Rachel River Valley, changes began to take place. The salmon still ran the river and humans still fed on the salmon, but the circumstances quickly changed. The settlers wanted more land, and Native Americans were forced to give way; they were either killed or forcibly moved to other places, such as reservations, while the settlers did all they could to erase Native American beliefs and cultural inheritance. The salmon still ran the streams.

After the settlers drove out the Native Americans, the salmon continued to run for a while, but more non-Native Americans continued to pour into the area. As the area became more crowded, the salmon still ran, but by now their home, their habitat, the Rachel River, had begun to show the effects of modern civilization. The prevailing philosophy was, "If we don't want it any more, we can just throw it away." The river provided a seemingly endless dump—out of the way, out of sight, out of mind. And they threw their trash, all the mountains of trash they could manufacture, into the river. The salmon still ran.

More time passed. More people moved in, and the more people that came into the area, the bigger their demands. In its natural course, sometimes the river flooded, creating problems for the settler populations. Also, everyone wanted power to maintain their modern lifestyles, and hydropower was constantly pouring down the Rachel River to the ocean. So the people built flood control systems and a dam to convert hydropower to hydroelectric power. (Funny … the Native Americans didn't have a problem with flood control. When the river

---

* From Spellman, F.R., *Environmental Science and Technology*, 2nd ed., Government Institutes, Rockville, MD, 2006.

rose, they broke camp and moved to higher ground. Hydroelectric power? If you don't build your life around things, you don't need electricity to make them work. With the sun, the moon, and the stars and healthy, vital land at hand, who would want hydroelectric power?)

The salmon still ran.

Building dams and flood control systems takes time, but humans, although impatient, have a way of conquering and using time (and anything else that gets in the way) to accomplish their goals, including construction projects. As the years passed, the construction moved closer to completion, and finally ended. The salmon still ran—but in reduced numbers and size. Soon local inhabitants couldn't catch the quantity and quality of salmon they had in the past. They began to ask, "Where are the salmon?"

But no one seemed to know. Obviously, the time had come to call in the scientists, the experts. The inhabitants' governing officials formed a committee, funded a study, and hired some scientists to tell them what was wrong. "The scientists will know the answer. They'll know what to do," they said, and that was partly true. Notice that they didn't try to ask the Native Americans. They also would have known what to do. The salmon had already told them.

The scientists came and studied the situation, conducted tests, tested their tests, and decided that the salmon population needed to grow. They determined that an increased population could be achieved by building a fish hatchery, which would take the eggs from spawning salmon, raise the eggs to fingerling-sized fish, release them into specially built basins, and later release them to restock the river. A lot of science goes into the operation of a fish hatchery. It can't operate successfully on its own (although Mother Nature never has a serious problem with it when left alone) but must be run by trained scientists and technicians following a proven protocol based on biological studies of salmon life cycles.

When the time was right, the salmon were released into the river—meanwhile, other scientists and engineers realized that some mechanism had to be installed in the dam to allow the salmon to swim downstream to the ocean, and the reverse, as well. In the lives of salmon (anadromous species that spend their adult lives at sea but return to freshwater to spawn), what goes downstream must go upstream. The salmon would eventually need some way of getting back up past the dam and into their home water, their spawning grounds. So, the scientists and engineers devised and installed fish ladders in the dam (Figures SB6.1.1) so the salmon could climb the ladders, scale the dam, and return to their native waters to spawn and die.

After a few seasons, the salmon again ran strong in the Rachel River. The scientists had temporarily—and at a high financial expenditure—solved the problem. Nothing in life or in nature is static or permanent. All things change. They shift from static to dynamic, in natural cycles that defy human intervention, relatively quickly, without notice—like a dormant volcano, or the Pacific Rim tectonic plates.

After a few years, local Rachel River residents noticed an alarming trend. Studies over a 5-year period showed that no matter how many salmon were released into the river, fewer and fewer returned to spawn each season. So they called in the scientists again. And again they thought, "Don't worry. The scientists will know. They'll tell us what to do." The scientists came in, analyzed the problem, and came up with five conclusions:

1. The Rachel River is extremely polluted from both point and nonpoint sources.
2. The Rachel River Dam had radically reduced the number of returning salmon to the spawning grounds.
3. Foreign fishing fleets off the Pacific Coast were depleting the salmon.
4. Native Americans were removing salmon downstream, before they even got close to the fish ladder at Rachel River Dam.

FIGURE SB6.1.1   Fish ladder at Rocky Reach Dam on the Columbia River, Washington. (Photographs by Frank R. Spellman.)

5. A large percentage of water was being withdrawn each year from rivers for cooling machinery in local factories. Large rivers with rapid flow rates usually can dissipate heat rapidly and suffer little ecological damage unless their flow rates are sharply reduced by seasonal fluctuations. This was not the case, though, with the Rachel River. The large input of heated water from Rachel River area factories back into the slow-moving Rachel River was creating an adverse effect called *thermal pollution*. Thermal pollution and salmon do not mix. First and foremost, increased water temperatures lower the dissolved oxygen (DO) content by decreasing the solubility of oxygen in the river water. Warmer river water also causes aquatic organisms to increase their respiration rates and consume oxygen faster, increasing their susceptibility to disease, parasites, and toxic chemicals. Although salmon can survive in heated water—to a point—many other fish and organisms (the salmon's food supply) cannot. Heated discharge water from the factories also disrupts the spawning process and kills the young fry.

The scientists prepared their written findings and presented them to city officials, who read them and were initially pleased. "Ah!" they said. "Now we know why we have fewer salmon!" But what was the solution? The scientists looked at each other and shrugged. "That's not our job," they said. "Call in the environmental folks."

The salmon still ran, but not up the Rachel River to its headwaters.

Within days, the city officials hired an environmental engineering firm to study the salmon depletion problem. The environmentalists came up with the same causal conclusions as the scientists, but they also noted the political, economic, and philosophical implications of the situation. The environmentalists explained that most of the pollution constantly pouring into the Rachel River would be eliminated when the city's new wastewater treatment plant came online and that specific *point-source pollution* would be eliminated. They explained that the state agricultural department and their environmental staff were working with farmers along the lower river course to modify their farming practices and pesticide treatment regimes to help control the most destructive types of *nonpoint-source pollution*. The environmentalists explained that the Rachel River dam's current fish ladder was incorrectly configured but could be modified with minor retrofitting.

The environmentalists went on to explain that the overfishing by foreign fishing fleets off the Pacific Coast was a problem that the federal government was working to resolve with the governments involved. The environmentalists explained that the state of Washington and the federal government were also addressing a problem with the Native Americans fishing the downriver locations, before the salmon ever reached the dam. Both governmental entities were negotiating with the local tribes on this problem. Meanwhile, local tribes had litigation pending against the state and federal government to determine who actually owned fishing rights to the Rachel River and the salmon.

The final problem was thermal pollution from the factories, which was making the Rachel River unfavorable for spawning, decreasing salmon food supply, and killing off the young salmon fry. The environmentalists explained that to correct this problem, the outfalls from the factories would have to be changed and relocated. The environmentalists also recommended construction of a channel basin whereby the ready-to-release salmon fry could be released in a favorable environment, at ambient stream temperatures. This would give them a controlled one-way route to safe downstream locations where they could thrive until it was time to migrate to the sea. After many debates and newspaper editorials, the city officials put the matter to a vote and voted to fund the projects needed to solve the salmon problem in the Rachel River. Some short-term projects are already showing positive signs of change, long-term projects are underway, and the Rachel River is on its way to recovery. In short, scientists are professionals who study to find *the* answer to a problem through scientific analysis and study. Their interest is in pure science. The environmentalists (also scientists) can arrive at the same causal conclusions as general scientists, but they are also able to factor in socioeconomic, political, and cultural influences, as well.

But, wait! It's not over yet. Concerns over disruption of the wild salmon gene pool by hatchery trout are drawing attention from environmentalists, conservationists, and wildlife biologists. Hatchery- or farm-raised stock of any kind is susceptible to problems caused by, among other things, a lack of free genetic mixing and the spread of disease, infection, and parasites, as well as reinforcement of negative characteristics. When escaped hatchery salmon breed with wild salmon, the genetic strain is changed and diseases can be transmitted. Many problems can arise.

Yes, many problems arise and solutions are constantly sought. When nature's natural processes are interrupted, changed, or manipulated in any way, humans need to adjust to the changes, but so does Mother Nature. The question is are the human-made changes to natural surroundings a good or bad thing? It depends. On what? It depends on your point of view.

For many, the Rachel River case study probably generates more questions than answers, because there are a number of vignettes within the account, many of which could give rise to separate case studies of their own. However, if we focus on the dam only and its implications, not only for the human inhabitants but also for the natural resources involved, environmental scientists would study the construction of such a human-made structure based on facts, science, and the pros and cons. For example, let's consider the pros and cons (USGS, 2014a).

Pros to hydroelectric power (as compared to other power-producing methods) include the following:

- Fuel is not burned, so there is minimal pollution.
- Water to run the power plant is provided free by nature.
- Hydropower plays a major role in reducing greenhouse gas emissions.
- Operations and maintenance costs are relatively low.
- The technology is reliable and has been proven over time.
- It is renewable, because rainfall renews the water in the reservoir, so the fuel is almost always there.

Cons to hydroelectric power (as compared to other power-producing methods) include the following:

- Investment costs are high.
- Hydropower is dependent on precipitation.
- In some cases, there is an inundation of a wildlife habitat.
- In some cases, there is a loss or modification of fish habitat.
- Dams can cause fish entrainment or passage restriction (stranding).
- In some cases, there can be changes in reservoir and stream water quality.
- In some cases, local populations can be displaced.

FLOW ALTERATIONS, FLOW FLUCTUATIONS, UNREGULATED RIVERS, AND REGULATED RIVERS*

Hydropower plants can, to varying capacities, change instream flow patterns in rivers below the dams and powerhouses. These changes can be classified into two categories: *flow alterations* and *flow fluctuations*. Flow alterations are changes in flow over long periods of time (weeks, months, or seasons) resulting from the storage of water, irrigation diversions, municipal diversions, or the reactions of flow between dams and powerhouses. These changes in net flow usually change the availability of fish habitat and thus change the fish production potential of a river. Flow alterations are evaluated by studying the fish habitat requirements and estimating the changes in habitat area at different flows using a hydraulic model. The Instream Flow Incremental Methodology (IFIM) (Bovee, 1982) has become a standard method for estimating habitat changes resulting from flow alterations. The IFIM that was developed under the guidance of the U.S. Fish and Wildlife Service (USFWS) is a process utilizing various technical methodologies to evaluate changes in the amount of estimated usable habitat for various species or groups of species as flow changes (Stalnaker et al., 1995). The IFIM methodology is routinely used to facilitate negation of instream flow requirements, usually minimum flow requirements that meet the habitat needs of economically important or threatened fish species.

---

* Adapted from Hunter, M.A., *Hydropower Flow Fluctuations and Salmonids: A Review of the Biological Effects, Mechanical Causes, and Options for Mitigation*, Washington Department of Fisheries, Olympia, 1992.

DID YOU KNOW?

River stage is an important concept when analyzing how much water is moving in a stream at any given moment. *Stage* is the water level above some arbitrary point in the river, usually with the zero height being near the river bed, and is commonly measured in feet. For example, on a normal day when no rain has fallen for a while, a river might have a stage of 2 feet (base-flow conditions). If a big storm hits, the river stage could rise to 15 or 20 feet, sometimes very quickly. This is important because, from past records, we might know that when the stage hits 21 feet the water will start flowing over its banks and into the basements of houses along the river. How high and how fast a river will rise during a storm depends on many things. Most important, of course, is how much rain is falling, but we also have to look at other things, such as the stage of the river when the storm begins, what the soil is like in the drainage basin where it is raining (is the soil already saturated with water from a previous storm?), and how hard and in what parts of the watershed the rain is falling (USGS, 2014b).

Flow fluctuations are unnaturally rapid changes in the flow over periods of minutes, hours, and days. Flow fluctuations can be immediately lethal or have indirect and delayed biological effects. Flow fluctuations can be measured either by changes in *flow*, which is the volume of water passing a specific river transect, or by changes in *stage*, which is the water surface elevation or gauge height. Both units are needed to understand the problem, and the terms are used interchangeably in this text. Hydrologists and engineers require flow measurements for many applications; however, the biological impact of flow fluctuations is best measured by stage. These two units do not have a simple functional relationship, thus *rating tables* or *rating curves* are used to define the flow at each stage for a specific river transect.

Flows in *unregulated rivers* respond to changes in precipitation and snow melt. West of the Cascade Range, the peak flows occur from heavy rain storms in November, December, and January. A lesser but more sustained peak occurs from a combination of rain and snow melt in the spring. The lowest flows coincide with the dry season that occurs in late summer and early fall. Glacial streams and streams on the east side of the Cascades have a somewhat different pattern. Here, the highest flows often occur in the spring and extend into the early summer. The lowest flows in some years occur during cold periods in the winter. In either case, periods of heavy rainfall or dry weather can create flows that are above or below seasonal averages. These natural flow variations indirectly affect fish production as a result of changes in the quantity and quality of instream habitats.

On a shorter time scale, individual storms can rapidly increase the river stage in less than a day. After the storm, the stage declines to a relatively stable level over a longer period of time, usually days or weeks. In addition to storm events, limited daily stage changes sometimes occur during sunny weather as a result of snow-melt runoff.

Flows in *regulated rivers* respond to measures taken to improve river channels so that they can be used more efficiently in the national economic interest. When properly engineered and constructed, river regulation ensures the creation of favorable conditions for navigation and timber rafting, the

DID YOU KNOW?

With the advent of modern computer and satellite technology, the U.S. Geological Survey can monitor the stage of many streams almost instantly. Because some streams, especially those in the normally arid western United States, can rise dramatically in a matter of minutes during a major storm, it is important to be able to remotely monitor how fast water is rising in real time in order to warn people who might be affected by a dangerous flood (USGS, 2014b).

> **DID YOU KNOW?**
>
> Salmonids are a family of ray-finned fish that includes salmon, trout, chars, freshwater white-fishes, and graylings.

maintenance of the necessary water levels at water intake works, the protection of populated areas and agricultural land from flooding during spring floods and high water, the slow movement of river sediment, and the smooth flow of water toward the openings of hydraulic engineering structures such as dams.

BIOLOGICAL IMPACTS OF FLOW FLUCTUATIONS

INCREASES IN FLOW

Evidence of biological impacts from rapid flow increases is scarce. Some impacts associated with rapid flow increases might be more appropriately associated with high flows. Eggs and alevins (fry) can be killed when gravel scour occurs, and juvenile fish may be physically flushed down the river (Rochester et al., 1984). Some species of aquatic insects that swim in pools can be physically flushed downstream from a sudden increase in flow (Trotzky and Gregory, 1974). The biological effects of unnatural flow increases are usually irrelevant in regulating hydropower operations because public safety concerns justify more stringent regulations than biological concerns. Flow increases can strand and occasionally drown fishermen and other people located on bars, rocks, or in confined canyons. Boaters might also be at risk under some circumstances.

STRANDING

Stranding is the separation of fish from flowing surface water as a result of declining river stage. Stranding can occur during any drop in stage. It is not exclusively associated with complete or substantial dewatering of a river. Stranding can be classified into two categories: *Beaching* is when fish flounder out of the water on the substrate. *Trapping* is the isolation of fish in pockets of water with no access to the free-flowing surface water. Stranding cannot always be neatly classified as beaching or trapping. Thus, in this text we use the term stranding unless a more specific term is appropriate. Salmonid stranding associated with hydropower operations has been widely documented in Washington and Oregon (e.g., Bauersfeld, 1977, 1978a; Becker et al., 1981; Fiscus, 1977; Olson, 1990; Phinney, 1974a,b; Satterthwaite, 1987; Thompson, 1970; Witty and Thompson, 1974). Stranding can occur many miles downstream of the powerhouse (Phillips, 1969; Woodin, 1984). The estimated numbers of fish stranded in flow fluctuation events range from negligible to 120,000 fry (Phinney, 1974a). Stranding mortality is difficult or impossible to estimate. Estimates are usually very conservative or highly variable. Stranding can also occur as a result of other events, including natural declines in flow (author's observation of Colorado River over time), ship wash (Bauersfeld, 1977), municipal water withdrawals, and irrigation withdrawals. Many factors affect the incidence of stranding. A recurrent theme in much of the following discussion is the high vulnerability of small salmonid fry.

Juvenile salmonids are more vulnerable to stranding than adults. Salmonid fry that have just absorbed the yolk sac and have recently emerged from the gravel are by far the most vulnerable. They are poor swimmers and settle along shallow margins of rivers (Phinney, 1974a; Woodin, 1984), where they seek refuge from currents and larger fish. Once Chinook attain the size of 50 to 60 mm in length, their vulnerability drops substantially. For steelhead, vulnerability drops significantly when the fry reach 40 mm (Beck Associates, 1989). Larger juveniles are more inclined to inhabit pools, glides, overhanging banks, and midchannel substrates, where they are less vulnerable

FIGURE 6.10 Series of potholes in an ancient river bed in Red Rock Canyon, Sedona, Arizona. (Photograph by Frank R. Spellman.)

to stranding; however, many juveniles still inhabit shoreline areas and remain vulnerable to stranding until they emigrate to saltwater (Hamilton and Buell, 1976). Adult stranding as a result of hydropower fluctuations has been documented (Hamilton and Buell, 1976).

The river channel configuration is a major factor in the incidence of stranding. A river channel with many side channels, potholes, and low gradient bars will have a much greater incidence of stranding than a river confined to a single channel with steep banks. Large numbers of small fry die from beaching on gravel bars when unnatural flow fluctuations occur (Phillips, 1969; Phinney, 1974a; Woodin, 1984). Bauersfeld (1978a) observed beaching primarily on bars with slopes less than 4%. Beck Associates (1989) determined that beaching occurred primarily on bars with slopes less than 5%. Under laboratory conditions, Monk (1989) determined that Chinook fry stranded in significantly larger numbers on 1.8% slopes than on 5.1% slopes, but the results were not significant for steelhead. Stranding on steep gravel bars (>5% slope) has not been thoroughly studied.

Long side channels with intermittent flows are notorious for trapping juvenile fish. Substantial trapping can occur even with unregulated flows (Hunter, 1992). Side channels are valuable rearing habitats, and juveniles of several species prefer side channels over the main channel; however, unnatural fluctuations will repeatedly trap fish, eventually killing some or all of them (Hamilton and Buell, 1976; Olson, 1990; Witty and Thompson, 1974; Woodin, 1984). Side channels can trap substantial numbers of fingerlings and smolts (up to 150 cm) as well as fry.

As water recedes from river margins, juvenile salmonids may become trapped in deep pools called *potholes* (Woodin, 1984). River potholes are formed at high flows from scouring (a process called *corrasion*; see Figure 6.10) around boulders and rootwads and where opposing flows meet. Potholes may remain watered for hours or months depending on the depth of the pothole and the river stage. Beck Associates (1989) extensively studied pothole stranding in the Skagit River (Washington State). Among the conclusions were that (1) only a small fraction of the potholes in a river channel posed a threat to fish if fluctuations are limited in range, (2) the incidence of stranding is independent of the rate of stage decrease, and (3) the incidence of stranding is inversely related to the depth of water over the top of each pothole at the start of the decline in flow.

Most documented observations of stranding have occurred on gravel; however, stranding has also occurred in mud (Becker et al., 1981) and vegetation (Phillips, 1969; Satterthwaite, 1987). Under laboratory conditions, Monk (1989) found significantly different rates of stranding on different types of gravel. In fact, substrate was statistically the most significant factor contributing to the stranding of Chinook and steelhead fry. On cobble substrate, fry (especially steelhead fry) were inclined to maintain a stationary position over the streambed (i.e., rheotaxis); whereas, over small gravel, fry swam around, often in schools. When the water surface dropped, fry maintaining their position became trapped in pockets of water between cobbles, whereas mobile fish were more inclined to retreat with the water margin. When beaching became imminent, fry over cobble substrate retreated into inter-gravel cavities, where they became trapped. The difference in stranding rate was facilitated by the flow of water along a receding margin of the stream. On cobble substrate, the water drained into the substrate, whereas on finer substrates a significant portion of the water flowed off on the surface.

Fry of some species are more vulnerable to stranding than others. In Washington State, stranding of Chinook and steelhead fry has been frequently observed. Although pink salmon fry and chum salmon fry occur in the same rivers, they strand in lower numbers than Chinook fry and steelhead fry (Woodin, 1984). However, Beck Associates (1989) determined that the rate of chum and pink fry stranding per the available fry was substantially higher than for Chinook. The low numbers of pink and chum salmon stranding is a result of the short freshwater residency; they emigrate to saltwater shortly after emergence, whereas Chinook and steelhead remain in the river for months or years.

Hamilton and Buell (1976) observed extensive coho stranding in the Campbell River (British Columbia), and coho stranding has been observed in incidental numbers in other studies (Olsen, 1990). The overall incidence of coho stranding is rather low in the studies conducted to date. The likely reason for this is that coho prefer streams for spawning and breeding, whereas the formal research and evaluation have taken place in large and medium rivers. Juvenile coho rear for a full year in freshwater; thus, it is reasonable to assume that stranding would occur at rates similar to those for Chinook and steelhead.

The total drop in stage from an episode of flow fluctuation (known as *ramping range*) affects the incidence of stranding by increasing the gravel bar area exposed. In addition, it increases the number of side channels and potholes that become isolated from surface flow (Beck Associates, 1989). Stranding increases dramatically when flow drops below a certain water level, defined as the *critical flow* (Bauersfeld, 1978a; Phinney, 1974a; Thompson, 1970; Woodin, 1984). In hydropower mitigation settlements, the critical flow is defined as the minimum operating discharge, or as an upper end of a flow range where more restrictive operation criteria are applied. The factors that likely account for this response have been discussed above. The exposure of the lowest gradient gravel bars often occurs in a limited range of flows. The exposure of spawning gravel from which fry are emerging may also account for the higher incidence of stranding.

DID YOU KNOW?

To measure instantaneous discharge, the USGS sends hydrologic technicians to go out and stand in a stream, on a bridge, or in a boat to measure the depth and how fast the water is moving at many places in the stream. By doing this many times and at many stream stages, over the years the USGS has identified the relationships between stream state and discharge.

In rivers with seasonal side channels and off-channel sloughs (slews, slues), even a natural flow reduction can trap fry and smolts. Under normal circumstances, the natural population can sustain a small loss several times a year; however, when a hydropower facility causes repeated flow fluctuations, these small losses can accumulate to a very significant cumulative loss (Bauersfeld, 1978a).

The *ramping rate* is the rate of change in stage resulting from regulated discharges. Unless otherwise noted, it refers to the rate of state decline. The faster the ramping rate, the more likely fish are to be stranded (Bauersfeld, 1978a; Phinney, 1974). Ramping rates less than 1 inch per hour were needed to protect steelhead fry on the Sultan River. Olson (1990) determined that a ramping rate of 1 inch per hour was adequate to protect steelhead fry; however, the ramping rate was measured at a confined river transect, whereas the stranding was observed on lower gradient bars farther downstream. Thus, the effective ramping rate at these bars was less than 1 inch per hour.

Although many hydropower mitigation settlements specify ramping rates, some research has indicated that ramping rates cannot always protect fish from stranding. Woodin (1984) determined that *any* daytime ramping stranded Chinook fry. Beck Associates (1989) could not find any correlation between ramping and the incidence of pothole trapping, nor was there any correlation between the ramping rate and steelhead fry stranding during the summer. In both cases, stranding occurred regardless of the ramping rate.

Small fry are highly vulnerable to stranding and are present in streams only at certain times of the year. Chinook, coho, pink, and chum fry emerge during late winter and early spring, while steelhead emerge in late spring through early fall (Olson, 1989). Fingerlings, smolts, and adults are vulnerable to stranding in other seasons; however, less restrictive ramping criteria are often sufficient to protect them.

For at least some species, the incidence of stranding is influenced by the time of day. Chinook fry are less dependent on substrate for cover at night and thus are less vulnerable to stranding at night (Woodin, 1984). Two studies (Olson, 1990; Stober et al., 1982) concluded that steelhead fry are less vulnerable during the day, presumably because this species feeds during the day. However, two other studies (Beck Associates, 1989; Monk, 1989) found no difference in the rate of steelhead fry stranding relative to day and night.

Salmonids respire using their gills and do not survive out of water for more than 10 minutes; thus, beaching is always fatal. Juvenile salmonids trapped in side channels and potholes can survive for hours, days, or, under favorable circumstances, for months (Hunter, 1992). However, many trapped fish die from predation, temperature shock, or oxygen depletion. Survivors that are rescued by higher flows are probably in poorer condition than fish in the free-flowing channel.

Some observations suggest that a highly stable flow regime for a week or more prior to a flow fluctuation will increase the incidence of fry stranding (Phinney, 1974b). Two hypotheses might explain this observation. One hypothesis states that after long periods of stable flow more fry are available for stranding. In other words, a major flow reduction after a week of stable flows strands seven daily cohorts of emerging fry at once, rather than one cohort when fluctuations occur daily. An alternative hypothesis is that juveniles become accustomed to residing and feeding along the margins of a stream, either as a behavioral response to stable flows or in response to aquatic invertebrate populations that thrive along the water's edge under stable flows. These hypotheses should be thoroughly tested before they are applied to mitigation practices.

Juvenile Emigration (Salmonid Drift)

Flow fluctuations in an experimental stream channel caused juvenile Chinook to emigrate downstream (McPhee and Brusven, 1976). The pre-test rate of emigration under stable flows was about 1% a day. Severe flow fluctuations (from 51 L/s to 17 to 3 to 51, with each flow being held for 24 hours) caused 60% of the Chinook to emigrate. A high rate of emigration continued even after initial flows were reestablished. A less severe daily fluctuation in flow (between 51 and 17 L/s for four 24-hour periods) caused 14% of the Chinook to emigrate. Alternating flows between 51 L/s and 17 L/s every 24 hours cause a greater rate of emigration than alternating the same flows every 12 hours. Most of the emigration occurred at night, a behavior observed in aquatic invertebrates. The behavioral response to flow fluctuations and how this may affect the juvenile salmonid rearing capacity is not well understood. Under conservative ramping requirements, flow fluctuations may cause downstream emigration, driving many fish to habitats that may be less desirable or overcrowded and leaving upstream rearing habitats underutilized. This could be a particular concern in a stream with a falls or other barrier that prevents juveniles from returning upstream.

Increased Predation

It has been suggested that juvenile fish forced from the river margins as a result of declining flows suffer from predation by larger fish (Phillips, 1969). This effect does not appear to have been documented anywhere, but it is a credible hypothesis under some circumstances.

Aquatic Invertebrates

Like fish, aquatic invertebrates are not necessarily adapted to unnatural drops in flow. Cushman (1985) extensively reviewed the effects of flow fluctuations on aquatic life, especially aquatic invertebrates. Interested readers are encouraged to refer to this review. Research on the effects of flow fluctuations on aquatic invertebrates in the Pacific Northwest is limited, although more information is available elsewhere in North America. The studies that have been done suggest that aquatic invertebrates can be severely impacted by flow fluctuations. Fluctuations substantially reduce invertebrate diversity and total biomass and change the species composition under most circumstances. One study from the Skagit River found that flow fluctuations had a greater adverse impact on the aquatic invertebrate community than a substantial reduction in average flow (Gislason, 1985). The reduction in aquatic invertebrate production can impact salmonid production as a result of reduced feeding (Cushman, 1985).

Additional research is needed on the effects of flow fluctuations on aquatic invertebrates in the Pacific Northwest; however, a thorough study would be a formidable task. It would involve many species with different life cycles, behavioral responses, lethal responses, and contributions as prey to salmonids. Populations of some species may change rapidly under normal conditions, thus it may be difficult to associate cause and effect. Flow fluctuations can impact the aquatic invertebrates in the following ways:

DID YOU KNOW?

Stream stages are not always cooperative, so it's not uncommon for someone to have to go measure a stream at 2:00 in the morning during a storm, sometimes in freezing conditions. Also, the stream can be uncooperative in that it changes—a big storm may come along and scour out bottom material of a creek or lodge a big log sideways in the creek, or sometimes do both at the same time. These kind of changes result in changes in the relationship between stage and discharge (USGS, 2014b).

- *Stranding*—Flow fluctuations can strand many species of aquatic invertebrates, much in the same way fish can become stranded (Gislason, 1985; Phillips 1969). Death may result from suffocation, desiccation, temperature shock, or predation.

- *Increased drift*—Many aquatic invertebrates are sensitive to reductions in flow and respond by leaving the substrate and floating downstream. This floating behavior is drift. Nighttime drift is normal; however, drift becomes highly elevated under unnatural fluctuations in flow (McPhee and Brusven, 1975; Cushman, 1985). This elevated drift may be an emergency response to avoid stranding, a response to overcrowding of the inter-gravel habitat, or a response by aquatic species that are adapted to a narrow range of water velocity. This response may temporarily increase fish food supply (McPhee and Brusven, 1975), but when repeated fluctuations occur many species are flushed out of the river and the aquatic invertebrate biomass usually declines, often substantially (Cushman, 1985; Gislason, 1985). Elevated drift also occurs in response to sudden increases in flow, capturing terrestrial insects from the river banks and scouring some aquatic invertebrates from the river substrate (Mundie and Mounce, 1976).

- *Detritus feeders*—Under stable flow conditions, floating detritus (leaves, woody debris) accumulates on the shores of the river as a result of current and wind action on sand or gravel substrate. This detritus remains close to the river margin and often remains damp for days or weeks at a time. Under fluctuating flows, this organic detritus becomes suspended (Mundie and Mounce, 1976) and is flushed out of the river or redeposited at the high waterline where it desiccates during low flow periods. As a result the invertebrate detritus community is less capable of exploiting this resource.

- *Herbivorous invertebrates*—Impacts are similar to those on the detritus community. Algae grows on exposed rock surfaces on which herbivorous aquatic invertebrates graze. Fluctuations desiccate and disrupt the growth of the exposed algae (Gislason, 1985) and reduce access by herbivores.

REDD DEWATERING

Research has extensively documented the lethal impact of redd dewatering on salmonid eggs and alevins (i.e., larval fish) (Frailey and Graham, 1982; Fraser, 1972; Fustish et al., 1988; Satterthwaite et al., 1985). Salmonid eggs can survive for weeks in dewatered gravel (Becker and Neitzel, 1985; Neitzel et al., 1985; Reiser and White, 1983; Stober et al., 1982), if they remain moist and are not subjected to freezing or high temperatures. The necessary moisture may originate from subsurface river water or from groundwater. If the subsurface water level drops too far, the inter-gravel spaces will dry out, and the eggs will desiccate and die. Thus, redd dewatering is not always lethal or even harmful to eggs; however, site conditions, weather, and duration of exposure will all affect survival. Because alevins rely on gills to respire, dewatering is lethal (Neitzel et al., 1985; Stober et al., 1982). Alevins can survive in subsurface, inter-gravel flow from a river or groundwater source. If inter-gravel spaces are not obstructed with pea gravel, sand, or fines, some alevins will survive by descending through inter-gravel spaces with the declining water surface (Stober et al., 1982). Both alevins and eggs may die from being submerged in stagnant water. Standing inter-gravel water may lose its oxygen to biotic decay, and metabolic wastes may build up to lethal levels. A redd can be dewatered between spawning and hatching without harm to the eggs under some circumstances, and in one situation a hydropower facility is operated to allow limited redd dewatering (Neitzel et al., 1985). However, in most Pacific Northwest rivers, anadromous fish spawn over an extended period. Different species spawn in different seasons and individual species may spawn over a range of 2 to 6 months. As a result, when eggs are present, alevins and fry are also present, both of which are highly vulnerable to flow fluctuations.

Spawning Interference

Bauersfeld (1978b) found that repeated dewatering caused Chinook salmon to abandon attempts to spawn and move elsewhere, often to less desirable or crowded locations. Hamilton and Buell (1976) performed a highly detailed study using observation towers situated over spawning beds to track activity on the spawning bed and to observe individual tagged fish. They observed that spawning Chinook were frequently interrupted by flow fluctuations. Females repeatedly initiated redd digging and then abandoned the redd sites when flows changed. The authors concluded that flow fluctuations decrease viability due to untimely release of eggs, failure to cover eggs once they were released, and a failure of males to properly fertilize eggs laid in incomplete redds. Other researchers offered conflicting conclusions. Stober et al. (1982) noted that Chinook salmon successfully spawned in an area that was dewatered several hours a day, and Chapman et al. (1986) found that 8 hours a day of dewatering still permitted successful spawning.

Hydraulic Response to Flow Fluctuations

The ramping rate attenuates as a function of the distance downstream from the source of a fluctuation event (Nestler et al., 1989). The characteristics of the river greatly influence this attenuation. A fluctuation in flow passing through a narrow bedrock river channel will experience little or no attenuation. Pools, side-channels, and gravel bars attenuate the ramping rate by storing water from higher flows and release this water gradually. Tributary inflow will attenuate the ramping rate and the ramping range. Hydraulic equations (e.g., unsteady flows; see Chow, 2009, for one of the best texts written on open-channel hydraulics) exist to describe these responses. The time it takes for a fluctuation to pass from one place to another on a river is known as *lag time*. The river channel configuration, gradient, and flow all influence the speed at which the fluctuation travels downstream. Lag time can be determined by field observations as several flows. For projects with long penstocks, the term *bypass lag time* refers to the time that flow fluctuations take to pass down the natural stream channel from the dam to the powerhouse tailrace.

TYPES OF HYDROPOWER ACTIVITY THAT CAUSE FLOWS TO FLUCTUATE

Hydropower facilities cause flow fluctuations in a variety of ways. A brief overview of mechanical causes is provided in the following.

- *Load following*—Load following occurs when a hydropower plant follows daily changes in power demand and adjusts its power output as demand for electricity fluctuates throughout the day.
- *Peaking*—A peaking hydropower plant operates only during times of peak demand. Peaking is the most widely documented source of fish stranding. Biologists and fishermen have observed major fish kills from peaking (Bauersfeld, 1977, 1978a; Becker et al., 1981; Graybill et al., 1979; Phinney, 1974a; Thompson, 1970).
- *Low-flow shutdowns*—When turbine flow is below the level for practical turbine operation, the plant must take various turbines offline. In addition, a minimum flow is usually required to maintain the aquatic habitat in the bypass reach. Dam facilities with seasonal storage can operate for years without a low-flow shutdown.
- *Low-flow startups*—Run-of-river plants will cause a drop in flow in the bypass and downstream reaches during powerhouse start-ups. In these situations, operators must ramp flows at the start of power generation to reduce stranding.

- *Powerhouse failures*—Powerhouse failures are rare disruptions of the penstock flow originating from the powerhouse. These disruptions usually result from mechanical problems but can also occur due to load rejection, which is the inability of the utility line to receive power generated from the turns. Powerhouse failures can occur at any facility. This type of failure can cause a sudden drop in flow level in the downstream reach.
- *Flow continuation*—Flow continuation is the mechanical capacity to maintain flow through the penstock during powerhouse failures.
- *Intake failures*—Intake failures are all penstock flow disruptions that occur at the intake structure. Intake failures are less frequent than powerhouse failures; they usually occur as a result of the accumulation of debris, the failure of fish screen cleaning equipment, or failure of the dam and associated gates to deliver water into the intake.
- *Cycling*—A way to generate power when flow is not enough for continuous or efficient operation; it is not an attempt to follow load demands. Cycling will normally occur at low stream flows when the salmonids would be most vulnerable to fluctuations.
- *Multiple turbine operations*—If a powerhouse has two or more turbines, operators can cause abrupt changes in flow when changing the number of turbines in operation.
- *Forebay surges*—Forebay surges occur when the powerhouse of some run-of-the-river plants start generation and are probably caused by a drop in head at the intake during start-up.
- *Reservoir stranding*—In large reservoirs, stranding is routinely anticipated as a consequence of drawdowns, and it is sometimes employed as a method of eradicating undesirable fish.
- *Tailwater maintenance and repair activities*—All hydropower plants will eventually require inspection, maintenance, and repair. However, it is often impossible to inspect or repair the structure or equipment submerged in the tailwater without completely or substantially disrupting flow to the river.

LOW WATER LEVELS AND EVAPORATION OF RESERVOIRS

They say a picture is worth a thousand words. Well, from the photographs shown in Figure 6.11, it should be obvious to even the most casual observer that Lake Mead, the reservoir formed by the Hoover Dam and the Colorado River, has a very distinctive bathtub ring along the channel cliffs that hold the reservoir water in place until the water passes through the dam's penstocks to its turbines and then back into the Colorado River below. The obvious conclusion to be drawn from the photographs is that Lake Mead is much lower than it has been in the past. This conclusion is correct, of course. As of October 2010, Lake Mead was only about 40% full. The lake has dropped 130 ft since 1999 and in 2010 was at 1084 ft, depths not seen since 1956.

Observers, hydrologists, critics, and laypersons (i.e., non-scientists) have differing opinions as to the causes of Lake Mead's current low water level. Most of these opinions have some merit. For example, the USGS (2013) observed that the bathtub rings and lower water level in Lake Mead are the result of a "stealth disaster." Drought is the culprit. Drought is a stealthy incremental disaster that is much more costly to the national economy than most people suspect. A prolonged dry spell, lasting over a decade, is steadily draining the water sources that power Hoover Dam's turbines. The Southwest and vast stretches of the High Plains region have suffered lengthy drought conditions for the past several years.

In addition to drought, heavy usage of the Colorado and Green Rivers has contributed to Lake Mead's low water level. For example, residents along the river use the water for showers, meals, hydration, and lawns. Farms use the water for their crops, livestock, and cleaning. Industrial facilities use the water for production, for cleaning their facilities, and for their employees.

In arid regions such as the Hoover Dam on the Arizona/Nevada border there is another reason for Lake Mead's low water levels: evaporation. This is a major issue and concern of the USGS and others. If water totals are at a premium to begin with (and they are), the evaporation rate of the remaining water is important and must be closely monitored. As a case in point, consider the following USGS 1997–1999 report on the impact of evaporation on Lake Mead.

FIGURE 6.11 Lake Mead bathtub ring on the steep, barren, rocky cliffs surrounding the lake in June 2013. By 2015, the width of the bathtub ring had increased by several feet. (Photographs by Frank R. Spellman.)

ESTIMATING EVAPORATION FROM LAKE MEAD[*]

Lake Mead is one of a series of large Colorado River reservoirs operated and maintained by the Bureau of Reclamation. The Colorado River system of reservoirs and diversions is an important source of water for millions of people in seven Western states and Mexico. The U.S. Geological Survey, in cooperation with the Bureau of Reclamation, conducted a study from 1997 to 1999 to estimate evaporation from Lake Mead. For this study, micrometeorological and hydrologic data were collected continually from instrumented platforms deployed at four locations on the lake: open-water areas of Boulder Basin, Virgin Basin, and Overton Arm and a protected cove in Boulder Basin. Data collected at the platforms were used to estimate Lake Mead evaporation by solving an energy-budget equation. The average annual evaporation rate at open-water stations from January 1998 to December 1999 was 7.5 ft. Because the spatial variation of monthly and annual evaporation rates was minimal for the open-water stations, a single open-water station in Boulder Basin would provide data that are adequate to estimate evaporation from Lake Mead.

---

[*] Adapted from Westennberg, C.I. et al., *Estimating Evaporation from Lake Mead (1997–99),* USGS Scientific Investigation Report 2006-5252, U.S. Geological Survey, Washington, DC, 2006.

Lake Mead is the largest reservoir by volume in the United States and was formed when Hoover Dam was completed in 1935. It took until July 1941 for water to fill Lake Mead, which has a maximum surface elevation of 1229 ft, a maximum surface area of 162,700 acres, and a maximum available capacity of 27,377,000 acre-ft (Bureau of Reclamation, 1967). At the maximum elevation of 1229 ft, Lake Mead extends 65.9 miles upstream of Hoover Dam and has a maximum width of 9.3 miles (LaBounty and Horn, 1997). The average lake elevation from 1942 (first complete calendar year of full pool) to 1995, based on monthly end-of-month elevations, is 1169.9 ft, which corresponds to a lake surface area of 125,600 acres.

The drainage area of Lake Mead at Hoover Dam is about 171,700 m^2 (Tadayon et al., 2000). Ninety-seven percent of the inflow into Lake Mead is from the Colorado River (13.12 million acre-ft/yr (Tadayon et al., 2000), with the remaining 2% coming from the combined flow of the Las Vegas Wash (148,000 acre-ft/yr) and the Virgin River (176,000 acre-ft/yr) (Jones et al., 2000), as well as ephemeral streams. The average annual release from Hoover Dam from 1935 to 1999 was about 10.1 million acre-ft (Tadayon et al., 2000). Flow of the Colorado River at Diamond Creek (130 miles upstream of Hoover Dam) in calendar year 1999 was 12.69 million acre-ft, while the release from Hoover Dam was 11.04 million acre-ft (Tadayon et al., 2001). Retention time for Lake Mead averages 3.9 years, depending on release and flow patterns.

Lake Mead is in an environment with a warm, arid climate. From 1961 to 1990, the average maximum air temperature was 105.9°F for Las Vegas, Nevada, and 107.8°F for Overton, Nevada, whereas the average minimum air temperatures were 33.6°F and 28.0°F, respectively. Average annual precipitation (1961–1990) was 4.13 in. for Las Vegas and 3.31 in. for Overton. This warm, arid environment is conducive to high rates of evaporation. Sparsely vegetated, gentle to moderately sloping alluvial fans, and steep, barren, rocky cliffs surround the lake. Generally, the adjacent hills rise to low or moderate height above the lake surface. The vast majority of lake is exposed to winds from the southwest to southeast.

Computation Methods

Lake Mead evaporation was computed using micrometeorological data collected at four floating instrumented platforms deployed in Water Barge Cove near Callville Bay, Boulder Basin near Sentinel Island, the Virgin Basin, and in the Overton Arm near Echo Bay. The platform in Water Barge Cove was in a relatively shallow cove protected from prevailing winds. The platforms in Boulder Basin near Sentinel Island, Virgin Basin, and Overton Arm were in different areas of the lake that are representative of open-water conditions in each basin. The depth of water at the four platforms fluctuated during the study due to water-level changes of Lake Mead. Lake elevation ranged from less than 1196 ft in February 1997 to almost 1216 ft in September 1998. Each floating platform was equipped with instruments to measure and record meteorological data and water temperature. Air temperature and relative humidity were measured with a temperature–humidity probe (THP), wind speed and direction were measured with an anemometer and wind monitor, net radiation was measured with a net radiometer, and water temperature was measured at various depths with temperature probes. A two-point mooring system was used to secure each barge to prevent drifting and maintain the directional aspect of the wind monitor and net radiometer.

DID YOU KNOW?

The amount of heat necessary to change 1 kilogram of a liquid into a gas is called the *latent heat of vaporization*. When this point is reached, the entire mass of substance is in the gas state. The temperature of the substance at which this change from liquid to gas occurs is known as the *boiling point*.

Energy-Budget Method to Measure Evaporation

The energy-budget method was used to measure evaporation from Lake Mead. This method also was used to quantify evaporation from western reservoirs during the 1950s (Anderson, 1954), and it has been used by the USGS on a few long-term lake studies (Rosenberry et al., 1993; Sturrock and Rosenberry, 1992; Swancar et al., 2000). The energy-budget method is the most accurate method for measuring lake evaporation (Winter, 1981). Use of the energy-budget method requires a large amount of data collection, but the effort is important because accurate measurements of lake evaporation are rare. An energy budget is similar to a water budget in that the change in stored energy is equal to the fluxes in and out of the system. Energy drives the process of evaporation because the water at the surface of the lake must absorb a certain amount of energy (known as the latent heat of vaporization; approximately 580 calories per gram) before the water will evaporate. After some derivation, the following equation can be used to calculate evaporation from the lake (Anderson, 1954):

$$E = \frac{Q_s - Q_r - Q_b - Q_v - Q_\theta}{\rho L(R+1)} \tag{6.19}$$

where

E = Evaporation rate.

Q_s = Solar radiation incident to the water surface.

Q_r = Reflected solar radiation.

Q_b = Net energy lost by the body of water through the exchange of long-wave radiation between the atmosphere and the body of water.

Q_v = Net energy advected into the body of water.

Q_θ = Change in energy stored in the body of water.

ρ = Density of evaporated water.

L = Latent heat of vaporization.

R = Bowen ratio.

The Bowen ratio (R) can be expressed as

$$R = \gamma \left[\frac{T_0 - T_a}{e_0 - e_a} \right] P \tag{6.20}$$

where

γ = Empirical constant.

T_0 = Temperature of the water surface.

T_a = Temperature of the air.

e_0 = Vapor pressure of saturated air at the temperature (T_0).

e_a = Vapor pressure of the air above the water surface.

P = Atmospheric pressure.

Net radiation (Q_n) was measured at Lake Mead evaporation platforms and replaces three energy terms (Q_s, Q_r, and Q_b) in Equation 6.19. Net advected energy (Q_v) is disregarded based on the assumption that advected energy is negligible during a 20-minute evaporation period. Thus, after modifying Equation 6.19 by substituting for net radiation, removing the net advected energy term, and replacing R with Equation 6.20, evaporation can be estimated from measured meteorological and hydrological parameters:

$$E = \frac{Q_n - Q_\theta}{\rho L \gamma_c \left(\dfrac{T_0 - T_a}{e_0 - e_a} \right) + 1} \tag{6.21}$$

where γ_c is a psychrometric constant, a product of γ and P (Laczniak et al., 1999).

Meteorological Data

Meteorological and water temperature data were used to compute evaporation rates for Lake Mead. Measurements were made every 10 to 30 seconds and were averaged for 20-minute periods. Some of the missing or incorrect 20-minute data were estimated or computed to maximize the amount of data available for the evaporation computation. Where data were missing for short periods, they were estimated from trends of the data before and after the periods of missing or incorrect data. Where data were missing or incorrect for longer periods, they were computed from other available data at that station, data from another station, or data for another year. Typically, a regression was developed with two sets of measurements for a different period, but for similar environmental conditions, using a complete set of data. The regression was then used to estimate the missing incorrect data from a complete set of data.

Monthly values of meteorological data and water temperatures were computed from 20-minute averaged data collected at each station for 1997, 1998, and 1999. Average daily air temperature was about the same at all four platforms, whereas relative humidity and water temperature were similar at the open-water platforms. Generally, water temperature was higher and relative humidity was lower at the sheltered cove platform (Water Barge Cove) than at the open-water platforms.

Daily air temperature, water temperature, and relative humidity at the Sentinel Island station were compared for 1997, 1998, and 1999. Air temperature varied from year to year, but the seasonal pattern was consistent. The maximum daily air temperature occurred about mid-July, with some daily average air temperatures exceeding 95°F, and the minimum daily air temperature occurred from late December through February. Water temperature did not vary much from year to year. Maximum daily water temperature occurred in late July to early August, and the minimum daily water temperature occurred in late February. Relative humidity fluctuated from day to day and differed greatly from year to year; however, there was a seasonal pattern of high relative humidity in January that gradually decreased to a low at the end of June, followed by a gradual increase to higher relative humidity in December.

Daily average wind speed for 1999 was 5 mph. Wind direction was predominately from the southeast to the southwest, occasionally from the northwest, and rarely from the northeast. Most daily wind speeds were less than 10 mph. The Virgin Basin location experienced more daily wind greater than 10 mph than other locations.

Evaporation Rates

Evaporation rates were computed at 20-minute intervals to evaluate diurnal fluctuations of lake evaporation. The 20-minute period evaporation rates were also used to identify periods of poor or missing energy-budget data. Daily evaporation rates are the sum of 20-minute periods, monthly rates are the sum of daily evaporation, and annual rates are the sum of monthly evaporation.

Daily Rates

The daily evaporation rates at the four evaporation stations were compared for calendar year 1999, and they showed similar daily fluctuations. However, the magnitude of fluctuations in daily evaporation was greater at the Virgin Basin and Sentinel Island than at Overton Arm or Water Barge Cove. Daily evaporation was greatest from late June through early July and was least from mid-December through late January.

Monthly Rates

To evaluate the temporal (size of) variation in monthly evaporation at each station, monthly evaporation rates for each station were averaged and compared to the total evaporation rate for each month of data collection. Temporal data were available for Water Barge Cove, Sentinel Island, and Virgin Basin stations (see Table 6.5); however, monthly data were insufficient for the Overton Arm station to compute average monthly evaporation. Total monthly evaporation rates at three stations, with some exceptions, generally were within 10% of average monthly rates; consequently, annual variation in monthly evaporation typically was minimal between 1997 and 1999. Some months, however, exhibited significant differences between average and total evaporation. These differences were as great as 31% at Water Barge Cove (February 1997), 14% at Sentinel Island (January 1998 and 1999), and 22% at Virgin Basin (December 1998 and 1999). Some of the difference in monthly evaporation rates from year to year may be due to errors in meteorological and water-temperature data collected at each station, but year-to-year differences for most months likely are due to actual differences in evaporation. To evaluate the spatial variation in evaporation, total monthly evaporation rates at all four stations were averaged for every month of data collection, and the average was compared to the total monthly evaporation rate at each station. For each station, total monthly evaporation rates compared well to average monthly rates with correlation coefficients of 0.96 or higher. However, total monthly evaporation at Water Barge Cove generally was less than total evaporation rates at the open-water stations when rates were less the 6.5 in. Total monthly evaporation rates for all three open-water stations were nearly equal and compared well to average evaporation rates; the correlation coefficient for Sentinel Island and for Overton Arm was 0.98, and for Virgin Basin it was 0.96. This evaluation suggests that the spatial variation in evaporation is minimal for open-water areas of Lake Mead. The monthly volume of evaporated water was computed using the average monthly open-water evaporation rate, in feet, and the average monthly surface area of Lake Mead for July 1997 through December 1999, in acres. The volume of water evaporated in 1 month ranged from 46,000 acre-ft in February 1998 to 126,000 acre-ft in July 1998.

Annual Rates

Average monthly rates for the Lake Mead open-water evaporation stations were computed for 1998 and 1999. For open-water stations, the sum of the average monthly rates for 1998 was 88.9 in. (7.4 ft), and for 1999 it was 90.7 in. (7.6 ft). For these 2 years, the average annual Lake Mead evaporation rate was 89.9 in. (7.5 ft). Monthly evaporation rates were available for only the Sentinel Island station for January to March 1998; those rates are used instead of an average rate. Evaporation rates at the Sentinel Island station are generally representative of evaporation of the lake as a whole. For April 1998 through November 1999, total evaporation at the Sentinel Island station was 161.3 in., whereas the total average monthly evaporation for the open-water stations was 159.9 in. The difference of 1.4 in. is less than 1% of the total average open-water evaporation. The annual volume of water evaporated from Lake Mead exceeded 1.1 million acre-ft in 1998 and 1999 and is probably higher than a long-term average annual evaporation due to higher-than-normal lake elevations and corresponding larger-than-normal surface area for the period. For example, the average surface area of Lake Mead was 125,000 acres from 1942 to 1995 and the computed average annual evaporation rate was 7.5 ft from 1997 to 1998, which would equal a long-term average annual volume of 937,500 acre-ft of evaporated water.

HYDROPOWER SECURITY

When I say that terrorism is war against civilization, I may be met by the objection that terrorists are often idealists pursuing worthy ultimate aims—national or regional independence, and so forth, I do not accept this argument. I cannot agree that a terrorist can ever be an idealist, in that the objects sought can ever justly terrorism. The impact of terrorism, not merely on individual nations, but on humanity as a whole, is intrinsically evil, necessarily evil and wholly evil.

Netanyahu (1981)

TABLE 6.5
Evaporation from Lake Mead at Four Floating Instrumented Platforms, Arizona and Nevada, 1999

Site	Jan	Feb	Mar	Apr	May	Jun	Jul	Aug	Sep	Oct	Nov	Dec
Total Monthly Evaporation, 1997 (inches)												
Water Barge Cove	—	5.4	9.1	7.9	—	10.8	10.8	9.2	8.3	7.6	4.1	2.6
Sentinel Island	—	—	—	—	—	—	9.1	9.1	8.8	8.7	6.4	4.6
Virgin Basin	—	—	—	—	—	—	—	—	—	—	—	—
Overton Arm	—	—	—	—	—	—	—	—	—	—	—	—
Total Monthly Evaporation, 1998 (inches)												
Water Barge Cove	2.9	3.2	6.6	8.2	8.9	10.2	10.1	8.8	8.5	6.7	3.7	2.6
Sentinel Island	3.7	3.6	7.1	7.4	9.9	9.4	10.0	8.9	8.9	8.5	7.3	6.0
Virgin Basin	—	—	—	8.6	9.7	9.1	9.9	8.7	9.7	7.1	5.9	4.1
Overton Arm	—	—	—	—	—	—	—	—	—	—	—	—
Average Evaporation, 1998 (inches)												
All available sites	3.3	3.4	6.9	8.0	9.5	9.6	10.0	8.8	9.0	7.4	5.6	4.2
Open-water sites	3.7	3.6	7.1	8.0	9.8	9.2	9.9	8.8	9.3	7.8	6.6	5.0
Total Monthly Evaporation, 1999 (inches)												
Water Barge Cove	3.0	3.8	7.4	6.8	9.2	10.5	9.4	8.7	9.4	8.0	4.3	3.1
Sentinel Island	5.0	4.4	8.1	6.9	9.0	9.2	8.4	8.6	9.9	8.8	6.7	—
Virgin Basin	4.0	5.3	7.9	7.1	9.6	9.3	8.8	8.7	9.1	8.0	7.3	6.5
Overton Arm	—	—	7.8	6.9	9.5	9.5	9.4	8.6	9.5	8.7	7.0	4.1
Average Evaporation, 1999 (inches)												
All available sites	4.0	4.5	7.8	6.9	9.3	9.6	9.0	8.6	9.5	8.4	6.3	4.5
Open-water sites	4.5	4.8	8.0	7.0	9.4	9.4	8.9	8.6	9.5	8.5	7.0	5.3

Source: USGS, *Estimating Evaporation from Lake Mead,* U.S. Geological Survey, Washington, DC, 2006 (http://pubs.usgs.gov/sir/2006/5252/section3.html).

FIGURE 6.12   North face of Mount St. Helens where the lateral explosion occurred. (Photograph by Frank R. Spellman.)

MOUNT ST. HELENS

That humankind cannot predict, at the present time, exactly when natural disasters such as floods, volcanic eruptions, earthquakes, tsunamis, or other geologic processes will occur is a given; it is a reality. It is important to note, however, that none of these natural occurrences or events can be classified as disastrous unless there is a vulnerable population residing within the event area. Now, the very definition of a "population" can be argued by some. As an example, consider the Mount St. Helens catastrophic North Slope lateral eruption that occurred on May 18, 1980 (see Figure 6.12). It killed 57 people and destroyed 200 houses, 27 bridges, 15 miles (24 km) of railways, and 185 miles (298 km) of highway. Could these humans and material things be considered to be a population or part of population? Do we just count humans? Or were the more than 4,000,000,000 board feet (9,400,000 m³) of timber destroyed or damaged during the eruption part of the population? What about downwind of the volcano, where many agricultural crops, such as wheat, apples, potatoes, and alfalfa, were destroyed by a thick ash accumulation? Are these crops part of a population?

How about the estimated 1500 elk and 5000 deer that were killed, in addition to an estimated 12 million Chinook and coho salmon fingerlings that died when their hatcheries were destroyed (Tilling et al., 1990; USGS, 2015)? And how about the billions of insects killed? Are they part of the population, too? Would you consider their demise disastrous? Additionally, consider that the level of Spirit Lake at the foot of Mount St. Helens was raised to its current position of 200 feet above original level, and that the 24-megaton thermal energy explosion (1600 times the size of the atomic bomb dropped on Hiroshima) rerouted parts of the North and South Toutle Rivers (USGS, 2014c). How exactly do we define the components of a population?

Depends.

Depends on what?

It depends on whether the bridges, houses, fisheries, apple farms, or timber areas contained or were occupied by humans who were killed. Understand that when a bridge is destroyed, it is bad. On the other hand, if that same bridge is destroyed and humans in a van or RV die as a result, then we are talking disaster. At least, that is how the media will play it up and how affected relatives and loved ones will view it.

At this point in the presentation the reader is probably wondering what the 1980 disastrous eruption of Mount St. Helens and other natural disasters have to do with dam infrastructure protection. Well, consider that when Mount St. Helens erupted an estimated 40,000 young salmon were lost when they swam through turbine blades of hydroelectric generators when reservoir levels were lowered along the Lewis River to accommodate possible mudflow and flood waters. The hydroelectric dams near Mount St. Helens were not damaged but it was necessary to adjust the flow and kill swarms of fish to prevent further downstream damage caused by the eruption (Tilling et al., 1990; USGS, 2015).

The Mount St. Helens eruption occurred somewhat recently, but the Johnstown Flood and the St. Francis Dam catastrophic failure come to mind when considering dam disasters. Before discussing these two disasters, which killed lots of people, it is important to point out right up front that neither event was the result of natural causes; instead, they were (in the author's opinion) the result of flawed decision-making, poor design, faulty engineering, and dysfunctional management.

JOHNSTOWN FLOOD

On May 31, 1889, the Johnstown Flood, also known as the Great Flood of 1889, occurred as a result of the catastrophic failure of the South Fork Dam on the Little Conemaugh River 14 miles (23 km) upstream of the town of Johnstown, Pennsylvania. After several days of extremely heavy rainfall, the dam broke and unleashed 20 million tons of water (18 million m³) from Lake Conemaugh. The flood killed 2209 people and caused $450 million (in 2015 dollars) (Gibson, 2006; Perkins, 2009; World Digital Library, 2015). What caused the dam to break? The exact cause of the dam break has been the subject of ongoing debate for decades. The courts eventually ruled that the dam break was an Act of God, and the survivors were granted no compensation. Notwithstanding God's influence, let's get back to the basic question of what caused the dam to break. Over the decades, some have blamed the managers of the dam for the break. The managers supposedly modified the dam in ways to make the area more attractive to hunters and fishermen who belonged to a select club in the area which was composed of members who were responsible for maintaining the dam in good condition. Two things are certain. First, when the man-made dam broke it killed 2209 people and damaged numerous properties. Second, if the dam had not been constructed in the first place, the killer flood could not have occurred as it did.

ST. FRANCIS DAM

The name Mulholland is familiar to many people for various reasons. Maybe the name is familiar because of the song and movie titled *Mulholland Drive*. The name is also well known to those who have traveled or live within the Los Angeles area, in the San Fernando Valley, or Los Angeles and Ventura counties. The name is a salute to the pioneering Los Angeles civil engineer William Mulholland. What is not well known, especially to those outside the Los Angeles area, is that William Mulholland is also known for the St. Francis Dam, which was built between 1924 and 1926 under the direction of the Bureau of Water Works and Supply and was managed by William Mulholland. The dam was designed to store and regulate water for the City of Los Angeles and was an integral part of the aqueduct water supply system. The dam itself was located about 10 miles north of present-day Santa Clara and about 40 miles from downtown Los Angeles.

Just before midnight on March 12, 1928, the dam catastrophically failed, and the resulting flood took the lives of an estimated 600 people (an event that occupies second place for greatest loss of life in California after the 1906 San Francisco earthquake and fire) (Pollack, 2010). The disaster marked the end of Mulholland's career (Mulholland, 1995). As you might imagine, the collapse of the St. Francis Dam and the resulting death and destruction were investigated by multiple teams of engineers and geologists, committees, and agencies and is still being studied in college-level civil engineering classes today. The initial investigations were conducted as quickly as possible because construction of the much larger Hoover Dam was set to begin, and many people were worried about the possibility of a similar catastrophe occurring. With so many different groups investigating the dam catastrophe, it is to be expected that several different opinions and findings would be rendered. That was the case, but in general the findings were pretty consistent. The investigators reached consensus on the inadequacy of the dam foundation. One report stated that, "The foundation under the entire dame left very much to desired. … With such a [geologic] formation, the ultimate failure of this dam was inevitable, unless water could have been kept from reaching the foundation" (Jackson and Hundley, 2004). The report of a committee appointed by the City Council of Los Angeles to investigate the cause of the failure of the St. Francis Dam offered the following conclusions:

1. The type and dimensions of the dam were amply sufficient if based on suitable foundations.
2. The concrete of which the dam was built was of ample strength to resist the stresses to which it would normally be subjected.
3. The failure cannot be laid to movement of the Earth's crust (i.e., not an Act of God).
4. The dam failed as a result of defective foundations.
5. This failure reflects in no way the stability of a well-designed gravity dam properly founded on suitable bedrock.

WHEN DISASTERS ARE WAITING TO HAPPEN

Natural disasters occur. Nonetheless, some people choose to ignore them or shrug off the possibility of them occurring in their own backyards. Take, for example, those who choose to live in a known floodplain. The question is why? They might answer that they have lived there all of their lives; they like it there; they are used to it and have boats ready to go; they have flood insurance; or maybe they just think the taxpayers will bail them out each and every time (and we do). And then there are those who live in areas underlined by earthquake faults, just waiting for the "big one," just waiting for disaster to occur.

The point being made here is that we all know that some disasters cannot be avoided, no matter what. We are human, and our self-protective mechanisms are limited to body armor, staying away from bad places, staying away from drugs/alcohol abuse, and living in so-called safe zones (if there are any). When a natural disaster does occur, such as the aforementioned eruption of Mount St. Helens, the Johnston Flood, or the St. Francis Dam disaster, can these events fall within the parameters detailed earlier by Benjamin Netanyahu? That is, was the eruption of Mount St. Helens, the Johnston Flood, and the St. Francis Dam collapse intrinsically evil, necessarily evil, and wholly evil?

Depends.

Depends on what?

It depends on one's point of view. It depends on the victims' and survivors' points of view. Again, common sense tells us that Mother Nature is in charge, and if she wants to shake us up a bit here and there, she will. Woe be to those who think they can prevent this. When any natural disaster occurs that results in damage to dam infrastructure, we have to classify such an occurrence as a non-intentional event—an event not caused intentionally by any human hand, that is. However, in the current global, anti-Western political climate, it is the events best described by Benjamin Netanyahu that are now a concern to us with regard to hydropower and associated operations: the intentional damage or destruction of our dam infrastructure, including hydropower plants, navigation locks, levees, dikes, hurricane barriers, mine tailings and other industrial waste impoundments, and other water retention facilities, just to mention a few. Thus, unfortunately, it is the human element—the one laced with evil destructive intent—that we must consider, must be aware of, and must guard against.

HYDROPOWER IMPACTS ON HUMAN HEALTH AND SAFETY

Possible impacts on human health and safety during operations include exposure to electromagnetic fields (EMFs) and accidental worker injury or death during operation and maintenance activities. Working around, with, or on hydropower facilities and equipment presents many of the same hazards to the installers, operators, and maintenance personnel as for wind turbine and solar operators and maintenance personnel. Hydropower safety issues include the following:

- Falls
- Confined space entry
- Fire
- Lockout/tagout
- Medical and first aid
- Crane, derrick, and hoist safety
- Electrical safety
- Machine guarding
- Respiratory protection

In addition, worker health and safety issues also include working in potential weather extremes and possible exposure to the hazards of nature, such as uneven terrain and dangerous plants, animals, and insects. Risk to the public of accidental death or injury is unlikely, as the facilities are fenced and guarded from unlawful entry; however, access to the impoundment or tailrace would be a potential source for accidents.

BOTTOM LINE ON HYDROPOWER

Hydropower offers advantages over the other energy sources but presents unique environmental challenges. As mentioned, the advantages of using hydropower begin with that fact that hydropower does not pollute the air like power plants that burn fossil fuels, such as coal and natural gas. Moreover, hydropower does not have to be imported into the United States like foreign oil does; it is produced in the United States. Because hydropower relies on the water cycle, driven by the sun, it is a renewable resource that will be around for at least as long as humans. Hydropower is controllable; that is, engineers can control the flow of water through the turbines to produce electricity on demand. Finally, hydropower impoundment dams create huge lake areas for recreation, irrigation of farm lands, reliable supplies of potable water, and flood control.

Hydropower, though, also has some disadvantages; it is these disadvantages or environmental impacts that have been the focus of this text. For example, fish populations can be impacted if fish cannot migrate upstream past impoundment dams to spawning grounds or if they cannot migrate downstream to the ocean. Many dams have installed fish ladders or elevators to aid upstream fish passage. Downstream fish passage is aided by diverting fish from turbine intakes using screens or racks or even underwater lights and sounds, and by maintaining a minimum spill flow past the turbine. Hydropower can also impact water quality and flow. Hydropower plants can cause low dissolved oxygen levels in the water, a problem that is harmful to riparian habitats and is addressed using various aeration techniques that oxygenate the water. Maintaining minimum flows of water downstream of a hydropower installation is also critical for the survival of riparian habitats. Hydropower is also susceptible to drought. When water is not available, the hydropower plants cannot produce electricity. Typical activities during operation of a hydropower plant include operation of the facility, power generation, and associated maintenance activities that would require vehicular access and heavy equipment operation when components are being replaced. Finally, construction of new hydropower facilities impacts investors and others by competing with other uses of the land. Preserving local flora and fauna and historical or cultural sites is often more highly valued than electricity generation.

DISCUSSION AND REVIEW QUESTIONS

1. A few environmentalist groups advocate tearing down existing hydroelectric dams in the United States to allow affected rivers to run free. Do you support or oppose this argument? Why?
2. Do we have any more water on Earth today than yesterday or 1000 years ago?

3. According to the Rachel River account, the scientists and environmental scientists devised several recommendations to alleviate the downward trend in salmon population. What additional recommendations would you make to mitigate the situation?

4. In the Rachel River account, explain the meaning of the statement: "Notice they didn't try to ask the Native Americans. They also would have known what to do. The salmon had already told them."

5. Explain why an increase in water temperature affects salmon populations.

6. When you compare the pros and cons of hydroelectric power it sounds great—so why don't we use it to produce all of our power?

7. Does the Rachel River account and the indicated mitigation procedures indicate to you that science and scientists can solve all problems?

8. Do you think hydropower plants and facilities are ripe targets for terrorists?

REFERENCES AND RECOMMENDED READING

Anderson, E.R. (1954). Energy-budget studies. In: *Water-Loss Investigations: Lake Hefner Studies, Technical Report*. Washington, DC: U.S. Geological Survey, pp. 71–119 (pubs.usgs.gov/pp/0269/report.pdf).

Bauersfeld, K. (1977). *Effects of Peaking (Stranding) of Columbia River Dams on Juvenile Anadromous Fish Below the Dalles Dam, 1974 and 1975*. Olympia: Washington Department of Fisheries.

Bauersfeld, K. (1978a). *Stranding of Juvenile Salmon by Flow Reductions at Mayfield Dam on the Cowlitz River, 1976*. Olympia: Washington Department of Fisheries.

Bauersfeld, K. (1978b). *The Effect of Daily Flow Fluctuations on Spawning Fall Chinook in the Columbia River*. Olympia: Washington Department of Fisheries.

Beck Associates. (1989). *Skagit River Salmon and Steelhead Fry Stranding Studies*, prepared for Seattle City Light Environmental Affairs Division, Seattle, WA.

Becker, C.D. and Neitzel, D.A. (1985). Assessment of intergravel conditions influencing egg and alvein survival during salmonid redd dewatering. *Environmental Biology of Fishes*, 12: 33–46.

Becker, C.D., Fickeison, D.H., and Montgomery, J.C. (1981). *Assessments of Impacts from Water Level Fluctuations on Fish in the Hanford Reach, Columbia River*. Richland, WA: Pacific Northwest Laboratory, Batelle Memorial Institute.

Bender, M.D., Hauser, G.E., and Shiao, M.S. (1991). *Modeling Boone Reservoir to Evaluate Cost-Effectiveness of Point and Nonpoint Source Pollutant Controls*, Report No. WR28-1-31-107. Norris, TN: Engineering Laboratory, Tennessee Valley Authority.

Berryman, A.A. (1999). *Principles of Population Dynamics and Their Application*. London: Garland Science.

Berryman, A.A. (2002). *Population Cycles: The Case for Trophic Interactions*. London: Oxford University Press.

Bovee, K.D. (1982). *A Guide to Stream Habitat Analysis Using the Instream Flow Incremental Methodology*, Instream Flow Information Paper 12. Washington, DC: U.S. Fish and Wildlife Service, Office of Biological Services.

Brahic, C. (2009). Fish "an ally" against climate change. *New Scientist*, January 16 (http://www.newscientist.com/article/dn16432-fish-an-ally-against-climate-change.html).

Brookfield, H.C. (1989). *Sensitivity to Global Change: A New Task for Old/New Geographers*. Reading, UK: Norma Wilkinson Memorial Lecture, University of Reading.

Brundtland, G.H. (1987). *Our Common Future*. New York: Oxford University Press.

Bureau of Reclamation. (1967). *Lake Mead Area and Capacity Tables*. Washington, DC: Bureau of Reclamation.

Bureau of Reclamation. (2003). Compilation of Records in Accordance with Article V of the Decree of the Supreme Court of the United States in *Arizona vs. California et al.*, Dated March 9, 1964. Boulder City, NV: Bureau of Reclamation.

Caldeira, K. and Wickett, M.E. (2003). Anthropogenic carbon and ocean pH. *Nature*, 425(6956): 365.

Calder, W.A. (1983). An allometric approach to population cycles of mammals. *Journal of Theoretical Biology*, 100: 275–282.

Calder, W.A. (1996). *Size, Function and Life History*. Mineola, NY: Dover Publications.

Chapman, D.W., Weitkamp, D.E., Welsh, T.L., Dell, M.B., and Schadt, T.H. (1986). Effects of river flow on the distribution of Chinook salmon redds. *Transactions of the American Fisheries Society*, 115: 537–547.

Chow, V.T. (2009). *Open Channel Hydraulics*. Caldwell, NJ: Blackburn Press.

Christensen, C. (1982). In: *Making It Happen: A Positive Guide to the Future*, Richardson, J.M., Ed. Washington, DC: U.S. Association for the Club of Rome.

Commoner, B. (1971). *The Closing Circle: Nature, Man, and Technology*. New York: Knopf.

Cushman, R.M. (1985). Review of ecological effects of rapidly varying flows downstream from hydroelectric facilities. *North American Journal of Fisheries Management*, 5: 330–339.

Damuth, J. (1991). Of size and abundance. *Nature*, 351: 268–269.

EERE. (2008). *History of Hydropower*. Washington, DC: Office of Energy Efficiency & Renewable Energy, U.S. Department of Energy (http://www1.eere.energy.gov/windandhydro/hydro_history.html).

Enger, E., Kormelink, J.R., Smith, B.F., and Smith, R.J. (1989). *Environmental Science: The Study of Interrelationships*. Dubuque, IA: William C. Brown.

Fiscus, G. (1977). Cedar River Fish Damage Observations and Reports on February 18, 1977; Report of Cedar River Fish Kills on March 1, 1977; Report of Cedar River Fish Kill on March 8, 1977; and Investigation of Cedar River Flow Fluctuation on May 21, 1977 [internal memos]. Olympia: Washington Department of Fish and Wildlife.

Frailey, J.J. and Graham. P.J. (1982). *The Impact of Hungry Horse Dam on the Fishery in the Fathead River*, Final Report. Boise, ID: Bureau of Reclamation.

Fraser, J.C. (1972). Regulated discharge and the stream environment. In: *River Ecology and Man*, Oglesby, R.T., Carlson, C.A., and McCann, J.A., Eds., pp. 263–286. New York: Academic Press.

Fustish, C.A., Jacobs, S.E., McPherson, B.P. and Frazier, P.A. (1988). *Effects of the Applegate Dam on the Biology of Anadromous Salmonids in the Applegate River*, DAWCW5-77-C-0033, prepared by Research and Development Section, Oregon Department of Fish and Wildlife, for U.S. Army Corps of Engineers.

Gibilisco, S. (2007). *Alternative Energy Demystified*. New York: McGraw-Hill.

Gibson, C. (2006). Our 10 greatest natural disasters. *American Heritage*, 57(4): 26–37.

Ginzburg, L.R. (1986). The theory of population dynamics. 1. Back to first principles. *Journal of Theoretical Biology*, 122: 385–399.

Ginzburg, L.R. and Colyvan, M. (2004). *Ecological Orbits: How Planets Move and Populations Grow*. New York: Oxford University Press.

Ginzburg, L.R. and Jensen C.X.J. (2004). Rules of thumb for judging ecological theories. *Trends in Ecology and Evolution*, 19: 121–126.

Gislason, J.C. (1985). Aquatic insect abundance in a regulated stream under fluctuating and stable diel flow patterns. *North American Journal of Fisheries Management*, 5: 39–46.

Goldfarb, T.D., Ed. (1989). *Taking Sides: Clashing Views on Controversial Environmental Issues*. Guilford, CT: Duskin Publishing.

Graybill, J.P., Burgner, R.L., Gislason, J.C., Huffman, P.E., Wyman, K.H. et al. (1979). *Assessment of Reservoir Related Effects of the Skagit River Project on Downstream Fishery Resources of the Skagit River, Washington*, Final Report to Seattle City Light Environmental Affairs Division, Seattle, WA.

Hamilton, R. and Buell, J.W. (1976). *Effects of Modified Hydrology on Campbell River Salmonids*, Technical Report Series No. Pac/T-76-20. Vancouver, BC: Canada Department of the Environment, Fisheries and Marine Service.

Hardin, G. (1986). Cultural carrying capacity: a biological approach to human problems. *BioScience*, 36: 599–606.

Hickman, C.P., Roberts, L.S., and Hickman, F.M. (1990). *Biology of Animals*. St. Louis, MO: Times Mirror/Mosby College Publishing.

Holdren, J.P. (1991). Population and the energy problem. *Population and Environment*, 12: 231–255.

Hubbell, S.P. and Johnson, L.K. (1977). Competition and nest spacing in a tropical stingless bee community. *Ecology*, 58: 949–963.

Hunter, M.A. (1992). *Hydropower Flow Fluctuations and Salmonids: A Review of the Biological Effects, Mechanical Causes, and Options for Mitigation*. Olympia: Washington Department of Fisheries.

Jackson, D.C. and Hundley, N. (2004). Privilege and responsibility: William Mulholland and the St. Francis dam disaster. *California History*, Fall, pp. 8–47.

Jarvie, M. and Solomon, B. (1998). Point–nonpoint effluent trading in water sheds: a review and critique. *Environmental Impact Assessment Review*, 18: 135–157.

Jones, C.Z., Row, T.G., Sexton, R.J., and Tanko, D.J. (2000). *Water Resources Data, Nevada, Water Year 1999*, USGS Water-Data Report NV-99-1. Washington, DC: U.S. Geological Survey.

Keller, E.A. (1988). *Environmental Geology*. Columbus, OH: Merrill.

Krebs, R.E. (2001). *Scientific Laws, Principles and Theories*. Westport, CT: Greenwood Press.

Kuhlken, R. (2002). Intensive agricultural landscapes of Oceania. *Journal of Cultural Geography*, 19(2): 161–195.

LaBounty, J.F. and Horn, M.J. (1997). The influence of drainage from the Las Vegas Valley on the limnology of Boulder Basin, Lake Mead, Arizona–Nevada: *Journal of Lake and Reservoir Management*, 13(2): 95–108.

Laczniak, R.J., DeMeo, G.A., Reiner, S.R., Smith, J.L., and Nyland, W.E. (1999). *Estimates of Ground-Water Discharge as Determined from Measurements of Evapotranspiration, Ash Meadows Area, Nye County, Nevada*, USGS Water-Resources Investigations Report 99-4079. Washington, DC: U.S. Geological Survey.

Leopold, A. (1949). *A Sand County Almanac: And Sketches Here and There*. New York: Oxford University Press, Selection 6.

Library of Congress. (2009). *The World's First Hydroelectric Power Plant Began Operation September 30, 1882*. Washington, DC: Library of Congress (http://www.americaslibrary.gov/jb/gilded/jb_gilded_hydro_1.html).

Liebig, J. (1840). *Chemistry and Its Application to Agriculture and Physiology*. London: Taylor & Walton.

Lund, J.W. (2007). Characteristics, development and utilization of geothermal resources. *Geo-Heat Centre Quarterly Bulletin*, 28(2): 1–9.

March, P.A., Brice, T.A., Mobley, M.H., and Cybularz, J.M. (1992). Turbines for solving the DO dilemma. *Hydro Review*, 11(1), 30–36.

Masters, G.M. (1991). *Introduction to Environmental Engineering and Science*. Englewood Cliffs, NJ: Prentice Hall.

McPhee, C. and Brusven. M.A. (1976). *The Effect of River Fluctuations Resulting from Hydroelectric Peaking on Selected Aquatic Invertebrates and Fish*. Moscow, ID: Office of Water Research and Technology, Idaho Water Resources Research Institute, University of Idaho.

Miller, G.T. (1988). *Environmental Science: An Introduction*. Belmont, CA: Wadsworth.

Monk, C.L. (1989). Factors That Influence Stranding of Juvenile Chinook Salmon and Steelhead Trout, master's thesis, University of Washington, Seattle.

Mulholland, C. (1995). *William Mulholland and the St. Francis Dam: St. Francis Dam Disaster Revisited*. Santa Clara, CA: Historical Society of Southern California.

Mundie, J.H. and Mounce, D.E. (1976). Effects of changes in discharge in the Lower Campbell River on the transport of food organisms of juvenile salmon, appendix report. In: *Effects of Modified Hydrology on Campbell River Salmonids*, Technical Report Series No. Pac/T-76-20, Hamilton, R. and Buell, J.W., Eds. Vancouver, BC: Canada Department of the Environment, Fisheries and Marine Service.

Nathanson, J.A. (1997). *Basic Environmental Technology: Water Supply, Waste Management, and Pollution Control*, 2nd ed. Upper Saddle River, NJ: Prentice Hall, pp. 21–22.

Neitzel, D.A., Becker, C.D., and Abernathy, C.S. (1985). Laboratory tests to assess water-level fluctuations at Vernita Bar, Washington. In: *Proceedings of the Symposium on Small Hydropower and Fisheries*, Olson, F.W., White, R., and Hamre, R.H., Eds. Bethesda, MD: American Fisheries Society.

Nestler, M.N., Milhous, R.T., and Layzer, J.B. (1989). Instream habitat modeling techniques. In: *Alternatives in Relative River Management*, Gore, J.A. and Petts, G.E., Eds., pp. 295–315. Boca Raton, FL: CRC Press.

Netanyahu, B. (1981). *International Terrorism: Challenge and Response*. New Brunswick, NJ: Transaction.

NREL. (2009). *Concentrating Solar Power*. Washington, DC: National Renewable Energy Laboratory (http://www.nrel.gov/learning/re_csp.html).

Olson, F.W. (1990). *Downramping Regime for Power Operations to Minimize Stranding of Salmon Fry in the Sultan River*, contract report by CH2M Hill (Bellevue, WA) for Snohomish County PUD 1.

Perkins, S. (2009). Johnstown flood matched volume of Mississippi River. *Science News*, 176(11): 21.

Phillips, R.W. (1969). *Effects of an Unusually Low Discharge from Pelton Regulating Reservoir, Deschutes River, on Fish and Other Aquatic Organisms*, Special Report 1. Portland: Basin Investigation Section, Oregon State Game Commission.

Phinney, L.A. (1974a). *Further Observations on Juvenile Salmon Stranding in the Skagit River, March 1973*. Olympia: Washington Department of Fisheries and Wildlife.

Phinney, L.A. (1974b). *Report on the 1972 Study of the Effect of River Flow Fluctuations Below Merwin Dam on Downstream Migrant Salmon*. Olympia: Washington Department of Fisheries and Wildlife.

Phinney, L.A., Rothfus, L.O., Hamilton J.A.R., and Weiss, E. (1973). *The Study of the Effect of River Flow Fluctuations Below Merwin Dam on Downstream Migrant Salmon*, FPC Project No. 935. Olympia: Washington Department of Fisheries and Wildlife.

Podar, M.K., Crossman, J.C., Burmaster, D.E., Ruane, R.J., Jaksch, J.A., Hauser, G., and Sessions, S.L. (1985). *Optimizing Point/Nonpoint Source Tradeoff in the Houston River Near Kingsport, Tennessee*, paper presented at the National Conference on Nonpoint Sources of Pollution, Kansas City, MO, May 19–22.

Pollack, A. (2010). President's Message. *The Heritage Junction Dispatch*, 36(3): 1 (scvhs.org/news/dispatch36-3.pdf).

Reid, W. et al. (1988). *Bankrolling Successes*. Washington, DC: Environmental Policy Institute and National Wildlife Federation.

Reiser, D.W. and White. R.G. (1983). Effects of complete redd dewatering on salmonid egg-hatching and development of juveniles. *Transactions of the American Fisheries Society*, 112: 532–540.

Rochester, Jr., H., Lloyd, T., and Farr, M. (1984). *Physical Impacts of Small-Scale Hydroelectric Facilities and Their Effects on Fish and Wildlife*. Washington, DC: U.S. Fish and Wildlife Service.

Rosenberry, D.O., Sturrock, A.M., and Winter, T.C. (1993). Evaluation of the energy budget method of determining evaporation at Williams Lake, Minnesota, using alternative instrumentation and study approaches. *Water Resources Research*, 29(8): 2473–2483.

Satterthwaite, T.D. (1987). *Effects of Lost Creek Dam on Spring Chinook in the Rouge River, Oregon: An Update*, DACW57-77-C-0027. Portland: Oregon Department of Fish and Wildlife.

Searchinger, T. et al. (2008). Use of U.S. croplands for biofuels increases greenhouse gases through emissions from land-use change. *Science*, 319(5867): 1238–1240.

Sharov, A. (1992). Life-system approach: a system paradigm in population ecology. *Oikos*, 63: 485–494.

Simon, J.L. (1980). Resources, population, environment: an oversupply of false bad news. *Science*, 208: 1431–1437.

Sinha, A. (2006). Cultural landscape of Pavagadh: the abode of Mother Goddess Kalika. *Journal of Cultural Geography*, 23(2): 89–103.

Spellman, F.R. (1996). *Stream Ecology and Self-Purification*. Lancaster, PA: Technomic.

Spellman, F.R. (2008). *The Science of Water*, 2nd ed. Boca Raton, FL: CRC Press.

Stalnaker, C., Lamb, B.L., Henriksen, J., Bovee, K.D., and Bartholow, J. (1995). *The Instream Flow Incremental Methodology: A Primer for IFIM*, Biological Report 29. Washington, DC: U.S. Geological Survey.

Stober, Q.J., Crumley, S.C., Fast, D.E., Killebrew, E.S., Woodin, R.M., Engman G.E., and Tutmark. G. (1982). *Effects of Hydroelectric Discharge Fluctuation on Salmon and Steelhead in the Skagit River, Washington*, Final Report to Seattle City Light Environmental Affairs Division, Seattle, WA.

Sturrock, A.M. and Rosenberry, D.O. (1992). Energy budget evaporation from Williams Lake: a closed lake in north central Minnesota. *Water Resources Research*, 28(6): 1605–1617.

Swancar, A., Lee, T.M., and O'Hare, T.M. (2000). *Hydrogeologic Setting, Water Budget, and Preliminary Analysis of Ground-Water Exchange at Lake Starr, a Seepage Lake in Polk County, Florida*, USGS Water-Resources Investigations Report 00-4030. Washington, DC: U.S. Geological Survey.

Tadayon, S., Duet, N.R., Fisk, G.G., McCormack, H.F., Partin, C.K., Poe, G.L., and Rigas, P.D. (2000). *Water Resources Data for Arizona, Water Year 1999*, USGS Water-Data Report AZ-99-1. Washington, DC: U.S. Geological Survey.

Tadayon, S, Duet, N.R., Fisk, G.G., McCormack, H.F., Partin, C.K., Pope, G.L., and Rigas, P.D. (2001). *Water Resources Data for Arizona, Water Year 2000*, USGS Water-Data Report AZ-00-1. Washington, DC: U.S. Geological Survey.

Thompson, J.S. (1970). *The Effect of Water Regulation at Gorge Dam on Stranding of Salmon Fry in the Skagit River, 1969–1970*. Olympia: Washington Department of Fisheries and Wildlife.

Tilling, R., Topinka, L., and Swanson, D.A. (1990). *Eruptions of Mount St. Helens: Past, Present, and Future*. Washington, DC: U.S. Geological Survey (http://pubs.er.usgs.gov/publication/7000008).

Tovey, N.K. (2005). *ENV-2E02 Energy Resources 2004–2005 Lecture*. Norwich, UK: University of East Anglia (http://www2.env.uea.ac.uk/gmmc/energy/env2e02/env2e02.htm).

Trotzky, H.M. and Gregory, R.W. (1974). The effects of water flow manipulation below a hydroelectric power dam on the bottom fauna of the Upper Kennebec River, Maine. *Transactions of the American Fisheries Society*, 103: 318–324.

Turchin, P. (2001). Does population ecology have general laws? *Oikos*, 94: 17–26.

Turchin, P. (2003). *Complex Population Dynamics: A Theoretical/Empirical Synthesis*. Princeton, NJ: Princeton University Press.

UCS. (2008). *Land Use Changes and Biofuels*. Cambridge, MA: Union of Concerned Scientists (ucsusa.org/assets/documents/clean_vehicles/Indirect-Land-Use-Factsheet.pdf).

University of Michigan. (2006). *Urbanization and Global Change*. Ann Arbor: University of Michigan (http://globalchange.umich.edu/globalchange2/current/lectures/urban_gc/).

USDOE. (2006). *Geothermal Technologies Program: A History of Geothermal Energy in the United States*. Washington, DC: U.S. Department of Energy (http://www1.eere.energy.gov/geothermal/history.html).

USDOE. (2008). *Advanced Hydropower Technology*. Washington, DC: U.S. Department of Energy (http://www1.eere.energy.gov/windandhydro/hydro_advtech.html).

USDOE. (2009a). *Fossil Fuels.* Washington, DC: U.S. Department of Energy (http://www.energy.gov/ energysources/fossilfuels.htm).

USDOE. (2009b). *Solar Energy Technologies Program.* Washington, DC: U.S. Department of Energy (http://www1.eere.Energy.gov/solar/printable_versions/about.html).

USEPA. (2009). *Ozone Science: The Facts Behind the Phase-Out.* Washington, DC: U.S. Environmental Protection Agency (http://www.epa.gov/ozone/science/sc_fact.html).

USGS. (1999). *Hawaiian Volcano Observatory.* Washington, DC: U.S. Geological Survey (http://hvo.wr.usgs.gov/).

USGS. (2006). *Estimating Evaporation from Lake Mead.* Washington, DC: U.S. Geological Survey (http://pubs.usgs.gov/sir/2006/5252/section3.html).

USGS. (2013). *Drought—The Stealth Disaster.* Washington, DC: U.S. Geological Survey (http://www.usgs.gov/blogs/features/usgs_top_story/drought-the-stealth-disaster/).

USGS. (2014a). *Hydroelectric Power Water Use.* Washington, DC: U.S. Geological Survey (http://water.usgs.gov/edu/wuhy.html).

USGS. (2014b). *What Does the Term "River Stage" Mean?* Washington, DC: U.S. Geological Survey (http://water.usgs.gov/edu/qa-measure-streamstage.html).

USGS. (2014c). *Mount St. Helens—From the 1980 Eruption to 2000,* Fact Sheet 036-00. Washington, DC: U.S. Geological Survey (http://pubs.usgs.gov/fs/2000/fs036-00/).

USGS. (2015). *The Climactic Eruption of May 18, 1980.* Washington, DC: U.S. Geological Survey (http://pubs.usgs.gov/gip/msh/climactic.html).

Vitousek, P.M., Ehrlich, A.H., and Matson, P.A. (1986). Human appropriation of the products of photosynthesis. *BioScience,* 36: 368–373.

Volterra, V. (1926). Variazioni e fluttuazioni del numero d'indivudui in specie animali conviventi. *Memorie dell'Accademia dei Lincei,* 2: 31–113.

Wetzel, R.G. (1975). *Limnology.* Philadelphia, PA: W.B. Saunders.

Wilson, E.O. (1989). Threats to biodiversity. *Scientific American,* 261(3): 108–116.

Wilson, E.O. (2001). *The Future of Life.* New York: Knopf.

Winter, T.C. (1981). Uncertainties in estimating the water balance of lakes. *Water Resources Bulletin,* 17(1), 82–115.

Witty, K. and Thompson, K. (1974). Fish stranding surveys. In: *Anatomy of a River: An Evaluation of Water Requirements of the Hell's Canyon Reach of the Snake River, Conducted March, 1973,* Bayha, K. and Koski, C., Eds., pp. 113–120. Vancouver, WA: Pacific Northwest River Basins Commission.

Woodin, R.M. (1984). *Evaluation of Salmon Fry Stranding Induced by Fluctuating Hydroelectric Discharge in the Skagit River, 1980–1983.* Olympia: Washington Department of Fisheries and Wildlife.

World Commission on Environment and Development. (1987). *Our Common Future.* New York: Oxford University Press.

World Digital Library. (2015). *Johnstown, Pennsylvania, 1904,* http://wdl.org/en/item/9571/.

Zelinsky, W. (2004). Globalization reconsidered: the historical geography of modern western male attire. *Journal of Cultural Geography,* 22(1): 83–134.

7 Bioenergy

The wretched and the poor look for water and find none,
their tongues are parched with thirst;
but I the Lord will give them an answer,
I, the God of Israel, will not forsake them.
I will open rivers among the sand-dunes
and wells in the valleys;
I will turn the wilderness into pools
and dry land into springs of water;
I will plant cedars in the wastes,
And acacia and myrtle and wild olive;
the pine shall grow on the barren heath
side by side with fir and box …

—Isaiah 41: 17–20

A nation that runs on oil can't afford to run short.

—Old oil industry slogan

I'm the Lorax who speaks for the trees, which you seem to be chopping as fast as you please.

—Dr. Seuss (*The Lorax*)

INTRODUCTION

In a recent article in *The New York Times*, Friedman (2010) stated that, "The fat lady has sung." Specifically, he was speaking about America's transition from the "greatest generation" to what novelist Kurt Anderson has referred to as the "grasshopper generation." According to Friedman, we are "going from the age of government handouts to the age of citizen givebacks, from the age of companions fly free to the age of paying for each bag." Our parents were the greatest generation, but we have changed course and become the grasshopper generation, which is "eating through the prosperity that was bequeathed us like hungry locusts." Emphasizing again the major theme of this text, what we are eating through, among other things, is our readily available, relatively inexpensive source of energy. The point is we can, like the grasshopper, gobble it all up until it is all gone or we can find alternatives—renewable alternatives of energy.

One of the promising forms of alternative energy is *bioenergy*, a general term for energy derived from materials such as straw, wood, or animal wastes, which, in contrast to fossil fuels, were living matter relatively recently. Such materials can be burned directly as solids (*biomass*) to produce heat or power, but they can also be converted into liquid biofuels. Interest is growing in *biofuels*, liquid fuels (biodiesel and bioethanol) that can be used for transport. At the moment, transport has taken center stage in our search for renewable, alternative fuels to eventually replace hydrocarbon fuels. Unlike biofuels, solid biomass fuel is used primarily for electricity generation or heat supply.

Even though we have stated that bioenergy is a promising source of energy for the future, it is rather ironic that the experts (or anyone else for that matter) frequently make this point without qualification. The qualification? The reality? Simply, keep in mind that only 100 yeas ago our economy was based primarily on bioenergy from biomass, or carbohydrates, rather than from hydrocarbons. In the late 1800s, the largest selling chemicals were alcohols made from wood and grain, the first plastics were produced from cotton, and about 65% of the nation's energy came from wood

(USDOE, 2004). By the 1920s, the economy started shifting toward the use of fossil resources, and after World War II this trend accelerated as technology breakthroughs were made. By the 1970s, fossil energy was established as the backbone of the U.S. economy, and all but a small portion of the carbohydrate economy remained (Morris, 2002). In 1989, in the industrial sector, plants accounted for about 16% of input, compared with 35% in 1925 (USDOE, 2006a).

Processing costs and the availability of inexpensive fossil energy resources continue to be driving factors in the dominance of hydrocarbon resources. In many cases, it is still more economical to produce goods from petroleum or natural gas than from plant matter. This trend is about to shift dramatically as we reach peak oil production and as the world continues to demand unprecedented amounts of petroleum supplies from an ever dwindling supply. Assisting in this trend shift are technological advances in the biological sciences and engineering, political change, and concern for the environment, which have begun to swing the economy back toward carbohydrates on a number of fronts. Consumption of biofuels in vehicles, for example, has risen from 0 in 1977 to nearly 1.5 billion gallons in 1999. The use of inks produced from soybeans in the United States increased fourfold between 1989 and 2000 and now represents more than 22% of total use (Morris, 2002).

Technological advances are also beginning to make an impact on reducing the costs of producing industrial products and fuels from biomass, making them more competitive with those produced from petroleum-based hydrocarbons. Developments in pyrolysis, ultracentrifuges, membranes, and the use of enzymes and microbes as biological factories are enabling the extraction of valuable components from plants at a much lower cost. As a result, industry is investing in the development of new bioproducts that are steadily gaining a share of current markets (USDOE, 2004).

New technologies are helping the chemical and food processing industries develop processes for more cost-effective production of all kinds of industrial products from biomass. One example is a plastic polymer derived from corn that is now being produced at a 300-million-pound-per-year plant in Nebraska, a joint venture between Cargill, one of the largest grain trading companies in the world, and Dow Chemical, one of the largest chemical producers (Fahey, 2001).

Other chemical companies are exploring the use of low-cost biomass processes to make chemicals and plastics that are now made from more expensive petrochemical processes (USDOE, 2006a). In this regard, new innovative processes such as biorefineries may become the foundation of the new bioindustry. A biorefinery is similar in concept to a petroleum refinery, except that it is based on conversion of biomass feedstocks rather than crude oil. Biorefineries in theory would use multiple forms of biomass to produce a flexible mix of products, including fuels, power, heat, chemicals, and materials (see Figure 7.1). In a biorefinery, biomass would be converted into high-value chemical products and fuels (both gas and liquid). Byproducts and residues, as well as some portion of the fuels produced, would be used to fuel onsite power generation or cogeneration facilities producing

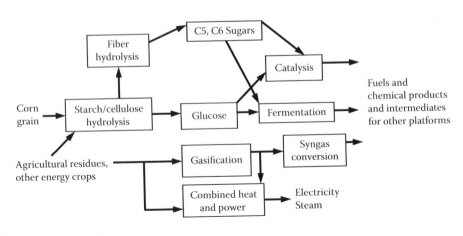

FIGURE 7.1 Biorefinery.

TABLE 7.1

U.S. Energy Consumption by Renewable Energy Source, 2014

Energy Source	Energy Consumption (quadrillion Btu)
Renewable total	9.656
Biomass (total)	**4.812**
Biofuels (ethanol and biodiesel)	**2.067**
Waste	**0.516**
Wood and wood-derived fuels	**2.230**
Geothermal	0.215
Hydroelectric power	2.475
Solar	0.421
Wind	1.733

Source: EIA, *Primary Energy Consumption by Source*, Table 10.1, Energy Information Administration, Washington, DC, 2016 (https://www.eia.gov/totalenergy/data/monthly/index.cfm#renewable).

heat and power. The biorefinery concept has already proven successful in U.S. agricultural and forest products industries, where such facilities now produce food, feed fiber, or chemicals, as well as heat and electricity to run plant operations. Biorefineries offer the most potential for realizing the ultimate opportunities of the bioenergy industry. Table 7.1 highlights bioenergy's quadrillion Btu ranking in current renewable energy source use. As pointed out in the table, the energy consumption by energy source computations are from the year 2014. Thus, the 4.812 bioenergy-biomass (biofuels, wastes, wood and wood-derived) quadrillion Btu figure is expected to steadily increase to an increasingly higher level and this should be reflected in future figures when they are released.

KEY TERMS

To assist in understanding the material presented in this chapter, the key components and operations related to bioenergy, biofuels, and biomass production and processing are defined in the following:

Algal biofuels—Utilization of primarily microalgae to produce high quantities of biomass per unit land area. The lipids in the microalgae can be used to produce biodiesel.

Annual removals—The net volume of growing stock trees removed from the inventory during a specified year by harvesting, cultural operations (such as timber stand improvement), or land clearing.

Asexual reproduction—The naturally occurring ability of some plant species to reproduce asexually through seeds, meaning the embryos develop without a male gamete. This ensures that the seeds will produce plants identical to the mother plant.

DID YOU KNOW?

In 2004, it was estimated that in the United States alone more than 100 billion gallons of light fuel oil per year could be produced from waste and biomass, including municipal solid waste, municipal sewage sludge, hazmat waste, agricultural crop waste, feedlot manure, plastics, tires, heavy oil or tar sands, forestry waste, restaurant grease, and biomass crops (switchgrass) grown on idle land and cropland.

Barrel (bbl)—A barrel of oil is 42 gallons.

Billion gallons per year—Abbreviated BGY.

Biobased product—As defined by the Farm Security and Rural Investment Act, a product determined by the U.S. Secretary of Agriculture to be a commercial or industrial product (other than food or feed) that is composed, in whole or in significant part, of biological products or renewable domestic agricultural materials (including plant, animal, and marine materials) or forestry materials.

Biodiesel—Fuel derived from vegetable oils or animal fats. It is produced when a vegetable oil or animal fat is chemically reacted with an alcohol, typically methanol. It is mixed with petroleum-based diesel.

Bioenergy—Useful renewable energy produced from organic matter through the conversion of the complex carbohydrates in organic matter to energy. This energy may be used directly as a fuel, may be processed into liquids and gases, or may be a residual of processing and conversion.

Biofuels—Fuels made from biomass resources or their processing and conversion derivatives. Biofuels include ethanol, biodiesel, and methanol.

Biomass—Any organic matter that is available on a renewable or recurring basis, including agriculture crops and trees, wood and wood residues, plants (including aquatic plants), grasses, animal manure, municipal residues, and other residue material. Biomass is generally produced in a sustainable manner from water and carbon dioxide by photosynthesis. There are three main categories of biomass: primary, secondary, and tertiary.

Biopower—The use of biomass feedstock to produce electric power or heat through direct combustion of the feedstock, through gasification and then combustion of the resultant gas, or through other thermal conversion processes. Power is generated with engines, turbines, fuel cells, or other equipment.

Biorefinery—A facility that processes and converts biomass into value-added products. These produces can range from biomaterials to fuels, such as ethanol or important feedstocks for the production of chemicals and other materials. Biorefineries can be based on a number of processing platforms using mechanical, thermal, chemical, and biochemical processes.

Black liquor—Solution of lignin residue and the pulping chemicals used to extract lignin during the manufacture of paper.

C4 species—C3 and C4 are the two main photosynthetic pathways in plants. C3 plants (e.g., beans, rice, wheat, potatoes, most temperate crops, all wood trees) fix carbon dioxide (CO_2) through photorespiration and require stomatal openings to acquire CO_2. C4 plants (e.g., corn, sugarcane, amaranth, most grasses, some shrubs) acquire CO_2 from malate (i.e., a source of CO_2 obtained via the Calvin cycle) and do not require open stomata; as a result, they provide higher water use efficiency and produce more biomass in hotter, drier climates. Under conditions of moderate temperatures and available soil water, C3 plants typically have an advantage in CO_2 fixation and thus overall growth.

Coarse materials—Wood residues suitable for chipping, such as slabs, edgings, and trimmings.

Commercial species—Tree species suitable for industrial wood products.

Component ration method (CRM)—Introduced by the U.S. Department of Agriculture Forest Service in 2009, a method for estimating non-merchantable volume from merchantable trees.

Composite integrated operations—Simultaneous production of both commercial (merchantable) wood products and thinnings.

Construction and demolition debris (C&D)—Wood waste generated during the construction of new buildings and structures, the repair and remodeling of existing buildings and structures, and the demolition of existing buildings and structures.

Conventionally sourced wood—Wood that has commercial uses other than fuel (e.g., pulpwood) but is used for energy because of market conditions. This would most likely only include smaller diameter pulpwood-sized trees.

Coppice—Regrowing from a (tree) stump after harvest. This would be equivalent to a ratoon (from the Spanish for "sprout") crop for sugarcane or sorghum, which would regrow from the stem after harvest.

Cotton gin trash—Residue available at a processing site, including seeds, leaves, and other material.

Cotton residue—Cotton stalks available for collection after cotton harvest.

Crop residues—The portion of a crop remaining after the primary product is harvested. Crop residues include corn stover and wheat, barley, oats, and sorghum straw. Other residues are rice field residue (straw), cotton field residues, and sugarcane residues (trash-leaves, tops, and remaining stalk after primary harvest of the stalk).

Cropland—Total cropland includes five components: cropland harvested, crop failure, cultivated summer fallow, cropland used only for pasture, and idle cropland.

Cropland pasture—Land used for long-term crop rotation; however, some cropland is marginal for crop uses and may remain in pasture indefinitely. This category also includes land that was used for pasture before crops reached maturity, and some land used for pasture that could have been cropped without additional improvement.

Cropland used for crops—Cropland used for crops includes cropland harvested, crop failure, and cultivated summer fallow.

Cropland harvested includes row crops and closely sown crops; hay and silage crops; tree fruits, small fruits, berries, and tree nuts; vegetables and melons; and miscellaneous other minor crops. In recent years, farmers have double-cropped about 4% of this acreage.

Crop failure primarily consists of the acreage on which crops failed because of weather, insects, and diseases but includes some land not harvested due to lack of labor, low market prices, or other factors. The acreage planted to cover and soil improvement crops not intended for harvest is excluded from crop failure and is considered idle.

Cultivated summer fallow refers to cropland in sub-humid regions of the West cultivated for one or more seasons to control weeds and accumulate moisture before small grains are planted. This practice is optional in some areas, but it is a requirement for crop production in the drier cropland areas of the West. Other types of fallow—such as cropland planted with soil improvement crops but not harvested, and cropland left idle all year—are not included in cultivated summer fallow, but they are included as idle cropland.

Cull tree—A live tree, 5 inches in diameter at breast height or larger, that is non-merchantable for saw logs now or prospectively because of rot, roughness, or species.

Diameter at breast height (dbh)—The common measure of wood volume approximated by the diameter of trees measured at approximately breast height from the ground.

Energy cane—Related to sugarcane. It can be a high-fiber sugarcane variety or a hybrid between sugarcane and wild relatives of sugarcane. Energy cane is designed to give higher biomass yields than sugarcane but to have a lower sugar concentration.

Energy Independence and Security Act of 2007 (EISA)—Act designed to increase the production of clean renewable fuels; protect consumers; increase the efficiency of products, buildings, and vehicles; promote research on and deploy greenhouse gas capture and storage options; and improve the energy performance of the federal government.

Ethanol—Also known as ethyl alcohol or grain alcohol. It is a volatile, flammable, and colorless liquid with the chemical formula C_2H_6O. It is produced by the fermentation of sugars into ethanol. Its primary uses are for drinking and fuel. In the United States, most fuel ethanol is currently produced by fermentation of glucose from the starch in corn.

Feedstock—A product used as the basis for the manufacture of another product.

Fiber products—Products derived from fibers of herbaceous and woody plant materials. Examples include pulp, composition board products, and wood chips for export.

Fine materials—Wood residues not suitable for chipping, such as planer shavings and sawdust.

Forest land—Land at least 10% stocked by forest trees of any size, including land that formerly had such tree cover and that will be naturally or artificially regenerated. Forest land includes transition zones, such as areas between heavily forested and non-forested lands that are at least 10% stocked with forest trees and forest areas adjacent to urban and built-up lands. Also included are pinyon–juniper and chaparral areas in the West and afforested areas. The minimum area for classification of forest land is 1 acre. Roadside, streamside, and shelterbelt strips of trees must have a crown width of a least 120 feet to qualify as forest land. Unimproved roads and trails, streams, and clearings in forest areas are classified as forest if less than 120 feet wide.

Fuelwood—Wood used for conversion to some form of energy, primarily for residential use.

Grassland pasture and range—All open land used primarily for pasture and grazing, including shrub and brush land types of pasture; grazing land with sagebrush and scattered mesquite; and all tame and native grasses, legumes, and other forage used for pasture or grazing. Due to the diversity in vegetative composition, grassland pasture and range are not always clearly distinguishable from other types of pasture and range. At one extreme, permanent grassland may merge with cropland pasture, or grassland may often be found in transitional areas with forested grazing land.

Growing stock—A classification of timber inventory that includes live trees of commercial species meeting specified standards of quality or vigor. Cull trees are excluded. When associated with volume, this classification only includes trees 5 inches in diameter at breast height and larger.

Harvest index (HI)—For traditional crops, the ratio of residue to grain.

Idle cropland—Land in cover and soil improvement crops and cropland on which no crops were planted. Some cropland is idle each year for various physical and economic reasons. Acreage diverted from crops to soil-conserving uses (if not eligible for and used as cropland pasture) under federal farm programs is included in this component. Cropland enrolled in the Federal Conservation Reserve Program is included in idle cropland.

Industrial wood—All commercial roundwood products except for fuelwood.

Live cull—A classification that includes live cull trees. When associated with volume, it is the net volume in live cull trees that are 5 inches in diameter at breast height and larger.

Logging residues—The unused portions of growing-stock and non-growing-stock trees that have been cut or killed by logging and left in the woods.

Mill residues—Bark and woody materials that are generated in primary wood-using mills when roundwood products are converted to other products. Examples are slabs, edgings, trimmings, sawdust, shavings, veneer cores and clippings, and pulp screenings. Includes bark residues and wood residues (both coarse and fine materials) but excludes logging residues. May include both primary and secondary mills.

Municipal solid waste (MSW)—Wastes (garbage) collected from municipalities consisting mainly of yard trimmings and paper products.

Nonforest land—Land that has never supported forests and lands formerly forested where use of timber management is precluded by development for other uses. (Note that this includes area used for crops, improved pasture, residential areas, city parks, improved roads of any width and adjoining clearings, powerline clearings of any width, and 1- to 4.5-acre areas of water classified by the Census Bureau as land. If intermingled in forest areas, unimproved roads and nonforest strips must be more than 120 feet wide, and clearings, etc., must be more than 1 acre in area to qualify as nonforest land.)

Nonindustrial private—An ownership class of private lands where the owner does not operate wood-using processing plants.

Other forest land—Forest land other than timberland, as well as reserved forest land. It includes available forest land that is incapable of annually producing 20 cubic feet per acre of industrial wood under natural conditions because of adverse site conditions such as sterile soils, dry climate, poor drainage, high elevation, steepness, or rockiness.

Other removals and residues—Unutilized wood volume from cut or otherwise killed growing stock, from cultural operations such as precommercial thinnings, or from timberland clearing for other uses (i.e., cropland, pastureland, roads, and urban settlement). Does not include volume removed from inventory through reclassification of timberland to productive reserved forest land.

Other wood sources—Sources of roundwood products that are not growing stock. These include salvable dead, rough and rotten trees, trees of noncommercial species, trees less than 5 inches in diameter at breast height, tops, and roundwood harvested from nonforest land (for example, fence rows).

Perennial—A crop that lives for more than 2 years. Well-established perennial crops have a good root system and provide cover that reduces erosion potential. They generally have reduced fertilizer and herbicide requirements compared to annual crops.

Poletimber trees—Live trees at least 5 inches in diameter at breast height but smaller than sawtimber trees.

Primary agricultural resources—Primary agricultural resources include energy feedstocks (annual energy crops, coppice and non-coppice woody crops, perennial grasses), crop residues (barley straw, corn stover, oat straw, sorghum stubble, wheat straw), and conventional crops (barely, corn, cotton, hay, oats, rice, sorghum, soybeans, wheat). The projections included for this category of feedstocks are two baseline scenarios—one with no energy crops (e.g., feed price of zero) and another including energy crops—and four high-yield scenarios with estimated biomass prices ranging between $40 and $80 at $5 increments.

Primary wood-using mill—A mill that converts roundwood products into other wood products. Common examples are sawmills that convert saw logs into lumber and pulp mills that convert pulpwood roundwood into wood pulp.

Pulpwood—Roundwood, whole-tree chips, or wood residues that are used for the production of wood pulp.

Rotten tree—A live tree of commercial species that does not contain a saw log now or prospectively primarily because of rot (that is, when rot accounts for more than 50% of the total cull volume).

Rough tree—(1) A live tree of commercial species that does not contain a saw log now or prospectively primarily because of roughness; that is, when sound cull, due to such factors as poor form, splits, or cracks, accounts for more than 50% of the total cull volume. (2) A live tree of non-commercial species.

Roundwood products—Logs and other round timber generated from harvesting trees for industrial or consumer use.

Salvable dead tree—A downed or standing dead tree that is considered currently or potentially merchantable by regional standards.

Saplings—Live trees 1.0 to 4.9 inches in diameter at breast height.

Secondary wood-processing mills—Mills that use primary wood products in the manufacture of finished wood products, such as cabinets, moldings, and furniture.

Soil conditioning index (SCI)—An index indicating the impact of crop management activities on soil organic matter.

Sound dead—The net volume in salvable dead trees.

Stand density index (SDI)—A measure of stocking of trees per unit area based on the number of trees per unit area and the average of diameter at breast height in the area. It is usually well correlated with the stand volume.

Starch—A carbohydrate consisting of many glucose units. It is the most common carbohydrate in the human diet.

Stumpage value—The sale value of the products that can be obtained from a stand of trees. This is the value of the wood products at a processing or end-use facility minus transport and harvest costs and a profit for the harvester.

Sugarcane trash—Tops and branches of sugarcane plants left on the field available for collection.

Thinnings (other forest land treatment thinnings)—Thinnings can come from operations to reduce fuel load (i.e., removal of small trees to reduce the fire danger) and from composite integrated operations on forest land (activities to harvest merchantable commercial wood and low-quality wood for bioenergy applications simultaneously). Thinnings can also come from pre-commercial operations and from other forest land to improve forest health.

Timberland—Forest land that is producing or is capable of producing crops of industrial wood and that is not withdrawn from timber utilization by statute or administrative regulations. Areas qualifying as timberland are capable of producing more than 20 cubic feet per acre per year of industrial wood in natural stands. Currently inaccessible and inoperable areas are included.

Urban wood wastes—These come from municipal solid waste (MSW) and construction and demolition debris. In the MSW portion, there is a wood component in containers, packaging and discarded durable goods (e.g., furniture), and yard and tree trimmings.

Wheat dust—The portion of wheat left after processing, known as dust and chaff.

BIOMASS

Biomass (all Earth's living matter) consists of the energy from plants and plant-derived organic-based materials; it is essentially stored energy from the sun. Biomass can be biochemically processed to extract sugars, thermochemically processed to produce biofuels or biomaterials, or combusted to produce heat or electricity. Biomass is also an input into other end-use markets, such as forestry products (pulpwood) and other industrial applications. This complicates the economics of biomass feedstock and requires that we differentiate between what is technically possible from what is economically feasible, taking into account relative prices and intermarket competition.

Biomass has been used since people began burning wood to cook food and keep warm. Trees have been the principal fuel for almost every society for over 5000 years, from the Bronze Age until the middle of the 19th century (Perlin, 2005). Wood is still the largest biomass energy resource today, but other sources of biomass can also be used. These include food crops, grassy and woody plants, residues from agriculture or forestry, and the organic component of municipal and industrial wastes. Even the fumes from landfills (which are methane, a natural gas) can be used as a biomass energy source. This category excludes organic material that has been transformed by geological processes into substances such as coal or petroleum. The biomass industry is one of the fastest-growing industries in the United States.

FEEDSTOCK TYPES

A variety of biomass feedstocks can be used to produce transportation fuels, biobased products, and power. Feedstocks refer to the crops or products, such as waste vegetable oil, that can be used as or converted into biofuels and bioenergy. With regard to the advantages or disadvantages of one type of feedstock as compared to another, this is gauged in terms of how much usable material they yield, where they can grow, and how energy and water intensive they are. Feedstock types are categorized as first-generation or second-generation feedstocks. First-generation feedstocks include those that are already widely grown and used for some form of bioenergy or biofuel production, which means that food vs. fuel conflicts could arise. First-generation feedstocks include sugars (sugar beets, sugarcane, sugar palm, sweet sorghum, and *Nypa* palm), starches (cassava, corn, milo, sorghum, sweet potato, and wheat), waste feedstocks such as whey and citrus peels, and oils and fats (coconut oil, oil palm, rapeseed, soy beans sunflower seed, castor beans, *Jatropha*, jojoba,

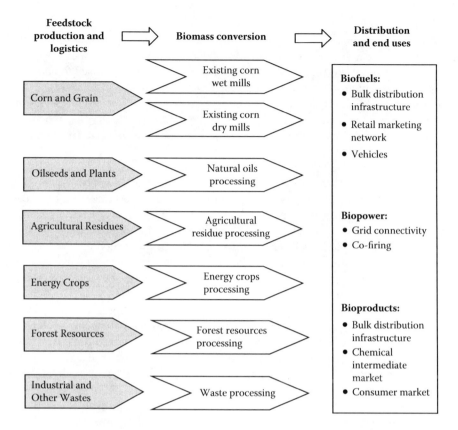

FIGURE 7.2 Resource-based biorefinery pathways.

karanj, waste vegetable oil, and animal fat). Second-generation feedstocks include crops that offer high potential yields of biofuels but are not widely cultivated or not cultivated as an energy crop. Examples are cellulosic feedstocks or conventional crops such as *Miscanthus* grasses, prairie grass and switchgrass, and willow and hybrid poplar trees. Algae and halophytes (saltwater plants) are other second-generation feedstocks.

Currently, a majority of the ethanol produced in the United States is made from corn or other starch-based crops (see Figure 7.2). The current focus, however, is on the development of cellulosic feedstocks—non-grain, non-food-based feedstocks such as switchgrass, corn stover, and wood material—and on technologies to convert cellulosic material into transportation fuels and other products. Using cellulosic feedstocks not only can alleviate the potential concern of diverting food crops to produce fuel but can also offers a variety of environmental benefits (EERE, 2008). Because such a wide variety of cellulosic feedstocks can be used for energy production, potential feedstocks are grouped into categories—or pathways. Figure 7.2 shows some of the specific feedstocks in each of these areas.

COMPOSITION OF BIOMASS

The ease with which biomass can be converted to useful products or intermediates is determined by the composition of the biomass feedstock. Biomass contains a variety of components, some of which are readily accessible and others that are much more difficult and costly to extract. The composition and subsequent conversion issues for current and potential biomass feedstock compounds are listed and described below:

- *Starch* (glucose) is readily recovered and converted from grain (corn, wheat, rice) into products. Starch from corn grain provides the primary feedstock for today's existing and emerging sugar-based bioproducts, such as polylactide, as well as the entire fuel ethanol industry. Corn grain serves as the primary feedstock for starch used to manufacture today's biobased products. Wet corn mills use a multistep process to separate starch from the germ, gluten (protein), and fiber components of corn grain. The starch streams generated by wet milling are very pure, and acid or enzymatic hydrolysis is used to break the glycosidic linkages of starch to yield glucose. Glucose is then converted into a multitude of useful products.

- *Lignocellulosic biomass* is the non-grain portion of biomass (e.g., cobs, stalks), often referred to as agricultural stover or residues. Energy crops such as switchgrass also contain valuable components, but they are not as readily accessible as starch. These lignocellulosic biomass resources (also called *cellulosic*) are comprised of cellulose, hemicellulose, and lignin. Generally, lignocellulosic material contains 30 to 50% cellulose, 20 to 30% hemicellulose, and 20 to 30% lignin. Some exceptions to this are cotton (98% cellulose) and flax (80% cellulose). Lignocellulosic biomass is perceived as a valuable and largely untapped resource for the future bioindustry; however, recovering the components in a cost-effective way presents a significant technical challenge.

- *Cellulose* is one of nature's polymers and is composed of glucose, a six-carbon sugar. The glucose molecules are joined by glycosidic linkages, which allow the glucose chains to assume an extended ribbon conformation. Hydrogen bonding between chains leads to formation of the flat sheets that lie on top of one another in a staggered fashion, similar to the way staggered bricks add strength and stability to a wall. As a result, cellulose is very chemically stable and insoluble and serves as a structural component in plant walls.

- *Hemicellulose* is a polymer containing primarily five-carbon sugars such as xylose and arabinose, with some glucose and mannose dispersed throughout. It forms a short-chain polymer that interacts with cellulose and lignin to form a matrix in the plant wall, strengthening it. Hemicellulose is more easily hydrolyzed than cellulose. Much of the hemicellulose in lignocellulosic material is solubilized and hydrolyzed to pentose and hexose sugars.

- *Lignin* helps bind the cellulosic/hemicellulose matrix while adding flexibility to the mix. The molecular structure of lignin polymers is very random and disorganized and consists primarily of carbon ring structures (benzene rings with methoxyl, hydroxyl, and propyl groups) interconnected by polysaccharides (sugar polymers). The ring structures of lignin have great potential as valuable chemical intermediates; however, separation and recovery of the lignin are difficult.

- *Oils* and *proteins* are obtained from the seeds of certain plants (e.g., soybeans, castor beans) and have great potential for bioproducts. These oils and proteins can be extracted in a variety of ways. Plants raised for this purpose include soy, corn, sunflower, safflower, rapeseed, and others. A large portion of the oils and proteins recovered from oilseeds and corn is processed for human or animal consumption, but they can also serve as raw materials for lubricants, hydraulic fluids, polymers, and a host of other products.

- *Vegetable oils* are composed primarily of triglycerides, also referred to as triacylglycerols. Triglycerides contain a glycerol molecule as the backbone with three fatty acids attached to glycerol's hydroxyl groups.

- *Proteins* are natural polymers with amino acids as the monomer unit. They are incredibly complex materials and their functional properties depend on molecular structure. There are 20 amino acids, each differentiated by their side chain or R-group, and they can be classified as nonpolar and hydrophobic, polar uncharged, or ionizable. The interactions among the side chains, the amide protons, and the carbonyl oxygen help create the three-dimensional shape of the protein.

TABLE 7.2

Comparison of Plants and Animals

Plants	Animals
Plants contain chlorophyll and can make their own food.	Animals cannot make their own food and are dependent on plants and other animals for food.
Plants give off oxygen and take in carbon dioxide given off by animals.	Animals give off carbon dioxide, which plants need to make food, and take in oxygen, which they need to breathe.
Plants generally are rooted in one place and do not move on their own.	Most animals have the ability to move fairly freely.
Plants have either no or a very basic ability to sense.	Animals have a much more highly developed sensory and nervous system.

PLANT BASICS*

To optimize plant biomass for more efficient processing requires a better understanding of plants and plant cell-wall structure and function. The plant kingdom ranks second in importance only to the animal kingdom (at least from the human point of view). The importance of plants and plant communities to humans, bioenergy production, and their environment cannot be overstated. Some of the important things plants provide are listed below:

Aesthetics—Plants add to the beauty of the places where we live.
Medicine—80% of all medicinal drugs originate in wild plants.
Food—90% of the world's food comes from only 20 plant species.
Industrial products—Plants are very important for the goods they provide (e.g., plant fibers provide clothing), and wood is used to build homes.
Recreation—Plants are the basis for many important recreational activities, including fishing, nature observation, hiking, and hunting.
Air quality—The oxygen in the air we breathe comes from the photosynthesis of plants.
Water quality—Plants aid in maintaining healthy watersheds, streams, and lakes by holding soil in place, controlling stream flows, and filtering sediments from water.
Erosion control—Plant cover helps to prevent wind or water erosion of the top layer of soil that we depend on.
Climate—Regional climates are impacted by the amount and type of plant cover.
Fish and wildlife habitat—Plants provide the necessary habitat for wildlife and fish populations.
Ecosystem—Every plant species serves an important role or purpose in its community.
Feedstock for bioenergy production—Some important fuel chemicals come from plants, such as ethanol from corn and soy diesel from soybeans.

Although both are important kingdoms of living things, plants and animals differ in many important aspects. Some of these differences are summarized in Table 7.2. Before discussing the basic specifics of plants, it is important to first define a few key plant terms.

PLANT TERMINOLOGY

Apical meristem consists of meristematic cells located at the tip (apex) of a root or shoot.
Cambium is the lateral meristem in plants.
Chloroplasts are disk-like organelles with a double membrane that are found in eukaryotic plant cells.
Companion cells are specialized cells in the phloem that load sugars into the sieve elements.

---

* Adapted from Spellman, F.R., *Biology for the Non-Biologist*, Government Institutes Press, Lanham, MD, 2009.

TABLE 7.3
Main Phyla (Divisions) of Plants

Phylum (Division)	Examples
Bryophyta	Mosses, liverworts, and hornworts
Coniferophyta	Conifers such as redwoods, pines, and firs
Cycadophyta	Cycads, sago palms
Gnetophyta	Shrub trees and vines
Ginkophyta	*Ginkgo* (the only genus)
Lycophyta	Lycopods (look like mosses)
Pterophyta	Ferns and tree-ferns
Anthophyta	Flowering plants, including oak, corn, maize, and herbs

Cotyledons are leaf-like structures (sometimes referred to as a *seed leaf*) present in the seeds of flowering plants.

Dicots are one of the two main types of flowering plants; they are characterized by having two cotyledons.

Diploid refers to having two of each kind of chromosome (2*n*).

Guard cells are specialized epidermal cells that flank stomata and whose opening and closing regulate gas exchange and water loss.

Haploid refers to having only a single set of chromosomes (*n*).

Meristem is a group of plant cells that can divide indefinitely, providing new cells for the plant.

Monocots are one of two main types of flowering plants; they are characterized by having a single cotyledon.

Periderm is a layer of plant tissue derived from the cork cambium and secondary tissue; it replaces the epidermis and acts as a protective coating.

Phloem is a complex vascular tissue that transports carbohydrates throughout the plant.

Sieve cells are conducting cells in the phloem of vascular plants.

Stomata are pores on the underside of leaves that can be opened or closed to control gas exchange and water loss.

Thallus is the main plant body, not differentiated into a stem or leaves.

Tropism is the plant behavior that controls the direction of plant growth.

Vascular tissue is found in the bodies of vascular plants that transport water, nutrients, and carbohydrates. The two major kinds are xylem and phloem.

Xylem is vascular tissue of plants that transports water and dissolved minerals from the roots upward to other parts of plant. Xylem often also provides mechanical support against gravity.

Although not typically acknowledged, plants are as intricate and complicated as animals. Plants evolved from photosynthetic protists and are characterized by photosynthetic nutrition, cell walls made from cellulose and other polysaccharides, lack of mobility, and a characteristic life cycle involving an alternation of generations. The phyla (divisions) of plants and some examples are shown in Table 7.3.

PLANT CELL

A brief summary of plant cells is provided here (see Figure 7.3).

- *Plants have all the same organelles that animal cells have* (e.g., nucleus, ribosomes, mitochondria, endoplasmic reticulum, Golgi apparatus).
- *Plants have chloroplasts*, the special organelles that contain chlorophyll and allow plants to carry out photosynthesis.

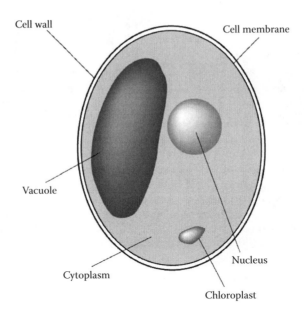

Cell wall

Cell membrane

Vacuole

Nucleus

Cytoplasm

Chloroplast

FIGURE 7.3   Plant cell.

- *Plant cells can sometimes have large vacuoles for storage.*
- *Plant cells are surrounded by a rigid cell wall made of cellulose,* in addition to the cell membrane that surrounds animal cells. These walls provide support.

VASCULAR PLANTS

Vascular plants, also called *tracheophytes*, have special vascular tissue for the transport of necessary liquids and minerals over long distances. Vascular tissues are composed of specialized cells that create "tubes" through which materials can flow throughout the plant body. These vessels are continuous throughout the plant, allowing for the efficient and controlled distribution of water and nutrients. In addition to this transport function, vascular tissues also support the plant. The two types of vascular tissue are xylem and phloem:

- *Xylem* consists of a tube or a tunnel (pipeline) in which water and minerals are transported throughout the plant to leaves for photosynthesis. In addition to distributing nutrients, xylem (wood) provides structural support. After a time, the xylem at the center of older trees ceases to function in transport and takes on a supportive role only.
- *Phloem* tissue consists of cells called *sieve tubes* and *companion cells.* Phloem tissue moves dissolved sugars (carbohydrates), amino acids, and other products of photosynthesis from the leaves to other regions of the plant.

The two most important tracheophytes are gymnosperms (*gymno*, naked; *sperma*, seed) and angiosperms (*angio*, vessel, receptacle, container).

- *Gymnosperms*—The plants we recognize as gymnosperms represent the sporophyte generation (i.e., the spore-producing phase in the life cycle of a plant that exhibits alternation of generation). Gymnosperms were the first tracheophytes to use seeds for reproduction. The seeds develop in protective structures called *cones.* A gymnosperm contains some cones that are female and some that are male. Female cones produce spores that, after fertilization, become eggs enclosed in seeds that fall to the ground. Male cones produce pollen, which is taken by the wind and fertilizes female eggs by that means. Unlike flowering plants, the gymnosperm does not form true flowers or fruits. Coniferous tress such as firs and pines are good examples of gymnosperms.

- *Angiosperms*—These flowering plants are the most highly evolved plants and currently the most dominant. They have stems, roots, and leaves. Unlike gymnosperms such as conifers and cycads, the seeds of angiosperms are found in the flowers. Angiosperm eggs are fertilized and develop into a seed in an ovary that is usually in a flower.

The two types of angiosperms are monocots and dicots:

- *Monocots*—These angiosperms start with one seed leaf (cotyledon), thus their name, which is derived from the presence of a single cotyledon during embryonic development. Monocots include grasses, grains, and other narrow-leaved angiosperms. The main veins of their leaves are usually parallel and unbranched, the flower parts occur in multiples of three, and a fibrous root system is present. Monocots include orchids, lilies, irises, palms, grasses, and wheat, corn, and oats.
- *Dicots*—Angiosperms in this group grow two seed leaves (two cotyledons). Most plants are dicots and include maples, oaks, elms, sunflowers, and roses. Their leaves usually have a single main vein or three or more branched veins that spread out from the base of the leaf.

LEAVES

The principal function of leaves is to absorb sunlight for the manufacturing of plant sugars in photosynthesis. The broad, flattened surfaces of the leaves gather energy from sunlight, while apertures on their undersides bring in carbon dioxide and release oxygen. Leaves develop as a flattened surface in order to present a large area for efficient absorption of light energy. On its two exteriors, the leaf has layers of epidermal cells that secrete a waxy, nearly impermeable cuticle (chitin) to protect against water loss (dehydration) and fungal or bacterial attack. Gases diffuse in or out of the leaf through *stomata*, small openings on the underside of the leaf. The opening or closing of the stomata occurs through the swelling or relaxing of *guard cells*. If the plant wants to limit the diffusion of gases and the transpiration of water, the guard cells swell together and close the stomata. Leaf thickness is kept to a minimum so that gases that enter the leaf can diffuse easily throughout the leaf cells.

Chlorophyll and Chloroplasts

The green pigment in leaves is *chlorophyll*. Chlorophyll absorbs red and blue light from the sunlight that falls on leaves; therefore, the light reflected by the leaves is diminished in red and blue and appears green. The molecules of chlorophyll are large. They are not soluble in the aqueous solution that fills plant cells. Instead, they are attached to the membranes of disc-like structures, called *chloroplasts*, inside the cells. Chloroplasts are the site of photosynthesis, the process in which light energy is converted to chemical energy. In chloroplasts, the light absorbed by chlorophyll supplies the energy used by plants to transform carbon dioxide and water into oxygen and carbohydrates. Chlorophyll is not a very stable compound; bright sunlight causes it to decompose. To maintain the amount of chlorophyll in their leaves, plants continuously synthesize it. The synthesis of chlorophyll in plants requires sunlight and warm temperatures; therefore, during summer chlorophyll is continuously broken down and regenerated in the leaves of trees.

Photosynthesis

Because our quality of life, and indeed our very existence, depends on photosynthesis, it is essential to understand it. In photosynthesis, plants (and other photosynthetic autotrophs) use the energy from sunlight to create the carbohydrates necessary for cell respiration. More specifically, plants take water and carbon dioxide and transform them into glucose and oxygen:

$$6CO_2 + 6H_2O + \text{Light energy} \rightarrow C_6H_{12}O_6 + 6O_2$$

This general equation of photosynthesis represents the combined effects of two different stages. The first stage is called the *light reaction* and the second stage is called the *dark reaction*. The light reaction is the photosynthesis process in which solar energy is harvested and ultimately converted into NADPH and adenosine triphosphate (ATP); the light reaction can only occur in light. In the dark reaction, the NADPH and ATP formed by the light reaction reduce carbon dioxide and convert it into carbohydrates; the dark reaction can occur in the dark as long as ATP is present.

ROOTS

Roots absorb nutrients and water, anchor the plant in the soil, provide support for the stem, and store food. They are usually below ground and lack nodes, shoots, and leaves. There are two major types of root systems in plants. *Taproot systems* have a stout main root with a limited number of side-branching roots. Examples of taproot system plants are nut trees, carrots, radishes, parsnips, and dandelions. Taproots make transplanting difficult. The second type of root system, *fibrous*, has many branched roots. Examples of fibrous root plants are most grasses, marigolds, and beans. Radiating from the roots is a system of root hairs, which vastly increase the absorptive surface area of the roots. Roots also anchor the plant in the soil.

GROWTH IN VASCULAR PLANTS

Vascular plants undergo two kinds of growth (growth is primarily restricted to meristems). *Primary growth* occurs relatively close to the tips of roots and stems. It is initiated by apical meristems and is primarily involved in the extension of the plant body. The tissues that arise during primary growth are called primary tissues, and the plant body composed of these tissues is called the *primary plant body*. Most primitive vascular plants are entirely made up of primary tissues. *Secondary growth* occurs in some plants; secondary growth thickens the stems and roots. Secondary growth results from the activity of lateral meristems, or *cambia*, of which there are two types:

1. *Vascular cambium* gives rise to secondary vascular tissues (secondary xylem and phloem). The vascular cambium gives rise to xylem on the inside and phloem on the outside.
2. *Cork cambium* forms the periderm (bark). The periderm replaces the epidermis in woody plants.

PLANT HORMONES

Plant growth is controlled by plant hormones, which influence cell differentiation, elongation, and division. Some plant hormones also affect the timing of reproduction and germination.

- *Auxins* affect cell elongation (tropism), apical dominance, and fruit drop or retention. Auxins are also responsible for root development, secondary growth in the vascular cambium, inhibition of lateral branching, and fruit development. Auxin is involved in the absorption of vital minerals and fall color. As a leaf reaches its maximum growth, auxin production declines. In deciduous plants this triggers a series of metabolic steps that cause the reabsorption of valuable materials (such as chlorophyll) and their transport into the branch or stem for storage during the winter months. When the chlorophyll is gone, the other pigments typical of fall color become visible.

- *Kinins* promote cell division and tissue growth in the leaf, stem, and root. Kinins are also involved in the development of chloroplasts, fruits, and flowers. In addition, they have been shown to delay senescence (aging), especially in leaves, which is one reason why florists use cytokinins on freshly cut flowers—when treated with cytokinins, they remain green, protein synthesis continues, and carbohydrates do not break down.
- *Gibberellins* are produced in the root growing tips and act as a messenger to stimulate growth, especially elongation of the stem, and can also end the dormancy period of seeds and buds by encouraging germination. Additionally, gibberellins play a role in root growth and differentiation.
- *Ethylene* controls the ripening of fruits. Ethylene may ensure that flowers are carpelate (female), while gibberellin confers maleness on flowers. It also contributes to the senescence of plants by promoting leaf loss and other changes.
- *Inhibitors* restrain growth and maintain the period of dormancy in seeds and buds.

TROPISMS

Tropism is the movement (including growth in plants) of an organism in response to an external stimulus. Tropisms controlled by hormones are a unique characteristic of sessile organisms such as plants that enable them to adapt to different features of their environment—gravity, light, water, and touch—so they can flourish. The three main types of tropisms are

- *Phototropism*—The tendency of plants to grow or bend (move) in response to light. Phototropism results from the rapid elongation of cells on the dark side of the plant, which causes the plant to bend in the opposite direction. For example, the stems and leaves of a geranium plant growing on a windowsill always turn toward the light.
- *Gravitropism*—The tendency of plants to grow toward or against gravity. A plant that displays positive gravitropism (plant roots) will grow downward, toward the center of earth. That is, gravity causes the roots of plants to grow down so that the plant is anchored in the ground and has enough water to grow and thrive. Plants that display negative gravitropism (plant stems) will grow upward, away from the earth. Most plants are negatively gravitropic. Gravitropism is also controlled by auxin. In a horizontal root or stem, auxin is concentrated in the lower half, pulled by gravity. In a positively gravitropic plant, this auxin concentration will inhibit cell growth on the lower side, causing the stem to bend downward. In a negatively gravitropic plant, this auxin concentration will inspire cell growth on that lower side, causing the stem to bend upward.
- *Thigmotropism*—The tendency of plants to grow or bend in response to touch. Some people notice that their houseplants respond to thigmotropism by growing better when they are touched and have attention paid to them. Touch causes parts of the plant to thicken or coil as they touch or are touched by environmental entities. For example, tree trunks grow thicker when exposed to strong winds and vines tend to grow straight until they encounter a substrate to wrap around.

PHOTOPERIODISM

Photoperiodism is the response of an organism, such as plants, to naturally occurring changes in light during a 24-hour period. The site of perception of photoperiod in plants is leaves. Sunflowers are particularly well known for their photoperiodism, or their ability to open and close in response to the changing position of the sun throughout the day. All flowering plants have been placed in one of three categories with respect to photoperiodism:

- *Short-day plants*—Flowering is promoted by day lengths shorter than a certain critical day length; includes poinsettias, chrysanthemums, goldenrod, and asters.
- *Long-day plants*—Flowering is promoted by day lengths longer than a certain critical day length; includes spinach, lettuce, and most grains.
- *Day-neutral plants*—Flowering response is insensitive to day length; includes tomatoes, sunflowers, dandelions, rice, and corn.

PLANT REPRODUCTION

Plants can reproduce both sexually and asexually. Each type of reproduction has its benefits and disadvantages. A comparison of sexual and asexual plant reproduction is provided in the following text.

Sexual Reproduction
- Sexual reproduction occurs when a sperm nucleus from the pollen grain fuses with an egg cell from the ovary of the pistil (the female reproductive structures in flowers, consisting of the stigma, style, and ovary).
- Each brings a complete set of genes and produces genetically unique organisms.
- The resulting plant embryo develops inside the seed and grows when the seed is germinated.

Asexual Reproduction
- Asexual reproduction occurs when the vegetative part of a plant, root, stem, or leaf gives rise to a new offspring plant whose genetic content is identical to the parent plant. An example would be a plant reproducing by root suckers, shoots that come from the root system. The breadfruit tree is an example.
- Asexual reproduction is also called *vegetative propagation*. It is an important way for plant growers to get many identical plants from one very quickly.
- By asexual reproduction plants can spread and colonize an area quickly (e.g., crabgrass).

PLANT CELL WALLS

Figure 7.3 shows the basic organelles within a standard plant cell, including the cell wall. Figure 7.4 shows plant cell walls. It should be pointed out, however, that plants can have two types of cell walls, primary and secondary (see Figure 7.4). Primary cell walls contain cellulose consisting of hydrogen-bonded chains of thousands of glucose molecules, in addition to hemicellulose and other

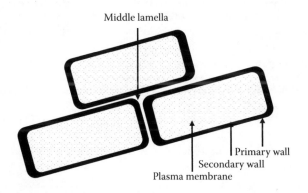

FIGURE 7.4   Plant cell walls.

DID YOU KNOW?

Cellulose microfibrils are composed of parallel and linear chains of glucose molecules that are tightly cross-linked by intermolecular hydrogen bonds.

materials all woven into a network. Certain types of cells, such as those in vascular tissues, develop secondary walls inside the primary wall after the cell has stopped growing. These cell-wall structures also contain lignin, which provides rigidity and resistance to compression. The area formed by two adjacent plant cells, the middle lamella, typically is enriched with pectin. Cellulose in higher plants is organized into microfibrils, each measuring about 3 to 6 nm in diameter and containing up to 36 glucan chains having thousands of glucose residues. Like steel girders stabilizing a skyscraper's structure, the mechanical strength of the primary cell wall is due mainly to the microfibril scaffold (i.e., crystalline cellulose core) (USDOE, 2007).

FEEDSTOCKS[*]

FIRST-GENERATION FEEDSTOCKS

For first-generation feedstocks, the type of major end-use product is easily categorized (see Table 7.4). Corn and sugarcane are the most commonly used feedstocks for ethanol production. Soybean and other vegetable oils and animal fats are used for the production of biodiesel (and bioproducts). Manure and landfill organic waste are used for methane production and the generation of electricity. Corn is used for ethanol production and currently is the leading feedstock used in the United States. Several factors favor a positive outlook for further near-term growth in corn ethanol production. Continued high oil prices will provide economic support for the expansion of all alternative fuel programs, including corn ethanol. Technology improvements that increase feedstock productivity and fuel conversion yields and positive spillovers from second-generation technologies (biomass gasification in ethanol refiners) will also help to lower production costs for corn ethanol. Among the factors likely to limit future growth of corn ethanol production are increased feedstock and other production costs, increased competition from unconventional liquid fossil fuels (e.g., oil sands, coal, heavy oil, shale), the emergence of cellulosic ethanol as a low-cost competitor, and new policies to reduce greenhouse gas (GHG) emissions that could favor advanced biofuels over corn ethanol.

Biodiesel is another type of biofuel experiencing expansion. Although its production costs are higher than those for ethanol, biodiesel has some environmental advantages, including biodegradability and lower sulfur and carbon dioxide emissions when burned. Biodiesel production in the United States increased rapidly from less than 2 million gallons in 2000 to about 500 million gallons in 2007. Policy incentives in the Energy Independence and Security Act of 2007 are expected to sustain demand for 1 billion gallons per year of this fuel after 2011. A variety of oil-based feedstocks are converted to biodiesel using a process known as *transesterification*. This is the process of exchanging the organic group R″ of an ester with the organic group R′ of an alcohol. These reactions are often catalyzed by the addition of an acid or base as shown below:

$$\text{Transesterification: Alcohol + Ester} \rightarrow \text{Different alcohol + Different ester}$$

The oil-based feedstocks include vegetable oils (mostly soy oil), recycled oils and yellow grease, and animal fats such as beef tallow. It takes 3.4 kg of oil/fat to produce 1 gallon of biodiesel (Baize, 2006). Biodiesel production costs are high compared to ethanol, with feedstocks accounting for 80% or more of total costs.

---

[*] Adapted from BR&Di, *The Economics of Biomass Feedstocks in the United States: A Review of the Literature*, Biomass Research and Development Initiative, Washington, DC, 2008.

TABLE 7.4
Biomass Feedstocks and Major Bioenergy End-Use Applications

Feedstocks		First-Generation Fuels			Second-Generation Fuels	
		Ethanol	Biodiesel	Methane	Cellulosic Ethanol	Thermoch Fuels[a]
First-generation						
Starch and sugar feedstock for ethanol	Corn	X				
	Sugarcane	X				
	Molasses	X				
	Sorghum	X				
Vegetable oil and fats and biodiesel	Vegetable oils		X			
	Recycled fats and grease		X			
	Beef tallow		X			
Second-generation (short-term)						
Agricultural residues and livestock byproducts	Corn stover				X	
	Wheat straw				X	
	Rice straw				X	
	Bagasse				X	
	Manure			X		
Forest biomass	Logging residues				X	X
	Fuel treatments				X	X
	Conventional wood				X	X
Urban woody waste and landfills	Primary wood products				X	X
	Secondary mill residues				X	X
	Municipal solid wastes			X	X	X
	Construction/ demolition wood					
	Landfills			X	X	
Second-generation (long-term)						
Herbaceous energy crops	Switchgrass				X	X
	Miscanthus				X	X
	Reed canary grass				X	X
	Sweet sorghum					
	Alfalfa				X	X
Short-rotation woody crops	Willow				X	X
	Hybrid poplar				X	X
	Cottonwood				X	X
	Sycamore pines				X	X
	Eucalyptus				X	X

[a] Ethanol, diesel, and butanol, for example.

The biodiesel industry consists of many small plants that are highly dispersed geographically. Decisions about plant location are primarily determined by local availability and access to the feedstock. Recent expansion in biodiesel production is affecting the soybean market. Achieving a nationwide target of 2% biodiesel blend in diesel transportation fuel, for example, would require 2.8 million metric tons (MT) of vegetable oil, or about 30% of current U.S. soybean oil production (USDA, 2012).

One of the key biodiesel byproducts is glycerin. At the present time there is concern about producing glycerin mainly because there are no existing markets for the product; however, recent technological developments include an alternative chemical process that would produce biodiesel without glycerin. In addition, new processes are being tested that further transform glycerin into propylene glycol, which is used in the manufacture of antifreeze.

SIDEBAR 7.1. FIRST-GENERATION FEEDSTOCK EXAMPLES*

Sugarcane and Sorghum

Sugarcane

As shown in Table 7.4, sugarcane is a first-generation biomass feedstock. It is a large-stature, jointed grass that is cultivated as a perennial row crop, primarily for its ability to store sucrose in the stem, in approximately 80 countries in tropical, semitropical, and subtropical regions of the world (Tew, 2003). It is one of the most efficient C4 grasses in the world, with an estimated energy in/energy out (I/O) ratio of 1:8 when grown for 12 months under tropical conditions and processed for ethanol instead of sugar (Bourne, 2007; Macedo et al., 2004; Muchow et al., 1996). Under more temperate environments, where temperature and sunlight are limited, I/O ratios of 1:3 are easily obtainable with current sugarcane cultivars if ethanol production from both sugar and cellulosic biomass is the goal (Tew and Cobill, 2008). In addition, sugarcane ethanol cuts GHGs at least 60% compared to gasoline—better than any other biofuel produced today. The U.S. Environmental Protection Agency (USEPA) has confirmed the superior environmental performance of sugarcane ethanol by designating it as an advanced renewable fuel. Most of the sugarcane grown in the United States is dedicated to the production of sugar. Although an energy cane industry has not developed to date, one can assume that the production process of the mature sugarcane industry can be modified to ensure the sustainable production of energy cane, as well.

Let's take a closer look at sugarcane. What is it, exactly? Sugarcane (*Saccharum* spp.) (see Figure SB7.1.1) is a genetically complex crop, with a genomic makeup resulting from successful interspecific hybridization efforts primary involving *S. officinarum* and *S. spontaneum* (Tew and Cobill, 2008). Improvement of sugarcane for increased energy efficiency and adaptability to a wide range of environments is considered by many geneticists as synonymous with "genetic base broadening" (i.e., utilization of wild *Saccharum* germplasm), particularly *S. spontaneum*, in sugarcane breeding programs (Ming et al., 2008). *S. spontaneum*, considered a noxious weed in the United States, can be found in the continents of Africa, Asia, and Australia in environments ranging from the Equator to the foothills of the Himalayas. This makes it an excellent source of a number of valuable genes (Mukherjee, 1950; Panje, 1970; Panje and Babu, 1960; Roach, 1978). The U.S. Department of Agriculture Economic Research Service (USDA ERS) Sugarcane Research Unit (SRU) at Houma, Louisiana, in the 1960s took on the role of introgressing desirable genes from sugarcane's wild and near relatives (*Miscanthus* and *Erianthus*) to build new parents for utilization in its commercial sugarcane varietal development program. Because these wild accessions contain only small amounts of sugar, three or four rounds of backcrossing to elite sugarcane varieties must be done to obtain a commercially acceptable sugarcane variety that has high sugar yields with minimum amounts of fiber.

Sugarcane is grown as a monoculture (i.e., a single crop) with fields being replanted every 4 or 5 years. This type of culture is conducive to the development of perennial weeds such as johnsongrass and bermudagrass, as the row top (i.e., raised bed) remains relatively

---

* Adapted from USDOE, *U.S. Billion-Ton Update: Biomass Supply for a Bioenergy and Bioproducts Industry*, ORNL/ TM-2011/224, Oak Ridge National Laboratory, Oak Ridge, TN, 2011.

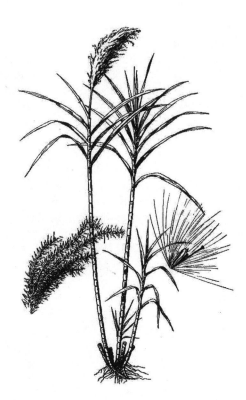

FIGURE SB7.1.1   Sugarcane (*Saccharum officinarum*). (From Hitchcock, A.S., *Manual of the Grasses of the United States*, 2nd ed. (revised by A. Chase), USDA Miscellaneous Publication No. 200, U.S. Government Printing Office, Washington, DC, 1950.)

undisturbed for the 5-year crop cycle. The selective control of these weeds within the crop is difficult with currently registered herbicides once these weeds produce rhizomes (i.e., the belowground horizontal stem of the plant). To minimize this risk, growers disk the old stubble fields in the winter or early spring and fallow the fields until they are planted to sugarcane again. Frequent disking and/or the use of multiple applications of glyphosate (a systemic herbicide) are ways to deplete the soil of weed seed and rhizomes during the fallow period.

With regard to actual planting, because of the need to accommodate mechanical harvesting, sugarcane is vegetatively planted by laying 6- to 8-foot long stakes end-to-end in a planting furrow along widely spaced rows. Each stalk is planted 5 to 6 feet apart and covered with 2 to 4 inches of soil. One acre of seedcane can plant 6 to 10 acres of sugarcane, depending on the length and number of stalks at the time of harvesting the seedcane for planting and the number of stalks per foot of row being planted. When harvesting seedcane for planting, stalks are cut at the soil surface and at the last mature node at the top of the stem. An alternative method of planting is to plant 12- to 18-inch stalk pieces (billets) that can be harvested with the same chopper harvester used to harvest sugarcane for delivery to sugar mills. Once the stalks are planted, a broad-spectrum pre-emergence herbicide is applied to control seedling weeds. New plants emerge from the axillary buds located in the nodal regions along the stalk. Growers produce most of their own seedcane for planting; hence, planting is generally done a few weeks prior to the beginning of the harvest season to ensure that stalks are plentiful and tall.

It is important to point out that vegetative planting of sugarcane is often considered a drawback by growers who are accustomed to planting large areas of seeded crops relatively quickly with one tractor and one plant. It is an expensive process, as it requires considerable labor and equipment, is relatively slow, and requires that the grower plant sugarcane that would normally be sent to the raw sugar factory for processing. However, with energy cane, 20% to 30% higher planting ratios can be expected, because stalk numbers and height are higher. In addition, at least two additional harvests per planting can be expected, significantly lowering the number of acres requiring planting each year. It is estimated that the cost to plant one acre of sugarcane is about $500 when the grower uses seedcane from the farm. Because planting costs are spread over four annual harvests, the annual cost would be approximately $125 per acre. With energy cane, one acre of seedcane would plant about 13 additional acres, reducing per-acre planting costs to $346. If spread over the anticipated six annual fall harvests, annual planting costs would be $58 per acre.

Vegetative planting also has advantages, especially when planting must be done under conditions of less than idle seedbed preparation. The crop emerges 14 to 21 days later and continues to grow until the first heavy frost of the fall. The first production year (plant-cane crop) actually begins in the spring following planting with the emergence of the crop from winter dormancy. Herbicides are applied each spring to the subsequent ratoon crops (first-through third-ratoon crop) to minimize early weed competition. Nitrogen is applied at rates of 70 to 90 pounds per acre of the plant-cane crop (first growing season) and 90 to 120 pounds per acre to the subsequent ratoon crops. These applications are generally made in the spring about 2 months into the growing season.

The sugarcane crop is susceptible to the rapid spread of a number of bacterial, fungal, and viral pathogens that can be spread easily by machinery and wind currents. These pathogens can affect the yield and ratooning ability (number of yearly harvests per planting) of the crop. Insects, primarily stalk borers, grubs, and aphids, also plague the industry. The compactness of the industry and the fact that the crop is grown continuously as a monoculture makes sugarcane especially vulnerable to the rapid spread of diseases and insects. The planting of resistant varieties is the predominant means of managing diseases in sugarcane. For insects, mainly the stalk borer, an effective integrated pest management program is used that involves field scouting, the use of tolerant varieties, and insecticides when established infestation thresholds are exceeded. Additional research is exploring the use of multiple crop-production systems for year-round delivery of feedstocks and to respond to biotic stresses (McCutchen and Avant, 2008).

Crop yields tend to decline with each successive fall harvest to the point that it is not practical to keep the old stubble for another year. The milling season in more subtropical climates such as Louisiana lasts about 100 days, beginning in late October and ending in later December or early January. Given that the crop emergence in the spring depends on the date of the last killing frost, it is obvious that sugarcane harvested in December will produce higher yields than sugarcane harvested in September. For this reason, growers try to have a balance in ratoon crop ages from 0 (plant cane that was never harvested) to 3 (third ratoon that was harvested three times previously) and begin the harvest with the third-ratoon fields that would lack the vigor of the plant-cane crops. In addition, early harvested crops tend not to yield as much the following year. A similar scenario is anticipated for energy cane with the exception that fourth- and fifth-ratoon crops would also be harvested.

In Louisiana, some energy cane varieties have produced average yields of 56 green tons per acre, with 8 to 14 tons per acre being fiber and 4 to 6 tons per acre being brix (bx), which is the sugar content of an aqueous solution. One degree brix is 1 gram of sucrose in 100 grams of solution and represents the strength of the solution as a percentage by mass on a dry weight basis.

With regard to potential yield and production costs, the theoretical maximum for above-ground sugarcane biomass (total solids) yield is estimated to be 62 green tons per acre annually (Loomis and Williams, 1963). This is dependent on temperature and sunlight and would probably occur under tropical conditions. Sugarcane breeding programs have reported sugar yield gains on the order of 1 to 2% per year (Edme et al., 2005). The economic sustainability of growing energy cane in non-traditional cane growing regions will require yearly biomass yield gains of this magnitude or greater, with a goal of ensuring that the I/O ratio of 1:8 projected for tropical countries can be met and ultimately exceeded under the subtropical cane growing conditions of the southeastern United States.

The "sun-dried crop" concept of allowing the crop to desiccate in the field and perhaps devoid itself of some of its leaves and moisture, as is proposed for many of the perennial grasses being considered for biofuels, is not an option for energy cane, as the stalks are thick with a waxy coating on them and the new growing season should begin as soon after harvest as possible. Consequently, energy cane, like sugarcane, will have to be harvested green and dewatered if the fiber is to be stored and processed later in the year. The value of this liquid is in question because it will add to transportation costs. However, if water is needed for the digestion of the fiber or maintenance of the bagasse under anaerobic conditions to minimize deterioration during outside storage, it would be present at no additional charge. What is also overlooked is the fact that the water contains sugar that is easily and much more inexpensively converted to ethanol. Furthermore, in some conversion processes the yeast used in fermentation requires a substrate to grow and multiply on, and sucrose is an ideal substrate. Conceivably, the biorefinery would have two processes for the production of biofuel, with one having sugar (brix) obtained from dewatering at the biorefinery as the feedstock and the other the fiber (bagasse). Economics would have to be considered with these options especially because of the very short cut-crush interval needed for sugar recovery and to prevent spoilage.

It is interesting to note that, in addition to the potential sustainability of energy cane, sustainability can also be provided by the fibrous extraneous matter generated during harvest. Green cane harvesting of sugarcane deposits 2.7 to 3.6 tons per acre of a mixture of brown and green leafy material and fragments of the stalks (Richard, 1999; Viator et al., 2006, 2009a,b). The fibrous extraneous matter generated has an energy value; however, the greatest value may be as a mulch to limit soil erosion, depress weed development, conserve moisture, and to serve as a means to recycle nutrients. All are potential contributors to the sustainable production of energy cane (de Resende et al., 2006). It is estimated that the extraneous residue generated during the green cane harvesting of sugarcane contains 0.7% nitrogen, 0.07% phosphorus, and 0.7% potassium by weight. Using 2009 USDA economic data (USDA ERS, 2010), 2.7 tons of residues would equate to a savings of approximately $50 per acre.

The critical amount of post-harvest residue that can be removed from the field has not been determined for energy cane. Sugarcane is harvested mechanically with a chopper harvester. These harvesters chop the sugarcane stalks into small 6- to 8-inch-long pieces (billets) and use wind currents from an extractor fan to remove the leaves attached to the stalks. When they are used to harvest energy cane, the speed of the extractor fans could be adjusted to deposit a percentage of the extraneous matter on the soil surface while the remainder is collected with the stalks and used as an additional source of fiber.

Predicting production costs has proven difficult and complicated because of first-year production of seedcane crop and the number of ratoon crops before re-establishment. For example, different costs are associated with fallow field and seedbed preparation, seedcane planting and harvest, plant cane, and ratoon operations and harvest over multiple years (Salassi and Deliberto, 2011). Using the most common sugarcane assumptions from the production costs from Salassi and Deliberto, an extrapolated cost to produce and harvest the energy cane is estimated to be about $34 per dry ton. The assumed yield is about 14 dry tons per acre.

Energy cane will be grown as a perennial. Commonly discussed disadvantages of perennial feedstocks include the difficulty of establishing a perennial crop from seed and rhizomes, the control of weeds during the establishment period, and the fact that an economic return will not be realized the first year. Energy cane, like sugarcane, is vegetatively planted by placing the stalks in a planting furrow in mid- to late summer. Herbicides are labeled for at-planting pre-emergence applications in sugarcane; presumably, these herbicides can also be applied to energy cane with similar results. These herbicides are reapplied each spring during the course of a 3- to 4-year sugarcane reproduction cycle. Because of the vigor (i.e., early spring emergence and high stalk population) of energy cane, use of these herbicides beyond the first spring after planting is not anticipated. The vigor of energy cane is also advantageous due to the fact that, when the crop is planted in the summer, it emerges and produces a uni-form stand quickly. The crop continues to grow until the aboveground portion is winter-killed in the fall. This aboveground material, which can equate to 2 to 4 dry tons per acre, could be harvested and converted to fuel in the first year of establishment.

The final consideration in the sustainability of energy cane production is the utilization of biorefinery byproducts to supply nutrients and reduce the impact of crop removal on soil health. These byproducts can include vinasse (a byproduct of the sugar industry) from the fermentation process, biochar from pyrolysis, and filter press mud and boiler fly ash from the squeezing of stalks. A positive synergistic response was observed when the application of fertilizer was combined with an application of filter press mud in Florida (Gilbert et al., 2008). In Brazil, vinasse from the ethanol distillery is typically returned to the recently har-vest fields to supplement fertilizer requirements.

The bottom line on energy cane becoming a suitable cellulosic feedstock for the future is dependent on several factors. These factors include extending energy cane's range of geo-graphic distribution outside of the traditional sugarcane growing areas by increasing its cold tolerance outside of tropical areas; drought and flood (saturated soil) tolerance, because this crop will probably be grown on marginal soils that may be prone to flooding or where irriga-tion is difficult; insect and disease resistance; and a further exploitation of some varieties of sugarcane that encourages symbiotic relationships with nitrogen-fixing bacteria.

The success of the sugarcane industry has been, and continues to be, dependent on the development of new hybrids with superior yields and increased resistance to many of the abiotic and biotic stresses previously mentioned. This formula will not change if the crop is grown as a dedicated feedstock for the production of liquid biofuels or electricity. Successful hybridization begins with the introgression of desirable traits from the wild relative of sugar-cane, *Saccharum spontaneum*. Early-generation progeny from these crosses with elite sugar-cane clones exhibit high levels of hybrid vigor, which translates into increased cold tolerance, greater ratooning ability, enhanced levels of tolerance to moisture extremes, increased insect and disease tolerance, and more efficient nutrient utilization (Legendre and Burner, 1995). Much of the vigor of these early-generation hybrids is lost in a conventional breeding pro-gram for sugar, as progeny from these crosses must be backcrossed with elite high-sugar-pro-duction clones three to four times before a commercial sugarcane variety can be produced. These early-generation hybrids would be considered ideal candidates as dedicated cellulosic biomass crops like energy canes. Most of these varieties can average over 15 dry tons per acre annually over at least four annual fall harvests.

Sorghum

Of the crops recently indentified for their potential bioenergy production (and as a sinkhole for greenhouse gases), sorghum has historically had the most direct influence on human development (see Figure SB7.1.2). Sorghum was domesticated in arid areas of northwest-ern Africa over 6000 years ago (Kimber, 2000). Sorghum is traditionally known for grain production; it is the fifth most widely grown and produced cereal crop in the world (FAO, 2007). However, in many regions of the world, sorghum is just as important (if not more) as a forage crop. In addition to forage and grain, sorghum types high in stalk sugar content and extremely lignified types (for structural building) have been grown throughout the world.

FIGURE SB7.1.2   Sorghum. (From Hitchcock, A.S., *Manual of the Grasses of the United States*, 2nd ed. (revised by A. Chase), USDA Miscellaneous Publication No. 200, U.S. Government Printing Office, Washington, DC, 1950.)

Given the demands for renewable fuel feedstocks, sorghum is now being developed as a dedicated bioenergy crop. This designation is not new; sorghum was mentioned prominently as a bioenergy crop over 20 years ago (Burton, 1986). The interest in the crop is justifiable based on several independent factors that separately indicate good potential, but when combined they clearly designate sorghum as a logical choice for bioenergy production. These factors include yield potential and composition, water-use efficiency and drought tolerance, established production systems, and the potential for genetic improvement by using both traditional and genomic approaches.

With regard to biology and adaptation, whether measured in grain yield or total biomass yield, sorghum is a highly productive C4 photosynthetic species that is well adapted to warm and dry growing regions. Although sorghum is technically a perennial in tropical environments, it is planted from seed, then grown and managed as an annual crop. Sorghum has a long-established breeding history through which the productivity, adaptation, and utilization of the crop have continually been improved. These efforts have resulted in numerous cultivars and hybrids of sorghum that are used for various purposes.

The optimum type of sorghum to be grown for biofuels production is dependent on the type of conversion process that will be used. Sorghums are divided into distinct types based on the amount of different carbohydrates they produce. Grain sorghum hybrids produce large quantities of grain (approximately 50% of total biomass); the grain is composed primarily of starch (approximately 75%) and may be used as a food grain, feed grain, or ethanol substrate via starch hydrolysis and fermentation. If mechanically harvested, these hybrids are usually less than 6.6 feet (2 meters) tall. Residue is typically returned to the soil, but it is used as forage under drought conditions, and it could be used as a biomass feedstock as well.

Forage sorghums are usually one of two types. Sorghum–sudangrass hybrids are tall, leafy, thin-stalked hybrids used for grazing or hay production. Silage–sorghum hybrids are typically taller and thicker stalked with high grain yields (25% of total biomass). They are chopped and ensiled (i.e., stored in a silo for preservation) for animal feeding. Both of these types of sorghum have been highly selected for optimum forage production, palatability, and conversion in an animal system. There have been significant breeding efforts to enhance the forage quality of this material by incorporating the brown-midrib trait into many forage hybrids. These brown-midrib hybrids have lower lignin and are more palatable, which increases conversion efficiency and consumption rates in ruminant feeding programs (Aydin et al., 1999; Oliver et al., 2004). These types may have application in certain energy sorghum applications where lower lignin content is desirable.

Sweet sorghum is a unique type of sorghum that accumulates high concentrations of soluble sugars. Traditionally, these sorghums were grown for the stalk, which was milled to extract the juice. The juice was then cooked down, and the resulting syrup was used as sweetener. Although these types of sorghum continue to be grown for syrup on an artisan level, there has been significant interest in the development of sweet sorghum as a dedicated bioenergy crop using a sugarcane system model. In the mid-1970s, significant research was conducted to explore the development of sweet sorghum as a bioenergy source for biofuels and energy production, and breeding programs were initiated to develop high-yielding sorghum specifically for ethanol production (McBee et al., 1987).

Dedicated biomass sorghums are the most recent class of sorghum that has been developed in response to interest in bioenergy crops. These sorghums are highly photoperiod sensitive, meaning that they do not initiate reproductive growth until well into the fall season of the year; consequently, in temperature environments such as those found in most of the United States these sorghums will not mature. This absence of reproductive growth reduces sensitivity to periods of drought and allows the crop to effectively photosynthesize throughout the entire growing season. This results in higher yields of primarily lignocellulosic biomass completed in a single annual season. Although phenotypically similar to forage sorghums, these biomass sorghums are distinctly different in that they are not selected for animal palatability, resulting in plants with large culms (stems) and flexible harvest schedules, which minimize nitrogen extraction at the end of the season.

With regard to production and agronomics, the biomass yield potential of sorghum is strongly influenced by both genetic and environmental factors. For example, grain sorghum is commonly grown in more arid regions of the country, and the plant itself is genetically designed to be shorter to facilitate mechanical harvesting. Alternatively, specific dedicated biomass sorghums are very efficient at producing large amounts of lignocellulosic biomass. Finally, both sweet sorghum and forage sorghum are prolific when environmental conditions allow the plants to reach full genetic potential. Hallam et al. (2001) compared perennial grasses with annual row crops and found that sweet sorghum had the highest yield potential, averaging over 17 tons per acre (dry-weight basis) and also performing well when intercropped with alfalfa. Rooney et al. (2007) reported biomass yields of energy sorghum in excess of 44.6 tons per acre (fresh weight) and 13.4 tons per acre (dry weight). They reported that potential improvements could extend the potential of these types of hybrids to a wide range of environments. Under irrigation in the Texas panhandle, McCollum et at. (2005) reported yields of commercial photoperiod-sensitive sorghum hybrids as high as 36 tons per acre (65% moisture) from a single harvest. In subtropical and tropical conditions, single cut yields are generally lower, which is likely due to increased night temperatures, but cumulative yields are higher due to the ratoon potential of the crop. Total biomass yields as high as 13.4 tons per acre (dry-weight basis) were reported near College Station, Texas (Blumenthal et al., 2007).

The composition of sorghum is highly dependent on the type that is produced, such as grain sorghum, sweet sorghum, forage, or cellulosic (high biomass) sorghum. Sorghum grain is high in starch, with lower levels of protein, fat, and ash (Rooney, 2004). Significant variation in the composition of grain is controlled by both genetic and environmental components, making consistency in composition a function of the environment at the time of production; consequently, these factors influence ethanol yield (Wu et al., 2007). Juice extracted from sweet sorghum is predominantly sucrose with variable levels of glucose and fructose; in some genotypes, small amounts of starch are detectable (Billa et al., 1997; Clark, 1981). In forage and dedicated biomass sorghums, the predominant compounds that are produced are structural carbohydrates (lignin, cellulose, and hemicellulose) (McBee et al., 1987; Monk et al., 1984). Amaducci et al. (2004) reported that the environment influences sucrose, cellulose, and hemicellulose concentrations, while lignin content remains relatively constant.

With regard to potential yield and production costs, sorghum has a long history as a grain and forage crop, and production costs range from $200 to $320 per acre (USDA ERS, 2010). This history provides an excellent basis for estimating crop production costs for energy sorghums with only a few modifications. Seed costs, planting costs, and production costs will be similar to grain and/or forage sorghum. Fertilizer rates will likely be less than forage sorghum on a production dry-ton basis (due to the reduced nitrogen content in the mature culm), but it is expected that yields will be higher, so total nitrogen requirements will be equalized. Production is expected under rain-fed conditions; therefore, no additional costs are added for irrigation. On a dry ton basis, given an average production of 10 dry tons per acre and assuming $400 per acre cost of production for dedicated biomass sorghums, biomass sorghum will cost $40 per dry ton at the farmgate.

Production practices for dedicated biomass and sweet sorghum are similar to traditional sorghum crops with some minor modifications. For both types of energy sorghums, it is expected that plant populations will be lowered relative to grain and certainly compared to forage sorghum. This drop will allow the plants to produce larger culms and reduce the potential for lodging and interplant competition. Pests and diseases of sorghum are well known and described, and there are some that will require management plans and effective deployment of host-plant resistance for control. Of particular note is the disease anthracnose (caused by *Colletotrichum graminicola*), which is prevalent in the southeastern United States and is capable of killing susceptible sorghum genotypes. Fortunately, there are many sources of genetic resistance to the disease, and effective control relies on effective integration of these anthracnose resistance genes.

Harvesting and preprocessing of energy sorghums represent an area of significant research, and both processes will likely require the greatest amount of modification compared to grain sorghum. Sweet sorghum will be harvested and moisture extracted for soluble sugars at a centralized location. For dedicated biomass sorghums, the forage harvest systems work very well, but there is a need to reduce moisture content to minimize transportation and storage costs.

The range of sorghum production varies with the type being produced. Both sweet sorghum and dedicated biomass sorghums grow well throughout the eastern and central United States as far north as 40° latitude, and the range of dedicated biomass sorghums is considered to be composed of most of the eastern and central United States. In the western United States, productivity will be directly related to available moisture from rainfall or irrigation. It is unlikely that the crop (or any crop) will be economically viable as a biomass crop in regions with less than 20 inches of available moisture annually. Dedicated biomass sorghums have shown yields of 7 to 11 dry tons per acre in the northern areas of the United States, with even higher yield potential possible in a southern environment due to the longer growing seasons. Therefore, dedicated biomass sorghum should find wide adaptation throughout most of the country that is suitable for herbaceous biomass production from an annual crop.

Although sweet sorghum is productive at northern latitudes, the logistics of processing make the production of the crop unlikely in more temperate latitudes. Because soluble sugars are not stable for long periods, a processor requires a long harvest and processing window for effective use of capital equipment. The farther north the production, the shorter the growing season; hence, the harvest season is further reduced and the ability to consistently grow the high-yield potential sweet sorghum varieties is limited (Wortmann et al., 2010). Consequently, the areas of the United States that process sugarcane are also ideal locations for the production of sweet sorghum. Production in other regions will be dependent on detailed economic analyses of the cost of processing vs. the length of the processing season.

It is interesting to note that most of the bioenergy crops are perennial, but sorghum is unique because it is an annual crop, even though the general opinion is that bioenergy crops should be perennial for sustainability purposes. Although most of the bioenergy crops are perennial, there are several reasons why annual bioenergy crops are necessary. First, annual crops deliver large yields in the first year, as compared to most perennial crops, which typically increase annual yield in subsequent years following establishment. Given the challenges of propagating and establishing perennial crops, annual crops can provide insurance and production stability to industrial processors in the early phases of bringing a new processing facility online if perennial crop stand failures or establishment problems are encountered. Second, for the most part, the U.S. farming system has been based on annual crop production systems. Farmers, bankers, and processors are much more familiar and accepting of these systems, and while this will eventually be overcome, annual energy crops will be needed for that transition. Finally, annual crops are much more tractable to genetic improvements through breeding due to the simple fact that breeding is accelerated by multiple generations per year.

There are several traits of specific importance to sorghum improvement relative to bioenergy production. These include, but are not limited to, maturity and height, drought tolerance, pest tolerance and/or resistance, and composition and/or quality. Improvements in these areas will increase yield potential, protect existing yield potential, and enhance conversion efficiency during processing.

The reason for producing bioenergy feedstock is to produce renewable fuel, but one of the critical components in their production will be water. Thus, both drought tolerance and water-use efficiency are critical, as many of these feedstocks will be produced in marginal environments where rainfall is limited and irrigation is either too expensive or would deplete water reserves. Sorghum is more drought tolerant than many other biomass crops. Depending on the type of biomass production in sorghum, both pre- and post-flowering drought tolerance mechanisms will be important. In sweet sorghum, both traits are important, but there has been little research regarding the impact of drought stress on sweet sorghum productivity.

For high-biomass, photoperiod-sensitive sorghums, preflowering drought tolerance is critical because, in most environments, this germplasm does not transition to the reproductive phase of growth. Each type of tolerance is associated with several phenotypic and physiological traits; these relationships have been used to fine map quantitative trait loci (QTL) associated with both pre-and post-flowering drought tolerance. Traits that have been associated with drought resistance include heat tolerance, osmotic adjustment (Basnayake et al.,

1995), transpiration efficiency (Muchow et al., 1996), rooting depth and patterns (Jordan and Miller, 1980), epicuticular wax (Maiti et al., 1984), and staying green (Rosenow et al., 1983). Combining phenotypic and marker-assisted breeding approaches should enhance drought tolerance breeding in energy sorghums.

Unlike perennial bioenergy crops, sorghum will require crop rotation to maintain high yields and soil conditioning. Continuous cropping studies of sorghum have confirmed that yields will drop in subsequent years unless additional nitrogen is provided to maintain yields (Peterson and Varvel, 1989). Therefore, it is critical to consider rotations when accounting for potential land area needs in energy sorghum production. The exact rotation sequence and time frame will vary with locale, but sorghum production once every 2 or 3 years will be acceptable in most regions. Failure to rotate may result in reduced yields and quality, as well as increased weed, insect, and disease problems.

The bottom line on the efficacy and use of sorghum as a bioenergy feedstock is based on its long-established production history, existing research infrastructure, and a relatively simple genetic system, all of which allow for the rapid modification of the crop and delivery of specific sorghum types, developed specifically for bioenergy production.

Second-Generation Feedstocks: Short-Term Availability

Agricultural residues, a second-generation biomass feedstock, offer a potentially large and readily available biomass resource, but sustainability and conservation constraints could place much of it out of reach. Given current U.S. cropland use, corn and wheat offer the most potentially recoverable residues; however, these residues play an important role in recycling nutrients in the soil and maintaining long-term fertility and land productivity. Removing too much residue could aggravate soil erosion and deplete the soil of essential nutrients and organic matter.

Safe removal rate methodologies, based on soil erosion, have been developed. Methodologies to determine removal rates while safeguarding soil fertility and meeting conservation objectives still need to be developed. Studies have shown that under current tillage practices the national average safe removal rate based on soil erosion for corn stover is less than 30%. Actual rates vary widely depending on local conditions. In other words, much of the generated crop residues may be out of reach for biomass use if soil conservation goals are to be achieved (Graham et al., 2007).

The estimated delivery costs for agricultural residues vary widely depending on crop type, load resource density, storage and handling requirements, and distance and transportation costs. Moreover, existing estimates are largely derived from engineering models, which may not account for economic conditions. Agricultural residue feedstocks (such as corn stover) have a significant advantage in that they can be readily integrated into the expanding corn ethanol industry; however, dedicated energy crops (such as switchgrass) may have more benign environmental impacts.

Another significant biomass source is *forest biomass*, which can be immediately available should the bioenergy market develop. Logging residues are associated with timber industry activities and constitute significant biomass resources in many states, particularly in the Northeast, North Central, Pacific Northwest, and Southeast. In the western states, the predominance of public lands and environmental pressure reduce the supply potential for logging residues, but there is a vast potential for biomass from thinning undertaken to reduce the risk of forest fires. However, the few analyses that have examined recoverability of logging residues cite the need to account for factors such as the scale and location of biorefineries and biopower plants, as well as regional resource density.

The potential for forest residues may be large but the actual quantities available for biomass conversion may be low due to the economics of harvesting, handling, and transporting the residues from forest areas to locations where they could be used. It is not clear how these residues compete with fossil fuels in the biopower and co-firing industries. In addition, there are competing uses for these products in the pulp and paper industry, as well as different bioenergy end uses. Economic studies of logging residues suggest a current lack of competitiveness with fossil fuels (coal, gas), but logging residues could become more cost competitive with further improvements in harvesting and transportation technologies and with policies that require a fuller accounting of the social and environmental benefits from converting forest residues to biopower or biofuels.

Another source of forest residues that could be recovered in significant quantities is biomass from fuel treatments and thinnings. Fuel treatment residues are the byproduct of efforts to reduce the risk of loss from fire, insects, and disease and therefore present substantially different challenges than do logging residues. The overall value of forest health benefits such as clean air and water is generally believed to exceed the cost of treatment. However, treated forests are often distant from end-use markets, resulting in high transportation costs to make use of the harvested material. Road or trail access, steep terrain, and other factors commonly limit thinning operations in western forests.

Transportation costs can be a significant factor in the cost of recovering biomass. As much as half the cost of the material delivered to a manufacturing facility may be attributed to transportation. The offset to the high-cost transportation of forest thinnings is onsite densification of the biomass. This could entail pelletization, fast pyrolysis (to produce bio-oil), or baling. The economics of transporting thinned woody residues vs. onsite densification depend on the distance to end-use markets. Densification may be more economical if power generation facilities are far away. In addition to co-firing or co-generation facilities, improvements in thermochemical conversion efficiency and establishment of small-scale conversion facilities using gasification and/or pyrolysis may favor the use of forest residues for biofuel production (Polagye et al., 2007).

A third major category of immediately available second-generation biomass is wood residues from *secondary mill products* and *urban wood waste*. Urban wood waste provides a relatively cheap feedstock to supplement other biomass resources (Wiltsee, 1998). Urban wood waste encompasses the biomass portion of commercial, industrial, and municipal solid waste (MSW), while secondary mill residues include sawdust, shavings, wood trims, and other byproducts generated from processing lumber, engineered wood products, or wood particles. Both urban wood waste and secondary mill residues have several primary uses and disposal methods. Urban wood waste not used in captive markets (such as the pulpwood industry) could be used as a biomass either to generate electricity or to produce cellulosic ethanol when it becomes commercially viable.

The amount of urban wood wastes produced in the United States is significant and their use as biomass could be economically viable, particularly in large urban centers (Wiltsee, 1998). Several national availability estimates exist for various types of urban wood wastes, but estimates vary depending on methodology, product coverage, and assumptions about alternative uses (McKeever, 2004; Wiltsee, 1998). One of the challenges with regard to the potential availability of urban woody waste is to sort out the portion that is available (not currently used) and determine alternative uses, including those used by captive markets (not likely to be diverted to bioenergy). One assessment of urban wood waste finds that 36% of total biomass generated is currently sold to noncaptive markets, and 50% of the unused residues are not available due to contamination, quality, or recoverability.

One source of recurring and potentially available carbon feedstock is municipal solid waste. The U.S. Environmental Protection Agency (USEPA) estimated that in 2005 245.7 million tons of MSW were generated in the United States, of which 79 million tons were recycled, 33.4 millions tons were diverted to energy recovery, and 133.3 million tons were disposed of in landfills. As such, landfilled material represents a potentially significant source of renewable carbon that could be used for fuel/energy production or in support of biofuel production.

SECOND-GENERATION FEEDSTOCKS: LONG-TERM AVAILABILITY

Large-scale biofuels production, in the long run, will require other resources, including dedicated energy crops. Dedicated feedstocks are perennial grasses and trees grown as crops specifically to provide the required raw materials to bioenergy producers. A steady supply of low-cost, uniform,

DID YOU KNOW?

With regard to establishing switchgrass, no single method can be suggested for all situations (Parrish et al., 2008). No-till and conventional tillage can be used to establish the crop. When seeded as part of a diverse mixture, planting guidelines for warm-season grass mixtures for conservation plantings should be followed. Several key factors can increase the likelihood of success for establishing switchgrass, including the following:

- Plant switchgrass after the soil is well warmed during the spring.
- Use seeds that are highly germinable and plant 0.6 to 1.2 cm deep, or up to 2 cm deep in sandy soils.
- Pack or firm the soil both before and after seeding.
- Provide no fertilization at planting to minimize competition.
- Control weeds with chemical or cultural control methods.

and consistent-quality biomass feedstock will be critical for the economic viability of cellulosic ethanol production. During the late 1980s, the Department of Energy sponsored research on perennial herbaceous (grassy) biomass crops, particularly switchgrass, which is considered a model energy crop because of its many perceived advantages: (1) native to North America, (2) high biomass yield per acre, (3) wide regional coverage, and (4) adaptability to marginal land conditions. An extensive research program on switchgrass in the 1990s generated a wealth of information on high-yielding varieties, regional adaptability, and management practices. Preliminary field trials show that the economic viability of switchgrass cultivation depends critically on the initial establishment success. During this phase, seed dormancy and seedling sensitivity to soil and weed conditions require that recommended practices be closely followed by growers. Viable yields require fertilization rates at about half the average for corn.

Switchgrass (*Panicum virgatum*) is a hardy, deep-rooted, perennial, rhizomatous, warm-season grass native to North America. It is believed to be most suitable for cultivation in marginal lands, low-moisture lands, and lands with lower opportunity costs such as pastures, including lands under the Conservation Reserve Program (CRP) where the federal government pays landowners annual rent for keeping land out of production (McLaughlin and Kzos, 2005). Additionally, a large amount of highly erodible land in the Corn Belt is unsuitable for straw or stover removal but is potentially viable for dedicated energy crops such as switchgrass. Factors favoring adoption of switchgrass include selection of suitable lands, environmental benefits (carbon balances, improved soil nutrients and quality), and the use of existing hay production techniques to grow the crop. Where switchgrass is grown on CRP lands, payments help to offset production costs. Factors discouraging switchgrass adoption include no possibility for crop rotation; farmers' risk aversion for producing a new crop because of lack of information, skills, and know-how; potential conflict with on-farm and off-farm scheduling activities; and a lack of compatibility with long-term land tenure. Overall, production budget and delivery cost assessments suggest that switchgrass is a high-cost crop (undercurrent technology and price conditions) and may not compete with established crops, except in areas with low opportunity costs (e.g., pasture land, marginal lands).

DID YOU KNOW?

The main advantage of using switchgrass over corn as an ethanol feedstock is that its cost of production is generally about half that of grain corn and more biomass energy per hectare can be captured in the field (Samson et al., 2008).

Substantial variability is apparent in the economics of switchgrass production and assessments of production budgets and delivered costs. Factors at play include methods for storage and handling, transport distances, yields, and types of land used (cropland vs. grassland). When delivered costs of switchgrass are translated into break-even prices (compared with conventional crops), it becomes apparent that cellulosic ethanol or biopower plants would have to offer relatively high prices for switchgrass to induce farmers to grow it (Rinehart, 2006). However, the economics of switchgrass could improve if growers benefited from CRP payments and other payments tied to environmental services (such as carbon credits). In the long run, the viability of an energy crop such as switchgrass hinges on continued reductions in cellulosic ethanol conversion costs and sustained improvements in yield and productivity through breeding, biotechnology, and agronomic research.

Although it is an important biofuels crop, switchgrass does have limitations. Switchgrass is not optimally grown everywhere. For example, in the upper Midwest under wet soils, reed canarygrass (*Phalaris arundinacea*) is more suitable, while semitropical grass species are better adapted to the Gulf Coast region. State and local efforts are testing alternatives to switchgrass, including reed canarygrass, a tall, perennial grass that forms extensive single-species stands along the margins of lakes and streams; *Miscanthus* (often confused with "elephant grass"), which is a rapid-growth, low-mineral-content, high-biomass-yield plant; and other species.

Miscanthus x giganteus Greef et Deuter (hereafter referred to as Giant Miscanthus; Greef and Deuter, 1993) feedstock has become a mainstay of increased biomass production in the United States. Giant Miscanthus is a sterile triploid hybrid resulting from the cross of the diploid *M. sinensis* and tetraploid *M. sacchariflorus* (Scally et al., 2001). Originally discovered in Japan, Giant Miscanthus was thereafter introduced into the Unites States as landscape plant (Scally et al., 2001). Giant Miscanthus exhibits many characteristics that allow it to meet or exceed the criteria for desirable biomass crops. As a perennial, it typically requires fewer yearly agronomic inputs than annual row crops. After establishment, time spent in the field is usually limited to a single annual harvest. In some years in some locations, additional field time may be spent applying fertilizer, but applications have not been shown to be required every year nor in every location. For example, Christian et al. (2008) reported no yield response to nitrogen applications to a 14-year-old stand in England, whereas Ercoli et al. (1999) did see a nitrogen response when nitrogen was applied to Giant Miscanthus in Italy. Thus, Giant Miscanthus crops have not required annual planting, pest controls, or fertilization in ongoing studies. Its perennial growth also controls soil erosion. As it becomes established and grows, Giant Miscanthus develops an extensive layer of rhizomes and mass of fibrous roots that can hold soil in place. Finally, the belowground growth can contribute soil organic carbon levels as shown in Germany (Schneckenberger and Kuzyakov, 2007) and Denmark (Foereid et al., 2004).

With regard to long-run sustainability, the ecology of perennial grassy crops favors a multiplicity of crops or even a mix of species within the same area. Both ecological and economic sustainability favor the development of a range of herbaceous species for optimal use of local soil and climatic conditions. A mix of several energy crops in the same region would help reduce the risk of epidemic pests and disease outbreak and optimize the supply of biomass to an ethanol or biopower plant because different grasses mature and can be harvested at different times. Moreover, development of future energy crops must be evaluated from the standpoint of their water use efficiency, impact on soil nutrient cycling, effect on crop rotations, and environmental benefit (improved energy use efficiency and reduced greenhouse gas emissions, nutrient runoff, pesticide runoff, and land-use impacts). In the long run, developing a broad range of grassy crops for energy use is compatible with both sustainability and economic viability criteria.

Short-rotation woody crops (SRWCs) represent another important category of future dedicated energy crops. Among the SRWCs, hybrid poplar, willow, American sycamore, sweetgum, and loblolly have been extensively researched for their very high biomass yield potential. Breeding programs and management practices continue to be developed for these species. SRWCs are based on a high-density plantation system and more frequent harvesting (every 3 to 4 years for willow and 7 years for hybrid poplar). Following is a summary of short-rotation crops:

- *Hybrid poplar*—This species is very site specific and has a limited growing niche. It requires an abundant and continuous supply of moisture during the growing season. Soils should be moist but not continually saturated and should have good internal drainage. It prefers damp, well-drained, fine sandy-loam soils located near streams, where coarse sand is first deposited as flooding occurs. Hybrid poplars are among the fastest-growing trees in North America—in just six growing seasons, hybrid poplars can reach 60 feet or more in height. They are well suited for the production of bioenergy, fiber, and other biobased products.
- *Willow species (Salix)*—In folklore and myth, a willow tree is believed to be quite sinister, capable of uprooting itself and stalking travelers. The reality is that willows are used for biofiltration, constructed wetlands, ecological wastewater treatment systems, hedges, land reclamation, landscaping, phytoremediation, stream bank stabilization (bioengineering), slope stabilization, soil erosion control, shelterbelt and windbreak, soil building, soil reclamation, tree bog composting toilets, and wildlife habitat. Willow is grown for biomass or biofuels in energy forestry systems as a consequence of its high energy-in/energy-out ratio, large carbon mitigation potential, and fast growth (Aylott, 2008). Willow is also grown and harvested for making charcoal.
- *American sycamore*—Sycamores prefer alluvial soils along streams in bottomlands. Sycamore growth and yield are less than those of poplars and willows.
- *Sweetgum (Liquidambar styraciflua)*—Sweetgum is a species tolerant of a variety of soils, but it grows best on rich, moist, alluvial clay and loam soils of river bottoms.
- *Loblolly pine (Pinus taeda)*—Loblolly pine is quite adaptable to a variety of sites. It performs well on both poorly drained bottomland flats and modestly arid uplands. Biomass for energy is currently being obtained from precommercial thinnings and from logging residues in loblolly pine stands. Utilization will undoubtedly increase, and loblolly pine energy plantations may become a reality.

In many parts of the country, plantations of poplar, willow, pines, and cottonwood have been established and are being commercially harvested. Willows are being planted in New York, particularly following enactment of state renewable portfolio standard (RPSs) and other incentives. Over 30,000 hectares of poplars are being grown in Minnesota, and several thousand hectares are also grown as part of a DOE-funded project to provide biomass for a power utility company in southern Minnesota. The Pacific Northwest has large plantations of hybrid poplars, estimated at 60,000 hectares as of 2007. Most of these plantations are currently used for pulp wood, with little volume being used for bioenergy. Because SRWCs can be used either for biomass or as feedstock for pulp and other products, pulp demand will influence the cost of using it for bioenergy production.

An important consideration for energy crops (e.g., switchgrass, poplar, willow) is the potential for increasing yields and developing other desirable characteristics. Most energy crops are unimproved or have been bred only recently for biomass yield, whereas corn and other commercial food crops have undergone substantial improvements in yield, disease resistance, and other agronomic traits. A more complex understanding of biological systems and application of the latest biotechnological advances would accelerate the development of new biomass crops with desirable attributes. These attributes include increased yields and processability, optimal growth in specific microclimates, better pest resistance, efficient nutrient use, and greater tolerance to moisture deficits and other sources of stress. Agronomic and breeding improvements of these new crops could provide a significant boost to future energy crop production.

BIOETHANOL PRODUCTION BY DRY CORN MILL PROCESS

Ethanol production from dry corn milling follows a seven-step process as shown in Figure 7.5. First, corn feedstock is cleaned and milled (ground into corn meal), and the milled corn is then converted to a slurry. The slurry is liquefied (slurried with water to form a mash), and the enzymes

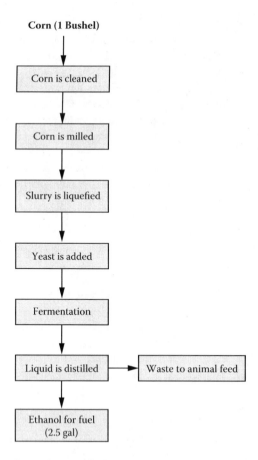

FIGURE 7.5 Flow diagram for production of ethanol from corn.

yield glucose. Yeast is then added to the mash to convert starch into a simple sugar (dextrose) in a process known as *saccharification*. After liquefaction, the mash is cooked in a saccharification tank to reduce bacterial levels and it then moves through several fermenters where fermentation takes place (sugar is converted to ethanol by yeast). The resulting "beer" containing 2 to 12% ethanol is then distilled into ethanol at 95% alcohol and 5% water. The remaining solids (stillage) are collected during distillation, dried, and sold as an animal feed known as dried distillers' grain (DDG). The final removal of water from the ethanol to less than 1% (dehydration) allows it to be blended with gasoline.

DID YOU KNOW?

Fermentation is a series of chemical reactions that convert sugars to ethanol. The fermentation reaction is caused by yeast or bacteria, which feed on the sugars. Ethanol and carbon dioxide are produced as the sugar is consumed. The simplified fermentation reaction equation for the 6-carbon sugar glucose is

$$C_6H_{12}O_6 \rightarrow 2\ CH_3CH_2OH + 2\ CO_2$$

Glucose   Ethanol   Carbon
                                          dioxide

TABLE 7.5
Gasoline Gallon Equivalent (GGE)

Fuel	GGE	Btu
Gasoline (base)	1.0000 gal	114,000 Btu/gal
Gasoline (conventional, summer)	0.9960 gal	114,500 Btu/gal
Gasoline (conventional, winter)	1.0130 gal	112,500 Btu/gal
Gasoline (reformulated gasoline, ethanol)	1.0190 gal	111,836 Btu/gal
Gasoline (reformulated gasoline, ETBE)	1.0190 gal	111,811 Btu/gal
Gasoline (reformulated gasoline, MTBE)	1.0200 gal	111,745 Btu/gal
Gasoline (10% MBTE)	1.0200 gal	112,000 Btu/gal
Gasoline (regular unleaded)	1.0000 gal	114,100 Btu/gal
Diesel #2	0.8800 gal	129,500 Btu/gal
Biodiesel (B100)	0.9600 gal	118,300 Btu/gal
Biodiesel (B20)	0.9000 gal	127,250 Btu/gal
Liquid natural gas (LNG)	1.5362 gal	75,000 Btu/gal
Compressed natural gas	1.26 ft^3 (3.587 m^3)	900 Btu/ft^3
Hydrogen at 101.325 kPa	357.37 ft^3	319 Btu/ft^3
Hydrogen by weight	0.997 kg (2.198 lb)	119.9 MJ/kg (51,500 Btu/lb)
Liquefied petroleum gas (LPG)	1.3500 gal	84,300 Btu/gal
Methanol fuel (M100)	2.0100 gal	56,800 Btu/gal
Ethanol fuel (E100)	1.5000 gal	76,100 Btu/gal
Ethanol fuel (E85)	1.3900 gal	81,800 Btu/gal
Jet fuel (naphtha)	0.9700 gal	118,700 Btu/gal

Source: USEPA, *Fuel Economy Impact Analysis of RFG*, U.S. Environmental Protection Agency, Washington, DC, 2007.

Note: ETBE, ethyl *tert*-butyl ether; MTBE, methyl *tert*-butyl ether.

GASOLINE GALLON EQUIVALENT

Before beginning our discussion about alternative renewable fuels used for transportation purposes, it is important that the reader has a fundamental understanding of the difference between conventional gasoline and diesel fuel energy output as compared to non-conventional renewable products. Typically, this comparison is made utilizing a standard engineering parameter known as the *gasoline gallon equivalent* (GGE), which is the ratio of the number of British thermal units (Btu) available in 1 U.S. gallon (1 gal) of gasoline to the number of British thermal units available in 1 gal of the alternative substance in question. NIST (2012) defined a gasoline gallon equivalent as 5660 pounds of natural gas. The GGE parameter allows consumers to compare the energy content of competing fuels against a commonly known fuel—gasoline. Table 7.5 provides GGE and Btu/unit value comparisons for various fuels.

PROS AND CONS OF BIOETHANOL

With regard to the use of ethanol for transportation (propulsion), it can be used as an extender and octane enhancer in conventional gasoline. Its primary advantage is that its feedstock (plants) is renewable. It can also be used as a primary fuel (in E85), thus reducing dependence on petroleum products. Ethanol is an important player in the ongoing effort to reduce environmental pollution from carbon monoxide (CO) gas and global carbon dioxide (CO_2). In some locations, E85 and gasohol are less expensive per gallon than conventional gasoline. As the production and use of ethanol as a fuel increase, farmers will benefit from the increased demand for their products. From the practical and safety and health point of view, ethanol can be used to prevent gasline freeze in extremely cold weather, and because it is not as flammable as gasoline it is less likely to cause accidental explosions.

Ethanol use for transportation purposes does have a few negatives. Probably the most obvious shortcoming is its lack of availability at local gas stations. Gas stations selling E85 as part of their normal products are not as abundant as conventional gas stations. In some locations, E85 and gasohol are more expensive per gallon than conventional gasoline. E85 can damage vehicles that are not designed to burn it as a fuel. Also, ethanol contains less energy than gasoline. This means that a car will not go as far on a gallon of E85, and fuel economy will decrease by 20 to 30%. Ethanol is produced from plant matter that could otherwise be consumed as human and animal food products. Critics argue that food products used for fuel instead of to feed people and animals will contribute indirectly to mass world hunger. Moreover, it will lead to higher food prices. However, even though higher corn prices increase animal feed and ingredient costs for farmers and food manufacturers, these costs are passed through to retail at a rate less than 10% of the change in corn price.

SIDEBAR 7.2. CORN: ETHANOL PRODUCTION OR FOOD SUPPLY?[*]

In 2007, record U.S. trade driven by economic growth in developing countries and favorable exchange rates, combined with tight global grain supplies, resulted in record or near-record prices for corn, soybeans, and other food and feed grains. For corn, these factors, along with increased demand for ethanol, helped push prices from under $2 per bushel in 2005 to $3.40 per bushel in 2007. By the end of the 2006–2007 crop year, over 2 billion bushels of corn (19% of the harvested crop) were used to produce ethanol, a 30% increase from the previous year. Higher corn prices motivated farmers to increase corn acreage at the expense of other crops, such as soybeans and cotton, raising their prices as well.

The pressing and pertinent question becomes: What effect do these higher commodity costs have on retail food prices? In general, retail food prices are much less volatile than farm-level prices and tend to rise by a fraction of the change in farm prices. The magnitude of response depends on both the retailing costs beyond the raw food ingredients and the nature of competition in retail food markets. The impact of ethanol on retail food prices depends on how long the increased demand for corn drives up farm corn prices and the extent to which higher corn prices are passed through to retail.

Retail food prices adjust as the cost of inputs into retail food production change and the competitive environment in a given market evolves. Strong competition among three to five retail store chains in most U.S. markets has had a moderating effect on food price inflation. Overall, retail food prices have been relatively stable over the past 20 years, with prices increasing an average of 3.0% per year from 1987 through 2007, just below the overall rate of inflation. Since then, food price inflation has averaged just 2.5% per year.

Field corn is the predominant corn type grown in the United States, and it is primarily used for animal feed. Currently, less than 10% of the U.S. field corn crop is used for direct domestic human consumption in corn-based foods such as corn meal, corn starch, and corn flakes; the remainder is used for animal feed, exports, ethanol production, seed, and industrial uses. Sweet corn, both white and yellow, is usually consumed as immature whole-kernel corn by humans and also as an ingredient in other corn-based foods, but it makes up only about 1% of the total U.S. corn production.

---

[*] Adapted from Leibtag, E., Corn prices near record high, but what about food costs? *Amber Waves*, February, 2008 (http://webarchives.cdlib.org/sw1vh5dg3r/http://ers.usda.gov/AmberWaves/February08/Features/CornPrices.htm).

Because U.S. ethanol production uses field corn, the most direct impact of increased ethanol production should be on field corn prices and on the price of food products based on field corn. However, even for those products heavily based on field corn, the effect of rising corn prices is dampened by other market factors. For example, an 18-ounce box of corn flakes contains about 12.9 ounces of milled field corn. When field corn is priced at $2.28 per bushel (the 20-year average), the actual value of corn represented in the box of corn flakes is about 3.3 cents (1 bushel = 56 pounds). (The remainder of the price represents packaging, processing, advertising, transportation, and other costs.) At $3.40 per bushel, the average price in 2007, the value is about 4.9 cents. The 49% increase in corn prices would be expected to raise the price of a box of corn flakes by about 1.6 cents, or 0.5%, assuming no other cost increases.

In the 1980s, Coca-Cola made the shift from sugar to corn syrup in most of its U.S.-produced soda, and many other beverage makers followed suit. Currently, about 4.1% of U.S.-produced corn is made into high-fructose corn syrup. A 2-liter bottle of soda contains about 15 ounces of corn in the form of high-fructose corn syrup. At $3.40 per bushel, the actual value of corn represented is 5.7 cents, compared with 3.8 cents when corn is priced at $2.28 per bushel. Assuming no other cost increases, the higher corn price in 2007 would be expected to raise soda prices by 1.9 cents per 2-liter bottle, or 1%. These are notable changes in terms of price measure and inflation, but relatively minor changes in the average household budget.

In addition to the impact on corn flakes and soda beverage prices, livestock prices are also impacted. This stands to reason when you consider that livestock rations traditionally contain a large amount of corn; thus, a bigger impact would be expected in meat and poultry prices due to higher feed costs than in other food products. Currently, 55% of corn produced in the United States is used as animal feed for livestock and poultry. However, estimating the actual corn used as feed to produce retail meat is a complicated calculation. Livestock producers have many options when deciding how much corn to include in a feed ration. For example, at one extreme, grass-fed cattle consume no corn, while other cattle may have a diet consisting primarily of corn. For hog and poultry producers, ration variations may be less extreme but can still vary quite a bit. To estimate the impact of higher corn prices on retail meat prices, it is necessary to make a series of assumptions about feeding practices and grain conversion rates from animal to final retail meat products. To avoid downplaying potential impacts, this analysis uses upper-bound conversion estimates of 7 pounds of corn to produce 1 pound of beef, 6.5 pounds of corn to produce 1 pound of pork, and 2.6 pounds of corn to produce 1 pound of chicken.

Using these ratios and data from the Bureau of Labor Statistics, a simple pass-through model provides estimates of the expected increase in meat prices given the higher corn prices. The logic of this model is illustrated by an example using chicken prices. Over the past 20 years, the average price of a bushel of corn in the United States has been $2.28, implying that a pound of chicken at the retail level uses 8 cents worth of corn, or about 4% of the $2.05 average retail price for chicken breasts. Using the average price of corn for 2007 ($3.40 per bushel) and assuming producers do not change their animal-feeding practices, retail chicken prices would rise 5.2 cents, or 2.5%. Using the same corn data, retail beef prices would go up 14 cents per pound, or 8.7%, while pork prices would rise 13 cents per pound, or 4.1%.

These estimates for meat, poultry, and corn-related foods, however, assume that the magnitude of the corn price change does not affect the rate at which cost increases are passed through to retail prices. It could be the case that corn price fluctuations have little impact on retail food prices until corn prices rise high enough for a long enough time to elicit a large price adjustment by food producers and notably higher retail food prices.

DID YOU KNOW?

Although higher commodity costs may have a relatively modest impact on U.S. retail food prices, there may be greater effect on retail food prices in lower-income developing countries. As a relatively low-priced food, grains have historically accounted for a large share of the diet in less developed countries. Even with incomes rising, consumers in such countries consume a less processed diet than is typical in the United States and other industrialized countries, so food prices are more closely tied to swings in both domestic and global commodity prices.

On the other hand, these estimates may be overstating the effect of corn price increases on retail food prices because they do not account for producers potentially switching from more expensive to less costly inputs. Such substitution would dampen the effect of higher corn prices on retail meat prices. Even assuming the upper-bound effects outlined above, the impact of rising corn commodity prices on overall food prices is limited. Given that less than a third of retail food contains corn as a major ingredient, these rising prices for corn-related products would raise overall U.S. retail food prices less than 1 percentage point per year above the normal rate of inflation.

Continuing elevated prices for corn will depend on the extent to which corn remains the most efficient feedstock for ethanol production and ethanol remains a viable source of alternative energy. Both of these conditions may change over time as other crops and biomass are used to produce ethanol and other alternative energy sources develop.

Even if these conditions do not change in the near term, market adjustments may dampen long-run impacts. In 1996, when field corn prices reached an all-time high of $3.55 per bushel due to drought-related tighter supplies in the United States and strong demand for corn from China and other parts of Asia, the effect on food prices was short lived. At that time, retail prices rose for some foods, including pork and poultry, but these effects did not extend beyond the middle of 1997. For the most part, food markets adjusted to the higher corn prices and corn producers increased supply, bringing down price.

Food producers, manufacturers, and retailers may also adjust to the changing market conditions by adopting more efficient production methods and improved technologies to counter higher costs. For example, soft drink manufacturers may consider substituting sugar for corn syrup as a sweetener if corn prices remain high, while livestock and poultry producers may develop alternative feed rations that minimize the corn needed for animal feed. Adjustments by producers, manufacturers, and retailers, along with continued strong retail competition, imply that U.S. retail food prices will remain relatively stable.

Brazilian Example[*]

Brazil has the world's second largest ethanol program and is capitalizing on plentiful soybean supplies to expand into biodiesel. More than half of the nation's sugarcane crop is processed into ethanol, which now accounts for about 20% of the country's fuel supply. Initiated in the 1970s after the OPEC oil embargo, Brazil's policy program was designed to promote the nation's energy independence and to create an alternative and value-added market for sugar producers. The government has spent billions to support sugarcane producers, develop distilleries, build up a distribution infrastructure, and promote production of pure-ethanol-burning and, later, flex-fuel vehicles able to run on gasoline, ethanol–gasoline blends, or pure

---

[*] Adapted from Coyle, W., The future of biofuels: a global perspective. *Amber Waves*, November, 2007 (http://www.thebioenergysite.com/articles/9/the-future-of-biofuels-a-global-perspective).

hydrous (wet; 7 to 4% water content) ethanol. Advocates contend that, while the costs were high, the program saved far more in foreign exchange from reduced petroleum imports. In the mid- to late 1990s, Brazil eliminated direct subsidies and price setting for ethanol. It pursued a less intrusive approach with two main elements—a blending requirement (now about 25%) and tax incentives favoring ethanol use and the purchase of ethanol-using or flex-fuel vehicles. Today, more than 80% of Brazil's newly produced automobiles have flexible fuel capability, up from 30% in 2004. With ethanol widely available at almost all of Brazil's 32,000 gas stations, Brazilian consumers currently choose primarily between 100% hydrous ethanol and a 2% ethanol–gasoline blend on the basis of relative prices.

Approximately 20% of current fuel use (alcohol, gasoline, and diesel) in Brazil is ethanol, but it may be difficult to raise the share as Brazil's fuel demand grows. Brazil is a middle-income economy with per capita energy consumption only 15% that of the United States and Canada. Current ethanol production levels in Brazil are not much higher than they were in the later 1990s. Production of domestic off- and on-shore petroleum resources has grown more rapidly than ethanol and accounts for a larger share of expanding fuel use than does ethanol in the last decade.

BIOMASS FOR BIOPOWER

In addition to advanced fuels, biomass can also be used for the production of biopower. This can be done in several ways, including direct combustion of biomass in dedicated power plants, co-firing biomass with coal, biomass gasification in a combined-cycle plant to produce steam and electricity, or via anaerobic digestion (EPRI, 1997). Combustion is the burning of biomass in air. This involves the conversion of chemical energy stored in biomass into heat, mechanical power, or electricity (McKendry, 2002). Although it is possible to use all types of biomass, combustion is preferable when the biomass is more than 50% dry. High-moisture biomass is better suited for biological conversion processes. Net bioenergy conversion efficiencies for biomass combustion power plants range from 20 to 40%. Higher efficiencies are obtained with combined heat and power (CHP) facilities and with large size power-only systems (over 100 megawatt-electrical, MWe), or when the biomass is co-fired with coal in power plants (McKendry, 2002).

Co-firing biomass with coal is a straightforward and inexpensive way to diversify the fuel supply, reduce coal plant air emissions (NO_x, SO_2, CO_2), divert biomass from landfills, and stimulate the biomass power industry (Hughes, 2000). Moreover, biomass is the only renewable energy technology that can directly displace coal. Given the dominance of coal-based power plants in U.S. electricity production, co-firing with biomass fuel is the most economical way to reduce greenhouse gas emissions. Possible biomass fuel for co-firing includes wood waste, short-rotation woody crops, switchgrass, alfalfa stems, various types of manure, landfill gas, and wastewater treatment gas (Tillman, 2000). In addition, agricultural residues such as straw can also be used for co-firing.

A promising technology development currently at the demonstration stage is biomass integrated gasification/combined cycle (BIG/CC), where a gas turbine converts the gaseous fuel to electricity with a high conversion efficiency, reaching 40 to 50% of the heating value of the incoming gas (McKendry, 2002). An important advantage of gasification is the ability to work with a wider variety of feedstocks, such as high-alkali fuels that are problematic with direct combustion. High-alkali fuels such as switchgrass, straw, and other agricultural residues often cause corrosion, but the gasification systems can easily remove the alkali species from the fuel gas before it is combusted. High silica, also a problem with grasses, can cause slagging in the reactor. The slagging problem is not unique to one form of biomass but instead is common among many different types of biomass fuels (Miles et al., 1993). Slagging deposits can reduce heat transfer, reduce combustion efficiency, and damage combustion chambers when large particles break off. Research has focused on two alkali metals, potassium and sodium, and on silica, all elements commonly found in living plants. In general, it appears that faster growing plants (or faster growing plant components such as seeds) tend to have higher concentrations of alkali metal and silica. Thus, materials such as straw, nut hulls, fruit pits, weeds, and grasses tend to create more problems when burned than does wood from a slow-growing tree.

Potassium and sodium metals, whether in the form of oxides, hydroxides, or metallo-organic compounds, tend to lower the melting point of ash mixtures containing various other minerals such as silica (SiO_2). The high alkali content (up to 35%) in the ash from burning annual crop residues lowers the fusion or "sticky temperature" of these ashes from 2200°F for wood ash to as low as 1300°F. This results in serious slagging on the boiler grate or in the bed and fouling of convection heat transfer surfaces. Even small percentages (10%) of some of these high-alkali residues burned with wood in conventional boilers will cause serious slagging and fouling in a day to two, necessitating combustion system shutdown. A method to predict slagging and fouling from combustion of biomass fuels has been adapted from the coal industry. The method involves calculating the weight in pounds of alkali ($K_2O + Na_2O$) per million Btu in the fuel as follows:

$$\frac{1 \times 10^6}{\text{Btu/lb}} \times \%\text{Ash} \times \%\text{Alkali of the ash} = \frac{\text{lb Alkali}}{\text{MMBtu}}$$

This method combines all the pertinent data into one index number. A value below 0.4 lb/MMBtu (MMBtu = thousand thousand Btu) is considered a fairly low slagging risk. Values between 0.4 and 0.8 lb/MMBtu will probably slag with increasing certainty of slagging as 0.8 lb/MMBtu is approached. Above 0.8 lb/MMBtu, the fuel is virtually certain to slag and foul (see Table 7.6).

TABLE 7.6
Alkali Content and Slagging Potential of Biofuels

Fuel	Total Alkali (lb/MMBtu)	Slagging Potential
Wood		
Pine chips	0.07	Minimal
White oak	0.14	·Minimal
Hybrid poplar	0.46	Probable
Urban wood waste	0.46	Probable
Clean tree trimmings	0.73	Probable
Pits, nuts, shells		
Almond shells	0.97	Certain
Refuse-derived fuel	1.60	Certain
Grasses		
Switch grass	1.97	Certain
Wheat straw (average)	2.00	Certain
Wheat straw (high-alkali)	5.59	Certain
Rice straw	3.80	Certain

Source: Adapted from Miles, T.R. et al., Alkali slagging problems with biomass fuels. In: *First Biomass Conference of the Americas: Energy, Environment, Agriculture, and Industry*, Vol. 1, National Renewable Energy Laboratory, Burlington, VT, 1993.

TABLE 7.7

Common Products from Biomass

Biomass Resource	Uses
Corn	Solvents, pharmaceuticals, adhesives, starch, resins, binders, polymers, cleaners, ethanol
Vegetable oils	Surfactants in soaps and detergents, pharmaceuticals (inactive ingredients), inks, paints, resins, cosmetics, fatty acids, lubricants, biodiesel
Wood	Paper, building materials, cellulose for fibers and polymers, resins, binders, adhesives, coatings, paints, ins, fatty acids, road and roofing pitch

Source: USDOE, *Industrial Bioproducts: Today and Tomorrow*, U.S. Department of Energy, Washington, DC, 2004.

Another process for biomass is the application of anaerobic digestion to produce biogas (methane) for electricity generation. Recall that anaerobic digestion involves the controlled breakdown of organic wastes by bacteria in the absence of oxygen. Major agricultural feedstocks for anaerobic digestion include food processing wastes and manure from livestock operations. The Energy Information Agency has also projected a significant increase in the generation of electricity from municipal waste and landfill gas—to about 0.5% of U.S. electricity consumption (EIA, 2006).

BIOMASS FOR BIOPRODUCTS[*]

Bioproducts are industrial and consumer goods manufactured wholly or in part from renewable biomass (plant-based resources). Today's industrial bioproducts are amazingly diverse, ranging from solvents and paints to pharmaceuticals, soaps, cosmetics, and building materials (see Table 7.7). Industrial bioproducts are integral to our way of life—few sectors of the economy do not rely in some way or another on products made from biomass. Corn, wood, soybeans, and plant oils are the primary resources used to create this remarkable diversity of industrial and consumer goods. In some cases, it is not readily apparent that a product is derived in part from biomass. Biomass components are often combined with other materials such as petrochemicals and minerals to manufacture the final product. Soybean oil, for example, is blended with other components to produce paints, toiletries, solvents, inks, and pharmaceuticals. Some products, such as starch adhesives, are derived entirely from biomass. The many derivatives of corn illustrate the diversity of products that can be obtained from a single biomass resource (see Figure 7.6). Besides being an important source of food and feed, corn serves as a feedstock for ethanol and sorbitol (a sweetish, crystalline alcohol), industrial starches and sweeteners, citric and lactic acid, and many other products. Biomass, which is comprised of carbohydrates, can be used to produce some of the products that are commonly manufactured from petroleum and natural gas, or hydrocarbons. Both resources contain the essential elements of carbon and hydrogen. In some cases, both resources have captured a portion of market share (see Table 7.8).

CLASSES OF BIOPRODUCTS

The thousands of different industrial bioproducts produced today can be categorized into five major areas, as follows:

---

[*] Adapted from USDOE, *Industrial Bioproducts: Today and Tomorrow*, U.S. Department of Energy, Washington, DC, 2004.

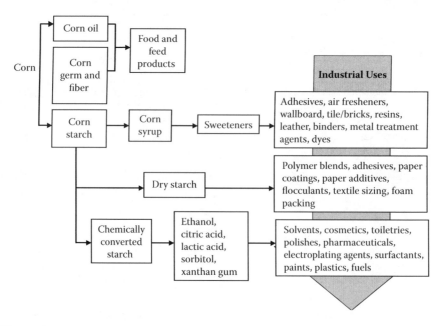

FIGURE 7.6   Bioproducts from corn grain.

1. *Sugar and starch bioproducts* derived through fermentation and thermochemical processes include alcohols, acids, starch, xanthum gum, and other products derived from biomass sugars. Primary feedstocks include sugarcane, sugarbeets, corn, wheat, rice, potatoes, barley, sorghum grain, and wood.
2. *Oil- and lipid-based bioproducts* include fatty acids, oils, alkyd resins, glycerine, and a variety of vegetable oils derived from soybeans, rapeseed, or other oilseeds.
3. *Gum and wood chemicals* include tall oil (liquid rosin), alkyd resins, rosins, pitch, fatty acids, turpentine, and other chemicals derived from trees.

TABLE 7.8
Products from Hydrocarbons vs. Carbohydrates

Product	Total Production (millions of tons)	% Derived from Plants
Adhesives	5.0	40.0
Fatty acids	2.5	40.0
Surfactants	3.5	35.0
Acetic acid	2.3	17.5
Plasticizers	0.8	15.0
Activated carbon	1.5	12.0
Detergents	12.6	11.0
Pigments	15.5	6.0
Dyes	4.5	6.0
Wall points	7.8	3.5
Inks	3.5	3.5
Plastics	30.0	1.8

Source: ILSR, *Accelerating the Shift to a Carbohydrate Economy: The Federal Role*, Executive Summary of the Minority Report of the Biomass Research and Development Technical Advisory Committee, Institute for Local Self-Reliance, Washington, DC, 2002.

4. *Cellulose derivatives, fibers, and plastics* include products derived from cellulose, including cellulose acetate (cellophane) and triacetate, cellulose nitrate, alkali cellulose, and regenerated cellulose. The primary sources of cellulose are bleached wood pulp and cotton linters.
5. *Industrial enzymes* are used as biocatalysts for a variety of biochemical reactions in the production of starch and sugar, alcohols, and oils. They are also used in laundry detergents, tanning of leathers, and textile sizing (Uhlig, 1998).

BIODIESEL

The diesel engine is the workhorse of heavy transportation and industrial processes; it is widely used to power trains, tractors, ships, pumps, and generators. Powering the diesel engine is conventional diesel fuel or biodiesel fuel. Biodiesel is a rather viscous liquid fuel made up of fatty acid alkyl esters, fatty acid methyl esters (FAMEs), and long-chain mono-alkyl esters. It is produced from renewable sources such as new and used vegetable oils and animal fats and is a cleaner-burning replacement for petroleum-based diesel fuel. It is nontoxic and biodegradable. This fuel is designed to be used in compression-ignition (diesel) engines similar or identical to those that burn petroleum diesel. Biodiesel has physical properties similar to those of petroleum diesel (see Table 7.9). In the United States, most biodiesel is made from soybean oil or recycled cooking oils. Animal fats, other vegetable oils, and other recycled oils can also be used to produce biodiesel, depending on the costs and availability. In the future, blends of all kinds of fats and oils may be used to produce biodiesel. Before providing a basic description of the vegetable and/or recycled grease to biodiesel process, it is important to define and review a few of the key technological terms associated with the conversion process.

- *Acid esterification*—Oil feedstocks containing more than 4% free fatty acids go through an acid esterification process to increase the yield of biodiesel. These feedstocks are filtered and preprocessed to remove water and contaminants and are then fed to the acid esterification process. The catalyst, sulfuric acid, is dissolved in methanol and then mixed with the pretreated oil. The mixture is heated and stirred, and the free fatty acids are converted to biodiesel. Once the reaction is complete, it is dewatered and then fed to the transesterification process.
- *Transesterification*—Oil feedstocks containing less than 4% free fatty acids are filtered and preprocessed to remove water and contaminants and then fed directly to the transesterification process along with any products of the acid esterification process. The catalyst, potassium hydroxide, is dissolved in methanol and then mixed with the pretreated oil. If an acid esterification process is used, then extra base catalyst must be added to neutralize the acid added in that step. Once the reaction is complete, the major co-products, biodiesel and glycerin, are separated into two layers.

TABLE 7.9
Biodiesel Physical Characteristics

Specific gravity	0.87 to 0.89
Kinematic viscosity at 40°C	3.7 to 5.8
Cetane number	46 to 70
Higher heating value (Btu/lb)	16,928 to 17,996
Sulfur (wt%)	0.0 to 0.0024
Cloud point (°C)	–11 to 16
Pour point (°C)	–15 to 13
Iodine number	60 to 135
Lower heating value (lb/lb)	15,700 to 16,735

- *Methanol recovery*—The methanol is typically removed after the biodiesel and glycerin have been separated to prevent the reaction from reversing itself. The methanol is cleaned and recycled back to the beginning of the process.
- *Biodiesel refining*—Once separated from the glycerin, the biodiesel goes through a clean-up or purification process to remove excess alcohol, residual catalyst, and soaps. This consists of one or more washings with clear water. It is then dried and sent to storage. Sometimes the biodiesel goes through an additional distillation step to produce a colorless, odorless, zero-sulfur biodiesel.
- *Glycerin refining*—The glycerin byproduct contains unreacted catalyst and soaps that are neutralized with an acid. Water and alcohol are removed to produce 50 to 80% crude glycerin. The remaining contaminants include unreacted fats and oils. In large biodiesel plants, the glycerin can be further purified, to 99% or higher purity, for sale to the pharmaceutical and cosmetic industries.

The most popular biodiesel production process is *transesterification* (production of the ester) of vegetable oils or animal fats, using alcohol in the presence of a chemical catalyst. About 3.4 kg of oil/fat are required for each gallon of biodiesel produced (Baize, 2006). The transesterification of degummed soybean oil produces ester and glycerin. The reaction requires heat and a strong base catalyst such as sodium hydroxide or potassium hydroxide. The simplified transesterification reaction is shown below:

Triglycerides + Free fatty acids (<4%) + Alcohol → Alkyl esters + Glycerin

Some feedstocks must be pretreated before they can go through the transesterification process. Feedstocks with less than 4% free fatty acids, which include vegetable oils and some food-grade animal fats, do not require pretreatment. Feedstocks with more than 4% free fatty acids, which include inedible animal fats and recycled greases, must be pretreated in an acid esterification process. In this step, the feedstock is reacted with an alcohol (such as methanol) in the presence of a strong acid catalyst (sulfuric acid), converting the free fatty acids into biodiesel. The remaining triglycerides are converted to biodiesel in the transesterification reaction:

Triglycerides + Free fatty acids (>4%) + Alcohol → Alkyl esters + Triglycerides

Figure 7.7 illustrates the basic technology for processing vegetable oils (such as soybeans) and recycled greases (used cooking oil and animal fat). When the feedstock is vegetable oil, the extracted oil is processed to remove all traces of water, dirt, and other contaminants. Free fatty acids are also removed. A combination of methyl alcohol and a catalyst, usually sodium hydroxide or potassium hydroxide, breaks the oil molecules apart in the esterification process. The resulting esters are then refined into usable biodiesel.

DID YOU KNOW?

Biodiesel is routinely blended with petroleum diesel. The percentage of biodiesel in the blend is written after an uppercase letter B to denote the proportion. For example, a mixture of 20% biodiesel and 80% petroleum diesel is called B20, a mixture of equal parts biodiesel and petroleum diesel is B50, and neat (or pure) biodiesel is B100. Most conventional diesel engines can burn blends from pure petroleum diesel up to B20 without modification. With minor modification, most diesel engines built since 1994 can use blends from B20 to B100. The diesel vehicle owner's manual and vehicle warranty should be checked before using any alternative fuel.

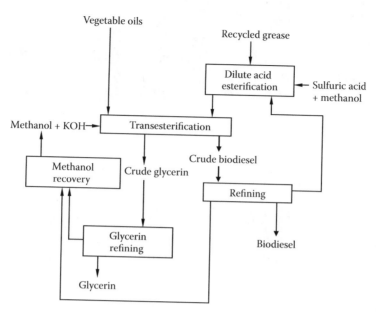

FIGURE 7.7 Biodiesel process.

When the feedstock is used-up cooking oil and animal fats refined to produce biodiesel, the process is similar to the way biodiesel is derived from vegetable oil, except there is an additional step involved (Figure 7.7). Methyl alcohol and sulfur are used in a process called *dilute acid esterification* to obtain a substance resembling fresh vegetable oil, which is then processed in the same way as vegetable oil to obtain the final product.

ALGAE TO BIODIESEL

The focus of our discussion of biofuels to this point has been on terrestrial sources of biomass fuels. This section is concerned with photosynthetic organisms that grow in aquatic environments—namely, macroalgae, microalgae, and emergents. Macroalgae, more commonly known as seaweed, are fast-growing marine and freshwater plants that can grow to considerable size (up to 60 m in length). Emergents are plants that grow partially submerged in bogs and marshes. Microalgae are, as the name suggests, microscopic photosynthetic organisms. With regard to the mass production of oil for alternative renewable energy purposes, the focus of this discussion is mainly on microalgae—organisms capable of photosynthesis that are less than 0.4 mm in diameter, including the diatoms and cyanobacteria—as opposed to macroalgae (seaweed). Microalgae are the preferred choice because they are less complex in structure, have a fast growth rate, and typically (for some species) contain a high concentration of oil. However, it is important to point out that recent research efforts also are focusing on using seaweeds for algae fuel (biofuels), probably due to their high availability (Lewis, 2005).

The following algal species are currently being studied for their suitability as a mass oil-producing crop across various locations worldwide (Anon., 2008):

Ulva
Botryococcus braunii
Chlorella
Dunaliella tertiolecta
Gracilaria
Pleurochrysis carterae
Sargassum (has 10 times the output volume of *Gracilaria*)

DID YOU KNOW?

Algae fuel, also called algal fuel, algaeoleum, or second-generation biofuel, is a biofuel derived from algae; that is, natural oils are used in the production of biodiesel.

The algae-to-biodiesel alternative renewable fuel source program got its start during the Carter Administration in response to the 1970s Arab fuel embargo. The Carter Administration consolidated all federal energy activities under the auspice of the newly established U.S. Department of Energy (USDOE). Among its various programs established to develop all forms of alternative energy (related to solar energy), the USDOE initiated research on the use of plant life as a source of transportation fuels (today known as the Biofuels Program). Before discussing the algae-to-biofuels process in detail, it is important for the reader to be well grounded in basic algal concepts. In many ways, the study of microalgae is a relatively limited field of study. Algae are not nearly as well understood as other organisms that have found a role in today's biotechnology industry. The study of microalgae represents an area of high risk and high gains; it also presents an opportunity for the curious and ambitious to break new ground in science by conducting research in this growing area of interest.

Algae can be both a nuisance and an ally; we might say they possess Jekyll and Hyde-like (good vs. bad) characteristics. Many ponds, lakes, rivers, streams, and bays (e.g., Chesapeake Bay) in the United States and elsewhere are undergoing *eutrophication*, which is basically the killing off (although in many cases very slowly) of calm water environments, especially ponds, marshes, and lakes, due to enrichment of an environment with inorganic substances (phosphorous and nitrogen). When eutrophication occurs, when filamentous algae such as *Caldophora* break loose in a pond, lake, stream, or river and wash ashore, algae makes it stinking, noxious presence known. More important than the offensive odor that dying algae give off is the fact that their deaths begin a biodegradation process whereby instead of adding oxygen to their watery environment they begin to use it up. Algae have a good side, too. For example, algae are allies in many wastewater treatment operations. In addition, they can be valuable in long-term oxidation ponds where they aid in the purification process by producing oxygen. Before discussing the specifics and different types of algae, it is important to be familiar with terminology associated with algae.

- *Algae*—Large and diverse assemblages of eukaryotic organisms that lack roots, stems, and leaves but have chlorophyll and other pigments for carrying out oxygen-producing photosynthesis.
- *Algology* or *phycology*—The study of algae.
- *Antheridium*—Special male reproductive structures where sperm are produced.
- *Aplanospore*—Nonmotile spores produced by sporangia.
- *Benthic*—Algae attached and living on the bottom of a body of water.
- *Binary fission*—Nuclear division followed by division of the cytoplasm.

DID YOU KNOW?

In the 1980s, the USDOE gradually shifted its focus to technologies that could have large-scale impacts on national consumption of fossil energy. Much of the USDOE's publications from this period reflect a philosophy of energy research that might, somewhat pejoratively, be called "the quads mentality." A quad is a short-hand name for the unit of energy often used by the USDOE to describe the amounts of energy that a given technology might be able to displace. Quad is short for "quadrillion Btus"—a unit of energy representing 10^{15} (1,000,000,000,000,000) Btus of energy. This perspective led the USDOE to focus on the concept of immense algae farms.

- *Chloroplasts*—Packets that contain chlorophyll *a* and other pigments.
- *Chrysolaminarin*—The carbohydrate reserve in organisms of division Chrysophyta.
- *Diatoms*—Photosynthetic, circular or oblong chrysophyte cells.
- *Dinoflagellates*—Unicellular, photosynthetic protistan algae.
- *Dry mass factor*—The percentage of dry biomass in relation to the fresh biomass; for example, if the dry mass factor is 5%, one would need 20 kg of wet algae (algae in the media) to get 1 kg of dry algae cells.
- *Epitheca*—The larger part of the frustule (diatoms).
- *Euglenoids*—Contain chlorophylls *a* and *b* in their chloroplasts; representative genus is *Euglena*.
- *Fragmentation*—A type of asexual algal reproduction in which the thallus breaks up and each fragmented part grows to form a new thallus.
- *Frustule*—The distinctive two-piece wall of silica in diatoms.
- *Hypotheca*—The small part of the frustule (diatoms).
- *Lipid content*—The percentage of oil in relation to the dry biomass needed to get it; for example, if the algae lipid content is 40%, one would need 2.5 kg of dry algae to get 1 kg of oil.
- *Neustonic*—Algae that live at the water–atmosphere interface.
- *Oogonia*—Vegetative cells that function as female sexual structures in the algal reproductive system.
- *Pellicle*—A *Euglena* structure that allows for turning and flexing of the cell.
- *Phytoplankton*—Made up of algae and small plants.
- *Plankton*—Free-floating, mostly microscopic aquatic organisms.
- *Planktonic*—Algae suspended in water as opposed to attached and living on the bottom (benthic).
- *Protothecosis*—A disease in humans and animals caused by the green algae, *Prototheca moriformis*.
- *Thallus*—The vegetative body of algae.

Algae are autotrophic, contain the green pigment chlorophyll, and are a form of aquatic plant. Algae differ from bacteria and fungi in their ability to carry out photosynthesis—the biochemical process requiring sunlight, carbon dioxide, and raw mineral nutrients. Photosynthesis takes place in the chloroplasts. The chloroplasts are usually distinct and visible. They vary in size, shape, distribution, and numbers. In some algal types, the chloroplast may occupy most of the cell space. They usually grow near the surface of water because light cannot penetrate very far through water. Although when they are *en masse* (multicellular forms such as marine kelp) the unaided eye easily sees them, many of them are microscopic. Algal cells may be nonmotile or motile by one or more flagella, or they may exhibit gliding *motility* as in diatoms. They occur most commonly in water (fresh and polluted water, as well as in saltwater), in which they may be suspended (planktonic) or attached and living on the bottom (benthic); a few algae live at the water–atmosphere interface (neustonic). Within the freshwater and saltwater environments, they are important primary producers, which means they are at the beginning of the food chain for other organisms. During their growth phase, they are important oxygen-generating organisms and constitute a significant portion of the plankton in water.

Algae belong to seven divisions (phylums), distributed between two different kingdoms (Plantae and Protista) in the standard biological five-kingdom system. Five divisions are discussed in this text:

- Chlorophyta—Green algae
- Euglenophyta—Euglenids
- Chrysophyta—Golden-brown algae, diatoms
- Phaeophyta—Brown algae
- Pyrrophyta—Dinoflagellates

The primary classification of algae is based on cellular properties. Several characteristics are used to classify algae, including: (1) cellular organization and cell wall structure; (2) the nature of chlorophyll(s) present; (3) the type of motility, if any; (4) the carbon polymers that are produced and stored; and (5) the reproductive structures and methods. Table 7.10 summarizes the properties of the five divisions discussed in this text.

Algae show considerable diversity in the chemistry and structure of their cells. Some algal cell walls are thin, rigid structures usually composed of cellulose modified by the addition of other polysaccharides. In other algae, the cell wall is strengthened by the deposition of calcium carbonate. Other forms have chitin present in the cell wall. Complicating the classification of algal organisms are the euglenids, which lack cell walls. In diatoms, the cell wall is composed of silica. The frustules (shells) of diatoms have extreme resistance to decay and remain intact for long periods of time, as the fossil records indicate.

The principal feature used to distinguish algae from other microorganisms (e.g., fungi) is the presence of chlorophyll and other photosynthetic pigments in algae. All algae contain chlorophyll *a*. Some, however, contain other types of chlorophylls. The presence of these additional chlorophylls is characteristic of a particular algal group. In addition to chlorophyll, other pigments encountered in algae include fucoxanthin (brown), xanthophylls (yellow), carotenes (orange), phycocyanin (blue), and phycoerythrin (red). Many algae have flagella (a threadlike appendage). The flagella are locomotor organelles that may be either single polar or multiple polar. The *Euglena* have a simple flagellate form with a single polar flagellum. Chlorophyta have either two or four polar flagella. Dinoflagellates have two flagella of different lengths. In some cases, algae are nonmotile until they form motile gametes (a haploid cell or nucleus) during sexual reproduction. Diatoms do not have flagella but have gliding motility. s

Algae can be either autotrophic or heterotrophic. Most are photoautotrophic; they require only carbon dioxide and light as their principal source of energy and carbon. In the presence of light, algae carry out oxygen-evolving photosynthesis; in the absence of light, algae use oxygen. Chlorophyll and other pigments are used to absorb light energy for photosynthetic cell maintenance and reproduction. One of the key characteristics used in the classification of algal groups is the nature of the reserve polymer synthesized as a result of utilizing carbon dioxide present in water.

Algae may reproduce either asexually or sexually. Three types of asexual reproduction occur: binary fission, spores, and fragmentation. In some unicellular algae, binary fission occurs where the division of the cytoplasm forms new individuals like the parent cell following nuclear division. Some algae reproduce through spores. These spores are unicellular and germinate without fusing with other cells. In fragmentation, the thallus breaks up and each fragment grows to form a new thallus. Sexual reproduction can involve the union of cells where eggs are formed within vegetative cells called *oogonia* (which function as female structures) and sperm are produced in male reproductive organs called *antheridia*. Algal reproduction can also occur through a reduction of chromosome number or the union of nuclei.

Characteristics of Algal Divisions

Chlorophyta (Green Algae)

The majority of algae found in ponds belong to this group; they also can be found in saltwater and soil. Several thousand species of green algae are known today. Many are unicellular; others are multicellular filaments or aggregated colonies. The green algae have chlorophylls *a* and *b*, along with specific carotenoids, and they store carbohydrates at starch. Few green algae are found at depths greater than 7 to 10 m, largely because sunlight does not penetrate to that depth. Some species have a holdfast structure that anchors them to the bottom of the pond and to other submerged inanimate objects. Green algae reproduce by both sexual and asexual means. Multicellular green algae have some division of labor, producing various reproductive cells and structures.

TABLE 7.10

Comparative Summary of Algal Characteristics

Algal Group	Common Name	Structure	Pigments	Carbon Reserve	Motility	Reproduction
Chlorophyta	Green algae	Unicellular to multicellular	Chlorophylls *a* and *b*, carotenes, xanthophylls	Starch, oils	Most are nonmotile	Asexual and sexual
Euglenophyta	Euglenoids	Unicellular	Chlorophylls *a* and *b*, carotenes, xanthophylls	Fats	Motile	Asexual
Chrysophyta	Golden-brown algae, diatoms	Multicellular	Chlorophylls *a* and *b*, special carotenoids, xanthophylls	Oils	Gliding by diatoms; others by flagella	Asexual and sexual
Phaeophyta	Brown algae	Unicellular	Chlorophylls *a* and *b*, carotenoids, xanthophylls	Fats	Motile	Asexual and sexual
Pyrrophyta	Dinoflagellated	Unicellular	Chlorophylls *a* and *b*, carotenes, xanthophylls	Starch	Motile	Asexual; sexual rare

Euglenophyta (Euglenoids)

The Euglenophyta are a small group of unicellular microorganisms that have a combination of animal and plant properties. Euglenoids lack a cell wall, possess a gullet, have the ability to ingest food, have the ability to assimilate organic substances, and, in some species, are absent of chloroplasts. They occur in fresh, brackish, and salt waters, and on moist soils. A typical *Euglena* cell is elongated and bound by a plasma membrane; the absence of a cell wall makes them very flexible in movement. Inside the plasma membrane is a structure called the *pellicle* that gives the organisms a definite form and allows the cell to turn and flex. Euglenoids are photosynthetic and contain chlorophylls *a* and *b*, and they always have a red eyespot (*stigma*) that is sensitive to light (photoreceptive). Some euglenoids move about by means of flagellum; others move about by means of contracting and expanding motions. The characteristic food supply for euglenoids is a lipopolysaccharide. Reproduction in euglenoids is by simple cell division.

> **Note:** Some autotrophic species of *Euglena* become heterotrophic when light levels are low.

Chrysophyta (Golden-Brown Algae)

The Chrysophyta phylum is quite large, having several thousand diversified members. They differ from green algae and euglenoids in that (1) chlorophylls *a* and *c* are present; (2) fucoxanthin, a brownish pigment, is present; and (3) they store food in the form of oils and leucosin, a polysaccharide. The combination of yellow pigments, fucoxanthin, and chlorophylls causes most of these algae to appear golden brown. The Chrysophycophyta division is also diversified in cell wall chemistry and flagellation. The division is divided into three major classes: golden-brown algae (Chrysophyceae), yellow–green algae (Xanthophycae), and diatoms (Bacillariophyceae). Some Chrysophyta lack cell walls; others have intricately patterned coverings external to the plasma membrane, such as walls, plates, and scales. The diatoms are the only group that has a hard cell wall, called a *frustule*, which is composed of pectin, cellulose, or silicon and consists of two valves: the epitheca and the hypotheca. Two anteriorly attached flagella are common among Chrysophyta; others have no flagella. Most Chrysophyta are unicellular or colonial. Asexual cell division is the usual method of reproduction in diatoms, but other forms of Chrysophyta can reproduce sexually. Diatoms have direct significance for humans. Because they make up most of the phytoplankton of the cooler ocean parts, they are the ultimate source of food for fish. Water and wastewater operators understand the importance of their ability to function as indicators of industrial water pollution. As water quality indicators, their specific tolerances to environmental parameters such as pH, nutrients, nitrogen, concentration of salts, and temperature have been determined.

> **Note:** Diatoms secrete a silicon dioxide shell (frustule) that forms the fossil deposits known as diatomaceous earth, which is used in filters and as abrasives in polishing compounds.

Phaeophyta (Brown Algae)

With the exception of a few freshwater species, all algal species of this division exist in marine environments as seaweed. They are a highly specialized group consisting of multicellular organisms that are sessile (attached, not free-moving). These algae contain essentially the same pigments seen in the golden-brown algae, but they appear brown because of the predominance and masking effect of a greater amount of fucoxanthin. Brown algal cells store food as the carbohydrate laminarin and some lipids. Brown algae reproduce asexually. Brown algae are used in foods, animal feeds, and fertilizers and as a source for alginate, a chemical emulsifier added to ice cream, salad dressing, and candy.

Pyrrophyta (Dinoflagellates)

The principal members of this division are the dinoflagellates. The dinoflagellates comprise a diverse group of biflagellated and nonflagellated unicellular eukaryotic organisms. The dinoflagellates occupy a variety of aquatic environments, with the majority living in marine habitats. Most of

DID YOU KNOW?

Cell division in dinoflagellates differs from most protistans, with chromosomes attaching to the nuclear envelope and being pulled apart as the nuclear envelope stretches. During cell division in most other eukaryotes, the nuclear envelope dissolves.

these organisms have a heavy cell wall composed of cellulose-containing plates. They store food as starch, fats, and oils. These algae have chlorophylls *a* and *c* and several xanthophylls. The most common form of reproduction in dinoflagellates is by cell division, but sexual reproduction has also been observed.

Algal Biomass

Algal biomass contains three main components:

- Carbohydrates
- Protein
- Natural oils

Biodiesel production applies exclusively to the natural oil fraction, the main product of interest to us in this section. The bulk of natural oil made by oilseed crops is in the form of triacylglycerols (TAGs), which consist of three long chains of fatty acids attached to a glycerol backbone. The algae species of concern can produce up to 60% of their body weight in the form of TAGs. (Recall that the species of concern, the oil producers, are *Ulva*, *Botryococcus braunii*, *Chlorella*, *Dunaliella tertiolecta*, *Gracilaria*, *Pleurochrysis carterae*, and *Sargassum*.) Thus, algae represent an alternative source of biodiesel, one that does not compete with the exiting oilseed market.

Algae can produce up to 300 times more oil per acre than conventional crops, such as rapeseed, palms, soybean, or *Jatropha* (Christi, 2007). Moreover, algae has a harvesting cycle of 1 to 10 days, permitting several harvests in a very short time frame, a strategy quite different from that for yearly crops. Algae can also be grown on land that is not suitable for other established crops, such as arid land, land with excessively saline soil, and drought-stricken land. This advantage minimizes the issue of taking away pieces of land from the cultivation of food crops (Schenk et al., 2008). Algae can grow 20 to 30 times faster than food crops (McDill, 2009). Algae can be produced and harvested for biofuel using various technologies. These technologies include photobioreactors (plastic tubes full of nutrients exposed to sunlight), closed-loop (not exposed to open air) systems, and open ponds. For the purposes of illustration (even though there are many objectors and dissenters to open-pond systems), the open-pond configuration of algae farms is discussed here, because the open-pond raceway system is a relatively low-cost system and is an easily understood process.

Open-Pond Algae Farms

Algae farms consist of open, shallow ponds in which some source of waste carbon dioxide (CO_2) can be bubbled into the ponds and captured by the algae. As shown in Figure 7.8, the ponds in an algae farm are "raceway" designs, in which the algae, water, and nutrients circulate around a racetrack. Paddlewheels provide the flow. The algae are thus kept suspended in water. Algae are circulated back up to the surface on a regular frequency. The ponds are kept shallow because of the need to keep the algae exposed to sunlight and the limited depth to which sunlight can penetrate the pond water. The ponds operate continuously; that is, water and nutrients are constantly fed to the pond while algae-containing water is removed at the other end. Some kind of harvesting system

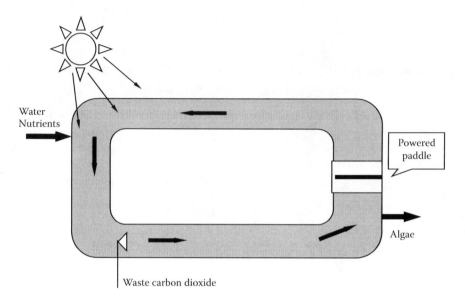

FIGURE 7.8 Raceway design algae pool.

is required to recover the algae, which contains substantial amounts of natural oil. Figure 7.9 illustrates the concept of an algae farm. The size of these ponds is measured in terms of surface area (as opposed to volume), because surface area is so critical to capturing sunlight. Their productivity is measured in terms of biomass produced per day per unit of available surface area. Even at levels of productivity that would stretch the limits of an aggressive research and development program, such systems will require acres of land. At such large sizes, it is more appropriate to think of these operations on the scale of a farm. Waste carbon dioxide is readily available from a number of sources.

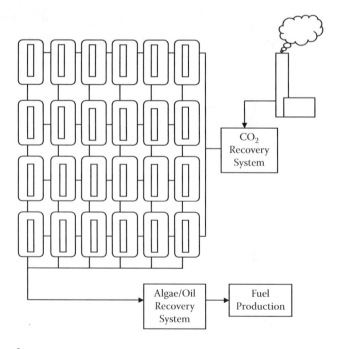

FIGURE 7.9 Algae farm.

Every operation that involves combustion of fuel for energy is a potential source. Generally, coal and other fossil fuel-fired power plants are targeted as the main sources of carbon dioxide, as typical coal-fired power plants emit flue gas from their stacks containing up to 13% carbon dioxide. This high concentration of carbon dioxide enhances the transfer and uptake of carbon dioxide in the ponds. The concept of coupling a coal-fired power plant with an algae farm provides a win–win approach for recycling the carbon dioxide from coal combustion into a usable liquid fuel.

JATROPHA TO BIODIESEL

The uninformed or misinformed might flinch when they discover that a byproduct of the plant *Jatropha curcas*—yes, the same plant known in some places as "black vomit nut" and in others as "bellyache bush" or "tuba-tub"—is being used as a product that has some credible value. *Jatropha* shrubs are inedible plants that grow mostly in countries such as the Philippines. *Jatropha* is resistant to drought and can easily be planted or propagated through seeds or cuttings. It starts producing seeds within 14 months, but reaches its maximum productivity level after 4 to 5 years. The *Jatropha* plant can produce an oil content of 30 to 58%, depending on the quality of the soil where it is planted. The seeds yield an annual equivalent of 0.75 to 2 tons or 1892 liters of biodiesel per hectare. The plant remains useful for around 30 to 40 years and can be planted in harsh climates where it would not compete for resources needed to grow food. Along with having the advantage of being a renewable fuel source, the *Jatropha* plant also reduces greenhouse gas emissions and our dependence on oil imports.

PROS AND CONS OF BIODIESEL

The greatest advantage of biodiesel over conventional petroleum diesel is that biodiesel comes from renewable resources. The supply can be grown, over and over again. Biodiesel combustion also produces fewer emissions (except for nitrous oxides) than combustion of an equal amount of petroleum diesel. The widespread use of biodiesel can also reduce the dependency on imported oil. From a safety standpoint, biodiesel is safer than petroleum diesel because it is less combustible. From an environmental standpoint, when accidentally spilled, biodiesel is not persistent within environmental media (air, water, soil) because it is biodegradable. Biodiesel can also be produced from waste products such as cooking oils and grease.

Probably the most pressing disadvantages or shortcomings of biodiesel, at least at the present time, are its lack of availability or accessibility and high cost relative to petroleum diesel. This trend in non-availability and non-accessibility is bound to change as the less expensive, more accessible petroleum diesel becomes more expensive and difficult to find. Additional disadvantages to consider are that biodiesel requires special handling, storage and transportation management as compared to petroleum diesel. With regard to environmental considerations, biodiesel produces more nitrous oxide emissions when combusted than an equal amount of petroleum diesel. Another potential problem with the production of biodiesel is its dependence on soybeans as its primary feedstock; there is some concern that the widespread use of biodiesel as fuel will contribute to higher food prices and indirectly to world hunger. There is a slight reduction in performance and mileage per gallon with biodiesel as compared to petroleum diesel. Biodiesel can also act as a solvent in some diesel engines, causing loosened deposits that may clog filters.

BIOGAS (METHANE)

Primarily known as a fuel for interior heating systems, methane or biogas can also be used as a replacement for natural gas—a fossil fuel for electricity generation and for cooking and heating—and as an alternative fuel to gasoline. Methane is a natural gas produced by the breakdown of organic material in the absence of oxygen in termite mounds and wetlands and by some animals.

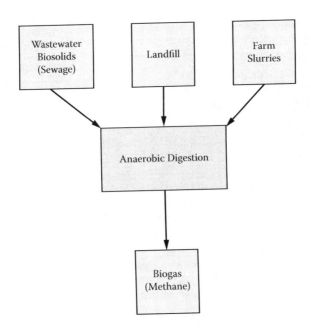

FIGURE 7.10   Production of biogas (methane, CH_4).

Humans are also responsible for the release of methane through biomass burning, rice production, cattle raising, and releases from gas exploration. Methane can also be obtained directly from the earth; however, other methods of production have been developed, most notably the fermentation or composting of plant and animal waste. The reasons for considering biogas (methane) as a possible biofuel include the following:

- It is viable because of its potential use as an alternative fuel source.
- It is a viable alternative fuel to use to improve air quality.
- It can be produced locally, reducing the need to use imported natural gas.

Methane is produced under anaerobic conditions where organic material is biodegraded or broken down by a group of microorganisms. The three main sources of feedstock material for anaerobic digestion are given in Figure 7.10 and are described in the following.

ANAEROBIC DIGESTION

Anaerobic digestion is the traditional method of managing waste, sludge stabilization, and releasing energy. It involves using bacteria that thrive in the absence of oxygen and is slower than aerobic digestion, but it has the advantage that only a small portion of the wastes is converted into new bacterial cells. Instead, most of the organics are converted into carbon dioxide and methane gas.

Cautionary Note: Allowing air to enter an anaerobic digester should be prevented because the mixture of air and gas produced in the digester can be explosive.

Stages of Anaerobic Digestion

Anaerobic digestion (see Figure 7.11) has four key biological and chemical stages (Spellman, 2009; USEPA, 1979, 2006):

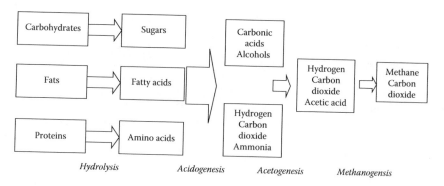

FIGURE 7.11   Key stages of anaerobic digestion.

1. *Hydrolysis*—Proteins, cellulose, lipids, and other complex organics are broken down into smaller molecules and become soluble by utilizing water to split the chemical bonds of the substances.
2. *Acidogenesis*—The products of hydrolysis are converted into organic acids (where monomers are converted to fatty acids).
3. *Acetogenesis*—The fatty acids are converted to acetic acid, carbon dioxide, and hydrogen.
4. *Methanogenesis*—Organic acids produced during the fermentation step are converted to methane and carbon dioxide.

The efficiency of each phase is influenced by the temperature and the amount of time the process is allowed to react. For example, the organisms that perform hydrolysis and volatile acid fermentation (often called *acidogenic bacteria*) are fast-growing microorganisms that prefer a slightly acidic environment and higher temperatures than the organisms that perform the methane formation step (*methanogenic bacteria*).

A simplified generic chemical equation for the overall processes outlined above is as follows:

$$C_6H_{12}O_6 \rightarrow 3CO_2 + 3CH_4$$

Biogas is the ultimate waste product of the bacteria feeding off the input biodegradable feedstock and is composed primarily of methane and carbon dioxide, with a small amount of hydrogen and trace hydrogen sulfide (see Table 7.11). Keep in mind that the ultimate output from a wastewater digester is water; biogas (methane) is more of an off-gas that can be used as an energy source. Wastewater digestion and the production of biogas are discussed in the next section.

TABLE 7.11

Typical Contents of Biogas

Matter	Percentage (%)
Methane (CH_4)	50–75
Carbon dioxide (CO_2)	25–50
Nitrogen (N_2)	0–10
Hydrogen (H_2)	0–1
Hydrogen sulfide (H_2S)	0–3
Oxygen (O_2)	0–2

DID YOU KNOW?

The primary purpose of a secondary digester is to allow for solids separation.

Anaerobic Digestion of Sewage Biosolids (Sludge)

Equipment used in anaerobic digestion typically includes a sealed digestion tank with either a fixed or a floating cover or an inflatable gas bubble, heating and mixing equipment, gas storage tanks, solids and supernatant withdrawal equipment, and safety equipment (e.g., vacuum relief, pressure relief, flame traps, explosion-proof electrical equipment).

Note: Biosolids are inherently dangerous as possible sources of explosive gases, and biosolids sites should never be entered without following OSHA's confined-space entry permit requirements. Only fully trained personnel should enter permit-required confined spaces.

In operation, the process residual (thickened or unthickened biosolids/sludge) is pumped into the sealed digester. The organic matter digests anaerobically by a two-stage process. Sugars, starches, and carbohydrates are converted to volatile acids, carbon dioxide, and hydrogen sulfide. The volatile acids are then converted to methane gas. This operation can occur either in a single tank (one stage) or in two tanks (two stages). In a single-stage system, supernatant and digested solids must be removed whenever flow is added. In a two-stage operation, solids and liquids from the first stage flow into the second stage each time fresh solids are added. Supernatant is withdrawn from the second stage to provide additional treatment space. Periodically, solids are withdrawn for dewatering or disposal. The methane gas produced in the process may be used for many plant activities.

Various performance factors affect the operation of the anaerobic digester. The percent volatile matter in raw sludge, digester temperature, mixing, volatile acids-to-alkalinity ratio, feed rate, percent solids in raw biosolids, and pH are all important operational parameters that the operator must monitor (see Table 7.12). Along with being able to recognize normal and abnormal anaerobic digester performance parameters, digester operators must also know and understand normal operating procedures. Normal operating procedures include biosolids additions, supernatant withdrawal, sludge withdrawal, pH control, temperature control, mixing, and safety requirements.

Note: Keep in mind that in fixed-cover operations additions must be balanced by withdrawals. If not, structural damage occurs.

TABLE 7.12
Sludge Parameters for Anaerobic Digesters

Raw Biosolids (Sludge) Solids	Impact
<4% solids	Loss of alkalinity
	Decreased sludge retention time
	Increased heating requirements
	Decreased volatile acids-to-alkalinity ratio
4–8% solids	Normal operation
>8% solids	Poor mixing
	Organic overloading
	Decreased volatile acids-to-alkalinity ratio

Sludge must be pumped (in small amounts) several times each day to achieve the desired organic loading and optimum performance, and supernatant withdrawal must be controlled for maximum sludge retention time. All drawoff points are sampled, and the level with the best quality is selected. Digested sludge is withdrawn only when necessary; at least 25% seed remains. A pH of 6.8 to 7.2 is maintained by adjusting the feed rate, sludge withdrawal, or alkalinity additions.

Note: The buffer capacity of an anaerobic digester is indicated by the volatile acids/alkalinity relationship. Decreases in alkalinity cause a corresponding increase in ratio.

If the digester is heated, the temperature must be controlled to a normal temperature range of 90 to 95°F. The temperature is never adjusted by more than 1°F per day. In digesters equipped with mixers, the mixing process ensures that organisms are exposed to food materials. Again, anaerobic digesters are inherently dangerous—several catastrophic failures have occurred. To prevent such failures, safety equipment, such as pressure relief and vacuum relief valves, flame traps, condensate traps, and gas collection safety devices, is installed. It is important that these critical safety devices be checked and maintained for proper operation.

Note: Because of the inherent danger involved with working inside anaerobic digesters, they are automatically classified as permit-required confined spaces. All operations involving internal entry must be made in accordance with OSHA's confined-space entry standard. Questions concerning safe entry into confined spaces of any type should be addressed by a Certified Safety Professional (CSP), Certified Industrial Hygienist (CIH), or Professional Engineer (PE).

Anaerobic digesters must be continuously monitored and tested to ensure proper operation. Testing is performed to determine supernatant pH, volatile acids, alkalinity, biochemical oxygen demand (BOD) or chemical oxygen demand (COD), total solids, and temperature. Sludge (in and out) is routinely tested for percent solids and percent volatile matter. Normal operating parameters are listed in Table 7.13.

TABLE 7.13
Anaerobic Digester: Normal Operating Ranges

Parameter	Normal Range
Sludge retention time	
Heated	30–60 days
Unheated	180+ days
Volatile solids loading	0.04–0.1 lb/day/ft^3
Operating temperature	
Heated	90–95°F
Unheated	Varies with season
Mixing	
Heated—primary	Yes
Unheated—secondary	No
Methane in gas	60–72%
Carbon dioxide in gas	28–40%
pH	6.8–7.2
Volatile acids-to-alkalinity ratio	≤0.1
Volatile solids reduction	40–60%
Moisture reduction	40–60%

Process control calculations involved with anaerobic digester operation include determining the required seed volume, volatile acids-to-alkalinity ratio, sludge retention time, estimated gas production, volatile matter reduction, and percent moisture reduction in digester sludge. Examples of how to make these calculations are provided in the following sections.

Required Seed Volume in Gallons

$$\text{Seed volume (gal)} = \text{Digester volume} \times \%\text{Seed} \qquad (7.1)$$

■ EXAMPLE 7.1

Problem: A new digester requires a 25% seed to achieve normal operation within the allotted time. If the digester volume is 266,000 gal, how many gallons of seed material will be required?
Solution:

$$\text{Seed volume} = 266,000 \times 0.25 = 66,500 \text{ gal}$$

Volatile Acids-to-Alkalinity Ratio

The volatile acids-to-alkalinity ratio can be used to control operation of an anaerobic digester:

$$\text{Ratio} = \frac{\text{Volatile acids concentration}}{\text{Alkalinity concentration}} \qquad (7.2)$$

■ EXAMPLE 7.2

Problem: A digester contains 240 mg/L volatile acids and 1860 mg/L alkalinity. What is the volatile acids-to-alkalinity ratio?
Solution:

$$\text{Ratio} = \frac{\text{Volatile acids concentration}}{\text{Alkalinity concentration}} = \frac{240 \text{ mg/L}}{186 \text{ mg/L}} = 0.13$$

Note: Increases in the ratio normally indicate a potential change in the operation condition of the digester, as shown in Table 7.14.

Sludge Retention Time

Sludge retention time (SRT) is the length of time the sludge remains in the digester:

$$\text{SRT (days)} = \frac{\text{Digester volume (gal)}}{\text{Sludge volume added per day (gpd)}} \qquad (7.3)$$

TABLE 7.14
Volatile Acids-to-Alkalinity Ratios

Operating Condition	Volatile Acids-to-Alkalinity Ratio
Optimum	≤0.1
Acceptable range	0.1–0.3
Increase in % carbon dioxide in gas	≥0.5
Decrease in pH	≥0.8

■ EXAMPLE 7.3

Problem: Sludge is added to a 525,000-gal digester at the rate of 12,250 gal per day. What is the sludge retention time?

Solution:

$$SRT = \frac{Digester\ volume}{Sludge\ volume\ added\ per\ day} = \frac{525,000\ gal}{12,250\ gpd} = 42.9\ days$$

Estimated Gas Production in Cubic Feet/Day

The rate of gas production is normally expressed as the volume of gas (ft^3) produced per pound of volatile matter destroyed. The total cubic feet of gas that a digester will produce per day can be calculated by

$$Gas\ production = VM_{in}\ (lb/day) \times \%VM\ reduction \times Production\ rate\ (ft^3/lb) \qquad (7.4)$$

■ EXAMPLE 7.4

Problem: A digester receives 11,450 lb of volatile matter per day. Currently, the volatile matter reduction achieved by the digester is 52%. The rate of gas production is 11.2 ft^3 of gas per pound of volatile matter destroyed.

Solution:

$$Gas\ production = 11,450\ lb/day \times 0.52 \times 11.2\ ft^3/lb = 66,685\ ft^3/day$$

Percent Volatile Matter Reduction

Because of the changes occurring during sludge digestion, the calculation used to determine percent volatile matter reduction is more complicated:

$$\%VM\ reduction = \frac{(\%VM_{in} - \%VM_{out}) \times 100}{\left[\%VM_{in} - (\%VM_{in} \times \%VM_{out})\right]} \qquad (7.5)$$

■ EXAMPLE 7.5

Problem: Using the data provided below, determine the percent volatile matter reduction for the digester:

Raw sludge volatile matter = 74%
Digested sludge volatile matter = 55%

Solution:

$$\%VM\ reduction = \frac{(0.74 - 0.55) \times 100}{\left[0.74 - (0.74 \times 0.55)\right]} = 57\%$$

Percent Moisture Reduction in Digested Sludge

$$\%Moisture\ reduction = \frac{(\%Moisture_{in} - \%Moisture_{out}) \times 100}{\left[\%Moisture_{in} - (\%Moisture_{in} \times \%Moisture_{out})\right]} \qquad (7.6)$$

■ **EXAMPLE 7.6**

Problem: Using the digester data provide below, determine the percent moisture reduction and percent volatile matter reduction for the digester.

Raw sludge percent solids = 6%
Digested sludge percent solids = 14%

Solution:

Note: Percent moisture = 100% − Percent solids.

$$\%\text{Moisture reduction} = \frac{(0.94 - 0.86) \times 100}{\left[0.94 - (0.94 \times 0.86) \right]} = 61\%$$

Anaerobic Digestion of Animal Wastes

Animal waste accounts for 10% of methane emissions in the United States. Ruminant animals, particularly cows and sheep, contain bacteria in their gastrointestinal systems that help to break down plant material. Some of these microorganisms use the acetate from the plant material to produce methane, and, because these bacteria live in the stomachs and intestines of ruminants, whenever the animal burps or defecates it emits methane as well (Spellman and Whiting, 2007). When not correctly managed, farm waste slurries can also seriously pollute local watercourses. Small anaerobic digesters have been installed on farms to treat excess animal slurries that cannot be placed on the land. The biogas formed is normally used for heat but can also be used to fuel engines and other onsite energy needs such as electricity and heating. Onsite biogas production and management also reduce offensive odors from overloaded or improperly managed manure storage facilities. These odors impair air quality and may be a nuisance to nearby communities. Anaerobic digestion of animal waste reduces these offensive odors because the volatile organic acids, the odor-causing compounds, are consumed by biogas-producing bacteria. In addition to biogas, another important byproduct of anaerobic digestion is ammonium, which is the major constituent of commercial fertilizer, which is readily available and utilized by crops. The bottom line on the production of biogas on the farm: Biogas recovery can improve profitability while improving environmental quality.

LANDFILL BIOGAS

Landfills can be a source of energy. Some wastewater treatment plants with anaerobic digesters located close to landfills harness the methane from the landfill and combine it with methane from their anaerobic digesters to provide additional power for their plant site. Landfills produce methane as organic waste decomposes in the same anaerobic digestion process used to convert wastewater and farm waste slurries into biogas. Most landfill gas results from the degradation of cellulose contained in municipal and industrial solid waste. Unlike animal manure digesters, which control the anaerobic digestion process, the digestion occurring in landfills is an uncontrolled process of biomass decay. To be technically feasible, a landfill must be at least 40 feet deep and have at least a million tons of waste in place for landfill gas collection.

The efficiency of the process depends on the waste composition and moisture content of the landfill, cover material, temperature, and other factors. The biogas released from landfills, commonly called *landfill gas*, is typically 50% methane, 45% carbon dioxide, and 5% other gases. The energy content of landfill gas is 400 to 550 Btu per cubic foot.

Figure 7.12 shows a landfill energy system. Such a system consists of a series of wells drilled into the landfill. A piping system connects the wells and collects the gas. Dryers remove moisture from the gas and filters remove impurities. The gas typically fuels an engine–generator set or gas turbine

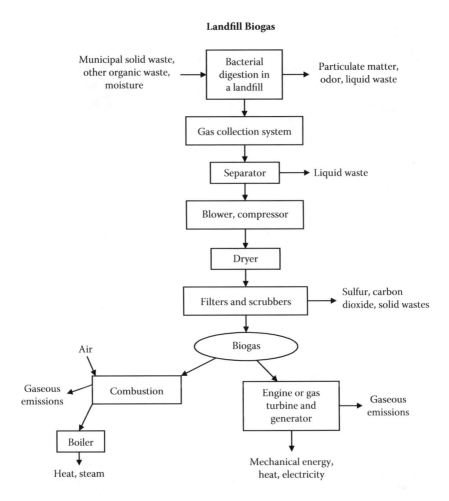

FIGURE 7.12   Landfill biogas system flow diagram.

to produce electricity. The gas can also fuel a boiler to produce heat or steam. Because waste-generated biogas is considered to be a dirty gas, as compared to natural gas, further gas cleanup is required to improve biogas to pipeline quality, the equivalent of natural gas. Reforming the gas to hydrogen would make possible the production of electricity using fuel cell technology.

ENVIRONMENTAL IMPACTS OF BIOMASS ENERY GENERATION

The combination of constructing biomass facilities (and the associated ancillaries), producing biomass feedstock, and operating biomass energy facilities may have environmental impacts. For example, the construction activities that could have an environmental impact include ground clearing, grading, excavation, blasting, trenching, drilling, facility construction, and vehicular and pedestrian traffic. Additionally, potential environmental impacts could result from biomass feedstock production activities such as the collection of waste materials and the growth and harvesting of woody and agricultural crops or algae, preprocessing, and transportation activities. Finally, operations activities that may have an environmental impact include operation of the biomass energy facility, power generation, biofuel production, and associated maintenance activities. Many of the environmental impacts resulting from these activities are discussed in the following text.

BIOMASS ENERGY CONSTRUCTION IMPACTS*

During the biomass energy facility construction phase, typical construction activities include ground clearing (removal of vegetative cover), grading, excavation, blasting, trenching, drilling, vehicular and pedestrian traffic, and construction and installation of facilities. Biomass power plants and some biogas plants that produce more electricity than required to operate the facility need transformers and transmission lines to deliver electricity to the power grid. Landfill gas production would require the drilling of wells for extraction of the gas and might require pipeline construction to deliver the gas to the user. Activities conducted in locations other than the facility site might include excavation and blasting for construction materials (e.g., sand, gravel) and access road construction. Potential impacts from these activities are presented below, by the type of affected resources.

Air Quality

Emissions generated during the construction phase include vehicle emissions; diesel emissions from large construction equipment and generators; release of volatile organic compounds (VOCs) from the storage and transfer of vehicle/equipment fuels; small amounts of carbon monoxide, nitrogen oxides, and particulates from blasting activities; and fugitive dust from any sources such as disturbing and removing soils (clearing, grading, excavating, trenching, backfilling, dumping, and truck and equipment traffic), mixing concrete, storage of unvegetated soil piles, and drilling and pile driving. Note that a permit is needed from the state or local air agency to control or mitigate these emissions; therefore, these emissions would not likely cause an exceedance of air quality standards nor have an impact on climate change. Moreover, a construction permit under the mandated prevention of significant deterioration (PSD) or air quality regulations might be required.

Cultural Resources

Direct impacts on cultural resources could result from construction activities, and indirect impacts might be caused by soil erosion and increased accessibility to possible site locations. Potential impacts include the following:

- Complete destruction of resources in areas undergoing surface disturbance or excavation
- Degradation or destruction of near-surface cultural resources on- and offsite resulting from topographic or hydrological pattern changes or from soil movement (removal, erosion, and sedimentation), although the accumulation of sediment could protect some localities by increasing the amount of protective cover
- Unauthorized removal of artifacts or vandalism to the site as a result of increases in human access to previously inaccessible areas, if significant cultural resources are present
- Visual impacts resulting from vegetation clearing, increases in dust, and the presence of large-scale equipment, machinery, and vehicles if the affected cultural resources have an associated landscape or other visual component that contributes to their significance, such as Native American sacred landscape or a historic trail

Ecological Resources

Ecological resources that could be affected include vegetation, fish, and wildlife, as well as their habitats. Vegetation and topsoil would be removed for the construction of the biomass energy facility, associated access roads, transmission lines, pipelines, and other ancillary facilities. This would lead to a loss of wildlife habitat, reduction in plant diversity, potential for increased erosion, and potential for the introduction of invasive or noxious weeds. The recovery of vegetation following interim and final reclamation would vary by the type of plant community desired. Dust settling on vegetation may alter or limit plants' abilities to photosynthesize or reproduce. Although the

---

* Adapted from Tribal Energy and Environmental Information Clearinghouse, http://teeic.anl.gov/er/biomass/impact/construct/index.cfm.

DID YOU KNOW?

Biomass power plants emit nitrogen oxides and a small amount of sulfur dioxide. The amounts emitted depend on the type of biomass that is burned and the type of generator used. Although the burning of biomass also produces carbon dioxide, the primary greenhouse gas, it is considered to be part of the natural carbon cycle of the Earth. The plants take up carbon dioxide from the air while they are growing and then return it to the air when they are burned, thereby causing no net increase.

potential for an increase in the spread of invasive and noxious weeds would occur during the construction phase due to increasing traffic and human activity, the potential impacts could be reduced by interim reclamation and implementation of mitigation measures. Adverse impacts on wildlife could occur during construction from

- Erosion and runoff
- Fugitive dust
- Noise
- Introduction and spread of invasive vegetation
- Modification, fragmentation, and reduction of habitat
- Mortality of biota (i.e., death of plants and animals)
- Exposure to contaminants
- Interference with behavioral activities

Wildlife would be most affected by habitat reduction within the project site, access roads, and gas and water pipeline rights-of-way. Wildlife within surrounding habitats might also be affected if the construction activity (and associated noise) disturbs normal behaviors, such as feeding and reproduction. Depletion of surface waters from perennial streams could result in a reduction of water flow, which could lead to habitat loss or degradation of aquatic species.

Water Resources

With regard to water resources (surface water and groundwater), water would be used for dust control when clearing vegetation and grading and for road traffic; for making concrete for foundations and ancillary structures; and for consumptive use by the construction crew. Water is likely to be obtained from nearby surface water bodies or aquifers, depending on availability, but could be trucked in from offsite. The bottom line on water for potable use always comes down to the Q and Q factors: quantity and quality. The quantity of water used would be small relative to water availability. Water quality could be affected by

- Activities that cause soil erosion
- Weathering of newly exposed soils that could cause leaching and oxidation, thereby releasing chemicals into the water
- Discharges of waste or sanitary water
- Untreated groundwater used to control dust could deposit dissolved salts on the surface, allowing the salts to enter surface water systems
- Chemical spills
- Pesticide applications

Surface water and groundwater flow systems could be affected by withdrawals made for water use, wastewater and stormwater discharges, and the diversion of surface water flow for access road construction or stormwater control systems. A stormwater discharge permit might be required.

Excavation activities and the extraction of geological materials could affect surface and groundwater flow. The interaction between surface water and groundwater could also be affected if the surface water and groundwater were hydrologically connected, potentially resulting in unwanted dewatering or recharging of water resources.

Land Resources

Impacts on land use could occur during construction if there were conflicts with existing land use plans and community goals; conflicts with existing recreational, educational, religious, scientific, or other use areas; or conversion of the existing commercial land use for the area (e.g., agriculture, grazing, mineral extraction). Existing land use during construction would be affected by intrusive impacts such as ground clearing, increased traffic, noise, dust, and human activity, as well as by changes in the visual landscape. In particular, these impacts could affect recreationists seeking solitude or recreation opportunities in a relatively pristine landscape. Ranchers or farmers could be affected by the loss of available grazing or crop lands, the potential for the introduction of invasive plants that could affect livestock forage availability, and possible increases in livestock/vehicle collisions. An expanded access road system could increase the numbers of off-highway vehicle users, hunters, and other recreationists in the surrounding area.

Impacts on aviation could be possible if the project is located within 20,000 feet (6100 meters) or less of an existing public or military airport, or if proposed construction involves objects greater than 200 feet (61 meters) in height. The Federal Aviation Administration (FAA) must be notified if either of these two conditions occurs, and the FAA would be responsible for determining if the project would adversely affect commercial, military, or personal air navigation safety. Similarly, impacts on military operations could occur if a project was located near a military facility, if that facility conducts low-altitude military testing and training activities.

Soils and Geologic Resources

Sands, gravels, and quarry stone for construction access roads, making concrete for foundations and ancillary structures, and improving ground surface for laydown areas and crane staging areas would either be brought in from offsite sources or would be excavated on site. Depending upon the extend of excavation and blasting required to install access roads and support facilities, there is a limited risk of triggering geological hazards (e.g., landslides). Altering drainage patterns could also accelerate erosion and create slope instability. Disturbed soil surfaces (crusts) now cover vast areas in the western United States as a result of ever-increasing recreational and commercial uses of these semi-arid and arid areas. Based on the findings of several studies (Belnap and Gillette, 1997; McKenna-Neumann et al., 1996; Williams et al., 1995), the tremendous land area currently affected by human activity may lead to significant increases in regional global wind erosion rates. Surface disturbance, heavy equipment traffic, and changes to surface runoff patterns resulting from biomass energy construction activities could cause soil erosion and impacts on special soils (e.g., cryptobiotic soil crusts; discussed below). Impacts of soil erosion could include soil nutrient loss and reduced water quality in nearby surface water bodies.

Cryptobiotic Soils Crust

With regard to disturbance of cryptobiotic soil crusts, this is an important but often overlooked and not fully appreciated or understood soil disturbance problem, especially within the western and southwestern United States. Whether the renewable energy source is solar, wind, hydro, or biomass, the western and southwestern states are key players in harnessing and processing these energy sources. Cryptobiotic soil crusts, consisting of soil cyanobacteria, lichens, and mosses, play an important ecological role in the arid Southwest. In the cold deserts of the Colorado Plateau region (parts of Utah, Arizona, Colorado, and New Mexico), these crusts are extraordinarily well developed, often representing over 70% of the living ground cover. Cryptobiotic crusts increase the stability of otherwise easily eroded soils, increase water infiltration in regions that receive little precipitation, and increase fertility in soils often limited in essential nutrients such as nitrogen and carbon (Belnap, 1994; Belnap and Gardener, 1993; Harper and Marble, 1988; Johansen, 1993; Metting, 1991; Williams et al., 1995).

Cyanobacteria occur as single cells or as filaments. The most common type found in desert soils is the filamentous type. The cells or filaments are surrounded by sheaths that are extremely persistent in these soils. When moistened, the cyanobacterial filaments become active, moving through the soils and leaving a trail of the sticky, mucilaginous sheath material behind. This sheath material sticks to surfaces such as rock or soil particles, forming an intricate webbing of fibers in the soil. In this way, loose soil particles are joined together, and otherwise unstable and highly erosion-prone surfaces become resistant to both wind and water erosion. The soil-binding action is not dependent on the presence of living filaments. Layers of abandoned sheaths, built up over long periods of time, can still be found clinging tenaciously to soil particles at depths greater than 15 cm in sandy soils. This provides cohesion and stability in these loose sandy soils even at depth.

Cyanobacteria and cyanolichen components of these soil crusts are important contributors of fixed nitrogen (Mayland and McIntosh, 1966; Rychert and Skujins, 1974). These crusts appear to be the dominant source of nitrogen in cold-desert pinyon–juniper and grassland ecosystems over much of the Colorado Plateau (Evans and Belnap, unpub. data; Evans and Ehleringer, 1993). Biological soil crusts are also important sources of fixed carbon on sparsely vegetated areas common throughout the arid West (Beymer and Klopatek, 1991). Plants growing on crusted soil often show higher concentrations and/or greater total accumulation of various essential nutrients when compared to plants growing in adjacent, uncrusted soils (Belnap and Harper, 1995; Harper and Pendleton, 1993).

Cryptobiotic soil crusts are highly susceptible to soil-surface disturbance such as trampling by hooves or feet, or driving of off-road vehicles, especially in soils with low aggregate stability such as areas of sand dunes and sheets in the Southwest, in particular over much of the Colorado Plateau (Belnap and Gardner, 1993; Gillette et al., 1980; Webb and Wilshire, 1983). When crusts in sandy areas are broken in dry periods, previously stable areas can become moving sand dunes in a matter of only a few years.

Cyanobacterial filaments, lichens, and mosses are brittle when dry, and crush easily when subjected to compressional or shear forces by activities such as trampling or vehicular traffic. Many soils in these areas are thin and are easily removed without crust protection. As most crustal biomass is concentrated in the top 3 mm of the soil, even very little erosion can have profound consequences for ecosystem dynamics. Because crustal organism are only metabolically active when wet, re-establishment time is slow in arid systems. Although cyanobacteria are mobile and can often move up through disturbed sediments to reach needed light levels for photosynthesis, lichens and mosses are incapable of such movement and often die as a result. On newly disturbed surfaces, mosses and lichens often have extremely slow colonization and growth rates. Assuming adjoining soils are stable and rainfall is average, recovery rates for lichen cover in southern Utah have been most recently estimated at a minimum of 45 years, while recovery of moss cover was estimated at 250 years (Belnap, 1993).

Because of such slow recolonization of soil surfaces by the different crustal components, underlying soils are left vulnerable to both wind and water erosion for at least 20 years after disturbance (Belnap and Gillette, 1997). Because soils take 5000 to 10,000 years to form in arid areas such as in southern Utah (Webb, 1983), accelerated soil loss may be considered an irreversible loss. Loss of soil also means loss of site fertility through loss of organic matter, fine soil particles, nutrients, and microbial populations in soils (Harper and Marble, 1988; Schimel et al., 1985). Moving sediments further destabilize adjoining areas by burying adjacent crusts, leading to their death, or by providing material for "sandblasting" nearby surfaces, thus increasing wind erosion rates (Belnap, 1995; McKenna-Neumann et al., 1996).

Soil erosion in arid lands is a global problem. Beasley et al. (1984) estimated that in the rangelands of the United States alone, 3.6 million hectares have undergone some degree of accelerated wind erosion. Relatively undisturbed biological soil crusts can contribute a great deal of stability to otherwise high erodible soils. Unlike vascular plant cover, crustal cover is not reduced in drought, and unlike rain crusts, these organic crusts are present year-round; consequently, they offer stability over time and under adverse conditions that is often lacking in other soil surface protectors.

Paleontological Resources

Impacts on paleontological resources could occur directly from the construction activities or indirectly from soil erosion and increased accessibility to fossil locations. Potential impacts include the following:

- Complete destruction of resources in areas undergoing surface disturbance or excavation
- Degradation or destruction of near-surface fossil resources on- and offsite caused by changes in topography, changes in hydrological patterns, and soil movement (removal, erosion, and sedimentation), although the accumulation of sediment could serve to protect some locations by increasing the amount of protective cover
- Unauthorized removal of fossil resources or vandalism to the site as a result of increased human access to previously inaccessible areas, if significant paleontological resources are present

Transportation

Short-term increases in the use of local roadways would occur during the construction period. Heavy equipment likely would remain at the site. Shipments of materials are unlikely to affect primary or secondary road networks significantly, but this would depend on the location of the project site relative to material source. Oversized loads could cause temporary transportation disruptions and could require some modifications to roads or bridges (such as fortifying bridges to accommodate the size or weight). Shipment weight might also affect the design of access roads for grade determinations and turning clearance requirements.

Visual Resources

The magnitude of visual impacts of construction of a biomass facility is dependent upon the distance of the construction activities from the viewer, the view duration, and the scenic quality of the landscape. Possible sources of visual impacts during construction include the following:

- Ground disturbance and vegetation removal could result in visual impacts that produce contrasts of color, form, texture, and line. Excavation for foundations and ancillary structures, trenching to bury pipelines, grading and surfacing roads, cleaning and leveling staging areas, and stockpiling soil might be visible to viewers in the vicinity of the site. Soil scars and exposed slope faces would result from excavation, leveling, and equipment movement.
- Road development (new roads or expansion of existing roads) and parking areas could introduce strong visual contrasts in the landscape, depending on the route relative to surface contours, and the width, length, and surface treatment of the roads.
- Conspicuous and frequent small-vehicle traffic for worker access and frequent large-equipment (trucks, graders, excavators, and cranes) traffic for road construction, site preparation, and biomass facility construction could produce visible activity and dust in dry soils. Suspension and visibility of dust would be influenced by vehicle speeds and road surface materials.
- There would be a temporary presence of large equipment producing emissions while in operation and creating visible exhaust plumes. Support facilities and fencing associated with the construction work would also be visible.
- Night lighting would change the nature of the visual environment in the vicinity.

Socioeconomics

Direct impacts would include the creation of new jobs for construction workers and the associated income and taxes generated by the biomass facility. Indirect impacts would occur as a result of the new economic development and would include new jobs at businesses that support the expanded workforce or provide project materials, and associated income and taxes. Proximity to biomass facilities could potentially affect property values, either positively from increased employment effects or negatively from proximity to residences or local businesses and any associated or perceived environmental effect (noise, visual, etc.). Adverse impacts could occur if a large in-migrant workforce, culturally different from the local indigenous group, is brought in during construction. This influx of migrant workers could strain the existing community infrastructure and social services.

Environmental Justice

If significant impacts occurred in any resource areas, and these impacts disproportionately affected minority or low-income populations, then there could be an environmental justice impact. Issues of potential concern during construction are noise, dust, and visual impacts from the construction site and possible impacts associated with the construction of new access roads. Additional impacts include limitations on access to the area for recreation, subsistence, and traditional activities.

BIOMASS FEEDSTOCK PRODUCTION IMPACTS

The impacts of biomass production are essentially the same as those of farming and forestry. The biomass production phase can be broken down into feedstock production and feedstock logistics. Feedstock production is the cultivation of crops such as corn, soybeans, or grasses and the collection of crop residues and wood residues from forests. These can be further categorized as primary, secondary, and tertiary resources:

- Primary feedstock includes grain and oilseed crops, such as corn or soybeans, that are grown specifically to make biofuels; crop residues such as corn stover and straw; perennial grasses and woody crops; algae; logging residue; and excess biomass from forests.
- Secondary feedstock consists of manure from farm animals, food residue, wood processing mill residue, and pulping liquors.
- Tertiary feedstock includes municipal solid waste, municipal sanitary waste sludge, landfill gases, urban wood waste, construction and demolition debris, and packaging waste.

Feedstock logistics consisting of harvesting or collecting the feedstock from the production area, processing it for use in a biomass facility, storing it to provide for a steady supply, and delivering it to the plant. The following potential impacts may result from biomass energy production activities.

Air Quality

Emissions generated during the feedstock production phase include vehicle emissions, diesel emissions from large equipment, emissions from storage and dispensing of fuels, and fugitive dust from many sources. The level of emissions would vary with the scale of operations and may be greater for agriculture operations than forestry operations. For feedstocks that do not require annual replanting (e.g., switchgrass, hybrid poplars) and cultivation (e.g., mill residues) or for algae, which is grown in enclosed aquaculture facilities, potential air emissions would be greatly reduced. If all vehicles and equipment have emission control devices and dust control measures are implemented, air emissions are unlikely to cause an exceedance of air quality standards. The removal of biomass from forests can reduce the potential for major forest fires and limit the need for prescribed burns, thereby eliminating some air pollution sources. However, from a climate change perspective, large reductions in forest mass (clear cutting) can remove biomass that served to capture carbon dioxide.

Carbon dioxide, a greenhouse gas, is considered a major contributor to climate change. Mechanisms that can capture or contain carbon dioxide, such as forests, are considered to be a viable mitigation measure against climate change.

Cultural Resources

Any cultural material present on the surface or buried below the surface of existing agricultural areas has already been disturbed, some for many decades. The conversion of uncultivated land to agricultural use to produce feedstock for biomass facilities would disturb previously undisturbed land and could affect cultural resources on or buried below the surface. Harvesting and collecting biomass from the forests could also affect cultural resources on or buried below the surface associated with the harvesting. If new access roads were required, this construction could also affect cultural resources. These agricultural and forestry activities could affect areas of interest to Native Americans depending on their physical placement and level of visual intrusion. Surveys conducted prior to the commencement of farming uncultivated land or harvesting in the forest to evaluate the presence and significance of cultural resources in the area would assist developers in properly managing cultural resources so they can plan their project to avoid or minimize impacts on these resources.

Ecological Resources

Vegetation and wildlife, including threatened, endangered, and sensitive species and their critical habitats, have been displaced from years of crop production. Converting uncultivated or fallow land to agriculture crops would result in additional displacement of native vegetation and wildlife. Forest stand thinning improves the growth of the remaining trees and reduces fire hazards; however, some native wildlife populations may decline as a result of habitat loss, fragmentation, and disturbance due to forest openings resulting from road construction and biomass collection. The presence of workers could increase human disturbance to wildlife. Limiting work activities in the vicinity of any known active nesting sites would help protect wildlife. Habitat alteration, including canopy cover and soil compaction, can degrade habitat for native plant populations and provide for the establishment of invasive plant species. After clearing a given area of biomass, additional seeding of highly disturbed soils with native grasses and taking steps to prevent the spread of noxious weeds could minimize impacts. However, a lower abundance of birds is sometimes found in reforested areas compared with natural forest or grassland.

Water Resources

Agricultural land use can degrade water quality where it results in runoff or the migration of nutrients, pesticides, and other chemicals into surface water and groundwater. Conversion of idle land to agriculture use would add to the degradation. Converting annual crops to perennial crops reduces the requirement for pesticides and fertilizers. If the conversion of idle land requires irrigation, then water would have to be withdrawn from surface water or groundwater sources. Large withdrawals could affect the water availability for other uses. Sedimentation from road construction and other ground-disturbing activities in forested regions could increase sedimentation levels in streams.

Land Use

Demand for increasing amounts of agricultural biomass feedstock would convert the land and cropland pasture to cultivation of perennial crops such as grass and wood crops. No change in land use in forests would occur as a result of clearing and thinning to remove biomass.

Soils and Geologic Resources

Crop residue left in the field and tree residue left in the forest help to maintain soil moisture, soil organic matter content, and soil carbon levels and to limit wind erosion. Removing too much residue would be detrimental to the soil. Soil compaction in agricultural operations would result from multiple passes of equipment for crop residue collection. These impacts can be partially mitigated by converting land from annual crop to perennial biomass crop production. This would increase the organic matter content of the soils and maximize the potential benefits listed above. The application of pesticides affects soil quality by adding toxic chemicals to the soil. Converting annual crops to perennial crops reduces the requirement for pesticides and fertilizers. Proper management of the type and quantities of pesticides can reduce the impact. Soil compaction, erosion, and topsoil loss result from logging operations. Biomass removal would utilize the same footprint as commercial harvesting activities and would not add to the amount of compacted or disturbed soil.

Paleontological Resources

Any paleontological resources present on the surface or buried below the surface of existing agricultural areas have already been disturbed. The conversion of uncultivated land to agricultural use to produce feedstock for biomass facilities would disturb previously undisturbed land and affect paleontological resources on or buried below the surface. Harvesting and collecting biomass from the forests could also affect paleontological resources on or buried below the surface associated with the harvesting. If new access roads were required, this construction could also affect these resources. Surveys conducted prior to the commencement of farming uncultivated land or harvesting in the forest to evaluate the presence and/or significance of paleontological resources in the area would assist developers in properly managing paleontological resources so they can plan their project to avoid or minimize impacts on these resources.

Transportation

Increased road congestion from agricultural vehicles, logging trucks, and workers would occur for the duration of the activities in a given area. Transportation of collected biomass from the point of generation to storage facilities or to biomass energy production facilities could also result in impacts on the transportation system.

Visual Resources

Converting idle land to agricultural use would change the visual aspect of an area and probably would be noticeable by any nearby residents or travelers that are familiar with the area. Major timber harvests change the look and character of hillsides and the views that local residents and tourists have of the forest. Collection of the residue biomass would not add to the visual degradation. Changes in the character of the forest resulting from the collection of biomass in forest thinning projects would most likely not be observed from a distance.

Socioeconomics

Direct impacts would include the creation of new jobs for farmworkers and the associated income and taxes generated by increased production of crops and grasses and new markets for crop residue. Increased biomass collection in forested regions would also create new jobs. Indirect impacts would include new jobs at businesses that support the expanded workforce or provide farm and logging equipment materials, and associated income taxes.

Environmental Justice

If significant impacts occurred in any resource areas, and these impacts disproportionately affect minority or low-income populations, then there could be an environmental justice impact. Issues of potential concern are noise, dust, and visual impacts from biomass production and harvesting and potential construction of new access roads. Additional impacts include limitations on access to the area for recreation, subsistence, and traditional activities.

Biomass Energy Operations Impacts

Operations activities that may cause environmental impacts include operation of the biomass energy facility, power generation, biofuel production, and associated maintenance activities. Typical activities during biomass facility operation include power generation or production of biofuels and associated maintenance activities that would require vehicular access and heavy equipment operation when components are being replaced. Biomass power plants require pollution control devices to reduce emissions from combustion and large cooling systems. Water requirements vary greatly among the various biomass facilities. Potential impacts for these activities are presented below, by the type of affected resource.

Air Quality

Operation of biomass facilities results in emissions of criteria air pollutants and hazardous air pollutants (HAPs). Criteria air pollutants include particulate matter, carbon monoxide, sulfur oxides, nitrogen oxides, lead, and volatile organic compounds (VOCs). HAPs are 189 toxic chemicals, known or suspected to be carcinogens, which are regulated by the U.S. Environmental Protection Agency as directed by the 1990 Clean Air Act. If the facility is in an area designated as "attainment" for all state and national ambient air quality standards (NAAQS), then emissions from operation, when added to the natural background levels, must not cause or contribute to ambient pollution levels that exceed the ambient air quality standards.

In particular, combustion of municipal solid wastes could result in trace quantities of mercury, other heavy metals, and dioxins in the air emissions. The use of best available control technology (BACT) would minimize the potential for adverse air quality impacts from biomass facilities. A gas-fired regenerative thermal oxidizer would reduce VOCs by 95%. Baghouses, which are a type of dust collector using fabric filters, control particulate matter. Enclosing the processing equipment in a slight negative pressure envelope in addition to the use of baghouses could minimize fugitive dust emissions from milling operations.

The use of cultivated biomass fuel (i.e., fuel specifically grown for energy production) in place of possible fuels such as coal, oil, and natural gas can result in a reduction in the amount of carbon dioxide that accumulates in the atmosphere only if the carbon released by combustion of biomass fuels is effectively recaptured by the next generation of feedstock plants. If the biomass source is not replaced by growing more plants, the carbon released in biomass combustion is not recaptured; therefore, these forms of biomass energy can only be considered to be carbon free if the energy production cycle includes replacing the feedstock. Using perennial or fast-growing biomass plants, such as switchgrass or poplar hybrids, can increase the rate of carbon recapture. Although the combustion of biomass fuels under these conditions can be considered to be carbon free, in practice any gains in terms of reduced carbon dioxide emissions are offset by carbon dioxide emissions associated with the use of fossil fuels in the cultivation, harvesting, and transportation of the biomass feedstock. Certain agricultural practices (e.g., no-till agriculture, use of perennial feedstock crops) produce fewer carbon dioxide emissions than conventional practices. Biomass energy derived from waste product fuels (e.g., residues from forestry operations, construction wastes, municipal wastes) is not considered to be carbon free, as the energy production cycle does not involve any cultivation of new biomass.

Cultural Resources

Impacts during the operations phase would be limited to unauthorized collection of artifacts and visual impacts. The threat of unauthorized collection would be present once the access roads are constructed in the construction phase, making remote lands accessible to the public. Visual impacts resulting from the presence of a biomass facility and transmission lines could affect some cultural resources, such as sacred landscapes or historic trails.

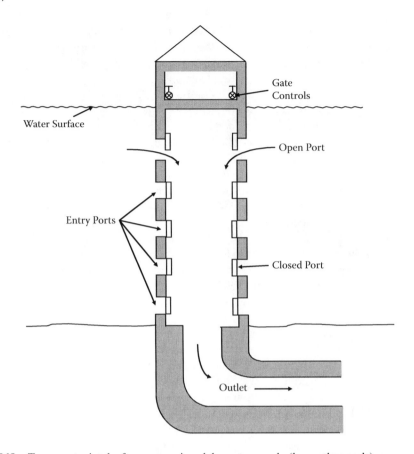

FIGURE 7.13   Tower water intake for a reservoir or lake water supply (larger than scale).

Ecological Resources

During operation, adverse ecological effects could occur from (1) disturbance of wildlife by equipment noise and human activity, (2) exposure of biota to chemical spills and other contaminants, and (3) mortality of wildlife due to increased vehicular traffic and collisions with or electrocution by transmission lines. Disturbed wildlife would be expected to eventually acclimatize to facility operations. Deposition of water and salts from the operation of mechanical-draft cooling towers has the potential to impact vegetation. Water intake structures (see Figure 7.13) for withdrawal of water from lakes or rivers would result in impingement and entrainment of aquatic species. Proper design of these structures can minimize these impacts. Discharge of heated cooling water into water bodies could be beneficial or adverse, depending on the design of the discharge structure and the temperature of the effluent.

Water Resources

Withdrawals of surface water and/or groundwater are expected to continue during the operations phase of both biomass power plants and biofuel production and refinery facilities. The amount of water needed depends on the type of facility. In a typical biomass power plant, the primary consumptive use of water will be to support the cooling system used to condense spent steam for reuse. Once-through cooling systems require large quantities of water to be withdrawn from and returned to a surface body of water. Wet recirculating cooling systems recycle cooling water through cooling towers where some portion of water is allowed to evaporate and must be continuously replenished. Wet recirculating cooling systems also periodically discharge small volumes of water as blowdown and replace that amount with freshwater to control chemical and biological contaminants to acceptable levels. A third type of cooling system, the dry cooling system, condenses and cools steam using only ambient air and requires no water to operate; however, some dry cooling systems can also be hybridized into wet/dry systems that use minimal amounts of water, which is allowed to evaporate to improve performance. Other consumptive uses of water at a biomass power plant include the initial filling and maintenance of the steam cycle, sanitary applications to support the workforce, and a wide variety of incidental maintenance-related industrial applications.

Most uses of water at a biomass power plant will ultimately result in the generation of some wastewater. Blowdown from both the steam cycle and the wet recirculating cooling system will represent the largest wastewater stream and, because water in both the steam cycle and the cooling system undergoes some chemical treatment, the discharge will contain chemical residuals. Its temperature will also be elevated. Water discharged from once-through systems does not undergo chemical treatment, but the temperature of the discharge will be elevated. All wastewater discharges from biomass power plants can be directed to a holding pond for evaporation, cooling, and further treatment but are likely to be eventually discharged to surface waters. The Clean Water Act requires any facility that discharges from a point source into water of the United States to obtain a National Pollutant Discharge Elimination System (NPDES) permit. The NPDES permit ensures that the state's water quality standards are being met.

Water is used in a wide variety of applications for biofuel production and refining facilities and can be consumed at rates as high as 400 gallons per minute (gpm). Some water used in production and refining activities can be recovered and recycled to reduce the demand on the water source. Algae production ponds can be large but are very shallow (about 12 inches). Only a small volume of water would need to be added to replace any evaporation. Bioreactors for algae production are closed systems and require very little additional water. As much as 100 gpm of wastewater can be discharged from a biofuel production and refining plant. The effluent discharge temperature would be at or slightly above ambient temperature and would often contain small amounts of chemicals. As with wastewaters from biomass power plants, such discharges can be directed to lined holding ponds for further treatment or discharged directly to a surface water body under the authority of an USEPA-issued NPDES permit.

Land Use

Any land use impacts would occur during construction, and no further impacts would be expected to result from biomass facility operation.

Soils and Geologic Resources

During operation, the soil and geologic conditions would stabilize with time. Soil erosion and soil compaction are both likely to continue to occur along access roads. Within the project footprint, soil erosion, surface runoff, and sedimentation of nearby water bodies will continue to occur during operation, but to a lesser degree than during the construction phase.

Paleontological Resources

Impacts during the operations phase would be limited to unauthorized collection of fossils. This threat is present once the access roads are constructed in the construction phase, making remote lands accessible to the public.

Transportation

Increases in the use of local roadways and rail lines would occur during operations. Biomass fuels for boilers and power plants would arrive daily by truck or rail. Feedstock for biofuels facilities, such as corn, soybeans, wood products, manure, and sludge, would also arrive by truck or rail, and ethanol and biodiesel produced would most likely be trucked to the end user, who would blend or sell the product. Depending upon the size and function of the facility, truck traffic could be on the order of 250 trucks per day. Biogas facilities would either combust the gas at the production plant or send it by pipeline to the user. Landfill gas would either be used to produce electricity near the point of collection or be sent by pipeline to the user.

Visual Resources

The magnitude of visual impacts from operation of a biomass facility is dependent on the distance of the facility from the viewer, the view duration, and the scenic quality of the landscape. Facility lighting would adversely affect the view of the night sky in the immediate vicinity of the facility. Plumes from stacks of cooling towers might be visible, particularly on cold days. Additional visual impacts would occur from the increase in vehicular traffic.

Socioeconomics

Direct impacts would include the creation of new jobs for operation and maintenance workers and the associated income and taxes paid. Indirect impacts are those impacts that would occur as a result of the new economic development and would include new jobs at businesses that support the workforce or that provide project materials, and associated income and taxes. The number of project personnel required during the operation and maintenance phase would be fewer than during construction; therefore, socioeconomic impacts related directly to jobs would be smaller than during construction.

Environmental Justice

Possible environmental justice impacts during operation include the alteration of scenic quality in areas of traditional or cultural significance to minority or low-income populations and disruption of access to those areas. Noise impacts, health and safety impacts, and water consumption are also possible sources of disproportionate effect.

BIOENERGY IMPACTS ON HUMAN HEALTH AND SAFETY*

As demand for low-carbon-impact, domestically sourced fuels has increased, biofuels have become a fast-growing part of the energy sector. Biofuels are produced from renewable resources, such as grains, plants biomass, vegetable oils, and treated municipal and industrial wastes. They are flammable or combustible, and their manufacture can involve potentially dangerous chemical reactions. Employers must protect workers from the hazards of these fuels and their production processes. There are currently two major types of biofuels being produced in the United States:

- *Ethanol* is a flammable liquid that is readily ignited at ordinary temperatures. Renewable ethanol is produced by the fermentation of grains, or, using advanced technologies, from cellulosic material such as waste paper, wood chips, and agricultural wastes. The production process can involve other hazardous materials, such as acids, bases, and gasoline (to denature the alcohol or for blending). Up to 10% ethanol is blended with gasoline in most automotive fuel currently sold in the United States. Higher ethanol blends, up to E85 (85% ethanol blended with gasoline), are also available in some parts of the country.
- *Biodiesel* is a combustible liquid that burns readily when heated; blending it with petroleum-based diesel fuel or contamination by materials used in manufacturing can increase its flammability. Biodiesel is produced by reacting organic materials such as vegetable oils with an alcohol, typically methanol, using a strong base, such as a caustic, as a catalyst. Glycerin, a combustible liquid, is produced as a byproduct. The caustic is neutralized with acid, typically sulfuric acid. All of these materials may require careful management to protect workers. Biodiesel blended with petroleum-based diesel is widely available.

---

* Adapted from OSHA's *Green Job Hazards: Biofuels*, https://www.osha.gov/dep/greenjobs/biofuels.html.

Biofuels Job Hazards

Potential hazards in biofuels production and handling include the following:

- Fire/explosion hazards
- Chemical reactivity hazards
- Toxicity hazards

Fire and Explosion Hazards[*]

Employers producing biofuels may expose workers to potential fire and explosion hazards, and they must protect them from these hazards by preventing releases, avoiding ignition of spills, and having appropriate fire protection systems and emergency response procedures in place. Engineering controls that should be used include the following:

- Good facility layout
- Proper design of vessels and piping systems
- Proper selection of electrical equipment for use in hazardous (classified) areas
- Adequate instrumentation with alarms, interlocks, and shutdowns

Administration controls that should be used include the following:

- Operating procedures
- Good maintenance practices
- Safe work practices/procedures

Facilities processing more than 10,000 pounds of flammable liquids or flammable mixtures may be covered by 29 CFR 1910.119 (Process Safety Management of Highly Hazardous Chemicals).

Chemical Reactivity Hazards[†]

Biofuels manufacturing processes can present reactive hazards. Although ethanol production by fermentation involves biological reactions that do not present an explosive "run-away reaction" hazard, some processes for making ethanol from materials such as waste paper and wood chips use concentrated acids and bases that can react vigorously with many materials. Also, the gases produced during ethanol fermentation need to be properly vented to avoid overpressuring equipment and piping. Biodiesel is produced by the chemical reaction of organic oils with an alcohol, typically using a strong base as a catalyst. The glycerin that is co-produced with the biodiesel is then often treated with acid. These reactions need to be carefully controlled. Failure to control potentially dangerous chemical reactions can lead to the rupture of equipment and piping, explosions, fires, and exposures to hazardous chemicals. Employers need to protect their workers from these hazards. Engineering and administrative controls to keep the process within safe limits include controlling the rate and order of chemical addition, providing robust cooling, segregating incompatible materials to prevent inadvertent mixing, and the use of detailed operating procedures. Several OSHA standards address potential reactive hazards:

- 29 CFR 1910.119, Process Safety Management of Highly Hazardous Chemicals
- 29 CFR 1910.1200, Hazard Communication
- 29 CFR 1910.147, The Control of Hazardous Energy (Lockout/Tagout)

---

[*] Adapted from OSHA's *Green Job Hazards: Biofuels—Fire and Explosion Hazards of Biofuels*, https://www.osha.gov/dep/greenjobs/bio_fireexplosion.html.
[†] Adapted from OSHA's *Green Job Hazards: Biofuels—Chemical Reactivity Hazards in Biofuel Manufacturing*, https://www.osha.gov/dep/greenjobs/bio_chemical.html

DID YOU KNOW?

Hazards, including toxic hazards that are not addressed by specific OSHA standards, still need to be controlled. Under Section 5(a)(1) of the Occupational Safety and Health Act, the "General Duty Clause," employers are required to provide workers with a place of employment that is free from recognized hazards that are causing or are likely to cause death or serious physical harm to employees.

Toxicity Hazards[*]

Biofuels and the chemicals used in their manufacture present toxic exposure hazards that need to be carefully controlled to protect workers. Material Safety Data Sheets (MSDSs) should be consulted to determine the potential for toxic exposures to feedstocks, products, and other chemicals used in biofuel processes, including, but not limited to, methanol, caustic, sulfuric acid, ethanol, and biodiesel, as well as hydrocarbons used for blending and alcohol denaturing. Engineering and administrative controls should be used to control hazards, including, but not limited to, good engineering, design, fabrication, and maintenance practices to prevent releases, ventilation and drainage to reduce exposures, and appropriate use of personal protective equipment, when needed. A number of OSHA standards address these potential hazards:

- 29 CFR 1910.119, Process Safety Management of Highly Hazardous Chemicals
- 29 CFR 1910.1200, Hazard Communication
- 29 CFR 1910, Subpart I, Personal Protective Equipment
- 29 CFR 1910.134, Respiratory Protection
- 29 CFR 1910.146, Permit-Required Confined Spaces
- 29 CFR 1910.147, The Control of Hazardous Energy (Lockout/Tagout)

BOTTOM LINE ON BIOFUELS

When oil prices are high, the future of biofuel—made from plant material (biomass)—is of keen interest worldwide. Using biofuels to power industry, private vehicles, and personal appliances offers several benefits over the use of conventional fuels. For example, environmental benefits include the use of biomass energy to greatly reduce greenhouse gas emissions. Burning biomass releases about the same amount of carbon dioxide as burning fossil fuels. However, fossil fuels release carbon dioxide captured by photosynthesis millions of years ago—an essentially "new" greenhouse gas. Biomass, on the other hand, releases carbon dioxide that is largely balanced by the carbon dioxide captured in its own growth (depending on how much energy was used to grow, harvest, and process the fuel).

Another benefit of biomass use for fuel is that it can reduce dependence on foreign oil because biofuels are the only renewable liquid transportation fuels available. Moreover, biomass energy supports U.S. agricultural and forest-product industries. The main biomass feedstocks for power are paper mill residue, lumber mill scrap, and municipal waste. For biomass fuels, the feedstocks are corn (for ethanol) and soybeans (for biodiesel), both surplus crops. In the near future—and with developed technology—agricultural residues such as corn stover (the stalks, leaves, and husks of the plant) and wheat straw will also be used. Long-term plans include growing and using dedicated energy crops, such as fast-growing trees and grasses that can grow sustainably on land that will not support intensive food crops.

---

[*] Adapted from OSHA's *Green Job Hazards: Biofuels—Toxicity Hazards in Biofuel Manufacturing*, https://www.osha. gov/dep/greenjobs/bio_toxicity.html.

The preceding lists many of the benefits of using biomass fuel; that is all well and good, but the reality is that the future role of biofuels depends on profitability and new technologies. Technological advances and efficiency gains—higher biomass yields per acre and more gallons of biofuel per ton of biomass—could steadily reduce the economic cost and environmental impact of biofuel production. Biofuel production will likely be most profitable and environmentally benign in tropical areas where growing seasons are longer, per-acre biofuel yields are higher, and fuel and other input costs are lower. For example, Brazil uses *bagasse*, which is a byproduct from sugar production, to power ethanol distilleries, whereas the United States uses natural gas or coal.

Biofuels will most likely be part of a portfolio of solutions to high oil prices, including conservation and the use of other alternative fuels. The role of biofuels in global fuel supplies is likely to remain modest because of their land intensity. In the United States, replacing all current gasoline consumption with ethanol would require more land in corn production than is currently used for all agricultural production. Technology will be central to boosting the role of biofuels. If the energy of widely available, cellulose materials could be economically harnessed around the world, biofuel yields per acre could more than double, reducing land requirements significantly (USDA, 2012).

DISCUSSION AND REVIEW QUESTIONS

1. Do you agree with Kurt Anderson's assertion that the current generation is the grasshopper generation? Explain.
2. Can you think of any other biomass feedstocks that might be considered for biofuel production?
3. Why does the author characterize algae as Jekyll and Hyde-like organisms?
4. Do you think using food corn as a feedstock for ethanol production is a good practice? Explain.
5. Is the production of biomass carbon neutral? Explain.
6. Is deforestation worth the production of biofuel? Explain.
7. In folklore and myth, why is the willow tree believed to be quite sinister?
8. Do you think using food corn as a feedstock for ethanol production is a good practice? Explain.

REFERENCES AND RECOMMENDED READING

Allen, H.L., Fox, T.R., and Campbell, R.G. (2005). What's ahead for intensive pine plantation silviculture? *Southern Journal of Applied Forestry*, 29: 62–69.

Amaducci, S., Monti, A., and Venturi, G. (2004). Non-structural carbohydrates and fiber components in sweet and fiber sorghum as affected by low and normal input techniques. *Industrial Crops and Products*, 20(1): 111–118.

Anon. (2008). Algae eyed as biofuel alternative. *Taipei Times*, January 12, p. 2 (http://www.taipeitimes.com/News/taiwan/archives/2008/01/12/2003396760).

Ashford, R.D. (2001). *Ashford's Dictionary of Industrial Chemicals*, 2nd ed. London: Wavelength Publications.

Aydin, G.R., Grant, R.H., and O'Rear, J. (1999). Brown midrib sorghum in diets for lactating dairy cows. *Journal of Dairy Science*, 82(10): 2127–2135.

Aylott, M.J. (2008). Yield and spatial supply of bioenergy poplar and willow short-rotation coppice in the U.K. *New Phytologist*, 178(2): 358–370.

Baize, J. (2006). *Bioenergy and Biofuels*, Agricultural Outlook Forum, February 17, Washington, DC.

Baker, J.B. and Broadfoot, W.M. (1979). *A Practical Field Method of Site Evaluation for Commercially Important Southern Hardwoods*, General Technical Report SO-26. New Orleans, LA: U.S. Department of Agriculture, Forest Service, South Forest Experiment Station.

Basnayake, J., Cooper, M., Ludlow, M.M., Henzell, R.G., and Snell, P.J. (1995). Inheritance of osmotic adjustment to water stress in three grain sorghum crosses. *Theoretical and Applied Genetics*, 90(5): 675–682.

Beasley, R.P., Gregory, M., and McCarty, T.R. (1984). *Erosion and Sediment Pollution Control*, 2nd ed. Des Moines: Iowa State University Press.

Belnap, J. (1993). Recovery rates of cryptobiotic crusts: inoculant use and assessment methods. *Great Basin Naturalist*, 53: 80–95.

Belnap, J. (1994). Potential value of cyanobacterial inoculation in revegetation efforts. In: *Proceedings— Ecology and Management of Annual Rangelands*, Technical Report INT-GRR-313, Monsen, S.B. and Kitchen, S.G., Eds., pp. 179–185. Ogden, UT: U.S. Department of Agriculture, Forest Service.

Belnap, J. (1995). Surface disturbances: their role in accelerating desertification. *Environmental Monitoring and Assessment*, 37: 39–57.

Belnap, J. and Gardener, J.S. (1993). Soils microstructure in soils of the Colorado Plateau: the role of the cyanobacterium *Microcoleus vaginatus*. *Great Basin Naturalist*, 53: 40–47.

Belnap, J. and Gillette, D.A. (1997). Disturbance of biological soil crusts: impacts on potential wind erodibility of sandy desert soils in southeastern Utah. *Land Degradation and Development*, 8(4): 355–362.

Belnap, J. and Harper, K.T. (1995). The influence of cryptobiotic soil crusts on elemental content of tissue of two desert seed plants. *Arid Soil Research and Rehabilitation*, 9: 107–115.

Bender, M. (1999). Economic feasibility review for community-scale farmer cooperatives for biodiesel. *Bioresource Technology*, 70: 81–87.

Beymer, R.J. and Klopatek, J.M. (1991). Potential contribution of carbon by macrophytic crusts in pinyon–juniper woodlands. *Arid Soil Research and Rehabilitation*, 5: 187–198.

Billa, E., Koullas, D.P., Monties, B., and Koukios, E.G. (1977). Structure and composition of sweet sorghum stalk components. *Industrial Crops and Products*, 6(3-4): 297–302.

Blumenthal, J.B., Rooney, W.L., and Wang, D. (2007). Yield and ethanol production in sorghum genotypes. In: *Abstracts, Annual Meeting of ASA-CSSA-SSSA*, New Orleans, November 4–8.

Bourne, Jr., J.K. (2007). Green dreams. *National Geographic*, October, pp. 38–59.

Burton, G.W. (1986). Biomass production from herbaceous plants. In: *Biomass Energy Development*, Smith, W.H., Ed., pp. 163–175. New York: Plenum Press.

Christi, Y. (2007). Biodiesel from microalgae. *Biotechnology Advances*, 25: 294–306.

Christian, D.G., Riche, A.B., and Yates, N.E. (2008). Growth, yield and mineral content of *Miscanthus x giganteus* grown as a biofuel for 14 successive harvests. *Industrial Crops and Products*, 28(3): 320–327.

Clark, J. (1981). The inheritance of fermentable carbohydrates in stems of *Sorghum bicolor* (L.) Moench, doctoral dissertation, Texas A&M University, College Station.

Coyle, W. (2007). The future of biofuels: a global perspective, *Amber Waves*, November (http://www.thebio-energysite.com/articles/9/the-future-of-biofuels-a-global-perspective).

de Resende, A.S., Xavier, R.P., de Oliveira, O.C., Urquiaga, S., Alves, B.J.R., and Boddey, R.M. (2006). Long-term effects of pre-harvest burning and nitrogen and vinasse applications on yield of sugarcane and soil carbon and nitrogen stocks on a plantation in Pernambuco, NE Brazil. *Plant and Soil*, 281(1–2): 339–351.

Dickman, D. (2006). Silviculture and biology of short-rotation wood crops in temperate regions: then and now. *Biomass and Bioenergy*, 30: 696–705.

Edme, S.H., Miller, J.D., Glaz, B., Tai, P.Y.P, and Comstock, J.C. (2005). Generation contributions to yield gains in the Florida sugarcane industry across 33 years. *Crop Science*, 45(1): 92–97.

EERE. (2008). *Biomass Program*. Washington, DC: Office of Energy Efficiency & Renewable Energy, U.S. Department of Energy (http://www1.eere.energy.gov/biomass/feedstocks_types.html).

EIA. (2006). *Annual Energy Outlook 2007 with Projections to 2030*, DOE/EIA-0383(2007). Washington, DC: U.S. Department of Energy, Energy Information Administration, Office of Integrated Analysis and Forecasting.

EIA and EPRI. (1997). *Renewable Energy Technology Characterizations*, TR-109496. Washington, DC: U.S. Department of Energy, Energy Efficiency and Renewable Energy (http://www1.eere.energy.govba/pba/tech_cahraterizaations.html).

Elliott, D.C., Fitzpatrick, S.W., Bozell, J.J. et al. (1999). Production of levulinic acid and use as a platform chemical for derived products. In: *Biomass: A Growth Opportunity in Green Energy and Value-Added Products, Proceedings of the Fourth Biomass Conference of the Americas*, Overend, R.P. and Chornet, E., Eds., pp. 595–600. Oxford: Elsevier Science.

EPRI. (1997). *Renewable Energy Technology Characterizations*, TR-109496. Washington, DC: U.S. Department of Energy, Electric Power Research Institute.

Ercoli, L., Mariotti, M., Masoni, A., and Bonan, E. (1999). Effect of irrigation and nitrogen fertilization on biomass yield and efficiency of energy use in crop production of *Miscanthus*. *Field Crops Research*, 63(1): 3–11.

Evans, R.D. and Ehleringer, J.R. (1993). Broken nitrogen cycles in arid lands: evidence from ^{15}N of soils. *Oecologia*, 94: 314–317.

Fahey, J. (2001). Shucking petroleum. *Forbes Magazine*, 168(13): 206–208.

FAO. (2007). *State of the World's Forests 2007*. Rome: Food and Agriculture Organization of the United Nations (http://www.fao.org/docrep/009/a0773e/a0773e00.htm).

Farrell, A.E. and Gopal, A.R. (2008). Bioenergy research needs for heat, electricity, and liquid fuels. *MRS Bulletin*, 33: 373–387.

Foereid, B., de Neergaard, A., and Hogh-Jensen, H. (2004). Turnover of organic matter in a *Miscanthus* field: effect of time in *Miscanthus* cultivation and inorganic nitrogen supply. *Soil Biology and Biochemistry*, 36(7): 1075–1085.

Friedman, T. (2010). The fat lady has sung. *The New York Times*, December 12, p. WK8 (http://www.nytimes.com/2010/02/21/opinion/21friedman.html).

Gentille, S.B. (1996). *Reinventing Energy: Making the Right Choices*. Darby, PA: Diane Publishing.

Gilbert, R.A., Morris, D.R., Rainbolt, C.R., McCray, J.M., Perdomo, R.E., Eiland, B., Powell, G., and Montes, G. (2008). Sugarcane response to mill mud, fertilizer, and soybean nutrient sources on a sand soil. *Agronomy Journal*, 100(3): 845–854.

Gillette, D.A., Adams, J., Endo, A., Smith, D., and Kihl, R. (1980). Threshold velocities for input of soil particles into the air by desert soils. *Journal of Geophysical Research*, 85: 5621–5630.

Graham, R., Nelson, R., Sheehan J., Perlack, R., and Wright L. (2007). Current and potential U.S. corn stover supplies. *Agronomy Journal*, 99: 1–11.

Greef, J.M. and Deuter, M. (1993). *Miscanthus x giganteus*. *Angewandte Botanik*, 67: 87–90.

Haas, M., McAloon, A., Yee, W., and Foglia, T. (2006). A process model to estimate biodiesel production costs. *Bioresource Technology*, 97: 671–678.

Hallam, A.I., Anderson, C., and Buxton, D.R. (2001). Cooperative economic analysis of perennial, annual and intercrops for biomass production. *Biomass and Bioenergy*, 21(6): 407–424.

Harper, K.T. and Marble, J.R. (1988). A role for nonvascular plants in management of arid and semiarid rangelands. In: *Vegetation Science Applications for Rangeland Analysis and Management*, Tueller, P.T., Ed., pp. 135–169. Dordrecht: Kluwer Academic.

Harper, K.T. and Pendleton, R.L. (1993). Cyanobacteria and cyanolichens: can they enhance availability of essential minerals for higher plants? *Great Basin Naturalist*, 53: 89–95.

Harrar, E.S. and Harrar, J.G. (1962). *Guide to Southern Trees*, 2nd ed. New York: Dover.

Hileman, B. (2003). Clashes over agbiotech. *Chemical & Engineering News*, 81: 25–33.

Hoffman, J. (2001). BDO outlook remains healthy. *Chemical Market Reporter*, 259(14): 5.

Hughes, E. (2000). Biomass cofiring: economics, policy and opportunities. *Biomass and Bioenergy*, 19: 457–465.

ILSR. (2002). *Accelerating the Shift to a Carbohydrate Economy: The Federal Role*, Executive Summary of the Minority Report of the Biomass Research and Development Technical Advisory Committee, Institute for Local Self-Reliance, Washington, DC.

Johansen, J.R. (1993). Cryptogamic crusts of semiarid and arid lands of North America. *Journal of Phycology*, 29: 140–147.

Jordan, W.R. and Miller, F.R. (1980). Genetic variability in sorghum root systems: implications for drought tolerance. In: *Adaptation of Plants to Water and High Temperature Stress*, Turner, P.C. and Kramer, P.J., Eds., pp. 383–399. New York: John Wiley & Sons.

Kantor, S.L., Lipton, K., Manchester, A., and Oliveira, V. (1997). Estimating and addressing America's food losses. *Food Review*, 20(1): 3–11.

Karthick, N. (2010). Biomass energy costs. *Buzzle*, August 9 (http://www.buzzle.com/articles/biomass-energy-costs.html).

Kimber, C. (2000). Origins of domesticated sorghum and its early diffusion to India and China. In: *Sorghum: Origin, History, Technology, and Production*, Smith, C.W. and Frederiksen, R.A., Eds., pp. 3–98. New York: John Wiley & Sons.

Klass, D.I. (1998). *Biomass for Renewable Energy, Fuels, and Chemicals*. San Diego, CA: Academic Press.

Lee, S. (1996). *Alternative Fuels*. Boca Raton, FL: Taylor & Francis.

Legendre, B.L. and Burner, D.M. (1995). Biomass production of sugarcane cultivars and early-generation hybrids. *Biomass and Bioenergy*, 8(2): 55–61.

Lewis, L. (2005). Seaweed to breathe new life into fight against global warming. *London Times*, May 14.

Little, A. (2001). *Aggressive Growth in the Use of Bioderived Energy and Products in the United States by 2010: Final Report*. Washington, DC: U.S. Department of Energy.

Loomis, R.S. and Williams, W.A. (1963). Maximum crop productivity: an estimate. *Crop Science*, 3(1): 67–72.

Macedo, I.C., Leal, M.R.L.V., and da Silva, J. (2004). *Assessment of Greenhouse Gas Emissions in the Production and Use of Fuel Ethanol in Brazil*. Sao Paulo, Brazil: Government of the State of San Paulo (https://www.wilsoncenter.org/sites/default/files/brazil.unicamp.macedo.greenhousegas.pdf).

Maiti, R.K., Rao, K.E., Raju, P.S., House, L.R., and Prasada-Rao, K.E. (1984). The glossy trait in sorghum: its characteristics and significance in crop improvement. *Field Crops Research*, 9: 279–289.

Markarian, J. (2003). New additives and basestocks smooth way for lubricants. *Chemical Market Reporter*, April 28.

Masters, G.M. (1991). *Introduction to Environmental Engineering and Science*. Englewood Cliffs, NJ: Prentice Hall.

Mayland, H.F. and McIntosh, T.H. (1966). Availability of biologically fixed atmosphere nitrogen-15 to higher plants. *Nature*, 209: 421–422.

McBee, G.G., Miller, F.R., Dominy, R.E., and Monk, R.L. (1987). Quality of sorghum biomass for methanogenesis. In: *Energy from Biomass and Waste*, Klass, D.L., Ed., pp. 251–260. London: Elsevier.

McCollum, T., McCuistion, K., and Bean, B. (2005). Brown midrib and photoperiod sensitive forage sorghums. In: *Proceedings of the 2005 Plains Nutrition Council Spring Conference*, San Antonio, TX, April 14–15.

McCutchen, B.F. and Avant, Jr., R.V. (2008). High-tonnage dedicated energy crops: the potential of sorghum and energy cane. In *Proceedings of the Twentieth Annual Conference of the National Agricultural Technology Council*, Columbus, OH, June 3–5.

McDill, S. (2009). Can algae save the world—again? *Reuters*, February 20 (http://www.reuters.com/article/2009/02/10/us-biofuels-algae-idUSTRE5196HB20090210?pageNumber=2&virtualBrandChannel=0).

McGraw, L. (1999). Three new crops for the future. *Agricultural Research Magazine*, 47(2): 17.

McKeever, D.B. (1998). Wood residual quantities in the United States, *BioCycle: Journal of Composting and Recycling*, 39(1): 65–68; as cited in Antares Group, Inc., *Assessment of Power Production at Rural Utilities Using Forest Thinnings and Commercially Available Biomass Power Technologies*, prepared for the U.S. Department of Agriculture, U.S. Department of Energy, and National Renewable Energy Laboratory, Washington, DC, 2003.

McKeever, D.B. (2004). Inventories of woody residues and solid wood waste in the United States, 2002. In: *Proceedings of the Ninth International Conference, Inorganic-Bonded Composite Materials*, Vancouver, BC, October 10–13.

McKendry, P. (2002). Energy production from biomass. Part I. Overview of biomass. *Bioresource Technology*, 83: 37–46.

McKenna-Neumann, C., Maxwell, C., and Bolton, J.W. (1996). Wind transport of sand surface with photoautotrophic microorganisms. *Catena*, 27: 229–247.

McLaughlin, S.B. and Kzos, L.A. (2005). Development of switchgrass (*Pancium virgatum*) as a bioenergy feedstock in the United States. *Biomass and Bioenergy*, 28, 515–535.

Metting, B. (1991). Biological surface features of semiarid lands and deserts. In: *Semiarid Lands and Deserts: Soil Resource and Reclamation*, Skujins, J., Ed., pp. 256–293. New York: Marcel Dekker.

Miles, T.R., Miles, Jr., T.R., Baxter, L.L., Jenkins, B.M., and Oden, L.L. (1993). Alkali slagging problems with biomass fuels. In: *First Biomass Conference of the Americas: Energy, Environment, Agriculture, and Industry*, Vol. 1, pp. 406–421. Burlington, VT: National Renewable Energy Laboratory.

Miller, G.T. (1988). *Environmental Science: An Introduction*. Belmont, CA: Wadsworth.

Ming, R., Moor, P.H., Wu, K.K., D'Hont, A., Tew, T.L. et al. (2008). Sugarcane improvement through breeding and biotechnology. *Plant Breeding Reviews*, 27: 17–118.

Monk, R.L., Millier, F.R., and McBee, G.G. (1984). Sorghum improvement for energy production. *Biomass*, 6(1–2): 145–185.

Morey, R.V., Tiffany, D.G., and Hartfield D.L. (2006). Biomass for electricity and process heat at ethanol plants. *Applied Engineering in Agriculture*, 22: 723–728.

Morris, D. (2002). *Accelerating the Shift to a Carbohydrate Economy: The Federal Role*. Washington, DC: Institute for Local Self-Reliance.

Muchow, R.C., Cooper, M., and Hammer, G.L. (1996). Characterizing environmental challenges using models. In: *Plant Adaptation and Crop Improvement*, Cooper, M. and Hammer, G.L., Eds., pp. 349–364. London: CABI.

Mukherjee, S.K. (1950). Search for wild relatives of sugarcane in India. *International Sugar Journal*, 52: 261–262.

NIST. (1994). *Handbook 44 Appendix D Definitions*. Washington DC: U.S. Department of Commerce.

NIST. (2012). *Specifications, Tolerances, and Other Technical Requirements for Weighing and Measuring Devices*, as adopted by the 96th National Conference on Weights and Measures 2011, Handbook 44. Washington DC: National Institute of Standards and Technology.

NREL. (1993). *Alkali Content and Slagging Potential of Various Biofuels*. Washington, DC: National Renewable Energy Laboratory (http://cta.ornl.gov/bedb/biopower/Alkali_Content_and_Slagging_Potential_of_Various_Biofuels.xls).

NREL. (2002). *The Biomass Economy*, NREL/JA-810031967. Washington, DC: National Renewable Energy Laboratory (http://www.afdc.energy.gov/pdfs/6748.pdf).

Oliver, A.L., Grant, R.J., Pedersen, J.F., and O'Rear, J. (2004). Comparison of Brown Midrib-6 and -18 forage sorghum with convention sorghum and corn silage in diets of lactating dairy cows. *Journal of Dairy Science*, 87(3): 637–644.

Panje, R.R. (1972). The role of *Saccharum spontaneum* in sugarcane breeding. *Proceedings of the International Society of Sugarcane Technology*, 14: 217–223.

Panje, R.R. and Babu, C.N. (1960). Studies in *Saccharum spontaneum*. Distribution and geographic association of chromosome numbers. *Cytologia*, 25: 152–172

Parrish, D.J., Fike, J.H., Bransby, D.I., and Samson, R. (2008). Establishing and managing switchgrass as an energy crop. *Forage and Grazinglands*, February 20.

Perlin, J. (2005). *A Forest Journey: The Story of Wood and Civilization*. Woodstock, VT: Countryman Press.

Peterson, T.A. and Varvel, G.E. (1989). Crop yield as affected by rotation and nitrogen rate. II. Grain sorghum. *Agronomy Journal*, 81(5): 731–734.

Polagye, B., Hodgson, K., and Malte, P. (2007). An economic analysis of bioenergy options using thinning from overstocked forests. *Biomass and Bioenergy*, 31: 105–125.

Richard, Jr., E.P. (1999). Management of chopper harvester-generated green cane trash blankets. A new concern for Louisiana. *Proceedings of the International Society of Sugarcane Technology*, 23: 52–62.

Rinehart, L. (2006). *Switchgrass as a Bioenergy Crop*. Butte, MT: National Center for Appropriate Technology.

Roach, B.T. (1978). Utilization of *Saccharum spontaneum* in sugarcane breeding. *Proceedings of the International Society of Sugarcane Technology*, 16: 43–58.

Rooney, W.L. (2004). Sorghum improvement—integrating traditional and new technology to produce improved genotypes. *Advances in Agronomy*, 83: 37–109.

Rooney, W.L., Blumenthal, J., Bean, B., and Mullet, J.E. (2007). Designing sorghum as a dedicated bioenergy feedstock. *Biofuels, Bioproducts and Biorefining*, 1(2): 147–157.

Rosenow, D.T., Quisenberry, J.E., Wendt, C.W., and Clark, L.E. (1983). Drought tolerant sorghum and cotton germplasm. *Agricultural Water Management*, 7(1–3): 207–222.

Rossell, J.B. and Pritchard, J.L.R., Eds. (1991). *Analysis of Oilseeds, Fats and Fatty Foods*. Elsevier, London.

Rychert, R.C. and Skujins, J. (1974). Nitrogen fixation by blue–green algae–lichens crusts in the Great Basin desert. *Soil Science Society of American Proceedings*, 38: 768–771.

Salassi, M.E. and Deliberto, M.A. (2011). *Sugarcane Production in Louisiana—2011 Projected Commodity Costs and Returns*, Information Series No. 267. Baton Rouge: Department of Agricultural Economics and Agribusiness, Louisiana State University.

Samson, R. et al. (2008). Developing energy crops for thermal applications: optimizing fuel quality, energy security and GHG mitigation. In: *Biofuels, Solar and Wind as Renewable Energy Systems: Benefits and Risks*, Pimental, D., Ed., pp. 395–424. Berlin: Springer.

Sauer, P. (2000). Domestic spearmint oil producers face flat pricing. *Chemical Market Reporter*, 258(18): 14.

Scally, L., Hodkinson, T., and Jones, M.B. (2001). Origins and taxonomy of *Miscanthus*. In: *Miscanthus for Energy and Fibre*, Jones, M.B. and Walsh, M., Eds., pp. 1–9. London: James and James, Ltd.

Schenk, P., Thomas-Hall, S., Stephens, R., Marx U., Mussgnug, J., Posten, C., Kruse, O., and Hankamer, B. (2008). Second generation biofuels: high-efficiency microalgae for industrial production. *BioEnergy Research*, 1(1): 20–43.

Schimel, D.S., Kelly, E.F., Yonker, C., Aquilar, R., and Heil, R.D. (1985). Effects of erosional processes on nutrient cycling in semiarid landscapes. In: *Planetary Ecology*, Caldwell, D.E., Brierley, J.A., and Brierley, C.L., Eds., pp. 571–580. New York: Van Nostrand Reinhold.

Schneckenberger, K. and Kuzyakov, Y. (2007). Carbon sequestration under *Miscanthus* in sandy and loamy soils estimated by natural ^{13}C abundance. *Journal of Plant Nutrition and Soil Science*, 170(4): 538–542.

Sedjo, R. (1997). The economics of forest-based biomass supply. *Energy Policy*, 25(6): 559–566.

Silva, B. (1998). Meadowfoam as an alternative crop. *AgVentures*, 2(4): 28.

Sioru, B. (1999). Process converts trash into oil. *Waste Age*, 30(11): 20.

Spellman, F.R. (2009). *Handbook of Water and Wastewater Treatment Plant Operations*, 2nd ed. Boca Raton, FL: CRC Press.

Spellman, F.R. and Whiting, N. (2007). *Concentrated Animal Feeding Operations (CAFOs)*. Boca Raton, FL: CRC Press.

Tew, T.L. (2003). World sugarcane variety census—year 2000. *Sugarcane International*, March/April, pp. 12–18.

Tew, T.L. and Cobill, R.M. (2008). Genetic improvement of sugarcane (*Saccharum* spp.) as an energy crop. In: *Genetic Improvement of Bioenergy Crops*, Vermerris, W., Ed., pp. 249–272. New York: Springer-Verlag.

Tillman, D. (2000). Biomass cofiring: the technology: the experience, the combustion consequences. *Biomass and Bioenergy*, 19: 365–384.

Uhlig, H. (1998). *Industrial Enzymes and Their Applications*. New York: John Wiley & Sons.

USDA. (2010). *The Forest Inventory and Database: Database Description and User's Guide Version 6.0.2 for Phase 2*. Washington, DC: U.S. Department of Agriculture (http://www.fia.fs.fed.us/library/database-documentation/.

USDA ERS. (2010). *Commodity Costs and Returns*. Washington, DC: U.S. Department of Agriculture, Economic Research Service (http://www.ers.usda.gov/data-products/commodity-costs-and-returns.aspx).

USDA. (2012). *World Agricultural Supply and Demand Estimates*. Washington, DC: U.S. Department of Agriculture (http://www.usda.gov/oce/commodity/wasde/).

USDOE. (2004). *Industrial Bioproducts: Today and Tomorrow*. Washington, DC: U.S. Department of Energy.

USDOE. (2006a). *The Bioproducts Industry: Today and Tomorrow*. Washington, DC: U.S. Department of Energy.

USDOE. (2006b). *Breaking the Biological Barriers to Cellulosic Ethanol: A Joint Research Agenda*, Report from the December 2005 Workshop, DOE-SC-0095. Washington, DC: U.S. Department of Energy.

USDOE. (2007). *Understanding Biomass: Plant Cell Walls*. Washington, DC: U.S. Department of Energy.

USDOE. (2011). *U.S. Billion-Ton Update: Biomass Supply for a Bioenergy and Bioproducts Industry*, ORNL/TM-2011/224. Oak Ridge, TN: Oak Ridge National Laboratory.

USDOE. (2014a). *Genomic Science Program: Systems Biology for Energy and Environment*. Washington, DC: U.S. Department of Energy (http://genomicscience.energy.gov/).

USDOE. (2014b). *Non-Hydroelecric Renewable Energy*. Washington, DC: U.S. Department of Energy (http://www.epa.gov/cleanenergy/energy-and-you/affect/non-hydro.html).

USEPA. (1979). *Process Design Manual: Sludge Treatment and Disposal*, EPA/625/625/1-79-011. Washington, DC: U.S. Environmental Protection Agency.

USEPA. (2004). Overview of biogas technology. In: *AgSTAR Handbook*, Roos, K.F., Martin, Jr., J.B., and Moser, M.A., Eds., Chap. 1. Washington, DC: U.S. Environmental Protection Agency (http://www.epa.gov/agstar/documents/chapter1.pdf).

USEPA. (2006). *Biosolids Technology Fact Sheet: Multi-Stage Anaerobic Digestion*, EPA/832-F-06-031. Washington, DC: U.S. Environmental Protection Agency.

USEPA. (2007). *Fuel Economy Impact Analysis of RFG*. Washington, DC: U.S. Environmental Protection Agency (http://www.epa.gov/oms/rfgecon.htm).

Valigra, L. (2000). Tough as soybeans. *The Christian Science Monitor*, January 20.

Viator, R.P, Johnson, R.M., Grimm, C.C., and Richard Jr. E.P. (2006). Allelopathic, autotoxic, and hermetic effects of postharvest sugarcane residue. *Agronomy Journal* 98(6):1526-1531.

Viator, R.P., Johnson, R.M., Boykin, D.L., and Richard, Jr., E.P. (2009a). Sugarcane post-harvest residue management in a temperate climate. *Crop Science*, 49(3): 1023–1028.

Viator, R.P., Johnson, R.M., and Richard, Jr., E.P. (2009b). Mechanical removal and incorporation of post-harvest residue effects on sugarcane ratoon yields. *Sugarcane International*, 24(4): 149–152.

Webb, R.H. (1983). Compaction of desert soils by off-road vehicles. In: *Environmental Effects of Off-Road Vehicles: Impacts and Management in Arid Regions*, Webb R.H. and Wilshire, H.G., Eds., pp. 31–80. New York: Springer-Verlag.

Webb, R.H. and Wilshire, H.G. (1983). *Environmental Effects of Off-Road Vehicles: Impacts and Management in Arid Regions*. New York: Springer-Verlag.

Wilhelm, W.W., Johnson, J.M.F., Karlen, D.L., and Lightle, D.T. (2002). Corn stover to sustain soil organic carbon further constrains biomass supply. *Agronomy Journal*, 99: 1665–1667.

Williams, J.D., Dobrowolski, J.P., West, N.E., and Gillette, D.A. (1995). Microphytic crust influences on wind erosion. *Transactions of the American Society of Agricultural Engineers*, 38: 131–137.

Wiltsee, G. (1998). *Urban Wood Waste Resource Assessment*, NREL/SR-570-25918. Golden, CO: National Renewable Energy Laboratory.

Wood, M. (2002). Desert shrub may help preserve wood. *Agricultural Research Magazine*, 50(4): 10–11.

Wortmann, C.S., Liska, A.J., Ferguson, R.B., Lyon, D.J., Klein, R.N., and Dweikat, I. (2010). Dryland performance of sweet sorghum and grain crops for biofuel in Nebraska. *Agronomy Journal*, 102(1): 319–326.

Wu, X., Zhao, R., Bean, S.R., Seib, P.A., McLearne, J.S., Madl, R.L., Tuinstra, M.R., Lenz, M.C., and Wang, D. (2007). Factors impacting ethanol production from grain sorghum in the dry-grind process. *Cereal Chemistry*, 84(2): 130–136.

Zeman, N. (2007). The preventive chemist. *Biodiesel Magazine*, February.

8 Geothermal Energy

The U.S. Geological Survey has calculated that the heat energy in the upper 10 kilometers of the Earth's crust in the U.S. is equal to over 600,000 times the country's annual non-transportation energy consumption. Probably no more than a tiny fraction of this energy could ever be extracted economically. However, just one hundredth of 1% of the total is equal to half the country's current non-transportation energy needs for more than a century, with only a fraction of the pollution from fossil-fueled energy sources.

—**McLarty et al. (2000)**

Geothermal heat is the only renewable energy source created naturally by the Earth itself.

—**Kimberly K. Smith, Carlton College**

If we utilize waste biomass, solar (passive and thermal), wind (on shore), photovoltaic, geothermal, and other renewable resources available to us in the United States, we would exceed the demand (what we need) by at least five times as much energy as we need, all from clean, renewable sources.

—**Frank R. Spellman**

INTRODUCTION*

Approximately 4000 miles below the Earth's surface is the core, where temperatures can reach 9000°F. This heat—geothermal energy (*geo* meaning "earth" and *thermos* meaning "heat")—flows outward from the core, heating the surrounding area, which can form underground reservoirs of hot water and steam. These reservoirs can be tapped for a variety of uses, such as to generate electricity or heat buildings. The geothermal energy potential in the uppermost 6 miles of the Earth's crust amounts to 50,000 times the energy of all oil and gas resources in the world. In the United States, most geothermal reservoirs are located in the western states, Alaska, and Hawaii; however, geothermal heat pumps (GHPs), which take advantage of the shallow ground's stable temperature for heating and cooling buildings, can be used almost anywhere. Again, it is important to point out that there is nothing new about renewable energy. From solar power to burning biomass (wood) in caves and elsewhere, humans have taken advantage of renewable resources from time immemorial. Hot springs have been used for bathing since Paleolithic times or earlier (USDOE, 2010), and the early Romans used hot springs to supply public baths and for underfloor heating systems. The world's oldest geothermal district heating system has been operating in France since the 14th century (Lund, 2007). The history of geothermal energy use in the United States is interesting and lengthy; following is a brief chronology of major geothermal events in this country (EERE, 2014b).

GEOTHERMAL TIMELINE

8000 B.C. (and Earlier)

Paleo-Indians used hot springs for cooking and for refuge and respite. Hot springs were neutral zones where members of warring nations would bathe together in peace. Native Americans have a history with every major hot spring in the United States.

---

* Based on information from USDOE, *Renewable Energy: An Overview*, National Renewable Energy Laboratory, U.S. Department of Energy, Washington, D.C., 2001.

> **DID YOU KNOW?**
> Scientists estimate that geothermal potential could be as large as 100 million kW.

1807

As European settlers moved westward across the continent, they gravitated toward these springs of warmth and vitality. In 1807, the first European visited the Yellowstone area; John Colter (c. 1774–1813), widely considered to be the first mountain man, probably encountered hot springs, leading to the designation "Colter's Hell." Also that year, settlers founded the city of Hot Springs in Arkansas, where, in 1830, Asa Thompson charged $1 each for the use of three spring-fed baths in a wooden tub—the first known commercial use of geothermal energy.

1847

William Bell Elliot, a member of John C. Fremont's survey party, stumbled upon a steaming valley just north of what is now San Francisco, California. Elliot called the area *The Geysers*—a misnomer—and thought he had found the gates of Hell.

1852

The Geysers was developed into a spa called The Geysers Resort Hotel. Guests include J. Pierpont Morgan, Ulysses S. Grant, Theodore Roosevelt, and Mark Twain.

1862

At springs located southeast of The Geysers, businessman Sam Brannan poured an estimated half million dollars into an extravagant development dubbed "Calistoga," replete with hotel, bathhouse, skating pavilion, and racetrack. Brannan's spa was one of many spas reminiscent of those of Europe.

1864

Homes and dwellings had been built near springs through the millennia to take advantage of their natural heat, but the construction of the Hot Lake Hotel near La Grande, Oregon, marked the first time that the energy from hot springs was used on a large scale.

1892

Boise, Idaho, provided the world's first district heating system as water was piped from hot springs to town buildings. Within a few years, the system was serving 200 homes and 40 downtown businesses. Today, there are four district heating systems in Boise that provide heat to over 5 million square feet of residential, business, and governmental space. The United States has 17 district heating systems, and dozens more can be found around the world.

1900

Hot springs water was piped to homes in Klamath Falls, Oregon.

1921

John D. Grant drilled a well at The Geysers with the intention of generating electricity. This effort was unsuccessful, but a year later Grant met with success across the valley at another site, and the United States' first geothermal power plant went into operation. Grant used steam from the first well to build a second well, and, several wells later, the operation was producing 250 kilowatts, enough electricity to light the buildings and streets at the resort. The plant, however, was not competitive with other sources of power, and it soon fell into disuse.

1927

Pioneer Development Company drilled the first exploratory well at Imperial Valley, California.

1930

The first commercial greenhouse use of geothermal energy was undertaken in Boise, Idaho. The operation used a 1000-foot well drilled in 1926. In Klamath Falls, Charlie Lieb developed the first downhole heat exchanger (DHE) to heat his house. Today, more than 500 DHEs are in use around the country.

1940

The first residential space heating in Nevada became available in the Moan area in Reno.

1948

Geothermal technology moved east when Carl Nielsen developed the first ground-source heat pump for use at his residence. J.D. Krocker, an engineer in Portland, Oregon, pioneered the first commercial building use of a groundwater heat pump.

1960

The country's first large-scale geothermal electricity-generating plant began operation. Pacific Gas and Electric operated the plant located at The Geysers. The first turbine produced 11 megawatts (MW) of net power and operated successfully for more than 30 years. Today, 69 generating facilities are in operation at 18 resource sites around the country.

1978

Geothermal Food Processors, Inc., opened the first geothermal food-processing (crop-drying) plant in Brady Hot Springs, Nevada. The Loan Guaranty Program provided $3.5 million for the facility.

1979

The first electrical development of a water-dominated geothermal resource occurred at the east Mesa field in the Imperial Valley in California. The plant was named for B.C. McCabe, the geothermal pioneer who, with his Magma Power Company, did field development work at several sites, including The Geysers.

1980

TAD's Enterprises of Nevada pioneered the use of geothermal energy for the cooking, distilling, and drying processes associated with alcohol fuel production. UNOCAL built the country's first flash plant, generating 10 MW at Brawley, California.

1982

Economical electrical generation began at California's Salton Sea geothermal field through the use of crystallizer-clarifier technology. The technology resulted from a government/industry effort to manage the high-salinity brines at the site.

1984

A 20-MW plant began generating power at Utah's Roosevelt Hot Springs. Nevada's first geothermal electricity was generated when a 1.3-MW binary power plant went into operation.

1987

Geothermal fluids were used in the first geothermal-enhanced heap leaching project for gold recovery near Round Mountain, Nevada.

1989

The world's first hybrid (organic Rankine cycle/gas engine) geopressure–geothermal power plant began operation at Pleasant Bayou, Texas, using both the heat and the methane of a geopressured resource.

1992

Electrical generation began at the 25-MW geothermal plant in the Puna field of Hawaii.

1993

A 23-MW binary power plant was completed at Steamboat Springs, Nevada.

1995

In Empire, Nevada, Integrated Ingredients dedicated a food-dehydration facility capable of processing 15 million pounds of dried onions and garlic per year. A DOE low-temperature resource assessment of 10 western states identified nearly 9000 thermal wells and springs and 271 communities collocated with a geothermal resource greater than 50.

2002

Organized by GeoPowering the West, geothermal development working groups were active in five states—Nevada, Idaho, New Mexico, Oregon, and Washington. Group members represented all stakeholder organizations. The working groups began identifying barriers to geothermal development in their states and bringing together all interested parties to arrive at mutually beneficial solutions.

2003

The Utah Geothermal Working Group was formed.

GEOTHERMAL ENERGY AS A RENEWABLE ENERGY SOURCE

Table 8.1 shows geothermal energy's ranking as a renewable energy source. The 0.215 geothermal quadrillion Btu figure is expected to steadily increase, which should be reflected in later figures when they are released.

TABLE 8.1
U.S. Energy Consumption by Renewable Energy Source, 2014

Energy Source	Energy Consumption (quadrillion Btu)
Renewable total	9.656
Biomass (total)	4.812
Biofuels (ethanol and biodiesel)	2.067
Waste	0.516
Wood and wood-derived fuels	2.230
Geothermal	**0.215**
Hydroelectric power	2.475
Solar	0.421
Wind	1.733

Source: EIA, *Primary Energy Consumption by Source*, Table 10.1, Energy Information Administration, Washington, DC, 2016 (https://www.eia.gov/totalenergy/data/monthly/index.cfm#renewable).

GEOTHERMAL ENERGY: THE BASICS

Geothermal energy processes utilize the natural heat of the Earth for beneficial purposes when the heat is collected and transported to the surface. To gain a proper understanding of geothermal energy as it is used at present, it is important to define *enthalpy*, the heat content of a substance per unit mass. Temperature alone is not sufficient to define the useful energy content of a steam/water mixture. A mass of steam at a given temperature and pressure can provide much more energy than the same mass of water under the same conditions. Enthalpy is a function of

Pressure + Volume + Temperature = Enthalpy

Geothermal practitioners usually classify geothermal resources as high enthalpy (water and steam at temperatures above about 180 to 200°C), medium enthalpy (about 100 to 180°C), and low enthalpy (<100°C). For the purpose of this text, though, it is sufficient to think of temperature and enthalpy as the same.

KEY TERMS

To assist in understanding material presented in this chapter, the key components and operations related to geothermal energy and its production are defined in the following.

Ambient—Natural condition of the environment at any given time.

Aquifer—Water-bearing stratum of permeable sand, rock, or gravel.

Baseload plants—Electricity-generating units operated to meet the constant or minimum load on the system; the cost of energy from such units is usually the lowest available to the system.

Binary-cycle plant—A geothermal electricity generating plant employing a closed-loop heat-exchange system in which the heat of the geothermal fluid ("primary fluid") is transferred to a lower-boiling-point fluid ("secondary" or "working" fluid), which is vaporized and used to drive a turbine/generator set.

Brine—A geothermal solution containing appreciable amounts of sodium chloride or other salts.

Cap rocks—Rocks of low permeability that overlie a geothermal reservoir.

Cascading heat—A process that uses a stream of geothermal hot water or steam to perform successive tasks requiring lower and lower temperatures.

Condensate—Water formed by condensation of steam.

Condenser—Equipment that condenses turbine exhaust steam into condensate.

Cooling tower—A structure in which heat is removed from hot condensate.

Crust—Earth's outer layer of rock; also called the *lithosphere*.

Direct use—Use of geothermal heat without first converting it to electricity, such as for space heating and cooling, food preparation, industrial processes, etc.

District heating—A type of direct use in which a utility system supplies multiple users with hot water or steam from a central plant or well field.

Drilling—Boring into the Earth to access geothermal resources, usually with oil and gas drilling equipment that has been modified to meet geothermal requirements.

Dry steam—Very hot steam that does not occur with liquid.

Efficiency—The ratio of the useful energy output of a machine or other energy-converting plant to the energy input.

Enhanced geothermal systems—Rock fracturing, water injection, and water circulation technologies to sweep heat from the unproductive areas of existing geothermal fields or new fields lacking sufficient production capacity.

Fault—A fracture or fracture zone in the Earth's crust along which slippage of adjacent Earth material has occurred at some time.

Flash steam—Steam produced when the pressure on a geothermal liquid is reduced; also called *flashing*.

Fumarole—A vent or hole in the Earth's surface, usually in a volcanic region, from which steam, gaseous vapors, or hot gases issue.

Geothermal—Of or relating to the Earth's interior heat.

Geothermal energy—The Earth's interior heat made available by extracting it from hot water or rocks.

Geothermal gradient—The rate of temperature increase in the Earth as a function of depth; temperature increases an average of 1°F for every 75 feet in descent.

Geothermal heat pumps—Devices that take advantage of the relatively constant temperature of the Earth's interior, using it as a source and sink of heat for both heating and cooling. When cooling, heat is extracted from the space and dissipated into the Earth; when heating, heat is extracted from the Earth and pumped into the space.

Geyser—A spring that shoot jets of hot water and steam into the air.

Heat exchanger—A device for transferring thermal energy from one fluid to another.

Heat flow—Movement of heat from within the Earth to the surface, where it is dissipated into the atmosphere, surface water, and space by radiation.

Hot dry rock (HDR)—Subsurface geologic formations of abnormally high heat content that contain little or no water.

Hydrothermal resource—Underground systems of hot water and/or steam.

Injection—The process of returning spent geothermal fluids to the subsurface; sometimes referred to as *reinjection*.

Known geothermal resource area (KGRA)—A region identified by the U.S. Geological Survey as containing geothermal resources.

Load—The simultaneous demand of all customers required at any specified point in an electric power system.

Mantle—The Earth's inner layer of molten rock, lying beneath the Earth's crust and above the Earth's core of liquid iron and nickel.

Multiplier effect—Sometimes called the *ripple effect* because a single expenditure in an economy can have repercussions throughout the entire economy; the multiplier is a measure of how much additional economic activity is generated from an initial expenditure.

Peaking plants—Electricity generating plants that are operated to meet the peak or maximum load on the system. The cost of energy from such plants is usually higher than from baseload plants.

Permeability—The capacity of a substance (such as rock) to transmit a fluid. The degree of permeability depends on the number, size, and shape of the pores and/or fractures in the rock and their interconnections. It is measured by the time it takes a fluid of standard viscosity to move a given distance. The unit of permeability is the darcy.

Plate tectonics—A theory of global-scale dynamics involving the movement of many rigid plates of the Earth's crust. Tectonic activity is evident along the margins of the plates where bulking, grinding, faulting, and vulcanism occur as the plates are propelled by the forces of deep-seated mantle convection currents.

Porosity—The ratio of the aggregate volume of pore spaces in rock or soil to its total volume, usually stated as a percent.

Subsidence—A sinking of an area of the Earth's crust due to fluid withdrawal and pressure decline.

Thermal gradient—The rate of increase or decrease in the Earth's temperature relative to depth.

Total dissolved solids (TDS)—Term used to describe the amount of solid materials in water.

Transmission lines—Structures and conductors that carry bulk supplies of electrical energy from power-generating units.

Vapor-dominated—A geothermal reservoir system in which subsurface pressures are controlled by vapor rather than by liquid; sometimes referred to as a *dry-steam reservoir*.

Well logging—Assessing the geologic, engineering, and physical properties and characteristics of geothermal reservoirs with instruments placed in the wellbore.

EARTH'S LAYERS

Earth is made up of three main compositional layers: crust, mantle, and core (see Figure 8.1). The crust has variable thickness and composition; the continental crust is 25 to 70 km thick, while the oceanic crust is 7 to 10 km thick. The elements silicon, oxygen, aluminum, and ion make up the Earth's crust. Like the shell of an egg, the Earth's crust is brittle and can break. Based on seismic (earthquake) waves that pass through the Earth, we know that below the crust is the mantle, a dense, hot layer of semisolid rock approximately 2900 km thick. The mantle might be thought of as the white of a boiled egg. It contains silicon, oxygen, aluminum, and more iron, magnesium, and calcium than the crust. The mantle is hotter and more dense than the crust because temperature and pressure inside the Earth increase with depth. The temperature at the top of the mantle is 870°C; the temperature at the bottom is 2200°C. The 30-km thick transitional layer between the mantle and crust is called the Moho layer.

At the center of the Earth lies the core, which is nearly twice as dense as the mantle because its composition is metallic (comprised of iron–nickel alloy) rather than stony. Unlike the yolk of an egg, however, the Earth's core is actually made up of two distinct parts: a 2200-km-thick liquid outer core and a 1250-km-thick solid inner core. As the Earth rotates, the liquid outer core spins, creating the Earth's magnetic field. Several important points related to Earth's structure and geothermal properties include the following:

- Heat flows outward from the center as a result of radioactive decay.
- The crust insulates us from the interior heat.
- The temperature at the base of crust is about 1000°C and increases slowly deeper into the core.
- Hot spots are located 2 to 3 km from the surface.

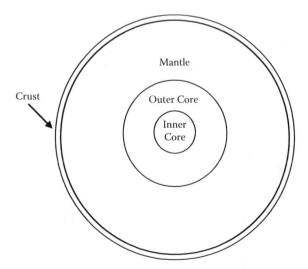

FIGURE 8.1 Layers of the Earth.

DID YOU KNOW?

Geothermal power plants are compact with a small footprint, using less land per GWh than coal, wind, or solar photovoltaics with a center station.

CRUSTAL PLATES

Geologists have developed the theory of *plate tectonics* (from the Greek word for "builder"). The theory of plate tectonics deals with the formation, destruction, and large-scale motions of great segments of the Earth's surface (crust), called *plates*. This theory relies heavily on the older concept of continental drift (developed during the first half of the 20th century) and newer understanding of seafloor spreading, both of which help to explain earthquakes and volcanic eruptions, as well as the origin of fold mountain systems.

Large quantities of heat (that are economically extractable) tend to be concentrated in places where hot or even molten (magma) rock exists at relatively shallow depths in the Earth's outmost layer (the crust). Such hot zones generally are near the boundaries of the dozen or so slabs of rigid rock (or plates) that form the Earth's lithosphere. These crustal plates are composed of great slabs of rock (lithosphere) about 100 km thick and cover many thousands of square miles (they are thin in comparison to their length and width). Geologists recognize at least eight main plates:

- African Plate covering Africa (Continental Plate)
- Antarctic Plate covering Antartica (Continental Plate)
- Australian Plate covering Australia (Continental Plate)
- Eurasian Plate covering Asia and Europe (Continental Plate)
- Indian Plate covering Indian subcontinent and a part of the Indian Ocean (Continental Plate)
- Pacific Plate covering the Pacific Ocean (Oceanic Plate)
- North American Plate covering North America and northeast Siberia (Continental Plate)
- South American Plate covering south America (Continental Plate)

as well as several minor plates, including

- Arabian Plate
- Caribbean Plate
- Cocos Plate
- Juan de Fuca Plate
- Nazea Plate
- Philippine Plate
- Scotia Plate

The plates literally ride on the *asthenosphere*, which is the ductile, soft, plastic-like zone in the upper mantle. Crustal plates move in relation to one another at one of three types of plate boundaries: convergent (collision) boundaries, divergent (spreading) boundaries, and transform boundaries. These boundaries between plates are typically associated with deep-sea trenches, large faults, fold mountain ranges, and mid-oceanic ridges.

Convergent Boundaries

Convergent boundaries (or active margins) develop where two plates slide toward each other, commonly forming either a subduction zone (if one plate subducts or moves underneath the other) or a continental collision (if the two plates contain continental crust). To relieve the stress created by the colliding plates, one plate deforms and slips below the other.

Divergent Boundaries

Divergent boundaries occur where two plates slide apart from each other. Oceanic ridges, which are examples of divergent boundaries, occur where new oceanic, melted lithosphere materials well up, resulting in basaltic magmas, which intrude and erupt at the oceanic ridge, in turn creating new oceanic lithosphere and crust (new ocean floor). Along with volcanic activity, the mid-oceanic ridges are also areas of seismic activity.

Transform Boundaries

Transform boundaries do not separate or collide; rather, they slide past each other in a horizontal manner with a shearing motion. Most transform boundaries occur where oceanic ridges are offset on the sea floor. The San Andreas Fault in California is an example of a transform fault.

ENERGY CONVERSION

The conversion of heat to electricity is common to most power plants. This is the case whether the energy source is coal, gas, nuclear power, wind power, solar power, water power, or geothermal power. Powering the non-transportation section of our economy is important, of course, so converting any fuel source to electrical power for industrial use is one of our premium objectives in our constant and insatiable appetite for energy. Because of our increasing need for more electricity generation one might surmise that we ought to focus solely on production of electricity not only to power our economy and industrial complex but also to power everything else—one genie in one bottle to accomplish everything. Liquid fuels are the fuels of choice now because they are accessible, available, and relatively inexpensive. Thus, though our ongoing research for other energy sources persists, we still do not have that absolute pressing need to come up with a liquid fuel replacement—at least not yet. Anyway, when our energy-needs focus shifts due to necessity, absolute or otherwise, geothermal will be available, and we need to continue our research in this important area. Moreover, the energy conversion occurring today utilizes all forms of energy. Geothermal conversion is nothing new, only the procedures and methodologies differ.

Geothermal energy conversion refers to the power-plant technology that converts the hot geothermal fluids into electric power. Although geothermal power plants have much in common with traditional power-generating stations—turbines, generators, heat exchangers, and other standard power-generating equipment—there are important differences between geothermal and other power-generating technologies. Each geothermal site, for example, has its own unique set of characteristics and operating conditions that must be taken into account. The fluid produced from a geothermal well can be steam, brine, or a mixture or the two, and the temperature and pressure of the resource can vary substantially from site to site. The chemical composition of the resource can contain dissolved minerals, gases, and other substances that are difficult to manage. These site-specific conditions can have a profound affect on efficiency, productivity, and economic viability, so engineers must fine-tune geothermal conversion technology, precisely matching plant design to the site-specific conditions.

GEOTHERMAL POWER PLANT TECHNOLOGIES

Geothermal power plants fall into one of three conversion categories: *dry steam*, *flash steam*, or *binary cycle*. The type of conversion used depends on the state of the fluid (whether steam or water) and its temperature. The first geothermal power generation plant built was a dry steam power plant; this type of plant uses the steam from the geothermal reservoir as it comes from wells and routes it directly through turbine/generator units to produce electricity. Flash steam plants, the most common type of geothermal power generation plants in operation today, pump water at temperatures greater than 360°F (182°C) and under high pressure to generation equipment at the surface. Binary-cycle geothermal power generation plants differ from dry-steam and flash-steam systems in that the water or steam from the geothermal reservoir never comes in contact with the turbine/generator units.

Dry Steam Power Plants

Location-specific, dry-steam geothermal power plants take advantage of subterranean rocks that are so hot that the water vaporizes on its way up through the production well (see Figure 8.2). This type of geothermal power plant is able to use steam directly from the ground to drive the turbine. This is the oldest type of geothermal power plant. It was first used in Italy in 1904 and is still very effective. These geothermal plants emit only excess steam and very minor amounts of gases (EERE, 2015a).

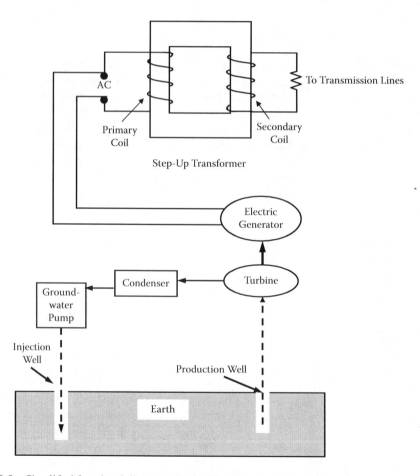

FIGURE 8.2   Simplified functional diagram of a dry steam geothermal power plant.

DID YOU KNOW?

The temperature of the Earth increases, on average, by about 28°C for every kilometer (80°F for every mile) of depth below the surface for the first several kilometers down.

FLASH STEAM POWER PLANTS

Figure 8.3 illustrates a flash steam geothermal power plant. Fluid is sprayed into a tank held at a much lower pressure than the fluid, causing some of the fluid to rapidly vaporize, or "flash." The vapor then drives a turbine, which drives a generator. If any liquid remains in the tank, it is returned to the groundwater pump to be forced down into the Earth again so it can be flashed again to extract more energy (EERE, 2015a).

BINARY-CYCLE POWER PLANTS

In a binary-cycle geothermal power plant (see Figure 8.4), water is pumped into the Earth and comes back up hot, just as it does in the flash-steam system. Instead of going into a flash tank, however, the hot water enters a *heat exchanger*, where most of its energy is transferred to another fluid, the binary liquid. This fluid can be water, but more often it is a volatile liquid resembling refrigerant that boils

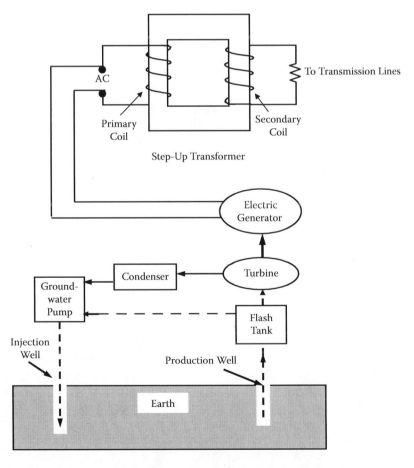

FIGURE 8.3 Simplified functional diagram of a flash steam geothermal power plant.

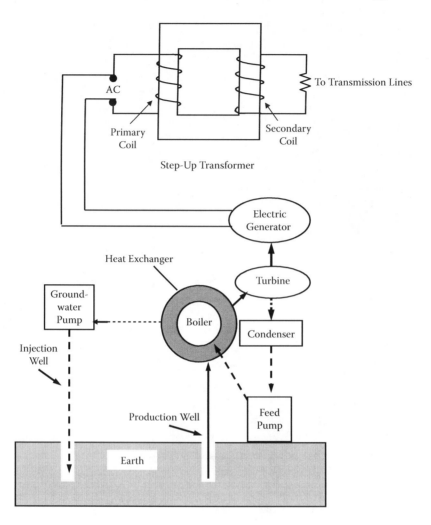

FIGURE 8.4 Simplified functional diagram of a binary-cycle geothermal power plant.

easily into vapor at a lower temperature than the water. The liquid-to-vapor conversion occurs in a special low-temperature boiler. Because this is a closed-loop system, virtually nothing is emitted to the atmosphere. Moderate temperature water is by far the more common geothermal resource, and most geothermal power plants in the future will be binary-cycle plants (EERE, 2015a).

ENHANCED GEOTHERMAL SYSTEMS

The great potential for dramatically expanding the use of geothermal energy can be realized by using enhanced geothermal systems (EGSs), also sometimes called *engineered geothermal systems*. Current geothermal power generation comes from hydrothermal reservoirs and is somewhat limited in geographic application to specific ideal places in the western United States. This represents the "lower hanging fruit" of geothermal energy potential (EERE, 2015b). Enhanced geothermal systems offer the ability to extend the use of geothermal resources to large areas of the western United States, as well as into new geographic areas of the entire country. More than 100,000 MWe (megawatt electrical) of economically viable capacity may be available in the continental United States, representing a 40-fold increase over current geothermal power generating capacity. This potential is about 10% of the overall U.S. electrical capacity today, and such systems represent a domestic energy source that is clean, reliable, and proven. The EGS concept is to extract heat by creating a subsurface fracture system, a reservoir, to which water can be added from injection wells, which are drilled into hot basement rock that has limited permeability and fluid content. This type of geothermal resource is sometimes referred to as "hot dry rock" and represents an enormous potential energy resource. Creating an enhanced, or engineered, geothermal system requires improving the natural permeability of rock. Rocks are permeable due to minute fractures and pore spaces between mineral grains. Water is injected at sufficient pressure to ensure fracturing; the water is heated by contact with the rock and returns to the surface through production wells, as in naturally occurring hydrothermal systems. Additional production wells are drilled to extract heat from large volumes of rock mass to meet power generation requirements. A previously unused but large energy resource is now available for clean, geothermal power generation.

GEOTHERMAL HEAT PUMPS

Geothermal heat pumps (Figure 8.5), sometimes referred to as geoexchange, earth-coupled, ground-source, or water-source heat pumps, have been in use since the later 1940s. Geothermal heat pumps use the constant temperature of the Earth as the exchange medium instead of the outside air

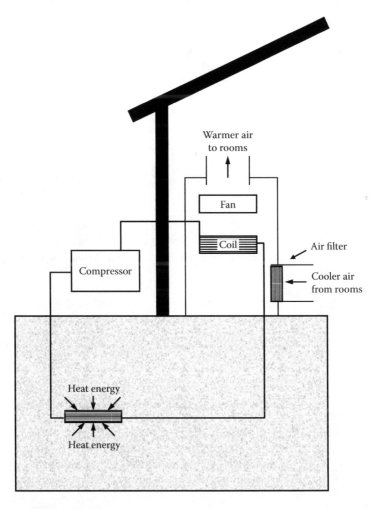

FIGURE 8.5 Geothermal heat pump system. (Adapted from Gibilisco, S., *Alternative Energy Demystified*, McGraw Hill, New York, 2007, p. 54.)

DID YOU KNOW?

Geothermal system life is estimated at 25 years for the inside components and 50+ years for the ground loop. Approximately 50,000 geothermal heat pumps are installed in the United States each year. Even with the large number of annual installations, it is important to keep in mind that heat pumps are relatively expensive to install new. This is especially true of the deep ground source type. It may take a long time for a new system to pay for itself.

temperature. This allows the system to reach fairly high efficiencies (300 to 600%) on the coldest of winter nights, compared to 175 to 250% for air-source heat pumps on cool days. Geothermal heat pumps are used for space heating and cooling, as well as water heating. Their great advantage is that they work by concentrating naturally existing heat, rather than by producing heat through the combustion of fossil fuels. The system includes three principal components:

- *Geothermal earth connection subsystems*—Using the Earth as a heat source (sink), a series of pipes, commonly called a loop, is buried in the ground near the building to be conditioned. The loop can be buried either vertically or horizontally. It circulates a fluid (water, or a mixture of water and antifreeze) that absorbs heat from, or relinquishes heat to, the surrounding soil, depending on whether the ambient air is colder or warmer than the soil.
- *Geothermal heat pump subsystem*—For heating, a geothermal heat pump removes the heat from the fluid in the Earth connection, concentrates it, and then transfers it to the building. For cooling, the process is reversed.
- *Geothermal heat distribution subsystem*—Conventional ductwork is generally used to distribute heated or cooled air from the geothermal heat pump throughout the building.

In addition to space conditioning, geothermal heat pumps can be used to provide domestic hot water while the system is operating. Many residential systems are now equipped with *desuperheaters*, which transfer excess heat from the compressor of the geothermal heat pump compressor to the hot water tank. A desuperheater provides no hot water during the spring and fall when the geothermal heat pump system is not operating; however, because the geothermal heat pump is so much more efficient than other means of water heating, manufacturers are beginning to offer "full demand" systems that use a separate heat exchanger to meet all of a household's hot water needs. These units cost-effectively provide hot water as quickly as any competing system (EERE, 2015c).

TYPES OF GEOTHERMAL HEAT PUMPS

There are four basic types of ground-loop geothermal heat pump systems. Three of these—horizontal, vertical, and pond/lake—are closed-loop systems. The fourth type of system is an open-loop option. Which one of these is best depends on the climate, soil conditions, available land, and local installation costs at the site. All of these approaches can be used for residential and commercial building applications (EERE, 2015c).

Closed-Loop Systems
- *Horizontal*—This type of installation is generally most cost effective for residential installations, particularly for new construction where sufficient land is available. It requires trenches at least 4 feet deep. The most common layouts use either two pipes, one buried at 6 feet and the other at 4 feet, or two pipes placed side-by-side 5 feet in the ground in a 2-foot-wide trench. The Slinky™ method of looping pipe allows more pipe in a short trench, which cuts down on installation costs and makes horizontal installation possible in areas where conventional horizontal applications would not be possible.

- *Vertical*—Large commercial buildings and schools use vertical systems when the land area required for horizontal loops would be prohibitive. Vertical loops are also used where the soil is too shallow for trenching, and they minimize the disturbance to existing landscaping. For a vertical system, holes (approximately 4 inches in diameter) are drilled about 20 feet apart and 100 to 400 feet deep. Into these holes go two pipes that are connected at the bottom with a U-bend to form a loop. The vertical loops are connected with horizontal pipe (i.e., manifold), placed in trenches, and connected to the heat pump in the building.
- *Pond/lake*—If the site has an adequate water body, this may be the lowest cost option. A supply line pipe is run underground from the building to the water and coiled into circles at least 8 feet under the surface to prevent freezing. The coils should only be placed in a water source that meets minimum volume, depth, and quality criteria.

Open-Loop System

This type of system uses well or surface body water as the heat-exchange fluid that circulates directly through the geothermal heat pump system. Once it has circulated through the system, the water returns to the ground through the well, a recharge well, or surface discharge. This option is obviously practical only where there is an adequate supply of relatively clean water and all local codes and regulations regarding groundwater discharge are met.

ENVIRONMENTAL IMPACTS OF GEOTHERMAL POWER DEVELOPMENT[*]

There are several potential environmental impacts from any geothermal power development:

- *Gaseous emissions*—Gaseous emissions result from the discharge of noncondensable gases (NCGs) that are carried in the source stream to the power plant. For hydrothermal installations, the most common NCGs are carbon dioxide (CO_2) and hydrogen sulfide (H_2S), although species such as methane, hydrogen, sulfur dioxide, and ammonia are often encountered in lower concentrations.
- *Water pollution*—Liquid streams from well drilling, stimulation, and production may contain a variety of dissolved minerals, especially for high-temperature reservoirs (>230°C). The amount of dissolved solids increases significantly with temperature. Some of these dissolved minerals (e.g., boron and arsenic) could poison surface or ground waters and also harm local vegetation. Liquid streams may enter the environment through surface runoff or through breaks in the well casing.
- *Solids emissions*—There is practically no chance for contamination of surface facilities or the surrounding area by the discharge of solids *per se* from the geofluid. The only conceivable situation would be an accident associated with a fluid treatment or minerals recovery system that somehow failed in a catastrophic manner and spewed removed solids onto the area.
- *Noise pollution*—Noise pollution from geothermal operations is typical of many industrial activities (DiPippo, 1991). The highest noise levels are usually produced during the well drilling, stimulation, and testing phases when noise levels ranging from about 80 to 115 decibels A-weighted (dBA) may occur at the plant fence boundary. During normal operations of a geothermal power plant, noise levels are in the range of 71 to 83 decibels at a distance of 900 m (DiPippo, 2005).
- *Land use*—Land footprints for geothermal power plants vary considerably by site because the properties of the geothermal reservoir fluid and the best options for wastestream discharge (usually reinjection) are highly site specific. Typically, the power plant is built at

---

[*] Adapted from MIT, *The Future of Geothermal Energy*, Massachusetts Institute of Technology, Cambridge, MA, 2006 (https://www1.eere.energy.gov/geothermal/pdfs/egs_chapter_8.pdf); Tribal Energy and Environmental Information Clearinghouse, http://teeic.anl.gov/er/biomass/impact/construct/index.cfm.

or near the geothermal reservoir because long transmission lines degrade the pressure and temperature of the geofluid. Although well fields can cover a considerable area, typically 5 to 10 km^2 or more, the well pads themselves will only cover about 2% of the area.

- *Land subsidence*—If geothermal fluid production rates are much greater than recharge rates, the formation may experience consolidation, which will manifest itself as a lowering of the surface elevation (i.e., surface subsidence). This was observed early in the history of geothermal power at the Wairakei field in New Zealand where reinjection was not used. Subsidence rates in one part of the field were as high as 0.45 m per year (Allis, 1990). Wairakei used shallow wells in a sedimentary basin. Subsidence in this case is very similar to mining activities at shallow depths where raw minerals are extracted, leaving a void that can manifest itself as subsidence on the surface.
- *Induced seismicity*—In normal hydrothermal settings, induced seismicity has not been a problem because the injection of waste fluids does not require very high pressures. However, the situation in the case of many EGS reservoirs will be different and requires serious attention. Induced seismicity continues to be under active review and evaluation by researchers worldwide.
- *Induced landslides*—There have been instances of landslides at geothermal fields. The cause of the landslide is often unclear. Many geothermal fields are in rugged terrain that is prone to natural landslides, and some fields actually have been developed atop ancient landslides.
- *Water use*—Geothermal projects, in general, require access to water during several stages of development and operation.
- *Disturbance of natural hydrothermal manifestations*—Although numerous cases can be cited of the compromising or total destruction of natural hydrothermal manifestations, such as geysers, hot springs, or mud pots, by geothermal developments (Jones, 2006; Keam et al., 2005), EGS projects will generally be located in non-hydrothermal areas and will not have the opportunity to interfere with such manifestations.
- *Disturbance of wildlife habitat, vegetation, and scenic vistas*—It is undeniable that any power generation facility constructed where none previously existed will alter the view of the landscape.
- *Catastrophic events*—Accidents can occur during various phases of geothermal activity, including well blowouts, ruptured steam pipes, turbine failures, and fires.
- *Thermal pollution*—Although thermal pollution is currently not a specially regulated quantity, it does represent an environmental impact for all power plants that rely on a heat source for their motive force.

Specific environmental impacts related to geothermal energy exploration, drilling, construction, operation, and maintenance activities are discussed in the following sections.

GEOTHERMAL ENERGY EXPLORATION AND DRILLING IMPACTS

Activities during the resource exploration and drilling phase are temporary and are conducted at a smaller scale than those during the construction, operations, and maintenance phases. The impacts described for each resource would occur from typical exploration and drilling activities, such as localized ground clearing, vehicular traffic, seismic testing, positioning of equipment, and drilling. Most impacts during the resource exploration and drilling phase would be associated with the development (improving or constructing) of access roads and exploratory and flow testing wells. Many of these impacts would be reduced by implementing good industry practices and restoring disturbed areas once drilling activities have been completed.

Air Quality

Emissions generated during the exploration and drilling phase include exhaust from vehicular traffic and drilling rigs, fugitive dust from traffic on paved and unpaved roads, and the release of geothermal fluid vapors (especially hydrogen sulfide, carbon dioxide, mercury, arsenic, and boron, if present in the reservoir). Initial exploration activities such as surveying and sampling would have minimal air quality impacts. Activities such as site clearing and grading, road construction, well pad development, sump pit construction, and the drilling of production and injection wells would have more intense exhaust-related emissions over a period of 1 to 5 years. Impacts would depend on the amount, duration, location, and characteristics of the emissions and the meteorological conditions (e.g., wind speed and direction, precipitation, relative humidity). Emissions during this phase would not have a measurable impact on climate change. State and local regulators may require permits and air monitoring programs.

Cultural Resources

Cultural resources could be impacted if additional roads or routes are developed across or within the historic landscape of a cultural resource. Additional roads could lead to increased surface and subsurface disturbances that could increase illegal collection and vandalism. The magnitude and extent of impacts would depend on the current state of the resource and their eligibility for the National Register of Historic Places. Drilling activities could result in long-term impacts on archeological artifacts and historic buildings or structures, if present. Surveys conducted during this phase to evaluate the presence and/or significance of cultural resources in the area would assist developers in locating sensitive resources and siting project facilities in such a way as to avoid or minimize impacts on these resources.

Ecological Resources

Most impacts on ecological resources (vegetation, wildlife, aquatic biota, special status species, and their habitats) would be low to moderate and localized during exploration and drilling (although impacts due to noise could be high). Activities such as site clearing and grading, road construction, well drilling, ancillary facility construction, and vehicle traffic have the potential to affect ecological resources by disturbing habitat, increasing erosion and runoff, and creating noise at the project site. Impacts on vegetation include the loss of native species and species diversity, increased risk of invasive species, increased risk of topsoil erosion and seed bank depletion, increased risk of fire, and alteration of water and seed dispersal. Exploration and drilling activities have the potential to destroy or injure wildlife (especially species with limited mobility); disrupt the breeding, migration, and foraging behavior of wildlife; reduce habitat quality and species diversity; disturb habitat (e.g., causing loss of cover or food source); and reduce the reproductive success of some species (e.g., amphibians). Accidental spills could be toxic to fish and wildlife. The noise from seismic surveys and drilling has a high potential to disturb wildlife and affect breeding, foraging, and migrating behavior. If not fenced or covered in netting, sump pits containing high concentrations of minerals and chemicals from drilling fluids could adversely impact animals (e.g., birds, wild horses and burrows, grazing livestock). Surveys conducted during this phase to evaluate the presence and significance of ecological resources in the area would assist developers in locating sensitive resources and siting project facilities in such a way as to avoid or minimize impact to these resources.

Water Resources

Impacts on water resources during the exploration and drilling phase would range from low to high. Survey activities would have little or no impact on surface water or groundwater. Exploration drilling would involve some ground-disturbing activities that could lead to increased erosions and surface runoff. Drilling into the reservoir can create pathways for geothermal fluids (which are under high pressure) to rise and mix with shallower groundwater. Impacts of these pathways may include alteration of the natural circulation of geothermal fluids and the usefulness of the resource. Geothermal fluids may also degrade the quality of shallow aquifers. Best management practices based on stormwater pollution prevention requirements and other industry guidelines would ensure that soil erosion and surface runoff are controlled. Proper drilling practices and closure and capping of wells can reduce the potential for drilling-related impacts.

Temporary impacts on surface water may also occur as a result of the release of geothermal fluids during any testing, if they are not contained. Geothermal fluids are hot and highly mineralized and, if released to surface water, could cause thermal changes and changes in water quality. Accidental spills of geothermal fluids could occur due to well blowouts during drilling, leaks in piping or well heads, or overflow from sump pits. Proper well casing and drilling techniques would minimize these risks.

Extracting geothermal fluids could also cause drawdowns in connected shallower aquifers, potentially affecting connected springs or streams. The potential for these types of adverse effects is moderate to high, but may be reduced through extensive aquifer testing and selection combined with compliance with the state and federal regulations that protect water quality and the limitations of water rights as issued. During the exploration and drilling phase, water would be required for dust control, making concrete, consumptive use by the construction crew, and drilling of wells. Depending on availability, it may be trucked in from offsite or obtained from local groundwater wells of nearby municipal supplies.

Land Use

Temporary and localized impacts on land use would result from exploration and drilling activities. These activities could create a temporary disturbance in the immediate vicinity of a surveying or drilling site (e.g., recreational activities, livestock grazing). The magnitude and extent of impacts from constructing additional roads would depend on the current land use in the area; however, long-term impacts on land use would be minimized by reclaiming all roads and routes that are not needed when exploration and drilling activities have been completed. All other land uses on land under well pads would be precluded as long as they are in operation. Exploration activities are unlikely to affect mining and energy development activities, military operations, livestock grazing, or aviation on surrounding lands. Activities affecting resources and values identified for protection areas would likely be prohibited.

Soils and Geologic Resources

Impacts on soils and geologic resources would be proportional to the amount of disturbance. The amount of surface disturbance and use of geologic materials during exploration would be minimal. Surface effects from vehicular traffic could occur in areas that contain special soils. The loss of biological or desert crusts can substantially increase water and wind erosion. Also, soil compaction due to development activities at the exploratory well pads and along access roads would reduce aeration, permeability, and the water-holding capacity of the soils and cause an increase in surface runoff, potentially causing increased sheet, rill, and gully erosion. The excavation and reapplication of surface soils could cause the mixing of shallow soil horizons, resulting in a blending of soil characteristics and types. This blending would modify the physical characteristics of the soils, including structure, texture, and rock content, that could lead to reduced permeability and increased runoff from these areas. Soil compaction and blending could also impact the viability of future vegetation. Any geologic resources within the areas of disturbance would not be accessible during the life of the development. Possible geological hazards (earthquakes, landslides, and subsidence) could be activated by drilling and blasting. Altering drainage patterns could also accelerate erosion and create slope instability.

Paleontological Resources

Paleontological resources are nonrenewable resources. Disturbance to such resources, whether through mechanical surface disturbance, erosion, or paleontological excavation, irrevocably alters or destroys them. The potential for impacts on paleontological resources is high where grading for access roads and drilling sites intercept geologic units with important fossil resources. Seismic surveys, ground clearing, and vehicular traffic have the potential to impact the fossil resources at the surface. The disturbance caused by all these activities could increase illegal collection and vandalism. Surveys conducted during this phase to evaluate the presence and significance of paleontological resources in the area would assist developers in locating significant resources so they can be studied and collected or so that project facilities can be sited in other areas.

Transportation

No impacts on transportation are anticipated during the exploration and drilling phase. Transportation activities would be temporary and intermittent and limited to a low volume of light utility trucks and personal vehicles.

Visual Resources

Impacts on visual resources would be considered adverse if the landscape were substantially degraded or modified. Exploration and drilling activities would have only temporary and minor visual effects, resulting from the presence of drill rigs, workers, vehicles, and other equipment (including lighting for safety), as well as from vegetation damage, scarring of the terrain, and altering landforms or contours. Reclamation following exploration and drilling to restore visual resources to pre-disturbance conditions would lessen these impacts.

Socioeconomics

As the activities conducted during the exploration and drilling phase are temporary and limited in scope, they would not result in significant socioeconomic impacts on equipment, local services, or property values.

Environmental Justice

Exploration activities are limited and would not result in significant long-term impacts in any resource area; therefore, environmental justice is not expected to be an issue during this phase.

GEOTHERMAL ENERGY CONSTRUCTION IMPACTS

Activities that may cause environmental impacts during construction include site preparation (e.g., clearing and grading), facility construction (e.g., geothermal power plant, pipelines, transmissions lines), and vehicular and pedestrian traffic. The construction of the geothermal power plant would disturb about 15 to 25 acres of land. Transmission line construction would disturb about 1 acre of land per mile of line. Impacts would be similar to but more extensive than those addressed for the exploration and drilling phase; however, many of these impacts would be reduced by implementing good industry practices and restoring disturbed areas when construction activities have been completed.

Air Quality

Emissions generated during the construction phase include exhaust from vehicular traffic and construction equipment, fugitive dust from traffic on paved and unpaved roads, and the release of geothermal fluid vapors (especially hydrogen sulfide, carbon dioxide, mercury, arsenic, and boron, if present in the reservoir). Activities such as site clearing and grading, power plant and pipeline system construction, and transmission line construction would have more intense exhaust-related emissions over a period of 2 to 10 years. Impacts would depend on the amount, duration, location, and characteristics of the emissions and the meteorological conditions (e.g., wind speed and direction, precipitation, relative humidity). Emissions during this phase would not have a measurable impact on climate change. State and local regulators may require permits and air monitoring programs.

Cultural Resources

Potential impacts on cultural resources during the construction phase could occur due to land disturbance related to the construction of the power plant and transmission lines. Impacts include destruction of cultural resources in areas undergoing surface disturbance and unauthorized removal of artifacts or vandalism as a result of human access to previously inaccessible areas (resulting in lost opportunities to expand scientific study and education and interpretive uses of these resources). In addition, for cultural resources that have an associated landscape component that contributes to their significance (e.g., sacred landscapes, historic trails), visual impacts could result from large areas of exposed surface, increases in dust, and the presence of large-scale equipment, machinery, and vehicles. Although the potential for encountering buried sites is relatively low, the possibility that buried sites would be disturbed during construction does exist. Unless the buried site is detected early in the surface-disturbing activities, the impact to the site can be considerable. Disturbance that uncovers cultural resources of significant importance that would otherwise have remained buried and unavailable could be viewed as a beneficial impact, provided the discovery results in study, curation, or recording of the resource. Vibration, resulting from increased traffic and drilling/development activities may also have effects on rock art and other associated sites (e.g., sites with standing architecture).

Ecological Resources

Most impacts on ecological resources (vegetation, wildlife, aquatic biota, special status species, and their habitats) would be low to moderate and localized during the construction phase (although impacts due to noise could be high). Activities such as site clearing and grading, road construction, power plant construction, ancillary facility construction, and vehicle traffic have the potential to affect ecological resources by disturbing habitat, increasing erosion and runoff, and creating noise at the project site. Impacts on vegetation include loss of native species and species diversity, increased risk of invasive species, increased risk of topsoil erosion and seed bank depletion, increased risk of fire, and alteration of water and seed dispersal. Construction activities have the potential to destroy or injure wildlife (especially species with limited mobility); disrupt the breeding, migration, and foraging behavior of wildlife; reduce habitat quality and species diversity; disturb habitat (e.g., causing loss of cover or food source); and reduce the reproductive success of some species (e.g., amphibians). Accidental spills could be toxic to fish and wildlife. The noise from construction and vehicle traffic has a high potential to disturb wildlife and affect breeding, foraging, and migrating behavior. Wild horses, burros, and grazing livestock could be adversely affected by the loss of forage, reduced forage palatability (due to dust settlement on vegetation), and restricted movement around the development area.

Water Resources

Impacts on water resources during the construction phase would be moderate because of ground-disturbing activities (related to road, well pad, and power plant construction) that could lead to an increase in soil erosion and surface runoff. Impacts on surface water would be moderate but temporary and could be reduced by implementing best management practices based on stormwater pollution-preventing requirements and other industry guidelines. During the construction phase, water would be required for dust control, making concrete, and consumptive use by the construction crew. Depending on availability, it may be trucked in from offsite or obtained from local groundwater wells or nearby municipal supplies.

Land Use

Temporary and localized impacts on land use would result from construction activities. These activities could create a temporary disturbance in the immediate vicinity of a construction site (e.g., to recreational activities or livestock grazing). The magnitude and extent of impacts from constructing power plants and pipeline systems would depend on the current land use in the area; however, long-term impacts on land use would be minimized by reclaiming all roads and routes that are not needed when construction has been completed. All other land uses on land under well pads, buildings, and structures would be precluded as long as they are in operation. Construction activities are unlikely to affect mining and energy development activities, military operation, livestock grazing, or aviation on surrounding lands. Activities affecting resources and values identified for protection areas would likely be prohibited.

Soils and Geologic Resources

Impacts on soils and geologic resources would be greater during the construction phase than for other phases of development because of the increased footprint and would be particularly significant if biological or desert crusts are disturbed. Construction of additional roads, well pads, the geothermal power plant, and structures related to the power plant (e.g., pipeline system, transmission lines) would occur during this phase. Construction of well pads, the geothermal power plant, and structures related to the power plant, pipeline system, access roads, and other project facilities could cause topographic changes. These changes would be minor but long term. Soil compaction due to construction activities would reduce aeration, permeability, and water-holding capacity of the soils and cause an increase in surface runoff, potentially causing increased sheet, rill, and gully erosion. The excavation and reapplication of surface soils could cause the mixing of shallow soil horizons, resulting in a blending of soil characteristics and types. This blending would modify the physical characteristics of the soils, including structure, texture, and rock content, which could lead to reduced permeability and increased runoff from these areas. Soil compaction and blending could also impact the viability of future vegetation. Any geologic resources within the areas of disturbance would not be accessible during the life of the development. It is unlikely that construction activities would activate geologic hazards; however, altering drainage patterns or building on steep slopes could accelerate erosion and create slope instability. It is unlikely that construction activities would activate geologic hazards; however, altering drainage patterns or building on steep slopes could accelerate erosion and create slope instability.

Paleontological Resources

The potential for impacts on paleontological resources is high where grading and excavation intercept geologic units with important fossil resources. Ground clearing and vehicular traffic have the potential to impact the fossil resources at the surface. The disturbance caused by all these activities could increase illegal collection and vandalism. Disturbance that uncovers paleontological resources of significant importance that would otherwise have remained buried and unavailable could be viewed as a beneficial impact, provided the discovery results in study, collection, or recording of the resource.

Transportation

Geothermal development would result in the need to construct or improve access roads and would result in an increase in industrial traffic. Overweight and oversized loads could cause temporary disruptions and could require extensive modifications to roads or bridges (e.g., widening roads or fortifying bridges to accommodate the size or weight of truck loads). An overall increase in heavy truck traffic would accelerate the deterioration of pavement, requiring local government agencies to schedule pavement repair or replacement more frequently than under the existing traffic conditions. Increased traffic would also result in a potential for increased accidents within the project area.

The locations at which accidents are most likely to occur are intersections used by project-related vehicles to turn onto or off of highways from access roads. Conflicts between industrial traffic and other traffic are likely to occur, especially on weekends, holidays, and seasons of high use by recreationists. Increased recreational use of the area could contribute to a gradual increase in traffic on the access roads.

Visual Resources

Impacts on visual resources would be considered adverse if the landscape were substantially degraded or modified. Construction activities would have only temporary and minor visual effects, resulting from the presence of workers, vehicles, and construction equipment (including lighting for safety) and from vegetation damage, dust generation, scarring of the terrain, and altering landforms or contours. Reclamation following construction to restore visual resources to pre-disturbance conditions would lessen these impacts.

Socioeconomics

Construction-phase activities would contribute to the local economy by providing employment opportunities, monies to local contractors, and recycled revenues through the local economy. The magnitude of these benefits would vary depending on the resource potential. Construction of a typical 50-megawatt (MW) power plant and related transmission lines would provide an estimated 387 jobs and $22.5 million in income, which would vary depending on the community. Job availability would vary with different stages of construction. Expenditures for equipment, materials, fuel, lodging, food, and other needs would stimulate the local economy over the duration of construction. Economic impacts may occur if other land use activities (e.g., recreation, grazing, hunting) are altered by geothermal development. Constructing facilities will alter the landscape and could affect the nonmarket values of the immediate area. Many of these land uses may be compatible; however, it is possible that some land uses will be displaced by geothermal development.

Environmental Justice

Environmental justice impacts occur only if significant impacts in other resource areas disproportionately affect minority or low-income populations. It is anticipated that the development of geothermal energy could benefit low-income, minority, and tribal populations by creating job opportunities and stimulating local economic growth via project revenues and increased tourism. However, noise, dust, visual impacts, and habitat destruction could have an adverse effect on traditional tribal life ways and religious and cultural sites. Development of wells and ancillary facilities could affect the natural character of previously undisturbed areas and transform the landscape into a more industrialized setting. Development activities could impact the use of cultural sites for traditional tribal activities (hunting and plant-gathering activities), as well as areas in which artifacts, rock art, or other significant cultural sites are located.

GEOTHERMAL ENERGY OPERATIONS AND MAINTENANCE IMPACTS

Typical activities during the operations and maintenance phase include operation and maintenance of production and injection wells and pipeline systems, operation and maintenance of the power plant, waste management, and maintenance and replacement of facility components.

Air Quality

Emissions generated during the operations and maintenance phase include exhaust from vehicular traffic and fugitive dust from traffic on paved and unpaved roads, most of which would be generally limited to worker and maintenance vehicle traffic. In addition, emission could include the release of geothermal fluid vapors (especially hydrogen sulfide, carbon dioxide, mercury, arsenic, and boron, if present in the reservoir). Impacts would depend on the amount, duration, location, and characteristics of the emissions and the meteorological conditions (e.g., wind speed and direction, precipitation, relative humidity). Carbon dioxide emissions would be considerably less than for comparable power plants using fossil fuel. State and local regulators may require permits and air monitoring programs.

Cultural Resources

During the operations and maintenance phase, impacts on cultural resources could occur primarily from unauthorized collection of artifacts and from visual impacts. In the latter case, the presence of the aboveground structures could impact cultural resources with an associated landscape component that contributes to their significance, such as a sacred landscape or historic trail. The potential for indirect impacts (e.g., vandalism, unauthorized collection) would be greater during the operations and maintenance phase compared to prior phases due to its longer duration.

Ecological Resources

Most impacts on ecological resources (vegetation, wildlife, aquatic biota, special status species, and their habitat) would be less during the operations and maintenance phase than for the exploration and drilling and construction phases because no new drilling or construction activities would take place. However, operations and maintenance activities have the potential to affect ecological resources mainly by reducing the acreage for foraging and migrating animals, fragmenting habitat, and creating noise at the project site during the life cycle of the project (which could last up to 50 years). Some of these impacts could be significant. Increased human activity also increases the risk of fire, especially in arid or semiarid areas. Application of herbicides to control vegetation along access roads, buildings, and power plant structures would increase the risk of wildlife exposure to contaminants.

Water Resources

Impacts on water resources during the operations and maintenance phase result mainly from the water demands associated with operating a geothermal power plant. Water resources during operations would be needed for replenishment of the geothermal reservoir through reinjection. However, because some water would be consumed by evaporation, additional water would have to be added to the system from another source. Makeup water to replace the evaporative losses and blowdown in a water-cooled power plant system would also be needed, depending on the type of power plant used (e.g., flash steam facilities can lose up to 20% of its cooling water due to evaporation, but binary plants are nonconsumptive because they use a closed-loop system). Water can also be lost due to pipeline failures or surface discharge for monitoring and testing the geothermal reservoir. The availability of water resources could be a limiting factor in siting or expanding a geothermal development at a given location. Cooling water or water from geothermal wells that is discharged to the ground or to an evaporation pond could affect the quality of shallow groundwater if allowed to percolate through the ground. However, the potential for this type of impact is considered minor or negligible because the facility would have to comply with the terms of the discharge permit required by the state.

Land Use

Impacts on land use during the operations and maintenance phase are an extension of those that occurred during the exploration and drilling and construction phases. Although, to some extent, land use can revert to its original uses (e.g., livestock grazing), many other uses (e.g., mining, farming, hunting) would be precluded during the life span of the geothermal development. Mineral resources would remain available for recovery, and operations and maintenance activities are unlikely to affect mining and energy development activities, military operations, livestock grazing, or aviation on surrounding lands.

Soils and Geologic Resources

Impacts on soils and geologic resources would be minimal during the operations and maintenance phase. The initial areas disturbed during the construction phase would continue to be used during standard operation and maintenance activities, but no additional impacts would occur unless new construction projects or drill sites are needed. Impacts associated with new construction projects or drill sites would be similar to those described for the exploration and drilling construction phases.

Paleontological Resources

The potential for impacts on paleontological resources would be limited primarily to the unauthorized collection of fossils. This threat is present when access roads have been constructed, making remote areas more accessible to the public. Damage to locations caused by off-highway vehicle use could also occur. The potential for indirect impacts (e.g., vandalism, unauthorized collection) would be greater during the production phase compared to the drilling/development phase, due to the longer duration of the production phase.

Transportation

Daily traffic levels, particularly heavy truck traffic, would be expected to be lower during the operations and maintenance phase compared to other phases of geothermal development. For the most part, heavy truck traffic would be limited to periodic monitoring and maintenance activities at the well pads and power plant.

Visual Resources

Adverse impacts on visual resources would occur during the 10- to 30-year life of the geothermal development. Impacts during the operations and maintenance phase would result from the presence of facility structures and roads (where undeveloped land once stood), increased vehicular traffic to the site, and releases of steam plumes from the geothermal power plant. Periodic construction projects occurring through the life of the development would have impacts similar to those described for the construction phase.

Socioeconomics

Activities during the operations and maintenance phase would contribute to the local economy by providing employment opportunities, monies to local contractors, and recycled revenues through the local economy. The magnitude of these benefits would vary depending on the resource potential. Operation of a typical 50-MW power plant and related transmission line would provide an estimated 93 jobs and $8 million in income, but would vary depending on the community. Job availability would vary with different stages of construction. Expenditures for equipment, materials, fuel, lodging, food, and other needs would stimulate the local economy over the duration of the project, which could last up to 50 years. Economic impacts may occur if other land use activities (e.g., recreation, grazing, hunting) are altered by geothermal development. Constructing facilities will alter the landscape and could affect the nonmarket values of the immediate area during the life of the geothermal development.

Environmental Justice

Possible environmental justice impacts during the operations and maintenance phase include alteration of the scenic quality in areas of traditional or cultural significance to minority populations. Noise, water, and health and safety impacts are also potential sources of disproportionate effects to minority or low-income populations.

GEOTHERMAL ENERGY IMPACTS ON HUMAN HEALTH AND SAFETY*

The construction of geothermal energy production sites and the use of geothermal energy can be a potentially dangerous undertaking or occupation. Geothermal systems use the heat from the Earth to create electricity and to heat and cool buildings. Some geothermal systems pump water underground through piping, allow it to be heated by the Earth, and then use the hot water to create electricity or heat and cool buildings. Other systems drill directly into the Earth's natural geothermal reservoirs, using the resulting hot water and steam to create electricity. Some geothermal systems use a brine or saltwater solution while others use glycol. These solutions may pose hazards of their own to workers.

HAZARDS AND CONTROLS

The hazards that are associated with this growing industry include some very familiar safety issues for which the Occupational Safety and Health Administration (OSHA) already has standards and information available. The hazards that workers in the geothermal industry may face include the following:

- Trenching and excavations
- Silica
- Personal protective equipment
- Electrical
- Welding and cutting
- Fall protection

OSHA and other regulatory requirements are contained in 29 CFR 1910 and 1926, along with detailed explanations of each standard, regulation, or requirement.

GEOTHERMAL ENERGY: THE BOTTOM LINE

The supply of geothermal energy is vast and can be considered renewable, as long as each site is properly engineered and is operated in such a way as to ensure that excessive water is not pumped into the earth in one location in too short a time. Geothermal energy can be, and already is, accessed by drilling water or steam wells in a process similar to that for drilling for oil. Geothermal energy is an enormous, underused heat and power resource that is clean (emits little or no greenhouse gases), reliable (average system availability of 95%), and home grown (making us less dependent on foreign oil). Geothermal resources range from shallow ground to hot water and rock several miles below the Earth's surface, and even farther down to the extremely hot molten rock called magma. Wells that are a mile or more deep can be drilled into underground reservoirs to tap steam and very hot water that can be brought to the surface for use in a variety of applications. In the United States, most geothermal reservoirs are located in the western states, Alaska, and Hawaii; however, before geothermal electricity can be considered a key element of the U.S. energy infrastructure, it must become more cost competitive compared with traditional forms of energy.

---

* Adapted from OSHA's *Green Job Hazards: Geo-Thermal Energy*, http://www.osha.gov/dep/greenjobs/geothermal.html.

DISCUSSION AND REVIEW QUESTIONS

1. Geothermal power plants can possibly cause groundwater contamination when drilling wells and extracting hot water or steam. How would you prevent this type of contamination? Explain.
2. Is geothermal energy a viable alternative renewable energy source that is capable of replacing fossil fuels in the United States? Explain.
3 Why do you think there has not been more extensive development of geothermal energy? Explain.

REFERENCES AND RECOMMENDED READING

Allis, R.G. (1990). Subsidence at Wairakei field, New Zealand. *Transactions—Geothermal Resources Council*, 14: 1081–1087.

Anon. (1996). Heat from the earth: geothermal heat pumps. *The Family Handyman*, 46(9): 9–12.

Atkinson, L. and Sancetta, C. (1993). Hail and farewell. *Oceanography*, 6(34).

Bonner, J.T. (1965). *Size and Cycle*. Princeton, NJ: Princeton University.

Cataldo, S. (1999). Ground-source heat pumps dig in. *Home Energy Magazine Online*, September/October.

DiPippo, R. (1991). Geothermal energy: electricity production and environmental impact: a worldwide perspective. In: *Energy and the Environment in the 21st Century*, Tester, J.W., Ed., pp. 741–754. Cambridge, MA: MIT Press.

DiPippo, R. (2005). *Geothermal Power Plants: Principles, Applications and Case Studies*. Oxford: Elsevier.

EERE. (2010). *Types of Geothermal Heat Pump Systems*. Washington, DC: Office of Energy Efficiency & Renewable Energy, U.S. Department of Energy (http://www.energysavers.gov/your_home/space_heating_cooling/index.cfm/mytopic=12650).

EERE. (2014a). *History of Hydropower*. Washington, DC: Office of Energy Efficiency & Renewable Energy, U.S. Department of Energy (http://energy.gov/eere/water/history-hydropower).

EERE. (2014b). *A History of Geothermal Energy in the United States*. Washington, DC: Office of Energy Efficiency & Renewable Energy, U.S. Department of Energy (http://www1.eere.energy.gov/geothermal/history.html).

EERE. (2015a). *Geothermal Energy at the U.S. Department of Energy*. Washington, DC: Office of Energy Efficiency & Renewable Energy, U.S. Department of Energy (http://www1.eere.energy.gov/geothermal/powerplants.html).

EERE. (2015b). *How an Enhanced Geothermal System Works*. Washington, DC: Office of Energy Efficiency & Renewable Energy, U.S. Department of Energy (http://energy.gov/eere/geothermal/how-enhanced-geothermal-system-works).

EERE. (2015c). *Geothermal Heat Pumps*. Washington, DC: Office of Energy Efficiency & Renewable Energy, U.S. Department of Energy (http://energy.gov/eere/geothermal/geothermal-heat-pumps).

Energy Savers. (2009). *Energy 101: Geothermal Heat Pumps*. Washington, DC: Office of Energy Efficiency & Renewable Energy (http://energy.gov/videos/energy-101-geothermal-heat-pumps).

Enger, E., Kormelink, J.R., Smith, B.F., and Smith, R.J. (1989). *Environmental Science: The Study of Interrelationships*. Dubuque, IA: William C. Brown.

Gibilisco, S. (2007). *Alternative Energy Demystified*. New York: McGraw-Hill.

Holmes, A. (1978). *Principles of Physical Geology*, 3rd ed. New York: John Wiley & Sons.

Jones, G.L. (2006). *Geysers/Hot Springs Damaged or Destroyed by Man*, http://www.wyojones.com/destroye.htm.

Keam, R.R., Luketina, K.M., and Pipe, L.Z. (2005). Definition and listing of significant geothermal feature types in the Waikato region, New Zealand. In: *Proceedings of World Geothermal Congress 2005*, Antalya, Turkey, April 24–29.

Lund, J.W. (2007). Characteristics, development and utilization of geothermal sources. *Geo-Heat Centre Quarterly Bulletin*, 28(2): 1–9.

Lyman, J. and Fleming, R.H. (1940). Composition of seawater. *Journal of Marine Research*, 3: 134–146.

McKnight, T. (2004). *Geographica: The Complete Illustrated Atlas of the World*. New York: Barnes & Noble.

McLarty, L., Grabowski, P., Entingh, D., and Robertson-Tait, A. (2000). Enhanced geothermal systems R&D in the United States. In: *Proceedings of World Geothermal Congress 2000*, Beppu-Morioka, Japan, May 28–June 10.

Oreskes, N., Ed. (2003). *Plate Tectonics: An Insiders History of the Modern Theory of the Earth.* New York: Westview.

Rafferty, K. (2001). *An Information Survival Kit for the Prospective Geothermal Heat Pump Owner.* Portland, OR: Oregon Institute of Technology, Geo-Heat Center.

Stanley, S.M. (1999). *Earth System History.* New York: W.H. Freeman, pp. 211–222.

Sverdrup, H.U., Johnson, M.W., and Fleming, R.H. (1942). *The Oceans: Their Physics, Chemistry, and General Biology.* New York: Prentice-Hall.

Turcotte, D.L. and Schubert, G. (2002). *Geodynamics,* 2nd ed. New York: John Wiley & Sons.

USDOE. (2010). *Fossil Fuels.* Washington, DC: U.S. Department of Energy (http://www.energy.gov/science-innovation/energy-sources/fossil).

USEPA. (1993). *Space Conditioning: The Next Frontier.* Washington, DC: U.S. Environmental Protection Agency.

9 Blue Energy

One day King Canute told his followers to carry him and his throne to the seashore at low tide. "Set me down right there at the water's edge," he said. "And I will command the sea to stay away. I will order the tide not to rise." "Oh, wow!" said his followers as they obeyed the King. "This ought to be something to see!" King Canute sat. He shouted: "Sea, stay away! Tide, do not rise!" But slowly, slowly, the tide came in and the sea rose, over his feet, past his knees, up to his waist. The crowd of followers pulled back a little to keep their own feet dry. They were puzzled. Why did the sea not obey the King? The water rose higher and higher, up to the King's shoulders, over his chin. "Loyal [glub] subjects," he gurgled. "Now you see that [glub] there are some things no man, be he [glub] king or commoner, can [glub] do! Now pull me the [glub] out of here!"

—Based on Viking legend

INTRODUCTION

As King Canute reportedly discovered, the rise and fall of the seas represent a vast and relentless natural phenomenon—certainly beyond the absolute control of all earthly subjects. The ocean can produce two types of energy: *thermal energy* from the sun's heat and *mechanical energy* from the tides and waves. Generating technologies for deriving electrical power from the ocean include tidal power, wave power, ocean thermal energy conversion, ocean currents, ocean winds, and salinity gradients. Of these, the three most well-developed technologies are *tidal power*, *wave power*, and *ocean thermal energy conversion* (OTEC). Ocean thermal energy conversion is limited to tropical regions, such as Hawaii, and to a portion of the Atlantic coast. Ocean thermal energy can be used for many applications, including electricity generation, which utilizes either warm surface water or boiled seawater to turn a turbine, which activates a generator. Tidal power requires large tidal differences which, in the United States, occur only in Maine and Alaska. Wave power has a more general application. The western coastline of the United States has the highest wave potential; in California, the greatest potential is along the state's northern coast. Ocean thermal energy can be used for many applications, including electricity generation. Electricity conversion systems either use the warm surface water or boil the seawater to turn a turbine, which activates a generator.

The electricity conversion of both tidal and wave energy usually involves mechanical devices. It is important to distinguish tidal energy from hydropower. Basically, for example, hydropower dams block the flow of water, but hydrokinetic devices do not disturb natural water flow. Moreover, hydropower is derived from the hydrological climate cycle, powered by solar energy, which is usually harnessed via hydroelectric dams. In contrast, tidal energy is the result of the interaction of the gravitational pull of the moon and, to a lesser extent, the sun on the seas. Processes that use tidal energy rely on the twice-daily tides and the resultant upstream flows and downstream ebbs in estuaries and the lower reaches of some rivers, as well as, in some cases, tidal movement out at sea.

A dam is typically used to convert tidal energy into electricity by forcing the water through turbines that activate a generator. The mechanical power created from these systems either directly activates a generator or transfers to a working fluid, water, or air, which then drives a turbine or generator. Before discussing the thermal and mechanical energy potential of the ocean, we will provide a basic understanding of oceans and especially their margins, where most, if not all, ocean energy is harnessed using current technology. The following section provides a foundation for better understanding the ocean energy concepts presented later in this chapter.

Key Terms

To assist in understanding material presented in this chapter, the key components and operations related to marine and hydrokinetic energy and its production are defined in the following:

Attenuator—Wave energy capture device with the principal axis oriented parallel to the direction of the incoming wave that converts the energy due to the relative motion of the parts of the device as the wave passes along it.

Axial flow turbine—Typically has two or three blades mounted on a horizontal shaft to form a rotor; the kinetic motion of the water current creates lift on the blades, which causes the rotor to turn and drives a mechanical generator. These turbines must be oriented in the direction of flow. There are shrouded and open rotor models.

Blue energy—Ocean energy.

Closed-cycle—Systems that use fluid with a low-boiling point, such as ammonia, to rotate a turbine to generate electricity. Warm surface seawater is pumped through a heat exchanger, where the low-boiling-point fluid is vaporized. The expanding vapor turns the turbo-generator. Cold deep seawater pumped through a second heat exchanger condenses the vapor back into a liquid, which is then recycled through the system.

Cross-flow turbine—Typically has two or three blades mounted along a vertical shaft to form a rotor; the kinetic motion of the water current creates lift on the blades, which causes the rotor to run and drives a mechanical generator. These turbines can operate with flow from multiple directions without reorientation. There are shrouded and open rotor models.

Hybrid—Systems that combine the features of both closed-cycle and open-cycle systems. In a hybrid system, warm seawater enters a vacuum chamber where it is flash-evaporated into steam, similar to the open-cycle evaporation process. The steam vaporizes a low-boiling-point fluid (in a closed-cycle loop) that drives a turbine to produce electricity.

Magnus effect—Named after the German scientist H.G. Magnus, who first (1853) experimented with sidewise force generation (i.e., departing from a straight path) on a spinning cylindrical or spherical solid immersed in a fluid (liquid or gas) when there was relative motion between the spinning body and the fluid. The Magnus effect is responsible for the curve of a served tennis ball or a driven golf ball.

Open-cycle—Systems that use the tropical oceans' warm surface water to make electricity. When warm seawater is placed in a low-pressure container, it boils. The expanding steam drives a low-pressure turbine attached to an electrical generator. The steam, which has left its salt behind in the low-pressure container, is almost pure freshwater. It is condensed back into a liquid by exposure to the cold temperatures of deep ocean water.

Oscillating hydrofoil—Similar to an airplane wing but in water; yaw control systems adjust their angle relative to the water stream, creating lift and drag forces that cause device oscillation; mechanical energy from this oscillation feeds into a power conversion system.

Oscillating water column—Partially submerged structure that encloses a column of air above a column of water. A collector funnels waves into the structure below the waterline, causing the water column to rise and fall; this alternately pressurizes and depressurizes the air column, pushing or pulling it through a turbine. There are shore-based and floating models.

Oscillating wave surge converter—Any of several devices that capture wave energy directly without a collector by using relative motion between a float/flap/membrane and a fixed reaction point; the float/flap/membrane oscillates along a given axis dependent on the device, and mechanical energy is extracted from the relative motion of the body part relative to its fixed reference.

Overtopping device—Partially submerged structure. A collector funnels waves over the top of the structure into a reservoir; water runs back out to the sea from the reservoir through a turbine. There are shore-based and floating models.

Point absorber—Wave energy capture device, with principal dimension relatively small compared to the wavelength; it is able to capture energy from a wave front greater than the
physical dimension of the device. There are floating and submerged models.

Reciprocating device—Uses the flow of water to produce the lift or drag of an oscillating part
transverse to the flow direction. This behavior can be induced by a vortex, the Magnus
effect, or by flow flutter.

Submerged pressure differential—Wave energy capture device that can be considered a fully
submerged point absorber; a pressure differential is induced within the device as the wave
passes, driving a fluid pump to create mechanical energy.

OCEANS AND THEIR MARGINS*

Oceans are a principal component of the hydrosphere and the storehouse of Earth's water. Oceans
cover about 71% of Earth's surface. The average depth of the oceans is about 3800 m, but the greatest ocean depth of 11,036 m was recorded in the Mariana Trench. The volume of all of the oceans is
about 1.35 billion cubic kilometers, representing 96.5% of Earth's total water supply; however, the
volume fluctuates with the growth and melting of glacial ice. The composition of ocean water has
remained constant throughout geologic time, with the major constituents dissolving in ocean water
from rivers and precipitation, as well as from weathering and degassing of the mantle by volcanic
activity. Seawater is approximately 3.5% salt and 96.5% water, by weight. The dissolved salts include
chloride (55.07%), sodium (30.62%), sulfate (7.72%), magnesium (3.68%), calcium (1.17%), potassium
(1.10%), bicarbonate (0.40%), bromine (0.19%), and strontium (0.02%). The most significant factor
related to ocean water that everyone is familiar with is the salinity of the water—how salty it is.
Salinity, a measure of the amount of dissolved ions in the water, ranges between 33 and 37 parts per
thousand. The concentration is the amount (by weight) of salt in water, usually expressed as parts per
million (ppm); 1 ppm is equivalent to a shotglass of water taken from an Olympic-sized swimming
pool. Water is saline if it has a concentration of more than 1000 ppm of dissolved salts; ocean water
contains about 35,000 ppm of salt (USGS, 2007). Chemical precipitation, absorption onto clay minerals, and plants and animals prevent seawater from having higher salinity concentrations; however,
salinity does vary in the oceans because surface water evaporates, rain and stream water is added,
and ice forms or thaws. Another important property of seawater is temperature. The temperature of
surface seawater varies with latitude, from near 0°C close to the poles to 29°C near the equator. Some
isolated areas can have temperatures as high as 37°C. Temperature decreases with ocean depth.

OCEAN FLOOR

The bottoms of the ocean basins (ocean floors) are marked by mountain ranges, plateaus, and other
relief features similar to (although not as rugged as) those on the land. As shown in Figure 9.1, the
floor of the ocean can be divided into four divisions: continental shelf, continental slope, continental
rise, and deep-sea floor or abyssal plain:

- The *continental shelf* is the flooded, nearly flat true margins of the continents. Varying in
  width to about 40 miles and a depth of approximately 650 feet, continental shelves slope
  gently outward from the shores of the continents. Continental shelves occupy approximately 7.5% of the ocean floor.
- The *continental slope* is a relatively steep slope descending from the continental shelf; it
  descends rather abruptly to the deeper parts of the ocean. These slopes represent about
  8.5% of the ocean floor.

---

* Material in this section is adapted from Spellman, F.R., *Geology for Nongeologists*, Government Institutes Press,
Lanham, MD, 2009.

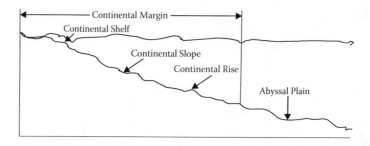

FIGURE 9.1 Cross-section of ocean floor showing major elements of topography.

- The *continental rise* is a broad gentle slope below the continental slope containing sediment that has accumulated along parts of the continental slope.
- The *abyssal plain* is a sediment-covered deep-sea plain about 12,000 to 18,000 feet below sea level. This plain accounts for about 42% of the ocean floor.

The deep ocean floor does not consist exclusively of the abyssal plain. Some areas have considerable relief, which can be described as follows:

- *Seamounts* are isolated mountain-shaped elevations more than 3000 feet high.
- *Mid-oceanic ridges* are submarine mountains, extending more than 37,000 miles through the oceans, generally 10,000 feet above the abyssal plain.
- *Trenches* are deep, steep-sided troughs in an abyssal plain.
- *Guyots* are flat-topped volcanic mountains that rise from the ocean bottom and usually are covered by 3000 to 6000 feet of water.

OCEAN TIDES, CURRENTS, AND WAVES

Water is the master sculptor of Earth's surfaces. The ceaseless, restless motion of the sea is an extremely effective geologic agent. Besides shaping inland surfaces, water sculpts the coast. Coasts include sea cliffs, shores, and beaches. Seawater set in motion erodes cliffs, transports eroded debris along shores, and dumps it on beaches. Most coasts retreat or advance over time. In addition to the unceasing causes of motion—wind, density of seawater, and rotation of the Earth—the chief agents in this process are tides, currents, and waves.

TIDES

Tides are the periodic rise and fall of the sea, once every 12 hours and 26 minutes. The gravitational pull of the moon and, to a lesser extent, the sun causes water on Earth to bulge toward them. It is interesting to note that a similar bulge occurs on the opposite side of the Earth due to inertial forces (further explanation is beyond the scope of this text). The effect of the tides is not too noticeable in the open sea, as the difference between high and low tide amounts to only about 2 feet. The tidal range may be considerably greater near shore, however. It may range from less than 2 feet to as much as 50 feet. The tidal range will vary according to the phase of the moon and the distance of the moon from the Earth. The type of shoreline and the physical configuration of the ocean floor will also affect the tidal range.

CURRENTS

Ocean currents are localized movements of masses of seawater. They are the result of drift of the upper 50 to 100 m of the ocean due to drag by wind. Thus, surface ocean currents generally follow the same patterns as atmospheric circulation, with the exception that atmospheric currents continue over the land surface while ocean currents are deflected by the land. Along with wind action, currents may also be caused by tides, variations in the salinity of the water, rotation of the Earth, and concentrations of turbid or muddy water. In addition to this surface circulation, seawater also circulates vertically due to differences in temperature and salinity that affect water density.

WAVES

Waves, varying greatly in size, are produced by the friction of wind on open water. Wave height and power depend on wind strength and fetch, the amount of unobstructed ocean over which the wind has blown. In a wave, water travels in loops (essentially up-and-down movements), with the diameter of the loops decreasing with depth. The diameter of loops at the surface is equal to the wave height. Breakers are formed when waves come into shallow water near the shore. The lower part of the wave is retarded by the ocean bottom, and the top, having greater momentum, is hurled forward, causing the wave to break. These breaking waves may do great damage to coastal property as they race across coastal lowlands driven by high winds or gales of hurricane velocities.

COASTAL EROSION, TRANSPORTATION, AND DEPOSITION

The geologic work of the sea consists of erosion, transportation, and deposition. The sea accomplishes its work of coastal landform sculpting largely by means of waves and wave-produced currents; their effect on the seacoast may be quite pronounced. The coast and accompanying coastal deposits and landform development represent a balance between wave energy and sediment supply.

WAVE EROSION

Waves attack shorelines and erode by a combination of several processes. The resistance of the rocks composing the shoreline and the intensity of wave action to which they are subjected are factors that determine how rapidly the shore will erode. Wave erosion works chiefly by hydraulic action, corrasion, and attrition. As waves strike a sea cliff, *hydraulic action* crams air into rock crevices, putting tremendous pressure on the surrounding rock; as waves retreat, the explosively expanding air enlarges cracks and breaks off chunks of rock (known as *scree*). Chunks hurled by waves against the cliff wear off more scree in a process called *corrasion*. In a similar process known as *attrition*, rocks collide with each other in the waves and become smaller and smoother until they are reduced to pebbles and sand grains (Lambert, 2007). Several coastal features are formed by marine erosion due to various combinations of wave action, rock types, and rock beds:

- *Sea cliffs* or *wave-cut platforms* are formed by wave erosion of underlying rock followed by caving-in of the overhanging rocks. As waves eat farther back inland, they leave a wave-cut beach or platform. Such cliffs are essentially vertical and are common at certain localities along the New England and Pacific coasts of North American.
- *Wave-cut benches* are the result of wave action not having enough time to lower the coastline to sea level. Because of the resistance to erosion, a relatively flat wave-cut bench develops. If subsequent uplift of the wave-cut bench occurs, it may be preserved above sea level as a marine terrace.
- *Headlands* are finger-like projections of resistant rock extending out into the water. Indentations between headlands are termed *coves*.
- *Sea caves*, *sea arches*, and *stacks* are formed by continued wave action on a sea cliff. Wave action hollows out cavities or caves in the sea cliffs. Eventually, waves may cut completely through a headland to form a sea arch; if the roof of the arch collapses, the rock left separated from the headland is called a *stack*.

Marine Transportation

Waves and currents are important transporting agents. Rip currents and undertow carry rock particles back to the sea, and long-shore currents will pick up sediments (some of it in solution), moving them out from shore into deeper water. Materials carried in solution or suspension may drift seaward for great distances and eventually be deposited far from shore. During the transportation process sediments undergo additional erosion, becoming reduced in size.

Marine Deposition

Marine deposition takes place whenever the velocity of currents and waves is reduced. Some rocks are thrown up on the shore by wave action. Most of the sediments thus deposited consist of rock fragments derived from the mechanical weathering of the continents, and they differ considerably from terrestrial or continental deposits. Due to the input of sediments from rivers, deltas may form; due to beach drift such features as spits and hooks, bay barriers, and tombolos may form. Depositional features along coasts are discussed below.

- *Beaches* are transitory coastal deposits of debris that lie above the low-tide limit in the shore zone.
- *Barrier islands* are long, narrow accumulations of sand lying parallel to the shore and separated from the shore by a shallow lagoon.
- *Spits and hooks* are elongated, narrow embankments of sand and pebble extending out into the water but attached by one end to the land.
- *Tombolos* are bars of sand or gravel connecting an island with the mainland or another island.
- *Wave-built terraces* are structures built up from sediments deposited in deep water beyond a wave-cut terrace.

WAVE ENERGY*

Waves are caused by the wind blowing over the surface of the ocean. In many areas of the world, the wind blows with enough consistency and force to provide continuous waves. Wave energy does not have the tremendous power of tidal fluctuations, but the regular pounding of the waves should not be underestimated because there is tremendous energy in the ocean waves. The total power of waves breaking on the world's coastlines is estimated at up to 2 terawatts. In optimal wave areas (such as the Pacific Northwest), 40 to 70 kilowatts of electricity could be produced per meter of coastline (EERE, 2010a). Because the wind is originally derived from the sun, we can consider the energy in ocean waves to be a stored, moderately high-density form of solar energy. According to certain estimates, wave technologies could feasibly fulfill 10% of the global electricity supply if fully developed (WEC, 2007). The west coasts of the United States and Europe and the coasts of Japan and New Zealand are good sites for harnessing wave energy.

Wave Energy: Facts, Parameters, and Equations

Three main processes create waves: (1) air flowing over the sea exerts a tangential stress on the water surface, resulting in the formation and growth of waves; (2) turbulent air flow close to the water surface creates rapidly varying shear stresses and pressure fluctuations (when these oscillations are in phase with existing waves, further wave development occurs); and (3) when waves have reached a certain size, the wind can exert a stronger force on the upwind face of the wave, resulting

---

* Adapted from USDOI, *Ocean Energy*, Minerals Management Service, U.S. Department of Interior, Washington, DC, 2010 (www.boemre.gov/mmsKids/PDFs/OceanEnergyMMS.pdf).

TABLE 9.1

World Meteorological Organization (WMO) Sea State Codes

WMO Sea State Code	Wave Height (meters)	Characteristics
0	0	Calm (glassy)
1	0–0.1	Calm (rippled)
2	0.1–0.5	Smooth (wavelets)
3	0.5–1.25	Slight
4	1.25–2.5	Moderate
5	2.5–4	Rough
6	4–6	Very rough
7	6–9	High
8	9–14	Very high
9	Over 14	Phenomenal

in additional wave growth. Waves located within or close to the areas where they are generated are called *storm waves*; waves known as *swell waves* can develop at great distances from their point of origin. The distance over which wind energy is transferred into the ocean to form waves is called the *fetch*. *Sea state* is the general condition of the free surface of a large body of water, with respect to wind waves and swell, at a certain location and moment (see Table 9.1).

The shape of a typical wave is described as *sinusoidal* (i.e., it has the form of a mathematical sine function) (see Figure 9.2). The difference in height between the peaks and troughs is height H, and the distance between successive peaks (or troughs) of the wave is wavelength λ. The time in seconds required for successive peaks (or troughs) to pass a given fixed point is period T. The *frequency* (ν) of the wave describes the number of peak-to-peak (or trough-to-trough) oscillations of the wave surface per second, as seen by a fixed observer and is the reciprocal of the period; that is, $\nu = 1/T$.

If a wave is traveling at velocity v past a given fixed point, it will travel a distance equal to its wavelength (λ) in a time equal to the wave period T (i.e., $v = \lambda/T$). The power (P) of an idealized ocean wave is approximately equal to the square of height H multiplied by wave period T. The exact expression is

$$P = \frac{g^2 H^2 T}{32}$$

where P is in units of watts per meter, and g is the acceleration due to gravity (9.81 m/s^2) (Phillips, 1977).

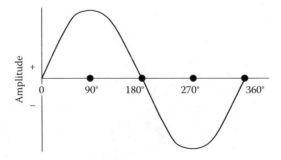

FIGURE 9.2 Sinusoidal wave showing wavelength and amplitude.

Wave Height

Wave height increases with increases in

- Wind speed
- Time duration of the wind blowing
- Fetch
- Water depth

Deep Water Waves

For deep water waves, the velocity of a long ocean wave can be shown to be proportional to the period (if the depth of water is greater than about half of wavelength λ) as follows:

$$v = \frac{gT}{2\pi}$$

The velocity in meters per second is approximately 1.5 times the wave period in seconds. It is interesting to note that in the deep ocean long waves travel faster than shorter waves. Moreover, if the above relationships hold, we can find the deep water wavelength (λ) for any given wave period:

$$\lambda = \frac{gT^2}{2\pi}$$

Intermediate Depth Waves

As the water becomes shallower, the properties of the waves become increasingly dominated by water depth. When waves reach shallow water, their properties are completely governed by the water depth, but in intermediate depths (i.e., between $d = \lambda/2\pi$ and $d = \lambda/4\pi$), the properties of the waves will be influenced by both water depth d and wave period T (Phillips, 1977).

Shallow Water Waves

As waves approach the shore, the seabed begins to have an effect on their speed, and it can be shown that if the water depth (d) is less than a quarter of the wavelength, the velocity is given by

$$v = \sqrt{gd}$$

Group Velocity

As waves propagate, their energy is transported. The energy transport velocity is the group velocity. The wave energy flux, through a vertical plane of unit width perpendicular to the wave propagation direction, is equal to

$$P = E \times c_g$$

where c_g is the group velocity (m/s).

DID YOU KNOW?

Hydro (water) kinetic energy is the energy possessed by a body of water due to its motion (kinetic energy = $1.2 \times m \times v^2$). Hydrostatic (at rest) energy is the energy possessed by a body of water due to its position or location at an elevation or height above a reference or datum (potential energy = $m \times g \times h$).

WAVE ENERGY CONVERSION TECHNOLOGY

In the early 1970s, the harnessing of wave power focused on using floating devices such as Cockerell rafts (a wave power hydraulic device), Salter's Duck (curved cam-like device that can capture 90% of waves for energy conversion), rectifiers (to convert AC to DC electricity), and the clam (a floating, rigid, doughnut-shaped device that converts wave energy to electrical energy). Wave energy converters can be classified in terms of their location: fixed to the seabed, generally in shallow water; floating offshore in deep water; or tethered in intermediate depths. These floating devices are not cost effective, and significant mooring problems remain to be solved, so current practice is to move closer to shore, sacrificing some energy. Fixed devices do have several advantages, including the following (Tovey, 2005):

- Easier maintenance
- Easier to land on device
- No mooring problems
- Easier power transmission
- Enhanced productivity
- Better design life

Wave energy devices can be classified by means of their reaction system, but it is often more instructive to discuss how they interact with the wave field. In this context, each moving body may be labeled as either a displacer or a reactor:

- A *displacer* is a body moved by the waves. It might be a buoyant vessel or a mass of water. If buoyant, the displacer may pierce the surface of the waves or be submerged.
- A *reactor* is a body that provides reaction to the displacer. It could a body fixed to the seabed, or the seabed itself. It could also be another structure or mass that is not fixed but moves in such a way that reaction forces are created (e.g., by moving by a different amount or at different times). A degree of control over the forces acting on each body or acting between the bodies (particularly stiffness and damping characteristics) is often required to optimize the amount of energy captured.

   In some designs, the reactor is actually inside the displacer, while in others it is an external body. Internal reactors are not subject to wave forces, but external reactors may experience loads that cause them to move in ways similar to a displacer; thus, some devices do not have dedicated reactors at all but rather a system of displacers whose relative motion creates a reaction system (INL, 2005). The three types of well-known wave energy conversion devices are point absorbers, terminators, and attenuators (see Figure 9.3):

- A *point absorber* is a floating structure that absorbs energy in all directions by virtue of its movements at or near the water surface (see Figure 9.4). It may be designed so as to resonate—that is, move with larger amplitudes than the waves themselves. This feature is useful for maximizing the amount of power that is available for capture. The power take-off system may take a number of forms, depending on the figuration of displacers and reactors.
- A *terminator* is also a floating structure that moves at or near the water surface, but a terminator absorbs energy in only a single direction (see Figure 9.5). The device extends in the direction normal to the predominant wave direction, so that as waves arrive the device restrains them. Again, resonance may be employed and the power take-off system may take a variety of forms.

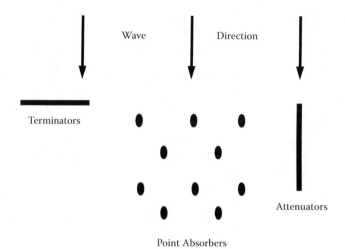

FIGURE 9.3 Types of wave energy converters. (Adapted from Tovey, N.K., *ENV-2E02 Energy Resources 2004–2005 Lecture*, University of East Anglia, Norwich, U.K., 2005; http://www2.env.uea.ac.uk/gmmc/energy/env2e02/env2e02.htm.)

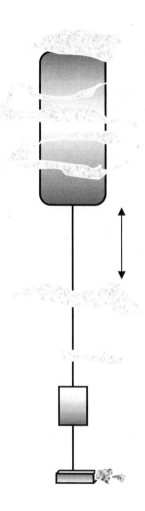

FIGURE 9.4 Point absorber.

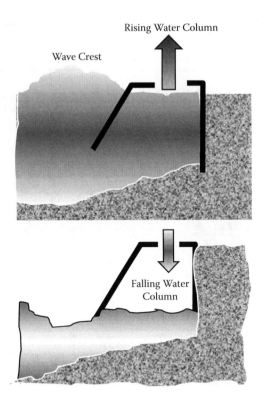

FIGURE 9.5   Terminator.

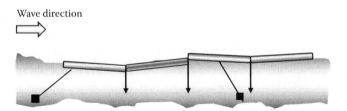

FIGURE 9.6   Attenuator.

- An *attenuator* is a long floating structure like the terminator, but it is orientated parallel to the waves rather than normal to them (see Figure 9.6). It rides the waves like a ship, and movements of the device at its bow and along its length can be restrained to extract energy. A theoretical advantage of the attenuator over the terminator is that its area normal to the waves is small so the forces it experiences are much lower.

TIDAL ENERGY

The tides rise and fall in eternal cycles. Tides are changes in the level of the oceans caused by the gravitational pull of the moon and sun and the rotation of the Earth. The relative motions of these cause several different tidal cycles, including a semidiurnal cycle (with a period of 12 hours and 25 minutes); a semimonthly cycle (spring or neap tides corresponding with the position of the moon, with the highest tides occurring in March and September); a semiannual cycle (with a period of about 178 days which is associated with the moon's orbit); and longer term cycles (e.g., a 19-year moon cycle). Nearshore water levels can vary up to 40 feet, depending on the season and local

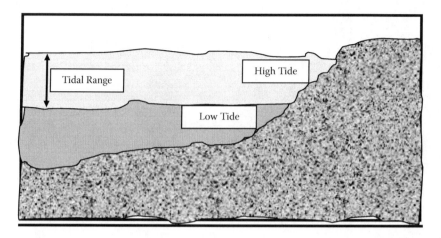

FIGURE 9.7   Tidal range is the difference between the high tide and the low tide.

factors. Only about 20 locations have good inlets and a large enough tidal range—about 10 feet—to produce energy economically (USDOI, 2010). The tide ranges (difference between high tide and low tide; see Figure 9.7) have been classified as follows (Masselink and Short, 1993):

- *Micromareal*, when the tidal range is less than 2 meters
- *Mesomareal*, when the tidal range is between 2 meters and 4 meters
- *Macromareal*, when the tidal range is higher than 4 meters

TIDAL ENERGY TECHNOLOGIES

Some of the oldest ocean energy technologies use tidal power. Tidal power is more predictable than solar power and wind energy. All coastal areas consistently experience two high and two low tides over a period of slightly greater than 24 hours. For those tidal differences to be harnessed into electricity, the difference between high and low tide must be at least 5 meters, or more than 16 feet. There only about 40 sites on the Earth with tidal ranges of this magnitude. Currently, there are no tidal power plants in the United States, although conditions are good for tidal power generation in both the Pacific Northwest and the Atlantic Northeast regions of the country. Tidal energy technologies include the following:

- *Tidal barrages*—A barrage is a simple generation system for tidal plants that involves installing a dam, or barrage, across an inlet. Sluice gates (gates commonly used to control water levels and flow rates) on the barrage allow the tidal basin to fill on the incoming high tides and to empty through the turbine system on the outgoing tide, also known as the ebb tide. Two-way systems generate electricity on both the incoming and outgoing tides. A potential disadvantage of a barrage tidal power system is the effect a tidal station can have on plants and animals in estuaries. Tidal barrages can change the tidal level in the basin and increase the amount of matter in suspension in the water (turbidity). They can also affect navigation and recreation.

DID YOU KNOW?

Spring tides have a range about twice that of neap tides; the other cycles can cause further variations of up to 15%. The tidal range is amplified in estuaries, and in some situations the shape of the estuary is such that near resonance occurs.

- *Tidal fences*—These look like giant turnstiles. A tidal fence has vertical-axis turbines mounted in a fence. All the water that passes is forced through the turbines. Some of these currents run at 5 to 8 knots (5.6 to 9 miles per hour) and generate as much energy as winds of much higher velocity. Tidal fences can be used in areas such as channels between two land masses. They are less expensive to install than tidal barrages and have less impact on the environment, although they can disrupt the movement of large marine animals.
- *Tidal turbines*—These are basically wind turbines in the water that can be located anywhere with strong tidal flow; they function best where coastal currents run at between 3.6 and 4.9 knots (4 to 5.5 mph). Because water is about 800 times more dense than air, tidal turbines have to be much sturdier than wind turbines. Tidal turbines are heaver and more expensive to build but capture more energy.

OCEAN THERMAL ENERGY CONVERSION

The most plentiful renewable energy source in our planet by far is solar radiation: 170,000 TW ($170,000 \times 10^{12}$ W) fall on Earth. Because of its dilute and erratic nature, however, it is difficult to harness. To capture this energy, we must employ the use of large collecting areas and large storage capacities, requirements satisfied on Earth by only the tropical oceans. We are all taught at an early age that water covers about 71% (or two-thirds) of Earth's surface. With regard to the vast oceans covering the majority of Earth, it is fitting that Ambrose Bierce (1842–1914) referred to them as "a body of water occupying about two-thirds of the world made for man who has no gills." True, we have no gills, so those who look out upon those vast bodies of water that cover the surface might ask: "What is their purpose?" This is a good question with several possible answers. With regard to renewable energy, we could look out upon those vast seas and wonder how we might use this massive storehouse of energy for our own needs; it is so vast and deep that it absorbs much of the heat and light that come from the sun. One thing seems certain: The secret to our origin, past, present, and future lies within those massive wet confines we call oceans.

Ocean Thermal Energy Conversion Process[*]

The ocean is essentially a gigantic solar collector. The energy from the sun heats the surface water of the ocean. In tropical regions, the surface water can be 40 or more degrees warmer than the deep water. This temperature difference can be used to produce electricity. Ocean thermal energy conversion (OTEC) has the potential to produce more energy than tidal, wave, and wind energy combined. The OTEC systems can be open or closed. In a closed system, an evaporator turns warm surface water into steam under pressure (see Figures 9.8 and 9.9). This steam spins a turbine generator to produce electricity. Water pumps bring cold deep water through pipes to a condenser on the surface. The cold water condenses the steam, and the closed cycle then begins again. In an open system, the steam is turned into freshwater, and new surface water is added to the system. A transmission cable carries the electricity to the shore. The OTEC systems must have a temperature difference of about 25°C to operate. This limits the use of OTEC to tropical regions where the surface waters are very warm and there is deep cold water. Hawaii, with its tropical climate, has experimented with OTEC systems since the 1970s. Because of the many challenges to their widespread use, no large or major OTEC systems are in operation today, but several experimental OTEC plants have been built. Pumping the water is a giant engineering challenge. Because of this, OTEC systems are not very energy efficient. It will probably be 10 to 20 years before the technology is available to produce and transmit electricity economically from OTEC systems.

---

[*] Adapted from USDOI, *Ocean Energy*, Minerals Management Service, U.S. Department of Interior, Washington, DC, 2010 (www.boemre.gov/mmsKids/PDFs/OceanEnergyMMS.pdf).

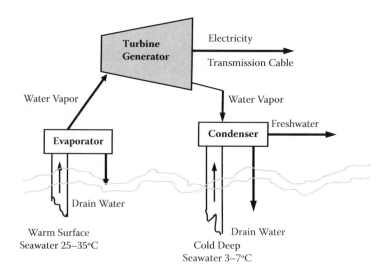

FIGURE 9.8   Schematic of ocean thermal energy conversion system.

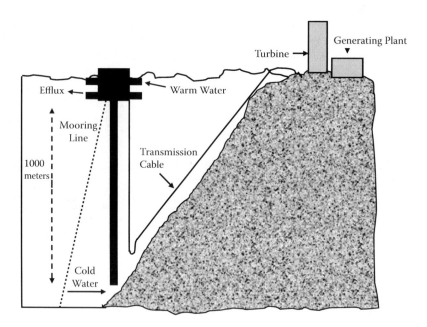

FIGURE 9.9   OTEC floating platform.

TYPES OF OTEC TECHNOLOGIES

The types of OTEC systems include the following:

- *Closed-cycle* systems use fluid with a low-boiling point, such as ammonia, to rotate a turbine to generate electricity. Warm surface seawater is pumped through a heat exchanger where the lower boiling point fluid is vaporized. The expanding vapor turns the turbogenerator. Cold, deep seawater pumped through a second heat exchanger condenses the vapor back into a liquid, which is then recycled through the system.

FIGURE 9.10 Free-flow water current turbine used in rivers and canals. (From NREL, *Image Gallery*, National Renewable Energy Laboratory, Washington, DC, 2014, http://images.nrel.gov/.)

- *Open-cycle* systems use the warm surface water of tropical oceans to make electricity. When warm seawater is placed in a low-pressure container, it boils. The expanding steam drives a low-pressure turbine attached to an electrical generator. The steam, which has left its salt behind in the lower pressure container, is almost pure fresh water. It is condensed back into a liquid by exposure to cold temperatures from deep ocean water.
- *Hybrid* systems combine the features of both the closed-cycle and open-cycle systems. In a hybrid system, warm seawater enters a vacuum chamber where it is flash-evaporated into steam, similar to the open-cycle evaporation process. The steam vaporizes a lower boiling point fluid (in a closed-cycle loop) that drives a turbine to produce electricity.

OCEAN ENERGY ENVIRONMENTAL IMPACTS[*]

This section summarizes the potential (generalized) environmental impacts of new ocean energy and hydrokinetic technologies. Environmental issues that apply to all technologies include the following:

- Alteration of river or ocean currents or waves
- Alteration of substrates and sediment transport and deposition
- Benthic organism habitat alterations
- Noise
- Electromagnetic fields
- Toxic effects of chemicals
- Interference with animal movements and migrations
- Interference with long-distance migrations of marine animals
- Potential for injury to aquatic organisms from strike or impingement due to designs that incorporate moving rotors or blades (see Figure 9.10)
- Impacts of ocean thermal energy conversion

---

[*] Adapted from USEPA, *Report to Congress on the Potential Environmental Effects of Marine and Hydrokinetic Energy Technologies*, U.S. Environmental Protection Agency, Washington, DC, 2009 (http://www1.eere.energy.gov/water/pdfs/doe_eisa_633b.pdf).

ALTERATION OF CURRENTS AND WAVES

The extraction of kinetic energy from river and ocean currents or tides will reduce water velocities in the vicinity (i.e., near field) of the project (Bryden et al., 2004). Large numbers of devices in a river will reduce water velocities, increase water surface elevations, and decrease flood conveyance capacity. These effects would be proportional to the number and size of structures installed in the water. Rotors, foils, mooring and electrical cables, and field structures will all act as impediments to water movement. The resulting reduction in water velocities could, in turn, affect the transport and deposition of sediment, organisms living on or in the bottom sediments, and plants and animals in the water column. Conversely, moving rotors and foils might increase mixing in systems where salinity or temperature gradients are well defined. Changes in water velocity and turbulence will vary greatly, depending on distance from the structure. For small numbers of units, the changes are expected to dissipate quickly with distance and are expected to be only localized; however, for large arrays, the cumulative effects may extend to a greater area. The alterations of circulation/mixing patterns caused by large numbers of structures might cause changes in nutrient inputs and water quality, which could in turn lead to eutrophication, hypoxia, and effects on the aquatic food web.

The presence of floating wave energy converters will alter wave heights and structures, both in the near field (within meters of the units or project) and, if installed in large numbers, potentially in the far field (extending meters to kilometers out from the project). The above-water structures of wave energy converters will act as a localized barrier to wind and thus reduce wind–wave interactions. Many of the changes would not directly relate to environmental impacts; for example, impacts on navigational conditions, wave loads on adjacent structures, and recreation on nearby beaches (e.g., surfing, swimming) might be expected (Michel et al., 2007). Reduced wave action could alter bottom erosion and sediment transport and deposition (Largier et al., 2008).

Wave measurements at operating wave energy conversion projects have not yet been made, and the data will be technology and project-size specific. The potential reductions in wave heights are probably smaller than those for wind turbines due to the low profiles of wave energy devices. For example, ASR, Ltd. (2007) predicted that operation of wave energy conversion (WEC) devices at the proposed Wave Hub (a wave power research facility off the coast of Cornwall, U.K.; http://www.wavehub.co.uk/) would reduce wave height at shorelines 5 to 20 kilometers away by 3 to 6%. Operation of six wave energy conversion buoys, a version of OPT's PowerBuoy®, in Hawaii was not predicted to impact oceanographic conditions (Department of the Navy, 2003). This conclusion was based on modeling analyses of wave height reduction due to both wave scattering and energy absorption. The proposed large spacing of buoy cylinders (51.5 m apart, compared to a buoy diameter of 4.5 m) resulted in predicted wave height reductions of 0.5% for a wave period (i.e., time between the passage of consecutive wave crests past a fixed point) of 9 seconds and less than 0.3% for a wave period of 15 seconds. Boehlert et al. (2008) summarized the changes in wave heights that were predicted in various environmental assessments. Recognizing that impacts will be technology and location specific, estimated wave height reductions range from 3 to 15%, with maximum effects closest to the installation and near the shoreline. Millar et al. (2007) used a mathematical model to predict that operation of the Wave Hub, with WECs covering an area 1 km by 3 km located 20 km from shore, could decrease average wave heights by about 1 to 2 centimeters at the coastline. This represents an average decrease in wave height of 1%; a maximum decrease in wave height of 3% was predicted to occur with a 90% energy transmitting wave farm (Smith et al., 2007). Other estimates in other environmental settings predict wave height reductions ranging from 3 to 13% (Nelson, 2008). Largier et al. (2008) concluded that height and incident angle are the most important wave parameters for determining the effects of reducing the energy supply to the coast.

The effects of reduced wave heights on coastal systems will vary from site to site. It is known that the richness and density of benthic organisms are related to such factors as relative tidal range and sediment grain size (Rodil and Lastra, 2004), so changes in wave height can be expected to alter benthic sediments and habitat for benthic organism. Coral reefs reduce wave heights and dissipate

DID YOU KNOW?

The PowerBuoy®, developed by Ocean Power Technologies (OPT), is one of the most widely deployed WEC device designs in the world. A 10-buoy test array of the PB150 PowerBuoy® has been proposed for deployment in Reedsport, Oregon. The PB150 is a utility-scale 150-kilowatt buoy that—in the initial design—contains hydraulic fluid, which is cycled as the buoy moves up and down with the waves. The moving fluid or mechanical parts are used to spin a generator, which produces electricity. The buoy is approximately 35 meters (115 feet) tall, of which approximately 9 meters (30 feet) project above the water's surface), and is 11 meters (36 feet) in diameter. It is held in place by a three-point mooring system (USDOE, 2012).

wave and tidal energy, thereby creating valuable ecosystems (Lugo-Fernandez et al., 1998; Roberts et al., 1992). In other cases, wave height reductions can have long-term adverse effects. Estuary and lagoon inlets may be particularly sensitive to changes in wave heights. For example, construction of a storm-surge barrier across an estuary in the Netherlands permanently reduced both the tidal range and mean high water level by about 12% from original values, and numerous changes to the affected salt marshes and wetlands soils were observed (de Jong et al., 1994).

Tidal energy converters can also modify wave heights and structure by extracting energy from the underlying current. It has been suggested that the effects of structural drag on currents would not be significant (Minerals Management Service, 2007), but few measurements of the effects of tidal/current energy devices on water velocities have been reported. Some tidal velocity measurements were made near a single, 150-kilowatt (kW) Stingray demonstrator in Yell Sound in the Shetland Islands (Engineering Business, Ltd., 2005). Acoustic Doppler current profilers were installed near the oscillating hydroplane (which travels up and down the water column in response to lift and drag forces) as well as upstream and downstream of the device. Too few velocity measurements were taken for firm conclusions to be made, but the data suggest that 1.5- to 2.0-m/s tidal currents were slowed by about 0.5 m/s downstream from the Stingray. In practice, multiple units will be spaced far enough apart to prevent a drop in performance (turbine output) caused by extraction of kinetic energy and localized water velocity reductions.

Modeling of the Wave Hub project in the United Kingdom suggested a local reduction in marine current velocities of up to 0.8 m/s, with a simultaneous increase in velocities of 0.6 m/s elsewhere (Michel and Burkhard, 2007). Wave energy converters are expected to affect water velocities less than submerged rotors and other, similar designs because only cables and anchors will interfere with the movements of tides and currents.

Tidal energy conversion devices will increase turbulence, which in turn will alter mixing properties, sediment transport, and, potentially, wave properties. In both the near field and far field, extraction of kinetic energy from tides will decrease tidal amplitude, current velocities, and water exchange in proportion to the number of units installed, potentially altering the hydrologic, sediment transport, and ecological relationships of rivers, estuaries, and oceans. For example, Polagye et al. (2008) used an idealized estuary to model the effects of kinetic power extraction on estuary-scale fluid mechanics. The predicted effects of kinetic power extraction included (1) reduction of the volume of water exchanged through the estuary over the tidal cycle, (2) reduction of the tidal range landward of the turbine array, and (3) reduction of the kinetic power density in the tidal channel. These impacts were strongly dependent on the magnitude of kinetic power extraction, estuary geometry, tidal regime, and nonlinear turbine dynamics.

Karsten et al. (2008) estimated that extracting the maximum of 7 gigawatts (GW) of power from the Minas Passage (Bay of Fundy) with in-stream tidal turbines could result in large changes in the tides of the Minas Basin (greater than 30%) and significant far-field changes (greater than 15%). Extracting 4 GW of power was predicted to cause less than a 10% change in tidal amplitudes, and 2.5 GW could be extracted with less than a 5% change. The model of Blanchfield et al. (2007) predicted that extracting the maximum value of 54 megawatts (MW) from the tidal current of Masset Sound (British Columbia) would decrease the water surface elevation within a bay and the maximum flow rate through the channel by approximately 40%. On the other hand, the tidal regime could be kept within 90% of the undisturbed regime by limiting extracted power to approximately 12 MW.

In the extreme far field (i.e., thousands of kilometers), there is an unknown potential for dozens or hundreds of tidal energy extraction devices to alter major ocean current such as the Gulf Stream (Michel et al., 2007). The significance of these potential impacts could be ascertained by predictive modeling and subsequent operational monitoring as projects are installed.

ALTERATION OF SUBSTRATES AND SEDIMENT TRANSPORT AND DEPOSITION

Operation of hydrokinetic or ocean energy technologies will extract energy from the water, which will reduce the height of waves or velocity of currents in the local area. This loss of wave/current energy could, in turn, alter sediment transport and the wave climate of nearby shorelines. Moreover, installation of many of the technologies will entail attaching the devices to the bottom by means of pilings or anchors and cables. Transmission of electricity to the shore will be through cables that are either buried in or attached to the seabed. Thus, project installation will temporarily disturb sediments, the significance of which will be proportional to the amount and type of bottom substrate disturbed. There have been few studies of the effects of burying cables from ocean energy technologies, but experience with other buried cables and trawl fishing indicate the possible severity of the impacts. For example, Kogan et al. (2006) surveyed the condition of an armored, 6.6-cm-diameter coaxial cable that was laid on the surface of the seafloor off Half Moon Bay, California. The cable was not anchored to the seabed. Whereas the impacts of laying the cable on the surface of the seabed were probably small, subsequent movements of the cable had continuing impacts on the bottom substrates. For example, cable strumming by wave action in shallower, nearshore areas created incisions in rocky siltstone outcrops ranging from superficial scrapes to vertical grooves and had minor effects on the habitats of aquatic organisms. At greater depths, there was little evidence of effects of the cable on the seafloor, regardless of exposure. Limited self-burial of the unanchored cable occurred over an 8-year period, particularly in deeper waters of the continental shelf.

During operation, changes in current velocities or wave heights will alter sediment transport, erosion, and sedimentation. Due to the complexity of currents and their interaction with structures, operation of the projects will likely increase scour and deposition of fine sediments on both localized and far-field scales. For example, turbulent vortices that are shed immediately downstream from a velocity-reducing structure (e.g., rotors, pilings, concrete anchor blocks) will cause scour, and this sediment is likely to be deposited further downstream. On average, extraction of kinetic energy from currents and waves is likely to increase sediment deposition in the shadow of the project (Michel et al., 2007), the depth and areal extent of which will depend on local topography, sediment types, and characteristics of the current and the project. Subsequent deposition of sediments is likely to cause shoaling and a shift to a finer sediment grain size on the lee side of wave energy arrays (Boehlert et al., 2008). Scour and deposition should be considered in project development, but many of the high-energy (high-velocity) river and nearshore marine sites that could be utilized for electrical energy production are likely to have substrates with few or no fine sediments. Changes in scour and deposition will alter the habitat for bottom-dwelling plants and animals.

Loss of wave energy may lead to changes in longshore currents, reductions in the width and energy of the surf zone, and changes in beach erosion and deposition patterns. Millar et al. (2007) modeled the wave climate near the Wave Hub electrical grid connection point off the north coast of Cornwall. The installation would be located 20 km off the coast, in water depths of 50 to 60 m.

Arrays of WECs connected to the Wave Hub would occupy a 1 km × 3 km site. The mathematical model predicted that an array of WECs would potentially affect the wave climate on the nearby coast on the order of 1 to 2 cm. It is unknown whether such small reductions in the average wave height would measurably alter sediment dynamics along the shore, given the normal variations in waves due to wind and storms.

Water quality will be temporarily affected by increased suspended sediments (turbidity) during installation and initial operation. Suspension of anoxic sediments may result in a temporary and localized decline in the dissolved oxygen content of the water, but dilution by oxygenated water current would minimize the impacts. Water quality may also be compromised by the mobilization of buried contaminated sediments during both construction and operation of the projects. Excavation to install the turbines, anchoring structures, and cables could release contaminants adsorbed to sediments, posing a threat to water quality and aquatic organisms. Effects on aquatic biota may range from temporary degradation of water quality (e.g., a decline in dissolved oxygen content) to biotoxicity and bioaccumulation of previously buried contaminants such as metals.

Benthic Organism Habitat Alterations

Installation and operation of hydrokinetic and marine energy projects can directly displace benthic (i.e., bottom-dwelling) plants and animals or change their habitats by altering water flows, wave structures, or substrate composition. Many of the designs will include a large anchoring system made of concrete or metal, mooring cables, and electrical cables that lead from the offshore facility to the shoreline. Electrical cables might simply be laid on the bottom, or they more likely will be anchored or buried to prevent movement. Large bottom structures will alter water flow, which may result in localized scour and/or deposition. Because these new structures will affect bottom habitats, changes to the benthic community composition and species interactions in the area defined by the project may be expected (Louse et al., 2008).

Displacement of Benthic Organisms by Installation of the Project

Bottom disturbances will result from the temporary anchoring of construction vessels, digging and refilling the trenches for power cables, and installation of permanent anchors, pilings, or other mooring devices. Motile organisms will be displaced and sessile organism destroyed in the limited areas affected by these activities. Displaced organisms may be able to relocate if similar habitats exist nearby and those habitats are not already at carrying capacity. That is, each population has an upper limit on size, called the *carrying capacity*. Carrying capacity can be defined as being the optimum number of individuals of a species that can survive in a specific area over time. A particular pond may be able to support only a dozen frogs based on the food resources for the frogs in the pond. If 30 frogs took up residency in the same pond, at least half of them would probably die because the pond environment would not have enough food for all of them to live. Carrying capacity, symbolized as K, is based on the quantity of food supplies, the physical space available, the degree of predation, and several other environmental factors.

Species with benthic-associated spawning or whose offspring settle into and inhabit benthic habitats are likely to be most vulnerable to disruption during project installation. Temporary increases in suspended sediments and sedimentation downcurrent from the construction area can be expected. The potential effects of suspended sediments and sedimentation on aquatic organisms are periodically reviewed (Newcombe and Jensen, 1996; Wilber and Clarke, 2001; Wilber et al., 2005; Wood and Armitage, 1997). When construction is completed, disturbed areas are likely to be recolonized by these same organisms, assuming that the substrate and habitats are restored to a similar state. For example, Lewis et al. (2003) found that numbers of clams and burrowing polychaetes (worms) fully recovered within a year after construction of an estuarine pipeline, although fewer wading birds returned to forage on these invertebrates during the same time period.

Alteration of Habitats for Benthic Organism during Operation

Installation of the project will alter benthic habitats over the longer term if the trenches containing electrical cables are backfilled with sediments of different size or composition than the previous substrate. Permanent structures on the bottom (ranging in size from anchoring systems to sea-bed-mounted generators or turbine rotors) will supplant the existing habitats. These new structures would replace natural hard substrates or, in the case of previously sandy areas, add to the amount of hard bottom habitat available to benthic algae, invertebrates, and fish. This could attract a community of rocky reef fish and invertebrate species (including biofouling organisms) that would not normally exist at that site. Depending on the situation, the newly created habitat could increase biodiversity or have negative effects by enabling introduced (exotic) benthic species to spread. Marine fouling communities developed on monopiles for offshore wind power plants are significantly different from the benthic communities on adjacent hard substrates (Wihelmsson and Maim, 2008; Wilhelmsson et al., 2006).

Changes in water velocities and sediment transport, erosion, and deposition caused by the presence of new structures will alter benthic habitats, at least on a local scale. This impact may be more extensive and long lasting than the effects of anchor and cable installation. Deposition of sand may impact seagrass beds by increasing mortality and decreasing the growth rate of plant shoots (Craig et al., 2008). Conversely, deposition of organic matter in the wakes of marine energy devices could encourage the growth of benthic invertebrate communities that are adapted to that substrate. Mussel shell mounts that slough off from oil and gas platforms may create surrounding artificial reefs that attract a large variety of invertebrates (e.g., crabs, sea stars, sea cucumbers, anemones) and fish (Love et al., 1999). Accumulation of shells and organic matter in the areas would depend on the wave and current energy, activities of biota, and numerous other factors (Widdows and Brinsley, 2002). Although the new habitats created by energy conversion structures may enhance the abundance and diversity of invertebrates, predation by fish attracted to artificial structures can greatly reduce the numbers of benthic organisms (Davis et al., 1982; Langlois et al., 2005).

Movements of mooring or electrical transmission cables along the bottom (sweeping) could be a continual source of habitat disruption during operation of the project. For example, Kogan et al. (2006) found that shallow water wave action shifted a 6.6-cm-diameter, armored coaxial cable that was laid on the surface of the seafloor. The strumming action caused incisions in rocky outcrops, but effects on seafloor organisms were minor. Anemones colonized the cable itself, preferring the hard structure over the nearby sediment-dominated seafloor. Some flatfishes were more abundant near the cable than at control sites, probably because the cable created a more structurally heterogeneous habitat. Sensitive habitats that may be particularly vulnerable to the effects of cable movements include macroalgae and seagrass beds, coral habitats, and other biogenic habitats such as worm reefs and mussel mounds.

Renewable energy projects may also have benefits for some aquatic habitats and populations. The presence of a marine energy conversion project will likely limit most fishing activities and other access in the immediate area. Bottom trawling can disrupt habitats, and benthic communities in areas that are heavily fished tend to be less complex and productive than in areas that are not fished in that way (Jennings et al., 2001; Kaiser et al., 2000). Blyth et al. (2004) found that cessation of towed-gear fishing resulted in significantly greater total species richness and biomass of benthic communities compared to sites that were still fished. The value of these areas in which fishing is precluded (or, at least limited to certain gear types) by the energy project would depend on the species of fish and their mobility. For relatively sedentary animals, reserves less than 1 km across have augmented local fisheries, and reserves in Florida of 16 km^2 and 24 km^2 have sustained more abundant and sizable fish than nearby exploited areas (Gell and Roberts, 2003). On the other hand, the protection of long-lived, late-maturing, or migratory marine fish species may require much larger marine protected areas (greater than 500 km^2) than those envisioned for most energy developments (Blyth-Skyrme et al., 2006; Kaiser, 2005; Nelson, 2008).

NOISE

Freshwater and marine animals rely on sound for many aspects of their lives including reproduction, feeding, predator and hazard avoidance, communication, and navigation (Popper, 2003; Weilgart, 2007). Consequently, underwater noise generated during installation and operation of a hydrokinetic or ocean energy conversion device has the potential to impact these organisms. Noise may interfere with sounds animals make to communicate or may drive animals from the area. If severe enough, loud sounds could damage their hearing or cause mortalities. For example, it is known from experience with other marine construction activities that the noise created by pile driving creates sound pressure levels high enough to impact the hearing of harbor porpoises and harbor seals (Thomsen et al., 2006). The effects are less certain for fish (Hastings and Popper, 2005), although fish mortalities have been reported for some pile-driving activities (Caltrans, 2001; Longmuir and Lively, 2001). Noise generated during normal operations is expected to be less powerful but could still disrupt the behavior of marine mammals, sea turtles, and fish at great distances from the source. Changes in animal behavior or physiological stresses could lead to decreased foraging efficiency, abandonment of nearby habitats, decreased reproduction, and increased mortality (National Research Council, 2005)—all of which could have adverse effects on both individuals and populations.

Construction and operation noise may disturb seabirds using the offshore and intertidal environment. Shorebirds will be disturbed by onshore construction and operations, causing them to abandon breeding colonies (Thompson et al., 2008). Pinnipeds such as seals, sea lions, and walruses may abandon onshore sites used for reproduction (rookeries) because of noise and other disturbing activities during installation. On the other hand, some marine mammals and birds may be attracted to the area by underwater sounds, lights, or increased prey availability.

There are many sources of sound and noise in the aquatic environment (National Research Council, 2003; Simmonds et al., 2003). Natural sources include wind, waves, earthquakes, precipitation, cracking ice, and mammal and fish vocalizations. Human-generated ocean noise comes from such diverse sources as recreational, military, and commercial ship traffic; dredging; construction; oil drilling and production; geophysical surveys; sonar; explosions; and ocean research (Johnson et al., 2008). Many of these sounds will be present in an area of new energy developments. Noises generated by marine and hydrokinetic energy technologies should be considered in the context of these background sounds. The additional noises from these energy technologies could result from installation and maintenance of the units, movements of internal machinery, waves striking the buoys, water flow moving over mooring and transmission cables, synchronous and additive nonsynchronous sound from multiple unit arrays, and environmental monitoring using hydroacoustic techniques.

Noise in the Aquatic Environment

There are many ways to express the intensity and frequency of underwater sound waves (Thomsen et al., 2006; Wahlberg and Westerberg, 2005). An underwater acoustic wave is generated by displacement of water particles; consequently, the passage of an acoustic wave creates local pressure oscillations that travel through water with a given sound velocity. These two parameters, pressure and velocity, are used to define the intensity of an acoustic field and therefore are useful for considering the effects of noise on aquatic animals. The intensity of the acoustic field is defined as the vector product of the local pressure fluctuations and the velocity of the particle displacement. A basic unit for measuring the intensity of underwater noise is the *sound pressure level* (SPL). The SPL of a sound, given in decibels (dB), is calculated by

$$\text{SPL (dB)} = 20 \log_{10} (p/p_0)$$

where p is a pressure fluctuation caused by a sound source, and p_0 is the reference pressure, defined in underwater acoustics as 1 µPa at 1 m from the source (Thomsen et al., 2006). Using the above formula, doubling the pressure of a sound (p) results in a 6-dB increase in SPL.

DID YOU KNOW?

According to several articles published in *Electrical Engineering* and the *Journal of the Acoustical Society of America*, the decibel suffers from the following disadvantages (Chapman, 2000; Clay, 1999; Hickling, 1999; Horton, 1954):

- The decibel creates confusion.
- The logarithmic form obscures reasoning.
- Decibels are more related to the area of slide rules than that of modern digital processing.
- Decibels are cumbersome and difficult to interpret.

Hickling (1999) concluded, "Decibels are a useless affectation, which is impeding the development of noise control as an engineering discipline."

The sound pressure of a continuous signal is often expressed by a root-mean-square (RMS) measure, which is the square root of the mean value of squared instantaneous sound pressures integrated over time (Madsen, 2005). Like SPL, the resulting integration of instantaneous sound pressure levels is also expressed in dB re 1 µPa (RMS). An RMS level of safe exposure to received noise has been established for marine mammals; the lower limits for concern about temporary or permanent hearing impairments in cetaceans and pinnipeds are currently 180 and 190 dB re 1 µPa (RMS), respectively (National Marine Fisheries Service, 2003; Southall et al., 2007). However, Madsen (2005) argued that RMS safety measures are insufficient and should be supplemented by other estimates of the magnitude of noise (e.g., maximum peak-to-peak SPL in concert with a maximum received energy flux level).

Sound intensity is greatest near the sound source and, in the far field, decreases smoothly with distance. As the acoustic wave propagates through the water, intensity is reduced by geometric spreading (dilution of the energy of the sound wave as it spreads out from the source over a larger and larger area) and, to a lesser extent, absorption, refraction, and reflection (Wahlberg and Westerberg, 2005). Attenuation of sound due to spherical spreading in deep water is estimated by $20 \log_{10} r$, where r is the distance in meters from the source (National Research Council, 2000). Assuming simple spherical spreading (no reflection from the sea surface or bottom) and the consequent transmission loss of SPL, a 190-dB source level would be reduced to 150 dB at 100 m. Close to the source, changes in sound intensity vary in a more complicated fashion, particularly in shallow water, as a result of acoustic interference from natural or man-made sounds or where there are reflective surfaces (seabed and water surface).

Sound exposure level (SEL) is a measure of the cumulative physical energy of the sound event which takes into account both intensity and duration. SELs are computed by summing the cumulative sound pressure squared (p^2) over time and normalizing the time to 1 second. Because calculation of the SEL for a given underwater sound source is a way to normalize to 1 second the energy of noise that may be much briefer (such as the powerful, but short impulses caused by pile driving), SEL is typically used to compare noise events of varying durations and intensities.

In addition to intensity, underwater noise will have a range of frequencies (Hz or cycles per second). For convenience, measurements of the potentially wide range of individual frequencies associated with noise are integrated into "critical bands" or filters; the width of a band is often given in 1/3-octave levels (Thomsen et al., 2006). Thus, sounds can be expressed in terms of the intensities (dB) at particular frequency (Hz) bands.

Four fundamental properties of sound transmission in water are relevant to consideration of the effects of noise on aquatic animals (National Research Council, 2000):

1. The transmission distance of sound in seawater is determined by a combination of geometric spreading loss and an absorptive loss that is proportional to the sound frequency. Thus, attenuation (weakening) of sound increases as its frequency increases.
2. The speed of a sound wave in water is proportional to the temperature.
3. The sound intensity decreases with distance from the sound source. Transmission loss of energy (intensity) due to spherical spreading in deep water is estimated by $20 \log_{10} r$, where r is the distance in meters from the source.
4. The strength of sound is measured on a logarithmic scale.

From these properties, it can be seen that high-frequency sounds will dissipate faster than low-frequency sounds, and a sound level may decrease by as much as 60 dB at 1 km from the source. Acoustic wave intensity of 180 dB is 10 times less intense than 190 dB, and 170 dB is 100 times less intense than 190 dB (National Research Council, 2000).

Noise Produced by Ocean Energy Technologies

There is very little information available on sound levels produced by construction and operation of ocean energy conversion structures (Michel et al., 2007). However, reviews of the construction and operation of European offshore wind farms provide useful information on the sensitivity of aquatic organisms to underwater noise. For example, Thomsen et al. (2006) reported that pile-driving activities generate brief, but very high, sound pressure levels over a broad band of frequencies (20 to 20,000 Hz). Single pulses are about 50 to 100 ms in duration and occur approximately 30 to 60 times per minute. The SEL at 400 m from the driving of a 1.5-m-diameter pile exceeded 140 dB re 1 µPa over a frequency range of 40 to 3000 Hz (Betke et al., 2004). It usually takes 1 to 2 hours to drive one pile into the bottom. Sounds produced by the pile-driving impacts above the water's surface enter the water from the air and from the submerged portion of the pile; they then propagate through the water column and into the sediments, from which they pass successively back into the water column. Larger diameter, longer piles require relatively more energy to drive into the sediments, which results in higher noise levels. For example, the SPL associated with driving 3.5-m-diameter piles is expected to be roughly 10 dB greater than for a 1.5-m-diameter pile (Thomsen et al., 2006). Pile-driving sound, while intense and potentially damaging, would occur only during the installation of some marine and hydrokinetic energy devices.

Some ocean energy technologies will be secured to the bottom by means of moorings and anchors drilled into rock. Like pile-driving, hydraulic drilling will occur during a limited time period, and noise generation will be intermittent. The Department of the Navy (2003) summarized underwater SPL measurements of three hydraulic rock drills; frequencies ranged from about 15 Hz to over 39 Hz, and SPLs ranged from about 120 to 170 dB re 1 µPa. SPLs were relatively consistent across the entire frequency range.

During operation, vibrations of the device's gearbox, generator, and other moving components are radiated as sound into the surrounding water. Noise during the operation of wind farms is of much lower intensity than noise during construction (Betke et al., 2004; Thomsen et al., 2006), and the same may be true for hydrokinetic and ocean energy farms; however, this source of noise will be continuous. Measurements of sound levels associated with the operation of hydrokinetic and ocean energy farms have not yet been published. One example of a wave energy technology, the WEC buoy (a version of OPT's PowerBuoy) which has been tested in Hawaii, has many of the mechanical parts contained within an equipment canister or mounted to a structure through mounting pads. Thus, the acoustic energy produced by the equipment is not well coupled to the seawater, which is expected to reduce the amount of radiated noise (Department of the Navy, 2003). Although no measurements had been made, it was predicted that the acoustic output from the WEC buoy system would probably be in the range of 75 to 80 dB re 1 µPa. This SPL is equivalent to light to normal density shipping noise, although the frequency spectrum of the WEC buoy is expected to be shifted to higher frequencies than typical shipping noise. By comparison, Thomsen et al. (2006) reported the ambient noise measured at five different locations in the North Sea. Depending on frequency, SPL values ranged from 85 to 115 dB, with most energy occurring at frequencies less than 100 Hz.

The Environmental Statement for the proposed installation of the Wave Dragon wave energy demonstrator off the coast of Pembrokeshire, U.K., predicted noise levels associated with instal-lation of a concrete caisson (gravity) block and steel cable mooring arrangement, installation of subsea cable, and support activity (Wave Dragon Wales, Ltd., 2007). The installation of gravity blocks is not expected to generate additional noise over and above that of the vessel conducting the operation. Vessel noise will depend on size and design of the ship but is expected to be up to 180 dB re 1 μPa at 1 m. Other predicted installation noise sources and levels stem from operation of the ship's echosounder (220 dB re 1 μPa at 1 m peak-to-peak), cable laying and fixing (159 to 181 dB re 1 μPa at 1 m), and directional drilling (129 dB re 1 μPa RMS at 40 m above the drill). There are no measurements available for the noise associated with operation of an overtopping device such as the Wave Dragon. Wave Dragon Wales, Ltd. (2007) predicted that operational noise would result from the Kaplan-style hydroturbines (an estimated 143 dB re 1 μPa at 1 m), as well as unknown levels and frequencies of sound from wave interactions with the body of the device, hydraulic pumps, and the mooring system.

In 2008, the Ocean Renewable Power Company (ORPC) made limited measurements of under-water noise associated with operation of their 1/3-scale working prototype in-stream tidal energy conversion device, its turbine generation unit (TGU). The TGU is a single horizontal-axis device with two advanced-design cross-flow turbines that drive a permanent magnet generator. An omnidi-rectional hydrophone, calibrated for a frequency range of 20 to 250 kHz, was used to take near-field measurements adjacent to the barge from which the turbine was suspended and at approximately 15 m from the turbine. Multiple far-field measurements were also made at distances out to 2.0 km from the barge. Noise measurements were made over one full tidal cycle, with supplemental mea-surements being taken later (USDOE, 2009). Sound pressure levels at 1/3-octave frequency bands were used to calculate RMS levels and SELs. During times when the turbine generator unit was not operating, background noise ranged from 112 to 138 dB re 1 μPa RMS, and SELs ranged from 120 to 140 dB re 1 μPa. A single measurement made when the turbine blades were rotating (at 52 rpm) resulted in an estimate of 132 dB re 1 μPa (RMS) and an SEL of 126 dB re 1 μPa at a horizontal distance of 15 m and a water depth of 10 m. These very limited readings suggest that the single 1/3-scale turbine generator unit did not increase noise above ambient levels.

In addition to the sound intensity and frequency spectrum produced by the operation of individ-ual machines, impacts of noise will depend on the geographic location of the project (water depth, type of substrate), the number of units, and the arrangement of multiple-unit arrays. For example, due to noise from surf and surface waves, noise levels in shallow, nearshore areas (≤100 m deep and within 5 km of the shore) are typically somewhat higher for low frequencies (≤1 kHz) and much higher for frequencies above 1 kHz.

Potential Effects of Noise on Aquatic Animals

Because of the complexity of describing underwater sounds, investigators have often used different units to express the effects of sound on aquatic animals and have not always reported precisely the experimental conditions. For example, acoustic signal characteristics that might be relevant to bio-logical effects include frequency content, rise time, pressure and particle velocity time series, zero-to-peak and peak-to-peak amplitude, mean-square amplitude, duration, integral of mean-square amplitude over duration, sound exposure level, and repetition rate (National Research Council, 2003; Thomsen et al., 2006). Each of these sound characteristics may differentially impact different species of aquatic animals, but the relationships are not sufficiently understood to specify which are the most important. Many studies of the effects of noise report the frequency spectrum and some measure of sound intensity (SPL, RMS, and/or SEL).

Underwater noise can be detected by fish and marine mammals if the frequency and intensity fall within the range of hearing for the particular species. An organism's hearing ability can be displayed as an audiogram, which plots sound pressure level (dB) against frequency (Hz). Nedwell et al. (2004) compiled audiograms for a number of aquatic organisms. If the pressure level of a generated sound is transmitted at these frequencies and exceeds the sound pressure level (i.e., above the line) on a given species' audiogram, the organism will be able to detect the sound. There is a wide range of sensitivity to sound among marine fish. Herrings (Clupeoidea) are highly sensitive to sound due to the structure of their swim bladder and auditory apparatus, whereas flatfish such as plaice and dab (Pleuronectidae) that have no swim bladder are relatively insensitive to sound (Nedwell et al., 2004). Possible responses to the received sound may include altered behavior (e.g., attraction, avoidance, interference with normal activities) (Nelson, 2008) or, if the intensity is great enough, hearing damage or mortality. For example, fish kills have been reported in the vicinity of pile-driving activities (Caltrans, 2001; Longmuir and Lively, 2001).

The National Research Council (2000) reviewed studies that demonstrated a wide range of susceptibilities to exposure-induced hearing damage among different marine species. The implications are that critical sound levels will not be able to be extrapolated from studies of a few species (although a set of representative species might be identified), and it will not be possible to identify a single sound level value at which damage to the auditory system will begin in all, or even most, marine mammals. Participants in a recent National Oceanic and Atmospheric Administration (NOAA) workshop (Boehlert et al., 2008) suggested that sounds that are within the range of hearing and "sweep" in frequency are more likely to disturb marine mammals than constant-frequency sounds. Thus, devices that emit a constant frequency may be preferable to ones that vary. They believed that the same may be true, although perhaps to a lesser extent, for sounds that change in amplitude.

Moore and Clarke (2002) compiled information on the reactions of gray whales (*Eschrichtius robustus*) to noise associated with offshore oil and gas development and vessel traffic. Gray whale responses included changes in swim speed and direction to avoid the sound sources, abrupt but temporary cessation of feeding, changes in calling rates and call structure, and changes in surface behavior. They reported a 0.5 probability of avoidance when continuous noise levels exceeded about 120 dB re 1 μPa and when intermittent noise levels exceeded about 170 dB re 1 μPa. They found little evidence that gray whales travel far or remain disturbed for long as a result of noises of this nature.

Weilgart (2007) reviewed the literature on the effects of ocean noise on cetaceans (whales, dolphins, porpoises), focusing on underwater explosions, shipping, seismic exploration by the oil and gas industries, and naval sonar operations. She noted that strandings and mortalities of cetaceans have been observed even when estimated received sound levels were not high enough to cause hearing damage. This suggests that a change in diving patterns may have resulted in injuries due to gas and fat emboli (a fat droplet that enters the blood stream). That is, aversive noise may prompt cetaceans to rise to the surface too rapidly, and the rapid decompression causes nitrogen gas supersaturation and the subsequent formation of bubbles (emboli) in their tissues (Fernandez et al., 2005). Other adverse (but not directly lethal) impacts could include increased stress levels, abandonment of important habitats, masking of important sounds, and changes in vocal behavior that may lead to reduced foraging efficiency or mating opportunities. Weilgart (2007) pointed out that responses of cetaceans to ocean noise are highly variable among species, age classes, and behavioral states, and many examples of apparent tolerance of noise have been documented.

Nowacek et al. (2007) reviewed the literature on the behavioral, acoustic, and physiological effects of anthropogenic noise on cetaceans and concluded that the noise sources of primary concern are ships, seismic exploration, sonars, and some acoustic harassment devices (AHDs) that are employed to reduce the by-catch of small cetaceans and seals by commercial fishing gear.

Two marine mammals whose hearing and susceptibility to noise have been studied are the harbor porpoise (*Phocoena phocoena*) and the harbor seal (*Phoca vitulina*). Both species inhabit shallow coastal waters in the north Atlantic and North Pacific. Harbor porpoises are found as far south as Central California on the West Coast. The hearing of the harbor porpoise ranges from below 1 kHz to around 140 kHz. In the United States, harbor seals range from Alaska to Southern California on the West Coast, and as far south as South Carolina on the East Coast. Harbor seal hearing ranges from less than 0.1 kHz to around 100 kHz (Thomsen et al., 2006). Sounds produced by marine energy devices that are outside of these frequency ranges would not be detected by these species.

Thomsen et al. (2006) compared the underwater noise associated with pile driving to the audiograms of harbor porpoises and harbor seals and concluded that pile-driving noise would likely be detectable at least 80 km away from the source. The zone of masking (the area within which the noise is strong enough to interfere with the detection of other sounds) may differ between the two species. Because the echolocation (sonar) used by harbor porpoises is in a frequency range (120 to 150 kHz) where pile-driving noises have little or no energy, they considered masking of echolocation to be unlikely. On the other hand, harbor seals communicate at frequencies ranging from 0.2 to 3.5 kHz, which is within the range of the highest pile-driving sound pressure levels; thus, harbor seals may have their communications masked at considerable distances by pile-driving activities.

The responses of green turtles (*Chelonia mydas*) and loggerhead turtles (*Caretta caretta*) to the sounds of air guns used for marine seismic surveys were studied by McCauley et al. (2000a,b). They found that above a noise level of 166 dB re 1 μPa RMS the turtles noticeably increased their swimming activity, and above 175 dB re 1 μPa RMS their behavior became more erratic, possibly indicating that the turtles were in an agitated state. On the other hand, Weir (2007) was not able to detect an impact on turtles of the sounds produced by air guns in geophysical seismic surveys. Caged squid (*Sepioteuthis australis*) showed a strong startle response to an air gun at a received level of 174 dB re 1 μPa RMS. When sound levels were ramped up (rather than a sudden nearby startup), the squid showed behavioral responses (e.g., rapid swimming) at sound levels as low as approximately 156 dB re 1 μPa RMS but did not display the startle response seen in the other tests.

Hastings and Popper (2005) reviewed the literature on the effects of underwater sounds on fish, particularly noises associated with pile driving. The limited number of quantitative studies found evidence of changes in the hearing capabilities of some fish, damage to the sensory structure of the inner ear, or, for fish close to the sources, mortality. They concluded that the body of scientific and commercial data is inadequate to develop more than the most preliminary criteria to protect fish from pile-driving sounds and suggested the types of studies that could be conducted to address the information gaps. Similarly, Viada et al. (2008) found very little information on the potential impacts on sea turtles of underwater explosives. Although explosives produce greater sound pressures than pile driving and are unlikely to be used in most ocean energy installations, studies of their effects provide general information about the peak pressures and distances that have been used to establish safety zones for turtles.

Wahlberg and Westerberg (2005) compared source level and underwater measurements of sounds from offshore windmills to information about the hearing capabilities of three species of fish: goldfish, Atlantic salmon, and cod. They predicted that these fish could detect offshore windmills at a maximum distance of about 0.4 to 25 km, depending on wind speed, type and number of windmills, water depth, and substrate. They could find no evidence that the underwater sounds emitted by windmill operation would cause temporary or permanent hearing loss in these species, even at a distance of a few meters, although sound intensities might cause permanent avoidance within ranges of about 4 m. They noted that shipping causes considerably higher sound intensities than operating windmills (although the noise from shipping is transient), and noises from installation may have much more significant impacts on fish than those from operation.

In the Environmental Assessment of the proposed Wave Energy Technology (WET) Project, the Department of the Navy (2003) considered the sounds made by hydraulic rock drilling to be detectable by humpback whales, bottlenose dolphins, Hawaiian spinner dolphins, and green sea turtles. Assuming a transmission loss due to spherical spreading, drilling sound pressure levels of 160 dB re 1 µPa would decrease by about 40 dB at 100 m from the source. They regarded a SPL of 120 dB re 1 µPa to be below the level that would affect these four species. In fact, they reported that other construction activities involving similar drilling attracted marine life (fish and sea turtles, in particular), perhaps because bottom organisms were stirred up by the drilling.

There are considerable information gaps regarding the effects of noise generated by marine and hydrokinetic energy technologies on cetaceans, pinnipeds, turtles, and fish. Sound levels from these devices have not been measured, but it is likely that installation will create more noise than operation, at least for those technologies that require pile driving. Operational noise from generators, rotating equipment, and other moving parts may have frequencies and magnitudes comparable to those measured at offshore wind farms; however, the underwater noise created by a wind turbine is transmitted down through the pilings, whereas noises from marine and hydrokinetic devices are likely to be greater because they are at least partially submerged. It is probable that noise from marine energy projects may be less than the intermittent noises associated with shipping and many other anthropogenic sound sources (e.g., seismic exploration, explosions, commercial, naval sonar).

The resolution of noise impacts will require information about the device's acoustic signature (e.g., sound pressure levels across the full range of frequencies) for both individual units and multiple-unit arrays, similar characterization of ambient (background) noise in the vicinity of the project, the hearing sensitivity (e.g., audiograms) of fish and marine mammals that inhabit the area, and information about the behavioral response to anthropogenic noise (e.g., avoidance, attraction, changes in schooling behavior or migration routes). Simmonds et al. (2003) described the types of *in situ* monitoring that could be carried out to develop information on the effects of underwater noise arising from a variety of activities. The studies include monitoring marine mammal activity in parallel with sound level monitoring during construction and operation. Baseline sound surveys would be needed against which to measure the added effects of energy generation. It will be important to measure the acoustic characteristics produced by both single units and multiple units in an array, due to the possibility of synchronous or asynchronous additive noise produced by the array (Boehlert et al., 2008). Minimally, the operational monitoring would quantify the sound pressure levels across the entire range of sound frequencies for a variety of ocean and river conditions in order to assess how meteorological, current strength, and wave height conditions affect sound generation and sound masking. The monitoring effort should consider the effects of marine fouling on noise production, particularly as it relates to mooring cables.

ELECTROMAGNETIC FIELDS

Underwater cables will be used to transmit electricity between turbines in an array (inter-turbine cables), between the array and a submerged step-up transformer (if part of the design), and from the transformer or array to the shore (CMACS, 2003). Ohman et al. (2007) categorized submarine electric cables into the following types: (1) telecommunications cables; (2) high-voltage, direct-current (HVDC) cables; (3) alternating-current, three-phase power cables; and (4) low-voltage cables. All types of cable will emit electromagnetic fields (EMFs) in the surrounding water. The electric current traveling through the cables will induce magnetic fields in the immediate vicinity, which can in turn induce a secondary electrical field when animals move through the magnetic fields (CMACS, 2003).

Nature of the Underwater Electromagnetic Field

In 1819, Hans Christian Oersted, a Danish scientist, discovered that a field of magnetic force exists around a single wire conductor carrying an electric current. The electromagnetic field created by electric current passing through a cable is composed of both an electric field (E field) and an induced magnetic field (B field). Although E fields can be contained within undamaged insulation surrounding the cable, B fields are unavoidable and will in turn induce a secondary electric field (iE field).

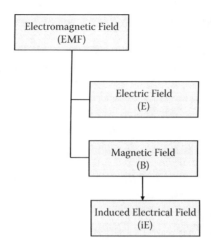

FIGURE 9.11 Simplified view of the fields associated with submarine power cables. (Adapted from Gill, A.B. et al., *COWRIE 1.5 Electromagnetic Fields Review: The Potential Effects of Electromagnetic Fields Generated by Sub-Sea Power Cables Associated with Offshore Wind Farm Developments on Electrically and Magnetically Sensitive Marine Organisms—A Review*, COWRIE–EM Field 2-06-2004, Cranfield University, Cranfield, U.K., 2005.)

Thus, it is important to distinguish between the two constituents of the EMF (E and B) and the induced field (iE) (Figure 9.11). Because the electric field is a measure of how the voltage changes when a measurement point is moved in a given direction, E and iE are expressed as volts/meter (V/m).

The intensity of a magnetic field can be expressed as magnetic field strength or magnetic flux density (CMACS, 2003). The magnetic field can be visualized as field lines, and the field strength (measured in amperes/meter [A/m]) corresponds to the density of the field lines. Magnetic flux density is a measure of the density of magnetic lines or force, or magnetic flux lines, passing through an area. Magnetic flux density (measured in teslas [T]) diminishes with increasing distance from a straight current-carrying wire. At a given location in the vicinity of a current-carrying wire, the magnetic flux density is directly proportional to the current in amperes. Thus, magnetic field B is directly linked to the magnetic flux density that is flowing in a given direction.

When electricity flows (electron flow) through the wire in a cable, every section of the wire has this field of force around it in a plane perpendicular to the wire, as shown in Figure 9.12. The strength of the magnetic field around a wire (cable) carrying a current depends on the current, because it is the current that produces the field. The greater the current flow in a wire, the greater the strength of the magnetic field. A large current will produce many lines of force extending far from the wire, whereas a small current will produce only a few lines close to the wire, as shown in Figure 9.13.

The EMFs associated with new marine and hydrokinetic energy designs have not been quantified; however, there is considerable experience with submarine electrical transmission cables, with some predictions and measurements of their associated electrical and magnetic fields. For example, the wave energy technology (WET) generator in Hawaii will be housed in a canister buoy and connected to shore by a 1190-m-long, 6.5-cm-diameter electrical cable. The cable is designed for three-phase AC transmission, can carry up to 250 kW, and has multiple layers of insulation and armoring to contain the electrical current. Depending on current flow (amperage), at 1 m from the cable, the magnetic field strength was predicted to range from 0.1 to 0.8 A/m and the magnetic flux density would range from 0.16 to 1.0 μT. The estimated strength of the electric field at the surface of the cable (apparently the iE) would range from 1.5 to 10.5 mV/m. The electric field strength, magnetic field strength, and magnetic flux density would all decrease exponentially with distance from the cable.

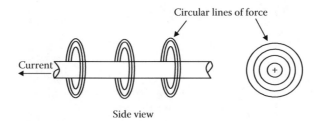

FIGURE 9.12   Circular fields of force around a wire carrying a current are in planes perpendicular to the wire.

FIGURE 9.13   The strength of the magnetic field around a wire carrying a current depends on the amount of current.

The Centre for Marine and Coastal Studies (CMACS, 2003) surveyed cable manufacturers and independent investigators to compile estimates of the magnitudes of E, B, and iE fields. Most agreed that the E field can be completely contained within the cable by insulation. Estimates of the B field strength ranged from 0 (by one manufacturer) to 1.7 and 0.61 µT at distances of 0 and 2.5 m from the cable, respectively. By comparison, the Earth's geomagnetic field strength ranges from approximately 20 to 75 µT (Bochert and Zettler, 2006). In another study cited by CMACS (2003), a 150-kV cable carrying a current of 600 A generated an induced electric field (iE) of more than 1 mV/m at a distance of 4 m from the cable; the field extended for approximately 100 m before dissipating. Lower voltage/amperage cables generated similarly large iE fields near the cable, but the fields dissipated much more rapidly with distance.

For short-distance undersea transmission of electricity, three-phase AC power cables are most common; HVDCs are used for longer distance, high-power applications (Ohman et al., 2007). In AC cables the voltage and current alternate sinusoidally at a given frequency (50 or 60 Hz); therefore, the E and B fields are also time varying. That is, like AC current, the magnetic field induced by a three-phase AC current has a cycling polarity, which is not like the natural geomagnetic fields. On the other hand, the E and B fields produced by a direct-current cable (e.g., HVDC) are static. Because the magnetic fields induced by DC and AC cables are different, they are likely to be perceived differently by aquatic organisms.

Because neither sand nor seawater has magnetic properties, burying a cable will not affect the magnitude of the magnetic (B) field; that is, the B fields at the same distance from the cable are identical, whether in water or sediment (CMACS, 2003). On the other hand, due to the higher conductivity of seawater compared to sand, the iE field associated with a buried cable is discontinuous across the sand/water boundary; the iE field strength is greater in water than in sand at a given distance from the cable. For example, for the three-phase AC cable modeled by CMACS (2003), the estimated iE field strengths at 8 m from the cable were 10 µV/m and 1 to 2 µV/m in water and sand, respectively.

The EMF generated by a multi-unit array of marine or hydrokinetic devices will differ from EMFs associated with a single unit or from the single cable sources that have been surveyed. Depending on the power generation device, a project may have electrical cable running vertically through the water column in addition to multiple cables running along the seabed or converging on a subsea pod. The EMF created by a matrix of cables has not been predicted or quantified.

Effects of Electromagnetic Fields on Aquatic Organisms

Electric Fields

Natural electric fields can occur in the aquatic environment as a result of biochemical, physiological, and neurological processes within an organism or as a result of an organism swimming through a magnetic field (Gill et al., 2005). Some of the elasmobranchs (e.g., sharks, skates, rays) have specialized tissues that enable them to detect electric fields (i.e., electroreception), an ability that allows them to detect prey and potential predators and competitors. Two species of Asian sturgeon have been reported to alter their behavior in changing electric fields (Basov, 1999, 2007). Other fish species (e.g., eels, cod, Atlantic salmon, catfish, paddlefish) respond to induced voltage gradients associated with water movement and geomagnetic emissions (Collin and Whitehead, 2004; Wilkens and Hofmann, 2005), but their electrosensitivity does not appear to be based on the same mechanism as sharks (Gill et al., 2005).

Balayev and Fursa (1980) observed the reaction of 23 species of marine fish to electric currents in the laboratory. Visible reactions occurred following exposure to electric fields ranging from 0.6 to 7.2 V/m and varied depending on the species and orientation to the field. They noted that changes in the fishes' electrocardiograms occurred at field strengths 20 times lower than those that elicited observable behavioral response. Enger et al. (1976) found that European eels (*Anguilla anguilla*) exhibited a decelerated heart rate when exposed to a direct current electrical field with a voltage gradient of about 400 to 600 μV/cm. In contrast, Rommel and McCleave (1972) observed much lower voltage thresholds of response (0.07 to 0.67 μV/cm) in American eels (*Anguilla rostrata*). The eels' electrosensitivity measured by Rommel and McCleave is well within the range of naturally occurring oceanic electric fields of at least 0.10 μV/cm in many currents in the Atlantic Ocean and up to 0.46 μV/cm in the Gulf Stream.

Kalmijn (1982) described the extreme sensitivity of some elasmobranchs to electric fields. For example, the skate (*Raja clavata*) exhibited cardiac response to uniform square-wave fields of 5 Hz at voltage gradients as low as 0.01 μV/cm. Dogfish (*Mustelus canis*) initiated attacks on electrodes from distances in excess of 38 cm and voltage gradients as small as 0.005 μV/cm.

Marra (1989) described the interactions of elasmobranchs with submarine optical communications cables. The cable created an iE field (1 μV/m at 0.1 m) when sharks crossed the magnetic field induced by the cable. The sharks responded by attacking and biting the cable. Marra (1989) was unable to identify the specific stimuli that elicited the attacks but suggested that at close range the sharks interpreted the electrical stimulus of the iE field as prey, which they then attacked.

The weak electric fields produced by swimming movements of zooplankton can be detected by juvenile freshwater paddlefish (*Polyodon spathula*). Wojtenek et al. (2001) used dipole electrodes to create electric fields that simulated those created by water flea (*Daphnia* sp.) swimming. They tested the effects of alternating current oscillations at frequencies ranging from 0.1 to 50 Hz and stimulus intensities ranging from 0.125 to 1.25 μA peak-to-peak amplitude. Paddlefish made significantly more feeding strikes at the electrodes at sinusoidal frequencies of 5 to 15 Hz compared to lower and higher frequencies. Similarly, the highest strike rate occurred at the intermediate electric field strength (stimulus intensity of 0.25 μA peak-to-peak amplitude). Strike rate was reduced at higher water conductivity, and the fish habituated (ceased to react) to repetitive dipole stimuli that were not reinforced by prey capture.

Gill and Taylor (2002) carried out a pilot study of the effects on dogfish of electric fields generated by a DC electrode in a laboratory tanks. They reported that the dogfish avoided constant electric fields as small as 1000 μV/m, which would be produced by 150 kV cables with a current of 600 A. Conversely, the dogfish were attracted to a field of 10 μV/m at 0.1 m from the source, which is similar to the bioelectric fields emitted by dogfish prey. The electrical field created by the three-phase, AC cable modeled by CMACS (2003) would likely be detectable by a dogfish (or other similarly sensitive elasmobranch) at a radial distance of 20 m. It is possible that the ability of fish to discriminate an electrical field is a function of not only the size and intensity but also the frequency (Hz) of the emitted field.

Like elasmobranchs, sturgeon (closely related to paddlefish) can utilize electroreceptor senses to locate prey and may exhibit varying behavior at different electric field frequencies (Basov, 1999). For this reason, electrical fields are a concern as they may impact migration or ability to find prey. The National Marine Fisheries Service (NMFS, 2008) proposed designating critical habitat for the Southern Distinct Population Segment of the threatened North American green sturgeon (*Acipenser medirostris*) along the coastline out to the 110-m isobath line. One of the principal elements in the proposal is safe passage along the migratory corridor. Green sturgeons migrate extensively along the nearshore coast from California to Alaska, and there is concern that these fish may be deterred from migration by either low-frequency sounds or electromagnetic fields created during operation of marine energy facilities.

Magnetic Fields

Many terrestrial and aquatic animals can sense the Earth's magnetic field and appear to use this magnetosensitivity for long-distance migrations. Aquatic species whose long-distance migrations or spatial orientation appear to involve magnetoreception include eels (Westerberg and Begout-Anras, 2000), spiny lobsters (Boles and Lohmann, 2003), elasmobranchs (Kalmijn, 2000), sea turtles (Lohmann and Lohmann, 1996), rainbow trout (Walker et al., 1997), tuna, and cetaceans (Lohmann et al., 2008a; Wiltschko and Wiltschko, 1995). Four species of Pacific salmon were found to have crystals of magnetite within them, and it is believed that these crystals serve as a compass that orients to the Earth's magnetic field (Mann et al., 1988; Walker et al., 1988). Because some aquatic species use the Earth's magnetic field to navigate or orient themselves in space, there is a potential for the magnetic fields created by the numerous electrical cables associated with offshore power projects to disrupt these movements.

Gill et al. (2005) placed magnetosensitive organisms into two categories: (1) those able to detect the iE field caused by movement through a natural or anthropogenic magnetic field, and (2) those with detection systems based on ferromagnetic minerals (i.e., magnetite or greigite). Johnsen and Lohmann (2005, 2008) added a third possible mechanism for magnetosensitivity—chemical reactions involving proteins known as crytochromes (i.e., a class of flavoproteins that are sensitive to blue light and are involved in circadian rhythm entrainment in plants, insects, and mammals). Those species using the iE mode may do it either passively (i.e., the animal estimates its drift from the electric fields produced by the interaction between tidal/wind-driven currents and the vertical component of the Earth's magnetic field) or actively (i.e., the animal derives its magnetic compass heading from its own interaction with the horizontal component of the Earth's magnetic field). For example, Kalmijn (1982) suggested that the electric fields that elasmobranchs induce by swimming through the Earth's magnetic field may allow them to detect their magnetic compass headings; the resulting voltage gradients may range from 0.05 to 0.5 µV/cm. Detection of a magnetic field based on internal deposits of magnetite occurs in a wide range of animals, including birds, insects, fish, sea turtles, and cetaceans (Bochert and Zettler, 2006; Gould, 1984). There is no evidence to suggest that seals are sensitive to magnetic fields (Gill et al., 2005).

Westerberg and Begout-Aranas (2000) studied the effects of a B field generated by a HVDC power cable on eels (*Anguilla anguilla*). The B field was on the same order of magnitude as the Earth's geomagnetic field and, coming from a DC cable, was also a static field. Approximately 60% of the 25 eels tracked crossed the cable, and the authors concluded that the cable did not appear to act as a barrier to the eel migration. In another behavioral study, Meyer et al. (2004) showed that conditioned sandbar and scalloped hammerhead sharks readily respond to localized magnetic fields of 25 to 100 µT, a range of values that encompasses the strength of the Earth's magnetic field.

Some sea turtles (see Sidebar 9.1) undergo transoceanic migrations before returning to nest on or near the same beaches where they were hatched. Lohmann and Lohmann (1996) showed that sea turtles have the sensory abilities necessary to approximate their global position of a magnetic map. This would allow them to exploit unique combinations of magnetic field intensity and field line inclination in the ocean environmental to determine direction and/or position during their long-distance migrations. Irwin and Lohmann (1996) found that magnetic orientation in loggerhead sea turtles (*Caretta caretta*) can be disrupted at least temporarily by strong magnetic pulses (i.e., five brief pulses of 40,000 μT with a 4-ms rise time). The impact of a changed magnetic environment would depend on the role of magnetic information in the hierarchy of cues used to orient and navigate (Wiltschko and Wiltschko, 1995). Juvenile loggerheads deprived of either magnetic or visual information were still able to maintain a direction of orientation, but when both cues were removed the turtles were disoriented (Avens and Lohmann, 2003). The magnetic map sense exhibited by hatchlings is also thought to allow female sea turtles to imprint upon the location of their natal beaches so that later in life they can return there to nest. This phenomenon is termed *natal homing* (Lohmann et al., 2008b), and it serves to drive genetic division among subpopulations of the same species. As a result, altering magnetic fields near nesting beaches could potentially result in altered nesting patterns. Given the important role of magnetic information in the movements of sea turtles, impacts of magnetic field disruption could range from minimal (i.e., temporary disorientation near a cable or structure) to significant (i.e., altered nesting patterns and corresponding demographic shifts resulting from large-scale magnetic field changes) and should be carefully considered when siting projects.

SIDEBAR 9.1. WITH REGARD TO TURTLES

Marine turtles have outlived almost all of the prehistoric animals with which they once shared the planet. Five species of marine turtles frequent the beaches and offshore waters of the southeastern United States:

- *Loggerhead turtles* are the most common turtles to nest in Florida. Over 50,000 loggerhead nests are recorded annually in Florida. This turtle is named for its disproportionately large head, and it feeds on crabs, mollusks, and jellyfish.
- *Green sea turtles* are the second most common turtles in Florida waters. Green sea turtles are the only herbivorous sea turtles. They feed on seagrasses in shallow areas through the Gulf. The lower jaw is serrated to help cut the seagrasses it eats.
- *Kemp's ridley sea turtles* are the rarest sea turtles in the world. They primarily nest on one beach on the Gulf Coast of Mexico and are the smallest species of sea turtle. Scientists have been trying to transplant Kemp's ridley eggs to Texas to establish a new nesting colony. They are the only species of sea turtle known to lay their eggs during the day.
- *Leatherback sea turtles* are the largest sea turtles in the world; they can be over 6 feet long and weigh 1400 pounds. The leatherback does not have a hard shell, but rather a leather-like carapace with bony ridges underneath the skin. The leatherback makes long migrations to and from its nesting beaches in the tropics as far north as Canada. Jellyfish are the favored prey for these turtles.
- *Hawksbill sea turtles* are usually found feeding primarily on sponges in the southern Gulf of Mexico and Caribbean. The hawksbill sea turtle was hunted to near extinction for its beautiful shell which features overlapping scales.

All five are reported to nest, but only the loggerhead and green turtles do so in substantial numbers. Most nesting occurs from southern North Carolina to the middle west coast of Florida, but scattered nesting occurs from Virginia through southern Texas. The beaches of Florida, particularly in Brevard and Indian River counties, host what may be the world's largest population of loggerheads (Dodd, 1995).

Marine turtles, especially juveniles and subadults, use lagoons, estuaries, and bays as feeding grounds. Areas of particular importance include Chesapeake Bay, Virginia (for loggerheads and Kemp's ridleys); Pamlico Sound, North Carolina (for loggerheads); and Mosquito Lagoon, Florida, and Laguna Madre, Texas (for greens). Offshore waters also support important feeding grounds such as Florida Bay and the Cedar Keys, Florida (for green turtles), and the mouth of the Mississippi River and the northeast Gulf of Mexico (for Kemp's ridleys). Offshore reefs provide feeding and resting habitat (for loggerheads, greens, and hawksbills), and offshore currents, especially the Gulf Stream, are important migratory corridors (for all species, but especially leatherbacks).

Note: Raccoons destroy thousands of sea turtle eggs each year and are the single greatest cause of sea turtle mortality in Florida.

Most marine turtles spend only part of their lives in U.S. waters. For example, hatchling loggerheads ride oceanic currents and gyres (giant circular oceanic surface currents) for many years before returning to feed as subadults in southeastern lagoons. They travel as far as Europe and the Azores, and even enter the Mediterranean Sea, where they are susceptible to longline fishing mortality. Adult loggerheads may leave U.S. waters after nesting and spend years in feeding grounds in the Bahamas and Cuba before returning. Nearly the entire world population of Kemp's ridleys uses a single Mexican beach for nesting, although juveniles and subadults, in particular, spend much time in U.S. offshore waters (Dodd, 1995).

The biological characteristics that make sea turtles difficult to conserve and manage include a long life span, delayed sexual maturity, differential use of habitats among species and over the turtles' life stages, adult migratory travel, high egg and juvenile mortality, concentrated nesting, and vast areal dispersal of young and subadults. Genetic analyses have confirmed that females of most species return to their natal beaches to nest (Bowen et al., 1992, 1993). Nesting assemblages contain unique genetic markers showing a tendency toward isolation from other assemblages (Bowen et al., 1993); thus, Florida green turtles are genetically different from green turtles nesting in Costa Rica and Brazil (Bowen et al., 1992). Nesting on warm sandy beaches puts the turtles in direct conflict with human beach use, and their use of rich offshore waters subjects them to mortality from commercial fisheries (National Research Council, 1990).

Marine turtles have suffered catastrophic declines since the European discovery of the New World (National Research Council, 1990). In a relatively short time, the huge nesting assemblages in the Cayman Islands, Jamaica, and Bermuda were decimated. In the United States, commercial turtle fisheries once operated in south Texas (Doughty, 1984) and in Cedar Keys, the Florida Keys, and Mosquito Lagoon in Florida; these fisheries collapsed from overexploitation of the mostly juvenile green turtle populations. Today, marine turtle populations are threatened worldwide and are under intense pressure in the Caribbean basin and Gulf of Mexico, including Cuba, Mexico, Hispaniola, Bahamas, and Nicaragua. Marine turtles can be conserved only through international efforts and cooperation (Dodd, 1995).

A number of interesting questions related to turtle migration remain unanswered. For example, how do turtles find their way precisely back to their natal beach over their vast travel distance? Do turtles imprint, as salmon do, on olfactory features in the water or is the location pinpointed using geomagnetic information?

Sea turtles have migration patterns somewhat similar to that of salmon. After hatching and entering the sea and facing and surviving the tribulations presented by the elements and predators, and after spending time in their sea feeding grounds, the females return to their natal grounds. Adult females lay eggs in the sand. Turtles may use the geomagnetic field to tell them their location and to lead them to their natal grounds (Goff et al., 1998).

The emphasis of most of these studies is on the value of magnetoreception for navigation; marine and hydrokinetic energy technologies are unlikely to create magnetic fields strong enough to cause physical damage. For example, Bochert and Zettler (2006) summarized several studies on the potential injurious effects of magnetic fields on marine organisms. They subjected several marine benthic species (i.e., flounder, blue mussel, prawn, isopods, and crabs) to static (DC-induced) magnetic fields of 3700 µT for several weeks and detected no differences in survival compared to controls. In addition, they exposed shrimp, isopods, echinoderms, polychaetes, and young flounder to a static, 2700 µT magnetic field in laboratory aquaria where the animals could move away from or toward the source of the field. At the end of the 24-hour test period, most of the test species showed a uniform distribution relative to the source, not significantly different from controls. Only one of the species, the benthic isopod *Saduria entomon*, showed a tendency to leave the area of the magnetic field. The oxygen consumption of two North Sea prawn species exposed to both static (DC) and cycling (AC) magnetic fields were not significantly different from controls. Based on these limited studies, Bochert and Zettler (2006) could not detect changes in marine benthic organisms' survival, behavior, or physiological response parameters (e.g., oxygen consumption) resulting from magnetic flux densities that might be encountered near an undersea electrical cable.

The current state of knowledge about the EMF emitted by submarine power cables is too variable and inconclusive to make an informed assessment of the effects on aquatic organisms (CMACS, 2003). Following a thorough review of the literature related to EMFs and extensive contacts with the electrical cable and offshore wind industries, Gill et al. (2005) concluded that there are significant gaps in our knowledge regarding the sources and effects of electrical and magnetic fields in the marine environment. They recommended developing information about likely electrical and magnetic field strengths associated with existing sources (e.g., telecommunications cables, power cables, electrical heating cables for oil and gas pipelines), as well as the generating units, offshore substations and transformers, and submarine cables that are a part of offshore renewable energy projects. They cautioned that networks of cables in close proximity to each other (as would be substations) are likely to have overlapping, and potentially additive, EMF fields. These combined EMF fields would be more difficult to evaluate than those emitted from a single, electrical cable. The small, time-varying B field emitted by a submarine three-phase AC cable may be perceived differently by sensitive marine organisms than the persistent, static, geomagnetic field generated by the Earth (CMACS, 2003).

Toxic Effects of Chemicals

Chemicals that are accidentally or chronically released from hydrokinetic and ocean energy installations could have toxic effects on aquatic organisms. Accidental releases include leaks of hydraulic fluids from a damaged unit or fuel from a vessel due to a collision with the unit; such events are unlikely but could potentially have a high impact (Boehlert et al., 2008). On the other hand, chronic releases of dissolved metals or organic compounds used to control biofouling in marine applications would result in low, predictable concentrations of contaminants over time. Even at low concentrations that are not directly lethal, some contaminants can cause sublethal effects on the sensory systems, growth, and behavior of animals; they may also be bioaccumulated (i.e., filter feeders such as limpets, oysters, and other shellfish concentrate heavy metals or other stable compounds present in dilute concentrations in seawater or freshwater).

Toxicity of Paints, Antifouling Coatings, and Other Chemicals

Biofouling (growth on external surfaces by algae, barnacles, mussels, and other marine organisms) will occur rapidly in ocean applications (Langhamer, 2005; Wilhelmsson and Malm, 2008). Sundberg and Langhamer (2005) observed that a 3-m-diameter buoy may accumulate as much as 300 kg of biomass on the buoy and mooring cables, whereas siting devices in deeper water with even slight currents will exhibit reduced biofouling. The encrustation of biofouling organisms could cause undesirable mechanical wear or changes in the weight, shape, and performance of energy conversion devices that would require increased maintenance or the application of antifouling measures. Encrustation by barnacles and other organisms could increase corrosion and fatigue and decrease electrical generating efficiency.

There are three options for removing marine biofouling (Michel et al., 2007): (1) use of antifouling coatings, (2) *in situ* cleaning using a high-pressure jet spray, and (3) removal of the device from the water for cleaning on a floating platform or onshore (Michel et al., 2007). Antifouling coatings hinder the development of marine encrustations by slowly releasing a biocide such as tribuyltin (TBT), copper, or arsenic. As the coatings wear away, they must be reapplied periodically. There are concerns about the immediate toxicity of these biocides to other, non-targeted organisms, and numerous countries and organizations have called for the ban of TBT as an antifouling coating (Antizar-Ladislao, 2008). As a result, alternative coatings are being explored. The release of toxic contaminants from a single unit may be relatively minor, but the cumulative impacts of persistent toxic compounds from dozens or hundreds of units may be considerable (Boehlert et al., 2008). Accumulations of biofouling organisms (e.g., barnacles) removed from the project structures may alter nearby bottom substrates and habitats.

Accidental releases of hydraulic fluids and lubricating oils from inside the energy conversion device or from vessels used to install and service the equipment could have toxic effects. At the least, leaks of inert (non-toxic) oils could cause physical/mechanical effects by coating organisms and blanketing the sediments.

INTERFERENCE WITH ANIMAL MOVEMENTS AND MIGRATION

Energy developments will add new structures to rivers and oceans that may affect the movements and migrations of aquatic organisms. Hydrokinetic devices, and their associated anchors and cables in a river, could attract or repel animals or interfere with their movements. In addition to seabed structures (e.g., anchors, turbines), many of the ocean energy devices would use mooring lines to attach a floating generator to the ocean bottom and electrical transmission lines to connect multiple devices to each other and to the shoreline. For example, the Minerals Management Service (2007) estimated that wave energy facilities may have as many as 200 to 300 mooring lines securing the wave energy devices to the ocean floor (based on 2 to 3 mooring lines per device and a 100-device facility). Mooring and transmission lines that extend from a floating structure to the ocean floor will create new fish attraction devices in the pelagic zone (i.e., the entire water column of the water body), pose a threat of collision for entanglement to some organisms, and potentially alter both local movements and long-distance migrations of marine animals (Nelson, 2008; Thompson et al., 2008). Because the transport of planktonic (drifting) life stages is affected by water velocity (DiBacco et al., 2001; Epifanio, 1988), localized reduction of water velocities by large, multi-unit projects could influence recruitment of some species. Various aquatic organisms use magnetic, chemical, and hydrodynamic cues for navigation (Cain et al., 2005; Loghmann et al., 2008a). Thus, in addition to mechanical obstructions, the electrical and magnetic fields and current and wave alterations produced by energy technologies could interfere with local movement or long-distance migrations.

Alteration of Local Movement Patterns

Anchors and other permanent structures on the bottom will create new habitats and thus may act as artificial reefs (Wilhelmsson et al., 2006). Artificial reefs are often constructed in order to increase fish production, but some studies suggest that they may be less effective than natural reefs (Carr and Hixon, 1997) and that they may even have deleterious effects on reef fish populations by stimulating overfishing and overexploitation (Grossman et al., 1997).

Similarly, new structures in the pelagic zone (e.g., pilings or mooring cables for floating devices) will create habitat that may act as fish aggregation/attraction devices (FADs). These devices are extremely effective in concentrating fish and making them susceptible to harvest (Dempster and Tacquet, 2004; Michel et al., 2007; Myers et al., 1986). Sea turtles are also known to be attracted to floating objects (Arenas and Hall, 1992). Fish are attracted to the devices as physical structures/shelter, and they may feed on organisms attached to the structures (Boehlert et al., 2008). Artificial lighting used to distinguish structures at night may also attract aquatic organisms.

The aggregation of predators near FADs may adversely affect juvenile salmonids or Dungeness crabs moving through the project area. Wilhelmsson et al. (2006) found that fish abundance in the vicinity of monopiles that supported wind turbines was greater than in surrounding areas, although species richness and diversity were similar. Most of the fish they observed near the structure were small (juvenile gobies), which may in turn attract commercially important fish looking for prey. Dempster (2005) observed considerable temporal variability in the abundance and diversity of fish associated with FADs moored between 3 and 10 km offshore. The variability was often related to the seasonal appearance of large schools of juvenile fish. Fish assemblages differed between times when predators were present or absent; few small fishes were observed near the FADs when predators were present, regardless of the season. Using FADs as an experimental tool, Nelson (2003) found that fish formed larger, more species-rich assemblages around large FADs compared to small ones, and they formed larger assemblages around FADs with fouling biota. Devices enriched with fish accumulated additional recruits more quickly than those in which fish were removed.

It is likely that floating wave energy devices will act as FADs, but the effect on fish populations may be difficult to determine. FADs are attractive to fish because they provide food and shelter (Castro et al., 2002); subsequently, they also attract predators (Dempster, 2005) that can in turn attract commercial and sport fisheries. Without well-designed monitoring, it will be difficult to determine whether an energy park will enhance populations of aquatic organisms (by providing more habitat to support more fish), will have no overall effect (because it simply draws fish from other, nearby areas), or will decrease fish populations (by facilitating harvest by predators and fishermen). The determination of the effects of FADs at a particular location is complicated by the influence of non-independent factors: proximity of other FADs (e.g., other wave energy units), interconnection of multiple FADs to provide routes for the movement of associated fishes, and temporal dependence (the number of fish present at one sampling date influencing the number at the next sampling date due to fish becoming residents) (Kingsford, 1999). Statistical approaches that could be applied to experiments on the effects of FADs on fish populations and solutions to the independent factor problems were also described.

Because anchoring systems and mooring lines will likely exclude fishing activities, energy parks could serve as marine protected areas. The Pacific Fisheries Management Council (2008) expressed concerns related to the prohibition of commercial fishing at wave energy test areas and suggested that there may be either a reduction in total fishing effort and lost productivity or a displacement of fishing effort to areas outside the areas closed to fishing. Displaced fishermen would likely concentrate their efforts in areas immediately outside the wave park boundaries, resulting in increased pressures on fish and habitats in those nearby areas.

Floating offshore wave energy facilities could create artificial haul-out sites for marine mammals (pinnipeds). Devices with a low profile above the waterline (desirable for aesthetic reasons) may enable seals and sea lions to use them as a haul-out site, particularly if the installations attract the marine mammals by acting as fish-concentrating devices. NOAA considers the creation of such artificial haul-outs as undesirable and recommends the use of deterrents to discourage use by marine mammals.

Floating devices could potentially impede movements of floating marine habitat communities, such as *Sargassum* communities. Masses of floating *Sargassum* algae form unique communities of organisms that serve as important habitats for hatchling sea turtles and juvenile fish (Coston-Clements et al., 1991). Strong current from the Sargasso Sea in the middle of the Atlantic Ocean carry these *Sargassum* communities around the world.

Floating devices with above-water structures may attract seabirds by creating artificial roosting sites or encouraging predation on fish near the FAD (Michel et al., 2007). There is particular concern about collision injuries to marine birds that are attracted to lighted structures at night or in inclement weather (Boehlert et al., 2008; Thompson et al., 2008). Peterson et al. (2006) monitored the interactions of birds and above-water structures at a Danish offshore wind farm from 1999 to 2005 and found that birds generally avoided the wind farms by flying around them, although there were considerable differences among species. The monitoring data suggested that avoidance was reduced at night. The authors obtained few data under conditions of poor visibility because bird migrations slowed or ceased during such times. Birds typically showed avoidance responses to the rotating wind turbine blades. A stochastic model predicted very low rates of Eider collisions with the offshore wind turbines, and the predictions were confirmed by subsequent monitoring (Petersen et al., 2006). Desholm (2006) provided a series of papers that describe techniques for predicting and monitoring interactions of birds and wind turbine structures at sea.

INTERFERENCE WITH LONG-DISTANCE MIGRATIONS OF MARINE ANIMALS

The numerous floating and submerged structures, mooring lines, and transmission cables associated with large ocean energy facilities could interfere with the long-distance migrations of marine animals (e.g., juvenile and adult salmonids, Dungeness crabs, green sturgeon, elasmobranchs, sea turtles, marine mammals, birds) if they are sited along migration corridors. On the U.S. Pacific Coast, effects on gray whales (*Eschrichtius robustus*) may be of particular concern because they migrate within 2.8 km of the shore line. Boehlert et al. (2008) noted that buoys attached to commercial crab pots already represent a major existing risk to gray whales off the coast of Oregon. Lines associated with lobster pots and other fishing gear are a source of injury and mortality to endangered North Atlantic right whales (*Eubalaena glacialis*) on the East Coast of the United States (Caswell et al., 1999; Kraus et al., 2005). Many marine fish species drift or actively migrate long distances in the sea and may interact with ocean energy developments. Anadromous fish (e.g., green sturgeon, salmon, steelhead) and catadromous fish (e.g., eels) migrate through both rivers and oceans and therefore may encounter buoy hydrokinetic devices in the rivers and ocean energy projects (Dadswell et al., 1987).

Entanglement of large, planktonic jellyfish with long tentacles (as well as actively swimming sea turtles and marine mammals) is a potential issue for energy technologies with mooring lines in the pelagic zone. Thin mooring cables are expected to be more dangerous than thick ones because they are more likely to cause lacerations and entanglements, and slack cables are more likely to cause entanglements than taut ones (Boehlert et al., 2008).

Michel et al. (2007) observed that smaller dolphins and pinnipeds could easily move around mooring cables, but larger whales may have difficulty passing through an energy facility with numerous, closely spaced lines. Marine species with proportionately large pectoral fins or flippers may be relatively more vulnerable to mooring lines, based on information from humpback whale entanglements with pot and gill net lines (Johnson et al., 2005). Boehlert et al. (2008) suggested that whales probably do not sense the presence of mooring cables and as a result could strike them or become entangled. In addition, they believed that if the cable density is sufficiently great and spacing is close, cables could have a "wall effect" that could force whales around them, potentially changing their migration routes. Whales and dolphins traveling or feeding together may be at a greater risk than solitary individuals because "ground response" may lead some individuals to follow others into danger (Faber Maunsell and Metoc, 2007).

Wave energy converters deployed near sea turtle nesting beaches have the potential to interfere with the offshore migration of hatchlings. Interference with migration could occur if the energy project acts as a physical barrier or alters wave action, which has been demonstrated to guide hatchlings away from the beach toward the open ocean (Goff et al., 1998; Lohmann et al., 1995; Wang et al., 1998).

Some marine fish species form spawning aggregations at specific sites or times (Coleman et al., 1996; Crawford and Carey, 1985; Cushing, 1969; Domeier and Colin, 1997; Sinclair and Tremblay, 1984). Smith (1972) reported a spawning aggregation consisting of 30,000 to 100,000 Nassau groupers (*Epinephelus striatus*) in the Bahamas. Because spawning success is important to the viability of populations, the siting and operation of ocean energy facilities would have to avoid interfering with these activities.

COLLISIONS AND STRIKES

Submerged structures present a collision risk to aquatic organisms and diving birds, and the above-water components of floating structures may be a risk to flying animals. Wilson et al. (2007) defined "collision" as physical contact between a device or its pressure field and an organism that may result in an injury to the organism. They noted that collisions can occur between animals and fixed submerged structures, mooring equipment, surface structures, horizontal- and vertical-axis turbines, and structures that, by their individual design or in combination, may form traps. Harmful effects to animal populations could occur directly (e.g., from strike mortality) or indirectly (e.g., if the loss of prey species to strike reduces food for predators). Attraction of marine mammals and other predators to fish congregations near structures may also expose them to increased risk of collision or blade strike. In an attempt to define the risk of collisions from marine renewable energy devices, Wilson et al. (2007) reviewed information from other industrial and natural activities: power plant cooling intakes, shipping, fishing gear, fish aggregation devices, and wind turbines. They concluded that, although animals may strike any of the physical structures associated with marine renewable energy devices (i.e., vertical or horizontal support piles, duct, nacelles, anchor locks, chains, cables, and floating structures), turbine rotors are the most intuitive sources of significant collision risks with marine vertebrates.

Effects of Rotor Blade Strike on Aquatic Animals

Many of the hydrokinetic and ocean current technologies extract kinetic energy by means of moving/rotating blades. A wide variety of swimming and drifting organisms (e.g., fish, sea turtles, driving birds, cetaceans, seals, otters) may be struck by the blades and suffer injury or mortality (Wilson et al., 2007). Mortality is a function of the probability of strike and the force of the strike. The seriousness of strike is related to the swimming ability of the animal (i.e., ability to avoid the blade), water velocity, number of blades, blade design (i.e., leading edge shape), blade length and thickness, blade spacing, blade movement (rotation) rate, and the part of the rotor that the animals strikes. A vertical axis turbine will have the same leading edge velocity along the entire length of the blade. On the other hand, blade velocity on a horizontal axis turbine will increase from the hub out to the tip. The rotor blade tip has a much higher velocity than the hub because of the greater distance that is covered in each revolution. For example, on a rotor spinning at 20 rpm, the leading edge of the blade 1 m from the center point will be traveling at about 2 m/s—a speed that is likely to be avoidable or undamaging to most organisms. However, a 20-m-diameter rotor spinning at 20 rpm would have a tip velocity of nearly 21 m/s. Fraenkel (2006, 2007b) described a horizontal-axis turbine with a maximum rotation speed of 12 to 15 rpm, resulting in a maximum blade tip velocity of 12 m/s. Wilson et al. (2007) suggested that rotor blades tips will likely move at or below 12 m/s because greater speeds will incur efficiency losses through cavitation.

The force of the strike is expected to be proportional to the strike velocity; consequently, the potential for injury from a strike would be greatest at the outer periphery of the rotor. Unfortunately, little is known about the magnitude of impact forces that cause injuries to most marine and freshwater organisms (Cada et al., 2005, 2006) or the swimming behavior (e.g., burst speeds) that organisms may use to avoid strikes. Although the blade tip will be moving at the highest velocity and exhibit the greatest strike force, animals may be able to avoid the tip of an unducted rotor. Relatively safe areas of passage through the rotor would be nearest the hub (because of low velocities) and potentially nearest the tip (because of the opportunity for the animal to move outward to avoid strike). The central zone of relatively high blade velocity and relatively less opportunity to avoid strike may be the most dangerous area (Coutant and Cada, 2005). For rotors contained in housings, there would be no opportunity for an organism entrained in the intake flow to escape strike by moving outward from the periphery; safe passage would depend on sensing and evading the intake flow or passing through the rotor between the blades. This suggestion of relatively high- and low-risk passage zones has not been tested and remains speculative until the phenomenon is investigated in field applications.

There have been several studies to estimate the potential of fish strike by rotating blades (e.g., Deng et al., 2005), but all involve conventional hydroelectric turbines that are enclosed in turbine housings and afford little opportunity for flow-entrained organisms to avoid strike. It is likely that both the probability and consequences of organisms striking the rotor blade are greater for a conventional turbine than for an unducted current energy turbine, due to the greater opportunities for organisms to avoid approaching the turbine rotor or moving outward from the periphery. However, passage through a conventional turbine poses only a single exposure to the rotor, whereas passage through a project consisting of large numbers of hydrokinetic energy turbines represents a larger risk of strike that has not been investigated.

Wilson et al. (2007) described a simple model to estimate the probability of aquatic animals entering the path of a marine turbine. The mode is based on the density of the animals and the water volume swept by the rotor. The volume swept by the turbine can be estimated from the radius of the rotor and the velocity of the animals and the turbine blades. They emphasized that their model predicts the probability of an animal entering the region swept by a rotor, not collisions. Entry into the path toward the rotor may lead to a collision but only if the animal does not take evasive action or has not already sensed the presence of the turbine and avoided the encounter. Applying this simplified model (no avoidance or evasive action) to a hypothetical field of 100 turbines, each with a two-bladed rotor 16 m in diameter, they predicted that 2% of the herring population and 3.6 to 10.7% of the porpoise population near the Scottish coast would encounter a rotating blade. At this time, there is no information about the degree to which marine animals may sense the presence of turbines, take appropriate evasive maneuvers, or suffer injury in response to a collision. Wilson et al. (2007) suggested that marine vertebrates may see or hear the device at some distance and avoid the area, or they may evade the structure by dodging or swerving when in closer range.

The potential injurious effects of turbine rotors have been compared to those of ship propellers, which are common in the aquatic environment. Fraenkel (2007a) pointed out that in contrast to ship propellers the rotors of hydrokinetic and current energy devices are much less energetic. He estimated that a tidal turbine rotor at a good site will absorb about 4 kW/m^2 of swept area from the current, whereas typical ship propellers release over 100 kW/m^2 of swept area into the water column. In addition to the greater power density, a ship propeller and ship hull generate suction that can pull objects toward them, increasing the area of influence for strike (Fraenkel, 2006).

Effects of Water Pressure Changes and Cavitation

In addition to direct strike, there is a potential for adverse effects due to sudden water pressure changes associated with movement of the blade. For example, if the local water pressures immediately behind the turbine blades drop below the vapor pressure of water, cavitation will occur. Cavitation is the process of forming water vapor bubbles in areas of extreme low pressure within liquids. As a turbine blade rotates, cavitation can occur in areas of low pressure (i.e., downstream surface of blades) causing increased local velocities, abrupt changes in the direction of flow, and roughness or surface irregularities (USACE, 1995). Once formed, cavitation bubbles stream from the area of formation and flow to regions of higher pressure where they collapse. The violent collapse of cavitation bubbles creates shock waves, the intensity of which depends on bubble size, water pressure in the region of collapse, and dissolved gas content. Within enclosed, conventional hydroelectric turbines, forces generated by cavitation bubble collapse may reach tens of thousands of kilopascals at the instant and point of collapse (Hamilton, 1983; Rodrigue, 1986). Cavitation is an undesirable condition that will reduce the efficiency of the turbine and damage blades as well as nearby organisms. Properly operating turbines would not cavitate, and the zone of low pressure that might be injurious to organisms would be relatively small. The pressure drops associated with the blades of hydrokinetic turbines have not been measured in field applications, but experimental evidence suggests that tidal turbines may experience strong and unstable sheet and cloud cavitation, as well as tip vortices at a shallow depth of submergence (Wang et al., 2007). If this occurs, aquatic organisms passing near the cavitation zones in the immediate blade area may be injured. The likelihood of cavitation-related injuries would depend on the extent of cavitation and the ability of aquatic organisms to avoid the area—the collapse of cavitation vapor bubbles creates noise which may act as a deterrent.

IMPACTS OF OCEAN THERMAL ENERGY CONVERSION

An OTEC technology operates a low-temperature heat engine based on the temperature differences between warm surface water and cold deep water (Holdren et al., 1980). This type of project consists of pumps and ducts for transferring large volumes of water (several times more flow than is needed for a once-through cooling system of a comparably sized steam electric power plant), large heat exchangers, and a working fluid that can be vaporized and recondensed (i.e., ammonia, propane, Freon®, or water). Electrical energy could be transported from offshore systems via subsea cables or alternatively could be converted to chemical energy *in situ* (e.g., hydrogen, ammonia, methanol) and transported to shore in tankers (Pelc and Fujita, 2002).

Effects on Ocean Ecosystems

Impacts of construction of an OTEC facility will depend on whether the project is located onshore or offshore. An offshore facility would require the installation of large, long water conduits on the seabed to access deep water. Alternatively, OTEC projects located on offshore platforms would depend on subsea cables to transfer electricity to shore. The installation and maintenance of pipelines and electrical cables would disturb bottom habitats and generate EMFs. Structures could become colonized with marine organisms and attract fish. Depending on the location of the warm water intake and discharges, these fish might be more susceptible to entrainment, impingement, or contact with the discharge plume.

The potential environmental effects of OTEC operation have been considered by a number of authors (Abbasi and Abbasi, 2000; Harrison, 1987; Holdren et al., 1980; Myers et al., 1986; Pelc and Fujita, 2002). Myers et al. (1986) provided the most comprehensive assessment of the possible effects on the marine environment resulting from operation of the types of OTEC facilities that were contemplated in the early 1980s. Most of the likely effects were expected to be physical and chemical changes in the ocean surface waters arising from the transfer of large volumes of cool, deep water. Abbasi and Abbasi (2000) suggested that OTEC plants will displace about $4 \ m^3/s$ of water per megawatt of electrify output from both the surface layer and the deep ocean layer, and then discharge the water at some intermediate depth. The warm water intake would be located at a depth of about 10 to 20 m, and the cold water intake might extend to a depth of 750 to 1000 m (Myers et al., 1986). The large transfer of water may disturb the thermal structure of the ocean near the plant, change salinity gradients, and change the amounts of dissolved gases, dissolved minerals, and turbidity. The transfer will result in an artificial upwelling of nutrient-rich deep water, which may increase marine productivity in the area. The stimulation of marine productivity may be especially strong in tropical waters, where nutrient levels are often low, and could have detrimental effects on nearby sensitive habitats such as coral reefs. Moreover, carbon dioxide will also be released when the deep water is warmed and subjected to lower pressures at the surface. The possible amounts of carbon dioxide released have not been rigorously quantified; some estimate that the quantities will be minute (Pelc and Fujita, 2002), and others suggest that the contribution will be relatively large (Holdren et al., 1980). The relatively high carbon dioxide and low dissolved oxygen content of the deep water may alter pH and dissolved oxygen concentrations in a surface mixing zone.

The large heat exchangers will have to be treated with biocides (e.g., chlorine or hypochlorite) in order to prevent the growth of bacterial slimes and other biofouling organisms; volumes of biocides would be proportional to the large volume of heating and cooling water. Degradation of the heat exchanger materials will result in chronic releases of metals (e.g., copper, nickel, aluminum).

Accidental release of the working fluid that is evaporated and condensed to drive the turbine could have toxic effects. The potential for acute and chronic toxicity and bioaccumulation of metals from deep ocean water will have to be considered (Fast et al., 1990).

Ocean thermal energy conversion projects would be sources of waterborne noise, arising from operation of ammonia turbines, seawater pumps, support systems associated with the energy-producing cycle, and in some cases propulsion machinery for dynamic positioning of the OTEC platform. Janota and Thompson (1983) measured noise from OTEC-1, a 1-MWe test facility that was moored near Keahole Point, Hawaii. The most significant sources of noise from the small project resulted from the interaction of inflow turbulence with the seawater pumps and from thrusters used for dynamic positioning. Based on their measurements, Janota and Thompson (1983) predicted that a 160-MWe OTEC plant would radiate less than 0.05 acoustic W of broadband sound in the frequency range of 10 to 1000 Hz, which is at least an order of magnitude less than that which is produced by a typical ocean-going freighter. Similarly, Rucker and Friedl (1985) predicted that pump noise (at 10 Hz) from a 40-MWe OTEC plant would be reduced from 136 dB to 78 dB at about 0.8 km; this is less than ambient noise at a sea state of 1 (very gentle sea with waves less than 0.3 m in height).

Large marine organisms may be impinged on the screens that protect the OTEC intakes, and smaller organisms (e.g., zooplankton, fish eggs, larvae) will pass through the screens and be entrained in the heat-exchanger system (Abbasi and Abbasi, 2000). The number of organisms entrained in the water will depend on their concentrations in the intake areas; more aquatic organisms are likely to be impinged and entrained at the surface water intake than from the deep water intake. Due to the large flow rates of water at the warm water intake, impingement and entrainment will especially need to be monitored there. As with steam electric power plants, the heat exchanger-entrained organisms will be susceptible to mechanical damage in the piping and to rapid changes in temperature, pressure, salinity, and dissolved gases that may cause mortality. For example, the temperature of cold, deep water is expected to increase by about 2 to 3°C after passing through the heat exchangers; likewise, the temperature of shallow, warm water is expected to decrease by the same amount. Myers et al. (1986) noted that there is insufficient information to judge the impacts of a 2 to 3°C temperature shock but assumed that most organisms will probably not be directly impacted by this amount of temperature change. However, secondary entrainment into the discharge plume will also expose marine organisms to chemical, physical, and temperature stresses. A mixed discharge of warm and cold water could subject organisms entrained from the warm surface waters to a drop of 10°C, which would likely cause lethal cold shock for some species. Few organisms are expected to be entrained in the deep, cold water flow, but those that do will be subjected to potentially lethal pressure decreases of 70 to 100 atmospheres (7100 to 10,100 kilopascals) (Myers et al., 1986).

ENVIRONMENTAL IMPACTS OF HYDROKINETIC ENERGY[*]

In the previous section, we provided a general discussion of potential environmental impacts of hydrokinetic energy technology. In this section, we provide a discussion of many of the specific impacts related to site evaluation, construction, and operations and maintenance (O&M) activities.

HYDROKINETIC ENERGY SITE EVALUATION IMPACTS

Site evaluation phase activities, such as monitoring and site characterization, are temporary and are conducted at a smaller scale than those during the construction and operation phases. Potential impacts from these activities are presented below, by the type of affected resource. The impacts described are for typical site evaluation activities, such as drilling to characterize the seabed or

---

[*] Adapted from Tribal Energy and Environmental Information Clearinghouse, http://teeic.indianaffairs.gov/er/hydrokinetic/impact/siteeval/index.htm.

riverbed. Onshore site characterization activities would be limited to a topographic survey to establish onshore site design and placement for an operations and maintenance facility, substation, and electric transmission lines. If road construction were necessary during this phase, potential impacts would be similar in character to those for the construction phase, but generally of smaller magnitude.

Air Quality

Impacts on air quality during site evaluation activities would be limited to barges conducting surveys and vehicular traffic to proposed sites for all hydrokinetic energy land-based facilities. These air pollutant emissions would be minor, of short duration, and intermittent.

Cultural Resources

Cultural material present within the project area could be impacted by any seafloor, riverbed, or ground disturbance. Such disturbance could result from drilling and sampling activities and, for land-based activities, vehicular and pedestrian traffic. These activities would be relatively limited in scope during this phase. Surveys conducted during this phase to evaluate the presence and significance of cultural resources in the area would assist developers in designing the project to avoid or minimize impacts on these resources.

Ecological Resources

Impacts on ecological resources would be minimal during site evaluation because of the limited nature of the activities. For offshore projects (e.g., tidal barrage, tidal turbine projects) and river projects, the potential effects of low-energy geological and geophysical surveys on marine mammals, sea turtles, and fish could include behavior responses such as avoidance and deflections in travel direction. A few individuals could be injured or killed by collisions with the survey vessels. Those individuals displaced because of avoidance behaviors during surveys are likely to return within relatively short periods following cessation of survey activities. Marine mammals, sea turtles, and fish could be exposed to discharges or accidental fuel releases from survey vessels and to accidentally released solid debris. Such spills would be small and would not be expected to measurably affect marine or river wildlife. Land-based activities could give rise to the introduction and spread of invasive vegetation as a result of vehicular traffic. Soil borings would destroy vegetation and disturb wildlife. Overall, site evaluations are not expected to cause significant impacts on terrestrial or aquatic biota. Surveys conducted during this phase to evaluate the presence and significance of ecological resources in the area would assist developers in properly locating the facility and its components.

Water Resources

Survey ships could contribute small amounts of fuel or oil to the ocean or river through bilge discharges or leaks. Anchoring of the ships can cause sediment from the seabed or riverbed to enter the water column. Negligible to minor impact on water quality would be expected. Relatively limited amounts of water would be used if drilling were required; this water could be obtained locally or it could be trucked in with the drilling equipment. Land-based site evaluation activities are anticipated to have minimal or no impact on water resources, local water quality, water flows, and surface water/groundwater interactions.

Land Use

Very few offshore and onshore site evaluation activities are expected; consequently, no impacts on existing land uses are anticipated.

Soils and Geologic Resources

Seabed, riverbed, and onshore ground disturbances would be minimal during the site evaluation phase and, as a result, impacts on seabed and riverbed sediments or soils are unlikely to occur. Site characterization activities would also be unlikely to activate geological hazards.

Paleontological Resources

Paleontological resources present within the project area could be impacted by any seafloor, riverbed, or ground disturbance. Such disturbance could result from drilling and sampling activities and, for land-based activities, vehicular and pedestrian traffic. These activities would be very limited in scope during this phase and would not be likely to affect paleontological resources. Surveys conducted during this phase to evaluate the presence and significance of paleontological resources in the area would assist developers in designing the project to avoid or minimize impacts on these resources.

Transportation

Impacts on transportation are anticipated to be insignificant during the site evaluation phase from the one or two survey vessels that might be deployed at any one time. Vehicular traffic would be temporary and intermittent and would be limited to very low volumes of heavy- and medium-duty equipment and personal vehicles.

Visual Resources

Site evaluation activities would have temporary and minor visual effects caused by the presence of survey vessels, workers, vehicles, and equipment.

Socioeconomics

Site evaluation activities are temporary and limited and would not result in socioeconomic impacts on employment, local services, or property values.

Environmental Justice

Site evaluation activities are limited and would not result in significant adverse impacts in any resource area; therefore, environmental justice impacts are not expected at this phase.

Acoustics (Noise)

Onshore and offshore drilling activities for all hydrokinetic energy facilities, if required, would generate the most noise during this phase, but impacts would be much lower than those that could occur during construction. Surveys using air-gun arrays may generate low-frequency noise that may be detected by marine mammals, sea turtles, and fish within the survey area. Other sea and river geophysical surveys and installation of wave-measuring devices equipped with recording equipment would generate some ship and boat noise.

Hazardous Materials and Waste Management

The only hazardous material associated with site evaluation activities would be the fuel for boats, barges, and vehicles. Impacts from operational discharges, accidental fuel releases, and accidentally released solid debris are expected to be small or nonexistent if appropriate management practices are followed.

HYDROKINETIC ENERGY FACILITY CONSTRUCTION IMPACTS

Typical activities during the wave or tidal turbine energy farm construction phase include assembling hydrokinetic units on shore, transporting each device to its designated location offshore, anchoring it to the seabed, connecting each device electrically to a central junction box, laying or burying submarine transmission and signal cable, and construction of onshore substation and electrical transmission lines to connect to the grid. Activities required for river in-stream facilities are essentially the same but are conducted in a river rather than offshore. For a barrage facility, a dam would be constructed across the inlet or estuary to contain the incoming tidal flow and a powerhouse would be constructed to produce hydroelectric energy. Onshore activities include ground clearing, grading, excavation, vehicular traffic, and construction of facilities.

Air Quality

Offshore activities that generate emissions include ship, boat, and barge traffic to and from the hydrokinetic energy facility site, installation of the hydrokinetic energy devices and their associated anchoring devices, and the laying of underwater cables. Air emissions result from the operation of ship engines and on-ship equipment such as cranes, generators, and air compressors. In most cases, an air quality permit would not be required for offshore facility construction. However, in areas of non-attainment for any criteria pollutants, the states have authority to regulate nearshore activities. Emissions generated during the construction phase of land-based facilities (including docks, equipment storage, and assembly area) include the following:

- Vehicle emissions
- Diesel emissions from large construction equipment and generators
- Volatile organic compound (VOC) releases from the storage and transfer of vehicle/equipment fuels
- Small amounts of carbon monoxide, nitrogen oxides, and particulates from blasting activities
- Fugitive dust from many sources, such as disturbing and moving soils (clearing, grading, excavation, trenching, backfilling, dumping, and truck and equipment traffic), mixing concrete, use of unvegetated soil piles, drilling, and pile driving

These emissions would also be expected during construction of a barrage facility. A permit may be required from the state or local air agency to control or mitigate these emissions, especially in non-attainment areas.

Cultural Resources

For offshore projects, trenching, dredging, and placement of hydrokinetic energy devices and associated components could impact shipwrecks or buried archeological artifacts. For onshore projects, impacts on cultural resources could occur from site preparation (e.g., clearing, excavation, grading) and construction of transmission-related facilities. For either offshore or onshore projects, visual impacts could also result from disruption of a historical setting that is important to the integrity of a historic structure, such as a lighthouse. Potential cultural resource impacts include the following:

- Complete destruction of the resource if present in areas undergoing surface disturbance or excavation
- Unauthorized removal of artifacts or vandalism to cultural resource sites resulting from increases in human access to previously inaccessible areas
- Visual impacts resulting from vegetation cleaning, increased industry, and the presence of large-scale equipment, machinery, and vehicles (if the affected cultural resources have an associated landscape or other visual component that contributes to their significance, such as a sacred landscape or historic trail)

Ecological Resources

Wave and Tidal Turbine Energy Farms

The potential effects of construction activities associated with the placement of wave and tidal turbine energy devices on marine mammals, sea turtles, and fish may include behavioral responses such as avoidance and deflections in travel direction. Noise and vibrations generated during the various activities could disturb the normal behaviors and mask sounds from other members of the same species or from predators. Coastal birds could be displaced from offshore feeding habitats; however, most birds would be likely to return within relatively short periods following cessation of construction activities. A few could be injured or killed by collisions with the survey vessels.

The movement and deposition of sediment during construction activities on the seafloor could kill benthic organisms, a source of food for fish. Effects to fish could potentially occur if spawning or nursery grounds are disturbed during construction or if resuspended sediments cause smothering of habitat. The area of seafloor disturbance from anchoring systems relative to the surface area occupied would be small.

Marine mammals, sea turtles, fish, and seabirds could be exposed to discharges or accidental fuel releases from construction vessels and to accidentally released solid debris. Such spills would be small and quickly diluted and would not be expected to measurably affect marine mammal or fish populations.

Onshore impacts from construction could affect terrestrial vegetation and wildlife, but the overall impact is anticipated to be minimal because permanent onshore facilities are expected to be small. Wildlife would be most affected by habitat reduction within the project site, access roads, and transmission line rights of way. Wildlife within surrounding habitats might also be affected if the construction activity (and associated noise) disturbs normal behaviors, such as feeding and reproduction. Impacts on wildlife are expected to be minor.

Turtles nest along the south Atlantic and Gulf coastlines. Nests containing eggs and emerging hatchlings could be affected by construction activities onshore. Lighting from the construction areas could disorient the hatchlings and increase their exposure to predators. The minimal amount of onsite construction would limit the impact to no more than a few nests.

River In-Stream Facilities

The potential effects of the placement of river in-stream energy devices and associated construction on fish may include behavioral responses such as avoidance and deflections in travel direction. Noise and vibrations generated during the various construction activities, especially placement of supporting structures and installation of submarine transmission lines, could disturb the normal behavior. Those displaced because of avoidance behaviors during construction are likely to return within relatively short periods following cessation of construction activities.

The movement and deposition of sediment during construction activities on the riverbed could kill benthic organisms, a source of food for fish. Effects to fish could potentially occur if spawning or nursery grounds are disturbed during construction or if resuspended sediments cause smothering of habitat. The area of riverbed disturbance would be very small relative to the availability of similar habitat in surrounding areas.

Terrestrial wildlife would be most affected by habitat reduction within the project site, access roads, and transmission line rights of way. Wildlife within surrounding habitats might also be affected if the construction activity (and associated noise) disturbs normal behaviors, such as feeding and reproduction. Impacts on wildlife are expected to be minor.

Barrage Facilities

Dam construction at a barrage facility would not increase the amount of wetted area inundated within the embayment, but it would alter the period of time that water is held in the embayment and could alter the aquatic environment of the embayment. These alterations could lead to habitat loss for terrestrial wildlife and bird species and/or degradation for aquatic species. Underwater habitat would be altered and marine species could be injured or killed during construction of the intake and dam. The ability of fish and marine mammals to enter and leave the embayment would be substantially altered. The significance of construction impacts on fish, marine mammals, and saltwater wetland-dependent birds and terrestrial species is likely to be site specific. Terrestrial wildlife also would be affected by habitat reduction caused by construction of land-based facilities within the project site, including access roads and transmission line rights of way. Wildlife within surrounding habitats might also be affected if the construction activity (and associated noise) disturbs normal behaviors, such as feeding and reproduction.

Water Resources

Water Use

Water would be used onshore for dust control when clearing vegetation and grading and for road traffic, for making concrete, and for domestic use by the construction crew. Water would likely be trucked in from offsite. The quantity of water required would be small.

Water Quality

Vessels used for transport and installation of hydrokinetic energy devices and components could contribute small amounts of fuel or oil to the ocean through bilge discharges or leaks. Anchoring of construction ships, and installation of anchoring devices and electrics cables, can cause sediment from the seabed to enter the water column. Onshore activities that cause soil erosion or discharges of waste or sanitary water could affect water quality. Negligible to minor impact on water quality would be expected.

Land Use

The wave and tidal turbine energy farm could occupy from 17 to 250 acres of ocean surface. The facility would exclude commercial shipping and, possibly, fishing activities. A river in-stream facility could occupy about 5 acres and could affect commercial shipping, recreation, and fishing activities. A barrage facility would impact the area behind the dam and would exclude commercial and recreational ships and boats from entering the previously accessible estuary unless a ship lock were constructed. Existing onshore land use during construction would be affected by intrusive impacts such as ground clearing, increased traffic, noise, dust, and human activity, as well as by changes in the visual landscape. In particular, these impacts could affect those seeking recreational opportunities on the shore or in the water. Generally, offshore and onshore impacts associated with these types of facilities are expected to be minor.

Soils and Geologic Resources

Seabed and riverbed disturbance would result from drilling or pile driving required for anchoring the hydrokinetic energy devices and from excavation required to bury electrical cable. The area disturbed is a small portion of the area occupied by the hydrokinetic energy facility. Construction activities would also be unlikely to activate geological hazards. Surface disturbance, heavy equipment traffic, and changes to surface runoff patterns during construction of onshore facilities could cause soil erosion. Impacts of soil erosion could include soil nutrient loss and reduce water quality in nearby surface water bodies.

Visual Resources

Viewers onshore and offshore would observe an increase in vessel traffic transporting hydrokinetic energy devices, components, and works to the site. The activity during installation would also be noticed. Wave, tidal, and in-river facilities are generally low-profile structures and, although visible, may not be as objectionable as larger, more visible structures. The overall effect on visual resources is also related to existing uses (especially land-based residential uses) that would have a view of the hydrokinetic energy farm areas and will require site-specific assessment. Construction of the dam at a barrage site would change the character of the water basin and could create visual concerns for nearby residents or recreational users. Possible sources of visual impacts during construction of onshore facilities include ground disturbance, construction of highly visible facilities, vegetation removal, road construction, and increased traffic. Increased truck and vessel traffic and human activity at the port facility supporting project construction would also be visible, although it is anticipated this would only be a short-term impact.

Paleontological Resources

For offshore projects, trenching, dredging, and placement of hydrokinetic energy devices and associated components could impact paleontological resources. For onshore projects, impacts on paleontological resources could occur directly from the construction activities and increased accessibility to fossil locations. Potential impacts include the following:

- Complete destruction of the resource if present in areas undergoing surface disturbance or excavation
- Unauthorized removal of paleontological resources or vandalism to the site as a result of increased human access to previously inaccessible areas, if significant paleontological resources are present

Transportation

Traffic at the port would increase as wave or tidal energy devices and components are delivered prior to assembly transport to the project site. Vessel traffic will increase during the construction phase. The same effect would occur with river in-stream projects although, because they are generally much smaller installations, the impacts on transportation would be less significant. Short-term increases in the use of local roadways would occur during the onshore construction period. Barrage projects would be relatively large projects and would likely require more labor for construction. This would cause an increase in traffic on local roads and potentially could disrupt local traffic use.

Socioeconomics

Direct impacts would include the creation of new jobs for construction workers and the associated income and taxes generated by the hydrokinetic energy facilities. Indirect impacts would occur as a result of the new economic development and would include new jobs at businesses that support the expanded workforce to provide project materials, and associated income and taxes. An influx of new workers could strain the existing community infrastructure and social services; however, because most hydrokinetic projects are relatively small, the workforce required is also expected to be relatively small and the impacts, therefore, are expected to be minor.

Environmental Justice

If significant impacts occurred in any resource areas, and these impacts disproportionately affected minority or low-income populations, then there could be an environmental justice impact. Issues of potential concern during construction are noise, dust, and visual impacts from the construction site and possible impacts associated with the construction of new access roads. Additional impacts could include limitations on access to the area for tribal recreation, subsistence, and traditional activities. Environmental justice impacts are dependent upon vulnerable populations being located within the area of influence of a project and are, therefore, site specific.

Acoustics (Noise)

Underwater and above-water noise sources include boat, ship, and barge activity associated with transporting workers, materials, and hydrokinetic energy devices to the offshore site, installing hydrokinetic facilities, and the laying of electrical and signal cables. Human receptors on the ocean shore likely would be far enough away for any impacts to be minor. Human receptors on the river shore could be close to the activities required for locating and anchoring river in-stream turbines. If pile driving is required for anchoring hydrokinetic energy devices or for construction of offshore power-gathering stations, the noise could be audible at the shoreline and might be annoying to populations. This impact would be intermittent. Techniques for laying cable could require use of air guns, rock cutters, or shaped explosive charges. The noise could be intense but would occur over a very short time period.

Onshore noise would result from preassembly of the hydrokinetic energy devices and from the construction of onshore facilities. The primary source of noise during construction of onshore facilities, transmission lines, and a barrage facility would be from equipment operation (e.g., rollers, bulldozers, diesel engines). Other sources of noise would include vehicular traffic, tree felling, and blasting. Whether the noise levels from these activities exceed U.S. Environmental Protection Agency (USEPA) guidelines or local ordinances would depend on the distance to the nearest residence and the effectiveness of any mitigating measures to reduce noise levels. If occurring near a residential area, noise levels from blasting and some equipment operation could exceed the USEPA guidelines but would be intermittent and extend for only a limited time.

Adverse impacts due to noise could occur if the site is located near a sensitive area, such as a park, wilderness, or other protected area. The primary impacts from noise would be localized disturbances to wildlife, recreationists, and residents.

Hazardous Materials and Waste Management

Hazardous materials associated with installation of hydrokinetic energy devices and construction of associated support components would include fuels, lubricants, and hydraulic fluids contained in the hydrokinetic energy devices or used in ships and construction equipment. Impacts from accidental spills, accidental fuel releases, and releases of solid debris are expected to be minor if appropriate management practices are followed. Garbage and sanitary waste generated onboard the vessels and barges would be returned to shore for disposal. Solid and industrial waste would be generated during onshore construction activities. The solid wastes would likely be nonhazardous and consist mostly of containers, packaging materials, and waste from equipment assembly and construction crews. Industrial wastes would include minor amounts of fuels, spent vehicle and equipment fluids (lubricating oils, hydraulic fluids, battery electrolytes, glycol coolants), and spent solvents. These materials would be transported offsite for disposal, but impacts could result if the wastes were not properly handled and were released to the environment. No impacts are expected from proper handling of all wastes.

HYDROKINETIC ENERGY FACILITY OPERATION AND MAINTENANCE IMPACTS

Typical activities during the hydrokinetic energy facility operational phase include operation of the hydrokinetic energy devices, power generation, and associated maintenance activities that would require operations from a vessel or barge when components are being maintained, repaired, or replaced.

Air Quality

There are no direct air emissions from the operation of hydrokinetic energy facilities. Air emissions result from the operation of maintenance ship engines and on-ship equipment such as cranes, generators, and air compressors. Onshore vehicular traffic will continue to produce small amounts of fugitive dust and tailpipe emissions during maintenance activities. These emissions would not likely exceed air quality standards and impacts on air quality would be minor.

Cultural Resources

The operation and maintenance of offshore facilities would have no direct impact on cultural resources unless previously undisturbed areas are disturbed. Potential indirect impacts associated with the operation and maintenance of onshore facilities would be limited to unauthorized collection of artifacts made possible by access roads if they make remote lands accessible to the public. Visual impacts resulting from the presence of a large wave, tidal turbine, river in-stream energy facility, or barrage facility and transmission lines could affect some cultural resources, such as sacred landscapes or historic trails.

Ecological Resources

Wave and Tidal Energy Farms

The presence of a wave or tidal turbine energy farm could cause some marine mammals to avoid the area. Collisions with maintenance vessels and underground structures are anticipated to be rare. Overtopping wave energy devices could trap hatching sea turtles and fish, resulting in injury or death. Impacts on the marine populations are expected to be minor to moderate depending on the specific site. Some whale species migrate along the Pacific coast from 1.5 to 2 miles (2.5 to 3 km) offshore. Any wave or tidal turbine facility located in this zone could impact whale migration.

Noise levels from wave energy devices would be similar to those from ship traffic but would be continuous for the life of the wave farm. This could result in long-term avoidance by wildlife, which could lead to abandonment of feeding or mating grounds. Noise from submerged tidal turbines would be low due to the low rotational speed of the turbine blades. Marine mammals, sea turtles, fish, and seabirds could be exposed to discharges or accidental fuel releases from maintenance vessels and to accidentally released solid debris. Such spills would be small and quickly diluted and would not be expected to measurably affect these wildlife populations. Depending on the design of the facilities, wave energy devices or any above-water portions of tidal energy devices could become the host for seabird colonies or may be used as haul-out areas for sea lions. Underwater structures may also create an artificial habitat for benthic species. This could complicate the maintenance and repair of wave energy devices. Electromagnetic fields from the transmission cable can be detected by some fish and might result in attraction or avoidance. Such impacts would be negligible to minor. Onshore impacts from operation could affect vegetation and wildlife by habitat reduction within the project site, access roads, and transmission line rights of way. Turtles nest along the south Atlantic and Gulf coastline, and nests and emerging hatchlings could be affected by maintenance activities onshore. Outdoor lighting from onshore facilities could disorient the hatchlings and increase their exposure to predators.

River In-Stream Facilities

Fish behavior is influenced primarily by the natural current in the river and only secondarily by the rotating mechanisms in the turbines. Proper location of the turbines would have minimal impact on fish movement or abundance. Allowance of sufficient turbine spacing in a turbine farm or in the river may minimize impact on fish. Noise generated by the turbines would have minimal effects on aquatic biota. The turbines are not expected to affect terrestrial wildlife since they are mostly, or completely, submerged. Wildlife would be most affected by habitat reduction within the onshore project site, access roads, and transmission line rights of way. Impacts on wildlife are expected to be minor.

Barrage Facilities

Dam construction at a barrage factory would not increase the amount of wetted area inundated within embayment, but it would alter the period of time that water is held in the embayment and could alter the aquatic environment of the embayment. These alterations could lead to habitat loss for terrestrial wildlife and bird species and/or degradation for aquatic species. Fish could be injured or killed during operation of the intake or during power generation. Because of a reduction in natural fishing of sediment, increases in sedimentation within the embayment may also adversely affect embayment ecosystems. The ability of fish and marine mammals to enter and leave the embayment would likely be substantially altered. The significance of operational impacts on fish, marine mammals, and saltwater wetland-dependent birds and terrestrial species would likely be site specific.

Water Resources

Water Use

Water use at the port would be required for normal operation, including fire protection, cleaning and maintenance of equipment, and consumptive use by personnel. Water would also be required for consumptive use on the vessels.

Water Quality

Vessels used for maintenance of hydrokinetic energy devices and components could contribute small amounts of fuel or oil to the ocean or river through bilge discharges or leaks. Damage to a hydrokinetic energy device, which may contain petroleum-based materials, could result in water contamination. Anchoring of the ships can cause sediment from the seabed to enter the water column. Onshore activities that could affect water quality are those that cause soil erosion or discharges of waste or sanitary water. Negligible or minor impact on water quality would be expected.

Land Use

All types of hydrokinetic facilities would likely exclude traditional uses of the areas where they are constructed. Commercial shipping, fishing, and recreation uses may be the most likely uses that could be affected. Depending on the nature and visibility of the facilities, the visual impacts of hydrokinetic development might also create conflicts with existing shore-based uses.

Soils and Geologic Resources

Seabed or riverbed disturbance would be minimal from maintenance activities, and sediments are unlikely to be affected. Maintenance activities would also be unlikely to activate geological hazards. A wave energy farm could cause a reduction in wave height of 10 to 15%, and this reduction could result in an interruption of the natural sediment transport along the shore, increasing erosion and drift. The impact is greater the closer the wave farm is to the shore and is greater for floating devices oriented parallel to the shore. Tidal and river in-stream turbines will cause turbulence downstream that might cause scour of the seabed or riverbed if the units are located near the bottom. During operation, the soil and geologic conditions would stabilize at onshore facilities. Soil erosion and soil compaction are both likely to continue to occur along access roads.

Paleontological Resources

The operation and maintenance of offshore facilities would have no direct impacts on paleontological resources unless additional undisturbed areas are developed. Potential indirect impacts associated with the operation and maintenance of onshore facilities would be limited to unauthorized collection of fossils made possible by access roads if they make remote lands accessible to the public.

Transportation

No noticeable impacts on transportation are likely during operations. Maintenance vessels would service the hydrokinetic energy devices at regular intervals, but the additional levels of activity resulting from this are anticipated to be small. Infrequent, but routine, truck shipments of component replacements to the dock during maintenance procedures are likely over the period of operation.

Visual Resources

Visual impacts of the operation of hydrokinetic energy devices would be the same as those identified for construction activities, with the exception of the increase in vessel and vehicle traffic associated with construction.

Socioeconomics

Direct impacts would include the creation of new jobs for operation and maintenance workers and the associated income and taxes paid. Indirect impacts are those impacts that would occur as a result of the new economic development and would include things such as new jobs at businesses that support the workforce or that provide project materials, and the associated income and taxes. However, the total number of operations and maintenance jobs likely would be small; therefore, the associated socioeconomic impacts are anticipated to be minimal.

Environmental Justice

If significant impacts occurred in any resource areas as a result of the operation of hydrokinetic facilities, and these impacts disproportionately affect minority or low-income populations, then there could be an environmental justice impact. Issues of potential concern during operations are noise, ecological, and visual impacts. Additional impacts include limitations on access to the area for tribal activities.

Acoustics (Noise)

Underwater and above-water noise sources include ship and barge noise associated with transporting workers for maintenance activities, which would require frequent (and possibly daily) trips to the wave or tidal turbine energy farm or to the river in-stream turbines. Wave energy device noise would result from the flexing action of attenuators and point absorbers, from compressed air released from oscillating water column turbines, and from the impact of waves on terminator and overtopping devices. Underwater noise from the operation of tidal or river in-stream turbines is expected to be low because the rotational speed of the turbine blades is low. Overall, noise from operation of the hydrokinetic energy devices is expected to be low. Onshore transformers would produce a humming noise and cooling fan noise. Sources of noise during operation of a barrage facility would be the turbines, generators, and transformers.

Hazardous Materials and Waste Management

Hazardous materials associated with the operation and maintenance of hydrokinetic energy devices and associated support compounds would include the fuel for boats, vessels, and barges, and lubricants and hydraulic fluids contained in the wave or tidal energy devices. Impacts from accidental spills, accidental fuel releases, and releases of solid debris are expected to be minor if appropriate management practices are followed. Garbage and sanitary waste generated onboard the vessels and barges would be returned to shore for disposal. Industrial wastes are generated during routine maintenance (used fluids, cleaning agents, and solvents). These wastes typically would be put in containers, characterized and labeled, possibly stored briefly, and transported by a licensed hauler to an appropriate permitted offsite disposal facility as a standard practice. Adverse impacts could result if these wastes were not properly handled and were released to the environment. Given current standards the impact of this is expected to be minor.

HYDROKINETIC ENERGY IMPACTS ON HUMAN HEALTH AND SAFETY

With regard to the site evaluation phase, the primary hazard is the risk of drowning while working on or above water. The potential health and safety risks that could result in injuries and fatalities include onboard accidents, collisions between the survey ship and marine vessels, and natural events such as hurricanes, earthquakes, tsunamis, and severe storms. Occupational and public health and safety risks normally associated with construction and outdoor activities (working in potential weather extremes and possible contact with natural hazards, such as uneven terrain and dangerous plant, animals, or insects) exist but are very limited during the site evaluation phase because of the limited range of activities and number of workers.

The primary hazard associated with hydrokinetic energy device installation and construction of associated components is the risk of drowning while working on or above water. The potential health and safety risks that could result in injuries and fatalities include onboard accidents, collisions between the vessel or barge and marine vessels, and natural events such as hurricanes, earthquakes, tsunamis, and severe storms. Potential onshore impacts on workers would be similar to those expected for any construction project involving earth moving, large equipment, and construction and installation of industrial facilities. Most accidents in the construction industry result from overexertion, falls, or being struck by equipment. Construction-related illnesses could also result from exposure to chemical substances from spills. In addition, health and safety issues include working in potential weather extremes and possible contact with natural hazards, such as uneven terrain and dangerous plants, animals, or insects. All personnel involved with the construction would utilize appropriate safety equipment and would be properly trained in required Occupational Safety and Health Administration (OSHA) practices.

The primary hazard associated with operation and maintenance of hydrokinetic energy devices and associated components is the risk of drowning while working on or above water. The potential health and safety risks that could result in injuries and fatalities include onboard accidents, collisions between the vessel or barge and marine vessels, and natural events, such as hurricanes, earthquakes, tsunamis, and severe storms. Also, other marine vessels could collide with the wave energy devices. Deploying navigational aids such as lighting and foghorns would minimize vessel collisions with wave energy devices. All personnel involved with the operations and maintenance activities would utilize appropriate safety equipment and would be properly trained in required Occupational Safety and Health Administration (OSHA) practices.

BLUE ENERGY: THE BOTTOM LINE

The three forms of blue (ocean) energy—tidal, wave, and ocean thermal energy conversion (OTEC) systems—are all renewable. This is a significant advantage over fossil fuels and other energy forms that pollute the environment. Although all three forms of ocean energy show promise for future development, it is the OTEC systems that appear to be most beneficial for use at the present time. OTEC systems provide both economic and noneconomic benefits. On the other hand, OTEC power plants have the potential to cause major adverse impacts on ocean water quality. Such plants would require entrainment and discharging enormous quantities of seawater. Sea surface temperatures in the vicinity of an OTEC plant could be lowered by the discharge of effluent from the cold-water pipe. Biocides, such as chlorine, may be irritating or toxic to organisms. OTEC plants could result in an upwelling of nutrients from the bottom and the subsequent growth of large algal blooms (Abbasi and Abbasi, 2000). The bottom line? Few marine and hydrokinetic renewable energy technologies have been tested at full scale, and it is therefore difficult to resolve all of the uncertainties about their specific environmental effects.

DISCUSSION AND REVIEW QUESTIONS

1   Is tidal energy a beneficial source of renewable energy? Explain your answer.
2   It has been said that tidal energy is expensive. Does it matter?
3   Tidal energy systems must be located near coastal areas. Is the tradeoff of lost land area worth it?

REFERENCES AND RECOMMENDED READING

Abbasi, S.A. and Abbasi, N. (2000). The likely adverse environmental impacts of renewable energy sources. *Applied Energy*, 65: 121–144.

Antizar-Ladislao, B. (2008). Environmental levels, toxicity, and human exposure to tributyltin (TBT)-contaminated marine environment: a review. *Environmental International*, 34(2008): 292–308.

Arenas, P. and Hall, M. (1992). The association of sea turtles and other pelagic fauna with floating objects in the eastern tropical Pacific Ocean. In: *Proceedings of the Eleventh Annual Workshop on Sea Turtle Biology and Conservation*, Jekyll Island, GA, February 26–March 2 (http://www.nmfs.noaa.gov/pr/pdfs/species/turtlesymposium1991.pdf).

ASR, Ltd. (2007). *Review of Wave Hub Technical Studies: Impacts on Inshore Surfing Beaches*, Final Report to South West of England Regional Development Agency, Sutton Harbour, Plymouth, U.K. (http://www.wavehub.co.uk/wp-content/uploads/2011/06/2007-April-Review-of-Wave-Hub-Technical-Studies.pdf).

Avens, L. and Lohmann, K.J. (2003). Use of multiple orientation cues by juvenile loggerhead sea turtles, *Caretta caretta. Journal of Experimental Biology*, 206: 4317–4325.

Balayev, L.A. and Fursa, N.N. (1980). The behavior of ecologically different fish in electric fields. I. Threshold of first reaction in fish. *Journal of Ichthyology*, 20(4): 147–152.

Basov, B.M. (1999). Behavior of sterlet *Acipenser ruthenus* and Russian sturgeon *A. gueldenstaedii* in low-frequency electric fields. *Journal of Ichthyology*, 39(9): 782–787.

Basov, B.M. (2007). On electric fields of power lines and on their perception by freshwater fish. *Journal of Ichthyology*, 47(8): 656–661.

Berinstein, P. (2001). *Alternative Energy: Facts, Statistics, and Issues*. Westport, CT: Oryx Press.

Betke, K., Schultz-von Glahn, M., and Matuschek, R. (2004). Underwater noise emission from offshore wind turbines. In: *Proceedings of the Joint Congress of CFA, DAGA '04*, Strasbourg, France, March 22–25.

Blanchfield, J., Rowe, A., Wild, P., and Garrett, C. (2007). The power potential of tidal streams including a case study for Masset Sound. In: *Proceedings of the 7th European Wave and Tidal Energy Conference*, Porto, Portugal, September 11–14.

Blyth, R.E., Kaiser, M.J., Edwards-Jones, G., and Hard, P.J.B. (2004). Implications of a zoned fishery management system for marine benthic communities. *Journal of Applied Ecology*, 41: 951–961.

Blyth-Skyrme, R.E., Kaiser, M.J., Hiddink, J.G., Edwards-Jones, G., and Hard, P.J.B. (2006). Conservation benefits of temperate marine protected areas: variation among fish species. *Conservation Biology*, 20(3): 811–820.

Bochert, R. and Zettler, M.I. (2006). Effect of electromagnetic fields on marine organisms. In: *Offshore Wind Energy*, Koller, J., Koppel, J., and Peters, W., Eds., pp. 223–234. Berlin: Springer-Verlag.

Boehlert, G.W., McMurray, G.R., and Tortorici, C.E., Eds. (2008). *Ecological Effects of Wave Energy Development in the Pacific Northwest*, NOAA Technical Memorandum NMFS-F/SPO-92. Washington, DC: U.S. Department of Commerce.

Boles, L.C. and Lohmann, K.J. (2003). True navigation and magnetic maps in spiny lobster. *Nature*, 421: 60–63.

Bowen, B.W., Meylan, A.B., Rose, J.P., Limpus, C.J., Balazs, G.H., and Avise, J.C. (1992). Global population structure and natural history of the green turtle in terms of matriarchal phylogeny. *Evolution*, 46: 865–881.

Bowen, F., Avise, J.C., Richardson, J.I., Meylan, A.B., Margaritoulis, D., and Hopkins-Humpy, S.R. (1993). Population structure of loggerhead turtles in the northwestern Atlantic Ocean and Mediterranean Seas. *Conservation Biology*, 7: 834–844.

Bryden, I.G., Grinstead, T., and Melville, G.T. (2004). Assessing the potential of a simple tidal channel to deliver useful energy. *Applied Ocean Research*, 26: 198–204.

Butti, K. and Perlin, J. (1980). *A Golden Thread: 2500 Years of Solar Architecture and Technology*. New York: Van Nostrand Reinhold.

Cada, G.F., Smith, J., and Busey, J. (2005). Use of pressure sensitive film to quantify sources of injury to fish. *North American Journal of Fisheries Management*, 25(2): 57–66.

Cada, G.F., Loar, J.M., Garrison, L., Fisher, R.K., and Neitzel, D. (2006). Efforts to reduce mortality to hydroelectric turbine-passed fish: locating and quantifying damaging shear stresses. *Environmental Management*, 37(6): 898–906.

Cada, G.F., Ahlgrimm, J., Bahleda, M., Bigford, T., Damiani Stavrakas, S., Hall, D., Moursund, R., and Sale, M. (2007). Potential impacts of hydrokinetic and wave energy conversion technologies on aquatic environments. *Fisheries*, 32(4): 174–181.

Cain, S.D., Boles, L.C., Wang, J.H., and Lohmann, K.J. (2005). Magnetic orientation and navigation in marine turtles, lobsters, and mollusks: concepts and conundrums. *Integrative and Comparative Biology*, 45: 539–546.

Caltrans. (2001). *Fisheries Impact Assessment*, Pile Installation Demonstration Project, San Francisco–Oakland Bay Bridge East Span Seismic Safety Project, PIDP EA 012081 (http://www.biomitigation. org/reports/files/PIDP_Fisheries_Impact_Assessment_0_ 1240.pdf).

Carr, M.H. and Hixon, M.A. (1997). Artificial reefs: the importance of comparisons with natural reefs. *Fisheries*, 22(4): 28–33.

Castro, J.J., Santiago, J.A., and Santana-Ortega, A.T. (2002). A general theory on fish aggregation to floating objects: an alternative to the meeting point hypothesis. *Reviews in Fish and Biology and Fisheries*, 11: 255–277.

Caswell, H., Fuiwara, M., and Brault, S. (1999). Declining survival probability threatens the North American right whale. *Proceedings of the National Academy of Sciences USA*, 96: 3308–3313.

Chapman, D.M.F. (2000). Decibels and SI units. *Journal of the Acoustical Society of America*, 106: 3048.

Clay, C.S. (1999). Underwater sound transmission and SI units. *Journal of the Acoustical Society of America*, 106: 3047.

CMACS. (2003). *A Baseline Assessment of Electromagnetic Fields Generated by Offshore Windfarm Cables*, COWRIE Report EMG-01-2002 66. Liverpool: Centre for Marine and Coastal Studies.

Coleman, F.C., Koenig, C.C., and Collins, L.A. (1996). Reproductive styles of shallow-water groupers (Pisces: Serranidae) in the eastern Gulf of Mexico and the consequences of fish spawning aggregations. *Environmental Biology of Fishes*, 47: 129–141.

Colin, P.L. (1992). Reproduction of the Nassau grouper, *Epinephrus striatus* (Pisces: Serranidae), and its relationship to environmental conditions. *Environmental Biology of Fishes*, 34: 357–377.

Collin, S.P. and Whitehead, D. (2004). The functional roles of passive electroreception in non-electric fishes. *Animal Biology*, 54(1): 1–25.

Coston-Clements, L., Settle, L.R., Hoss, D.E., and Cross, F.A. (1991). *Utilization of the Sargassum Habitat by Marine Invertebrates and Vertebrates: A Review*, NOAA Technical Memorandum NMFS-SEFSC-296. Beaufort, NC: U.S. Department of Commerce.

Coutant, C.C. and Cada, G.F. (2005). What's the future of instream hydro? *Hydro Review*, 24(6): 42–49.

Craig, C.I., Wyllie-Escheverria, S., Carrington, E., and Shafer, D. (2008). *Short-Term Sediment Burial Effects on the Seagrass Phyllospadix scouleri*, ERDC TN-EMRRP-EI-03. Vicksburg, MS: U.S. Army Engineer Research and Development Center.

Crawford, R.E. and Carey, C.G. (1985). Retention of winter flounder larvae with a Rhode Island salt pond. *Estuaries*, 8(2B): 217–227.

Cushing, D.H. (1969). The regularity of spawning season of some fishes. *Journal du Conseil—Conseil International pour l'Exploriation de la Mer*, 33(1): 81–92.

Dadswell, M.J.U., Klauda, R.J., Moffitt, C.M., Saunders, R.L., Rulifson, R.A., and Cooper, J.E. (1987). *Common Strategies of Anadromous and Catadromous Fishes*. Bethesda, MD: American Fisheries Society.

Davis, N., VanBlaricom, G.R., and Dayton, P.K. (1982). Man-made structures on marine sediments: effects on adjacent benthic communities. *Marine Biology*, 70: 295–303.

de Jong, D.J., de Jong, Z., and Mulder, J.P.M. (1994). Changes in area, geomorphology, and sediment nature of salt marshes in the Oosterschelde estuary (SW Netherlands) due to tidal changes. *Hydrobiologia*, 281/283: 303–316.

Dempster, T. (2005). Temporal variability of pelagic fish assemblages around fish aggregation devices: biological and physical influences. *Journal of Fish Biology*, 66: 1237–1260.

Dempster, T. and Taquet, M. (2004). Fish aggregation device (FAD) research: gaps in current knowledge and future directions for ecological studies. *Review in Fish Biology and Fisheries*, 14: 21–42.

Deng, D.L., Carlson, T.J., Ploskey, G.R., and Richmond, M.C. (2005). *Evaluation of Blade-Strike Models for Estimating the Biological Performance of Large Kaplan Turbines*, PNNL-15370. Richland, WA: Pacific Northwest National Laboratory.

Department of the Navy. (2003). *Environmental Assessment—Proposed Wave Energy Technology Project, Marine Corps Base Hawaii, Kaneohe Bay, Hawaii*. Washington, DC: U.S. Department of Defense.

Desholm, M. (2003). *Thermal Animal Detection System (TADS): Development of a Method for Estimating Collision Frequency of Migrating Birds at Offshore Wind Turbines*, NETI Technical Report No. 440. Roskilde: National Environmental Research Institute of Denmark.

Desholm, M. (2006). Wind Farm Related Mortality Among Avian Migrants: A Remote Sensing Study and Modeling Analysis, doctoral thesis, Department of Wildlife Ecology and Biodiversity, University of Copenhagen.

DiBacco, C., Sutton, D., and McConnico, L. (2001). Vertical migration behavior and horizontal distribution of brachyuran larvae in a low-inflow estuary: implications for bay–ocean exchange. *Marine Ecology Progress Series*, 217: 191–206.

Dodd, C.K. (1991). The status of the Red Hills salamander *Phaeognathus hubrichti*, Alabama, USA, 1976–1988. *Biological Conservation*, 55: 57–75.

Dodd, C.K. (1995). Marine turtles in the Southeast. In: *Our Living Resources: A Report to the Nation on the Distribution, Abundance, and Health of U.S. Plants, Animals, and Ecosystems*, LaRoe, E.T. et al., Eds., pp. 121–123. Washington, DC: National Biological Service.

Domeier, M.L. and Colin, P.L. (1997). Tropical reef fish spawning aggregations: defined and reviewed. *Bulletin of Marine Science*, 60(3): 698–726.

Doughty, R.W. (1984). Sea turtles in Texas: a forgotten commerce. *Southwestern Historical Quarterly*, 88: 43–70.

EERE. (2010a). *Ocean Wave Power*. Washington, DC: Office of Energy Efficiency & Renewable Energy, U.S. Department of Energy (http://www.energysavers.gov/renewable_energy/ocean/index.cfm/mytopic=50009).

EERE (2010b). *Ocean Thermal Energy Conversion*. Washington, DC: Office of Energy Efficiency & Renewable Energy, U.S. Department of Energy (http://www.energysavers.gov/renewable_energy/ocean/index.cfm/mytopic=50010).

EERE. (2014). *Water.* Washington, DC: Office of Energy Efficiency & Renewable Energy, U.S. Department of Energy (http://energy.gov/eere/renewables/water).

EERE. (2015). *Marine and Hydrokinetic Technology Glossary.* Washington, DC: Office of Energy Efficiency & Renewable Energy, U.S. Department of Energy (http://energy.gov/eere/water/marine-and-hydrokinetic-technology-glossary).

Enger, P.S., Kristensen, L., and Sand, O. (1976). The perception of weak electric D.C. currents by the European eel (*Anguilla anguilla*). *Comparative Biochemistry and Physiology*, 54A: 101–103.

Engineering Business, Ltd., The (2005). *Stingray Tidal Steam Energy Device—Phase 3*, Contract No. T/06/00230/00. Oxfordshire, U.K.: DTI New & Renewable Energy Programme (http://tethys.pnnl.gov/sites/default/files/publications/Stingray_Tidal_Stream_Energy_Device.pdf).

Epifanio, C.E. (1988). Transport of invertebrate larvae between estuaries and the continental shelf. *American Fisheries Society Symposium*, 3: 104–114.

Faber Maunsell and Metoc. (2007). *Scottish Marine Renewables SEA—Scoping Report.* Edinburgh, Scotland: Scottish Executive.

Fast, A.W., D'Itri, F.M., Barclay, D.K., Katase, S.A., and Madenjian, C. (1990). Heavy metal content of coho *Onchorhynchus kisutch* and Chinook salmon *O. tscharwytscha* reared in deep upwelled ocean waters in Hawaii. *Journal of the World Aquaculture Society*, 21(4): 271–276.

Fernandez, A., Edwards, J.F., Rodriguez, F., Espniosa de los Monteros, A., Harraez, P., Castro, P., Jaber, J.R., Martin, V., and Arbelo, M. (2005). Gas and fat embolic syndrome involving a mass stranding of beaked whales (Family Ziphiidae) exposed to anthropogenic sonar signals. *Veterinary Pathology*, 42: 446–457.

Flavin, C. (1990). *Beyond the Petroleum Age: Designing a Solar Economy.* Washington, DC: Worldwatch Institute.

Fraenkel, P.L. (2006). Tidal current energy technologies. *Ibis*, 148: 145–151.

Fraenkel, P.L. (2007a). Marine current turbines: pioneering the development of marine kinetic energy converters: proceedings of the Institution of Mechanical Engineers, Part A. *Journal of Power and Energy*, 221(2): 159–169.

Fraenkel, P.L. (2007b). Marine current turbines: moving from experimental test rigs to a commercial technology. In: *Proceedings of ASME 2007, 26th International Conference on Offshore Mechanics and Arctic Engineering*, San Diego, CA, June 10–15.

Garrett, C. and Cummins, P. (2004). Generating power from tidal currents. *Journal of Waterway, Port, Coastal and Ocean Engineering*, 130: 114–118.

Gell, F.R. and Roberts, C.M. (2003). Benefits beyond boundaries: the fishery effects of marine reserves. *Trends in Ecology and Evolution*, 18(9): 448–455.

Gill, A.B. (2005). Offshore renewable energy: ecological implications of generating electricity in the coastal zone. *Journal of Applied Ecology*, 42: 605–615.

Gill, A.B. and Taylor, H. (2002). *The Potential Effects of Electromagnetic Fields Generated by Cabling between Offshore Wind Turbines upon Elasmobranch Fishes*, CCW Contract Science Report No. 488. Bangor, Gwynedd, North Wales: Countryside Council for Wales.

Gill, A.B., Gloyne-Phillips, I., Neal, K.J., and Kimber, J.A. (2005). *COWRIE 1.5 Electromagnetic Fields Review: The Potential Effects of Electromagnetic Fields Generated by Sub-Sea Power Cables Associated with Offshore Wind Farm Developments on Electrically and Magnetically Sensitive Marine Organisms—A Review*, COWRIE–EM Field 2-06-2004. Cranfield, U.K.: Cranfield University.

Goda, Y. (2000). *Random Seas and Designs of Maritime Structures.* Singapore: World Scientific.

Goff, M., Salmon, M., and Lohmann, K.J. (1998). Hatchling sea turtles use surface waves to establish a magnetic compass direction. *Animal Behavior*, 55: 69–77.

Gould, J.L. (1984). Magnetic field sensitivity in animals. *Annual Reviews in Physiology*, 46: 585–598.

Gross, G.M. (1995). *Oceanography: A View of the Earth*, 7th ed. Englewood Cliffs, NJ: Prentice Hall.

Grossman, G.D., Jones, G.P., and Seaman, Jr., W.J. (1997). Do artificial reefs increase regional fish production? A review of existing data. *Fisheries*, 22(4): 17–23.

Hamilton, W.W. (1983). Preventing cavitation damage to hydraulic structures, Part One. *Water Power and Dam Construction*, November, pp. 48–53.

Harrison, J.T. (1987). *The 40 MWe OTEC Plant at Kahe Point, Oahu, Hawaii: A Case Study of Potential Biological Impacts*, NOAA Technical Memorandum NMFS, NOAA-TM-NMFS-SWFC-68. Honolulu, HI: Southwest Fisheries Center.

Hastings, M.C. and Popper, A.N. (2005). *Effects of Sound on Fish.* Sacramento: California Department of Transportation (http://www.dot.ca.gov/hq/env/bio/files/Effects_of_Sound_on_Fish23Aug05.pdf).

Hickling, R. (1999). Noise control and SI units. *Journal of the Acoustical Society of America*, 106: 3048.

Holdren, J.P., Morris, G., and Mintzer, I. (1980). Environmental aspects of renewable energy sources. *Annual Review of Energy*, 5: 241–291.

Horton, C.W. (1954). The bewildering decibel. *Electrical Engineering*, 73: 550–555.

IEA. (2016). International Energy Agency, http://www.iea.org/.

IEC. (2016). International Electrotechnical Commission, http://www.iec.ch/.

INL. (2005). *Hydropower: Hydrokinetic & Wave Technologies*. Idaho Falls: Idaho National Laboratory (http://hydropower.inl.gov/hydrokinetic_wave/).

Irwin, W.P. and Lohmann, K.J. (1996). Disruption of magnetic orientation in hatchling loggerhead sea turtles by pulsed magnetic fields. *Journal of Comparative Physiology A*, 191: 475–480.

Janota, C.P. and Thompson, D.E. (1983). Waterborne noise due to ocean thermal energy conversion plants. *Journal of the Acoustic Society of America*, 74(1): 256–266.

Jennings, S., Pinnegar, J.K., Polunin, N.V.C., and Warr, K.J. (2005). Impacts of trawling disturbance on the trophic structure of benthic invertebrate communities. *Marine Ecology Progress Series*, 213: 127–142.

Johnsen, S. and Lohmann, K.J. (2005). The physics and neurobiology of magnetoreception. *Neuroscience*, 6: 703–712.

Johnsen, S. and Lohmann, K.J. (2008). Magnetoreception in animals. *Physics Today*, 61(3): 29–35.

Johnson, A., Salvador, G., Kenney, J. et al. (2005). Fishing gear involved in entanglements of right and humpback whales. *Marine Mammal Science*, 21(4): 635–645.

Johnson, M.R., Boelke, C., Chiarella, L.A., Colosi, P.D., Green, K., Lellis-Dibble, K., Ludemann, H., Ludwig, M., McDermott, S., Ortiz, J., Rusanowsky, D., Scott, M., and Smith, J. (2008). *Impacts to Marine Fisheries Habitat from Nonfishing Activities in the Northeastern United States*, NOAA Technical Memorandum NMGS-NE-209. Gloucester, MA: U.S. Department of Commerce, National Marine Fisheries Service (http://nefsc.noaa.gov/nefsc/publications/tm/tm209/).

Kaiser, M.J. (2005). Are marine protected areas a red herring or fisheries panacea? *Canadian Journal of Fisheries and Aquatic Sciences*, 62: 1194–1199.

Kaiser, M.J., Spence, F.E., and Hart, P.J.B. (2000). Fishing-gear restrictions and conservation of benthic habitat complexity. *Conservation Biology*, 14(5): 1512–1525.

Kalmijn, A.T. (1982). Electric and magnetic field detection in elasmobranch fishes. *Science*, 218(4575): 916–918.

Kalmijn, A.T. (2000). Detection and processing of electromagnetic and near-field acoustic signals in elasmobranch fishes. *Philosophical Transactions of the Royal Society of London B*, 355: 1135–1141.

Karsten, R.H., McMillan, J.M., Lickley, M.J., and Haynes, R.D. (2008). Assessment of tidal current energy in the Minas Passage, Bay of Fundy: proceedings of the Institution of Mechanical Engineers, Part A. *Journal of Power and Energy*, 222(5): 493–507.

Kingsford, M.J. (1999). Fish attraction devices (FADs) and experimental designs. *Scientia Marina*, 63(3-4): 181–190.

Kogan, I., Paull, C.K., Kuhnz, L.A., Burton, E.J., Von Thun, S., Green, H.G., and Barry, J.P. (2006). ATOC/Pioneer Seamount cable after 8 years on the seafloor: observations, environmental impact. *Continental Shelf Research*, 26(2006): 771–787.

Krause, S.D., Brown, M.W., Caswell, H., Clark, C.W., Fujiwara, M., Hamilton, P.K., Keeney, R.D., Knowlton, A.R., Landry, S., Mayo, C.A., McLellan, W.A., Moore, M.J., Nowacek, D.P., Pabst, D.A., Read, A.J., and Rolland, R.M. (2005). North Atlantic right whales in crisis. *Science*, 309: 561–562.

Lambert D. (2007). *The Field Guide to Geology*. New York: Checkmark Books.

Langhamer, O. (2005). *Man-Made Offshore Installations: Are Marine Colonisers a Problem or an Advantage?* Uppsala, Sweden: Uppsala University (http://www.el.angstrom.uu.se/forskningsprojekt/Islandsberg_pek/Man-made%20offshore%20installations.pdf).

Largier, J., Behrens, D., and Robart, M. (2008). The potential impact of WEC development on nearshore and shoreline environments through a reduction in nearshore wave energy. In: *Developing Wave Energy in Coastal California: Potential Socio-Economic and Environmental Effects*, Nelson, P.A. et al., Eds., pp. 52–74. Sacramento: California Energy Commission.

Lewis, L.J., Davenport, J., and Kelly, T.C. (2003). A study of the impact of a pipeline construction on estuarine benthic invertebrate communities. Part 2. Recolonization by benthic invertebrates after 1 year and response of estuarine birds. *Estuarine, Coastal and Shelf Sciences*, 57(2003): 201–208.

Loch, S. and Norman, R.S. (1975). Osmotic power plants. *Science*, 189: 654–655.

Lohmann, K.J. and Lohmann, C.M.F. (1996). Detection of magnetic field intensity by sea turtles. *Nature*, 380: 59–61.

Lohmann, K.J., Swartz, A.W., and Lohmann, C.M.F. (1995). Perception of ocean wave direction by sea turtles. *Journal of Experimental Biology*, 198: 1079–1085.

Lohmann, K.J., Lohmann, C.M.F., and Endres, C.S. (2008a). The sensory ecology of ocean navigation. *Journal of Experimental Biology*, 211: 1719–1728.

Lohmann, K.J., Putman, N.F., and Lohmann, C.M.F. (2008b). Geomagnetic imprinting: a unifying hypothesis of long-distance natal homing in salmon and sea turtles. *Proceedings of the National Academy of Sciences USA*, 105(49): 19096–19101.

Longmuir, C. and Lively, T. (2001). Bubble curtain systems help protect the marine environment. *Pile Driver Magazine*, Summer, pp. 11–16.

Louse, D.P., Gaddam, R.N. and Raimondi, R.T. (2008). Predicted effects of wave energy conversion on communities in the nearshore environment. In: *Developing Wave Energy in Coastal California: Potential Socio-Economic and Environmental Effects*, Nelson, P.A. et al., Eds., pp. 75–99. Sacramento: California Energy Commission (http://www.energy.ca.gov/2008publications/CEC-500-2008-083/CEC-500-2008-083.PDF).

Love, M.S., Caselle, J., and Snook, L. (1999). Fish assemblages on mussel mounds surround seven oil platforms in the Santa Barbara Channel and Santa Maria Basin. *Bulletin of Marine Science*, 65(2): 497–513.

Lugo-Fernandez, A., Roberts, H.H., and Wiseman, Jr., W.J. (1998). Tide effects on wave attenuation and wave set-up on a Caribbean coral reef. *Estuarine, Coastal and Shelf Science*, 47: 385–393.

Madsen, P.T. (2005). Marine mammals and noise: problems with root mean square sound pressure levels for transients. *Journal of the Acoustic Society of America*, 117(6): 3952–3957.

Mann, S., Sparks, N.H.C., Walker, M.M., and Kirschvink, J.L. (1988). Ultrastructure, morphology and organization of biogenic magnetite from sockeye salmon, *Oncorhynchus nerka*: implications for magnetoreception. *Journal of Experimental Biology*, 140: 35–49.

Marra, L.J. (1989). Sharkbite on the SL submarine lightwave cable system: history, causes, and resolution. *IEEE Journal of Oceanic Engineering*, 14(3): 230–237.

Masselink, G. and Short, A.D. (1993). The effect of tidal range on beach morphodynamics: a conceptual beach model. *Journal of Coastal Research*, 9(3): 785–800.

McCabe, A., Bradshaw, A., Meadowcroft, J., and Aggidis, G. (2006). Developments in the design of the PS Frog Mk 5 wave energy converter. *Renewable Energy*, 31(2): 141–151.

McCauley, R.D., Fewtrell, J., Duncan, A.J., Jenner, C., Jenner, M.-N., Penrose, J.D., Prince, R.I.T., Adhitya, A., Murdoch, J., and McCabe, K. (2000a). Marine seismic surveys: a study of environmental implications. *Australian Petroleum Production Exploration Association Journal*, 2000: 692–708.

McCauley, R.D., Fewtrell, J., Duncan, A.J., Jenner, C., Jenner, M.-N., Penrose, J.D., Prince, R.I.T., Adhitya, A., Murdoch, J., and McCabe, K. (2000b). *Marine Seismic Surveys: Analysis and Propagation of Air-Gun Signals and Effects of Air-Gun Exposure on Humpback Whales, Sea Turtles, Fishes and Squid*, Report R99-15. Perth, Western Australia: Centre for Marine Science and Technology, Curtin University of Technology.

Meyer, C.G., Holland, K.N., and Papastamatiou, Y.P. (2004). Sharks can detect changes in the geomagnetic field. *Journal of the Royal Society Interface*, 2(2): 129–130.

Michel, J. and Burkhard, E. (2007). *Workshop to Identify Alternative Energy Environmental Information Needs—Workshop Summary*, OCS Report MMS 2007-057. Washington, DC: Mineral Management Service, U.S. Department of the Interior.

Michel, J., Dunagan, H., Boring, C., Healy, E., Evans, W., Dean, J., McGillis, A., and Hain, J. (2007). *Worldwide Synthesis and Analysis of Existing Information Regarding Environmental Effects of Alternative Energy Uses on the Outer Continental Shelf*, OCS Report MMS 2007-038. Washington, DC: Mineral Management Service, U.S. Department of the Interior.

Millar, D.L., Smith, H.C.M., and Reeve, D.E. (2007). Modeling analysis of the sensitivity of shoreline change to a wave farm. *Ocean Engineering*, 34(2007): 884–901.

Minerals Management Service. (2007). *Programmatic Environmental Impact Statement for Alternative Energy Development and Production and Alternate Uses of Facilities on the Outer Continental Shelf*, Final EIS, MMS 2007-046. Washington, DC: Minerals Management Service.

Moore, S.E. and Clarke, J.T. (2002). Potential impact of offshore human activities on gray whale (*Eschrichtius robustus*). *Journal of Cetacean Research and Management*, 4(1): 19–25.

Myers, E.P., Hoss, D.E., Peters, D.S., Matsumoto, W.M., Seki, M.P., Uchida, R.N., Ditmars, J.D., and Paddock, R.A. (1986). *The Potential Impact of Ocean Thermal Energy Conversion (OTEC) on Fisheries*, NOAA Technical Report NMFS 40. Seattle, WA: U.S. Department of Commerce.

National Marine Fisheries Service. (2003). Taking marine mammals incidental to conducting oil and gas exploration activities in the Gulf of Mexico. *Federal Register*, 68(41): 9991–9996.

National Marine Fisheries Service. (2008). Endangered and threatened wildlife and plants: proposed rule-making to designate critical habitat for the threatened southern distinct population segment of North American green sturgeon. *Federal Register*, 73(174): 52084–52110.

National Research Council. (1990). *Decline of the Sea Turtles: Causes and Prevention*. Washington, DC: National Academy Press.

National Research Council. (2000). *Marine Mammals and Low-Frequency Sound: Progress Since 1994*. Washington, DC: National Academy Press.

National Research Council. (2003). *Ocean Noise and Marine Mammals*. Washington, DC: National Academy Press.

National Research Council. (2005). *Marine Mammal Populations and Oceans Noise: Determining When Noise Causes Biologically Significant Effects*. Washington, DC: National Academy Press.

Nedwell, J.R., Edwards, B., Turnpenny, A.W.H., and Gordon, J. (2004). *Fish and Marine Mammal Audiograms: A Summary of Available Information*, Report 534R0214. Hampshire, U.K.: Subacoustech (http://www.subacoustech.com/wp-content/uploads/ 534r0214.pdf).

Nelson, P.A. (2003). Marine fish assemblages associated with fish aggregating devices: effects of fish removal, FAD size, fouling communities, and prior recruits. *Fishery Bulletin*, 101(40): 835–850.

Nelson, P.A. (2008). Ecological effects of wave energy conversion technology on California's marine and anadromous fishes. In: *Developing Wave Energy in Coastal California: Potential Socio-Economic and Environmental Effects*, Nelson, P.A. et al., Eds., pp. 100–122. Sacramento: California Energy Commission (http://www.energy.ca.gov/2008publications/CEC-500-2008-083/CEC-500-2008-083.PDF).

Newcombe, C.P. and Jensen, J.O.T. (1996). Channel suspended sediment and fisheries: a synthesis for quantitative assessment of risk and impact. *North American Journal of Fisheries Management*, 16(4): 693–727.

Nowacek, D.P., Thorne, L.H., Johnston, D.W., and Tyack, P.I. (2007). Responses of cetaceans to anthropogenic noise. *Mammal Review*, 37(2): 81–115.

Ocean Power Technologies. (2011). *Marine Energy Infrastructure*, http://www.oceanpowertechnologies.com.

Ohman, M.C., Sigray, P., and Westerberg, H. (2007). Offshore windmills and the effects of electromagnetic fields on fish. *Ambio*, 36(8): 630–633.

Pacific Fisheries Management Council. (2008). June 17, 2008, letter from D.O. McIsaac, Executive Director of PFMC, to Director Randall Luthi, Minerals Management Service, regarding Docket ID MMS-2008-OMM-0020.

Pelc, R. and Fujita, R.M. (2002). Renewable energy from the ocean. *Marine Policy*, 26: 471–479.

Petersen, I.K., Christensen, T.K., Kahlert, J., Desholm, M., and Fox, A.D. (2006). *Final Results of Bird Studies at the Offshore Wind Farms at Nysted and Horns Rev, Denmark*. Roskilde: National Environmental Research Institute of Denmark.

Phillips, O.M. (1977). *The Dynamics of the Upper Ocean*, 2nd ed. Cambridge, U.K.: Cambridge University Press.

Pinet, P.R. (1996). *Invitation to Oceanography*. St. Paul, MN: West Publishing.

Polagye, B., Malte, P., Kawase, M., and Durran, D. (2008). Effect of large-scale kinetic power extraction on time-dependent estuaries: proceedings of the Institution of Mechanical Engineers, Part A. *Journal of Power and Energy*, 222(5): 471–484.

Popper, A.N. (2003). Effects of anthropogenic sound on fishes. *Fisheries*, 28: 24–31.

Prato, T. (2003). Adaptive management of large rivers with special reference to the Missouri River. *Journal of the American Water Resources Association*, 39(4): 935–946.

Roberts, H.H., Wilson, P.A., and Lugo-Fernandez, A. (1992). Biologic and geologic responses to physical processes: examples from modern reef systems of the Caribbean–Atlantic region. *Continental Shelf Research*, 12(7/8): 809–834.

Rodil, I.F. and Lastra, M. (2004). Environmental factors affecting benthic macrofauna along a gradient of intermediate sandy beaches in northern Spain. *Estuarine, Coastal and Shelf Science*, 61: 37–44.

Rodrigue, P.R. (1986). *Cavitation Pitting Mitigation in Hydraulic Turbines*. Vol. 2. *Cavitation Review and Assessment*, EPRI AP-4719. Palo Alto, CA: Electric Power Research Institute.

Rommel, Jr., S.A. and McCleave, J.D. (1972). Oceanic electric fields: perception by American eels? *Science*, 176: 1233–1235.

Ross, D. (1995). *Power from Sea Waves*. Oxford: Oxford University Press.

Rucker, J.B. and Friedl, W.A. (1985). Potential impacts for OTEC-generated underwater sounds. *Oceans*, 17, 1279–1283.

Salter, S. (1974). Wave power. *Nature*, 249(5459): 720–724.

Shaw, R. (1982). *Wave Energy: A Design Challenge*. West Sussex, U.K.: Ellis Horwood.

Simmonds, M.P., Dolman, S.J., and Weilgart, L., Eds. (2003). *Oceans of Noise: A WDCS Science Report.* Wiltshire, U.K.: Whale and Dolphin Conservation Society (http://www.okeanos-foundation.org/assets/Uploads/OceansofNoise.pdf).

Sinclair, M. and Tremblay, M.J. (1984). Timing of spawning of Atlantic herring (*Clupea harengus harengus*) populations and the match–mismatch theory. *Canadian Journal of Fisheries and Aquatic Sciences*, 41: 1055–1065.

Smith, C.L. (1972). A spawning aggregation of Nassau grouper, *Epinephelus striatus* (Bloch). *Transactions of the American Fisheries Society*, 101(2): 257–261.

Smith, H.C.M., Millar, D.L., and Reeve, D.E. (2007). Generalization of wave farm impact assessment on inshore wave climate. In: *Proceedings of the 7th European Wave and Tidal Energy Conference*, Porto, Portugal, September 11–13.

Sorensen, B. (2000). *Renewable Energy: Its Physics, Engineering, Use, Environmental Impact, Economy and Planning Aspects.* San Diego, CA: Academic Press.

Southhall, B.L., Bowles, A.E., Ellison, W.T. et al. (2007). Marine mammal noise exposure criteria: initial scientific recommendations. *Aquatic Mammals*, 33(4): 411–521.

Sundberg, J. and Langhamer, O. (2005). Environmental questions related to point-absorbing linear wave-generators: impact, effects and fouling. In: *Proceedings of the 6th European Wave and Tidal Energy Conference*, Glasgow, Scotland, August 30–September 2.

Thompson, S.A., Castle, J., Mills, K.L., and Sydeman, W.J. (2008). Wave energy conversion technology development in coastal California: potential impacts on marine birds and mammals. In: *Developing Wave Energy in Coastal California: Potential Socio-Economic and Environmental Effects*, Nelson, P.A. et al., Eds., pp. 123–147. Sacramento: California Energy Commission (http://www.energy.ca.gov/2008publications/CEC-500-2008-083/CEC-500-2008-083.PDF).

Thomsen, F., Lüdemann, K., Kafemann, R., and Piper, W. (2006). *Effects of Offshore Wind Farm Noise on Marine Mammals and Fish.* Hamburg, Germany: Biola, on behalf of COWRIE, Ltd. (http://users.ece.utexas.edu/~ling/2A_EU3.pdf).

Tovey, N.K. (2005). *ENV-2E02 Energy Resources 2004–2005 Lecture.* Norwich, U.K.: University of East Anglia (http://www2.env.uea.ac.uk/gmmc/energy/env2e02/env2e02.htm).

USACE. (1995). *Proceedings: 1995 Turbine Passage Survival Workshop.* Portland, OR: U.S. Army Corps of Engineers.

USDOE. (2009). *Report to Congress on the Potential Environmental Effects of Marine and Hydrokinetic Energy Technologies.* Washington, DC: U.S. Department of Energy.

USDOI. (2010). *Ocean Energy.* Washington, DC: Minerals Management Service, U.S. Department of Interior (www.boemre.gov/mmsKids/PDFs/OceanEnergyMMS.pdf).

USGS. (2007). *The Water Cycle: Water Storage in Oceans.* Washington, DC: U.S. Geological Survey (http://ga.water.usgs.gov/edu/watercycleoceans.html).

Viada, S.T., Hammer, R.M., Racca, R., Hannay, D., Thompson, M.J., Balcom, B.J., and Phillips, N.W. (2008). Review of potential impacts to sea turtles from underwater explosive removal of offshore structures. *Environmental Impact Assessment Review*, 28(2008): 267–285.

Wahlberg, M. and Westerberg, H. (2005). Hearing in fish and their reactions to sounds from offshore wind farms. *Marine Ecology Progress Series*, 288: 295–309.

Walker, M.M., Quinn, T.P., Kirschvink, J.L., and Groot, C. (1988). Production of single-domain magnetite throughout life by sockeye salmon, *Onchorhynchus nerka*. *Journal of Experimental Biology*, 140: 51–63.

Walker, M.M., Diebel, C.E., Haugh, C.V., Pankhurst, P.M., Montgomery, J.C., and Green, C.R. (1997). Structure and function of the vertebrate magnetic sense. *Nature*, 390: 371–376.

Wang, D., Altar, M., and Sampson, R. (2007). An experimental investigation on cavitation, noise, and slip-stream characteristics of ocean stream turbines: proceedings of the Institution of Mechanical Engineers, Part A. *Journal of Power and Energy*, 221(2): 219–231.

Wang, J.H., Jackson, J.K., and Lohmann, K.J. (1998). Perception of wave surge motion by hatchling sea turtles. *Journal of Experimental Marine Biology and Ecology*, 229: 177–186.

Wave Dragon Wales, Ltd. (2007). *Wave Dragon Pre-Commercial Wave Energy Device, Environmental Statement*, http://www.wavedragon.co.uk/.

Wavebob. (2016). http://wavebob.com/home/.

Wavestar. (2016). http://wavestarenergy.com/.

WEC. (2010). *Survey of Energy Resources 2010.* London: World Energy Council (http://www.worldenergy.org/publications/3040.asp).

Weilgart, L.S. (2007). The impacts of anthropogenic ocean noise on cetaceans and implications for management. *Canadian Journal of Zoology*, 85: 1091–1116.

Weir, C.R. (2007). Observations of marine turtles in relation to seismic airgun sound off Angola. *Marine Turtle Newsletter*, 116: 17–20.

Westerberg, H. and Begout-Anras, M.L. (2000). Orientation of silver eel (*Anguilla anguilla*) in a disturbed geomagnetic field. In: *Advances in Fish Telemetry. Proceedings of the 3rd Conference on Fish Telemetry*, Moore, A. and Russell, I., Eds., pp. 149–158. Lowestoft, U.K.: Centre for Environment, Fisheries, and Aquaculture Studies.

Westerberg, H. and Lagenfelt, I. (2008). Sub-sea power cables and the migration behaviour of the European eel. *Fisheries Management and Ecology*, 15(5-6): 369–375.

Widdows, J. and Brinsley, M. (2002). Impact of biotic and abiotic processes on sediment dynamics and the consequences to the structure and functioning of the intertidal zone. *Journal of Sea Research*, 48: 143–156.

Wilber, D.H. and Clarke, D.G. (2001). Biological effects of suspended sediments: a review of suspended sediment impacts on fish and shellfish with relation to dredging activities in estuaries. *North American Journal of Fisheries Management*, 21(4): 855–875.

Wilber, D.H., Brostoff, W., Clarke, D.G., and Ray, G.L. (2005). *Sedimentation: Potential Biological Effects of Dredging Operations in Estuarine and Marine Environments*, ERDC TN-DOER-E20. Vicksburg, MS: US. Army Engineer Research and Development Center (http://el.erdc.usace.army.mil/elpubs/pdf/doere20.pdf).

Wilhelmsson, D. and Malm, T. (2008). Fouling assemblages on offshore wind power plants and adjacent substrata. *Estuarine, Coastal and Shelf Science*, 79(3): 459–466.

Wilhelmsson, D., Malm, T., and Ohman, M.C. (2006). The influence of offshore windpower on demersal fish. *ICES Journal of Marine Science*, 63: 775–784.

Wilkens, L.A. and Hoffman, M.H. (2005). Behavior of animals with passive, low-frequency electrosensory systems. In: *Springer Handbook of Auditory Research*. Vol. 21. *Electroreception*, Bullock, T.H., Hopkins, C.D., Popper, A.N., and Fay, R.R., Eds., pp. 229–263. New York: Springer.

Wilson, R., Batty, R.S., Daunt, F., and Carter, C. (2007). *Collision Risks Between Marine Renewable Energy Devices and Mammals, Fish and Diving Birds*, Report to the Scottish Executive. Oban, Scotland: Scottish Association for Marine Science.

Wiltschko, R. and Wiltschko, W. (1995). *Magnetic Orientation in Animals*. Berlin: Springer.

Wojtenek, W., Pei, X., and Wilkens, L.A. (2001). Paddlefish strike at artificial dipoles simulating the weak electric fields of planktonic prey. *Journal of Experimental Biology*, 204: 1391–1399.

Wood, P.J. and Armitage, P.D. (1997). Biological effects of fine sediment in the lotic environment. *Environmental Management*, 21(2): 203–217.

10 Fuel Cells

I believe fuel cell vehicles will finally end the hundred-year reign of the internal combustion engine as the dominant source of power for personal transportation. It's going to be a winning situation all the way around—consumers will get an efficient power source, communities will get zero emissions, and automakers will get another major business opportunity—a growth opportunity.

—William C. Ford, Jr., Ford chairman, International Auto Show, January 2000

I believe that water will one day be employed as a fuel, that hydrogen and oxygen which constitute it, used singly or together, will furnish an inexhaustible source of heat and light.

—Jules Verne (*Mysterious Island*, 1874)

INTRODUCTION

Depending on your education level or social, cultural, or economic background, the term *cell* may conjure up images as diverse in variety as the colors, sizes, and shapes of lightning bolts (which, by the way, are a huge source of renewable energy). Some may think of plant cells, animal cells, cell structure, cell biology, cell diagrams, cell membranes, human memory, cell theory, cell walls, cell parts, cell functions, honeycomb cells, prison cells, electrolytic cells (for producing electrolysis), aeronautic gas cells (contained in a balloon), ecclesiastical cells, or cell phones. You may have noticed that nowhere in this particular list are terms related to the subject of this book: renewable energy. Some would argue that any kind of biological cell mechanism and cell phones are types of energy devices, producers or consumers that can be renewed unless destroyed. It is not our intention to argue this point either way; instead, it is our point that an important type of cell was not included in the above list, but it should be. We are referring to the electric cell, electrochemical cell, galvanic cell, or voltaic cell—often referred to as a *battery*. This type of cell, no matter what we call it, is a device that generates electrical energy from chemical energy, usually consisting of two different conducting substances placed in an electrolyte. (Note that we could also include the solar cell in this discussion, but we discussed these earlier in the text.)

Before moving on to our basic discussion of fuel cells and their associated terminology and applications, it is important to point out that, although we do not consider fuel cells as frequently as we do other types of cells (e.g., cell phones), we predict that the day is coming when we will refer to our fuel cells just as commonly as we mention our cell phones—for it will be the fuel cell that will power our lives, just as the cell phone powers our communication.

Key Terms

To assist in understanding material presented in this chapter, the key components and operations related to fuel cells are defined in the following:

Alkaline fuel cell (AFC)—A type of hydrogen/oxygen fuel cell in which the electrolyte is concentrated potassium hydroxide (KOH) and the hydroxide ions (OH^-) are transported from the cathode to the anode.
Anion—A negatively charged ion; an ion that is attracted to the anode.

Anode—The electrode at which oxidation (a loss of electrons) takes place. For fuel cells and other galvanic cells, the anode is the negative terminal; for electrolytic cells (where electrolysis occurs), the anode is the positive terminal.

Bipolar plates—The conductive plate in a fuel cell stack that acts as an anode for one cell and a cathode for the adjacent cell. The plate may be made of metal or a conductive polymer (which may be a carbon-filled composite). The plate usually incorporates flow channels for the fluid feeds and may also contain conduits for heat transfer.

Catalyst—A chemical substance that increases the rate of a reaction without being consumed; after the reaction, it can potentially be recovered from the reaction mixture and is chemically unchanged. The catalyst lowers the activation energy required, allowing the reaction to proceed more quickly or at a lower temperature. In a fuel cell, the catalyst facilitates the reaction of oxygen and hydrogen. It is usually made of platinum powder very thinly coated onto carbon paper or cloth. The catalyst is rough and porous so the maximum surface area of the platinum can be exposed to the hydrogen or oxygen. The platinum-coated side of the catalyst faces the membrane in the fuel cell.

Catalyst poisoning—The process of impurities binding to a fuel cell's catalyst (fuel cell poisoning), lowering the catalyst's ability to facilitate the desired chemical reaction.

Cathode—The electrode at which reduction (a gain of electrons) occurs. For fuel cells and other galvanic cells, the cathode is the positive terminal, for electrolytic cells (where electrolysis occurs) the cathode is the negative terminal.

Cation—A positively charged ion.

Combustion—The burning fire produced by the proper combination of fuel, heat, and oxygen. In the engine, the rapid burning of the air–fuel mixture that occurs in the combustion chamber.

Combustion chamber—In an internal combustion engine, the space between the top of the piston and the cylinder head in which the air–fuel mixture is burned.

Composite—Material created by combining materials differing in composition or form on a macroscale to obtain specific characteristics and properties. The constituents retain their identity; they can be physically identified, and they exhibit an interface among one another.

Compressed hydrogen gas (CHG)—Hydrogen gas compressed to a high pressure and stored at ambient temperature.

Compressed natural gas (CNG)—Mixtures of hydrocarbon gases and vapors consisting principally of methane in gaseous form that has been compressed.

Compressor—A device used for increasing the pressure and density of gas.

Cryogenic liquefaction—The process through which gases such as nitrogen, hydrogen helium, and natural gas are liquefied under pressure at very low temperatures.

Current collector—The conductive material in a fuel cell that collects electrons (on the anode side) or disburses electrons (on the cathode side). Current collectors are microporous (to allow fluid to flow through them) and lie in between the catalyst/electrolyte surfaces and the bipolar plates.

Direct methanol fuel cell (DMFC)—A type of fuel cell in which the fuel is methanol (CH_3OH) in gaseous or liquid form. The methanol is oxidized directly at the anode instead of first being reformed to produce hydrogen. The electrolyte is typically a proton exchange membrane (PEM).

Dispersion—The spatial property of being scattered over an area or volume.

Electrode—A conductor through which electrons enter or leave an electrolyte. Batteries and fuel cells have a negative electrode (anode) and a positive electrode (cathode).

Electrolysis—A process that uses electricity, passing through an electrolytic solution or other appropriate medium, to cause a reaction that breaks chemical bonds (e.g., electrolysis of water to produce hydrogen and oxygen).

Electrolyte—A substance that conducts charged ions from one electrode to the other in a fuel cell, battery, or electrolyzer.

Endothermic—A chemical reaction that absorbs or requires energy (usually in the form of heat).

Energy content—Amount of energy for a given weight of fuel.

Energy density—Amount of potential energy in a given measurement of fuel.

Engine—A machine that converts heat energy into mechanical energy.

Ethanol (CH_3CH_2OH)—An alcohol containing two carbon atoms. Ethanol is a clear, colorless liquid and is the same alcohol found in beer, wine, and whiskey. Ethanol can be produced from cellulosic materials or by fermenting a sugar solution with yeast.

Exhaust emissions—Materials emitted into the atmosphere through any opening downstream of the exhaust ports of an engine, including water, particulates, and pollutants.

Exothermic—A chemical reaction that gives off heat.

Flammability limits—The flammability range of a gas is defined in terms of its lower flammability limit (LFL) and its upper flammability limit (UFL). Between the two limits is the flammable range in which the gas and air are in the proper proportions to burn when ignited. Below the lower flammability limit, there is not enough fuel to burn. Above the high flammability limit, there is not enough air to support combustion.

Flashpoint—The lowest temperature under very specific conditions at which a substance will begin to burn.

Flexible fuel vehicle—A vehicle that can operate on a wide range of fuel blends (e.g., blends of gasoline and alcohol) that can be put in the same fuel tank.

Fuel—A material used to create heat or power through conversion in such processes as combustion or electrochemistry.

Fuel cell—A device that produces electricity through an electrochemical process, usually from hydrogen and oxygen.

Fuel cell poisoning—The lowering of a fuel cell's efficiency due to impurities in the fuel binding to the catalyst.

Fuel cell stack—Individual fuel cells connected in a series. Fuel cells are stacked to increase voltage.

Fuel processor—Device used to generate hydrogen from fuels such as natural gas, propane, gasoline, methanol, and ethanol for use in fuel cells.

Gas—Fuel gas such as natural gas, undiluted liquefied petroleum gases (vapor phase only), liquefied petroleum gas–air mixtures, or mixtures of these gases.

- *Natural gas*—Mixtures of hydrocarbon gases and vapors consisting principally of methane (CH_4) in gaseous form.
- *Liquefied petroleum gas (LPG)*—Any material composed predominantly of any of the following hydrocarbons or mixtures of them: propane, propylene, butanes (normal butane or isobutane), and butylenes.
- *Liquefied petroleum gas–air mixture*—Liquefied petroleum gases distributed at relatively low pressures and normal atmospheric temperatures that have been diluted with air to produce the desired heating value and utilization characteristics.

Gas diffusion—Mixing of two gases caused by random molecular motions. Gases diffuse very quickly, liquids diffuse much more slowly, and solids diffuse at very slow (but often measurable) rates. Molecular collisions make diffusion slower in liquids and solids.

Graphite—Mineral consisting of a form of carbon that is soft, black, and lustrous and has a greasy feeling. Graphite is used in pencils, crucibles, lubricants, paints, and polishers.

Gravimetric energy density—Potential energy in a given weight of fuel.

Greenhouse effect—Warming of the Earth's atmosphere due to gases in the atmosphere that allow solar radiation (visible, ultraviolet) to reach the Earth's atmosphere but do not allow the emitted infrared radiation to pass back out of the Earth's atmosphere.

Greenhouse gas (GHG)—Gases in the Earth's atmosphere that contribute to the greenhouse effect.

Heat exchanger—Device (e.g., a radiator) designed to transfer heat from the hot coolant that flows through it to the air blown through it by the fan.

Heating value (total)—The number of British thermal units (Btu) produced by the combustion of 1 cubic foot of gas at constant pressure when the products of combustion are cooled to the initial temperature of the gas and air, when the water vapor formed during combustion is condensed, and when all of the necessary corrections have been applied.

- *Lower (LHV)*—The value of the heat of combustion of a fuel measured by allowing all products of combustion to remain in the gaseous state. This method of measure does not take into account the heat energy put into the vaporization of water (heat of vaporization).
- *Higher (HHV)*—The value of the heat of combustion of a fuel measured by reducing all of the products of combustion back to their original temperature and condensing all water vapor formed by combustion. This value takes into account the heat of vaporization of water.

Hybrid electric vehicle (HEV)—A vehicle combining a battery-powered electric motor with a traditional internal combustion engine. The vehicle can run on either the battery or the engine or both simultaneously, depending on the performance objectives for the vehicle.

Hydrides—Chemical compounds formed when hydrogen gas reacts with metals; used for storing hydrogen gas.

Hydrocarbon (HC)—An organic compound containing carbon and hydrogen, usually derived for fossil fuels, such as petroleum, natural gas, and coal.

Hydrogen (H_2)—Hydrogen (H) is the most abundant element in the universe, but it is generally bonded to another element. Hydrogen gas (H_2) is a diatomic gas composed of two hydrogen atoms and is colorless and odorless. Hydrogen is flammable when mixed with oxygen over a wide range of concentrations.

Hydrogen-rich fuel—A fuel that contains a significant amount of hydrogen, such as gasoline, diesel fuel, methanol (CH_3OH), ethanol (CH_3CH_2OH), natural gas, and coal.

Impurities—Undesirable foreign material(s) in a pure substance or mixture.

Internal combustion engine (ICE)—An engine that converts the energy contained in a fuel inside the engine into motion by combusting the fuel. Combustion engines use the pressure created by the expansion of combustion product gases to do mechanical work.

Liquefied hydrogen (LH_2)—Hydrogen in liquid form. Hydrogen can exist in a liquid state but only at extremely cold temperatures. Liquid hydrogen typically has to be stored at –253°C (–423°F). The temperature requirements for liquid hydrogen storage necessitate expending energy to compress and chill the hydrogen into its liquid state.

Liquefied natural gas (LNG)—Natural gas that is in liquid form at –162°C (–259°F) at ambient pressure.

Liquefied petroleum gas (LPG)—Any material that consists predominantly of any of the following hydrocarbons or mixtures of hydrocarbons: propane, propylene, normal butane, isobutylene, and butylenes. LPG is usually stored under pressure to maintain the mixture in the liquid state.

Liquid—A substance that, unlike a solid, flows readily but, unlike a gas, does not tend to expand indefinitely.

Membrane—The separating layer in a fuel cell that acts as electrolyte (an ion-exchanger) as well as a barrier film separating the gases in the anode and cathode compartments of the fuel cell.

Methanol (CH_3OH)—An alcohol containing one carbon atom. It has been used, together with some of the higher alcohols, as a high-octane gasoline component and is a useful automotive fuel.

Miles per gallon equivalent (MPGe)—Energy content equivalent to that of a gallon (114,320 Btu).

Molten carbonate fuel cell (MCFC)—A type of fuel cell that contains a molten carbonate electrolyte. Carbonate ions (CC_3^{-2}) are transported from the cathode to the anode. Operating temperatures are typically near 650°C.

Naflon®—Sulfonic acid in a solid polymer form that is usually the electrolyte of polymer electrolyte membrane (PEM) fuel cells.

Natural gas—A naturally occurring gaseous mixture of simple hydrocarbon components (primarily methane) used as a fuel.

Nitrogen (N_2)—A diatomic colorless, tasteless, odorless gas that constitutes 78% of the atmosphere by volume.

Nitrogen oxides (NO_x)—Any chemical compound of nitrogen and oxygen. Nitrogen oxides result from the high temperatures and pressure in the combustion chambers of automotive engines and other power plants during the combustion process. When combined with hydrocarbons in the presence of sunlight, nitrogen oxides form smog. Nitrogen oxides are basic air pollutants; automotive exhaust emission levels of nitrogen oxides are regulated by law.

Oxidant—A chemical, such as oxygen, that consumes electrons in an electrochemical reaction.

Oxidation—Loss of one or more electrons by an atom, molecule, or ion.

Oxygen (O_2)—A diatomic colorless, tasteless, odorless gas that makes up about 21% of air.

Partial oxidation—Fuel reforming reaction where the fuel is oxidized partially to carbon monoxide and hydrogen rather than fully oxidized to carbon dioxide and water. This is accomplished by injecting air with the fuel stream prior to the reformer. The advantage of partial oxidation over steam reforming of the fuel is that it is an exothermic reaction rather than an endothermic reaction and therefore generates its own heat.

Pascal (Pa)—The International System of Units (SI)-derived unit of pressure or stress. It is a measure of perpendicular force per unit area. The Pascal is equivalent to 1 newton per square meter.

Permeability—Ability of a membrane or other material to permit a substance to pass through it.

Phosphoric acid fuel cell (PAFC)—A type of fuel cell in which the electrolyte consists of concentrated phosphoric and (H_3PO_4). Protons (H^+) are transported from the anode to the cathode. The operating temperature range is generally 160 to 220°C.

Polymer—Natural or synthetic compound composed of repeated links of simple molecules.

Polymer electrolyte membrane (PEM)—A fuel cell incorporating a solid polymer membrane used as its electrolyte. Protons (H^+) are transported from the anode to the cathode. The operating temperature range is generally 60 to 100°C.

Polymer electrolyte membrane fuel cell (PEMFC or PEFC)—A type of acid-based fuel cell in which the transport of protons (H^+) from the anode to the cathode is through a solid, aqueous membrane impregnated with an appropriate acid. The electrolyte is a called a polymer electrolyte membrane (PEM). The fuel cells typically run at low temperatures (<100°C).

Reactant—A chemical substance that is present at the start of a chemical reaction.

Reactor—Device or process vessel in which chemical reactions (e.g., catalysis in fuel cells) take place.

Reformate—Hydrocarbon fuel that has been processed into hydrogen and other products for use in fuel cells.

Reformer—Device used to generate hydrogen from fuels such as natural gas, propane, gasoline methanol, and ethanol for use in fuel cells.

Reforming—A chemical process in which hydrogen-containing fuels react with steam, oxygen, or both to produce a hydrogen-rich gas stream.

Reformulated gasoline—Gasoline that is blended so that, on average, it reduces volatile organic compounds and air toxics emissions significantly relative to conventional gasolines.

Regenerative fuel cell—A fuel cell that produces electricity from hydrogen and oxygen and can use electricity from solar power or some other source to divide the excess water into oxygen and hydrogen fuel to be reused by the fuel cell.

Solid oxide fuel cell (SOFC)—A type of fuel cell in which the electrolyte is a solid, nonporous metal oxide, typically zirconium oxide (ZrO_2) treated with Y_2O_3 and O^{-2}, and is transported from the cathode to the anode. Any CO in the reformate gas is oxidized to CO_2 at the anode. Temperatures of operation are typically 800 to 1000°C.

Sorbent—Material that sorbs another (i.e., has the capacity or tendency to take it up either by adsorption or absorption).

Sorption—Process by which one substance takes up or holds another.

Steam reforming—The process for reacting a hydrocarbon fuel, such as natural gas, with steam to produce hydrogen as a product. This is a common method for bulk hydrogen generation.

Turbine—Machine for generating rotary mechanical power from the energy in a stream of fluid. The energy, originally in the form of head or pressure energy, is converted to velocity energy by passing through a system of stationary and moving blades in the turbine.

Turbocharger—A device used for increasing the pressure and density of a fluid entering a fuel cell power plant using a compressor driven by a turbine that extracts energy from the exhaust gas.

Turbocompressor—Machine for compressing air or other fluids (reactant if supplied to a fuel cell system) in order to increase the reactant pressure and concentration.

Volumetric energy density—Potential energy in a given volume of fuel.

Watt (W)—A unit of power equal to 1 Joule of work performed per second: 746 watts is the equivalent of 1 horsepower. The watt is named for James Watt, a Scottish engineer (1736–1819) and pioneer in steam engine design.

Weight percent (wt%)—Term that is widely used in hydrogen storage research to denote the amount of hydrogen stored on a weight basis; the term *mass percent* is also occasionally used. The term can be used for materials that store hydrogen or for the entire storage system (e.g., material or compressed/liquid hydrogen as well as the tank and other equipment required to contain the hydrogen such as insulation, valves, or regulators). For example, 6 wt% on a system basis means that 6% of the entire system by weight is hydrogen. On a material basis, the wt% is the mass of hydrogen divided by the mass of material plus hydrogen.

HYDROGEN PRODUCTION AND DELIVERY[*]

Containing only one electron and one proton, hydrogen, chemical symbol H, is the simplest element on earth. Hydrogen as is a diatomic molecule—each molecule has two atoms of hydrogen (which is why pure hydrogen is commonly expressed as H_2). Although abundant on Earth as an element, hydrogen combines readily with other elements and is almost always found as part of another substance, such as water hydrocarbons or alcohols. Hydrogen is also found in biomass, which includes all plants and animals.

- Hydrogen is an energy carrier, not an energy source. Hydrogen can store and deliver usable energy, but it does not typically exist by itself in nature; it must be produced from compounds that contain it. Its production is not inexpensive.
- Hydrogen can be produced using diverse, domestic resources including nuclear energy, natural gas and coal, and biomass and other renewables, including solar, wind, hydroelectric, and geothermal energy. This diversity of domestic energy sources makes hydrogen a

[*] Adapted from EERE, *Hydrogen Production*, Office of Energy Efficiency & Renewable Energy, U.S. Department of Energy, Washington, DC, 2016 (http://www1.eere.energy.gov/hydrogenandfuelcells/production/basics.html).

promising energy carrier and important to our nation's energy security. It is expected and desirable for hydrogen to be produced using a variety of resources and process technologies (or pathways).

- The U.S. Department of Energy (USDOE) focuses on hydrogen-production technologies that result in near-zero net greenhouse gas emissions and use renewable energy sources, nuclear energy, and coal (when combined with carbon sequestration). To ensure sufficient clean energy for our overall energy needs, energy efficiency is also important.
- Hydrogen can be produced via various process technologies, including thermal (natural gas reforming, renewable liquid and bio-oil processing, and biomass and coal gasification), electrolytic (water splitting using a variety of energy resources), and photolytic (splitting water using sunlight via biological and electrochemical materials).
- Hydrogen can be produced in large, central facilities (50 to 300 miles from point of use) or in smaller semi-central facilities (located within 25 to 100 miles of use), or it can be distributed (near or at the point of use).
- To be successful in the marketplace, hydrogen must be cost competitive with the available alternatives. In the light-duty vehicle transportation market, this competitive requirement means that hydrogen needs to be available untaxed at $2 to $3 per gasoline gallon equivalent (GGE). This price would result in hydrogen fuel cell vehicles having the same cost to the consumer on a cost-per-mile-driven basis as a comparable conventional internal-combustion engine or hybrid vehicle.
- The USDOE is engaged in the research and development of a variety of hydrogen production technologies. Some are further along in development than others—some can be cost competitive for the transition period (beginning in 2015), and others are considered long-term technologies (cost-competitive after 2030).

Infrastructure is required to move hydrogen from the location where it is produced to the dispenser at a refueling station or stationary power site. Infrastructure includes the pipelines, trucks, railcars, ships, and barges that deliver fuel, as well as the facilities and equipment required to load and unload them. Delivery technology for a hydrogen infrastructure is currently available commercially, and several U.S. companies are delivering bulk hydrogen today. Some of the infrastructure is already in place because hydrogen has long been used in industrial applications, but it is not sufficient to support widespread consumer use of hydrogen as an energy carrier. Because hydrogen has a relatively low volumetric energy density, its transportation, storage, and final delivery to the point of use represent significant costs and result in some of the energy inefficiencies associated with using it as an energy carrier.

Options and trade-offs for hydrogen delivery from the production facilities to the point of use are complex. The choice of a hydrogen production strategy greatly affects the cost and method of delivery; for example, larger, centralized facilities can produce hydrogen at relatively low cost due to economies of scale, but the delivery costs are higher than for smaller, localized production facilities. Although these smaller production facilities would have relatively lower delivery costs, their hydrogen production costs are likely to be higher, as lower volume production means higher equipment costs on a per-unit-of-hydrogen basis.

Key challenges to hydrogen delivery include reducing delivery costs, increasing energy efficiency, maintaining hydrogen purity, and minimizing hydrogen leakage. Further research is needed to analyze the trade-offs between the hydrogen production options and the hydrogen delivery options taken together as a system. Building a national hydrogen delivery infrastructure is a big challenge. Such an infrastructure will take time to develop and will likely include combinations of various technologies. Delivery infrastructure needs and resources will vary by region and type of market (e.g., urban, interstate, or rural). Infrastructure options will evolve as the demand for hydrogen grows and as delivery technologies develop and improve.

HYDROGEN STORAGE

Storing enough hydrogen on board a vehicle to achieve a driving range of greater than 300 miles is a significant challenge. On a weight basis, hydrogen has nearly three times the energy content of gasoline (120 MJ/kg for hydrogen vs. 44 MJ/kg for gasoline); however, on a volume basis, the situation is reversed (8 MJ/L for liquid hydrogen vs. 32 MJ/L for gasoline). On-board hydrogen storage in the range of 5 to 13 kg is required to encompass the full platform of light-duty vehicles. Hydrogen can be stored as either a gas or a liquid. Storage as a gas typically requires high-pressure tanks (5000- to 10,000-psi tank pressure). Storage of hydrogen as a liquid requires cryogenic temperatures, because the boiling point of hydrogen at 1 atmosphere pressure is –252.8°C. Hydrogen can also be stored on the surfaces of solids (by adsorption) or within solids (by absorption). In adsorption, hydrogen is attached to the surface of a material as either hydrogen molecules or hydrogen atoms. In absorption, hydrogen is dissociated into H atoms, and then the hydrogen atoms are incorporated into the solid lattice framework. Hydrogen storage in solids may make it possible to store large quantities of hydrogen in smaller volumes at low pressures and at temperatures close to room temperature. It is also possible to achieve volumetric storage densities greater than those of liquid hydrogen because the hydrogen molecule is dissociated into atomic hydrogen within a metal hydride lattice structure. Finally, hydrogen can be stored through the reaction of hydrogen-containing materials with water (or other compounds such as alcohols). In this case, the hydrogen is effectively stored in both the material and the water. The terms *chemical hydrogen storage* and *chemical hydride* are used to describe this form of hydrogen storage. It is also possible to store hydrogen in the chemical structures of liquids and solids.

HOW A HYDROGEN FUEL CELL WORKS

A fuel cell uses the chemical energy of hydrogen to cleanly and efficiently produce electricity. Fuel cells have a variety of potential applications; for example, they can provide energy for systems as large as utility power stations or as small as laptop computers. Fuel cells offer several benefits over conventional combustion-based technologies currently used in many power plants and passenger vehicles. They produce much smaller quantities of greenhouse gases and none of the air pollutants that create smog and cause health problems. If pure hydrogen is used as a fuel, fuel cells emit only heat and water as byproducts. A hydrogen fuel cell is a device that uses hydrogen (or hydrogen-rich fuel) and oxygen to create electricity through an electrochemical process. A single fuel cell consists of an electrolyte and two catalyst-coated electrodes (a porous anode and a cathode). The various types of fuel cells all work similarly:

- Hydrogen (or hydrogen-rich fuel) is fed to an anode, where a catalyst separates the negatively charged hydrogen electrons from positively charged ions.
- At the cathode, oxygen combines with electrons and, in some cases, with species such as protons or water, resulting in water or hydroxide ions, respectively.
- For polymer electrolyte membrane and phosphoric acid fuel cells, protons move through the electrolyte to the cathode, where they combine with oxygen and electrons to produce water and heat.

DID YOU KNOW?

Hydrogen fuel cell vehicles (FCVs) emit approximately the same amount of water per mile as vehicles using gasoline-powered internal combustion engines (ICEs).

- For alkaline, molten carbonate, and solid oxide fuel cells, negative ions travel through the electrolyte to the anode, where they combine with hydrogen to generate water and electrons.
- The electrons from the anode cannot pass through the electrolyte to the positively charged cathode; they must travel around it via an electrical circuit to reach the other side of the cell. This movement of electrons is an electrical current.

PROPERTIES OF HYDROGEN

Hydrogen has unique physical and chemical properties that present benefits and challenges to its successful widespread adoption as a fuel. Hydrogen is the lightest and smallest element in the universe. Hydrogen is 14 times lighter than air and rises at a speed of almost 20 m/s, 6 times faster than natural gas, which means that when released it rises and disperses quickly. Hydrogen is also odorless, colorless, and tasteless, making it undetectable by human senses. For these reasons, hydrogen systems are designed with ventilation and leak detection. Natural gas is also odorless, colorless, and tasteless, but a sulfur-containing odorant is added so people can detect it. There is no known odorant light enough to travel with hydrogen at an equal dispersion rate, so odorants are not used to provide a detection method. Many odorants can also contaminate fuel cells.

OTHER TYPES OF FUEL CELLS[*]

In addition to hydrogen fuel cells, there are other types of fuel cells. These other types are classified primarily by the kind of electrolyte they employ. The classification determines the kind of electrochemical reactions that take place in the cell, the kind of catalysts required, the temperature range in which the cell operates, the fuel required, and other factors. These characteristics, in turn, affect the applications for which these cells are most suitable. There are several types of fuel cells currently under development; those with the most promise, at the present time, are discussed below.

POLYMER ELECTROLYTE MEMBRANE FUEL CELLS

Polymer electrolyte membrane (PEM) fuel cells—also called proton exchange membrane fuel cells—deliver high power density and offer the advantages of low weight and volume compared with other fuel cells. PEM fuel cells use a solid polymer as an electrolyte and porous carbon electrodes containing a platinum or platinum alloy catalyst. They require only hydrogen, oxygen from the air, and water to operate. They are typically fueled with pure hydrogen supplied from storage tanks or reformers. PEM fuel cells operate at relatively low temperatures, around 80°C (176°F). Low-temperature operation allows them to start quickly (less warm-up time, which makes PEM fuel cells particularly suitable for use in passenger vehicles) and results in less wear on system components, resulting in better durability. However, the use of a noble-metal catalyst (typically platinum) is necessary to separate the hydrogen's electrons and protons, adding to system cost. The platinum catalyst is also extremely sensitive to carbon monoxide poisoning, making it necessary to employ an additional reactor to reduce carbon monoxide in the fuel gas if the hydrogen is derived from a hydrocarbon fuel. This reactor also adds cost. PEM fuel cells are used primarily for some stationary applications and transportation applications.

---

[*] Adapted from USDOE, *Types of Fuel Cells*, U.S. Department of Energy, Washington, DC, 2016 (http://energy.gov/eere/fuelcells/types-fuel-cells).

DIRECT METHANOL FUEL CELLS

Most fuel cells are powered by hydrogen, which can be fed to the fuel cell system directly or can be generated within the fuel cell system by reforming hydrogen-rich fuels such as methanol, ethanol, and hydrocarbon fuels. Direct methanol fuel cells (DMFCs), however, are powered by pure methanol, which is usually mixed with water and fed directly to the fuel cell anode. Direct methanol fuel cells do not have many of the fuel storage problems typical of some fuel cell systems because methanol has a higher energy density than hydrogen—though less than gasoline or diesel fuel. Methanol is also easier to transport and supply to the public using our current infrastructure because it is a liquid, like gasoline. DMFCs are often used to provide power for portable fuel cell applications such as cell phones or laptop computers.

ALKALINE FUEL CELLS

Alkaline fuel cells (AFCs) were one of the first fuel cell technologies developed, and they were the first type widely used in the U.S. space program to produce electrical energy and water onboard spacecraft. These fuel cells use a solution of potassium hydroxide in water as the electrolyte and can utilize a variety of non-precious metals as a catalyst at the anode and cathode. High-temperature AFCs operate at temperatures between 100 and 250°C (212 and 482°F); however, newer AFC designs operate at lower temperatures of roughly 23 to 70°C (74 to 158°F). In recent years, novel AFCs that use a polymer membrane as the electrolyte have been developed. These fuel cells are closely related to conventional PEM fuel cells, except that they use an alkaline membrane instead of an acid membrane. The high performance of AFCs is due to the rate at which electrochemical reactions take place in the cell. They have also demonstrated efficiencies above 60% in space applications.

The disadvantage of this fuel cell type is that it is easily poisoned by carbon dioxide (CO_2). In fact, even the small amount of CO_2 in the air can affect this cell's operation, making it necessary to purify both the hydrogen and oxygen used in the cell. This purification process is costly. Susceptibility to poisoning also affects the cell's lifetime (the amount of time before it must be replaced), further adding to cost. Alkaline membrane cells have lower susceptibility to CO_2 poisoning than liquid-electrolyte AFCs do, but performance still suffers as a result of CO_2 that dissolves into the membrane. Cost is less of a factor for remote locations, such as in space or under the sea; however, to compete effectively in most mainstream commercial markets, these fuel cells will have to become more cost effective. To be economically viable in large-scale utility applications, AFCs need to reach operating times exceeding 40,000 hours, something that has not yet been achieved due to material durability issues. This obstacle is possibly the most significant in commercializing this fuel cell technology.

PHOSPHORIC ACID FUEL CELLS

Phosphoric acid fuel cells (PAFCs) use liquid phosphoric acid as an electrolyte (the acid is contained in a Teflon®-bonded silicon carbide matrix) and porous carbon electrodes containing a platinum catalyst. The PAFC is considered the "first generation" of modern fuel cells. It is one of the most mature cell types and the first to be used commercially. This type of fuel cell is typically used for stationary power generation, but some PAFCs have been used to power large vehicles such as city buses. PAFCs are more tolerant of impurities in fossil fuels that have been reformed into hydrogen than PEM cells, which are easily poisoned by carbon monoxide because carbon monoxide binds to the platinum catalyst at the anode, decreasing the fuel cell's efficiency. PAFCs are more than 85%

efficient when used for the cogeneration of electricity and heat but they are less efficient at generating electricity alone (37 to 42%). PAFC efficiency is only slightly more than that of combustion-based power plants, which typically operate at around 33% efficiency. PAFCs are also less powerful than other fuel cells, given the same weight and volume. As a result, these fuel cells are typically large and heavy. PAFCs are also expensive. They require much higher loadings of expensive platinum catalyst than other types of fuel cells do, which raises the cost.

MOLTEN CARBONATE FUEL CELLS

Molten carbonate fuel cells (MCFCs) are currently being developed for natural gas and coal-based power plants for electrical utility, industrial, and military applications. MCFCs are high-temperature fuel cells that use an electrolyte composed of a molten carbonate salt mixture suspended in a porous, chemically inert ceramic matrix of lithium aluminum oxide. Because MCFCs operate at high temperatures of 650°C (roughly 1200°F), non-precious metals can be used as catalysts at the anode and cathode, reducing costs. Improved efficiency is another reason MCFCs offer significant cost reductions over phosphoric acid fuel cells. Molten carbonate fuel cells, when coupled with a turbine, can reach efficiencies approaching 65%, considerably higher than the 37 to 42% efficiencies of a phosphoric acid fuel cell plant. When the water heat is captured and used, overall fuel efficiencies can be over 85%. Unlike alkaline, phosphoric acid, and PEM fuel cells, MCFSs do not require an external reformer to convert fuels such as natural gas and biogas to hydrogen. At the high temperatures at which MCFCs operate, methane and other light hydrocarbons in these fuels are converted to hydrogen within the fuel cell itself by a process called *internal reforming*, which also reduces costs. The primary disadvantage of current MCFC technology is durability. The high temperatures at which these cells operate and the corrosive electrolyte used accelerate component breakdown and corrosion, decreasing cell life. Scientists are currently exploring corrosion-resistant materials for components as well as fuel cell designs that double cell life from the current 40,000 hours (~5 years) without decreasing performance.

SOLID OXIDE FUEL CELLS

Solid oxide fuel cells (SOFCs) use a hard, nonporous ceramic compound as the electrolyte. SOFCs are around 60% efficient at converting fuel to electricity. In applications designed to capture and utilize the system's waste heat (cogeneration), overall fuel use efficiencies could top 85%. SOFCs operate at very high temperatures—as high as 1000°C (1830°F). High-temperature operation removes the need for precious-metal catalysts, thereby reducing cost. It also allows SOFCs to reform fuels internally, which enables the use of a variety of fuels and reduces the cost associated with adding a reformer to the system. SOFCs are also the most sulfur-resistant fuel cell type; they can tolerate several orders of magnitude more sulfur than other cell types can. In addition, they are not poisoned by carbon monoxide, which can even be used as fuel. This property allows SOFCs to use natural gas, biogas, and gases made from coal. High-temperature operation has disadvantages. It results in a slow start-up and requires significant thermal shielding to retain heat and protect personnel, which may be acceptable for utility applications but not for transportation. The high operating temperatures also place stringent durability requirements on materials. The development of low-cost materials with high durability at cell operating temperatures is the key technical challenge facing this technology. Scientists are currently exploring the potential for developing lower temperature SOFCs operating at or below 700°C that have fewer durability problems and cost less. Lower temperature SOFCs have not yet matched the performance of the higher temperature systems, however, and stack materials that will function in this lower temperature range are still under development.

REVERSIBLE FUEL CELLS

Reversible fuel cells produce electricity from hydrogen and oxygen and generate heat and water as byproducts, just like other fuel cells. However, reversible fuel cell systems can also use electricity from solar power, wind power, or other sources to split water into oxygen and hydrogen fuel through a process called *electrolysis*. Reversible fuel cells can provide power when needed, but during times of high power production from other technologies (such as when high winds lead to an excess of available wind power), reversible fuel cells can store the excess energy in the form of hydrogen. This energy storage capability could be a key enabler for intermittent renewable energy technologies.

ENVIRONMENTAL IMPACTS OF FUEL CELLS

Beyond the expectation that hydrogen leakage from its use in fuel cells could greatly impact the hydrogen cycle and could, when oxidized in the stratosphere, cool the stratosphere and create more clouds, delaying the break-up of the polar vortex at the poles and making the holes in the ozone layer larger and longer lasting, little is understood about how hydrogen leakage would affect the environment. For example, much uncertainty exists over the extent of the impact of hydrogen emissions on soil absorption of hydrogen from the atmosphere. This concept is important because if we use extensive quantities of hydrogen for fuel cells, absorption of hydrogen by soils could have a compensatory effect on any possible anthropogenic emissions. Again, it is important to emphasize that little is understood about how hydrogen leakage would affect the environment.

DISCUSSION AND REVIEW QUESTION

1. Are hydrogen fuel cells a viable energy option? Explain.

REFERENCES AND RECOMMENDED READING

EERE. (2016a). *Hydrogen Production*. Washington, DC: Office of Energy Efficiency & Renewable Energy, U.S. Department of Energy (http://www1.eere.energy.gov/hydrogenandfuelcells/production/basics. html).
EERE. (2016b). *Types of Fuel Cells*. Washington, DC: Office of Energy Efficiency & Renewable Energy, U.S. Department of Energy (http://energy.gov/eere/fuelcells/types-fuel-cells).
NREL. (2009). *Ultracapacitors*. Washington, DC: National Renewable Energy Laboratory (http://www.nrel. gov/vehiclesandfuels/energystorage/ultracapacitors.html?print).

11 Carbon Capture and Sequestration

You cannot get through a single day without having an impact on the world around you—for better or for worse. What you do makes a difference.

—Jane Goodall

Human activities, especially the burning of fossil fuels such as coal, oil, and gas, have caused a substantial increase in the concentration of carbon dioxide (CO_2) in the atmosphere. This increase in atmospheric CO_2—from about 280 to more than 380 parts per million (ppm) over the last 250 years—is causing measureable global warming. Potential adverse impacts include sea-level rise; increased frequency and intensity of wildfires, floods, droughts, and tropical storms; changes in the amount, timing, and distribution of rain, snow, and runoff; and disturbance of coastal marine and other ecosystems. Rising atmospheric CO_2 is also increasing the absorption of CO_2 by seawater, causing the ocean to become more acidic, with potentially disruptive effects on marine plankton and coral reefs. Technically and economically feasible strategies are needed to mitigate the consequences of increased atmospheric CO_2.

USGS (2008)

Above all we should, in the century since Darwin, have come to know that man, while captain of the adventuring ship, is hardly the sole object of its quest, and that his prior assumptions to this effect arose from the simple necessity of whistling in the dark. These things, I say, should have come to us. I fear they have not come to many.

—Aldo Leopold (*A Sand County Almanac, and Sketches Here and There,* 1948)

INTRODUCTION TO CARBON CAPTURE AND SEQUESTRATION

The reader might wonder what carbon capture and sequestration (CCS) might have to do with the environmental impacts of renewable energy. Renewable energy has two pluses: (1) it is a possible source of energy now and in the future (it is renewable and sustainable) and will be called on to replace nonrenewable hydrocarbon energy sources as they are depleted; and (2) renewable energy produces little or no waste products such as carbon dioxide or other chemical pollutants, so it has minimal impact on the environment. It is the latter of these two pluses that is related to carbon capture and sequestration. That is, at the present time and in the near future we have (and will continue to have) an ongoing increase in atmospheric carbon dioxide. Many scientists agree that global climate change is occurring and that to prevent its most serious effects we must begin immediately to significantly reduce our greenhouse gas (GHG) emissions. One major contributor to climate change is the release of the greenhouse gas carbon dioxide (CO_2). This is the essence of the carbon capture and sequestration process: capturing and sequestering carbon dioxide. Further, to control atmospheric carbon dioxide requires deliberate mitigation with an approach that combines reducing emissions by utilizing renewable sources and by increasing capture and storage.

The term *carbon sequestration* is used to describe both natural and deliberate processes by which CO_2 is either removed from the atmosphere or diverted from emission sources and stored in the ocean, terrestrial environments (vegetation, soils, and sediments), and geologic formations. Before human-caused CO_2 emissions began to occur, the natural processes that make up the global carbon cycle (see Figure 11.1) maintained a near balance between the uptake of CO_2 and its release back to

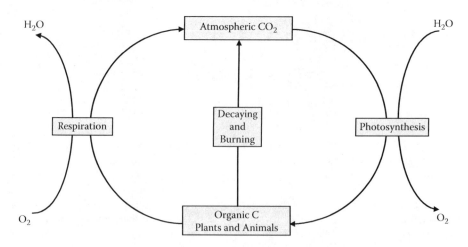

FIGURE 11.1 The global carbon cycle. Carbon naturally moves, or cycles, between the atmosphere and vegetation, soils, and the oceans over time scales ranging from years to millennia and longer. Human activities, primarily the burning of fossil fuels and clearing of forests, have increased the transfer of carbon as CO_2 to the atmosphere. Although some of the anthropogenic CO_2 is removed from the atmosphere by the natural uptake processes ("sinks") of the carbon cycle, much of it remains in the atmosphere and causes rising CO_2 concentrations. The goal of deliberate carbon sequestration is to decrease the net flux of CO_2 to the atmosphere by sequestering carbon in the oceans, vegetation, soils, and porous rock formations. (From USGS, *Carbon Sequestration to Mitigate Climate Change*, U.S. Geological Survey, Washington, DC, 2008.)

the atmosphere. However, existing CO_2 uptake mechanisms (sometimes called CO_2 or carbon sinks) are insufficient to offset the accelerating pace of emissions related to human activities. Annual carbon emissions from burning fossil fuels in the United States are about 1.6 gigatons (billion metric tons), whereas annual uptake amounts are only about 0.5 gigatons, resulting in a net release of about 1.1 gigatons per year (see Figure 11.2).

Scientists at the U.S. Geological Survey (USGS) and elsewhere are working to assess both the potential capacities and the potential limitations of the various forms of carbon sequestration and to evaluate their geologic, hydrologic, and ecological consequences. The USGS is providing information needed by decision makers and resource managers to maximize carbon storage while minimizing undesirable impacts on humans and their physical and biological environment.

DID YOU KNOW?

The world's oceans are the primary long-term sink for human-caused CO_2 emissions, currently accounting for a global net uptake of about 2 gigatons of carbon annually. This uptake is not a result of deliberate sequestration but instead occurs naturally through chemical reactions between seawater and CO_2 in the atmosphere. The absorption of atmospheric CO_2 causes the oceans to become more acidic. Ocean acidification (OA) is the term given to the chemical changes in the ocean as a result of carbon dioxide emissions. Many marine organisms and ecosystems depend on the formation of carbonate skeletons and sediments that are vulnerable to dissolution in acidic waters (see Figure 11.3). Laboratory and field measurements indicate that CO_2-induced acidification may eventually cause the rate of dissolution of carbonate to exceed its rate of formation in these ecosystems. The impacts of ocean acidification and deliberate ocean fertilization on coastal and marine food webs and other resources are poorly understood. Scientists are studying the effects of oceanic carbon sequestration on these important environments.

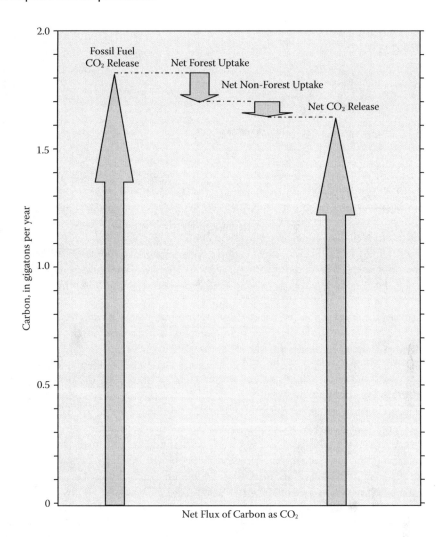

FIGURE 11.2 Estimated annual net CO_2 emissions and uptake in the United States in 2003, according to the U.S. Climate Change Science Program. U.S. fossil fuel CO_2 emissions accounted for more than 20% of the global total in 2003. Net uptake fluxes (photosynthesis minus oxidation) are shown for forest and non-forest areas. The largest net U.S. CO_2 uptake was associated with regrowing forests and harvesting wood products. Non-forest net carbon uptake—including accumulation in shrubs, agricultural soils, wetlands, rivers, and reservoirs—was smaller and more uncertain. (From USGS, *Carbon Sequestration to Mitigate Climate Change*, U.S. Geological Survey, Washington, DC, 2008.)

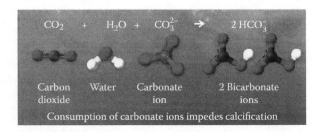

FIGURE 11.3 Ocean acidification. (From NOAA, *Ocean Acidification: The Other Carbon Dioxide Problem*, National Oceanic and Atmospheric Administration, Washington, DC, 2016, http://pmel.noaa.gov/co2/story/Ocean+Acidification.)

DID YOU KNOW?

The *carbon footprint* is the total amount of greenhouse gases that are emitted into the atmosphere each year by a person, family, building, organization, or company.

TERRESTRIAL CARBON SEQUESTRATION

Terrestrial sequestration (sometimes termed *biological sequestration*) is the removal of gaseous carbon dioxide from the atmosphere and binding it in living tissue by plants. Terrestrial sequestration is typically accomplished through forest and soil conservation practices that enhance the storage of carbon (such as restoring and establishing new forests, wetlands, and grasslands) or reduce CO_2 emissions (such as reducing agricultural tillage and suppressing wildfires). In the United States, these practices are implemented to meet a variety of land-management objectives (see Sidebar 11.1). Although the net terrestrial uptake fluxes shown in Figure 11.2 offset about 30% of U.S. fossil fuel CO_2 emissions, only a small fraction of this uptake results from activities undertaken specifically to sequester carbon. The largest net uptake is due primarily to ongoing natural regrowth of forests that were harvested during the 19th and early 20th centuries.

Existing terrestrial carbon storage is susceptible to disturbances such as fire, disease, and change in climate and land use. Boreal forests, also known as *taiga*, and northern peatlands, which store nearly half the total terrestrial carbon in North American, are already experiencing substantial warming, resulting in large-scale thawing of permafrost and dramatic changes in aquatic and forest ecosystems. USGS scientists have estimated that at least 10 gigatons of soil carbon in Alaska are stored in organic soils that are extremely vulnerable to fire and decomposition under warming conditions.

The capacity of terrestrial ecosystems to sequester additional carbon is uncertain. An upper estimate of potential terrestrial sequestration in the United States might be the amount of carbon that would be accumulated if U.S. forests and soils were restored to their historic levels before they were depleted by logging and cultivation. These amounts (about 32 and 7 gigatons for forests and soils, respectively) are probably not attainable by deliberate sequestration because restoration on this scale would displace a large percentage of U.S. agriculture and disrupt many other present-day activities. Decisions about terrestrial carbon sequestration require careful considerations of priorities and trade-offs among multiple resources. For example, converting farmlands to forests or wetlands may increase carbon sequestration, enhance wildlife habitat and water quality, and increase flood storage and recreational potential—but the loss of farmlands would decrease crop production. Converting existing conservation lands to intensive cultivation, while perhaps producing valuable crops (for example, for biofuels), may diminish wildlife habitat, reduce water quality and supply, and increase CO_2 emissions. Scientists are working to determine the effects of climate and land-use change on potential carbon sequestration and ecosystem benefits and to provide information about these effects for use in resource planning.

SIDEBAR 11.1. URBAN FORESTS AND CARBON SEQUESTRATION[*]

The urban environment presents important considerations for terrestrial carbon sequestration and global climate change. Over half of the world's population lives in urban areas (Population Reference Bureau, 2012). Because cities are more dense and walkable, urban per capita emissions of greenhouse gases (GHGs) are almost always substantially lower than average per capita emissions for the countries in which they are located (Cities Alliance, 2007; Romero-Lankao,

---

[*] Adapted from Safford, H. et al., *Urban Forests and Climate Change*, U.S. Department of Agriculture, Washington, DC, 2013.

2008). Urban areas are more likely than non-urban areas to have adequate emergency services and so may be better equipped to provide critical assistance to residents in the case of climate-related stress and events such as heat waves, floods, storms, and disease outbreaks (Myers et al., 2013). However, cities are still major sources of GHG emissions (Dodman, 2009). Studies suggest that cities account for 40 to 70% of all GHG emissions worldwide due to resource consumption and energy, infrastructure, and transportation demands (USEPA, 2009). Highly concentrated urban areas, especially in coastal regions and in developing countries, are dispro-portionately vulnerable to extreme weather and infectious disease.

The term *urban forest* refers to all trees within a densely populated area, including trees in parks, on streetways, and on private property. Although the composition, health, age, extent, and costs of urban forests vary considerably among different cities, all urban forests offer some common environmental, economic, and social benefits. Urban forests play an impor-tant role in climate change mitigation and adaptation. Active stewardship of a community's forestry assets can strengthen local resilience to climate change while creating more sustain-able and desirable places to live. Trees in a community help to reduce air and water pollu-tion, alter heating cooling costs, and increase real estate values. Trees can improve physical and mental health and strengthen social connections, and they are associated with reduced crime rates. Trees, community gardens, and other green spaces get people outside, helping foster active living and neighborhood pride.

Like any forest, urban forests help mitigate climate change by capturing and storing atmospheric carbon dioxide during photosynthesis and by influencing energy needs for heating and cooling buildings. Trees typically reduce cooling costs, but they can increase or decrease winter heating use depending on their location around a building and whether they are evergreen or deciduous. In the contiguous United States alone, urban trees store over 708 million tons of carbon (approximately 12.6% of annual carbon dioxide emissions in the United States) and capture an additional 28.2 million tons of carbon (approximately 0.05% of annual emissions) per year (Nowak et al., 2013; USEPA, 2014). The value of urban carbon sequestration is substantial: approximately $2 billion per year, with a total current carbon storage value of over $50 billion (Nowak et al., 2013). Shading and reduction of wind speed by trees can help to reduce carbon emissions by reducing summer air conditioning and win-ter heating demand and, in turn, the level of emissions from supplying power plants (Nowak et al., 2010). Shading can also extend the useful life of street pavement by as much as ten years, thereby reducing emissions associated with the petroleum-intensive materials and operation of heavy equipment required to repave roads and haul away waste (McPherson and Muchnick, 2005). Establishing 100 million mature trees around residences in the United States would save an estimated $2 billion annually in reduced energy costs (Akbari et al., 1992; Population Reference Bureau, 2012). However, this level of tree planting would only offset less than 1% of U.S. emissions over a 50-year period (Nowak and Crane, 2002).

The sustainable use of wood, food, and other goods provided by the local urban forest may also help mitigate climate change by displacing imports associated with higher levels of carbon dioxide emitted during production and transport. Urban wood is a valuable and underutilized resource. At current utilization rates, forest products manufactured from felled urban trees are estimated to save several hundred million tons of CO_2 over a 30-year period. Furthermore, wood chips made from low-quality urban wood may be combusted for heat and power to displace an additional 2.1 million tons of fossil fuel emissions per year (Sherrill and Bratkovitch, 2011).

Urban forests enable cities to better adapt to the effect of climate change on temperature patterns and weather events. Cities are generally warmer than their surroundings (typically by about 1 to 2°C, although this difference can be as high as 10°C under certain climatic condi-tions), meaning that average temperature increases caused by global warming are frequently amplified in urban areas (Bristow et al., 2012; Kovats and Akhtar, 2008). Urban forests help control this "heat island" effect by providing shade, by reducing urban albedo (the fraction of solar radiation reflected back into the environment), and through cooling evapotranspiration (Bristow et al., 2012; Novak et al., 2010; Romero-Lankao, 2008). Cities are also particularly susceptible to climate-related threats such as storms and flooding. Urban trees can help control runoff from these by catching rain in their canopies and increasing the infiltration rate of deposited precipitation. Reducing stormwater flow reduces stress on urban sewer systems by limiting the risk of hazardous combined sewer overflows (Fazio, 2010). Furthermore, well-maintained urban forests help buffer high winds, control erosion, and reduce drought (Cullington and Gye, 2010; Fazio, 2010; Nowak et al., 2010).

Urban forests provide critical social and cultural benefits that may strengthen community resilience to climate change. Street trees can hold spiritual value, promote social interaction, and contribute to a sense of place and family for local residents (Dandy, 2010). Overall, forested urban areas appear to have potentially stronger and more stable communities (Dandy, 2010). Community stability is essential to the development of effective long-term sustainable strategies for addressing climate change (Williamson et al., 2010). For example, neighborhoods with stronger social networks are more likely to check on the elderly and other vulnerable residents during heat waves and other emergencies (Klinenberg, 2002).

Urban forests help control the causes and consequences of climate-related threats; however, forests may also be negatively impacted by climate change. Although increased carbon dioxide levels and water temperature may initially promote urban tree growth by accelerating photosynthesis, too much warming in the absence of adequate water and nutrients stresses trees and retards future development (Tubby and Webber, 2010). Warmer winter temperatures increase the likelihood of winter kill, in which trees, responding to their altered environment, prematurely begin to circulate water and nutrients in their vascular tissue. If rapid cooling follows these unnatural warm periods, tissues will freeze and trees will sustain injury or death.

Warmer winter temperatures favor many populations of tree pest and pathogen species normally kept at low levels by cold winter temperatures (Tubby and Webber, 2010). Although climate change may reduce populations of some species, many others are better able than their arboreal host to adapt to changing environments due to their short lifecycles and rapid evolutionary capacity (Cullington and Gye, 2010; Tubby and Webber, 2010). The consequences of these population changes are compounded by the fact that hot, dry environments enrich carbohydrate concentrations in tree foliage, making urban trees more attractive to pests and pathogens (Tubby and Webber, 2010).

Climate change alters water cycles in ways that impact urban forests. Increased winter precipitation puts urban forests at greater risk from physical damage due to increased snow and ice loading (Johnston, 2004). Increased summer evaporation and transpiration create water shortages often exacerbated by urban soil compaction and impermeable surfaces. More frequent and intense extreme weather events increase the likelihood of severe flooding, which may uproot trees and cause injury or death to tree root systems if waterlogged soils persist for prolonged periods (Johnston, 2004). Especially cold regions may benefit from increased tourism, agricultural productivity, and ease of transport as a result of climate change (Romero-Lankao, 2008). The potential positive implications of climate change, however, are far eclipsed by the negative (Parry et al., 2007). Rising temperatures, increased pest and pathogen activity, and water-cycle changes impose physiological stresses on urban forests that compromise forest ability to deliver ecosystems services that protect against climate change. Climate change will also continue to alter species ranges and regeneration rates, further affecting the health and composition of urban forests (Nowak, 2010; Ordonez et al., 2010). Proactive management is necessary to protect urban forests against climate-related threats and to sustain desired urban forest structures for future generations.

GEOLOGIC CARBON SEQUESTRATION

Geologic sequestration begins with capturing carbon dioxide from the exhaust of fossil fuel power plants and other major sources. The captured carbon dioxide is piped 1 to 4 kilometers below the land surface and injected into porous rock formations. Compared to the rates of terrestrial carbon uptake shown in Figures 11.1 and 11.2, geologic sequestration is currently used to store only small amounts of carbon per year. Much larger rates of sequestration are envisioned to take advantage of the potential permanence and capacity of geologic storage.

The permanence of geologic sequestration depends on the effectiveness of several carbon dioxide trapping mechanisms. After carbon dioxide is injected underground, it will rise buoyantly until it is trapped beneath an impermeable barrier, or seal. In principle, this physical trapping mechanism, which is identical to the natural geologic trapping of oil and gas, can retain carbon dioxide for thousands to millions of years. Some of the injected carbon dioxide will eventually dissolve in groundwater, and some may be trapped in the form of carbonate minerals formed by chemical reactions with the surrounding rock. All of these processes are susceptible to change over time following carbon dioxide injection. Scientists are studying the permanence of these trapping mechanisms and developing methods to determine the potential for geologically sequestered carbon dioxide to leak back to the atmosphere. The capacity for geologic carbon sequestration is constrained by the volume and distribution of potential storage sites. According to the U.S. Department of Energy, the total storage capacity of physical traps associated with depleted oil and gas reservoirs in the United States is limited to about 38 gigatons of carbon and is geographically distributed in locations that are distant from most U.S. fossil fuel power plants. The potential U.S. storage capacity of deep porous rock formations that contain saline groundwater is much larger (estimated by the U.S. Department of Energy to be about 900 to 3400 gigatons of carbon) and more widely distributed, but less is known about the effectiveness of trapping mechanisms at these sites. Unmineable coal beds have also been proposed for potential carbon dioxide storage, but more information is needed about the storage characteristics and impacts of carbon dioxide injection in these formations. Scientists are developing methods to refine estimates of the national capacity for geologic carbon sequestration.

To fully assess the potential for geologic carbon sequestration, economic costs and environmental risks must be taken into account. Infrastructure costs will depend on the locations of suitable storage sites. Environmental risks may include seismic disturbances, deformation of the land surface, contamination of portable water supplies, and adverse effects on ecosystems and human health. Many of these environmental risks and potential environmental impacts are discussed in the sections that follow.

POTENTIAL IMPACTS OF TERRESTRIAL SEQUESTRATION

Potential environmental impacts associated with terrestrial sequestration include ground disturbance and the loss of soil resources due to erosion; equipment-related noise; visual impacts; air emissions; disturbance of ecological, cultural, and paleontological resources; and conflicts with current or proposed land uses. Establishing and managing a terrestrial sequestration plot could involve ground clearing (removal of vegetative cover) to prepare the ground for planting, grading, vehicular traffic, and pedestrian traffic. Management could require the use of water for dust control, and in some cases water could be required to establish and maintain seeds, seedlings, or crops. The addition of soil additives such as fertilizer and pesticides could have an impact on water quality. Equipment used to maintain a terrestrial sequestration plot could be a source of noise and air emissions and create a visual impact if frequent and conspicuous use was required.

Ecological, cultural, and paleontological resources could be impacted, especially if a terrestrial sequestration plot is going to replace an established ecological habitat or otherwise impact undisturbed land that hosts important cultural or paleontological resources. Impacts on land use could occur if there were conflicts with existing land use plans—for example, if land zoned for future commercial or housing development is used to establish a forest sequestration plot.

Soil resources can also be impacted by terrestrial sequestration. The careful management of a sequestration plot should result in an improvement of soil resources, but poor management practices could adversely impact soils and the viability of the sequestration project. Practices such as no-till cultivation and planting, crop rotation, and the use of cover crops should result in the maintenance of soil organic material and nutrients and an increase in the relative health of soil resources. Some management practices, however, could involve the use of hazardous materials such as herbicides to kill a cover crop before planting the terrestrial sequestration crop.

POTENTIAL IMPACTS OF GEOLOGIC SEQUESTRATION

The potential impacts of geologic sequestration, including the transportation of carbon, are discussed in this section. For this discussion, it is assumed that carbon capture would likely occur at a single power generating station. Because captured carbon may have to be transported for some distance away from the power station, transport, in general, has been evaluated. The significance of the impacts depends on factors such as the number and size of transport pipelines and injection wells, the amount of land disturbed by drilling and transport activities, the amount of land occupied by facilities over the life of the sequestration project, the project's location with respect to other resources (e.g., wildlife use, distance to surface water bodies), and so forth.

GEOLOGIC SEQUESTRATION EXPLORATION IMPACTS

Activities during the exploration phase (including seismic surveys, testing, and exploratory drilling) are temporary and are conducted at a smaller scale than those at the drilling/construction, sequestration, and decommissioning/reclamation phases. The impacts described for each resource would result from typical exploration activities, such as localized ground clearing, vehicular traffic, seismic testing, positioning of equipment, and exploratory drilling. Most impacts during the exploration phase would be associated with the development of access roads and exploratory wells. Impacts on resources would be similar in character, but lesser in magnitude, to those for the drilling phase. Potential impacts from these activities are presented below, by the type of affected resource.

Air Quality

Impacts on air quality during exploration activities would include emissions and dust from earth-moving equipment, vehicles, seismic surveys, well completion and testing, and drill rig exhaust. Pollutants would include particulates, oxides of nitrogen, carbon monoxide, sulfur dioxide, and volatile organic compounds (VOCs). Nitrogen oxides and VOCs may combine to form ground-level ozone. Impacts would depend on the amount, duration, location, and characteristics of the emission and the meteorological conditions (e.g., wind speed and direction, precipitation, relative humidity). Emissions during this phase would not have a measurable impact on climate change.

Cultural Resources

During the exploration phase, soil surface and subsurface disturbance is minimal. Cultural resources buried below the surface are unlikely to be affected; however, material present on the surface could be disturbed by vehicular traffic, ground clearing, and pedestrian activity (including collection of artifacts). Exploration activities could affect areas of interest to Native Americans, depending on the placement of equipment and level of visual intrusion. Surveys conducted during this phase to evaluate the presence and significance of cultural resources in the area would assist developers in siting project facilities, in order to avoid or minimize impacts on these resources.

Ecological Resources

Impacts on ecological resources (vegetation, wildlife, aquatic biota, special status species, and their habitats) would be minimal and localized during exploration due to the limited nature of the activities. The introduction or spread of some nonnative invasive vegetation could occur as a result of vehicular traffic but would be relatively limited in extent. Seismic surveys could disturb wildlife. Exploratory well establishment would destroy vegetation and impact wildlife. Surveys conducted during this phase to evaluate the presence and significance of ecological resources in the area would assist developers in siting project facilities to avoid or minimize impacts on these resources.

Water Resources

Minimal impact on water resources (water quality, water flows, and surface water/groundwater interactions) would be anticipated from exploration activities. Exploratory wellbores may provide a path for surface contaminants to come into contact with groundwater or for waters from subsurface formations to commingle. They may also decrease pressure in water wells and affect their quality. Very little produced water would likely be generated during the exploration phase. Most water needed to support drilling operations could be trucked in from offsite.

Land Use

Temporary and localized impacts on land use would result from exploration activities. These activities could create a temporary disturbance in the immediate vicinity of a surveying or monitoring site or an exploratory well (e.g., disturb recreational activities or livestock grazing). Exploration activities are unlikely to affect mining activities, military operations, or aviation.

Soils and Geologic Resources

Surface effects from vehicular traffic could occur in areas that contain special (e.g., cryptobiotic) soils. The loss of biological crusts can substantially increase water and wind erosion. Also, soil compaction due to development activities at the exploratory well pads and along access roads would reduce aeration, permeability, and water-holding capacity of the soils and cause an increase in surface runoff, potentially causing increased sheet, rill, and gully erosion. The excavation and reapplication of surface soils could cause the mixing of shallow soil horizons, resulting in a blending of soil characteristics and types. This blending would modify physical characteristics of the soils, including structure, texture, and rock content, which could lead to reduced permeability and increased runoff from these areas. Potential impacts on geologic and mineral resources would include depletion of hydrocarbons and sand and gravel resources. It is unlikely that exploration activities would activate geological hazards. Impacts on soils and geologic resources would be proportional to the amount of disturbance. The amount of surface disturbance and use of geologic materials during exploration would be minimal.

Paleontological Resources

Paleontological resources are nonrenewable resources. Disturbance of such resources, whether it is through mechanical surface disturbance, erosion, or paleontological excavation, irrevocably alters or destroys them. Direct impacts on paleontological resources would include surface disturbance during seismic surveys, the drilling of exploratory wells, and the construction of access roads and other ancillary facilities. The amount of subsurface disturbance is minimal during the exploration phase, and paleontological resources buried below the surface are unlikely to be affected. Fossil material present on the surface could be disturbed by vehicular traffic, ground clearing, and pedestrian activities (including collection of fossils). Surveys conducted during this phase to evaluate the presence and significance of paleontological resources in the area would assists developers in siting project facilities in order to avoid or minimize impacts on these resources.

Transportation

No impacts on transportation are anticipated during the exploration phase. Transportation activities would be temporary and intermittent and limited to low volumes of light utility trucks and personal vehicles.

Visual Resources

Impacts on visual resources would be considered adverse if the landscape were substantially degraded or modified. Exploration activities would have only temporary and minor visual effects, resulting from the presence of drill rigs, workers, vehicles, and other equipment.

Socioeconomics

Because the activities conducted during the exploration phase are temporary and limited in scope, they would not result in significant socioeconomic impacts on employment, local services, or property values.

Environmental Justice

Exploration activities are limited and would not result in significant adverse impacts in any resource area; therefore, environmental justice is not expected to be an issue during this phase.

Acoustics (Noise)

Primary sources of noise associated with exploration include earth-moving equipment, vehicle traffic, seismic survey, blasting, and drill rig operations.

Hazardous Materials and Waste Management

Seismic and exploratory well crews may generate waste (e.g., plastic, paper, containers, fuel leaks/spills, food, human waste). Wastes produced by exploratory drilling would be similar but would occur to a lesser extent than those produced during drilling and operation of injection wells. These would include drilling fluid and muds, used oil and filters, spilled fuel, drill cuttings, spent and unused solvents, scrap metal, solid waste, and garbage.

GEOLOGIC SEQUESTRATION DRILLING/CONSTRUCTION IMPACTS

Typical activities during the drilling/construction phase of a sequestration project include ground clearing and removal of vegetative cover, grading, drilling, waste management, vehicular and pedestrian traffic, and construction and installation of facilities. Activities conducted in locations other than at the injection well-pad site may include excavation and blasting for construction materials (sands, gravels), access road and storage area construction, and construction of pipelines, compressor stations, pumping stations, and other facilities (e.g., office buildings). Potential impacts from these activities are presented below, by the type of affected resource.

Air Quality

Emissions generated during the drilling/construction phase include vehicle emissions; diesel emissions from large construction equipment and generators, storage/dispensing of fuel, and, if installed at this stage, flare stacks; small amounts of carbon monoxide, nitrogen oxides, and particulates from blasting activities; and dust from many sources, such as disturbing and moving soils (clearing, grading, excavation, trenching, backfilling, dumping, and truck and equipment traffic), mixing concrete, and drilling. During windless conditions (especially in areas of thermal inversion), project-related odors may be detectable at more than a mile from the source. Excess increases in dust could decrease forage palatability for wildlife and livestock and increase the potential for dust pneumonia.

Cultural Resources

Potential impacts on cultural resources during the drilling/construction phase could include the following:

- Destruction of cultural resources in areas undergoing surface disturbance
- Unauthorized removal of artifacts or vandalism as a result of human access to previously inaccessible areas (resulting in lost opportunities to expand scientific study and educational and interpretive uses of these resources)
- Visual impacts resulting from large areas of exposed surface, increases in dust, and the presence of large-scale equipment, machinery, and vehicles if the cultural resources have an associated landscape component that contributes to their significance (e.g., sacred landscapes, historic trails)

Although the potential for encountering buried sites is relatively low, the possibility that buried sites would be disturbed during pipeline, access road, or well-pad construction does exist. Unless the buried site is detected early in the surface-disturbing activities, the impact on the site can be considerable. Disturbance that uncovers cultural resources of significant importance that would otherwise have remained buried and unavailable could be viewed as a beneficial impact. Vibration, resulting from increased traffic and drilling/development activities, may also have effects on rock art and other associated sites (e.g., sites with standing architecture).

Ecological Resources

Impacts on ecological resources would be proportional to the amount of surface disturbance and habitat fragmentation. Vegetation and topsoil would be removed for the development of well pads, access roads, pipelines, and other ancillary facilities. This would lead to a loss of wildlife habitat, reduction in plant diversity, potential for increased erosion, and potential for the introduction of invasive or noxious weeds. The recovery of vegetation following interim and final reclamation would vary by community (e.g., grasslands would recover before sagebrush or forest habitats). Indirect impacts on vegetation would include increased deposition of dust, spread of invasive and noxious weeds, and the increased potential for wildfires. Dust settling on vegetation may alter or limit a plant's ability to photosynthesize or reproduce. Over time, a composition of native and/or invasive vegetation would become established in areas disturbed by wildfire. Although injection field development would likely increase the spread of invasive and noxious weeds by increasing traffic and human activity, the potential impacts could be partially reduced by interim reclamation and implementation of mitigation measures. Adverse impacts on fish and wildlife could occur during the drilling/construction phase due to

- Erosion and runoff
- Dust
- Noise
- Introduction and spread on invasive nonnative vegetation
- Modification, fragmentation, and reduction of habitat
- Mortality of biota
- Exposure to contaminants
- Interference with behavioral activities
- Increased harassment and/or poaching

Depletion of surface waters from perennial streams could result in a reduction of water flow, which could lead to habitat loss and/or degradation of aquatic species.

Water Resources

Impacts on water resources could occur due to water quality degradation from increases in turbidity, sedimentation, and salinity; spills; cross-aquifer mixing; and water quantity depletion. During the drilling/construction phase, water would be required for dust control, making concrete, consumptive use by the construction crew, and drilling wells. Depending on availability, it may be trucked in from offsite or obtained from local groundwater wells or nearby surface water bodies. Where surface waters are used to meet drilling and construction needs, depletion of stream flows could occur. Drilling and well completion can require the use of drilling fluids that could, if not managed properly, contaminate soils and surface water features. Drilling activities may affect surface and groundwater flows. If a well is completed improperly, such that subsurface formations are not sealed off by the well casing and cement, aquifers can be impacted by other nonpotable formation waters. The interaction between surface water and groundwater may also be affected if the two are hydrologically connected, potentially resulting in unwanted dewatering or recharging. Soils compacted on existing roads, new access roads, and well pads generate more runoff than undisturbed sites. The increased runoff could lead to slightly higher peak storm flows into streams, potentially increasing erosion of the channel banks. The increased runoff could also lead to more efficient sediment delivery and increase turbidity during storm events. During drilling and construction, water quality can be affected by

- Activities that cause soil erosion or dust that can be washed into water bodies
- Weathering of newly exposed soils, causing leaching and oxidation that can release chemicals into the water
- Increased salinity levels resulting from increased sediment loading
- Discharges of waste or sanitary water
- Use of herbicides and dust suppressants (e.g., magnesium chloride)
- Contaminant spills

Land Use

Land use impacts would occur during the drilling/construction phase if there are conflicts with existing land use plans and community goals; existing recreational, educational, religious, scientific, or other use areas; or existing commercial land use (e.g., agriculture, grazing, mineral extraction). In general, the development of large-scale geologic sequestration facilities and transport pipelines is expected to change the character of the landscape from a rural to a more industrialized setting. Existing land use would be affected by intrusive impacts such as increased traffic, noise, dust, and human activity, as well as by changes in the visual landscape. In particular, these impacts could affect recreationists seeking solitude or recreational opportunities in a relatively pristine landscape. Ranchers or farmers could be affected by loss of available grazing or crop lands, the potential for the introduction of invasive and noxious plants that could affect livestock forage availability, and a possible increase in livestock/vehicle collisions. In forested areas, drilling could result in the long-term loss of timber resources. The expanded access road system could increase the numbers of off-highway vehicle (OHV) users, hunters, and other recreationists in the area. Although the change in landscape character could discourage hunters who prefer a more remote backcountry setting, the potential for illegal hunting activities could increase due to the expanded access road system. Construction and drilling noise could potentially be heard 20 miles (32 kilometers) or more from the project area. It would be barely audible at this distance, but it could affect residents' and recreationists' perception of solitude. Most land use impacts that occur during the drilling/construction phase would continue throughout the life of the sequestration project. Overall, land use impacts could range from minimal to significant depending on both the areal extent of the project, the density of injection wells and other ancillary facilities, and the compatibility of the project with the existing land uses.

Soils and Geologic Resources

Potential impacts on soils during the drilling/construction phase would occur due to the removal of vegetation, mixing of soil horizons, soil compaction, increased susceptibility of the soils to wind and water erosion, contamination of soils from spills of hazardous materials (e.g., drilling mud, fluids used to hydraulically fracture subsurface formations), loss of topsoil productivity, and disturbance of biological soil crusts. Impacts on soils would be proportionate to the amount of disturbance. Sands, gravels, and quarry stone could be excavated for use in the construction of access roads, foundations, and ancillary structures, in addition to well pads and storage areas. Construction of well pads, pipelines, compressors or pumping stations, access roads, and other project facilities could cause topographic changes. These changes would be minor but long term. Well pads located on canyon rims of the side slopes of canyons could result in bedrock disturbances. Additional bedrock disturbance could occur due to the construction of access roads, pipelines, rock borrow pits, and other ancillary facilities. Possible geological hazards (earthquakes, landslides, and subsidence) could be activated by drilling and blasting. Altering drainage patterns could also accelerate erosion and create slope instability.

Paleontological Resources

Impacts on paleontological resources can occur directly from construction and drilling activities or indirectly as a result of soil erosion and increased accessibility to fossil localities (e.g., unauthorized removal of fossil resources or vandalism to the resource). This would result in lost opportunities to expand scientific study and educational interpretive uses of these resources. Disturbance that uncovers paleontological resources of significant importance that would otherwise have remained buried and unavailable could be viewed as a beneficial impact. Direct impacts on unknown paleontological resources can be anticipated to be proportional to the total area impacted by construction and drilling activities.

Transportation

Development of a geologic sequestration project would result in the need to construct or improve access roads and would result in an increase in industrial traffic (e.g., hundreds of truck loads or more per well site). Overweight and oversized loads could cause temporary disruptions and could require extensive modifications to roads or bridges (e.g., widening roads or fortifying bridges to accommodate the size or weight of truck loads). An overall increase in heavy truck traffic would accelerate the deterioration of pavement, requiring local government agencies to schedule pavement repair or replacement more frequently than under the existing traffic conditions. Increased traffic would also result in a potential for increased accidents within the project area. The locations at which accidents are most likely to occur are intersections used by project-related vehicles to turn onto or off of highways from access roads. Conflicts between industrial traffic and other traffic are likely to occur, especially on weekends, holidays, and seasons of high use by recreationists. Increased recreational use of the area could contribute to a gradual increase in traffic on the access roads. Over 1000 truckloads per well could be expected during the drilling/construction phase.

Visual Resources

During the drilling/construction phase, impacts on visual resources would occur as a result of the addition of well pads, pipelines, access roads, and other facilities which would result in an industrial landscape throughout the project area. Additional components that would adversely affect the visual character of the landscape are pumping units, compressor stations, aggregate borrow areas, equipment storage areas, and, if needed, worker housing units and airstrips. Project facilities would introduce new elements of form, line, color, and texture into the landscape, which would dominate foreground views. In some instances, the facilities would also be visible from greater distances and could, occasionally, dominate the view. Vehicles and the dust they generate would also contribute to visual impacts. Because drilling activities typically take place 24 hours per day, visual impacts would include lighting of drill rigs during nighttime hours. Nighttime lighting on drill rigs would be visible from long distances.

Socioeconomics

Drilling/construction phase activities would contribute to the local economy by providing employment opportunities, monies to local contractors, and recycled revenues through the local economy. Additional revenues could be generated in the form of carbon avoidance-type emission credits sold by the sequestration facility operator in a commodity market. Taxes collected by federal, state, and local governments could also be involved for transportation, injection, and regulation of the sequestration project. Indirect impacts could occur as a result of the new economic development (e.g., new jobs at businesses that support the expanded workforce or that provide project materials). Depending on the source of the workforce, local increases in population could occur. Development of an injection well field also could potentially affect property values, either positively from increased employment effects or negatively from proximity to the field and any associated or perceived adverse environmental effects (e.g., noise of compressor stations, visual effects, air quality). Some economic losses could occur if recreationists (including hunters and fishermen) avoid the area. Increased growth of the transient population could contribute to increased criminal activities in the project area (e.g., robberies, drugs).

Environmental Justice

If significant impacts were to occur in any of the resource areas and these were to disproportionately affect minority or low-income populations, there could be an environmental justice impact. It is anticipated that the drilling/construction phase could benefit low-income, minority, and tribal populations by creating job opportunities and stimulating local economic growth via project revenues and increased tourism. However, noise, dust, visual impacts, and habitat destruction could have an adverse effect on traditional tribal life ways and religious and cultural sites. Development of wells and ancillary facilities could affect the natural character of previously undisturbed areas and transform the landscape into a more industrialized setting. Drilling and construction activities could impact the use of cultural sites for traditional tribal activities (hunting and plant-gathering activities and areas in which artifacts, rock art, or other significant cultural sites are located).

Acoustics (Noise)

Primary sources of noise during the drilling/construction phase would be equipment (bulldozers, drill rigs, and diesel engines). Other sources of noise include vehicular traffic and blasting. Blasting activities typically would be very limited, the possible exception being in areas where the terrain is hilly and bedrock shallow. With the exception of blasting, noise would be restricted to the immediate vicinity of the work in progress. Noise from blasting would be sporadic and of short duration but would carry for long distances. If noise-producing activities occur near a residential area, noise levels from blasting, drilling, and other activities could exceed U.S. Environmental Protection Agency (USEPA) guidelines. The movement of heavy vehicles and drilling could result in frequent or continuous noise. Drilling noise would occur continuously for 24 hours per day for one to two months or more depending on the depth of the formation. Exploratory wells that end up becoming injection wells would continue to generate noise during the sequestration phase.

Hazardous Materials and Waste Management

Solid and industrial wastes would be generated during the drilling/construction phase. Much of the solid wastes would be expected to be nonhazardous, considering of containers and packaging materials, miscellaneous wastes from equipment assembly and the presence of construction crews (food wrappers and scraps), and woody vegetation. Industrial wastes would include minor amounts of paints, coatings, and spent solvents. Most of these materials would likely be transported offsite for disposal.

In forested areas, commercial-grade timber could be sold, while slash may be spread or burned near the well site. Drilling wastes include hydraulic fluids, pipe dope, used oils and oil filters, rigwash, spilled fuel, drill cuttings, drums and containers, spent and unused solvents, paint and paint washes, sandblast media, scrap metal, solid waste, and garbage. Wastes associated with drilling fluids include oil derivatives, such as polycyclic aromatic hydrocarbons (PAHs), spilled chemicals, suspended and dissolved solids, phenols, cadmium, chromium, copper, lead, mercury, nickel, and drilling mud additives, including potentially harmful contaminants such as chromate and barite. Adverse impacts could result if hazardous wastes are not properly handled and are released to the environment.

GEOLOGIC SEQUESTRATION OPERATIONS IMPACTS

Typical activities during the operations phase include operation of wells and compressor stations or pump stations, waste management, and maintenance and replacement of facility components. Impacts could also result from the fact that a geologic sequestration project could be linked to an enhanced oil recovery or enhanced coalbed methane recovery project.

Air Quality

The primary emission sources during the operations phase would include compressor and pumping station operations, vehicle traffic, and operating wells. Venting of carbon dioxide may occur during injection and pipeline maintenance operations.

Cultural Resources

During the operations phase, impacts on cultural resources could occur primarily from the unauthorized collection of artifacts and from visual impacts. In the latter case, the presence of the aboveground structures could impact cultural resources with an associated landscape component that contributes to their significance, such as a sacred landscape or historic trail. Damage to localities caused by off-highway vehicle (OHV) use could also occur. The potential for indirect impacts (e.g., vandalism, unauthorized collecting) would be greater during the operations phase compared to the drilling/construction phase, due to the longer duration of the operations phase.

Ecological Resources

During the operations phase, adverse impacts on ecological resources could result from

- Disturbance of wildlife from noise and human activity
- Exposure of biota to contaminants
- Mortality of biota from colliding with aboveground facilities or vehicles

Ecological resources may continue to be affected by the reduction in habitat quality associated with habitat fragmentation due to the presence of operating wells, pipelines, ancillary facilities, and access roads. In addition, the presence of access roads may increase human use of surrounding areas, which, in turn, could impact ecological resources in the surrounding areas through

- Introduction and spread of invasive nonnative vegetation
- Fragmentation of habitat
- Disturbance of biota
- Increase in hunting (including poaching)
- Increased potential for fire

The presence of an injection well field could also interfere with migratory and other behaviors of some wildlife. In some coal bed methane production areas, methane gas or carbon dioxide gas could seep up into fields and create dead zones. High levels of carbon dioxide could asphyxiate wildlife in their burrows.

Water Resources

During the life of an injection well, the integrity of the well casing and cement will determine the potential for adverse impacts on groundwater. If subsurface formations are not sealed off by the well casing and cement, aquifers can be impacted by other nonpotable formation waters, hydraulic fracturing fluids, or the injected carbon dioxide. Other potential impacts on water availability and quality during the operations phase would include possible minor degradation of water quality resulting from vehicular traffic and machinery operations during maintenance (e.g., erosion, sedimentation) or herbicide contamination resulting from improper application. A spill or blowout could potentially cause extensive contamination of surface waters or a shallow aquifer. Contaminated groundwater could potentially be discharged into springs or as base flow into stream channels, leading to surface water contamination. Recovered waters used for hydraulic fracturing could cause altered surface water quality or an increase in flows in normally dry water bodies such as ephemeral drainages, if they are disposed of by discharge to the surface.

With regard to hydraulic fracturing wastewater discharge into the environment, a point of interest is that the Clean Water Act (CWA) made it unlawful to discharge any pollutant from a point source into the navigable waters of the United States unless done in accordance with a specific approved permit. The National Pollutant Discharge Elimination System (NPDES) permit program controls discharges from point sources that are discrete conveyances, such as pipes or man-made ditches. Industrial, municipal, and other facilities such as shale gas production sites or commercial facilities that handle the disposal or treatment of shale gas produced water must obtain permits if they intend to discharge directly into surface water. Large facilities usually have individual NPDES permits. Discharges from some smaller facilities may be eligible for inclusion under general permits that authorize a category of discharge under the CWA within a geographic area. A general permit is not specifically tailored to an individual discharger. Most oil and gas production facilities with related discharges are authorized under general permits because there are typically numerous sites with common discharges in a geographic area.

Land Use

Land use impacts during the operations phase would be an extension of those that occurred during the drilling/construction phase; however, to some extent, land can revert to its original uses after the major drilling/construction phase is over. For example, farmers can graze livestock or grow crops around the well sites. Other industrial projects would likely be excluded within the sequestration project area. Recreation activities (e.g., OHV use, hunting) are possible, although gun and archery restrictions would probably exist. Operations may conflict with livestock and farming operations.

Soils and Geologic Resources

Following construction and drilling, disturbed portions of well and ancillary facility sites not required for operations would be revegetated. This would help to stabilize soil and geologic conditions. Routine impacts on soils during the operations phase would be limited largely to soil erosion impacts caused by vehicular traffic. Any excavations required for maintenance would cause impacts similar to those from the drilling/construction phase, but to a lesser spatial and temporal extent. The accidental spill of product or other wastes would likely cause soil contamination. Except in the case of a large spill, soil contamination would be localized and limited in extent and magnitude. In areas where interim reclamation is implemented (e.g., reclamation of an individual well that is no longer needed), ground cover by herbaceous species could reestablish within 1 to 5 years following seeding of native plant species and diligent weed control efforts, thus reducing soil erosion. Operations might preclude or interfere with mineral development activities in the project area, including oil and gas development and mining activities. Possible geological hazards (earthquakes, landslide, and subsidence) could be activated by injection activities.

Paleontological Resources

Impacts on paleontological resources during the operations phase would be limited primarily to unauthorized collection of fossils. This threat is present when the access roads have been constructed, making remote areas more accessible to the public. Damage to localities caused by OHV use could also occur. The potential for indirect impacts (e.g., vandalism, unauthorized collecting) would be greater during the operations phase compared to the drilling/construction phase, due to its longer duration.

Transportation

Impacts on transportation during the operations phase would be similar to those for the drilling/construction phase. However, unless carbon dioxide is transported to the site by truck or rail, daily traffic levels, particularly heavy truck traffic, would be expected to be lower during the operations phase compared to the drilling/construction phase. For the most part, heavy truck traffic would be limited to periodic visits to a well site for workovers and formation treatment. The use of pipelines to convey carbon dioxide to the operating site would reduce the volume of traffic during the operations phase. If a pipeline is not used for the injection well field, multiple truckloads per day would be needed.

Visual Resources

When operating facilities have been installed, portions of well pads, access roads, and pipeline rights of way (ROWs) that are not needed for operations would be reclaimed; however, much of the disturbed area would continue to contrast with the natural form, line, color, and texture of the surrounding landscape. This would impact undisturbed vistas and areas of solitude. The aboveground portions of an injection well would be highly visible in rural or natural landscapes, many of which may have a few other comparable structures. The artificial appearance of an injection well may have visually incongruous "industrial" associations for some, particularly in a predominately natural landscape. Any nighttime lighting would be visible from long distances. During the operations phase, indirect impacts on visual resources would occur as a result of sequestration activities (e.g., industrial traffic, heavy equipment use, dust); however, human activity would be substantially lower than during the drilling/construction phase.

Socioeconomics

Direct socioeconomic impact would include the creation of new jobs and the associated royalties and taxes paid for carbon emission avoidance created by the sequestration project. Indirect impacts are those impacts that would occur as a result of the new economic development and would include new jobs at businesses that support the expanded workforce or that provide project materials, and associated taxes. Potential impacts on the value of residential properties located adjacent to an oil or gas field would continue during this phase.

Environmental Justice

Possible environmental justice impacts during the operations phase include the alteration of scenic quality in areas of traditional or cultural significance to minority populations. Noise and health and safety impacts are also potential sources of disproportionate impacts for minority or low-income populations.

Acoustics (Noise)

The main sources of noise during the operations phase would include compressor and pumping stations, producing wells (including occasional flaring), and vehicle traffic. Compressor stations produce noise levels ranging from 64 to 86 dBA at the station and from 58 to 75 dBA at about 1 mile

(1.6 kilometers) from the station. Use of remote telemetry equipment would reduce daily traffic and associated noise levels within the project area. The primary impacts from noise would be localized disturbances to wildlife, recreationists, and residents. Noise associated with cavitation is a major concern for landowners, livestock, and wildlife.

Hazardous Materials and Waste Management

Industrial wastes are generated during routine operations (lubricating oils, hydraulic fluids, coolants, solvents, and cleaning agents). These wastes are typically placed in containers, characterized and labeled, possibly stored briefly, and transported by a licensed hauler to an appropriate permitted offsite disposal facility as a standard practice. Impacts could result if these wastes are not properly handled and are released to the environment. Environmental contamination could result from accidental spills of herbicides or other chemicals. Chemicals in open pits used to store wastes may pose a threat to wildlife and livestock.

Should geologic sequestration become common, a wide diversity of geologic formations is likely to be encountered. Depending on the nature of the formation targeted for injection, it may be necessary to increase the injectivity of carbon dioxide by using hydraulic fracturing. Hydraulic fracturing fluids can contain innocuous constituents, such as sand and water, and potentially toxic substances, such as diesel fuel (which contains benzene, ethylbenzene, toluene, xylenes, naphthalene, and other chemicals), polycyclic aromatic hydrocarbons (PAHs), methanol, formaldehyde, ethylene glycol, glycol ethers, hydrochloric acid, and sodium hydroxide. Because some aspects of a hydraulic fracturing operation are considered proprietary, information about the specific constituents used in a given hydrofracturing operation may not be available, thus causing some concern over the risks presented by this practice. Some of the hydrofracture fluids used to increase the injectivity of a carbon dioxide injection well would probably be pumped out of the well and then be managed at the surface in tanks of impoundments; however, some of the fluids would remain in the underground formation. For example, in the process of producing hydrocarbons and produced water, about 20 to 40% of the fluids used for hydrofracturing may remain underground. Thus, should hydraulic fracturing be used in a carbon sequestration injection well, these fluids could have an impact on underground water sources that are close (horizontally or vertically) to the injection well.

During the operations phase, scale and sludge wastes can accumulate inside pipelines and storage vessels. They must be removed periodically from the equipment for disposal. These wastes may be transported to offsite disposal facilities. In some instances, they may be disposed of via landspreading, a practice that entails spreading the wastes over the surface of the disposal area and mixing it with the top few inches of soil.

CARBON SEQUESTRATION IMPACTS ON HUMAN HEALTH AND SAFETY

The potential impacts on human health and safety resulting from exploration activities could include occupational accidents and injuries, vehicle or aircraft accidents, exposure to weather extremes, wildlife encounters, trips and falls on uneven terrain, adverse health effects from dust generation and emissions, and contact with hazardous materials (e.g., from spills). The potential for these impacts to occur would be low due to the limited range of activities and small number of workers required during exploration.

Potential impacts on worker and public health and safety during the drilling phase would be similar to other projects that involve earth moving, use of large equipment, transportation of overweight and oversized materials, and construction and installation of industrial facilities. In particular, the risks would be very similar to those associated with oil and gas drilling activities. Statistical data on occupational accidents and fatalities for the oil and gas extraction labor category are available from the U.S. Bureau of Labor Statistics. In 2005, the oil and gas industry experienced a nationwide rate of 2.1 accidents per 100 full-time workers and 25.6 fatalities per 100,000 workers. The potential for occupational accidents and mortality would be highest during peak drilling periods. Drilling and construction activities also present the potential for well fires or explosions. Well blowouts are rare, but are typically caused by unsafe work practices and can be extremely dangerous (e.g., they can destroy rigs and kill nearby workers). If natural gas is in the blowout materials, the fluid may ignite from an engine spark or other source of flame. Blowouts may take days to months to cap and control. Also, increased human activity and increased public access could result in a higher potential for wildfires in the project area. Workers could also be exposed to air pollutants and could have body contact with product or other chemicals. Reckless driving by workers would also create safety hazards. In addition, health and safety issues include working in potential weather extremes and possible contact with natural hazards, such as uneven terrain and dangerous plants, animals, or insects.

Possible impacts on health and safety while operating an injection field include accidental injury or death to workers and, to a lesser extent, the public (e.g., from OHV collisions with project components, vehicle collisions with workers). Health impacts could result from water contamination, dust and other air emissions, noise, soil contamination, and stress (e.g., associated with living near an industrial zone). Potential fires and explosions would cause safety hazards. In addition, health and safety issues include working in potential weather extremes and possible contact with natural hazards, such as uneven terrain and dangerous plants, animals, or insects.

DISCUSSION AND REVIEW QUESTIONS

1. Do you think terrestrial carbon sequestration is an effective way to begin to reduce the global effect of atmospheric carbon dioxide? Explain.
2. Do you think geological carbon sequestration is an effective way to begin to reduce the global effect of atmospheric carbon dioxide? Explain.

REFERENCES AND RECOMMENDED READING

Akbari, H., Huang, J., Martien, P., Rainer, L., Rosenfeld, A., and Taha, H. (1988). The impact of summer heat islands on cooling energy consumption and global CO_2 concentrations. In: *Proceedings of 1988 ACEEE Summer Study on Energy Efficiency in Buildings*, Vol. 5, Diamond, R.C. and Goldman, C.A., Eds., pp. 5-11–5-21. Washington DC: American Council for an Energy-Efficient Economy.

Akbari, H., Davis, S., Dorsano, S., Huang, J., and Winnett, S. (1992). *Cooling our Communities: A Guidebook on Tree Planting and Light-Colored Surfacing*. Washington, DC: U.S. Environmental Protection Agency.

Bristow, R.S., Blackie, R., and Brown, N. (2012). Parks and the urban heat island: a longitudinal study in Westfield, Massachusetts. In: *Proceedings of the 2010 Northeastern Recreation Research Symposium*, Fisher, C.I. and Watts, Jr., C.E., Eds., pp. 224–230. Newtown Square, PA: U.S. Department of Agriculture.

Cities Alliance. (2007). *Liveable Cities: The Benefits of Urban Environmental Planning*. Washington, DC: Cities Alliance (http://www.citiesalliance.org/node/720).

Cullington, J. and Gye, J. (2010). *Urban Forests: A Climate Adaptation Guide*. British Columbia: British Columbia Ministry of Community, Sport and Cultural Development (http://www.cakex.org/virtual-library/urban-forests-climate-adaptation-guide).

Dandy, N. (2010). *Climate Change & Street Trees Project: Social Research Report. The Social and Cultural Values, and Governance, of Street Trees*. Surrey, U.K.: The Research Agency of the Forestry Commission.

Dodman, D. (2009). Blaming cities for climate change? An analysis of urban greenhouse gas emissions inventories. *Environment and Urbanization*, 21(1): 185–201.

Fazio, J.R., Ed. (2010). *How Trees Can Retain Stormwater Runoff*, Tree City USA Bulletin No. 55. Nebraska City, NE: Arbor Day Foundation.

Johnston, M. (2004). Impacts and adaptation for climate change in urban forests. In: *6th Canadian Urban Forest Conference Proceedings*, Kelowna, British Columbia, October 19–23.

Klinenberg, E. (2002). *Heat Wave: A Social Autopsy of Disaster in Chicago*. Chicago, IL: University of Chicago Press.

Kovats, S. and Akhtar, R. (2008). Climate, climate change and human health in Asian cities. *Environment and Urbanization*, 20: 165–175.

McPherson, G. and Muchnick, J. (2005). Effects of street tree shade on asphalt concrete pavement performance. *Journal of Arboriculture*, 31(6): 303–310.

Myers, S.R., Branas, C.C., French, B.C., Kallan, M.L., Wiebe, D.J., and Carr, H.G. (2013). Safety in numbers: are major cities the safest places in the United States? *Annals of Emergency Medicine*, 62(4): 408–418.

Nowak, D.J. (2010). Urban biodiversity and climate change. In: *Urban Biodiversity and Design*, Miller, N., Werner, P., and Kelcey, J.G., Eds., pp. 101–117. Hoboken, NJ: Wiley-Blackwell.

Nowak, D.J. and Crane, D.E. (2002). Carbon storage and sequestration by urban trees in the USA. *Environmental Pollution*, 116: 381–389.

Nowak, D.J., Stein, S.M., Randler, P.B. et al. (2010). *Sustaining America's Urban Trees and Forests: A Forests on the Edge Report*, Gen. Tech. Rep. NRS-62. Newtown Square, PA: U.S. Department of Agriculture.

Nowak, D.J., Greenfield, E.J., Hoehn, R., and LaPoint, E. (2013). Carbon storage and sequestration by trees in urban and community areas of the United States. *Environmental Pollution*, 178: 229–236.

Ordonez, C., Dunker, P.N., and Steenberg, J. (2010). *Climate Change Mitigation and Adaptation in Urban Forests: A Framework for Sustainable Urban Forest Management*, paper presented at the 18th Commonwealth Forestry Conference, Edinburgh, Scotland, June 28–July 2.

Parry, M.L., Canziani, O.F., Palutikof, J.P., van der Linden, P.J., and Hanson, C.E., Eds. (2007). *Climate Change 2007: Impacts, Adaptation, and Vulnerability*, Contribution of Working Group II to the Fourth Assessment Report of the Intergovernmental Panel on Climate Change. Cambridge, MA: Cambridge University Press.

Population Reference Bureau. (2012). *2012 World Population Data Sheet*, http://www.prb.org/Publications/Datasheets/2012/world-population-data-sheet.aspx.

Romero-Lankao, P. (2008). Urban areas and climate change: review of current issues and trends. In: *Global Report on Human Settlements 2011: Cities and Climate Change: Policy Directions*. Geneva: United Nations Human Settlements Programme.

Safford, H., Larry, E., McPherson, E.G., Nowak, D.J., and Westphal, L.M. (2013). *Urban Forests and Climate Change*. Washington, DC: U.S. Department of Agriculture.

Sherrill, S. and Bratkovitch, S. (2011). *Carbon and Carbon Dioxide Equivalent Sequestration in Urban Forest Products*. Minneapolis, MN: Dovetail Partners.

TEEIC. (2014). *Potential Impacts of Sequestration*. Washington, DC: Tribal Energy and Environmental Information Clearinghouse (http://www.teeic.anl.gov/er/carbon/impact/terrimp/index.cfm).

Tubby, K.V. and Webber, J.F. (2010). Pests and diseases threatening urban trees under a changing climate. *Forestry*, 83(4): 451–459.

USEPA. (2009). *Buildings and Their Impact on the Environment: A Statistical Summary*. Washington, DC: U.S. Environmental Protection Agency (http://www.epa.gov/greenbuilding/pubs/gbstats.pdf).

USEPA. (2013). *Inventory of U.S. Greenhouse Gas Emissions and Sinks: 1990–2011*. Washington, DC: U.S. Environmental Protection Agency (http://www3.epa.gov/climatechange/ghgemissions/usinventoryreport/archive.html).

USEPA. (2014). *National Greenhouse Gas Emissions Data*. Washington, DC: U.S. Environmental Protection Agency (http://www.epa.gov/climatechange/ghgemissions/usinventoryreport.html).

USGS. (2008). *Carbon Sequestration to Mitigate Climate Change*. Washington, DC: U.S. Geological Survey.

Williamson, T., Dubb, S., and Alperovitz, G. (2010). *Climate Change, Community Stability, and the Next 150 Million Americans*. College Park, MD: The Democracy Collaboration.

Afterword
Is It the Beginning of the End or the Beginning of the Beginning?

Like the science of global warming, the science of renewable energy will only be settled in hindsight.

—Frank R. Spellman

When we consider that we are probably at or near peak oil natural production and already absorbing the higher costs associated with it (i.e., if we are able to obtain a reliable and consistent supply of it in the first place), renewable energy alternatives begin to look a lot more viable and necessary to us. On the other hand, we are currently suffering severe economic conditions; if we are lucky enough to have reliable and consistent employment with accompanying benefits, the advantages of switching from oil or natural gas supplies to renewable energy sources may not be that readily apparent or pressing. Until the summer of 2008, when gasoline prices at the pump exceeded $4.00/gal, many of us had not given much thought to the issue. After the spike in gasoline prices, however, people (those with and without employment) began to look at hybrid cars and domestic renewable energy alternatives in a very new light. Using Interstate 40 as our example, it is not that difficult or unusual to pick out numerous hybrid automobiles along with those ubiquitous heavy trucks, RVs, and other standard vehicles sharing the roadway. In addition, along the interstate it is not unusual to find giant wind towers with their turbine blades slashing through the air, turning turbine-generator shafts and cranking out electricity. Along this same superhighway, especially at rest stops (the ones not closed due to budget cuts) and state visitor centers, small arrays of solar cells can be spotted here and there. These solar panels supply electricity to fans, signal lights, natural gas metering systems, alarm systems, pump monitoring equipment, digital signs, and many other small appliances and industrial devices.

It is not unusual for drivers and passengers traveling the interstates to overlook these wind turbines, solar panels, and hydroelectric dams, these generators of renewable energy, because they are focused on getting from point A to point B and maybe safely back again to point A. These folks are missing the point that someday we will be driving high-capacity, highly efficient electric or hydrogen-like fuel-cell vehicles that will be as common as the gas- and diesel-powered vehicles of today. Much of the electricity required to recharge these electricity- or fuel-cell-powered vehicles will be provided by these same wind turbines, solar panels, and hydroelectric dams.

In light of today's current economic situation, which is driven by many factors, including the volatile price of crude oil and its dwindling accessibility, we can ask ourselves this question: Is it the beginning of the end or the beginning of the beginning? And the reader might wonder what we mean by that. Is doomsday around the corner, or is it already here and we are too ignorant to recognize the signs? Maybe we just don't give a damn, or our judgment is clouded by what is going on in the world today. The beginning of the end or the beginning of the beginning—do we anticipate *the* end, or *an* end, whatever that end might be?

The end of what you might ask? (Geez, at least that's what I *hope* you are asking.) The author is referring to the end of the so-called good life as we know it—that is, the good life that our parents (the greatest generation) and we, the so-called grasshopper generation, knew very briefly, yesterday, in the not so distant past, and that, we can only hope, we will know again tomorrow.

Here, "yesterday" refers to the era just before that infamous dot.com bust in 1999 or thereabouts. Remember those days? When young computer whiz kids were driving around in all those BMWs and living the good life in million-dollar homes (mortgaged to the hilt, of course) with portfolios filled with promissory notes? Not to worry; the dot.com wizards were the new rich … until the bust, that is. Suddenly their paper was just paper and almost worthless, certainly not as utilitarian as toilet paper; it was just paper with print on it that promised untold fortunes that weren't meant to be. More recently, dysfunctional, asleep-at-the-switch government regulators and congressional officials allowed just about anyone and everyone to buy a home—even those with bad credit or no credit or no job or no future. These nonqualified buyers have brought us to the sharp edge of the abyss we are now straddling, and if we are not careful we will fall over.

So, is it all doom and gloom? Short answer: Don't have a clue. Long answer: Not necessarily, if we make the right moves soon. Right moves? What are they? How do we maintain the good life? How do we improve on our lives to the point where we are living as well as or better than our parents did?

The author does not profess to understanding all of the problems facing our country today nor do we know all the solutions. We know what we know from what we know—from what we see and feel. Our great country is at a crossroads. We can maintain our current lifestyles, fall on our swords, or progress through innovation. We hope that it is progression, moving forward to maintain the good life, that lies before us. This is our hope.

So, how do we progress—how do we maintain that so-called good life? Perhaps the first step is to achieve energy self-sufficiency through innovation and a fundamental shift from fossil fuels to renewable, nonpolluting energy supplies. Energy is the lifeblood of America. We must get off our knees and stop begging those who hate us to supply us with the energy we need. We must put America first. Using American genius we must innovate and come up with a solution to the pending energy crisis. Remember, every problem has a solution; we must find the solution to the energy crisis. Moreover, when we do innovate, we must make sure that our country holds onto its trade secrets regarding how to produce renewable energy sources; otherwise, other countries will simply steal our ideas and manufacture energy at rates against which we cannot compete. As an example, $9 of production costs (mainly for labor, raw materials, and taxes) in the United States cannot possibly compete with $1 production costs in other countries for the same product. If we are going to innovate, let's innovate for America. We have given away too much, and now we are paying for it.

Glossary*

Absorber: In a photovoltaic device, the material that readily absorbs photons to generate charge carriers (free electrons or holes).

Acid hydrolysis: A chemical process in which acid is used to convert cellulose or starch to sugar.

Albedo: The ratio of light reflected by a surface to the light falling on it.

Alcohol: A general class of hydrocarbons that contains a hydroxyl group (OH); the many types of alcohol include butanol, ethanol, and methanol.

Alcohol fuels: Alcohol can be blended with gasoline for use as transportation fuel. It may be produced from a wide variety of organic feedstock. The common alcohol fuels are methanol and ethanol. Methanol may be produced from coal, natural gas, wood, and organic waste; ethanol is commonly made from agricultural plants, primarily corn, containing sugar.

Alkaline fuel cell (AFC): A type of hydrogen/oxygen fuel cell in which the electrolyte is concentrated potassium hydroxide (KOH) and the hydroxide ions (OH) are transported from the cathode to the anode.

Alternating current (AC): An electric current that reverses its direction at regularly recurring intervals, usually 50 or 60 times per second.

Alternator: A device that turns the rotation of a shaft into alternating current (AC).

Ambient: Natural condition of the environment at any given time.

Amorphous silicon: An alloy of silica and hydrogen, with a disordered, noncrystalline internal atomic arrangement, that can be deposited in thin layers (a few micrometers in thickness) by a number of a deposition methods to produce thin-film photovoltaic cells on glass, metal, or plastic substrates.

Ampere (amp): A unit of electrical current flowing in a circuit; can be considered to be analogous to water flowing through a pipe being measured in liters per minute.

Ampere-hour: A measure of the flow of current (in amperes) over one hour; used to measure energy production over time and battery capacity.

Anaerobic digestion: A biochemical process by which organic matter is decomposed by bacteria in the absence of oxygen, producing methane and other byproducts.

Anemometer: Wind-speed measurement device used to send data to the controller; also used to conduct wind site surveys.

Angle of attack: In wind turbine operation, the angle of the airflow relative to the blade.

Angle of incidence: The angle that a ray of sun makes with a line perpendicular to the surface; for example, a surface that directly faces the sun has a solar angle of incidence of 0, but if the surface is parallel to the sun (e.g., sunrise striking a horizontal rooftop), the angle of incidence is 90°.

Anion: A negatively charged ion; an ion that is attracted to the anode.

Annual removals: Net volume of growing stock trees removed from the inventory during a specified year by harvesting, cultural operations such as timber land improvement, or land clearing.

Annualized growth rate: Calculated as follows: $(x_n/x_1)1/n$, where x is the value under consideration and n is the number of periods.

Anode: The electrode at which oxidation (a loss of electrons) takes place. For fuel cells and other galvanic cells, the anode is the negative terminal; for electrolytic cells, the anode is the positive terminal.

* Some definitions were retrieved from EERE, *Glossary of Energy-Related Terms*, Office of Energy Efficiency & Renewable Energy, U.S. Department of Energy, Washington, DC, 2013 (http://energy.gov/eere/energybasics/articles/glossary-energy-related-terms).

Aquifer: Water-bearing stratum of permeable sand, rock, or gravel.

Asexual reproduction: The naturally occurring ability of some plant species to reproduce asexually through seeds, meaning the embryos develop without a male gamete. This ensures that the seeds will produce plants identical to the mother plant.

Availability factor: A percentage representing the number of hours a generating unit is available to produce power (regardless of the amount of power) in a given period, compared to the number of hours in the period.

Azimuth angle: The angle between true south and the point on the horizon directly below the sun.

Bagasse: The fibrous material remaining after the extraction of juice from sugarcane; often burned by sugar mills as a source of energy.

Baseload plants: Electricity-generating units that are operated to meet the constant or minimum load on the system. The cost of energy from such units is usually the lowest available to the system.

Binary cycle: Combination of two power plant turbine cycles utilizing two different working fluids for power production. The waste heat from the first turbine cycle provides the heat energy for the operation of the second turbine, thus providing higher overall system efficiencies.

Biobased product: As defined by the Farm Security and Rural Investment Act (FSRIA), a product determined by the U.S. Secretary of Agriculture to be a commercial or industrial product (other than food or feed) that is composed, in whole or in significant part, of biological products or renewable domestic agriculture materials (including plant, animal, and marine materials) or forestry materials.

Biochemical conversion: The use of fermentation or anaerobic digestion to produce fuels and chemicals from organic sources.

Biodiesel: Fuel derived from vegetable oils or animals fats; it is produced when a vegetable oil or animal fat is chemically reacted with an alcohol.

Bioenergy: Useful, renewable energy produced from organic matter that may be used directly as a fuel or processed into liquids and gases.

Biofuels: Liquid fuels and blending components produced from biomass (plant) feedstocks; used primarily for transportation.

Biogas: Combustible gas derived from decomposing biological waste; normally consists of 50 to 60% methane.

Biomass: Any organic nonfossil material of biological origin constituting a renewable energy source; produced from organic matter that is available on a renewable or recurring basis, including agricultural crops and trees, wood and wood residues, plants (including aquatic plants), grasses, animal manure, municipal residues, and other residue materials. Biomass is generally produced in a sustainable manner from water and carbon dioxide by photosynthesis. The three main categories of biomass are primary, secondary, and tertiary.

Biomass gas (biogas): A medium-Btu gas containing methane and carbon dioxide; it is produced by the action of microorganisms on organic materials such as a landfill.

Biomaterials: Products derived from organic-based (as opposed to petroleum-based) products.

Bio-oil: Intermediate fuel derived from fast pyrolysis.

Biopower: The use of biomass feedstock to produce electric power through direct combustion of the feedstock, gasification, and then combustion of the resultant gas, or through other thermal conversion processes. Power is generated with engines, turbines, fuel cells, or other equipment.

Biorefinery: A facility that processes and converts biomass into value-added products, ranging from biomaterials to fuels such as ethanol or important feedstocks for the production of chemicals and other materials. Biorefineries can utilize mechanical, thermal, chemical, and biochemical processes.

Blackbody: An ideal substance that absorbs all radiation falling on it, reflecting nothing.

Black liquor (pulping liquor): The alkaline spent liquor removed from the digesters in the process of chemically pulping wood. After evaporation, the liquor is burned as a fuel in a recovery furnace that permits the recovery of certain basic chemicals.

Bone dry: Having zero moisture content.

Borehole breakouts: Failure of the borehole wall that occurs because of stress in the rock surrounding the borehole. The breakout is generally located symmetrically in the wellbore perpendicular to the direction of greatest horizontal stress on a vertical wellbore.

Brine: A geothermal solution containing appreciable amounts of sodium chloride or other salts.

British thermal unit (Btu): A basic measure of thermal (heat) energy. A Btu is defined as the amount of energy required to increase the temperature of 1 pound of water by $1°F$ at normal atmospheric pressure. 1 Btu = 1055 joules.

Bulk density: Weight per unit of volume, usually specified in pounds per cubic foot.

Cap rocks: Rocks of low permeability that overlie a geothermal reservoir.

Capacity factor: The ratio of the electrical energy produced by a generating unit for the period of time considered to the electrical energy that could have been produced at continuous full-power operation during the same period.

Capacity, gross: The full-load continuous rating of a generator, prime mover, or other electric equipment under specified conditions as designated by the manufacturer. It is usually indicated on a nameplate attached to the equipment.

Capitol cost: The cost of field development and plant construction and the equipment required for the generation of electricity.

Carbon dioxide (CO_2): A product of combustion; the most common greenhouse gas.

Carbon monoxide (CO): A colorless, odorless gas produced by incomplete combustion; carbon monoxide is poisonous when inhaled.

Carbon sequestration: The absorption and storage of carbon dioxide from the atmosphere by naturally occurring plants.

Carnot cycle: An ideal heat engine (conceived by Sadi Carnot) in which the sequence of operations forming the working cycle consists of isothermal expansion, adiabatic expansion, isothermal compression, and adiabatic compression back to its initial state.

Cascading heat: A process that uses a stream of geothermal hot water or steam to perform successive tasks requiring lower and lower temperatures.

Casing: Pipe placed in a wellbore as a structural interface between the wellbore and the surrounding formation. It typically extends from the top of the well and is cemented in place to maintain the diameter of the wellbore and provide stability.

Cast silicon: Crystalline silicon that is obtained by pouring pure molten silicon into a vertical mold and adjusting the temperature gradient along the mold during cooling to achieve slow, vertically advancing crystallization of the silicon. The resulting polycrystalline ingot is composed of large, relatively parallel, interlocking crystals. The cast ingots are sawed into wafers for further fabrication into photovoltaic cells. Cast-silicon wafers and ribbon-silicon sheets fabricated into cells are usually referred to as *polycrystalline photovoltaic cells.*

Cathode: The electrode at which reduction (a gain of electrons) occurs. For fuel cells and other galvanic cells, the cathode is the positive terminal; for electrolytic cells (where electrolysis occurs), the cathode is the negative terminal.

Cation: A positively charged ion.

Cellulose: The main carbohydrate in living plants; cellulose forms the skeletal structure of the plant cell wall.

Cellulosic ethanol: Ethanol derived from cellulosic and hemicellulosic portions of plant biomass.

Chips: Woody material cut into short, thin wafers. Chips are used as a raw material for pulping and fiberboard or as biomass fuel.

Closed-loop biomass: Crops grown in a sustainable manner for the purpose of optimizing their value for bioenergy and bioproduct uses. These include annual crops, such as maize and wheat, and perennial crops, such as trees, shrubs, and grasses (e.g., switchgrass).

Coarse materials: Wood residues suitable for chipping, such as slabs, edgings, and trimmings.

Co-firing: Practice of introducing biomass into the boilers of coal-fired power plants.

Combined cycle: An electric generating technology in which electricity is produced from otherwise lost waste heat exiting one or more gas (combustion) turbines. The exiting heat is routed to a conventional boiler or to a heat-recovery steam generator for utilization by a steam turbine in the production of electricity. Such designs increase the efficiency of the electric generating unit.

Combined heat and power (CHP) plant: A plant designed to produce both heat and electricity from a single heat source. This term has replaced the term "cogenerator." CHP better describes the facilities because some of the plants included do not produce heat and power in a sequential fashion; as a result, they do not meet the legal definition of cogeneration specified in the Public Utility Regulatory Polices Act (PURPA).

Commercial sector: An energy-consuming sector that consists of service-providing facilities and equipment of nonmanufacturing businesses; federal, state, and local governments; and other private and public organizations, such as religious, social, or fraternal groups. The commercial sector includes institutional living quarters, as well as sewage treatment facilities. Common uses of energy associated with this sector include space heating, water heating, air conditioning, lighting, refrigeration, cooking, and running a wide variety of other equipment. This sector includes generators that produce electricity or useful thermal output primarily to support the activities of the above-mentioned commercial establishments.

Commercial species: Tree species suitable for industrial wood products.

Compressed natural gas (CNG): Mixture of hydrocarbon gases and vapors, consisting principally of methane in gaseous form that has been compressed.

Concentrator: A reflective or refractive device that focuses incident insolation onto an area smaller than the reflective or refractive surface, resulting in increased insolation at the point of focus.

Condensate: Water formed by condensation of steam.

Condenser: Equipment that condenses turbine exhaust steam into condensate.

Conservation Reserve Program (CRP): Program that provides farm owners or operators with an annual per-acre rental payment and half the cost of establishing a permanent land cover in exchange for retiring environmentally sensitive cropland from production for 10 to 15 years. In 1996, Congress reauthorized CRP for an additional round of contracts, limiting enrollment to 36.4 million acres at any time. The 2002 Farm Act increased the enrollment limit to 39 million acres. Producers can offer land for competitive bidding based on an Environmental Benefits Index (EBI) during periodic signups, or they can automatically enroll more limited acreages in such practices as riparian buffers, field windbreaks, and grass strips on a continuous basis. CRP is funded through the Commodity Credit Corporation (CCC).

Cooling tower: A structure in which heat is removed from hot condensate.

Conventional hydroelectric (hydropower) plant: A plant in which all the power is produced from natural streamflow as regulated by available storage.

Core: A cylinder of rock recovered from a well by a special coring drill bit.

Crop failure: Acreage on which crops have failed due to weather, insects, or diseases, but can also include some land not harvested due to lack of labor, lower market prices, or other factors. Acreage planted with cover crops or soil improvement crops not intended for harvest is excluded and is considered idle.

Cropland: Total cropland includes cropland harvested, crop failure, cultivated summer fallow, cropland used only for pasture, and idle cropland.

Cropland harvested: Includes row crops and closely sown crops, hay and silage crops, tree fruits, small fruits, berries, tree nuts, vegetables, melons, and miscellaneous other minor crops and hay. In recent years, farmers have double-cropped about 4% of this acreage.

Cropland used for crops: Cropland used for crops includes cropland harvested, crop failure, and cultivated summer fallow.

Cropland used for pasture: Land used for long-term crop rotation; however, some cropland pasture is marginal for crop uses and may remain in pasture indefinitely. This category also includes land that was used for pasture before crops reached maturity and some land used for pasture that could have been cropped without additional improvement.

Crust: Earth's outer layer of rock; also called the *lithosphere*.

Cryogenic liquefaction: The process through which gases such as nitrogen, hydrogen, helium, and natural gas are liquefied under pressure at very low temperatures.

Cull tree: A live tree, 5 inches in diameter at breast height or larger, that is not merchantable for saw logs now or prospectively because of rot, roughness, or species.

Cut-in speed: The speed at which a shaft must turn to generate electricity and send it over a wire.

Daylighting: The use of direct, diffuse, or reflected sunlight to provide supplemental lighting for building interiors.

d.b.h: The diameter measured at approximately breast height from the ground.

Densification: A mechanical process to compress biomass (usually wood waste) into pellets, briquettes, cubes, or densified logs.

Depletion factor: Annual percentage of the depletion of a thermal resource.

Digester gas: Biogas that is produced using a digester, an airtight vessel or enclosure in which bacteria decomposes biomass in water to produce biogas.

Diode: A solid-state device that acts as a one-way valve for electricity.

Direct current (DC): An electric current that flows in a constant direction; the magnitude of the current does not vary or has only a slight variation.

Direct methanol fuel cell (DMFC): A fuel cell in which the fuel is methanol (CH_3OH) in gaseous or liquid form. The methanol is oxidized directly at the anode instead of first being reformed to produce hydrogen. The electrolyte is typically polymer electrolyte membrane (PEM).

Direct use: Use of geothermal heat without first converting it to electricity, such as for space heating and cooling, food preparation, industrial processes, etc.

Distributed generation (distributed energy resources): Refers to electricity provided by small, modular power generators (typically ranging in capacity from a few kilowatts to 50 megawatts) located at or near customer demand.

District heating: A type of direct use in which a utility system supplies multiple users with hot water or steam from a central plant or well field.

Downwind turbine: A turbine that does not face into the wind and whose direction is controlled directly by the wind.

Drag bit: Drilling bit that drills by scraping or shearing rock with fixed hard surfaces known as cutters.

Drilling: Boring into the ground to access geothermal resources, usually with oil and gas drilling equipment that has been modified to meet geothermal requirements.

Dry steam: Very hot steam that occurs without liquid.

Dump load: A device that allows excess energy to be safely disposed of.

E10: A mixture of 10% ethanol and 90% gasoline based on volume.

E85: A mixture of 85% ethanol and 15% gasoline based on volume.

Efficiency: The ratio of the useful energy output of a machine or other energy-converting plant to the energy input.

Electric power sector: An energy-consuming sector that consists of electricity-only and combined heat and power (CHP) plants whose primary business is to sell electrical, or electricity and heat, to the public (i.e., North American Industry Classification System 22 plants).

Electric utility: A corporation, person, agency, authority, or other legal entity or instrumentality aligned with distribution facilities for delivery of electric energy for use primarily by the public. Included are investor-owned electric utilities, municipal and state utilities, federal electric utilities, and rural electric cooperatives. A few entities that are tariff based and corporately aligned with companies that own distribution facilities are also included.

Emissions: Anthropogenic releases of gases to the atmosphere. In the context of global climate change, they consist of radiatively important greenhouse gases (e.g., the release of carbon dioxide during fuel combustion).

Endothermic: A chemical reaction that absorbs or requires energy (usually in the form of heat).

Energy: The ability to do work.

Energy crops: Crops grown specifically for their fuel value. They include food crops such as corn and sugarcane and nonfood crops such as poplar trees and switchgrass. Low-energy crops that are under development include short-rotation woody crops (fast-growing hardwood trees) harvested in 5 to 8 years and herbaceous energy crops (e.g., perennial grasses) harvested annually after taking 2 to 3 years to reach full productivity.

Enhanced geothermal systems (EGS): Engineered reservoirs that can extract economic amounts of heat from geothermal resources. Rock fracturing, water injection, and water circulation technologies sweep heat from unproductive areas of existing geothermal fields or new fields lacking sufficient production capacity.

Enthalpy: A thermodynamic property of a substance, defined as the sum of its internal energy plus the pressure of the substance times its volume, divided by the mechanical equivalent of heat. The total heat content of air; the sum of enthalpies of dry air and water vapor, per unit weight of dry air; measured in Btu per pound (or calories per kilogram).

Environmental Impact Statement: A document created from a study of the expected environmental effects of a new development or installation.

Ethanol (CH_3CH_2OH; also known as ethyl alcohol or grain alcohol): A clear, colorless flammable oxygenated hydrocarbon with a boiling point of 173.5°F in the anhydrous state. It readily forms a binary azetrope with water, with a boiling point of 172.67°F at a composition of 95.57% by weight ethanol. It is used in the United States as a gasoline octane enhancer and oxygenate (maximum 10% concentration). Ethanol can be used in higher concentrations (E85) in vehicles designed for its use. Ethanol is typically produced chemically from ethylene or biologically from the fermentation of various sugars from carbohydrates found in agricultural crops and cellulosic residues from crops or wood. The lower heating value, equal to 76,000 Btu per gallon, is assumed for most estimates in this text.

Evacuated tube: In a solar thermal collector, an absorber tube contained in an evacuated glass cylinder through which collector fluids flows.

Exothermic: A chemical reaction that gives off heat.

Externality: A cost or benefit not accounted for in the price of goods and services; often refers to the cost of pollution and other environmental impacts.

Fast pyrolysis: Thermal conversion of biomass by rapid heating to 450 to 600°C in the absence of oxygen.

Fault: A fracture in rock exhibiting relative movement between the adjoining surfaces.

Feedstock: A product used as the basis for the manufacture of another product.

Feller buncher: A self-propelled machine that cuts trees with giant shears near ground level and then stacks the trees into piles to await skidding.

Fenestration: Openings in a building (e.g., windows, doors, skylights). The whole-building design approach determines what type of fenestration products should be used. It is desirable to select products with characteristics that accommodate the local climate and address insulating, daylighting, heating and cooling, and natural ventilation needs.

Fermentation: Conversion of carbon-containing compounds by microorganisms for the production of fuels and chemicals such as alcohols, acids, or energy-rich gases.

Fiber products: Products derived from the fibers of herbaceous and woody plant materials. Examples include pulp, composition-board products, and wood chips for export.

Fine materials: Wood residues not suitable for chipping, such as planer shavings and sawdust.

Fischer–Tropsch fuels: Liquid hydrocarbon fuels produced by a process that combines carbon monoxide and hydrogen. The process is used to convert coal, natural gas, and low-value refinery products into high-value diesel substitutes.

Flash steam: Steam produced when the pressure on a geothermal liquid is reduced; also called *flashing.*

Flat plate pumped: A medium-temperature solar thermal collector that typically consists of a metal frame, glazing, absorbers (usually metal), and insulation and that uses a pump liquid as the heat-transfer medium; predominant use is in water heating applications.

Flexible-fuel vehicle: A vehicle with a single fuel tank designed to run on varying blends of unleaded gasoline with either ethanol or methanol.

Flow battery: An electrochemical energy storage device that utilizes tanks of rechargeable electrolyte to refresh the energy-producing reaction. Because its capacity is limited only by the size of its electrolyte tanks, it is useful for large-scale backup systems to supplement other forms of generation that may be intermittent in nature.

Forest land: Land at least 10% stocked by forest trees of any size, including land that formerly had such tree cover and that will be naturally or artificially regenerated. Forest land includes transition zones, such as areas between heavily forested and nonforested lands that are at least 10% stocked with forest trees and forested areas adjacent to urban or built-up lands. Also included are pinyon–juniper and chaparral areas in the West and afforested areas. The minimum area for classification of forest land is 1 acre. Roadside, streamside, and shelterbelt strips of trees must have a crown width of at least 120 feet to qualify as forest land. Unimproved roads and trails, streams, and clearings in the forest areas are classified as forest if less than 120 feet wide.

Fracture: Natural or induced break in rock.

Fracturing treatments: Fracturing treatments are performed by pumping fluid into the subsurface at pressures above the fracture pressure of the reservoir formation to create a highly conductive flow path between the reservoir and the wellbore.

Fuel cell: Cell capable of generating an electrical current by converting the chemical energy of a fuel directly into electrical energy. Fuel cells differ from conventional electrical cells in that the active materials such as fuel and oxygen are not contained within the cell but are supplied from outside.

Fuel cell poisoning: The lowering of a fuel cell's efficiency due to impurities in the fuel binding to the catalyst.

Fuel cell stack: Individual fuel cells connected in a series; fuel cells are stacked to increase voltage.

Fuel Treatment Evaluator (FTE): A strategic assessment tool capable of aiding the identification, evaluation, and prioritization of fuel treatment opportunities.

Fuelwood: Wood and wood products, possibly including coppices, scrubs, branches, etc., bought or gathered and used by direct combustion.

Full sun: The amount of power density in sunlight received at the Earth's surface at noon on a clear day (about 1000 W/m^2).

Fumarole: A vent or hole in the Earth's surface, usually in a volcanic region, from which steam, gaseous vapors, or hot gases issue.

Gallon: A volumetric measure equal to 4 quarts (231 cubic inches). One gallon equals 3785 liters; 1 barrel of fuel oil equals 42 gallons.

Gasification: A chemical or heat process to convert a solid fuel to a gaseous form.

Gasohol: A motor vehicle fuel that is a blend of 90% unleaded gasoline and 10% ethanol (by volume).

Gearbox: Increases the rpm of a low-speed shaft, transferring its energy to a high-speed shaft to provide enough speed to generate electricity.

Generation (electricity): The process of producing electric energy from other forms of energy; the amount of electric energy produced is expressed in watt-hours (Wh).

Geothermal: Of or relating to the Earth's interior heat.

Geothermal energy: The Earth's interior heat made available by extracting it from hot water or rocks. Electric power plants extract hot water or steam from geothermal reservoirs in the Earth's crust that is supplied to steam turbines to drive generators to produce electricity.

Geothermal gradient: The rate of temperature increase in the Earth as a function of depth. Temperature increases an average of 1°F for every 75 feet in descent.

Geothermal heat pumps: Devices that take advantage of the relatively constant temperature of the Earth's interior, using it as a source and sink of heat for both heating and cooling. For cooling purposes, heat is extracted from the space and dissipated into the Earth; for heating purposes, heat is extracted from the Earth and pumped into the space.

Geothermal plant: A plant in which a turbine is driven by either hot water or natural steam that derives its energy from heat found in rocks or fluids at various depths beneath the surface of the Earth. The fluids are extracted by drilling or pumping.

Geothermal resources: The natural heat of the Earth that can be used for beneficial purposes when the heat is collected and transported to the surface.

Geyser: A spring that shoots jets of hot water and steam into the air.

Glazing: Transparent or translucent material (glass or plastic) used to admit light or to reduce heat loss; used for windows, skylights, or greenhouses or for covering the aperture of a solar collector.

Global positioning system (GPS): A navigation system using satellite signals to fix the location of a radio receiver on or above the Earth's surface.

Grassland pasture and range: All open land used primarily for pasture and grazing, including shrub and brush land types of pasture, grazing land with sagebrush and scattered mesquite, and all tame and native grasses, legumes, and other forage used for pasture or grazing. Because of the diversity in vegetative composition, grassland pasture and range are not always clearly distinguishable from other types of pasture and range. At one extreme, permanent grassland may merge with cropland pasture, or grassland may often be found in transitional areas with forested grazing land.

Gravimetry: The use of precisely measured gravitational force to determine mass differences that can be correlated to subsurface geology.

Green pricing/marketing: In the case of renewable electricity, green pricing represents a market solution to the various problems associated with regulatory valuation of the non-market benefits of renewables. Green pricing programs allow electricity customers to express their willingness to pay for renewable energy development through direct payments on their monthly utility bills.

Greenhouse gas: Gases that trap the heat of the sun in the Earth's atmosphere, producing the greenhouse effect. The two major greenhouse gases are water vapor and carbon dioxide. Other greenhouse gases include methane, ozone, chlorofluorocarbons, and nitrous oxide.

Grid: The layout of an electrical distribution system.

Gross generation: The total amount of electric energy produced by the generating units at a generation station or stations, measured at the generator terminals.

Growing stock: A classification of timber inventory that includes live trees of commercial species meeting specified standards of quality or vigor. Cull tress are excluded. When associated with volume, includes only trees 5.0 inches in d.b.h. and larger.

Hardwoods: Usually broad-leaved and deciduous trees.

HDR (hot dry rock): Subsurface geologic formations of abnormally high heat content that contain little or no water.

Heat exchanger: A device for transferring thermal energy from one fluid to another.

Heat flow: Movement of heat from within the Earth to the surface, where it is dissipated into the atmosphere, surface water, and space by radiation.

Heat pump: A year-round heating and air-conditioning system employing a refrigeration cycle. In the refrigeration cycle, a refrigerant is compressed (as a liquid) and expanded (as a vapor) to absorb and reject heat. The heat pump transfers heat to a space to be heated during the winter period and, by reversing the operation, extracts (absorbs) heat from the same space to be cooled during the summer period. The refrigerant within a heat pump in the heating mode absorbs heat from an outside medium (air, ground, or groundwater); in the cooling mode, it absorbs heat from the space to be cooled and disperses it to the outside medium.

Heat pump, air-source: The most common type of heat pump; it absorbs heat from the outside air and transfers the heat to the space to be heated in the heating mode. In the cooling mode, the heat pump absorbs heat from the space to be cooled and rejects the heat to the outside air. When the outside air approaches 32°F or less, air-source heat pumps in the heating mode lose efficiency and generally require a back-up (resistance) heating system.

Heat pump efficiency: The electrical energy required to operate a heat pump is directly related to the temperatures between which it operates. Geothermal heat pumps are more efficient than conventional heat pumps or air conditioners that use the outdoor air, as the temperature of the ground or groundwater that is a few feet below the Earth's surface remains relatively constant throughout the year. It is more efficient in the winter to draw heat from the relatively warm ground than from the atmosphere where the air temperature is much colder, and in summer to transfer waste heat to the relatively cool ground rather than to hotter air. Geothermal heat pumps are generally more expensive to install than outside air heat pumps; however, depending on the location, geothermal heat pumps can reduce energy consumption (operating costs) and emissions by more than 20% compared to high-efficiency outside air heat pumps. Geothermal heat pumps also use the waste heat from air-conditioning to provide free hot water heating in the summer.

Heat pump, geothermal: A heat pump in which the refrigeration exchanges heat (in a heat exchanger) with a fluid circulating through an earth connection medium (ground or groundwater). The fluid is contained in a variety of loop (pipe) configurations depending on the temperature of the ground and the ground area available. Loops may be installed horizontally or vertically in the ground or submerged in a body of water.

Heating value: The maximum amount of energy available from burning a substance.

Herbaceous: Nonwoody type of vegetation, usually lacking permanent strong stems, such as grasses, cereals, and canola (rapeseed).

High-speed shaft: Transmits force from the gearbox to the generator.

High-temperature collector: A solar thermal collector designed to operate at a temperature of 180°F or higher.

Hub: The center part of the rotor assembly which connects the blades to the low-speed shaft.

Hydraulic stimulation: A stimulation technique performed using fluid.

Hydrothermal: Pertaining to hot water.

Hydrothermal reservoir: An aquifer, or subsurface water, that has sufficient heat, permeability, and water to be exploited without stimulation or enhancement.

Idle cropland: Land in cover and soil improvement crops and cropland on which no crops were planted. Some cropland is idle each year for various physical and economic reasons. Acreage diverted from crops to soil-conserving uses (if not eligible for and used as cropland pasture) under federal farm programs is included in this component; cropland enrolled in the Federal Conservation Reserve Program (CRP) is included in idle cropland.

Incident light: Light that shines onto the face of a solar cell or module.

Induced seismicity: Induced seismicity refers to typically minor earthquakes and tremors that are caused by human activity that alters the stresses and strains on the Earth's crust. Most induced seismicity is of an extremely low magnitude, and in many cases human activity is merely the trigger for an earthquake that would have occurred naturally in any case.

Industrial wood: All commercial roundwood products except fuelwood.

Injection: The process returning spent geothermal fluids to the subsurface; sometimes referred to as *reinjection.*

Interferometric synthetic aperture radar (InSAR): A remote sensing technique that uses radar satellite images to determine movement of the surface of the Earth.

Internal collector storage: A solar thermal collector in which incident solar radiation is absorbed by the storage medium.

Irradiance: The direct, diffuse, and reflected solar radiation that strikes a surface.

Joule: The basic energy unit for the metric system or, in a later more comprehensive formulation, the International System of Units (SI). It is defined in terms of the meter, kilogram, and second.

Kaplan turbine: A type of turbine that has two blades whose pitch is adjustable. The turbine may have gates to control the angle of the fluid flow into the blades.

Kilowatt (kW): One thousand watts of electricity.

Kilowatt-hour (kWh): One thousand watt-hours.

Landfill gas: Gas that is generated by the decomposition of organic material at landfill disposal sites; landfill gas is approximately 50% methane.

Langley (L): Unit of solar irradiance; 1 gram-calorie per square centimeter; $1 L = 85.93$ kWh/m^2.

Leeward: Away from the direction of the wind; opposite of windward.

Life-cycle analysis: Analysis focused on the environmental impact of a product during the entirety of its life cycle, from resource extraction to postconsumer waste disposal. It is a comprehensive approach to examining the environmental impacts of a product or package.

Lignin: Structural constituent of wood and (to a lesser extent) other plant tissues that encrusts the cell walls and cements the cells together.

Line shaft pump: Fluid pump that has the pumping mechanism in the wellbore and that is driven by a shaft connected to a motor on the surface.

Liner: A casing string that does not extend to the top of wellbore but instead is anchored or suspended from inside the bottom of the previous casing string.

Liquid collector: A medium-temperature solar thermal collector, employed predominately in water heating, which uses pumped liquid as the heat-transfer medium.

Lithology: The study and description of rocks, in terms of their color, texture, and mineral composition.

Live cull: A classification that includes live cull trees. When associated with volume, it is the net volume in live cull trees that are 5.0 inches in d.b.h. and larger.

Load: The simultaneous demand of all customers required at any specified point in an electric power system.

Load balancing: Keeping the amount of electricity produced (supply) equal to the consumption (demand). This is one of the challenges of wind energy production, which produces energy on a less predicable schedule than other methods.

Local solar time: A system of astronomical time in which the sun crosses the true north–south meridian at 12 noon and which differs from local time according to longitude, time zone, and equation of time.

Logging residues: The unused portions of growing stock and nongrowing stock trees cut or killed by logging and left in the woods.

Lost circulation: Zones in a well that imbibe drilling fluid from the wellbore, thus causing a reduction in the flow of fluid returning to the surface. This loss causes drilled rock particles to build up in the well and can cause problems in cementing the casing in place.

Low-speed shaft: Connects the rotor to the gearbox.

Low-temperature collectors: Metallic or nonmetallic solar thermal collectors that generally operate at temperatures below 110°F and use pumped liquid or air as the heat-transfer medium. They usually contain no glazing and no insulation, and they are often made of plastic or rubber, although some are made of metal.

Magma: Molten rock within the Earth from which igneous rock is formed by cooling.

Magnetic survey: Measurements of the Earth's magnetic field that are then mapped and used to determine subsurface geology.

Magnetotellurics: An electromagnetic method of determining structures below the Earth's surface using electrical currents and the magnetic field.

Mantle: The Earth's inner layer of molten rock, lying beneath the Earth's crust and above the Earth's core of liquid iron and nickel.

Matrix treatments: Treatments performed below the reservoir fracture pressure; they are generally designed to restore the natural permeability of the reservoir following damage to the near-wellbore area. Matrix treatments typically use hydrochloric or hydrofluoric acids to remove mineral material that reduces flow into the well.

Medium-temperature collectors: Solar thermal collectors designed to operate in the temperature range of 140 to 180°F but that can also operate at a temperature as low as 110°F. The collector typically consists of a metal frame, metal absorption panel with integral flow channels (attached tubing for liquid collectors or integral ducting for air collectors), and glazing and insulation on the sides and back.

Megawatt (WM): One million watts of electricity.

Methane: A colorless, flammable, odorless hydrocarbon gas (CH_4) that is the major component of natural gas. It is also an important source of hydrogen in various industrial processes. Methane is a greenhouse gas.

Methanol (CH_3OH; also known as methyl alcohol or wood alcohol): Methanol is usually produced by chemical conversion at high temperatures and pressures. Although normally produced from natural gas, methanol can be produced from gasified biomass.

Microseismicity: Small movements of the Earth causing fracturing and movement of rocks. Such seismic activity does not release sufficient energy for the events to be recognized except with sensitive instrumentation.

Mini-frac: A small fracturing treatment performed before the main hydraulic fracturing treatment to acquire stress data and to test prestimulation permeability.

MSW (municipal solid waste): Residential solid waste and some nonhazardous commercial, institutional, and industrial wastes.

MTBE (methyl *tert*-butyl ether): Fuel oxygenate produced by reacting methanol with isobutylene.

Multiplier effect: Sometimes called the *ripple effect* because a single expenditure in an economy can have repercussions throughout the entire economy. The multiplier is a measure of how much additional economic activity is generated from an initial expenditure.

Nacelle (or cowling): Contains and protects the gearbox and generator, sometimes large enough for an engineer or technician to stand in while doing maintenance.

Net metering: Arrangement that permits a facility (using a meter that reads inflows and outflows of electricity) to sell any excess power it generates over its load requirement back to the electrical grid to offset consumption.

Net photovoltaic cell shipment: The difference between photovoltaic cell shipments and photovoltaic cell purchases.

Net photovoltaic module shipment: The difference between photovoltaic cell shipments and photovoltaic cell purchases.

Net summer capacity: The maximum output, commonly expressed in megawatts, that generating equipment can supply to system load, as demonstrated by a multi-hour test, at the time of summer peak demand. This output reflects a reduction in capacity due to electricity use for station service or auxiliaries.

Nocturnal cooling: The effect of cooling by the radiation of heat from a building to the night sky.

Nonforest land: Land that has never supported forests and lands formerly forested where timber production is precluded by development for other uses; includes area used for crops, improved pasture, residential areas, city parks, improved roads of any width and adjoining clearings, powerline clearings of any width, and 1- to 4.5-acre areas of water classified by the Bureau of the Census as land. If intermingled in forest areas, unimproved roads and nonforest strips must be more than 120 feet wide, and clearings, etc., must be more than 1 acre in area to qualify as nonforest land.

Nonindustrial private: An ownership class of private lands where the owner does not operate wood-using processing plants.

Nonutility generation: Electric generation by nonutility power producers to supply electric power for industrial, commercial, and military operations or for sales to electric utilities.

Nonutility power producer: A corporation, person, agency, authority, or other legal entity or instrumentality that owns electric generating capacity and is not an electrical utility. Nonutility power producers include qualifying cogenerators, qualifying small power producers, and other nonutility generators without a designated, franchised service area that do not file forms listed in the Code of Federal Regulations, Title 18, Part 141.

Occupied space: The space within a building or structure that is normally occupied by people and that may be conditioned (heated, cooled, or ventilated).

Other biomass: Includes agricultural byproducts/crops (agricultural byproducts, straw), other biomass gas (digester waste alcohol), other biomass liquids (fish oil liquid acetonitrite, waste, tall oil, waste alcohol), and other biomass solids (medical waste, solid byproducts, sludge waste, tires).

Other forest land: Forest land other than timberland and reserved forest land; includes available forest land incapable of annually producing 20 ft³/acre of industrial wood under natural conditions because of adverse site conditions such as sterile soils, dry climate, poor drainage, high elevation, steepness, or rockiness.

Other removals: Wood volume not utilized from cut or otherwise killed growing stock, from cultural operations such as precommercial thinnings, or from timberland clearing; does not include volume removed from inventory through reclassification of timber land to productive reserved forest land.

Other sources of roundwood: Sources of roundwood products that are not growing stock. Include available dead, rough, and rotten trees; trees of noncommercial species; trees less the 5.0 inches d.b.h.; tops; and roundwood harvested from nonforest land (e.g., fence rows).

Packer: Device that can be placed in the wellbore to block vertical fluid flow so as to isolate zones.

Paper pellets: Paper compressed and bound into uniform diameter pellets to be burned in a heating stove.

Parabolic dish: A high-temperature (above 180°F) solar thermal concentrator, generally bow shaped, with two-axis tracking.

Passive solar: A system in which solar energy alone is used for the transfer of thermal energy. Pumps, blowers, or other heat-transfer devices that use energy other than solar are not used.

Peak watt: A manufacturer's unit indicating the amount of power a photovoltaic cell or module will produce at standard test conditions (normally 1000 W/m² and 25°C).

Peaking plants: Electricity generating plants that are operated to meet the peak or maximum load on the system. The cost of energy from such plants is usually higher than from baseload plants.

Peat: Peat consists of partially decomposed plant debris. It is considered an early stage in the development of coal. Peat is distinguished from lignite by the presence of free cellulose and a high moisture content (exceeding 70%). The heat content of air-dried peat (about 50% moisture) is about 9 million Btu/ton. Most U.S. peat is used as a soil conditioner.

Permeability: The ability of a rock to transmit fluid through its pores or fractures when subjected to a difference in pressure. Typically measured in darcies or millidarcies.

Photovoltaic (PV) cell: An electronic device consisting of layers of semiconductor materials fabricated to form a junction (adjacent layers of materials with different electronic characteristics) and electrical contacts and being capable of converting incident light directly into electricity (direct current).

Photovoltaic (PV) module: An integrated assembly of interconnected photovoltaic cells designed to deliver a selected level of working voltage and current at its output terminals, packaged for protection against environmental degradation, and suited for incorporation in photovoltaic power systems.

Plate tectonics: A theory of global-scale dynamics involving the movement of many rigid plates of the Earth's crust. Tectonic activity is evident along the margins of the plates where buckling, grinding, faulting, and vulcanism occur as the plates are propelled by the forces of deep-seated mantle convection current. Geothermal resources are often associated with tectonic activity that allows groundwater to come in contact with deep subsurface heat sources.

Poletimber trees: Live trees at least 5.0 inches in d.b.h. but smaller than sawtimber trees.

Polycrystalline diamond compact (PDC) drilling bit: A drilling bit that uses polycrystalline diamond compact cutters to shear rock with a continuous scraping motion.

Porosity: The ratio of the aggregate volume of pore spaces in rock or soil to its total volume, usually stated as a percent.

Primary wood-using mill: A mill that converts roundwood products into other wood products. Common examples are sawmills that convert logs into lumber and pulp mills that convert pulpwood roundwood into wood pulp.

Process heating: The direct process end use in which energy is used to raise the temperature of substances involved in the manufacturing process.

Proppant: Suspended particles in the fracturing fluid that are used to hold fractures open after a hydraulic fracturing treatment, thus producing a pathway that fluids can easily flow along.

Public Utility Regulatory Policies Act of 1978 (PURPA): Contains measures designed to encourage the conservation of energy, more efficient use of resources, and equitable rates. Principal among these are suggested retail rate reforms and new incentives for production of electricity by cogenerators and uses of renewable resources.

Pulpwood: Roundwood, whole tree chips, or wood residues that are used for the production of wood pulp.

Pumped-storage hydroelectric plant: A plant that usually generates electric energy during peak load periods by using water previously pumped into an elevated storage reservoir during off-peak periods when excess generating capacity is available to do so. When additional generating capacity is needed, the water can be released from the reservoir through a conduit to turbine generators located in a power plant at a lower level.

Pyrolysis: Thermal decomposition of biomass at high temperatures (greater than 400°F, or 200°C) in the absence of air. The end product is a mixture of solids (char), liquid (oxygenated oils), and gases (methane, carbon monoxide, and carbon dioxide), the proportions of which are determined by operating temperature, pressure, oxygen content, and other conditions.

Quadrillion Btu (quad): Equivalent to 10^{15} Btu.

Qualifying facility (QF): A cogeneration or small power production facility that meets certain ownership, operating, and efficiency criteria established by the Federal Energy Regulatory Commission (FERC) pursuant to the Public Utility Regulatory Polices Act of 1978 (PURPA).

Rankine cycle: The thermodynamic cycle that is an ideal standard for comparing the performance of heat engines, steam power plants, steam turbines, and heat pump systems that use a condensate vapor as the working fluid; efficiency is measured as work done divided by sensible heat supplied.

Recovery factor: The fraction of total resource that can be extracted for productive uses.

Renewable energy: Energy that is produced using resources that regenerate quickly or are inexhaustible. Wind energy is considered inexhaustible, because although it may blow intermittently it will never stop.

Renewable energy resources: Energy resources that are naturally replenishing but flow limited. They are virtually inexhaustible in duration but limited in the amount of energy that is available per unit of time. Renewable energy resources include biomass, hydro, geothermal, solar, wind, ocean thermal, wave action, and tidal action.

Renewable Portfolio Standard (RPS): A mandate requiring that renewable energy provide a certain percentage of total energy generation or consumption.

Reservoir: A natural underground container of liquids, such as water or steam (or, in the petroleum context, oil or gas).

Residues: Bark and woody materials that are generated in primary wood-using mills when roundwood products are converted to other products. Examples are slabs, edgings, trimmings, sawdust, shavings, veneer cores and clippings, and pulp screenings. Bark residues and wood residues (both coarse and fine materials) are included, but not logging residues.

Resistivity survey: Measurement of the ability of a material to resist or inhibit the flow of an electrical current, measured in ohm-meters. Resistivity is measured by inducing a current between two electrodes and measuring the resulting potential at other electrodes. Resistivity surveys can be used to delineate the boundaries of geothermal fields.

Resource base: All of a given material in the Earth's crust, whether its existence is known or unknown, and regardless of cost considerations.

Ribbon silicon: Single-crystal silicon derived by means of fabricating processes that produce sheets or ribbons of single-crystal silicon. These processes include edge-defined film-fed growth, dendritic web growth, and ribbon-to-ribbon growth.

Roller cone bit: Drill bit with studded rotating cones attached to it that crushes rock.

Rotten tree: A live tree of commercial species that does not contain a saw log now or prospectively primarily because of rot (that is, when rot accounts for more than 50% of the total cull volume).

Rough tree: A live tree of commercial species that does not contain a saw log now or prospectively primarily because of roughness (that is, when sound cull, due to such factors as poor form, splits, or cracks, accounts for more than 50% of the total cull volume), or a live tree of noncommercial species.

Roundwood: Wood cut specifically for use as a fuel.

Roundwood products: Logs and other round timber generated from harvesting trees for industrial or consumer use.

Salvable dead tree: A downed or standing dead tree that is considered currently or potentially merchantable by regional standards.

Salinity: A measure of the quantity or concentration of dissolved salts in water.

Saplings: Live trees 1.0 inch through 4.9 inches in d.b.h.

Secondary wood processing mills: A mill that uses primary wood products in the manufacture of finished wood products, such as cabinets, moldings, and furniture.

Seismic: Pertaining to, of the nature of, or caused by an earthquake or earth vibration, natural or manmade.

Seismicity: Earth movements; also, the frequency, distribution, and intensity of earthquakes.

Seismometer: Electrical device that is used on the surface and within wellbores to measure the magnitude and direction of seismic events.

Self-potential: Self-potential in geothermal systems measures currents induced in the subsurface because of the flow of fluids.

Silicon: A semiconductor material made from silica, purified for photovoltaic applications.

Single-crystal silicon, Czochralski: An extremely pure form of crystalline silicon produced by the Czochralski method of dipping a single crystal seed into a pool of molten silicon under high-vacuum conditions and slowly withdrawing a solidifying single-crystal boule rod of silicon. The boule is sawed into thin wafers and fabricated into single-crystal photovoltaic cells.

Skylight: A window located on the roof of a structure to provide interior building spaces with natural daylight, warmth, and ventilation.

Slim hole: Drill holes that have a nominal inside diameter less than about 6 inches.

Slotted liner: Liner that has slots or holes in it to let fluid pass between the wellbore and surrounding rock.

Sludge: A dense, slushy, liquid-to-semifluid product that accumulates as an end result of an industrial or technological process designed to purify a substance. Industrial sludges are produced from the processing of energy-related raw materials, chemical products, water, mined ores, sewerage, and other natural and manmade products. Sludges can also form from natural processes, such as the runoff produced by rainfall, and can accumulate on the bottom of bogs, streams, lakes, and tidelands.

Smart tracer: Tracer that is useful in determining the flow path between a well injecting fluid into the subsurface and a well producing fluid from an adjacent well; it can also be used to determine temperature along the flow path, the surface area contacted by the tracer, the volume of rock that the tracer interacts with, and the relative velocities of separate phases (gas, oil, and water in petroleum fields; steam and liquid water in geothermal systems).

Solar energy: The radiant energy of the sun which can be converted into other forms of energy, such as heat or electricity.

Solar heat gain coefficient (SHGC): Measures how well a product blocks heat caused by sunlight. The SHGC is expressed as a number between 0 and 1. The lower the SHGC, the less solar heat it transmits.

Solar thermal collector: A device designed to receive solar radiation and convert it into thermal energy. Normally, a solar thermal collector includes a frame, glazing, and an absorber, together with the appropriate insulation. The heat collected by the solar thermal collector may be used immediately or stored for later use.

Solar thermal collector, special: An evacuated-tube collector or a concentrating (focusing) collector; special collectors operate in the temperature range from just above ambient temperature (low concentration for pool heating) to several hundred degrees Fahrenheit (high concentration for air conditioning and specialized industrial processes).

Spent liquor: The liquid residue left after an industrial process; can be a component of waste materials used as fuel.

Spent sulfite liquor: End product of pulp and paper manufacturing processes that contains lignins and has a high moisture content; often reused in recovery boilers. Similar to black liquor.

Spinner survey: The use of a device with a small propeller that spins when fluid passes to measure fluid flow in a wellbore. The device is passed up and down the well, continuously measuring flow to establish where and how much fluid enters or leaves the wellbore at various depths.

Spinning reserve: Generation capacity that is online but unloaded and that can respond within 10 minutes to compensate for unexpected increases in demand or a decrease in supply.

Start-up speed: The wind speed at which a rotor begins to rotate.

Stimulation: A treatment performed to restore or enhance the productivity of a well. Stimulation treatments fall into two main groups: hydraulic fracturing treatments and matrix treatments.

Stress: The forces acting on rock. In the subsurface, the greatest force or stress is generally vertical and caused by the weight of overlying rock.

Structural discontinuity: A discontinuity of the rock fabric that can be a fracture, fault, intrusion, or differing adjacent rock type.

Submersible sump: Pump with both the pumping mechanism and a driving electric motor suspended together at depth in the well.

Subsidence: Sinking of an area of the Earth's crust due to fluid withdrawal and pressure decline.

Syngas: A synthesis gas produced through gasification of biomass; similar to natural gas. It can be cleaned and conditioned to form a feedstock for production of methanol.

TDS (total dissolved solids): Describes the amount of solid materials in water.

Tall oil: The oily mixture of rosin acids, fatty acids, and other materials obtained by acid treatment of the alkaline liquors from the digesting (pulping) of pine wood.

Thermal drawdown: Decline in formation temperature due to geothermal production.

Thermal gradient: The rate of increase in temperature as a function of depth into the Earth's crust.

Thermosiphon system: A solar collector system for water heating in which circulation of the collection fluid through the storage loop is provided solely by the temperature and density difference between the hot and cold fluids.

Thin-film silicon: A technology in which amorphous or polycrystalline material is used to make photovoltaic (PV) cells.

Tiltmeter: Device able to measure extremely small changes in its rotation from horizontal. The tilt measured by an array of tiltmeters placed over a stimulation allows delineation of inflation and fracturing caused by the stimulation.

Tip speed ratio (TSR): The ratio between the wind speed and the speed of the tips of the wind turbine blades.

Tipping fee: A fee for disposal of waste.

Tower: Steel structures that support a turbine assembly; higher towers allow for longer blades and the capture of faster moving air at higher altitudes.

Tracer: A chemical injected into the flow stream of a production or injection well to determine fluid path and velocity.

Transformer: Used to step up or step down AC voltage or AC current.

Transmission line: Structures and conductors that carry bulk supplies of electrical energy from power-generating units.

Transmission system (electric): An interconnected group of electric transmission lines and associated equipment for moving or transferring electric energy in bulk between points of supply and points at which it is transformed for delivery over the distribution system lines to consumers or is delivered to other electric systems.

Transportation sector: An energy-consuming sector that consists of all vehicles whose primary purpose is transporting people or goods from one physical location to another. Included are automobiles; trucks; buses; motorcycles; trains, subways, and other rail vehicles; aircraft; and ships, barges, and other waterborne vehicles. Vehicles whose primary purpose is not transportation (e.g., construction cranes and bulldozers, farming vehicles, warehouse tractors and forklifts) are classified in the sector of their primary use.

Turbine: A machine for generating rotary mechanical power from the energy of a stream of fluid (such as water, steam, or hot gas). Turbines convert the kinetic energy of fluids to mechanical energy through the principles of impulse and reaction, or a mixture of the two.

U-factor: Measures the rate of heat loss or how well a product prevents heat from escaping. It includes the thermal properties of the frame as well as the glazing. The insulating value is indicated by the R-value, which is the inverse of the U-factor. U-factor ratings generally fall between 0.20 and 1.20. The lower the U-factor, the greater a product's resistance to heat flow and the better its insulating value.

Under reamer: A heavy-duty drilling tool for enlarging a borehole significantly beyond the bit diameter. The device is placed about the drill bit; it can be opened during drilling and then closed to be brought back up through smaller diameter holes or casings.

Upwind turbine: A turbine that faces into the wind; it requires a wind vane and yaw drive to maintain proper orientation in relation to the wind.

Useful heat: Heat stored above room temperature (in a solar heating system).

Useful thermal output: The thermal energy made available for use in any industrial or commercial process or used in any heating or cooling application; the total thermal energy made available for processes and applications other than electrical generation.

Vapor dominated: A geothermal reservoir system in which subsurface pressures are controlled by vapor rather than by liquid. Sometimes referred to as a *dry-steam reservoir.*

Viewshed: The scenic characteristics of an area, when referred to as a resource.

Visible light transmittance: The amount of visible light that passes through the glazing material of a window, expressed as a percentage.

Voltage: The measure of electrical potential difference.

Watt (electric): The electrical unit of power; the rate of energy transfer equivalent to 1 ampere of electric current flowing under a pressure of 1 volt at unity power factor.

Watt-hour (Wh): The electrical energy unit of measure equal to 1 watt of power supplied to, or taken from, an electric circuit steadily for 1 hour.

Watt (thermal): A unit of power in the metric system, expressed in terms of energy per second, equal to the work done at a rate of 1 joule per second.

Well log: Logging includes measurements of the diameter of the well and various electrical, mass, and nuclear properties of the rock which can be correlated with physical properties of the rock. The well log is a chart of the measurement relative to depth in the well.

Wind energy: Energy present in wind motion that can be converted to mechanical energy for driving pumps, mills, and electric power generators. Wind pushes against sails, vanes, or blades radiating from a central rotating shaft.

Wind power plant: A group of wind turbines interconnected to a common utility system through a system of transformers, distribution lines, and (usually) one substation. Operation, control, and maintenance functions are often centralized through a network of computerized monitoring systems, supplemented by visual inspection. In Europe, this type of plant is called a *generating station.*

Wind rose: A diagram that indicates the average percentage of time that the wind blows from different directions on a monthly or annual basis.

Wind turbine or windmill: A device for harnessing the kinetic energy of the wind and using it to do work, or generate electricity.

Wind vane: Wind direction measurement device used to send data to the yaw drive.

Windward: Into, or facing the direction of, the wind; opposite of leeward.

Wood energy: Wood and wood products used as fuel, including roundwood (cord wood), limb wood, wood chips, bark, sawdust, forest residues, charcoal, pulp waste, and spent pulping liquor.

Wood pellets: Sawdust compressed into uniform-diameter pellets to be burned in a heating stove.

Wood/wood waste: Category of biomass energy that includes black liquor, wood/wood waste liquids (red liquor, sludge wood, spent sulfite liquor), and wood/wood waste solids (peat, paper pellets, railroad ties, utility poles, wood/wood waste).

Yaw: The rotation of a horizontal axis wind turbine around its tower or vertical axis.

Yaw drive: Motor that keeps an upwind turbine facing into the wind.

Zonal isolation: Various methods to selectively partition portions of the wellbore for stimulation, testing, flow restriction, or other purpose.

Zone: Area within the interior space of a building that is to be cooled, heated, or ventilated. A zone has its own thermostat to control the flow of conditioned air into the space.

Index